T0297757

CAMBRIDGE LIBRARY COLLECTION

Books of enduring scholarly value

Earth Sciences

In the nineteenth century, geology emerged as a distinct academic
discipline. It pointed the way towards the theory of evolution, as scientists
including Gideon Mantell, Adam Sedgwick, Charles Lyell and Roderick
Murchison began to use the evidence of minerals, rock formations and
fossils to demonstrate that the earth was older by millions of years than the
conventional, Bible-based wisdom had supposed. They argued convincingly
that the climate, flora and fauna of the distant past could be deduced from
geological evidence. Volcanic activity, the formation of mountains, and the
action of glaciers and rivers, tides and ocean currents also became better
understood. This series includes landmark publications by pioneers of the
modern earth sciences, who advanced the scientific understanding of our
planet and the processes by which it is constantly re-shaped.

Geological Manual

Henry De la Beche (1796–1855) was a geologist who published widely
on various aspects of this science and was elected to the Royal Society in
1823. He was involved with the Ordnance Survey maps of Britain, and
became president of the Geological Society in 1847. De la Beche was also
instrumental in the 1851 opening of two influential institutions: the Museum
of Practical Geology and the School of Mines and of Science Applied to the
Arts, which were housed on the same site in London. His *Geological Manual*,
first published in 1831, also had French, German and US editions. In this
expanded third edition, published in 1833, the work offers a comprehensive
survey of multiple aspects of geology. Topics covered include an overview
of the Earth, rivers, glaciers, rock types and fossils in France and England,
demonstrating the range and depth of geological knowledge in the early
nineteenth century.

Cambridge University Press has long been a pioneer in the reissuing of out-of-print titles from its own backlist, producing digital reprints of books that are still sought after by scholars and students but could not be reprinted economically using traditional technology. The Cambridge Library Collection extends this activity to a wider range of books which are still of importance to researchers and professionals, either for the source material they contain, or as landmarks in the history of their academic discipline.

Drawing from the world-renowned collections in the Cambridge University Library, and guided by the advice of experts in each subject area, Cambridge University Press is using state-of-the-art scanning machines in its own Printing House to capture the content of each book selected for inclusion. The files are processed to give a consistently clear, crisp image, and the books finished to the high quality standard for which the Press is recognised around the world. The latest print-on-demand technology ensures that the books will remain available indefinitely, and that orders for single or multiple copies can quickly be supplied.

The Cambridge Library Collection will bring back to life books of enduring scholarly value (including out-of-copyright works originally issued by other publishers) across a wide range of disciplines in the humanities and social sciences and in science and technology.

Geological Manual

Henry Thomas De La Beche

CAMBRIDGE UNIVERSITY PRESS

Cambridge, New York, Melbourne, Madrid, Cape Town,
Singapore, São Paolo, Delhi, Tokyo, Mexico City

Published in the United States of America by Cambridge University Press, New York

www.cambridge.org
Information on this title: www.cambridge.org/9781108072557

© in this compilation Cambridge University Press 2011

This edition first published 1833
This digitally printed version 2011

ISBN 978-1-108-07255-7 Paperback

A

GEOLOGICAL MANUAL.

BY

HENRY T. DE LA BECHE, F.R.S., V.P.G.S.

MEMB. GEOL. SOC. OF FRANCE, CORR. MEMB. ACAD. NAT. SCI. PHILADELPHIA, ETC.

THIRD EDITION,

CONSIDERABLY ENLARGED.

LONDON:

CHARLES KNIGHT,

22 LUDGATE STREET, AND 13 PALL-MALL EAST.

1833.

PRINTED BY RICHARD TAYLOR,
RED LION COURT, FLEET STREET.

PREFACE.

WHEN it is attempted, as in works of the following description, to sketch the actual state of a particular science, and at the same time to point out a few of the conclusions that may be hazarded from known facts, an author has always great difficulty in avoiding unnecessary and tedious detail on the one hand ; while, on the other, he must notice such facts as may convince a student that he is not wandering in a wilderness of crude hypotheses or unsupported assumptions.

The present edition contains so many additions to the greater part of the Work, that it would be tedious, and, indeed, somewhat difficult to enumerate them. The chief alteration consists in removing the various lists of organic remains to the end of the volume, where they can be more readily consulted. The Work has also, at the suggestion of friends, been printed in a larger form, and in a larger type, the small type of the former editions having only been retained for the lists of organic remains. Under the heads of Inferior Stratified, and Unstratified Rocks, calculations have been introduced respecting the substances of which such rocks are chemically composed; and it is hoped that these calculations may be found useful, as also some observations respecting geological maps and sections, and the geological examination of a country.

The Author has availed himself largely of the additions made by M. von Dechen, with the assistance of the celebrated Von Buch and other German geologists, to the German translation of this Work, more particularly as respects the geology of Germany and the lists of organic remains. He has not been able to avail himself of any additions to the French translation of the Manual, made under the superintendence of M. Brochant de Villiers, as it will not appear much before the present edition ; but the Author is informed that it will contain a further development, by M. Elie de Beaumont, of his theory of the elevation of mountain chains, as also additions to the lists of organic remains. He has not seen a copy of the American edition of this Work, and is therefore not aware that any additions have been made to it.

There can be little doubt that, from a strong desire to find similar organic remains in supposed equivalent deposits, even at great distances, and from an equally strong desire to discover new species, the same organic remains, particularly shells, often figure in our catalogues under two names, while different species are made to appear as one. Notwithstanding these difficulties, it will, however, be evident, from a glance of these catalogues, that a great mass of information has been gradually collected on this subject alone, from which the most important results must follow, even though the various lists may require very considerable correction.

While availing himself of these and similar catalogues, the student should be careful to recollect, that however great and valuable the aid of Zoology and Botany may be in geological investigations, Physics and Chemistry

are of still greater importance ; inasmuch as the former
can only be employed with advantage in explanation of
a portion of the phænomena observed, while the latter
are available to a very great extent in explanation of the
whole.

The Author has been particularly anxious to point out
his various sources of information, even when he himself
has visited the same countries ; that, independently of
the fundamental principle *suum cuique*, the student
should be enabled more fully to avail himself of the
labours of the various authors cited, by referring to their
published works for greater detail than could be admitted
into a volume of this description.

In a rapidly advancing science like Geology, to which
new facts are constantly added, and in which the chances
of new views by their combination are consequently multiplied, it is almost impossible to avoid hazarding certain
general conclusions, when the various known facts pass in
review before us. In those which the Author has ventured
to bring forward, he has endeavoured always to follow that
system of induction which can alone lead to exact knowledge; but as truth, and truth alone, is the object of all science, he can sincerely declare,—that if from the discovery
of new facts, or from more sound views respecting those
already known, his conclusions should not appear tenable,
he would not only be most ready to abandon them, but to
rejoice that an untenable hypothesis may have been the
means of leading to more exact knowledge, if it should
have fortunately so happened that it promoted the requisite inquiry. Essentially it is of little importance whose
or what theory may in the end be found most accurate; so
long as we approximate towards the truth, we accomplish
all that can be expected; and it is clear, that the greater
the amount of known facts, the greater the chance of accu-

racy, not only from the larger mass of information presented
to the mind, but also from the frequent checks offered to
hasty conclusions.

Happily facts have become so multiplied, that Geology
is daily emerging from that state when an hypothesis, provided it were brilliant or ingenious, was sure of advocates
and temporary success, even when it sinned against the
laws of physics and facts themselves. It is not difficult
to foresee, that this science, essentially one of observation,
instead of being, as formerly, loaded with ingenious speculations, will be divided into different branches, each investigated by those whose particular acquirements may
render them most competent to do so ; the various combinations of inorganic matter being examined by the
Natural Philosopher, while the Natural Historian will
find ample occupation in the remains of the various animals and vegetables which have lived at different periods
on the surface of the earth.

TABLE OF CONTENTS.

Section XII.

Section XIII.

APPENDIX.

ABBREVIATIONS

OF

AUTHORS' NAMES

IN

THE LISTS OF ORGANIC REMAINS.

Ag.	Agassiz.	Hœn.	Hœninghaus.
Bast.	Basterot.	Jäg.	Jäger.
Beaum.	Elie de Beaumont.	Lam.	Lamarck.
Blain.	Blainville.	Lam^x.	Lamouroux.
Blum.	Blumenbach.	Linn.	Linnæus.
Bobl.	Boblaye.	Lons.	Lonsdale.
Broc.	Brocchi.	L. & H.	Lindley and Hutton.
Al. Brong.	Alex. Brongniart.	Mant.	Mantell.
Ad. Brong.	Adolphe Brongniart.	Munst.	Munster.
Brug.	Bruguière.	Murch.	Murchison.
Buckl.	Buckland.	M. de S.	Marcel de Serres.
Conyb.	Conybeare.	Nils.	Nilsson.
Cuv.	Cuvier.	Park.	Parkinson.
De C., or De Cau.	De Caumont.	Pas.	Passy.
Defr.	Defrance.	Phil.	Phillips.
De la B.	De la Beche.	Raf.	Rafinesque.
Desh.	Deshayes.	Rein.	Reinecke.
Des M.	Des Moulins.	Sauv.	Sauveur.
Desm.	Desmarest.	Schlot.	Schlotheim.
Desn.	Desnoyers.	Sedg.	Sedgwick.
Dufr.	Dufrénoy.	Sow.	Sowerby.
Dum.	Dumont.	Sternb.	Sternberg.
Fauj. de St. F.	Faujas de St. Fond.	Thir.	Thirria.
Flem.	Fleming.	Thur.	Thurman.
Goldf.	Goldfuss.	Y. & B.	Young and Bird.
G. T.	German Transl. of Manual.	Wahl.	Wahlenberg.
Her.	Herault.	Weav.	Weaver.
His.	Hisinger.		

ERRATA.

Page 63, line 6, *for* adn *read* and
—— 141, —— 15, *for* hwich *read* which
—— 161, —— 9, for *Planobris* read *Planorbis*
—— 193, —— 34, *for* from places *read* from their places
—— 320, —— 15, *for* von Decken *read* von Dechen
—— 350, —— 15, *for* is *read* as
—— 351, —— 24, *for* beds deposited *read* beds were deposited

A

GEOLOGICAL MANUAL.

Section I.

Figure of the Earth.

IT has been concluded, both from astronomical and geo-
desical observations, that the figure of the earth is a spheroid.
This spheroid has been considered as one of rotation, or such
a figure as a fluid body would assume if possessed of rotatory
motion in space.

The amount of the flattening of the poles, or the difference
of the diameter of the earth from pole to pole, and its diameter
at the equator, has been variously estimated; but it is commonly received that the polar axis is to the equatorial diameter
as 304 to 305, the compression of the earth, or flattening at
the poles, being thus considered as $= \frac{1}{305}$.

The equatorial diameter about $= 7924$ miles.*
The polar axis . . . . . $= 7898$

Difference   26

Density of the Earth.

Various opinions have been entertained on this subject;
but it appears certain that the internal density is greater than

* Considering the flattening of the poles as $= \frac{1}{304}$, M. Daubuisson has
made the following calculations:—

Radius at the equator	6376851 metres.
Semi-terrestrial axis	6355943
Diff. or flattening of poles . . .	20908
Radius in lat. 45°	6366407
A degree at same lat.	111115
A degree of long. in same lat. .	78828
Surface of our earth	5098857 square myriametres.
The volume	1082634000 cubic myriametres.

Traité de Géognosie, ed. 2me, tom. i.

the solid superficial density. Daubuisson infers from the observations of Maskelyne, Playfair, and Cavendish, that " the
mean density of the earth is about five times greater than that
of water, and consequently, about double that of the mineral
crust of our globe*." Laplace considered the mean density
of our spheroid as = 1·55, the solid surface being 1. According to Baily, the density of the earth is 3·9326 times greater
than that of the sun, and is to that of water as 11 to 2†.

Superficial Distribution of Land and Water.

The relative proportion of dry land to the ocean, as it at
present exists, is such, that nearly three-fourths of the whole
surface of the globe may be assigned to the latter. Of the
former, the configuration is very various, presenting the
greatest surface in the Northern hemisphere. Although the
land sometimes rises high above the level of the sea, according
to our general ideas on such subjects, it is, in reality, but
slightly removed above that level, when considered, as it
should be, with reference to the radius of the earth ‡. The
superficies of the Pacific Ocean alone is estimated as somewhat greater than that of the whole dry land with which we
are acquainted. Dry land can only be considered as so much
of the rough surface of our globe as may happen, for the time,
to be above the level of the waters, beneath which it may again
disappear, as it has done at different previous periods. Laplace calculated that the mean depth of the ocean was a small
fraction of twenty-five miles, the difference produced in the
diameters of the earth by the flattening of the poles. It has been
variously estimated at between two and three miles. The mean
height of the dry land above the ocean-level does not exceed two
miles, but probably falls far short of it; therefore, assuming two
miles for the mean depth of the ocean, the waters occupying
three-fourths of the earth's surface, the present dry land might
be distributed over the bottom of the ocean, in such a manner
that the surface of the globe would present a mass of waters;
—an important possibility, for, with it at command, every variety of the superficial distribution of land and water may be
imagined, and consequently every variety of organic life, each
suited to the various situations and climates under which it
would be placed.

The surface of the globe's solid crust is so uneven, that the
ocean, preserving a general level, enters among the dry land

---

* Traité de Géognosie, ed. 2me, tom. i. p. 18.

† Baily, Astronomical Tables.

‡ See the diagram in my Sections and Views illustrative of Geological
Phænomena, pl. 40.

in various directions, forming what are commonly termed inland seas; such as the Baltic, Red, and Mediterranean Seas,
in which geological changes may be effected different from
those in the open ocean.

Masses of salt water are sometimes included in the dry land,
which have been termed Caspians, from the Caspian Sea, the
largest of them. These have no communication with the main
ocean; indeed the level of the Caspian is much lower than
that of the Black or Mediterranean Seas, the former body of
salt water occupying, with lake Aral and other minor lakes,
the lower part of an extensive depression in Western Asia,
(from 200 to 300 feet under the general ocean-level,) which
receives the waters of the Volga and other rivers. These
bodies of salt water have been variously accounted for; some
supposing that they have been left isolated by a change in the
relative level of land and water, while others imagine their
saltness to arise from their occurrence in countries impregnated with saline matter. It is stated, in support of the latter
opinion, that the Caspian, and the lakes Aral, Baikal, &c. are
situated where salt springs abound. Whatever may be their
origin, it will be obvious, that if the fresh water they receive
be not equal to their evaporation, they will become gradually
more saline, until, the water being saturated, the surplus salt
will be deposited at the bottom, and strata of it will be formed
of a size and depth proportioned to those of the lake or sea.

It would be out of place to attempt a general description of
all the various combinations of land and water, with which all
must be more or less familiar; but it may be useful to notice
that fresh-water lakes cover very considerable spaces, and
that thus very extensive deposits may now take place, which
can only envelope the remains of terrestrial or fresh-water
animals and vegetables.

Saltness and Specific Gravity of the Sea.

The whole body of the ocean is composed of salt water,
which does not vary very materially in composition, as far as
we can judge from the experiments made on it.

From evaporation and the fall of rain, the sea will be less
salt at the surface than at some little depth beneath it.

According to Dr. Murray, sea-water collected from the
Firth of Forth contained, in 10,000 parts,

Common salt	220·01
Sulphate of soda . . .	33·16
Muriate of magnesia . .	42·08
Muriate of lime . . .	7·84

303·09

Dr. Marcet states that 500 grains of sea-water, taken from the middle of the North Atlantic, contained,

Muriate of soda	. . .	13·3
Sulphate of soda	. . .	2·33
Muriate of lime	. . .	0·995
Muriate of magnesia	. .	4·955

21·580

According to the experiments of Dr. Fyfe (Edin. Phil. Journal, vol. i.), the waters of the ocean between 61° 52′ N. and 78° 35′ N. do not differ much in their saline contents, these being between 3·27 and 3·91 per cent.—The waters were obtained by Scoresby.

M. Eichwald informs us, that the waters of the Caspian Sea contain much sulphate of magnesia, in addition to the various other salts held in solution by them.

Dr. Marcet instituted a series of experiments on the specific gravity of water, of which the following are the results:

	Sp. Gr.		Sp. Gr.
Arctic Ocean	. . . . 1·02664	Sea of Marmora . .	1·01915
Northern Hemisphere	. 1·02829	Black Sea	1·01418
Equator	1·02777	White Sea	1·01901
Southern Hemisphere	. 1·02882	Baltic	1·01523
Yellow Sea	1·02291	Ice-Sea Water . .	1·00057
Mediterranean . . .	1·0293	Lake Ourmia . . .	1·16507

The same author concluded from his observations,

" 1. That the Southern Ocean contains more salt than the Northern Ocean in the ratio of 1·02919 to 1·02757.

" 2. That the mean specific gravity of sea-water near the equator is 1·02777, intermediate between that of the Northern and Southern hemispheres.

" 3. That there is no notable difference in sea-water under different meridians.

" 4. That there is no satisfactory evidence that the sea at great depths is more salt than at the surface *.

" 5. That the sea, in general, contains more salt where it is deepest and most remote from land; and that its saltness is always diminished in the vicinity of large masses of ice.

* The author of the abstract of Dr. Marcet's observations in the Edin. Phil Journal, cites the following observations of Mr. Scoresby in support of this conclusion.

		Sp. Gr.
Lat. 76° 16′ N.	Surface	1·0261
	At 738 feet . .	1·0270
	At 1380 feet . .	1·0269
Lat. 76° 34′ N.	Surface	1·0265
	At 120 feet . .	1·0264
	At 240 feet . .	1·0266
	At 360 feet . .	1·0268
	At 600 feet . .	1·0267

"6. That small inland seas, though communicating with the ocean, are much less salt than the ocean.

"7. The Mediterranean contains rather larger proportions of salt than the ocean *."

M. Lenz, who accompanied Kotzebue's expedition, inferred from numerous experiments that,

1. The Atlantic Ocean is salter than the South Sea; and the Indian Ocean, being the transition from the one to the other, is salter towards the Atlantic, on the west, than towards the South Sea, on the east.

2. In each of these three great oceans there exists a maximum of saltness towards the north, and another towards the south; the first being further from the equator than the second. The minimum between these two points is a few degrees south of the equator in the Atlantic, and probably also in the Pacific, though M. Lenz's observations did not extend sufficiently low in the Pacific.

3. In the Atlantic the western portion is more salt than the eastern. In the Pacific the saltness does not appear to alter with the longitude.

4. In proceeding north from the northern maximum, the specific gravity of the water diminishes constantly as the latitude increases.

The same author considers that, from the equator to 45° N., the water of the sea, to the depth of 1000 fathoms, possesses the same degree of saltness †.

The saltness of the sea, particularly that of its surface, would seem greatly to depend on the proximity of nearly permanent ice, and of large or numerous rivers. Thus, as is seen above, the Baltic, White, Black, and Yellow Seas are less salt than the main ocean, because they are supplied with comparatively large quantities of fresh water. From the small proportion of salt contained in the Black Sea and Sea of Azof, the bays of the former frequently contain ice, and the latter is stated to be frozen over during four months in the year.

The superior saltness of the Mediterranean, though an inland sea, is attributed to the evaporation of its surface, which is supposed greater than the quantity of fresh water with which it is supplied. In consequence, two great currents, one from the Black Sea and the other from the Atlantic, flow into it to supply the waste caused by evaporation.

The saline contents of the sea are important, as all chemical changes or deposits, taking place in it, will be more or less affected by them. The gravity and pressure of the sea are of

---

* Phil. Trans. 1819; and Edin. Phil. Journal, vol. ii.
† Edin. Journ. of Science, April 1832.

still greater consequence; for, as the pressure increases with the depth, effects, which would be possible at one depth, would be impossible at another. Thus, it is obvious from the ingenious experiments of Sir James Hall, that carbonate of lime may be fused by heat without the loss of its carbonic acid, if subjected to great pressure, such as exists at the bottom of the deep sea. The pressure of the sea must also have considerable influence on the kind of animal and vegetable life found at different depths ; and we may infer that beneath very deep seas such life does not exist, great pressure and the absence of the necessary light being as destructive to it as the cold and the rarity of the air are in the higher regions of the atmosphere.

The compressibility of water, which was for a long time doubted, has been proved by experiment. According to the observations of M. Œrsted, corrected for the pressure of part of the apparatus employed, this compressibility amounts to 46·65 millionths of its volume for a pressure equal to each atmosphere. The experiments of MM. Colladon and Sturm, corrected in the same manner, make the compressibility of water, not deprived of air, equal to 47·85 millionths for each atmosphere ; while that of water, deprived of air, is equal to 49·65 millionths under the same pressure *. M. Poisson estimates that it would require a pressure equal to 1100 atmospheres to reduce water six hundredths of its volume †. Water containing salts in solution is found to be somewhat less compressible. It follows, that at great depths, and beneath a great pressure of the ocean, a given quantity of water will occupy a less space than on the surface, and will, consequently, by this circumstance alone, have its specific gravity increased.

Temperature of the Earth.

The superficial temperature of our planet is certainly very materially influenced by, if it may not be entirely due to, solar heat. That the difference of seasons, and of the climates of various latitudes, originates in the greater or less exposure to the sun, is obvious. That local circumstances cause great variations of superficial temperature, is also well known ; yet the principle seems to prevail, that under equal circumstances, the temperature decreases from the tropics to the poles.

It would be useless to increase the size of this volume with a detail of the various temperatures that have been observed

---

* Pouillet, Élémens de Phys. Expérimentales, 2me ed. t. ii. p. 65.
† Poisson, Nouvelle Théorie de l'Action Capillaire, p. 277.

in different situations, or of the modifications arising from local causes; this will be found in various works devoted to the subject,—more particularly in Humboldt's Treatise on Isothermal Lines.

Respecting the temperature of our globe, M. Arago has made the following remarks :—" 1st, In no part of the earth on land, and in no season, will a thermometer raised from two to three metres above the ground, and protected from all reverberation, attain the 46th centigrade degree: 2ndly, In the open sea, the temperature of the air, whatever be the place and season, never attains the 31st centigrade degree: 3rdly, The greatest degree of cold which has ever been observed upon our globe, with the thermometer suspended in the air, is 50 centigrade degrees below zero : 4thly, The temperature of the water of the sea, in no latitude, and in no season, rises above + 30 centigrade degrees *."

Geologists have discovered that the superficial temperature of the earth has not always remained the same, and that there is evidence of a very considerable decrease.   This evidence will be found scattered over such parts of the following pages as treat of organic remains, and therefore need not be adduced here.   It may, however, be right to remark, that it rests on the discovery of vegetable and animal remains entombed in situations, where, from the want of a congenial temperature, such animals or vegetables would now be unable to exist. Undoubtedly this inference rests on the supposed analogy between animals and vegetables now existing, and those of a similar general structure found in various rocks, and at various depths beneath the earth's surface: but as we now find every animal and vegetable suited to the situations proper for them, we have a right to infer design at all periods, and under every possible state of our earth's surface; and therefore to consider, that similarly constituted animals and vegetables have, in general, had similar habitats.

This decrease in surface-temperature may arise either from external, superficial, or internal causes.

External Influence.—Heat, derived from the sun, producing such great effects at present, it has been supposed that a difference in the relative position of our planet and our great luminary would cause a corresponding change in the surface-temperature of the globe.   Theories have been invented which suppose such a change in the earth's axis as would render the present poles parts of the equator, and thus capable of having once supported a tropical vegetation, which has gradually disappeared, and been replaced by such plants as can exist

---

* Ann. de Phys. et de Chim. tom. xxvii. ; and Edin. Phil. Journ. 1825.

amid masses of ice and snow. Mr. Herschel, viewing this
subject with the eye of an astronomer, considers that a diminution of the surface-temperature might arise from a change
in the ellipticity of the earth's orbit, which, though slowly,
gradually becomes more circular. No calculations having yet
been made as to the probable amount of decreased temperature from this cause, it can at present be only considered as a
possible explanation of those geological phænomena which
point to considerable alterations in climates *.

Superficial Influence.—A decrease of temperature may arise
from such a variation in the relative position of land and water,
and in the elevation and form of land, as may cause the climate, in any given position on the earth's surface, so to change,
that a greater heat may precede a less heat, and the land be
capable of supporting the vegetables and animals of hot climates at one time, and be incapable of doing so at another.
For this ingenious theory we are indebted to Mr. Lyell †. It
supposes a combination of external and internal causes; the
latter raising or depressing the land in the proper situations,
the former supplying the necessary heat. It also supposes
the possible recurrence of a warm climate, so that the same
situations might alternately be placed under the influence of a
raised and a depressed temperature. We have so few data
for estimating the value of this theory, that it can only be considered as a possible explanation of a diminished temperature.
It must, however, be admitted, that, in every state of the
earth's surface, the relative disposition of land and water, and
the form or elevation of the land, would always have had, as
they now have, very considerable influence on climate.

Internal Influence.—From the earliest times an opinion has
existed among philosophers that a central heat exists;—an
opinion naturally arising from the phænomena of volcanos and
hot springs. But, notwithstanding this opinion, it was not
until a comparatively late period that direct experiments were
instituted, for the purpose of determining whether the temperature does, or does not, increase with the depth, or from
the surface downwards.

Various observations have been made on the temperature of
mines in Great Britain, France, Saxony, Switzerland, and
even Mexico. All those made previous to 1827 were collected,
arranged, and commented on by M. Cordier ‡. Experiments
on the temperature of mines have been made in various ways;

* Herschel, Geol. Trans., 2nd series, vol. iii.
† Principles of Geology.
‡ Essai sur la Température de l'Intérieur de la Terre: Mém. de l'Acad.
tom. vii.

sometimes by ascertaining the heat of air in the galleries, sometimes that of the stagnant water at various levels; at others, by observing the temperature of springs at different depths, or that of the waters pumped up from below; and sometimes, though rarely, by obtaining the temperature of the rock itself at various levels.

It soon suggested itself that, though these experiments pointed to an increase of temperature as we descended, the presence of the miners with their lamps or candles, and the explosions of gunpowder in some mines, would cause an increased heat of the air in galleries, sufficient to produce exceedingly grave errors. M. Cordier endeavours to assign to these and other objections their full value. It is calculated that a miner disengages, in an hour, a quantity of heat sufficient to raise the temperature of 542 cubic metres of air, one degree above a previous heat of 12° centigrade. It is also inferred that four miners' lamps will produce as much heat as three miners. It is further calculated that the presence of two hundred miners and two hundred lamps, properly separated from each other, would elevate the temperature of a gallery whose dimensions are one metre by two, and 93,000 metres long, about one degree (centigrade) in one hour. M. Cordier also mentions, that in the coal-mine of Carmeaux " nineteen lamps and twenty-four miners, scattered through two levels, and continually employed during six days in the week, produced, by the hour, a heat sufficient to raise the temperature of the air in the galleries by 1°·66 cent." The air in these galleries was estimated at 12,560 cubic metres.

Another source of error arises from the circulation of air in mines, and its introduction from the surface. This will vary according to the local distribution of the galleries in a mine; but there will always be a tendency to replace expanded and heated air by that which is more dense and cold; consequently, from whatever cause the heat of a mine may be derived, if the air in it be, as usually happens, warmer than that of the surface, the cold air will always strive to get into the mine, and the heated air to escape from it. It follows, that the entrance of air from the exterior surface tends to lower the temperature of the mine, and in some measure to check the heat caused by the workings. M. Cordier * observes, on this subject, that the mean temperature of *the mass of air*, introduced into a mine during a year, is lower than the mean temperature of the country for the same year, and estimates the difference

---

* Essai sur la Température de l'Intérieur de la Terre.

It has been supposed, the air in mines being under a greater pressure than that at the surface, and undergoing this change in a short time, that heat

between them at between 2° and 3° cent. for the greater part
of the mines in our climate.

The waters in mines may either give too high or too low a
temperature, as they may be either derived from beneath or
above. If waters descend from the surface into a mine, they
will carry with them their original temperature, modified by
the heat of the substances through which they pass; so that
their difference of temperature in the mine and on the surface
will depend on their abundance or scarcity, and on their
slowness or rapidity of motion. Moreover they will constantly
tend to reduce the surfaces of rock through which they percolate to their own temperature. The same remarks apply to
water derived from a lower level.

The temperature observed in the rock itself will be more or
less affected, according to circumstances, by that of the water
or air near it. So that the sides of a mine, to certain distances,
might possess a heat not common to the mass of rock at the
same level.

From these various sources of error, to which others might
be added, the observations made under circumstances that
might be influenced by them, can only be considered as approximations towards an estimate of the value of this mode of
inquiry. To render each set of observations available for what
they may be worth, M. Cordier has classed those made under
different circumstances under different heads. His tables,
thus formed, have also the great advantage of being reduced
to common measures of heat and depth.—From these the following have been selected as, perhaps, least liable to error*.

would be evolved sufficient to cause the appearance of an increase of temperature corresponding with an increased depth. But as the cold air will
become expanded by the heated air of the workings, and as the change of
pressure cannot be very sudden, this does not appear sufficient to account
for the phænomena observed. According to Mr. Ivory (Phil. Mag. and
Annals. of Phil. vol. i. p. 94), one degree of heat, of Fahrenheit's scale,
will be extricated from air when it undergoes condensation $= \frac{1}{180}$; and if a
mass of air were suddenly reduced to half its bulk, the heat evolved would
be $= 90°$.

* The temperature in these tables is marked in degrees of the centigrade
thermometer. When we consider the simplicity of this scale, and the facilities with which calculations can be made with it, it seems strange that its
use should not be generally adopted in this country, where we continue to
employ, from habit, the least philosophical of the three scales. The centigrade scale can easily be reduced to that of Fahrenheit, by considering that
the latter is to the former, between the freezing and boiling points of water,
as 180 to 100, or as 9 to 5. The degrees of Reaumur's scale are to those
of Fahrenheit's as 4 to 9. As the zero of Fahrenheit's scale is 32° of that
scale below the zero in the others, it is always necessary to make a proper
allowance for it.

Table of Observations made on the Springs in Mines.

Names, Authors, and Dates.	Mines.	Depth.	Temperature of the Springs.	mean of the Country.
		Metres.	Deg.	Deg.
Saxony. Daubu-isson. End of winter, 1802.	Lead and Silver of Junghohe-Birke.	78	9·4	8·
	. . .	217	12·5	8·
	Beschert Glück . .	256	13·8	8·
	Himmelfahrt . . .	224	14·4	8·
Brittany. Daubu-isson. 5th Sept., 1805.	Poullaouen	39	11·9	11·5
	———	75	11·9	11·5
	———	140	14·6	11·5
	Huelgoët	60	12·2	11·
	——	80	15·	11·
	———	120	15·	11·
	——	230	19·7	11·
Cornwall. Fox. Publ. 1821.	Dolcoath—Copper.	439	27·8	10·
Mexico. Humboldt.	Guanaxuato--Silver	522	35·8	16·

Tables of the Temperature of the Rock in Mines.

I. Thermometer placed in a niche cut in the rock, distant from the principal workings :—the bulb in the rock; the rest in a glass tube;—the whole covered by a glass door, closing the niche, and only opened for observation.

		Depth. Metres.	Temperature of Rock.	of Country.
Saxony. De Trébra. 1805, 1806, 1807. .	Mine of Beschert Glück; lead & sil.	180	11·25	8·
		260	15·	8·
Saxony. De Trébra. 1815	Mine of Alte Hoff-nung Gotes . . .	71·9	8·75	8·
		168·2	12·81	8·
		268·2	15·	8·
		379·54	18·75	8·

II. Thermometer plunged in the earthy matters at the bottom of galleries, which had been inundated two days*.

Cornwall. Fox. Published 1821. . .	United Mines	348	30·8	10·
		366	31·1	10·

III. Thermometer fixed in the rock of a gallery, for eighteen months at a yard deep.

Cornwall. Fox. Published 1822. . .	Dolcoath	421	24·2	10·

* M. Cordier remarks on the error that may, in this case, arise from the mixed temperature of the galleries, before inundation, produced by the usual causes in mines at work, and of the waters during inundation. On this subject he cites some observations of his own at Ravin, near Carmeaux, which show that the differences of temperature between the rubbish on the floor of the galleries, and that proper to the level, amounted to 2°·6, 2°·8, and even 3°·1 centigrade.

Table of the Temperatures of the Rock observed in the Coal-
mines at Carmeaux, Littry, and Decise.

Carmeaux.

	Depth. Metres.	Temperature. Deg.
Water of the well Vériac	6·2	12·9
Water of the well Bigorre	11·5	13·15
Rock at the bottom of Ravin Mine . . .	181·9	17·1
Rock at the bottom of Castellan Mine . .	192·	19·5

Littry.

	Depth. Metres.	Temperature. Deg.
Surface	0·	11·
Rock at the bottom of St. Charles Mine— mean of 2 obs. . . . . }	99·	16·135

Decise.

	Depth. Metres.	Temperature. Deg.
Water of the well Pélisson	8·8	11·4
Water of the Puits des Pavillons	16·9	11·67
Rock in the Jacobé Mine.	107·	17·78
————————————	171·	22·1

These observations were made with great care ; " the thermometer was loosely rolled in seven turns of silk paper, closed
at bottom, and tied by a string a little beneath the other extremity of the instrument, so that so much of the tube might
be withdrawn as might be necessary for an observation of the
scale, without fearing the contact of the air: the whole contained in a tin case." This was introduced into a hole from
24 to 26 inches in depth and $1\frac{1}{2}$ in diameter, inclined at
an angle of 10° or 15°; so that the air once entered into
the holes could not be renewed, because it became cooler,
and consequently heavier, than that of the galleries. The
thermometer was kept as nearly as possible at the temperature of the rock, by plunging it among pieces of rock or
coal freshly broken off, and by holding it a few instants at
the mouth of the hole, into which it was afterwards shut, a
strong stopper of paper closing the aperture. The thermometer generally remained in this hole about an hour [*].

Temperature of Water in Artesian Wells, and in neglected
Mines.

Artesian wells are well known as borings, by which water,
at different distances from the surface, rises to, and even above,
that surface, from its endeavour to escape. According to the

---

[*] Where the investigation of the increase or decrease of temperature, beneath such a depth as may be out of atmospheric influences, is so easy, with
a few necessary precautions, it is surprising, that in the British collieries,
which are so numerous, and many of which are very deep, so few direct experiments should have been made on the temperature of the rock itself.

observations of M. Arago, the greater the depth of these wells, the higher is the temperature of the waters that flow from them.

From experiments made by M. Fleuriau de Bellevue, in an Artesian well on the sea-side near Rochelle, the temperature increases with the depth. The well, at the time of the first experiment, was 3¼ inches in diameter, and 316 feet deep, and in it a column of brackish and stagnant water rose to the height of 294 feet. On February 14th, 1830, he found the temperature at the bottom, after the thermometer had remained there 24 hours, to be = 16°·25 centigrade; the external air being = 10°·6. At 11 feet beneath the surface of the water the temperature was found = 13°·12 cent. after the instrument had remained 17 hours. Common wells, varying in depth from 22 to 28 feet, afforded at the same time a mean temperature of 8°·75. On March 22nd, MM. Emy and Gon made further experiments on the same well, which was then sunk to the depth of 125·16 metres, or 369½ metrical feet. They found the temperature at the bottom, after the thermometer had remained there 25 hours, = 18°·12 cent. Fearful of some inaccuracy in this experiment, they repeated it the next day, when, after the instrument had remained at the bottom for 15 hours, they obtained exactly the same result. M.Fleuriau de Bellevue estimates the mean temperature of the country at 11°·87 cent.*

These experiments were conducted with great care, and seem highly illustrative of an increase of heat from the surface to the interior; for the column of water being subject to the usual laws, it would equalize its temperature by the descent of the cooler, and the ascent of the warmer water, if a constant source of comparatively considerable heat did not exist at the bottom.

In the waters of neglected mines also there are numerous observations tending to show that the waters do not follow the laws of their greatest specific gravity in such situations, but that the temperatures greatly increase with their depth. Certainly, in many situations, such as in recently flooded mines, the water would be heated by the galleries in which work had been carried on; but such influence could not continue for a long period, and there are numerous observations which show an increase of temperature in neglected mines. On a subject of this kind, however, great caution is necessary in obtaining the true temperature, and it is very desirable that many of the experiments should be repeated †.

* Fleuriau de Bellevue, Journal de Géologie, tom. i.
† A cold spring percolating rapidly from the surface to the deep waters of a neglected mine would tend to cool the waters at such depths.

Temperature of Springs.

The temperature of surface-springs has been supposed to
give nearly, if not altogether, the mean temperature of the
countries in which they appear. Their value in this respect
would depend on whether the waters which supply them be
derived from above or beneath, that is, whether they percolate
from the surface through porous strata until thrown out by impervious beds, or are forced by some means from comparatively
greater depths upwards. Many springs, we well know, come
within the first class; but many, we are also certain, come
within the second, for their temperatures are greatly above
what they could have acquired by mere percolation downwards.

At Paris, the oscillations of the temperature of the earth
do not quite cease at 28 metres. Professor Kupffer considers
that 25 metres from the surface will afford a depth beneath
which springs rise with a uniform temperature throughout the
year, being sufficiently removed from atmospheric influences.
Admitting this, it is clear that if surface-springs be small, and
rise slowly, they may have their temperatures somewhat
changed during their passage through the 25 metres; while if
they rise quickly, and their waters be copious, they will suffer
little change in their traverse through that thickness. The
question, however, of whence the waters may have been derived, remains the same.

Professor Kupffer has constructed the following Table,
principally from Von Buch's Treatise on the Temperature of
Springs, and from Humboldt's Treatise on Isothermal Lines,
with the view of corroborating the observations of Wahlenberg, that the temperature of springs in high latitudes is
greater than that of the air, and of those of Von Humboldt and Von Buch, who found that in low latitudes the temperature of springs was lower than that of the air;—showing
" that the temperature of the earth is sometimes very different
from the mean temperature of the air, and that its distribution
follows different laws*."

* Kupffer on the Mean Temperature of the Atmosphere and of the Earth
in some Parts of Russia: Edin. New Phil. Journ. vol. viii.: and Poggendorf's Annalen, 1829.

Places.	Latitude.	Height above sea.	Temp. of Earth. Fahr.	Temp. of Air. Fahr.	Observers.
	°	Metres.	°	°	
Congo	9 S.	45	72·95	78·12	Smith.
Cumana	10¼N	0	78·12	82·40	Humboldt.
St. Jago (Cape Verde Isles)	15 —	0	76·10	77·00	Hamilton.
Rock Fort (Jamaica) . . .	18 —	0	79·02	80·60	Hunter.
Havannah	23 —	0	74·30	78·12	Ferrier.
Nepaul	28 —	0?	73·85	77·00	Hamilton.
Teneriffe	28½—	0	64·40	70·92	Von Buch.
Cairo	30 —	0	72·5	72·5	Nouet.
Cincinnati	39 —	160	54·27	53·82	Mansfield.
Philadelphia	40 —	0	54·95	54·27	Warden.
Carmeaux	43 —	300?	55·40	57·87	Cordier.
Geneva	46 —	350	52·02	49·32	Saussure.
Paris	49 —	75	57·70	51·57	Bouvard.
Berlin	52½—	40	50·22	46·40	
Dublin	53 —	0	49·32	40·10	Kirwan.
Kendal	54 —	0	47·75	46·16	Dalton.
Keswick	54½—	0	48·65	47·97	
Konigsberg	54½—	0	46·62	43·25	Erman.
Edinburgh	56 —	0	47·75	47·75	Playfair.
Carlscrona	56¼—	0	47·30	47·30	Wahlenberg.
Upsal	60 —	0	43·70	42·12	———
Umeo	64 —	0	37·17	33·35	———
Giwartenfiäll	66 —	500	34·25	25·25	———

To this should be added Professor Kupffer's own observations in Russia.

Places.	Latitude.	Height.	Temp. of Earth.	Temp. of Air.
	°	Metres.	°	°
Kinekejewa	54½	300	39·87	34·7
Kasan	56	30	43·25	37·4
Nishney-tagilsk	58	200	37·17	31·55
Werchoturie	59	200	36·27	30·42
Bogoslowsk	60	200	35·37	29·30

The above tables, if correct, are sufficient to show that, though the terrestrial temperature, as deduced from springs, decreases from the equator to the poles, it does not decrease according to the mean temperature of the air above it. This seems to point out that there is some modifying cause in action independent of solar influence. Wahlenberg has noticed that many deep-rooted plants and trees only flourish because the temperature of the earth exceeds the mean temperature of the air; and Professor Kupffer remarks that he has often had occasion to confirm this observation in the northern Urals.

At the contact of the atmosphere and earth, we should expect, if they possessed different sources of temperature, that they would mutually act on each other, and that therefore the equal mean temperature of different parts of the earth's surface would, to a certain extent, correspond with equal terrestrial temperatures, as deduced from moderate depths. This may perhaps account for Professor Kupffer's conclusion, that " if we draw lines through all the points which have the same terrestrial temperature, these *isogeothermal lines* resemble the isothermal, as they are parallel to the equator, but diverge from it in several points*."

The temperature of the surface, as deduced from springs, is undoubtedly liable to many errors, as it rests on the assumption that they take the temperature of the earth at moderate depths. Those springs which percolate through porous strata, until thrown out, may take this temperature; but those which seem to come from beneath cannot be supposed, though cooled in their passage upwards, to do so.

The evidence that many springs rise from considerable depths, and possess a temperature independent of solar influence, rests on their great heat, which varies from the boiling point of water downwards to ordinary temperatures. It is impossible to account for this, otherwise than by supposing such heat communicated to the water in parts of the earth far beneath the surface, and removed from atmospheric influence.

The source of the heat in thermal waters has occupied the attention of Berzelius, Von Hoff, Keferstein, Bischoff, Daubeny, and others. The former remarks on those thermal springs which are charged with various salts of soda and carbonic acid, and attributes their origin to the percolation of atmospheric waters to volcanic regions, after which they are forced up to the surface, charged with the substances with which they have become combined in those situations. Von Hoff opposes the theory of a mere volcanic point supplying the necessary heat, and considers it much more probable that this is due to those processes in the interior of our globe which produce volcanos and earthquakes. Keferstein considers that hot vapours and springs are due to volcanic agency, which may be very deeply seated, even below the oldest formations. Bischoff, who details these various opinions†, does not appear to have adopted any decided one of his own on the subject, but directs attention to the possible increase of temperature in the waters by the internal heat of the earth at great depths, independent of

---

* Kupffer (memoir cited above).

† Uber die Vulchanischen Mineralquellen Deutschland und Frankreichs: and Edin. New Phil. Journal, 1830.

volcanic fires, and observes that if the channels through which
the waters flow upwards become once heated, their walls
would conduct little heat outwards, for rocks are bad conductors of heat, as is well shown in the case of lava streams, on
the outside of which the hand may sometimes be placed, while
the melted rock is still flowing inside*.

In support of the opinion that thermal waters may have
their high temperature caused by a general internal heat, and
not by mere volcanic points on the earth's surface, it may be
remarked that thermal springs occur in almost all situations,
some of which are far removed from any volcanic points on
the surface.

The immediate connexion of the Geysers and the volcanos
of Iceland is so obvious that few will be found to doubt it; yet
when hot springs have been found traversing cracks in strata
not volcanic, theories have been invented to explain their origin by chemical combinations at small depths. The salts,
however, usually held in solution in these waters do not afford
support to this view, and Berzelius has shown it to be untenable with respect to the Carlsbad waters.

To show the various rocks among which thermal springs
occur, we will select a few examples. In ranges of mountains
they would appear to be far from uncommon, a circumstance
which, supposing the ranges to have been elevated by a force
acting from beneath, lends additional probability to a general
heat beneath the surface. They have been observed in various
places in the range of the Himalaya. Captain Hodgson notices them in the course of the Jumna river, so hot that the
hand could not be kept in them many moments, and the temperature was too great to be measured by the short scaled
thermometer usually employed to ascertain atmospheric heat.
Again, at Jumnotri, very copious thermal springs rise through
crevices in the granite. The heat was estimated at nearly the
boiling point; the finger could not be kept in it two seconds.
As the height of Jumnotri is estimated at 10,483 feet above
the sea, the water would have the appearance of boiling at a
lower temperature than in the plains below: moreover, the
springs seem to evolve gas, for they rise with great ebullition;
still, however, the temperature of the waters would appear to
be very considerable†.

In the range of the Alps, there are also many thermal
springs, as has been already remarked by Bakewell. The
thermal waters of Bad-Gastein in the Salzburg country are
well known.

* Monticelli and Covelli.
† Hodgson, Asiatic Researches, vol. xiv. : and Edin. Phil. Journ vol. viii.

The following are Alpine warm springs noticed by Bakewell*: Naters, Haut Valais;—temperature = 86° Fahr. Leuk, Haut Valais,—twelve springs;—temperature varying from 117° to 126°. Bagnes, in the valley of the same name;—the baths, village, and one hundred and twenty inhabitants destroyed by the fall of part of a mountain in the year 1545;—temperature unknown. Thermal springs in the valley of Chamonix;—temperature unknown. St. Gervais, near the Mont Blanc;—temperature from 94° to 98°. Aix les Baines, Savoy;—two springs;—temperature from 112° to 117°. Montiers, Savoy;—temperature not noticed. Brida, Savoy;— temperature 93° to 97°. Saute de Pucelle, Savoy;—temperature not noticed. Thermal springs at Cormayeur and St. Didier, on the Italian side of the Pennine Alps;—temperature 94°. Warm springs in the Alps near Grenoble.

Many of these thermal waters are of recent discovery, although those of Aix were known to the Romans; therefore there may be many in other parts of the Alps which remain unnoticed.

There are also warm springs in the Caucasus, to the N.W. of the fortress of Constantinohor, with a temperature of from 110° to 114° F.; and there are, no doubt, numerous other thermal waters in great mountain ranges, with which we are as yet unacquainted.

In the Pyrenees, we have the two celebrated thermal waters of Barège and Bagnères; the former having a temperature of 120°, at the hottest spring, and the latter of 138°, also at the hottest spring.

The thermal springs at both these places are numerous. At the latter place there are no less than thirty of them, the temperature of the least hot of which is = 83¾° F.

There are also thermal waters in the valley of Barège, at St. Sauveur, = 98½°; as also several springs at Cautieres not far from the latter place, of which the temperatures vary from 98° to 131°. At Caberu, three leagues from Bagnères, there is a spring = 80°.

It would be tedious to give a long list of thermal springs; they occur in all parts of the world, as well remote from, as in the vicinity of, active volcanos. A great burst of hot springs takes place near the base of the south-eastern slope of the Ozark mountains, North America, and about six miles north from the Washita, from which they take their name. They are about seventy in number, and occur in a ravine between two slate hills. James states the temperature of these waters at 160° Fahr. Major Long gives that of several of them, as respectively, 122°, 104°, 106°, 126°, 94°, 92°, 128°, 132°,

---

* On the Thermal Waters of the Alps, Phil. Mag. and Annals, 1828.

151°, 148°, 132°, 124°, 119°, 108°, 122°, 126°, 128°, 130°, 136°, 140°. He also states that, "not only confervas and other vegetables grow in and about the hottest springs, but great numbers of little insects are seen constantly sporting about the bottom and sides*."

Another example of the existence of animals and vegetables in thermal springs is to be found at Gastein, where the *Ulva thermalis,* and a fresh-water shell, the *Limneus pereger* Drap., are found in waters at a temperature of 117° Fahr.

A very copious discharge of hot water takes place in an alluvial plain in a granitic district at Yom-Mack, about twenty miles from Macao, China. Three large springs have respectively the temperatures of 132°, 150°, and 186° Fahr. That with the temperature of 150° is described as in a state of active ebullition, about thirty feet in diameter, and discharging at least fifteen gallons in a minute†.

The temperature of the waters of Carlsbad is also considerable, being, according to Berzelius, 165° Fahr. Those of Aix-la-Chapelle are = 143°; and at Borset, near Aix-la-Chapelle, there are two springs, of which the temperatures are respectively 158° and 127° Fahr. At Balarue, department of Herault, there is one = 128° Fahr. The waters of Baden Baden are = 153° F.; of Bude, near Chemnitz, = 158° F.; of Ems, Nassau, = 122° F.; of Ofen, Hungary (in granite), = 147° F.; of Plombières, Vosges, = 140° F.; of Burtscheid, Prussian Provinces, = 171º F. Numerous thermal springs, varying from 118º to 122º F., rise amid a calcareous district near the Euphrates, about five wersts from Fort Diadine, in Armenia. The thermal springs of our own country are not very remarkable for their elevated temperature; for with the exception of those of Bath ‡, which are at 116° Fahr., the others can only be considered as tepid, the waters at Buxton being at 82º, those of the Hotwells, Bristol, at 74º, and those of Matlock 68º §.

In the volcanic districts of Italy, thermal springs, as might be expected, are numerous. The waters of the Bagni di Lucca are however sufficiently removed from a volcano to be here noticed: they rise on the sides of a hill, composed of a sandstone, the *macigno* of the Italians. The district is one of sandstone and limestone, and the hottest spring has a temperature of 131º F.

---

* James, Expedition to the Rocky Mountains.
† Livingstone, Edin. Phil. Journal, vol. vi.
‡ These rise through lias, traversing probably red sandstone, carboniferous limestone, &c.
§ The thermal springs of the Hotwells, Matlock, and Buxton, appear among carboniferous limestone.

It may not be altogether out of place to notice the thermal waters of Bath, St. Thomas in the East, Jamaica, to show how widely distributed these heated springs are. They rise at the base of the Blue Mountains, in a valley composed of trap, limestone, and slate. I observed their temperature to be = 127° F.*

The hot and cold springs of La Trinchera, three leagues from Valencia (America), may be cited to show, how differently derived waters may be, which make their appearance close to each other. According to Humboldt, there are two springs, only 40 feet asunder, the one cold, the other hot, the thermal waters having the great temperature of 90°·3 centigrade (194°·5 Fahr.). At Cannea, in Ceylon, a thermal spring is stated to exist which does not preserve a constant temperature, but varies from 38° to 41° cent. (100°·4 to 105°·8 F.).

Hot springs are common to the volcanic districts of different parts of the world, as also amid extinct volcanos, such as those of Central France; to enumerate them would be useless; but those of Iceland are so remarkable, that a short notice may not be unacceptable to the reader, particularly as they are the most extraordinary thermal springs with which we are acquainted.

Hot springs are numerous in Iceland, but those named the Geysers are the most singular. They are alternately in a state of rest and of violent activity, discharging, at intervals, immense quantities of hot water and steam.

Sir G. Mackenzie states that an eruption of the Great Geyser, which he witnessed, commenced with a sound resembling the distant discharge of a piece of ordnance. "The sound was repeated irregularly and rapidly; and I had just," observes this author, "given the alarm to my companions†, who were at a little distance, when the water, after heaving several times, suddenly rose in a large column, accompanied by clouds of steam, from the middle of the basin, to the height of ten or twelve feet. The column seemed as if it burst, and sinking down it produced a wave, which caused the water to overflow the basin in considerable quantity. After the first propulsion, the water was thrown up again to the height of about fifteen feet. There was now a succession of jets to the number of eighteen, none of which appeared to me to exceed fifty feet in height; they lasted about five minutes. Though the wind blew strongly, yet the clouds of vapour

* Although no active volcanos exist in Jamaica, there are the remains of an extinct one on the north side of the island; and earthquakes are, as is well known, sufficiently common.

† Dr. Bright and Dr. Holland.

were so dense, that after the first two jets I could only see the highest part of the spray, and some of it that was occasionally thrown out sideways. After the last jet, which was the most furious, the water suddenly left the basin, and sunk into the pipe in the centre*." The water sunk in the pipe to the depth of ten feet, but afterwards rose gradually; when sufficiently high, its temperature was observed, and found = 209° F.

A subsequent eruption of the same Geyser is thus described by the same author. After an alarm given of its approaching activity, "in an instant," he says, "we were within sight of the Geyser; the discharges continuing, being more frequent and louder than before, and resembling the distant firing of artillery from a ship at sea . . . . It raged furiously, and threw up a succession of magnificent jets, the highest of which was at least ninety feet†."

One of the other fountains, which was formerly an insignificant spring, and now known as the New Geyser, alternates in like manner. The eruption commences, as at the Great Geyser, by short jets, which increase in size. When a considerable mass of water is thrown out, the steam rushes forth furiously, accompanied by a loud thundering noise, carrying the water, when Sir G. Mackenzie observed it, to at least seventy feet. He describes it as continuing in this magnificent play for more than half an hour. "When stones are dropped into this pipe, while the steam is rushing out, they are immediately thrown up, and are commonly broken into fragments, some of which are projected to an astonishing height‡."

There are other alternating hot springs in Iceland, which are, however, of greatly inferior magnitude to the Geysers. The springs of Reikum, with a temperature of 212° Fahr., rise and fall, and dash up spray to the height of twenty or thirty feet. In the valley of Reikholt, there is a singular alternation of two boiling jets, one throwing the water up twelve feet, the other five§.

Temperature of the Sea and of Lakes.

This temperature will probably be in part derived from that of the atmosphere, and partly from the earth; but water being, under certain circumstances, able to communicate heat

---

* Mackenzie's Travels in Iceland.　　　　　† Ibid.
‡ Travels in Iceland: where views of these fountains in full operation will be found.
§ Waters actually at the boiling point seem exceedingly rare. The thermal waters of Urijino, in Japan, are stated to have a temperature of 212° Fahr., but it does not appear among what rocks they occur.

with great rapidity, the temperature will be more speedily equalized in it, than in the solid earth beneath. Water, moreover, at a given temperature possesses a greater specific gravity than when that temperature is either increased or diminished, and will consequently, at that given temperature, sink to the lowest depths. Even if it should be heated there, on the presumption of an internal heat in the earth, the water will still obey the same laws, the newly heated water will ascend, and be replaced by that which is cooler and of greater specific gravity. For, in order that the water should sink to these depths in the first instance, it must be of such a temperature, or specific gravity, as shall enable it to do so, and any change in that temperature, if it be that of the maximum density of water, will cause it to rise.

According to Dr. Hope, the maximum density of fresh water is at a temperature between 39¼° and 40° Fahr.*, and this determination has been confirmed by Professor Moll. According to the experiments of Professor Hällström, the maximum density of water occurs at the temperature of 4°·108 centigrade (39°·394 Fahr.).

It has been considered that the temperature of the maximum density of sea-water is not far removed from that of fresh water. On this head we have no very satisfactory experiments, but it may be supposed that the saline contents of sea-water would have considerable influence on its relative gravity at different temperatures.

In the years 1819 and 1820 I made numerous experiments, with great care, on the temperature of the Swiss Lakes at various depths, which are often considerable. The results of more than one hundred observations on the Lake of Geneva, in September and October 1819, were, that between the surface and a depth of 40 fathoms the temperature varied considerably. From 67° to 64° Fahr. was a common heat from one to five fathoms, and there was a general diminution of temperature downwards to the depth of 40 fathoms, whatever the surface-heat might be; in other words, there was a general increase of specific gravity downwards. From 40 fathoms to 90 fathoms the temperature was always 44°, with one exception near Ouchy, where 45° were observed at a depth of 40 fathoms. From 90 fathoms to the greatest depths, which amounted to 164 fathoms, between Evian and Ouchy, the temperature was invariably = 43°·5 Fahr. It will be observed, that in these experiments, made with a register thermometer constructed for the purpose, the water arranged itself according to the temperatures that would be expected,

* Trans. Roy. Soc. Edinburgh.

on the supposition of the maximum density of water being
between 39° and 40°*.

After the severe winter of 1819, I made some further experiments, and found that the temperature of the lake still
followed the same law.

In May, 1820, I tried the temperature of the lakes of Thun
and Zug, and obtained the following results†.

Lake of Thun.		Lake of Zug.	
Surface	60°	Surface	58°
At 15 fathoms	42	At 15 fathoms	42
At 50 fathoms	41·5	At 25 fathoms	41
At 105 fathoms	41·5	At 38 fathoms	41

In these experiments also, the results are in accordance
with the maximum density of water being between 39° and
40°, as was also the case in some which I made in the Lake
of Neufchatel, during very cold weather, so cold indeed, that
the water froze on the oars of the boat, when the temperature
increased towards the supposed maximum density of water.

We now turn to the experiments that have been made by
different navigators on the temperature of the sea at various
depths. The following observations by Scoresby show an increase of temperature from the surface downwards in cold
latitudes.

Situation.	Depth.	Deg. of Temp.	Situation.	Depth.	Deg. of Temp.
	Surface	29·0		Surface	28·8
	13 fathoms	31·0	Lat. 76° 16′ N.	50 fathoms	31·8
Lat. 79° 4′ N.	37 fathoms	33·8		123 fathoms	33·8
Long. 5° 4′ E.	57 fathoms	34·5		230 fathoms	33·3
	100 fathoms	36·0	Lat. 79° 4′ N.	Surface	29
	400 fathoms	36·0		730 fathoms	37

Again, in lat. 78° 2′ N. and long. 0° 10′ W. the same scientific navigator obtained 38° at 761 fathoms, the surface-water
being 32°. In one situation, indeed, in lat. 76° 34′ N. the
same observer obtained a temperature of 34° at 60 fathoms,
and 34°·7 at 100 fathoms, after having had 35° at 40 fathoms.
The experiments of Capt. Ross are, indeed, opposed to this
view, for they give a decrease down to 25° at 660 fathoms,
from 30° at 100 fathoms, 29° at 200 fathoms, and 28° at 400
fathoms; in lat. 60° 44′ N. and long. 59° 20′ W. According
also to Dr. Marcet, the maximum density of sea-water is not
at 40° Fahr. He states that this water increases in weight to
the freezing point, until actually congealed. In four experiments Dr. Marcet cooled sea-water down to between 18° and

---

* A detailed account of these experiments, with a chart of soundings in
the lake, were inserted in the Bibliothèque Universelle for 1819; from
whence it was copied, in part, into the Edin. Phil. Journal, vol. ii.
  † See also Bibliothèque Universelle for 1820.

19° Fahr., and found that it decreased in bulk till it reached 22°, after which it expanded a little, and continued to do so till the fluid was reduced to between 19° and 18°; when it suddenly expanded, and became ice with a temperature of 28°. According to M. Erman, salt water of the specific gravity of 1·027 diminishes in bulk down to 25° F., and does not reach its maximum density before congelation. It should always be recollected that a saturated solution of common salt does not become solid, or converted into ice at a less temperature than 4° Fahr.; and therefore if the sea should be, as is sometimes supposed, more saline at great depths, and as it appears to be in the Mediterranean from the experiments of Dr. Wollaston, ice could not be formed there at the same temperature as it could nearer the surface.

Kotzebue, in lat. 36° 9′ N. and long. 148° 9′ W. found the surface-water = 71°·9, the air being 73°; at 25 fathoms the water was at 57°·1; at 100 fathoms, 52° 8; and at 30 fathoms, 44°: showing a decrease of temperature towards 39° or 40°. In lat. 23° 3′ N. and 181° 56′ W. Krusenstern obtained, at the surface, 78°; at 25 fathoms, 75°; at 50 fathoms, 70°·5; and at 125 fathoms, 61°·5.

In latitudes south of the tropics, Kotzebue observed a temperature of 49°·5 at 35 fathoms, the surface being at 67°, the air at 68°, in lat. 30° 39′ S. The same navigator found the temperature at 196 fathoms to be = 38°·8, in lat. 44° 17′ S. and long. 57° 31′ W.; the surface-water being 54°·9, and the air at 57°·6.

The following are among the temperatures obtained by Captain Beechey* at various depths and situations. In lat. 47° 18′ S., and long. 53° 30′ W., the surface-water being at 49°·8, he found 44°·7 at 270 fathoms, 39°·2 at 603 fathoms, 40°·1 at 733 fathoms, and 39°·4 at 854 fathoms. In lat. 55° 58′ S., and long. 72° 10′ W., the surface-water being at 43°·5, he obtained 42°·5 at 100 fathoms, 42°·5 at 230 fathoms, 42·5 at 330 fathoms, and 41°·6 at 430 fathoms. In the South Pacific, he found in lat. 28° 40′ S., and long. 96° W., 71° at 100 fathoms, 53° at 200 fathoms, 49° at 300 fathoms, and 45° at 400 fathoms, the surface-water being at 74°. Among the observations made by the same navigator in the North Pacific are the following: in lat. 61° 10′ N., and long. 183° 28′ W., in July 1827, at 5 fathoms 41°·5′, at 10 fathoms 38°, at 20 fathoms 29°·5, at 20 fathoms 30°·5, (this is apparently a second observation at the same depth), at 30 fathoms 30°·5, at 52 fathoms 32°·5, at 100 fathoms 32°·5, and at 200 fathoms 32°·5, the surface-water being at 43°·5 and the air at 45°.

* Beechey, Voyage to the Pacific, &c.

Observations have been made at considerable depths in the tropics.   Capt. Sabine found in lat. 20° 30′ N., and long. 83° 30′ W., a temperature of 45°·5 at 1000 fathoms, the surface-water being at 83°.   Capt. Wauchope obtained in lat. 10° N. and long. 25° W., a temperature of 51° at 966 fathoms, the surface-water being at 80°: and the same observer also found, in lat. 3° 20′ S. and 7° 39′ E., a temperature of 42° at 1300 fathoms, the surface-water being at 73°.   M. Lenz, in lat. 21° 14′ N., and long. 196° 1′ W., found a temperature of 61°·4 at 150 fathoms; of 37°·7 at 440 fathoms; of 37°·2 at 709 fathoms; and of 36°·5 at 976 fathoms; the surface-water being at 79°·5 *. Other observations within the tropics, at inferior depths, show the same decrease of temperature downwards.   Thus Kotzebue in lat. 9° 21′ N. obtained 77° at 250 fathoms, the surface-water being at 83° and the air at 84°; and under the equator, in long. 177° 5′ W., 55° at a depth of 300 fathoms, the surface-water being at 82°·5 and the air at 83°.

M. Berard found, at a depth of 1200 fathoms, (without reaching bottom,) between the Balearic Isles and the coast of Algiers, a temperature of 53°·4, the surface-water being at 69°·8, and the air at 75°·2 F.   From other observations in the western part of the Mediterranean, at the respective depths of 600 and 750 fathoms, and another not stated, it was found that the water was still at 55°·4, though the temperature of the surface-water varied materially.   M. D'Urville remarks that these experiments accord with some made by himself, also in the western Mediterranean, at 300, 200, 250, 600, and 300 fathoms, when he obtained the respective temperatures of 54°·5, 54°·1, 57°·3, 54°·6, and 54°·8 F.   He hence infers that the waters of the western Mediterranean, beneath a depth of 200 fathoms, rest at a temperature of about 55° F.†

It will be observed, from what has been stated above, that the waters of lakes arrange themselves according to certain temperatures, which show that experiments made in the cabinet, and which fix the maximum density of fresh water at a temperature of between 39° and 40° Fahr., are correct.   With respect to the waters of the ocean, sea water evidently arranges itself, in the warmer climates of the globe, according to its density, supposing its maximum density to be at a temperature approaching that of its congelation; but in the colder regions there would appear to be some disturbing forces in action, which in many situations cause the temperature to increase with the depth, thus interfering with the densities, if it be true,

---

* The same observer obtained, in lat. 32° 6′ N., and in long. 136° 48′ W., a temperature of 56° at 96 fathoms; of 43°·7, at 228; of 38°·7, at 480; and of 35°·9, at 632; the surface-water being at 70°·6 F.

† D'Urville, Bul. de la Soc. de Géographie, t. xvii. p. 82.

as it seems to be, that the maximum density of sea-water is as
above stated.

The probability of a central heat would appear to rest, first,
on the experiments made in mines, which, notwithstanding
their liability to error from various sources, still seem to show,
particularly those made in the rock itself, an increase of temperature from the surface downwards; secondly, on thermal
springs, which are not only abundant among active and extinct volcanos, but also among all varieties of rocks, in various
parts of the world; thirdly, on the presence of volcanos themselves, which are distributed over the globe, and present such
a general resemblance to each other, that they may be considered as produced by a common cause, and that cause probably deep-seated; and fourthly, on the terrestrial temperature
at comparatively small depths, which does not coincide with
the mean temperature of the air above it.

The temperature at the bottom of seas and lakes is not at
variance with this probability, as the waters endeavour to arrange themselves according to their greatest specific gravity.
Indeed, so far from the temperature found at various depths
in the ocean being at variance with the hypothesis of a central
heat, they rather accord with it; more particularly when we
regard the increase of temperature with the depth in many
places in high latitudes.   We might assume that such increase
of temperature, interfering with the densities, was due to heat,
in the bottom beneath; which, though sufficient to produce a
visible effect upon waters so closely approaching their maximum density as those in high latitudes, was yet insufficient to
be apparent beneath warm climates, though it would still cause
an increase in the temperature of the waters in low latitudes.

Neither is the probability of internal heat at variance with
the figure of the earth or observed geological phænomena.
The figure of our planet being that which a fluid body would
assume if revolving in space, it is as probable that this fluidity
should be igneous as aqueous.   Geological phænomena attest
the eruptions of igneous matter from the interior at all periods;
as also elevations of mountains and great dislocations of the
earth's surface, caused by forces acting from beneath; and,
finally, a great decrease of surface temperature.   Should we
be inclined to build a theory on the probability of a central
heat, we may suppose, as has often been done, that our world
is a mass of igneous matter in the act of cooling.

Baron Fourier considered it as proved,—from the form of
our spheroid, the disposition of the internal strata (shown by
experiments with the pendulum) to increase in density with
their depth, and from other considerations,—that a very intense heat formerly penetrated all parts of our globe.   He

concluded that this temperature was dissipated into the surrounding planetary spaces, the temperature of which he considered, from the laws of radiant heat, to be $= -50°$ cent. ($-58°$ Fahr.). He moreover inferred that the earth had nearly reached its limit of cooling. The original heat contained in a spheroidal mass equal in magnitude to our globe, would diminish more rapidly at the surface than at great depths, where the elevated temperature would remain for a great length of time. He further inferred from these circumstances, and from the temperature of mines and springs, that there is an internal source of heat, raising the temperature of the surface above that which the action of the sun could alone give it*.

Temperature of the Atmosphere.

The gaseous compound termed the Atmosphere, which surrounds the earth, has been calculated, from its powers of refraction, to extend upwards about forty-five miles. Dr. Wollaston considered, from the laws of the expansions of gases, that it might reach to at least forty miles, with its properties uninjured by rarefaction. On this head Dr. Turner observes, "that the tension or elasticity of gaseous matter is lessened by two causes, diminution of pressure, and reduction of temperature." And he further remarks, that the former alone has been taken into account by Dr. Wollaston, while it appears to him that the extreme cold at great heights would also be sufficient to limit the extent of the atmosphere†.

Though no part of the solid earth is so elevated above the general surface as to be exposed to a very considerable depression of temperature, yet numerous mountains are of sufficient height to be covered, at their summits, with what has been termed perpetual snow, the prolific parent of innumerable rivers, without which many regions would be uninhabitable.

---

* M. Svanberg, calculating what might possibly be the temperature of the planetary spaces, proceeds upon another principle than that of the radiation of heat. He supposes that the planetary spaces never undergo any change of temperature, but that the capacity for elevation of temperature, above that which constantly reigns in the ethereal regions, exists only within the limits of the planetary atmosphere. He obtains for the result of his calculations a temperature $= -49°·85$ cent. Observing this near approach towards Baron Fourier's supposed temperature, he had the curiosity to calculate the temperature according to Lambert's statements, respecting the absorption which takes place in a ray of light passing from the zenith through the whole atmosphere, and found that he obtained $-50°·35$ for the result. A curious coincidence between the results of the three modes of calculation. —Berzelius. Annual Progress of Chemical and Physical Science. Edin. Journ. of Science, vol. iii. New Series.

† Turner, Elements of Chemistry, p. 221.

The line of perpetual snow differs generally according to the latitude, and is liable to very great variations from local causes. Some of these variations will be observed in the following Table, by Humboldt*, of the snow-line in certain mountain chains.

Mountains.	Latitude. Deg. Deg.	Height in English feet.
Cordillera of Quito	0 to 1½ S.	15,730
Cordillera of Bolivia	16 to 17¾ S.	17,070
Cordillera of Mexico	19 to 19¼ N.	15,020
Himalaya:		
Northern Flank	30¾ to 31 N.	16,620
Southern Flank.		12,470
Pyrenees	42½ to 43 N.	8,950
Caucasus	42½ to 43 N.	10,870
Alps.	45¾ to 46 N.	8,760
Carpathians	49 to 49¼ N.	8,500
Altai	49 to 51 N.	6,400
Norway:		
Interior	61 to 62 N.	5,400
Interior.	67 to 67¼ N.	3,800
Interior -	70 to 70¼ N.	3,500
Coast.	71 to 71½ N.	2,340

Among other variations from the theoretical line of perpetual snow which have been produced by a combination of physical circumstances, it will be observed that there is a difference between the northern and southern flanks of the Himalaya of more than 4000 feet in favour of the *former*, by which means a large surface is inhabited which would otherwise be unfit to support animal and vegetable life.

It has been supposed that the temperature of the atmosphere diminished equally upwards in different latitudes; but the following Table, also by Humboldt, will show that this is not the case, and that, on the contrary, the decrease is much more rapid in the temperate than in the equatorial zone.

Height in English feet.	Equatorial zone; from 0° to 10°.		Temperate zone; from 45° to 47°.	
	Mean Temp.	Differ- ence.	Mean Temp.	Differ- ence.
0	81·5	0	53·6	0
3,195	71·2	10·3	41·0	12·6
6,392	65·1	6·1	31·6	9·4
9,587	57·7	7·4	23·4	8·2
12,792	44·6	13·1		
15,965	34·7	9·9		

The curve, representing the line of perpetual snow, will not

* Fragmens Asiatiques, p. 549.

be equal in the northern and southern hemispheres, as the latter is found to be colder than the former.

From the variable height at which perpetual snow commences, it follows, all other circumstances being the same, that the extent of dry land capable of sustaining animal and vegetable life, will decrease from the equator to the poles, and, consequently, that there is a greater probability of an abundance of terrestrial remains being entombed in any deposit now taking place in the tropics, than in similar deposits in high latitudes*.

Valleys.

A classification of valleys cannot well be accomplished with out some violence, as the various depressions of land, to which the term valley has been much too generally applied, pass into each other in such a manner as to produce compounds of no easy arrangement. No great value is therefore attached to the following sketch.

Mountain Valleys.—These are both longitudinal and transverse; ranging either in the direction of the mountain chain, or across that direction. Their sides are generally rugged, crowned by lofty pinnacles and broken masses, and are, for the most part, steep. Atmospheric agents, far from producing a milder outline, generally add to their broken appearance. The melting of ice and snow, and the drain of rain-waters furrow their sides, bringing down detritus to the rivers, which, when levels are favourable, deposit it in situations well suited to vegetation; so that in mountain regions patches of verdure occur amid the wildest scenes, presenting a singular contrast to the broken forms of the surrounding heights. When levels are unfavourable, or the fallen blocks large, the masses accumulate in the water-courses, and produce innumerable cascades, adding to the desolate character of such regions.

Lowland Valleys.—These differ from the preceding in their rounded form, which would render a section of them an undulating line, the undulations varying in the proximity of the higher parts and in depth, so that the more elevated portions may even be many miles asunder, and the depth inconsiderable. From the comparatively gentle slopes of these valleys, atmospheric agents, though still able to decompose the rocks be-

---

* If we consider that animal and vegetable life decreases in proportion as the atmosphere becomes colder and less dense, and that marine life is less abundant as the pressure of the sea increases, and the necessary light diminishes, we obtain, if I may so express myself, two series of zones, one rising above the ocean-level, the other descending beneath it,—the terms of the two series, all other things remaining the same, affording the greater amount of animal and vegetable life, as they respectively approach the ocean-level.

neath, do not transport the detritus to any considerable
distance, except in climates and situations where heavy torrents of rain descend on land unfavourable to vegetation; yet,
even in this case, the general rounded outlines of the hills are
not very considerably impaired, though deep furrows are made
in their sides.

Ravines and Gorges.—These are bounded by more or less
perpendicular walls of rock, and are common both among
mountain and lowland valleys, but more particularly the
former. They frequently communicate between more open
spaces, and their edges may often be approached without any
suspicion that they exist, the country appearing as one continuous slope or level.

Broad Flat-bottomed Valleys.—Level plains of greater or
less extent, bounded by hills or mountains on either side;
such as the great valley of the Rhine below Basle, bounded on
one side by the Swartzwald, and on the other by the Vosges.

Such a diversity of form would seem to suggest a diversity
of origin. The mountain valleys for the most part resemble
large cracks, produced when the strata were suddenly elevated
and contorted, while the lowland valleys appear as if a large
body of water had passed over them, rounding the inequalities,
and acting on masses of strata in proportion to their power of
resistance. The gorges or ravines would seem due to the
cutting power of running waters, or to rifts in the rocks produced by violent convulsions. The flat-bottomed valleys have
the character of drained lakes, or situations where the rivers
or floods, not having any great velocity, deposit considerable
quantities of sediment over a flat surface.

As we may suppose hill and dale, mountain and valley, to
have existed from the earliest geological periods, and that
strata were by no means deposited in one even plane surface, we
have now a very complicated system of depressions; though as
a general fact it may be stated, that the superior stratified rocks
have filled up and covered over numerous inequalities of the
inferior stratified rocks, as is the case in Normandy, where the
oolite group covers over the uneven surface of slates, limestones
and grauwacke, the latter rocks here and there protruding
through the stratification of the former, and becoming visible
where rivers cut the superincumbent beds.

If we can imagine a violent disruption of strata, contorting
or throwing them on their edges, large rents and fractures
would be the natural consequences, producing longitudinal and
transverse fissures; but these would merely gape, and their
origin would appear clear, if not modified by some subsequent
action. If we suppose, with the advocates for no greater effects than we daily witness, that mountains have been raised

gradually by a multitude of earthquakes acting always in the same line, we shall have great difficulty in explaining the position of strata in high ranges, more particularly those (such are by no means uncommon in the calcareous Alps,) where whole mountains are contorted, and even appear as if thrown over, as at the Righi. Whereas, if we suppose that the elevations have been more violent, these difficulties would appear to vanish, and the upturned, overthrown, and contorted strata, the longitudinal and transverse cracks or valleys, would be more in harmony with each other.

If we should suppose a violent disruption of strata to take place beneath the waters of an ocean, these waters would be greatly agitated and react upon the land, rushing into the cracks; sweeping away pinnacles; driving blocks and loosely aggregated strata before them; rounding off angles; and accumulating detritus at the bottom of hollows. Should such a sudden elevation be effected, partly in the ocean, and partly out of it, the reaction of the sea would only reach the lower portion of the upraised strata, and these only would present rounded forms. Should the strata be elevated only in the atmosphere, the modification of the original cracks would be effected by atmospheric agency alone.

Although lowland valleys generally present rounded forms, the strata composing such districts are often far from undisturbed; on the contrary, they are often upturned, contorted, and fractured, the lines of valleys being frequently the same with those of the faults or fractures. Often, however, no appearances of fracture are visible in the hills, though these are traversed by faults in various directions. Of this fact the neighbourhood of Weymouth, in our own country, may be cited as affording good examples.

Valleys of Elevation are those which seem to have originated in a fracture of the strata, and a movement of the fractured part upwards, so that the strata dip from the valley on either side. Probably a very large proportion of mountain valleys might be arranged under this head; but at present geologists seem to have confined the application of the term to those which are bounded by hills of moderate height.

Prof. Buckland (in 1825) noticed valleys of this kind at New Kingsclere, Bower Chalk, near Shaftesbury, and Poxwell near Weymouth. The annexed diagram is a section of that of Kingsclere.

Fig. 1.

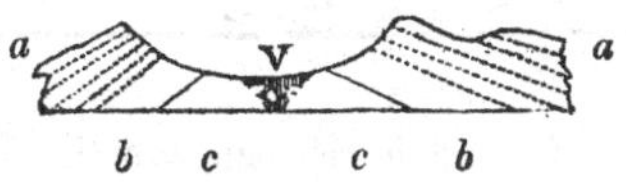

V. Valley of Kingsclere: *a a*, chalk with flints: *b b*, chalk without flints: *c c*, green sand.

It will at once be observed, that the strata on either side were once continuous, and that they have been upheaved, producing a fracture, which, by subsequent denudation, has been formed into the valley we now see.

Subsequently to the observations of Prof. Buckland, similar valleys in Germany have occupied the attention of M. Hoffman, who endeavours also to show that they are connected with springs impregnated with carbonic acid gas. In support of this opinion he cites the valley of Pyrmont, of which he gives the following section, which will be seen closely to correspond, in its general characters, with that of Kingsclere.

Fig. 2.

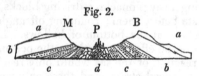

M, the Muhlberg, 1107 feet: B, the Bomberg 1136 feet: P, Pyrmont, the bottom of the valley being 250 feet: *a a*, keuper (red or variegated marl): *b b*, muschelkalk: *c c*, grès bigarré, broken into fragments at *d*, through which the acidulous waters are forced out.

As at Kingsclere, the strata have not been forced up to equal heights on either side. The grès bigarré rises to 850 feet on the Bomberg or north side; while on the Muhlberg or south side it only reaches 540 feet, with an inferior dip. The theoretical opinions connected with these appearances will be noticed in the sequel; at present it is only necessary to point out the existence of such valleys.

M. Hoffman also notices similar appearances, with acidulous springs, in the valley of Dribourg, on the left bank of the Weser, and several other combinations of the like kind *.

Valleys of Denudation.—Although the valleys of elevation above noticed, may also be termed valleys of denudation, this name seems given, in preference, to those valleys where the strata are not far removed from an horizontal position on either side, and of which also the former continuity cannot be doubted. Of these, the following section of the valley of Charmouth will afford an example.

Fig. 3.

* Journal de Géologie, tom. i.

a a, summits of the hills composed of angular flint and chert gravel, the remains of former superincumbent chalk and green sand which have been partially dissolved in place : *b b,* green sand, with an uneven upper surface resulting from the causes that have produced the gravel: *c c,* lias, in which the lower part of the valley has been excavated: *d,* small river Char flowing at the bottom. If proportions had been strictly attended to, the stream would have been invisible. On the sides of the hills, from *a* to *d,* much chert and flint gravel is distributed over the rocks *b* and *c,* and it may be questionable how much of it has, during a great lapse of time, descended from the heights, as has occurred on the slopes of similarly rounded hills, in the South Hams in Devon, and how much may have been left at the original formation of the valley. The advocates, indeed, of such excavations by no greater powers than those we daily witness, would consider this valley formed by the insignificant streamlet which now flows through it, aided by the rain-waters. This valley is, however, the sole channel of drainage for a district many miles in extent, in which the actual river, with every assistance from floods, has only effected a cut, varying from four to fifteen feet deep, bounded by perpendicular walls; these walls not composed, for the most part, of lias, but of gravel and drifted materials, such as are strewed over the valley of all heights, from the bed of the river to the tops of the hills. Such valleys are common in various parts of the world, and not unfrequently are without running waters in them, so that these could not have caused them. Even in Jamaica, where heavy tropical rains are sufficiently common, there are valleys, in which the waters are swallowed up by subterraneous cavities, or sink-holes, and no continuous streams are formed. In England we have examples of dry valleys, in our chalk districts, in the oolite of Yorkshire, and among the slates of the South Hams, Devon*; a covering of vegetation or turf most commonly protecting the surface from removal, even during heavy rains. On the west coast of Peru, where rain never falls, there are also some remarkable examples of dry valleys, which, judging from sketches, resemble many a lowland valley with rounded sides in Europe. The form of these valleys is also opposed to their production by running waters, for they are rounded and not bounded by perpendicular walls.

Sometimes the upper part of a hill being composed of harder materials than the lower portion, it advances with a somewhat bold escarpment.

* These latter are due to the highly inclined position of the strata, between the fissures of which the rain-water, after having been received in a porous superficial gravel, percolates.

D

The general form of these valleys would seem to suggest a mode of formation somewhat different from that of the mountain valleys; one which has permitted a very general removal of bold projecting points. There is scarcely a district of any considerable extent, composed of these valleys, which does not contain fissures, or faults, even when the strata are, as a mass, not far removed from an horizontal position. In other situations the strata are upheaved, contorted, and intruded by trap rocks, yet the general forms of the valleys are not considerably altered; the same rounded forms still prevail. From this general prevalence of the same character, it may not be unreasonable to conclude that some similar cause has produced them. They appear as if scooped out by large moving masses of waters; the least resisting parts first giving way. We might imagine this to have been effected by great disturbances beneath an ocean; such as would be caused by the elevation of long ranges of mountains near them, or a disruption of the strata of which they were actually composed;—in fact, by submarine earthquakes of much greater force than those which we now witness. Earthquakes of the present day frequently produce violent waves, which discharged on a coast ravage all within their reach. The sudden elevation of mountains to the height of several thousand feet would be accompanied by violent disturbance of the land, causing the waters of neighbouring seas to rise considerably, and overwhelm land within their reach; and these discharges of masses of water would have great scooping powers, more particularly if they acted on fractured strata, or small previously existing depressions. These valleys may also have been formed beneath agitated waters, in which currents moving with great velocity were produced; the land having been afterwards protruded above the level of the sea. These observations on the origin of lowland valleys should be considered as mere speculations, which future investigations may show to be either probable or improbable. One argument in their favour, in preference to the supposition that they have been scooped out by rivers, is, that in many instances the rivers quit the valleys, which would appear continuations of their natural channels, and pass through gorges or ravines cut in lands of considerable elevation on one side; the barrier to their natural passage onwards being merely a gentle rise of a few feet at the bottom of the valley, not easily observed.

Changes on the Surface of the Globe.

The present condition of our planet's surface is far from stable; on the contrary, if time enough could be allowed, a great change in the relations of land and water would be effected. This process is undoubtedly slow, but it is nevertheless certain, and so apparent, that many persons have been inclined to refer all geological phænomena to a continuance of those effects of existing causes which we daily witness. As far as we can judge from known facts, this opinion seems to have been somewhat hastily adopted, and not altogether in accordance with all those geological phænomena with which we are at present acquainted. As the student may, however, be supposed not to possess a knowledge of these phænomena, the consideration of their relative value must be waived until he becomes more familiar with the subject.

After geologists had ceased to amuse themselves by fabricating theories, without being at the trouble of examining the surface structure of that world which they made, modified, and broke to pieces at their own good-will and pleasure, and when it was thought that a knowledge of facts was somewhat necessary to a knowledge of the subject, it was soon observed that considerable changes had taken place on the world's surface. Facts being still few, hypotheses were easily formed, and were more or less plausible according to the knowledge of the day. These will be found in the various works which treat of the history of geology, and therefore need not be produced here; it will be sufficient to observe that the two prevailing theories of the present time are, 1st, That which attributes all geological phænomena to such effects of existing causes as we now witness; and, 2ndly, That which considers them referable to series of catastrophes or sudden revolutions. The difference in the two theories is in reality not very great; the question being merely one of intensity of forces, so that, probably, by uniting the two, we should approximate nearer to the truth.

Classification of Rocks.

The term Rock is applied by geologists, not only to the hard substances to which this name is commonly given, but also to those various sands, gravels, shales, marls, or clays, which form beds, strata, or masses*.

Rocks were first divided into two classes, Primitive and Secondary, it being considered that they originated under different circumstances; the latter only containing organic remains. To this Werner added a third class, which he named Transition, considering that it exhibited a passage from the primary into the secondary. Subsequently, from obser-

---

* For the terms used in geology, see Appendix A.

vations made by MM. Cuvier and Brongniart on the country round Paris, a fourth class was instituted, and called Tertiary, because the strata composing it occurred above the chalk, a rock considered as the highest of the secondary class.   These divisions or classes are more or less in use at the present time, though it seems somewhat generally admitted that they are insufficient, and not in accordance with the present state of science.   Numerous modifications and divisions have been proposed, which, though preferable to the preceding, have not been adopted, the force of habit, possibly, having prevailed.

To propose in the present state of geological science any classification of rocks which should pretend to more than temporary utility, would be to assume a more intimate acquaintance with the earth's crust than we possess.   Our knowledge of this structure is far from extensive, and principally confined to certain portions of Europe.   Still, however, a mass of information has gradually been collected, particularly as respects this quarter of the world, tending to certain general and important conclusions; among which the principal are,—that rocks may be divided into two great classes, the stratified and the unstratified;—that of the former some contain organic remains, and others do not; and that the nonfossiliferous stratified rocks, as a mass, occupy an inferior place to the fossiliferous* strata also taken as a mass.   The next important conclusion is, that among the stratified fossiliferous rocks there is a certain order of superposition, apparently marked by peculiar general accumulations of organic remains, though the mineralogical character varies materially.   It has even been supposed that in the divisions termed formations, there are found certain species of shells, &c. characteristic of each.   Of this supposition, extended observation can alone prove the truth; but it must not be supposed, as some now do, that in any accumulation of ten or twenty beds, characterized by the presence of distinct fossils in a given district, the organic remains will be found equally characteristic of the same part of the series at remote distances.

To suppose that all the formations, into which it has been thought advisable to divide European rocks, can be detected by the same organic remains in various distant points of the globe, is to assume that the vegetables and animals distributed over the surface of the world were always the same at the same time, and that they were all destroyed at the same moment, to be replaced by a new creation, differing specifically, if not generically, from that which immediately preceded

* The term *fossiliferous* is here confined to organic remains.

it. From this theory it would also be inferred that the whole surface of the world possessed an uniform temperature at the same given epoch.

It has been considered, but has not yet been sufficiently proved, that the lowest rocks in which organic remains are found entombed, show a general uniformity in their organic contents at points on the surface considerably distant from each other, and that this general uniformity gradually disappeared, until animal and vegetable life became as different in different latitudes, and even under various meridians, as it now is. How far this opinion may, or may not, be correct, can only be seen when geological facts shall have been sufficiently multiplied; but it is one which demands considerable attention, as the classification of fossiliferous rocks greatly depends upon it. Should it eventually be found to a certain degree correct, it would not be at variance with the theory of a central heat, which having diminished, permitted solar heat gradually to acquire an influence on the earth's surface.

Classifications of rocks should be convenient, suited to the state of science, and as free as possible from a leading theory. The usual divisions of Primitive, Transition, Secondary, and Tertiary, may perhaps be convenient, but they certainly cannot lay claim to either equality with the state of science, or freedom from theory.

In the accompanying Table, (pp. 38, 39,) rocks are first divided into Stratified and Unstratified, a natural division, or at all events one convenient for practical purposes, independent of the theoretical opinions that may be connected with either of these two great classes of rocks. The same may, perhaps, also be said of the next great division; namely, that of the stratified rocks into Superior or Fossiliferous, and Inferior or Non-fossiliferous. The superior stratified or fossiliferous rocks are divided into groups. We are yet well acquainted with so small a portion of the real structure of the earth's exposed surface, that all general classifications seem premature; and it appears useless to attempt others than those which are calculated for temporary purposes, and of such a nature as not to impede, by an assumption of more knowledge than we possess, the general advancement of geology.

Stratified Rocks. Group 1. (*Modern*) seems at first sight natural and easily determined; but in practice it is often very difficult to say where it commences. When we take into consideration the great depth of many ravines and gorges, which appear to originate in the cutting power of existing rivers, the cliffs even of the hardest rocks which more or less bound any extent of coast, and the immense accumulations of comparatively modern land, such as those which constitute the deltas

STRATIFIED ROCKS.	SUPERIOR STRATIFIED, or FOSSILIFEROUS.	**1. Modern Group ...** { Detritus of various kinds produced by causes now in action; Coral islands; Travertin, &c. } ...
		2. Erratic Block Group............ { Transported boulders and blocks; gravels on hills and plains, apparently produced by greater forces than those now in action. (*A provisional group.*) } ...
		3. Supracretaceous Group............ { Various deposits above the chalk, such as in England, the Crag, Isle of Wight beds, London and Plastic clays. In France, the freshwater and marine rocks of Paris, &c. } ...
		4. Cretaceous Group. { 1. Chalk. 2. Upper greensand. 3. Gault. 4. Lower greensand. To which may be added, for convenience, 1. Weald clay. 2. Hastings sands. 3. Purbeck beds. }
		5. Oolitic Group..... { The rocks usually known as the Oolite formation, including the Lias. } ...
		6. Red Sandstone Group............ { 1. Variegated or Red marl. 2. Muschelkalk. 3. Red sandstone. 4. Zechstein. 5. Red conglomerate. }
		7. Carboniferous Group............ { 1. Coal measures. 2. Carboniferous limestone. 3. Old red sandstone .... }
		8. Grauwacke Group. { Grauwacke, thick-bedded and schistose, sometimes red; Grauwacke limestones; Grauwacke clay slates, &c. Inferior beds, frequently mixed with stratified compounds resembling those of the unstratified rocks .... } ...
	INFERIOR STRATIFIED, or NONFOSSILIFEROUS.	**No determinate order of superposition ...** { Various schistose rocks, and many crystalline stratified compounds, such as Gneiss, Protogine, Mica Slate, &c. .... } ...
	UNSTRATIFIED ROCKS.	**Volcanic, Trappean, Serpentinous, and Granitic rocks .....** { Ancient and modern Lava, Trachyte, Basalt, Greenstone, Corneans, Augite, and Hornblende Porphyries, Serpentine, Diallage rock, Sienite, Quartziferous Porphyry, Granite, &c. } ...

Improved Wernerian.	Conybeare.	Omalius d'Halloy, 1830.	Brongniart, 1829.

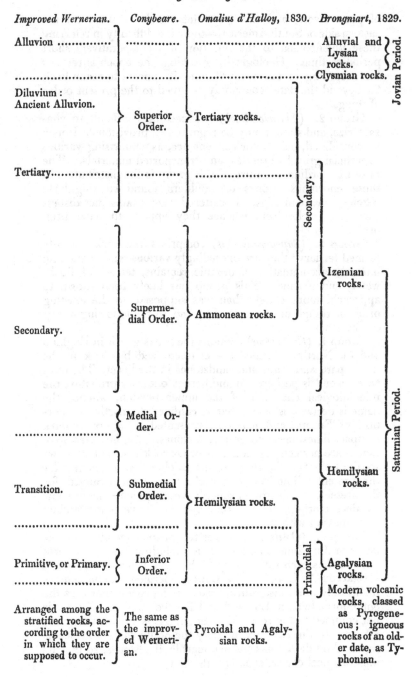

Alluvion .... Alluvial and Lysian rocks. — Clysmian rocks. } Jovian Period.

Diluvium: Ancient Alluvion.

Superior Order. } Tertiary rocks. } Secondary.

Tertiary.................

Secondary. — Supermedial Order. } Ammonean rocks. — Izemian rocks.

Medial Order.

Transition. — Submedial Order. } Hemilysian rocks. — Hemilysian rocks. } Saturnian Period.

Primitive, or Primary. } Inferior Order. — Agalysian rocks. } Primordial.

Arranged among the stratified rocks, according to the order in which they are supposed to occur. } The same as the improved Wernerian. } Pyroidal and Agalysian rocks. } Modern volcanic rocks, classed as Pyrogeneous; igneous rocks of an older date, as Typhonian.

of great rivers, and the great flats, such as those on the eastern side of South America,—there is a difficulty in referring these phænomena to the duration of a comparatively short period of time. Geologically speaking, the epoch is recent; but according to our ideas of time, it appears to reach back far beyond the dates commonly assigned to the present order of things.

Group 2. (*Erratic Block*) is exceedingly difficult to characterize, and should only be regarded as provisional. It may be considered, merely for convenience, as comprising various superficial gravels, breccias, and transported materials. The most extraordinary feature of this group is the distribution of those enormous blocks or boulders found so singularly perched on mountains, or scattered over plains, far distant from the rocks from whence they appear to have been broken.

Group 3. (*Supracretaceous*) comprises the rocks usually termed tertiary: they are exceedingly various, and contain an immense accumulation of organic remains, terrestrial, freshwater, and marine. This group has lately been shown to approach, more closely than was supposed, to the existing order of things on the one side, and to the following group on the other.

Group 4. (*Cretaceous*) contains the rocks which in England and the North of France are characterized by chalk in the upper part, and sands and sandstones in the lower. The term 'cretaceous' is perhaps an indifferent one; for probably, the mineralogical character of the upper portion, whence the name is derived, is local, that is, confined to particular portions of Europe, and may be represented elsewhere by dark compact limestones and even sandstones. As, however, geologists are perfectly agreed as to what rock is meant when we speak of the 'chalk,' there seems no objection to retain it for the present. The Wealden rocks have been arranged, for the present, in this group, though their organic remains show a different origin, because they may be conveniently studied in connexion with it.

Group 5. (*Oolitic*) comprises the various members of the oolite or Jura limestone formation, including lias. The term 'oolitic' has been retained upon the same principle as that of 'cretaceous.' In point of fact, this mineralogical character is found only in an insignificant part of the rocks known as the oolite formation in England and France; and moreover it is not confined to the rocks in question, but is common to many others. In the Alps and in Italy the oolite formation seems replaced by dark and compact marble limestones, so that its mineralogical character is of little value.

Group 6. (*Red Sandstone*) contains the red or variegated marls (marnes irisées, keuper), the muschelkalk, the new red or variegated sandstone (grès bigarré, bunter sandstein), the zechstein or magnesian limestone, and the red conglomerate (rothe todte liegende, grès rouge). The whole is considered as a mass of conglomerates, sandstones, and marls, generally of a red colour, but most frequently variegated on the upper parts. The limestones may be considered subordinate; sometimes only one occurs, sometimes the other, and sometimes both are wanting. There seems no good reason for supposing that other limestones may not be developed in this group in other parts of the world.

Group 7. (*Carboniferous.*) Coal-measures, carboniferous limestone, and old red sandstone of the English. The former would appear in the greater number of instances to be naturally divided from the group (6) above it; but the latter, though disconnected from the group (8) beneath in the North of England, is apparently so united with it in many other situations, that the old red sandstone may be considered as little else than the upper part of the grauwacke series in those places.

Group 8. (*Grauwacke.*) This may be considered as a mass of sandstones, slates, and conglomerates, in which limestones are occasionally developed. Sandstones which mineralogically resemble the old red sandstone of the English, not only occupy the upper part, but frequently also other situations in the series. In the lower portion of this group stratified compounds, resembling some of the unstratified rocks, are by no means unfrequent.

Inferior or *Non-Fossiliferous Stratified Rocks,*—comprising slates of different kinds, such as mica slate, chlorite slate, talcose slate, hornblende slate, &c., and various crystalline compounds arranged in strata, such as saccharine marble, in which other minerals may or may not be imbedded, gneiss, protogine, &c. From various circumstances, many rocks in the previous division so assume the mineralogical characters of those in this, as to be undistinguishable from them, except by geological situation; but it may be assumed, that, as a mass, the strata in this division are far more crystalline than in those of the superior stratified rocks, the origin of which seems chiefly mechanical.

Unstratified Rocks.—This great natural division is one of considerable importance in the history of our globe, as the rocks composing it seem to have caused, jointly with the forces that ejected them, very considerable changes on the earth's surface. They are very generally admitted to be of igneous origin; some of them indeed, those produced by active vol-

canoes, never could have been doubted. Their great characteristic is a tendency to a crystalline structure, though, in
many, this cannot be traced. Every gradation from the crystalline to the non-crystalline structure can frequently be observed in the same mass. The minerals, felspar, quartz, hornblende, mica, augite, diallage, and serpentine, enter largely
into the composition of these rocks, more particularly the
former.

In proposing this classification, I am fully aware that many
objections may be made to it; but it pretends to little beyond
convenience: and if geologists could be induced to use something of this kind, or any other that would better answer the
purpose of relieving us from the old theoretical terms, I cannot but imagine that the science would derive benefit from
the change.

In the following part of this volume, geological phænomena will be noticed in accordance with this classification.
But to enable those who prefer other arrangements to avail
themselves of any facts that may be brought forward, the
equivalents of the divisions or groups above noticed are given
in the Table (pages 38, 39), where the classifications of Conybeare, Brongniart, and Omalius D'Halloy, as also the improved Wernerian, are placed in parallel columns with it.

SECTION II.

MODERN GROUP.

Degradation of Land.

THERE is a constant tendency in all decomposed or disintegrated substances to be removed, by the agency of rains and
superficial waters, to a lower level than they previously occupied, and finally to be transported into the sea. There is no
rock, even the hardest, that does not bear some marks of what
has been termed weathering, or of the action of the atmosphere
upon it. The amount of surface-change, so produced, is exceedingly variable, depending much on local causes. Thus,
a rock may undergo complete disintegration in one situation,
though composed of nearly the same materials as that in
another, of which the change has been comparatively trifling.
When we contemplate the present surface of our continents
and islands, we cannot but be struck with the great effects
that have been produced upon them by the agents commonly
known as existing causes; and among these, the weathering
and degradation of land are very remarkable, attesting a lapse
of time far beyond the usual calculations. The tors of Dartmoor, Devon, may be referred to as excellent examples of the
weathering of a hard rock. These are composed of granite,
which, as Dr. MacCulloch has observed, are divided into
masses of a cubical or prismatic shape. " By degrees, surfaces which were in contact become separated to a certain distance, which goes on to augment indefinitely. As the wear-
ing continues to proceed more rapidly near the parts which
are most external, and therefore most exposed, the masses
which were originally prismatic acquire an irregular curvili-
near boundary, and the stone assumes an appearance resembling the Cheese-wring (Cornwall). If the centre of gravity
of the mass chances to be high and far removed from the perpendicular of its fulcrum, the stone falls from its elevation,
and becomes constantly rounder by the continuance of decomposition, till it assumes one of the spheroidal figures which
the granite boulders so often exhibit. A different disposition
of that centre will cause it to preserve its position for a greater
length of time, or, in favourable circumstances, may produce
a logging stone *." The weathering of these tors is so ex-

---

* MacCulloch. Geol. Trans. 1st series, vol. ii. ; where there are views of
Vixen Tor, the Cheese-wring, and Logan Rock: as also in Sections and
Views illustrative of Geological Phænomena, pl. 20.

ceedingly slow, that the life of man will scarcely permit him to
observe a change; therefore the period requisite to produce
their present appearances must have been very considerable.
The surface of the whole country round these districts attests
the same great lapse of time. Whatever may be the nature
of the rock, it is disintegrated to considerable depths; porphyries, slates, compact sandstones, trap rocks,—all have suffered;
but the valleys appear to have previously existed, and the general form of the land to have been much the same as it now is.
The following section will explain this decomposition of surface.

Fig. 4.

a a, decomposition of the rock *b b* following a line of previous elevation and depression, the accumulation being greatest at the bottom of the valley *c*, frequently cut through by a
river or rivulet, and sometimes exposing a stratified appearance, as if the disintegrated substances of the hill-sides had
slipped over each other to the bottom of the valley. The
maximum quantity of detritus so brought down to the bottom
of a valley, sometimes amounts to 25 or 30 feet. This detritus,
which is often very loosely aggregated, is now indeed protected
from removal, at least to a great extent, by grass and general
cultivation. The various appearances of this detritus are singular; for often larger pieces, perhaps of twenty or thirty
pounds weight, are included among small fragments and even
sand. Of this the following section, exhibited on the sea-
shore at Black Pool, Dartmouth, affords an example.

Fig. 5.

a a, detritus from the grauwacke slates *b b*, more thickly
accumulated at *e f. c c*, a high beach of small quartz shingles,
defending the bottom of the valley *d* (which is much lower than
the crown of the beach) and the cliffs on either side. The
drainage of the valley escapes in a serpentine manner by a
rivulet at *e*. At *e* and *f*, many large fragments are mixed with
the smaller.

The slates in the South Hams, Devon, are frequently surmounted by a superficial covering of fragments, which, at
their union with the undecomposed rock, appear as if some
force had been exercised at the commencement; the slates being
broken and turned back in the manner represented beneath.

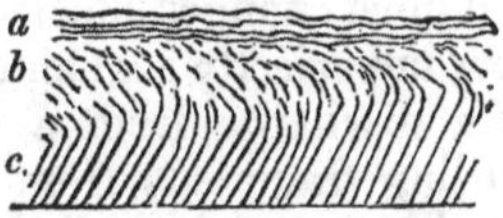

a, vegetable soil: *b*, small fragments of slate resting in various directions: *c*, portions of laminæ, turned backwards, sometimes without fracture.

Fig. 6.

If we proceed to the eastward from the South Hams, the same appearances present themselves, whatever may be the nature of the rock, though they become somewhat more complicated upon Haldon Hill, and on the coast of Sidmouth and Lyme Regis, as this decomposition of the surface seems mixed with a disintegration effected previous to the deposit of the supracretaceous rocks. A deep disintegration of surface, conforming to the undulations of the country, is well observed in Normandy, where it has been described by M. de Caumont and M. de Magneville, and seems due to the action of the same causes which have produced the decomposition of surface in the South of England.

This destruction of the surface is common to most countries; and if the rock so weathered be limestone, there is, not unfrequently, a reconsolidation of the parts by means of calcareous matter deposited by the water that percolates through the fragments, and which dissolves a portion of them. At Nice, the fractured surface thus reunited is so hard, that, if it occur on a line of road, it must be blasted by gunpowder for removal. There are some fine examples of this reconsolidation upon the limestone hills of Jamaica ; as for example near Rock Fort, and at the cliffs to the eastward of the Milk River's mouth.

The felspar contained in granite is often easily decomposed, and when this is effected the surface frequently presents a quartzose gravel. D'Aubuisson mentions that in a hollow way, which had been only six years blasted through granite, the rock was entirely decomposed to the depth of three inches. He also states that the granite country of Auvergne, the Vivarrais and the eastern Pyrenees, is frequently so much decomposed, that the traveller may imagine himself on large tracts of gravel*.

Some trap-rocks, from the presence of the same mineral, are so liable to decomposition that there is frequently much difficulty in obtaining a specimen. The depth to which some rocks of this nature are disintegrated in Jamaica is often very considerable.

This decomposition is attributed to the chemical as well as mechanical action of the atmosphere. The oxygen of the atmosphere produces considerable alteration in rocks, more particularly observed in those containing iron, which are thus

* Traité de Géognosie.

often reduced from a hard to a soft substance. With the slow and quiet changes effected by electricity on the surface we are very imperfectly acquainted; but all are familiar with the effects of a discharge from a thunder-storm, shivering rocks, and hurling fragments from the heights into the valleys beneath. In these electrical discharges the lightning often fuses the surface of rocks. Thus, De Saussure found a compound rock, on Mont Blanc, fused on the surface, white bubbles being on the felspar, and black bubbles on the hornblende. Similar observations have been made by other geologists in other parts of the world.

The effects of lightning upon loose sand were beautifully exhibited in the drifted sand hills between the sea and the embouchure of the Irt, near Drigg, in Cumberland. The sand consists of quartzose grains intermingled with a few grains of hornstone porphyry, and a few fragments of shells, and rests, at the depth of twenty-nine feet, upon a bed of pebbles. Two feet beneath these pebbles is a bed of wet sand containing small pebbles. Upon a single hillock, about forty feet above the sea, and within an area of fifteen yards, three hollow tubes were discovered, about an inch and a half in diameter. They consisted of the matter of the sand fused, and rendered vitreous. These tubes descended in a vertical manner to more than thirty feet, branching downwards in their descent. One tube coming in contact, at the depth of twenty-nine feet, with a fragment of hornstone porphyry, glanced off at an angle of 45°, and afterwards resumed its vertical course. As might have been expected, the tubes were very irregular in their passage into the wet sand.—The annexed sketch represents the termination of one of these vitreous tubes upon a granite pebble, and exhibits the manner in which they branched off in their descent, thus marking the course of the electrical discharge*.

Fig. 7.

At Peninis Point, St. Mary's, Scilly Islands, there is a curious example of that decomposition of granite, which antiquaries have termed *rock-basins*, and considered the work of the Druids. The Kettle and Pans, as these depressions are there named, occur in the large blocks of granite on the top of this promontory; they are generally three feet in diameter and about two feet deep; they are mostly circular and concave, but there are others much indented at the sides.

" Some have perpendicular sides and flat bottoms, some are

* Geol. Trans. vols. ii. and v. pl. 39.

of an oval form, and others of no regular figure. Many of the blocks are six or seven yards high, eight or nine yards square, and several of them have four, five, six or more of these cavities in them. A large rock near the extremity of this group has two basins, of an immense size, besides several smaller ones. The upper and larger one appears to have been formed by the junction of three or more large basins. It is irregularly shaped, and about eighteen feet in circumference and six feet deep. When the water in this basin has attained the height of three feet, it discharges itself by a lip into a lower basin, more regularly formed, the back of which is about five feet high, but which is incapable of containing more than a depth of two feet of water, owing to the declivity of the surface of the rock*." As a proof that similar decomposition sometimes takes place on the sides of a block, the author above cited mentions an oval cavity, six feet long, five wide, and nearly four feet deep, thus situated.—The following wood-cut will afford an idea of the Kettle and Pans†.

Fig. 8.

There is scarcely a substance, which having been exposed to the action of the atmosphere for a considerable time, does not exhibit marks of weathering. It will even be observed on the hardest siliceous rocks. The action of the atmosphere on cliffs of sandstone, in which the cement varies in induration or otherwise, produces the most grotesque forms, which must be more or less familiar to the least observing. Variations in temperature much assist the chemical decomposing power of the air.

Water may be considered as the principal mechanical agent in the great work of atmospheric destruction, uniting at the same time the character of a chemical agent. By infiltration it tends to separate the particles of which the rocks are composed, uniting chemically with the cementing matter in some cases, and in others forcing it away mechanically; in both instances leaving the particles not previously acted upon, more easily disturbed by a continuation of infiltration. In

---

* Rev. G. Woodley; View of the present State of the Scilly Islands, 1822.
† From a sketch by Mr. Holland.

those situations where the temperature descends sufficiently
low to produce frost, the mechanical action of the atmospheric water becomes much more considerable. Having entered
into the interstices of rocks when liquid, it assumes a greater
volume when it becomes solid from a sufficiently diminished
temperature, felt at greater or less depths in proportion to
the amount of decreased heat of the climates where the rocks
may be situate. Portions of rock are thus forced asunder,
and fine particles so separated, that the mere return of the
water to a liquid state, assisted by gravity, is sufficient to remove them. The large masses have their centres of gravity
often so altered relatively to rocks on which they rest, that
when no longer cemented by the ice, they fall from their situations to a lower level. The fall of rocks occasioned by
this means is common in lofty mountains, where considerable
heights are exposed to the alternations of frost and thaw.

By percolation through porous rocks the water attains strata
which are not so,—such as clays. The water thus stopped in
its course downwards, escapes as it best can to the sides of
hills and other situations, producing springs. At the places
where this discharge of water takes place, there is also a mechanical destruction of the parts through which the water de-
livers itself. Rocks are affected by this action of the water in
proportion to their composition; which, though not porous,
may still be acted on by the water. An argillaceous substra-
tum will get gradually moist at the surface, and in favourable
situations may become a wet clay. The stability of the mass
above will depend upon the relative position of the strata.
Thus in the wood-cut annexed, if
on the mountain *a*, water percolate through the porous strata *b*
to the impervious clay bed *c c*, the
surface of the latter would become
slippery, and the mass above be
launched into the valley *d*. Now this is precisely what happened in the case of the Ruffiberg in Switzerland. This
mountain, also known as the Rossberg, is 5196 feet above
the level of the sea, and rises opposite the well-known Righi.
Its upper part is composed of beds of a compound rock
formed from the debris of the Alps at a previous geological
epoch. These are to a certain extent porous, and the water
percolates through them to a clay stratum on which they
rest; the whole dipping at a considerable angle (about 45°).
The clay becoming soft by the action of the water, and the
thick superincumbent beds losing their support, the latter
were launched over the slippery and inclined surface beneath,
and the valley below was covered with their ruin.

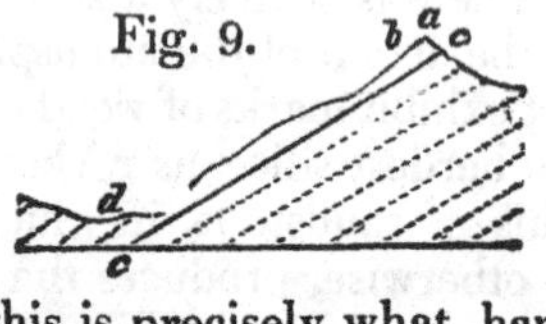

This slide took place on the 2nd of September 1806, and

Fig. 9.

covered a beautiful valley with rocks and mud. The villages of Goldau and Busingen, the hamlet of Huelloch, a large part of the village of Lowertz, the farms of Unter- and Ober-Rothen, and many scattered houses in the valley, were overwhelmed by the ruin. Goldau was crushed by masses of rocks, and Lowertz invaded by a stream of mud.

The torrent of rubbish and mud which rushed into the lake of Lowertz produced such a motion of the waters, that the village of Seven, situated at the other extremity, was inundated, and in great danger of being destroyed, two houses having been washed away. Live fish were found in the village of Steinen, thrown there by the flood. The lives lost were calculated from 800 to 900. Several travellers perished. It appears that there are traditionary accounts of former, though smaller, slides from the Rouffi or Rossberg*.

Large falls from mountains take place from the percolation of water to certain portions, which they mechanically loosen or chemically destroy without sliding over an inclined plane, as in the case of Rossberg, though the force of gravity still causes the fall. The Alps have afforded many examples of this fact, among others that of the great fall from the Diablerets in 1749.

Nothing is so common in mountainous regions as a talus of detritus brought to the foot of a cliff; this detritus composed of fragments of the rocks above, detached by decomposition from their surface, and brought down directly by their own gravity, or by the union of their own gravity and the force of surface-water, the latter derived from rains and the melting of snows. Avalanches of snow are great transporters of such fragments; and in the places where they fall there are always large accumulations of them, often borne from the greatest heights by the irresistible fury of the descending snow.

The under cliffs at Pinhay, near Lyme Regis, may be taken as an example of the destruction of a cliff by means of land springs, greater than that which is produced by the action of the sea in the same place.

Fig. 10.

---

* For a view of this fall, taken four days after the catastrophe,—see Sections and Views illustrative of Geological Phænomena, pl. 33.

a, gravel: *b,* chalk: *c,* green sand, both porous rocks
through which the water percolates to the clay bed *d,* composed of the lower part of the green sand beds *c* and the upper part of the lias beds *e*; being arrested in its progress
downwards, the water escapes by the easiest road, which is
that presented by the cliff originally formed by the sea. It
here gradually carries away the clay, first rendering it moist.
The chalk and green sand lose their support, give way, and
fall over into the sea. The lias *e* does not give way so fast
before the sea at the cliff *g,* as the superincumbent mass, affected by the land springs; therefore the latter retreats until
it has formed a great talus at *f*; but this talus tends constantly to move forward both by the destruction of the lias
cliff at *g,* and by the tendency of the land springs to loosen
its base, and to propel it into the sea. The chalk and green
sand containing hard substances, often of considerable size,
great protection is afforded to the cliff *g,* by their fall over
its top, the fury of the breakers being greatly spent upon
these masses.

Rivers.—These most frequently, though not always, take
their rise among hills and mountains, and are supplied either
by the melting of snows or glaciers, by the draining of rain-
waters, or by springs. They transport the detritus formed
either by the atmospheric agents, previously noticed, or by
themselves. The power of this transport depends upon their
velocities. Now, the velocity of a river current is greatest in
the centre, and least on the sides and bottom, being retarded
by friction; consequently, the transporting power of a river
is least where it comes in contact with the substances to be
transported. These substances are generally angular if detached from simple rocks for the first time, such as pieces of
limestone, granite, &c., and at the commencement present
great obstacles to transportation; for the velocity of a current
must be sufficient to move these angular fragments before
they can suffer attrition. Rocks composed of fragments which
have been previously rounded, such as conglomerates, will,
if they decompose easily, contribute ready-formed gravel to
the river, which might thus be able to carry them forward,
while its velocity was insufficient to transport angular fragments of equal weight. The transport of sandstones will
depend on their state of induration, and be easy where the
particles are slightly aggregated; difficult, when so compact
as to form angular fragments.

When the velocity of a river is sufficient to produce attrition
of the substances which it has either torn up, collected by un-
dermining its banks, or which have fallen into it, such substances gradually become more easy of transport, and would,

if the force of the current continued always the same, be forced forward until the river delivered itself into the sea; but as the velocity of a current greatly depends on the fall of the river from one level to another, the transport is regulated by the inclination of the river's bed. Now it is well known that this inclination varies materially, even in the same river; so that it may be able to carry detritus to one situation, but may be unable to transport it further, under ordinary circumstances, in consequence of diminished velocity. But this may be, and often is, so much increased further down, that its original transporting power may be, in a great measure, restored. It can now, however, only carry forward such detritus as it can receive or tear up in its course, and the pebbles which were left behind at the place of its first diminished velocity can only be brought within its power by floods, or, in other words, by extraordinary circumstances. As a general fact, it may be stated that rivers, where their courses are short and rapid, bear down pebbles into the seas near them, as is the case in the Maritime Alps, &c.; but that when their courses are long, and chan ed from rapid to slow, they deposit the pebbles where the force of the stream diminishes, and finally transport mere sand or mud to their mouths, as is the case with the Rhine, Rhone, Po, Danube, Ganges, &c.

It will follow that the form and weight of the detritus carried to the sea will depend upon the length and velocities of rivers, all other circumstances being the same.

If in its course the form of the land be such that lakes are produced, the detritus borne down by a river will be deposited in their beds, which have thus a tendency to be gradually filled up, the quality of the detritus depending on the velocity of the river. Such inequalities, producing small lakes, are common in mountain valleys, and have evidently been once much more so. The velocity of the stream issuing from the lake will greatly depend upon the fall of land over which it flows. The stream will endeavour to cut down the barrier which produced the lake, but if the velocity be trifling or the rocks hard, it will effect little; while if the stream be rapid or the rocks easily cut, it will traverse the natural bar, drain the lake, and permit the river to flow in an uninterrupted course. Should the lake, while it existed, have been partially filled up by the detritus from above, the river will cut through this also, and the part thus cut away will be transported to a lower level. The following diagram may assist the reader.

Fig. 11.

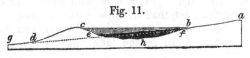

a b, course of a river flowing into the lake *b h c,* which is
filled with water to the level *b c,* the surplus falling over the
slope *c d,* and continuing its course in the direction *d g : e f,*
deposit of detritus derived from the river *a b,* at the bottom
of the lake *c h b : b d,* bed of the river formed by cutting
through the barrier *e c d,* and part of the detritus *e h f,* so as
to form a continuous course with *a b* on the one side and *d g*
on the other.

When lakes are large, such for instance as those of Geneva
and Constance, an immense lapse of time will be required to
fill them with detritus, so that, eventually, a continuous river
may traverse land occupying a space once filled by the water.
Lakes of this magnitude oppose great obstacles to the transport
of pebbles. The progress of a large proportion of detritus
from the Alps is arrested by lakes on their north and south
sides. Thus, on the north, the Rhine deposits its mountain
detritus in the lake of Constance, and the Rhone its transported pebbles and sands in the lake of Geneva. Between
these two great lakes, those of Zurich, Lucerne, &c., receive
the gravels of other Alpine rivers. On the south, the Lago
Maggiore receives the Alpine detritus of the Ticino; the lake
of Como, that of the Adda; and the lakes of Garda, &c.
perform the same office to other rivers. From these circumstances it will be evident, that the detritus of a large portion
of the Alps cannot travel, by the rivers, either into the ocean
or the Mediterranean. The Po receives the waters of a large
portion of the Alps, and carries sand and silt into the sea;
but the pebbles are arrested before it receives the Ticino,
which, though it transports rounded stones, does not bring
them directly from the Alps, but from its banks, after quitting
the Lago Maggiore, which banks contain the rounded Alpine
fragments of a previous epoch. The same with the Rhone
near Geneva, in which Alpine pebbles occur, and which could
not, in the actual state of things, be derived from the Alps,
because they would have been stopped in the lake of Geneva.
They are derived from its banks and bed immediately on
quitting the lake. Geological students, in examining river-
courses, should be very careful in distinguishing between
pebbles from the immediate banks of rivers, and those which
might be derived from a distance, but to the transport of
which, by the rivers, physical obstacles oppose themselves.
From a want of attention to this circumstance, many errors
have arisen. It has been considered that the mode in which
a river discharges itself into a lake, and pushes forward its
detritus, would be such that the deposit would assume a nearly
horizontal stratification. The angle of deposit must, however,
depend upon the depth of the lake and the quality of the

detritus discharged into it. Thus, if the detritus be composed of sand and mud, it will be propelled further into the body of the lake than if it consist of pebbles. Examples of both cases will be found in the lake of Geneva. The ordinary deposit from the Rhone is sandy and muddy, which sinks in clouds, from its greater specific gravity, beneath the clear waters of the lake; yet the initial velocity is sufficient to transport a part of it about a league and a quarter, for I found a portion of it at the depth of 90 fathoms, raising the bottom of the lake between St. Gingolph and Vevey *. This would give a very slight dip from the embouchure of the Rhone. Off the mouth of the Drance, a torrent rushing into the lake near Ripaille, the pebbles, forced down, must arrange themselves at a much more considerable angle; for 80 fathoms are obtained at a short distance from the shore. The same variations in dip will also be observed in the lake of Como, where the turbid waters of the Adda have deposited a considerable quantity of sand and mud, which slopes gradually at a gentle angle; while the torrent-borne detritus at Bellano, Mandello, Abbadia, and other places, arranges itself with a much more considerable inclination. It would seem to follow that the stratification of lake deposits derived from the land around them, would not be uniform, but would depend on local circumstances, rivers or torrents propelling detritus before them, which would be as various as the rocks they respectively traversed; each collection would have a mode of deposit of its own, independent of the others, and they would tend to approach, and finally to unite with each other.

The higher part of the lake of Como is nearly filled up by the detritus transported by the Adda and Mera†. The former has divided the lake into two; the smaller portion (known by the name of the Lago di Mesola,) being so shallow from the united deposit of the two rivers and some torrents, that aquatic plants grow through the water on the eastern part; while on the western, in which there is a greater depth, the process of filling up is hastened by means of stones, detached in such numbers, in particular seasons of the year, from the heights on that side, that a passage in a boat beneath the cliffs becomes exceedingly hazardous. Considering the many thousand revolutions of our planet round the sun, that must have taken place since the land assumed its present general form, we should expect to find the barriers even of considerable lakes cut through under favourable circumstances, and accordingly we do discover appearances which would seem to warrant this conclusion.

* For a map and sections of this lake, see Bibliothèque Universelle for 1819.
† See Sections and Views illustrative of Geological Phænomena, pl. 31.

It is by no means uncommon to find plains of greater or less extent, bounded on all sides by high land, and through which a principal river meanders, entering at one end by a valley, and passing out through a gorge at the other, augmented by tributary streams from the surrounding hills: sometimes these plains have no principal river passing through them; but many small streams, descending from the mountains, unite in the plain and pass out also through a gorge. In such cases the plain presents the appearance of a drained lake, such as we may suppose would be exhibited in many now existing, if passages for the waters were cut or broken through any part of the basins holding them. The gorge at Narni seems to have let out the waters of a lake supplied by the Nera, which now flows through the plain of Terni, the former bottom of the lake. The great fertile plain of Florence seems once to have been the bed of a lake, the drainage of which was effected by a cut through the high land that bounds it on the west. If this outlet were again closed, the waters of the Arno would again cover the plain and convert it into the bed of a lake. The period at which the break in the Jura was formed at the Fort de l'Ecluse, may perhaps be questionable; but if closed, it would stop the course of the Rhone, and convert the lake of Geneva into a much larger body of water.

These appearances are not confined to one part of the world; they would appear, from the descriptions of intelligent travellers, to exist very commonly. I have myself observed examples in Jamaica. The district named St. Thomas in the Vale is a marked one. Here we have low land bounded on all sides by hills, which would form the banks of a lake, were not the waters let out by the gorge through which the Rio Cobre flows.

It would therefore appear, though large lakes collect mountain detritus, which is distributed over a large surface, enveloping, probably, animal and vegetable remains, that the barriers of the lakes may be cut or broken through, and the rivers again act on a portion of the previous deposit.

The probability that many gorges originate from the cutting power of rivers discharged from lakes, is rendered stronger by examining those natural basins which are drained by subterraneous channels, and where gorges are not found. Thus Luidas Vale, in the island of Jamaica, is a district surrounded on all sides by high land, and would form a lake, were not the waters, derived from heavy tropical rains, carried off by sink-holes in the low grounds. A body of water, brought to turn the water-wheel of an estate's works, is swallowed up close to these works. A cavern, out of which water sometimes issues, near another estate, is speedily engulfed in a cave not far distant. In consequence of this escape of the waters, a gorge is not formed by means of a discharging river flowing over the lowest lip of the high land, as appears to have happened in the case of St. Thomas in the Vale, which adjoins Luidas Vale.

It is stated, "that a velocity of three inches per second at the bottom will just begin to work upon fine clay fit for pottery, and however firm and compact it may be, it will tear it up; yet no beds are more stable than clay when the velocities do not exceed this; for the water soon takes away the impalpable particles of the superficial clay, leaving the particles of sand sticking by their lower half in the rest of the clay, which they now protect, making a very permanent bottom, if the stream does not bring down gravel or coarse sand, which will rub off this very thin crust, and allow another layer to be worn off. A velocity of six inches will lift fine sand; eight inches will lift sand as coarse as linseed; twelve inches will sweep away fine gravel; twenty-four inches will roll along rounded pebbles an inch in diameter; and it requires three feet per second at the bottom to sweep along shivery angular stones of the size of an egg*."

The destructive power of rivers on solid rocks appears to act both chemically and mechanically. Chemically, by the affinity of water and of the air which it holds in solution for the various substances it encounters; and mechanically, by the friction of the detritus, independent of that of the water, upon the bottom and sides, but principally on the former. They may have thus effected a passage through the lake barriers previously noticed, and by these means they destroy the obstacles opposed to their courses. When a bank, a small hill, or the foot of a mountain, opposes their progress, they assail it, and form cliffs, the materials of which, if soft, fall into the stream, or make under cliffs, which are removed, and the work of destruction is slowly continued (Fig. 12. *a.*); or when

Fig. 12.

the cliff, thus formed, is of harder materials, blocks are accumulated in a talus at its base, and the cliff is secured, in a great measure, from attack, until this protecting mass is removed

* Encyclopædia Britannica, art. *River.*

(Fig. 12. *b*.). There is scarcely a river of any considerable length which does not afford examples of cliffs thus produced; very frequently they overhang flat or gently sloping land, on which the river has flowed while employed in cutting the cliff. It is not a little curious to trace, in countries where rivers wind considerably, the various obstacles which have determined the course of the stream, causing it to attack the original more or less rounded forms of the bases of moderately elevated hills.

Rivers appear to be constantly striving to arrange their beds in such a manner that they should suffer the least resistance in their courses, cutting down obstacles and filling up depressions which checked them. But the constant addition of new detritus from the neighbouring highlands embarrasses this operation, causing accumulations in one situation which direct the waters in another. Thus the fall of a considerable quantity of rocks on one side will throw the stream upon the opposite bank, which might previously have been little attacked. This again forces the current in a direction that it did not previously follow; the bottom becomes torn up by the new line of the principal stream, and the effect of such a fall is felt far down the course of the river. In consequence of this endeavour to avoid a new obstacle, continual changes in a river's bed take place, as also from the destruction of an old obstacle, which permits a new course in a direction that the river has been striving to follow.

D'Aubuisson observed two rocks at the falls of the Rhine, near Schafhausen, isolated at the head of the precipice over which the waters leap; these were observed corroded at their bases by the action of the pent-up current between them. By gradually diminishing their support, the rocks would finally be forced over the cataract, and the waters, having overcome this obstacle, would fall in a different manner on the bottom beneath, producing a different effect from that which they had previously caused.

As all rivers must vary in their cutting power, according to velocity, volume of water, and amount and quality of detritus in the act of transport, it becomes exceedingly difficult to generalize on the subject; but as barriers of even the hardest rocks have suffered, and as the destructive power of the same rivers on the same obstacles is so exceedingly small as to be scarcely perceptible during the life of man, it seems fair to infer that this also tends to confirm the opinion of the great age of the present general state of the world.

Mr. Lyell indeed produces, as an example of the comparatively quick cutting power of a river, a gorge in a lava-current at the foot of Etna, formed by the erosion of the Simeto. The

lava is considered modern, and Gamellaro is cited as supposing it thrown out in 1603. The lava is described as not
porous or scoriaceous, but as a compact homogeneous rock,
lighter than common basalt, and containing crystals of olivine
and glassy felspar. Though there are two waterfalls, each
about six feet, the general fall of the river's bed is stated as
not considerable. The gorge is cut in some places to the
depth of forty or fifty feet, and its breadth varies from fifty to
several hundred feet*. It is therefore inferred that this is a
good example of the speedy formation of gorges by running
water; and this inference cannot be denied, if the date of the
lava-current be correctly ascertained. It may be remarked
that the present fall in the bed of the Simeto does not give
that of the river during the great cutting operation. It must
once have occupied a different level, or else the gorge could
not have been commenced; and there must always have been
a rapid fall, or, in other words, a cascade into the low land off
the lava, equal to the height of the lava-current; the waters
being raised to the top of the lava, at this place, by the formation of a lake behind, produced by the bar of lava. It would
therefore follow, that the gorge in the lava-current has been
principally formed by the cutting back of rapids or a cataract.
Though this circumstance would facilitate the progress of
destruction, and render it less remarkable than if the Simeto,
with its present fall, had cut the gorge, it yet leaves this a
good example of a ravine formed in hard rock during the
course of two centuries, it being always understood that no
doubt exists of the period when the lava-current was ejected,
and crossed the previously existing valley.

The dates obtained by the well-known examples of the
Auvergne rivers are only relative; but they are sufficient to
show that a valley existed, through which a river kept its
course, conveying detritus in the usual way, and that the progress of the river was barred by a lava-current (as in the instance just cited), which descending from a neighbouring volcano traversed the valley, and formed a lake. This lake, when
full, discharged itself over the lower lip of its basin, which
happened to be in the direction of the valley, and over the
lava-current. This, by erosion, is cut down, not only to its
original bed, but through it into the rock which constituted
the bottom of the original valley.

Notwithstanding appearances, there are numerous gorges or
ravines through which rivers flow, which could not have been
cut out by them, at least during the existence of the present
general disposition of land; for the relative levels are such,

* Principles of Geology.

that the rivers must be supposed to have run over land of much
greater elevation towards their embouchures than they flowed
over from their sources; in other words, such rivers must be
supposed to have run up hill, if they be considered the agents
which have formed these gorges. As a striking example of
this fact, we may cite the course of the Meuse previous to, and
during its traverse through the Ardennes. M. Boblaye informs us, that previous to its passage through these mountains, the Meuse is only separated from the great basin of the
Seine by hills or low *cols*, not more than thirty or forty yards
above the present bed of the river; while the Ardennes,
through which it actually passes, rise to the height of several
hundred feet above the same level. Now, if all rivers had
really cut the beds or valleys through which they actually
flow, the Meuse must have run up hill, and have cut a narrow
channel about three hundred yards deep; while nothing prevented its flowing in the opposite direction into the Paris basin,
when it had effected a rise of not much more than a tenth part
of that height*.

At Clifton, near Bristol, we have also a striking example
of the same fact. The Avon here runs through a gorge or
ravine, which if closed would form a lake behind it; but this
lake would exert no action on the range of hill through which
the present channel passes; on the contrary, the lowest lip of
the basin, and consequently the drainage, would be found in
the direction of Nailsea, to the sea beyond which the Avon
would continue its course from Bristol. The real rise of land
between high water at Bristol and the sea beyond Nailsea is
trifling, and is bounded on the north by the high ridge through
which the Avon now finds its passage to the Severn.

M. v. Dechen remarks that these facts can nowhere be
better observed than at the confluence of the Nahe and Rhine
near Bingen. The Nahe flows from Kreutznach to Münster
and Büdesheim through a broad valley, bounded on the right
by low limestone hills. It then traverses the slate mountains
through a narrow defile, while the broad valley continues
south-east from the Rochusberg by Sponsheim and Okenheim
to the great valley of the Rhine at Gaulsheim. Nothing can
be more evident than that the Nahe was unable to cut this
defile between the Rochusberg and the heights of Weiler,
while such a broad and deep channel into the Rhine presented
itself so immediately to the passage of the waters†.

Other examples might easily be cited, but these are sufficient to point out the fact. There are many gorges through
which rivers pass, the formation of which remains question-

---

* Boblaye, Ann. des Sci. Nat. t. xvii. p. 37.
† Von Dechen, German Transl. of Manual.

able from our ignorance of the relative levels in their vicinity, and thus it becomes difficult to assign them any particular origin. They may be either due to the same causes which have produced the ravines of the Meuse in the Ardennes, and of the Avon near Bristol, or to the cutting power of rivers discharging the surplus waters of lakes. Under this head may be enumerated the celebrated Vale of Tempe, in Thessaly; the tortuous course of the Wye, between Monmouth and Chepstow; the famous Rheingau; the ravine by which the Potomack traverses the Blue Mountains in the United States; the Gates of Iron, through which the Danube escapes into Wallachia; &c.

The Falls of Niagara may be adduced as an example of a river discharging the surplus waters of a lake, and now cutting back a gorge to that lake, which may eventually be drained by it. This celebrated cataract is situated between the lakes Ontario and Erie. For some distance above the embouchure of the river into the former, the country is flat, and apparently alluvial, when suddenly a plateau rises above it and continues to lake Erie. Over this plateau the surplus waters of the latter lake have taken their course, and appear to have originally fallen over the face of the plateau fronting lake Ontario. By degrees they have cut back their passage about seven miles, leaving about eighteen more to be worn away by future ages. When this shall have been accomplished, the gorge or ravine will be similar to those previously noticed. The manner in which the river cuts its passage is singular, and perhaps somewhat different from what, at first sight, might have been expected. It will be best explained by the following diagram.

a b, original level of the plateau: *a h,* river flowing over the plateau, and falling over to the abyss *c,* forming the cascade *h c,* after which the waters take their course in the direction *c g :* *d,* beds of limestone resting on beds of shale *e,* both being surmounted, in the neighbouring flat country, by a

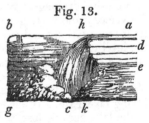

Fig. 13.

mass of transported substances, varying from ten to one hundred and forty feet in depth, and containing large blocks. The rush of waters from *h* to *c* occasions violent gusts of wind, charged with water, to be driven against the shale *e* at *f.* The continued action of these water-charged whirlwinds displaces the shale, and throws it down in a talus at *k.* From the removal of this shale, the superincumbent limestone loses its support, falls from the combined gravity of itself and the water above, is dashed into the abyss beneath, and thus the falls are

cut back so rapidly that they have considerably receded within
the memory of man. The same operations are again renewed,
and again the same results follow. So that unless some extraordinary circumstance should arrest their retreat, these falls
will discharge the waters of lake Erie; but not suddenly, as is
sometimes supposed, so as to produce a violent deluge over
the lower country further down the river, but much more
gradually; for the lake waters will only be lowered in proportion to the depth of the draining channel, as may be illustrated
by the annexed wood-cut, in which *a b* represents the level of

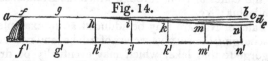

Fig. 14.

the lake and of the plateau, rising but little above it: *h e*, the
slope (exaggerated) of the lake bed from *h*, the spot where the
surplus waters are delivered over the plateau, *f' n'* the level of
the river below the falls. Supposing *g g'* to represent the falls
which have approached the lake by gradually cutting back the
channel from *f f'* to *g g'*, it will appear that the same kind of
retreat may be effected to *h h'* without discharging more water
than now passes down the river. But the falls being once at
h h', the retreat of every succeeding yard will occasion more
water to pass over them, by draining the waters of the lake
down to the point which now becomes its lowest lip; so that
when the falls have cut their way back to *i i'*, the surface of the
lake will sink to the horizontal line *i c*, and the mass of water
above the new level will have passed over the falls in addition
to the usual drainage. Such an addition must add greatly to
the velocity and cutting power of the falls, which will now retreat more rapidly and effect their passage to *k k'*, reducing
the level to *k d* in less time than it reduced it from *a b* to *i c*.
After a certain time the water forced over the falls would become less, because the superficies of the lake would be smaller,
if it were not for the diminished evaporation of the waters supplying lake Erie. It has been remarked * that as the surface
of the lake became less, the waters carried off by evaporation
would also decrease, and that consequently a greater body of
water would be carried over the falls of Niagara, thus accele-
rating their retrograde movement. Such would no doubt be
the effect of the union of the streams, now supplying lake Erie,
into one considerable river, without being exposed to that
great evaporation which they now suffer in the lake; but this
would by no means assist in producing a debacle or deluge

* Lyell, Principles of Geology, 2nd ed. vol. i. p. 209.

from the supposed sudden discharge of the waters of lake
Erie. The addition to the waters passing over the falls of
Niagara could only take place as the superficies of the lake
became less from the gradual drainage of lake Erie above noticed, and consequently the additional force could only come
into action and increase in intensity in proportion as the
possibility of a sudden discharge of waters would become less.
Finally, from the operation of these various causes, a river,
fed by a prolongation of the streams now discharging themselves into lake Erie, would traverse the ancient bed of the
lake, and run through the ravine, cut by the falls of Niagara,
into lake Ontario.

The waters of a lake with a rocky barrier can only be suddenly let out, and produce a debacle, when the hard barrier
separating it from the land at a lower level presents a perpendicular face to the whole depth of the lake, which, even then,
must be suddenly thrown down, in its whole height, to produce the effect required. Such rocky barriers must be exceedingly rare; and it must be still more rare, that where
they existed they were not cut down, to a certain extent, by
degrees. The common character of lakes, as respects the
inclination from their bottoms to the discharging outlet, varies
materially, but in general the slope is very gradual, particularly in lakes of considerable magnitude.

The often cited debacle caused by the bursting of a lake in
the Val de Bagnes was produced from a very different state
of things from that attending the drainage of a lake existing
in a depression of land, with a rocky barrier.

The Val de Bagnes, in the Vallais, is drained by the
Dranse, which, when unobstructed, is joined by the waters
from the valley of Entremont, leading to the Grand St. Bernard, and runs into the great valley of the Rhone, near
Martigny. In a part of the valley near the bridge of Mauvoisin, the channel is precipitous and much contracted. Mont
Pleureur and Mont Getroz rise near this spot on the north,
and Mont Mauvoisin on the south. Between the two former
there is a ravine communicating with the Val de Bagnes, having a considerable glacier at its upper extremity. Through
this ravine blocks of ice and avalanches of snow descend into
the Val de Bagnes, and more or less obstruct the channel
of the Dranse, which is able, under ordinary circumstances,
to remove the greater part, if not the whole, of such ob-
structions. When however the blocks of ice are numerous, and the avalanches are heavy, the force of the torrent is unable to contend with them, and they accumulate.
" For several years previous to 1818," says M. Escher de
la Linth, " the progress of the Dranse had begun to be
obstructed by the blocks of ice and avalanches of snow that

descended from the glacier of Getroz; and as soon as this accumulation was able to resist the heats of summer, it acquired new magnitude during every succeeding winter, till it became an homogeneous mass of ice of a conical form. The waters of the Dranse, however, still found their way beneath the icy cone till the month of April, when they were observed to have been dammed up, and to have formed a lake about half a league in length*.

The danger that threatened now became apparent, and accordingly the gradual drainage of the lake was attempted by means of a gallery through the ice. This reduced its contents from about 800,000,000 cubic feet to 530,000,000 cubic feet. Finally, the discharging waters attacked the debris at the foot of Mauvoisin, and excavating a passage between the rocks and the ice, rushed furiously out, carrying houses, trees, large blocks of rock, &c. before it. Escaping from the narrow valley it desolated a large portion of Martigny †, and passed with gradually diminished velocity down the Rhone into the lake of Geneva. As might be expected, the velocity of the torrent varied materially in different parts of its course. M. Escher de la Linth calculates that from the glacier to Le Chable, a distance of 70,000 feet, the velocity was 33 feet per second; from Le Chable to Martigny, 60,000 feet, at the rate of 18 feet; from Martigny to St. Maurice, 30,000 feet, at 11½; and from St. Maurice to the lake of Geneva, 80,000 feet, with a diminished velocity of 6 feet per second ‡. The lake was drained in half an hour.

As has been noticed by Mr. Yates §, lakes are produced in mountainous countries by the fall of rocky masses across narrow valleys, the waters being thus arrested in their progress down such valleys. Mr. Yates cites the Öschenen-see in the canton of Berne, as a good example of lakes thus formed ‡; and M. De Gasparin mentions a recent example (November, 1829) of the formation of such a lake, in the department of the Drome, by the fall of a mountain mass across the river Oule near Lamothe Chalancon. The lake produced in the latter case was 500 or 600 yards long, 60 broad, and 3 or 4 yards deep‖. It will be obvious that the possibility of the sudden discharge of waters, thus pent up, will depend upon the nature of the materials composing the dam or barrier: if

* Edin. Phil. Journ. vol. i. p. 188.

† Among the debris transported to Martigny were many trees, resting upright on their roots, the attached gravel and soil having kept them in a position with the branches upwards.

‡ Edin. Phil. Journ. vol. i. p. 191.

§ Yates, Remarks on Alluvial Deposits, Edin. New Phil. Journal, July, 1831.

‖ De Gasparin, Ann. des Sci. Nat., Avril 1830.

these be of such a form, quality, and magnitude, that the body of water is unable to overcome their resistance, and that they do not give way before the cutting power of the surplus waters discharged over the lower lip of the dam, the barrier will remain, adn be clothed with wood and other vegetation, as that of the Öschenen-see now is. Should, however, the dam be composed of soft materials, which might either suddenly give way before the force of the pent-up water, or be rapidly cut down when a discharge of the surplus water took place, a debacle somewhat analogous to that of the Val de Bagnes might be produced, the effects of which would depend upon the body of waters let out, the suddenness with which this was accomplished, and other obvious circumstances *.

Lakes may be suddenly drained, if but a thin perpendicular partition divides them from an inferior level; for this barrier maybe rendered soft by the percolation of water, and suddenly give way; but such cases must be of very rare occurrence; and the lakes are not likely to be of such magnitude as to cause appearances, by their sudden discharge, that may be equal to those producible by the passage of a more general mass of waters over land.

Mr. Strangways notices the bursting, or sudden considerable drainage, of the lake Souvando, on the north of St. Petersburg. Previous to 1818 this lake was separated from that of Ladoga by the little isthmus of Taipala. The lake discharged its waters into the Voxa at Keognemy, and so passed into the Ladoga at Kexholm. In the spring of 1818, the water broke down the isthmus and changed the direction of the discharging waters, by presenting a lower lip in another direction. The water has been lowered considerably, and continues to run through its new channel into the lake of Ladoga, having deserted the Voxa †.

The same author describes the falls or rapids of Imatra, about six wersts below the point where the surplus waters of the lake Saima first drain off by the Voxa. This river suddenly contracts itself above the rapids, over which it runs, with great noise and impetuosity, through a gorge that it has evidently cut for itself. According to Mr. Strangways, we

---

* The same observations apply to those cases also noticed by Mr. Yates, in the memoir above cited, where from various circumstances a torrent may bring with it from a transverse or tributary valley such a mass of detritus into the main valley, as to arrest the progress of water flowing down it. In these cases, however, the barrier, from the nature of things, is not likely to be permanent, but, on the contrary, to be removeable with greater or less rapidity by the main river or torrent.

† Strangways, Geol. Trans. First Series, vol. v. p. 344.

may consider the water to have originally passed over a platform between two ranges of hills, forming the bottom of a
valley. The platform is composed of gneiss, in very highly
inclined strata ; and into this the river has cut a channel.
" The surface of this platform is apparently now about fifty
feet above the level of the water, at the lower extremity of the
rapids. Its surface is in many parts quite bare and deeply
channelled in a direction parallel to the river. It is covered
with heaps of pebbles and boulders of great size, some of which
are hollowed and scooped into the most fanciful shapes. One
of the largest of the blocks now left dry, standing nearly in the
middle of the elevated platform, is worn through perpendicularly with a cylindrical hole *." It is stated that the level of
the lake Saima and its discharging river fall gradually.

Freshets.—These take place, more or less, in all rivers,
greatly augmenting their velocities and transporting power,
carrying forward substances that could not have been moved
under ordinary circumstances. They are also important, as
they surprise terrestrial animals in low situations, hurry them
on with trees and other matters into the sea, where they may
be entombed entire with estuary and marine animals in mud
and silt.

It has been observed that, during freshets, a river tends
chiefly to widen its bed, " without greatly deepening it : for the
aquatic plants, which have been growing and thriving during the
peaceable state of the river, are now laid along, but not swept
away, by the freshes, and protect the bottom from their attacks ; and the stones and gravel, which must have been left
bare in a course of years, working on the soil, will also collect in the bottom, and greatly augment its power of resist-
ance† ." During these freshes, low lands on the sides of the
river are frequently under water, and a deposit takes place ;
but notwithstanding all checks, a large quantity of detritus
passes onwards to the sea.

We should be careful, in our estimates of the effects of a
flood in a cultivated country, not only to separate the loss of
lives and the destruction of property, which may affect the
feelings, from the real physical change produced in the
country; but also to remember, that the works of man greatly
aid the destructive power of a flood. Instead of a body of
water rushing into a plain, where from its diffusion over a
more considerable space its velocity and transporting power
are both diminished, all cross hedges and bridges, though
they may check the waters for the moment, are the means of

---

* Strangways, Geol. Trans. First Series, vol. v. p. 341.
† Encyc. Brit. art. *River.*

producing innumerable debacles, when they give way before the pressure exerted upon them. Suppose a bridge arrests the progress of the flood downwards, and, as very frequently happens on small plains, a causeway connects the bridge with the hills on either side, the waters will accumulate, and will finally burst through the least resisting part of the barrier, which will most probably be the bridge. Having once found a vent, the pent-up waters will issue forth with a velocity proportioned to the difference of the level and the mass of water, and a debacle will be produced, whose transporting power will be much greater than that of the general force of the flood if no such barrier had existed. It must also be recollected, that man, by his contrivances of ditches and drains, prevents the rain-water from remaining the time that it would otherwise do on the slopes of hills, conducting it as he does by numerous free channels into the valleys below; so that, in a given time, a much greater body of water is collected than could happen in an uncultivated country. He moreover, by dams and banks, often confines a body of river water within narrower channels than it would naturally take; and thus its dispersion over a larger surface being prevented during a freshet, its ordinary velocity is greatly increased, and with this its transporting power.

Glaciers.—These are large bodies of ice or indurated snow, formed upon land in the cold regions of the atmosphere, which descend into the valleys of mountainous countries; thus frequently presenting the singular appearance of desolation amid fertility, of ice amid vegetation. The levels to which glaciers descend depend greatly on the latitude of the place. Thus, in the arctic regions, where the line of perpetual snow approaches very nearly to the level of the sea, glaciers are produced in lower hills than could be the case in the Alps, where the line of perpetual congelation is much more elevated. So again in the Himalaya range the line of perpetual congelation being higher than in the Alps, the glaciers form at higher levels. Glaciers are instruments of the degradation of land, inasmuch as they drive before them and transport such substances as they may have the power to move. In front of glaciers there is usually a pile of rubbish composed of pieces of rock, earth, and trees, which they have forced forward, known in Switzerland by the name of *moraine.* If there be a line of moraine some distance from the front of the glacier, it is considered that the glacier has retreated to the amount of that distance; but if there be no other than that which the glacier immediately drives before it, it is considered to be on the increase. Glaciers assist the degradation of land by transporting blocks, often of very large dimensions, into

F

lower regions than they could otherwise attain in so short a
time. Many glaciers, particularly where they pass beneath
precipices, are charged with fallen rubbish, which, as the ice
constantly advances, are carried on with it; and should a precipice occur in the front of the moving mass, they are hurled
over with it into the ravines beneath. Such falls are common
in the high regions of the Alps, producing, with the rents
suddenly formed in the glacier itself, the few interruptions to
the dead silence which reigns in those lofty and wild regions.
The velocity with which a glacier advances depends on the
angle that it makes with the horizon, of course increasing with
the steepness of the declivity.

A ladder, left by M. de Saussure at the upper end of a
glacier, when he first visited the Col du Géant, has lately
been discovered in the Mer de Glace, the continuation of the
same glacier, and nearly opposite the aiguille named Le Moine.
It must therefore have advanced about three leagues since the
year 1787 *. From some experiments by Chamonix guides,
mentioned by Capt. Sherwill, we learn that this rapid progress ceases, as might have been expected, where the declivity becomes less in the Mer de Glace itself; for it was there
found that a block of rock advanced about two hundred yards
in a twelvemonth†. No better proofs could be afforded of
the advance of a glacier, the amount of which corresponds
with the declivity. It hence appears to follow, that as the
declivity remains nearly the same for a long period, the advance or retreat of the lower part of a glacier will correspond
with the local variations in climate, which shall produce more
or less ice in the higher, or destroy more or less of the glacier
in the lower regions.

Almost all glacier waters are charged with detritus, the
larger portions of which are deposited near the ice, but the
lighter particles are transported to considerable distances; as
is, for example, the case with the Arve, which having deposited its heavier burden in the valley of Chamonix, carries
the lighter parts to its junction with the Rhone, near Geneva.
Not unfrequently the turbid glacier waters are carried on,
and deposit the detritus in some lake, as is the case with the
Rhone, which transports silt, mud, and occasionally pebbles,
into the lake of Geneva. The grinding of the glacier against
the bottom over which it passes, may perhaps mechanically
assist in the work of destruction.

In the northern regions glaciers have sometimes such a
short distance to pass over before they reach the sea, that
they project into it, as has been observed by northern navi-

---

gators. The mass so forced into the sea will have a constant
tendency to float, from its inferior specific gravity, and therefore when detached by any force from the glacier behind, it
will be carried away;—thus, forming those icebergs, so well
known and so dangerous, in the Northern Atlantic Ocean.

Delivery of Detritus into the Sea.

We have seen above, that from the action of the atmosphere, the melting of snows and glaciers, landslips, and the
cutting power of rivers, considerable destruction of dry land
is effected. Local circumstances arrest a considerable portion
of this detritus; lakes are filled up, and again cut through;
low lands are occasionally flooded, and considerable deposits
left upon them; the velocity of the streams diminishes, and
with it the power of transport; so that, as previously observed,
rivers when short and rapid may carry a large portion of their
detritus forward, while, when long, they leave a considerable
part of it in their courses. In favourable situations, such as
in plains, they will raise their beds, if confined within bounds,
that do not either permit a change of course, or a deposit in
a new channel. This fact is well observed in Italy, where
many plains have been under cultivation for a long period,
during which it was always necessary to restrain the rivers
within artificial banks, to prevent their range over the cultivated land, which would otherwise have been devastated by
them; so that, in travelling in that country, the road frequently passes up hill, over high artificial ridges, upon which
the rivers hold their course at a higher level than that of the
surrounding country. These artificial ridges are particularly
striking on the little plain of Nice, which has been under cultivation since the country was settled from the Phocæan co-
lony of Marseilles. The height of the latter elevated river-
courses is not only due to their antiquity, but to the loose
nature of the conglomerate hills behind, which permits an
easy transport of the pebbles.

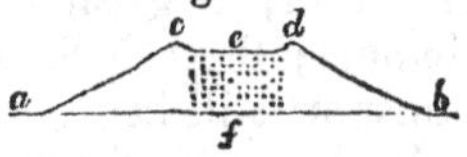

The annexed diagram will illustrate
this fact: *a b*, the level of the country,
now cultivated, upon which the artificial banks have been gradually raised

Fig. 15.

to *c d*, in order to protect the cultivated lands from being invaded by the detritus of the river or torrent *e*, which is thus
accumulated from *f* to *e*. There is a very general system of
endeavouring to check this accumulation, and consequent rise
of bed, by throwing, when the waters are low, the transported
detritus out of the bed *e*, upon the protecting banks *c d*.

The Po affords a well-known example of this rise of bed,

so that it becomes higher than the houses in the city of Fer-
rara. In Holland also the same phænomenon is observable,
though not on so great a scale; and may always be expected
where artificial banks prevent detritus-bearing rivers from
changing their beds on plains.

Although rivers, in certain situations, raise their beds, in
others they deepen them. This arises from two or more
streams uniting into one river, when the water does not expose
a surface equal to the two previous surfaces, but one very
considerably less, the action of the united waters being to
deepen their channel; so that even with a diminished general
inclination of the bed, the velocity continues the same, or is
even increased.

This deepening of beds by the union of rivers is well exhibited by the following facts observed in the Po :—

" About the year 1600, the waters of the Panaro, a very considerable river, were added to the Po Grande ; and although
it brings along with it in its freshes a vast quantity of sand
and mud, it has greatly deepened the whole Tronco di Venezia
from the confluence to the sea. This point was clearly ascertained by Manfredi about the year 1720, when the inhabitants
of the valleys adjacent were alarmed by the project of bringing in the waters of the Rheno, which then ran through the
Ferrarese. Their fears were overcome, and the Po Grande
continues to deepen its channel every day with a prodigious
advantage to the navigations; and there are several extensive
marshes which now drain off by it, after having been for ages
under water : and it is to be particularly remarked, that the
Rheno is the foulest river in its freshes of any river in that
country *."

It might be supposed that all rivers would, by means of
freshes, propel pebbles into the sea. They certainly accomplish by these means a greater transport than could be effected
in the same channels under ordinary circumstances; but
during freshes rivers can only be considered as of greater
magnitude, and are therefore still subject to the general laws
of rivers; a greater body of water tending to deepen the
channel; the velocities, inclinations of beds, and the power of
transport still being in proportion to each other.

In the beds of torrents, dry, or nearly dry, for the greater
part of the year, we see examples of the deepening of river
beds in proportion to the volume of water which passes through
them, to the inclination of the beds, and to the resisting power
of the bottoms and sides. The transport of detritus will also
be observed greater or less in proportion to these circum-

* Encyc. Brit., art. *River.*

stances : the finer particles being more easy of transport, there
are few rivers which, during freshes, do not convey a great
quantity of such detritus into the sea : other kinds of detritus
will be also transported, if levels permit ; if not, they remain
in the interior.   Consequently, according to the circumstances
already noticed will be the nature of the detritus conveyed to
the mouths of rivers.   But as circumstances vary in the same
river, a deposit of such detritus in these situations also varies,
and there may be alternations of clay or marl, and of sand or
gravel.

If the mouths of rivers be tidal, the river detritus is com-
mitted to the charge of the estuary tides, and is dealt with
according to the laws by which these are governed.   If they
be tideless, the whole mass of transported matter will be propelled without check into the seas at the embouchures.   Between the extremes of great resistance and non-resistance the
variations are so great and depend so much on local circumstances, as to be of exceedingly difficult classification.   The
principal variations are produced by the difference in the
volume of the discharging rivers, their velocities, and the
quantity and quality of the substances they may transport.
As a general fact, however, it may be stated that rivers tend
to form deltas in tideless, or nearly tideless, seas, or where
they can overcome the resistance of tides, currents, and the
destructive action of the breakers ; thus increasing the land
by their deposit, and splitting into several channels ; the superficial increase being in proportion to the depth of water
into which the rivers discharge themselves.

In calculations of the advance of deltas, care has not always
been taken to show the general depth of water into which
they may have been protruded ; so that a less quantity of
transported detritus might expose a larger surface when
thrown on a shallow bottom, than a larger quantity in deeper
water.

The Nile, Danube, Volga, Rhone, and Po, afford us examples of deltas thrown forward into seas, which may, in
common terms, be called tideless.   As the Nile receives little
atmospheric water from Egypt, on which rain seldom falls,
the detritus which it brings down must be principally derived
from above.   This river begins to rise in June, attains its
maximum of height—namely, twenty-four or twenty-eight
feet—in August, and then falls till the next May.   During a
succession of ages, the Nile has transported a great mass of
detritus into the Mediterranean, which has accumulated in a
delta at the mouth, and is constantly on the increase.   It has
been calculated, that, as the sea deepens at the rate of a fathom
in a mile, and supposing that the deposit is the same as in the

Thebais, the addition would amount to a mile and a quarter since the time of Herodotus. According to Girard, the Nile has raised the surface of Upper Egypt about six feet four inches since the commencement of the Christian æra. The quantity of water discharged per annum by this river is estimated at 250 times that of the Thames *. The delta is traversed by two main streams, which separate a few miles below Cairo; one descending to Rosetta, the other to Damietta. The present position of the latter city has led to very exaggerated ideas respecting the rapid increase of this delta. It was supposed that the present town was the same with that which during the first crusade of St. Louis was situated on the sea. Now, as Damietta is two leagues from the sea, it was calculated that this distance had been produced by deposits from the Nile within about 600 years. It now, however, appears, from the labours of M. Renaud, that after the departure of St. Louis, the Egyptian Emirs, wishing to prevent a new invasion on the same side, destroyed Damietta, and founded a new city in the interior, the present Damietta†. From the effect of the waves and currents, banks are thrown up on the outer edge of the delta, forming lakes, of which those of Menzalen, Bourlos, and that behind Alexandria, are the largest.

The delta of the Po advances at a rapid rate, in consequence of the shallow sea into which it is protruded. We are indebted to M. Prony for a very interesting collection of facts, which authorize him to conclude, " First, that at some ancient period, the precise date of which cannot now be ascertained, the waves of the Adriatic washed the walls of Adria. Secondly, that in the twelfth century, before a passage had been opened for the Po at Ficarrolo, on its left or northern bank, the shore had already been removed to the distance of nine or ten thousand metres from Adria. Thirdly, that the extremities of the promontories formed by the two principal branches of the Po, before the excavation of the Taglio di Porto Viro, had extended by the year 1600, or in four hundred years, to a medium distance of 18,500 metres beyond Adria; giving from the year 1200 an average yearly increase of the alluvial land of 25 metres. Fourthly, that the extreme point of the present single promontory, formed by the alluvions of the existing branches, is advanced to between thirty-two and thirty-three thousand metres beyond Adria; whence the average yearly progress is about seventy metres during the last two hundred years, being a greatly more rapid proportion than in former times‡."

* Supplement to Encyc. Brit., art. *Physical Geography.*
† Extraits des Historiens Arabes relatifs aux Guerres des Croisades.
‡ Prony, as quoted by Cuvier. Dis. sur les Rev. du Globe.

The Mississippi, the great drain of so large a portion of North America, may be considered as delivering its waters into a nearly tideless sea. Its delta is very considerable, and little raised above the level of the ocean. During the greatest heights of flood, the fall of the river from New Orleans to the sea, a distance of about one hundred miles, has been calculated at only one inch and a half in a mile. When the waters are low, the fall is scarcely perceptible, the level of the sea being then nearly that of the river at New Orleans*.

This river affords a good example of a flood being higher at a distance from the embouchure of a river than at the mouth itself; for the rise of water, during the great freshets, is fifty feet at Natchez, three hundred and eighty miles inland, while at New Orleans it is only thirteen†.

Darby has furnished us with a mass of information respecting a large portion of the Mississippi's course, and of its delta, from whence very important geological information may be obtained‡. It would appear that the Atchafalaya, which now, at a distance of about two hundred and fifty miles from the sea, conducts a large part of the Mississippi's waters into the Gulf of Mexico, did not always form a drain from that river, but that it once constituted a continuation of the Red River, which now flows into the Mississippi. During the autumns of 1807, 1808, 1809, Mr. Darby had frequent opportunities of examining the bed of the Atchafalaya, the waters in which were then at a low state. He found that " the upper stratum invariably consisted of a blueish clay common to the banks of the Mississippi. This is usually followed by a stratum of red ochreous earth peculiar to the Red River, under which the blue clay of the Mississippi was again to be perceived§." From this we may infer, not only that the Red River flowed through the channel of the Atchafalaya, previous to the present course of the Mississippi, but that the latter river preceded the former, and that there have been alternations.

From the form of the Mississippi, where the Atchafalaya detaches itself, an immense quantity of trees brought down by the former are thrown into the latter. About fifty-two years since, these trees began to accumulate and form the "raft." "This mass of timber rises and falls with the water in the river, and at all seasons maintains an equal elevation above the surface. The tales that have been narrated respecting this phænomenon, its having timber of large size, and in many places being compact enough for horses to pass, are entirely

* Hall's Travels in North America.   † *Ibid.*
‡ Darby's Geographical Description of the State of Louisiana.
§ *Ibid.*

void of truth. The raft is, in fact, subject to continual change
of position, which, superadding its recent formation, renders either the solidity of its structure, or the growth of
large timber, impossible. Some small willows and other
aquatic bushes are frequently seen among the trees, but are too
often destroyed by the shifting of the mass to acquire any considerable size. In the fall season, when the waters are low,
the surface of the raft is perfectly covered by the most beautiful flora, whose varied dyes, and the hum of the honey-bee,
seen in thousands, compensate to the traveller for the deep
silence and lonely appearance of nature at this remote spot*.

   Mr. Darby estimated the cubic contents of the raft, from
observations made in 1808, at 286,784,000 cubic feet, considering the breadth of the river = 220 yards, the length of the
raft = 10 miles, and the depth = 8 feet. The distance between the extremities of the raft was actually more than twenty
miles; but, as the whole distance was not filled up by timber,
he assumed ten miles as near the truth.

   Rafts of this description, but of less size, occur in other
parts of the Mississippi or its great tributaries. The banks are
destroyed by the currents, and large collections of trees are
suddenly hurled into the stream. Captain Hall was present
when a large mass of earth, loaded with trees, suddenly fell
into the Missouri, and a larger mass had been detached a
short time previous to his arrival†.

   There are few rivers whose course is more instructive than
the Mississippi, as man has not yet effected many changes on
its banks; and we thus contemplate great natural operations,
such as cannot be so well observed in those which have been
more or less under his dominion for a series of ages. Its course
is so long, and through such various climates, that the freshets
or floods produced in one tributary are over before they commence in another: hence arise those frequent deposits of detritus at the mouths of the tributaries. These latter have their
waters forced back, and rendered, to a certain distance, stagnant by the rush of the flood across their embouchures, and
the consequence is a deposit, which remains until the annual
floods in the tributary remove it‡. When the Ohio is in
flood, it stagnates the waters of the Mississippi for many leagues;
when the Mississippi is in flood, it dams up the waters of the
Ohio for seventy miles §.

   Darby remarks that the Mississippi, in its long course from

* Darby's Geographical Description of the State of Louisiana, p. 65.
† Hall's Travels in North America.
‡ James, Exp. to Rocky Mountains.
§ Hall's Travels in North America, vol. iii. p. 370. The same author
notices the curious mixture of the Missouri waters with those of the Missis-
sippi, the former charged with detritus and wood, the latter beautifully clear.

the embouchure of the Ohio to Baton Rogue, washes the
eastern bluffs, which it tends to carry away and destroy, and
that, even to the sea, it does not come in contact with the
western side of the valley through which it flows. He attributes this, with great probability, to the deposits brought
down by the great tributaries, which all enter the Mississippi
from the west, and thus accumulate detritus on that side.

Notwithstanding the general tendency of the river to the
eastward, innumerable smaller changes of channel take place.
Thus winding courses shorten themselves, by cutting through
isthmuses, the tendency of the winding currents being to destroy the barriers between them, as may be observed in numerous rivers flowing through plains. New obstacles present
themselves; new sinuosities of channel are produced; trees
growing upon old alluvial deposits of the river are carried
away; and new vegetation springs up upon the recent alluvium,
to be again removed by a new change of channel. During
these various minor changes of bed, the degradation of the
higher lands supplies a great abundance of detritus, which not
only tends to raise the general level of the valley, by deposits
over the low lands at floods, but is carried forward towards
the sea, and forms an immense delta, composed of clay, mud,
and silt, mixed with a large proportion of drifted trees and
other vegetable substances.

The delta is divided into innumerable lakes, marshes, and
streams, inhabited by a multitude of alligators. The main
stream of the Mississippi will be observed to project forward,
on all good maps, in a singular manner. The detritus brought
down by it produces constant alterations, which require all
the attention of the pilots. According to Captain Hall, millions of logs, or trunks of trees, are brought down during
freshets, and carried several miles into the sea, so that it is
difficult to navigate among them. When not carried to sea,
these logs are bound together by a kind of cane, which retards
the river and collects mud. The same author considers " that
a belt of uninhabitable country, from fifty to one hundred
miles in width, fringes the edge of the whole of that part of
the coast *."

It has been supposed that the mud of the Mississippi, sinking through one foot of water in an hour, would be carried by
the Gulf Stream a distance of fifteen hundred miles before it fell
to the depth of five hundred feet, estimating the velocity of the
current at three miles per hour †. This has been employed as an
argument in favour of the great extent over which river de-

---

* Hall's Travels in North America, vol. iii. p. 340.
† Babbage, Economy of Manufactures, 2nd edition, p. 51.

tritus may be transported; and there could be little doubt of
the correctness of the calculation if the data on which it is
founded were also correct.   There is, however, every reason
to conclude that the detritus of the Mississippi never finds its
way into the Gulf Stream, but that it is deposited on the
western shores of Florida, and in the northern portion of the
Gulf of Mexico, generally.   There is in fact no current to
carry the detritus of the Mississippi into the Gulf Stream, as is
well known to those who navigate the northern portion of the
Mexican Gulf.   The waters, moreover, of the Gulf Stream,
even so far southward as the Florida reefs, are among the
clearest of sea waters yet discovered; indeed, the depth at
which objects can be seen through them has always been remarked with surprise by those who have visited that part of
the world for the first time.   This great clearness continues
to the Bahama Bank and Islands.   It might also be shown
that three miles per hour is an over-estimate of the mean annual velocity of the Gulf Stream.

The mouth of the Ganges will afford us an example of the
power of rivers to force forward deltas where no violent currents run across their embouchures, and where the body of
water, particularly during freshets, is very considerable, even
when such rivers are opposed to considerable tides.   Major
Rennel described this delta in 1781, so that probably, since
his account was written, very material changes have been
effected; yet as all these changes are likely to have been made
in the same manner, Major Rennel's description will always
be valuable, as showing the mode in which they have been
carried on.   The delta of the Ganges commences about two
hundred and twenty miles from the sea in a direct line; or
nearly three hundred, if the distance be reckoned along the
windings of the river.   The Ganges makes frequent windings,
like many other rivers, and thus considerable changes of its
bed take place, the opposing bends cutting through the isthmus
between them, as in the Mississippi. During the eleven years
which Major Rennel remained in India, the head of the Jel-
linghy river was gradually removed three-quarters of a mile
further down.   He also states, that "there are not wanting
instances of a total change of course in some of the Bengal
rivers.   The Cosa (equal to the Rhine) once ran by Purneah,
and joined the Ganges opposite Rajenal.   Its junction is now
nearly forty-five miles higher up.   Gour, the ancient capital
of Bengal, once stood on the Ganges."   It seems probable
that the Ganges once ran in the line now occupied by the lakes
and morasses between Nattore and Jaffiergunge*.

* Rennel, Phil. Trans. 1781.

This delta is constantly on the increase. The quantity of detritus must be abundant, for the sea into which it is borne is by no means shallow, the depths being considerable. The amount of detritus held in mechanical suspension by the Ganges has, however, been greatly exaggerated ; for instead of being equal to one fourth of the volume of water discharged, as was supposed by Major Rennel, it appears that $2\frac{1}{7}$ per cent. is the utmost that can be allowed.* The usual checks are produced by the tide, but during the freshets the ebb and flow are little felt, except near the sea. During these times, therefore, the advance of the delta is most considerable, the quantity of transported detritus being then greatest, and the resistance of the sea at its minimum. The sea may ravage the new lands, and apparently remove them for a time ; but eventually they must gain, even from the accumulative power of the breakers themselves, which also equalize the depths, by conveying the detritus to a short distance : thus rendering the sea more shallow, and consequently more easily filled up by river-borne detritus.

Coarse gravel transported by the Ganges does not approach the sea within four hundred miles, and consequently does not occur within one hundred and eighty miles of the commencement of the delta ; therefore it would appear that during the present order of things the Ganges has not transported coarse gravel into the sea at its present relative level. A great portion of the periodical inundations, represented as flowing on the level lands at the rate of half a mile per hour, has been attributed to the rains which fall on the low lands of India, as it has a blackish tint, from being long almost stagnant among vegetables of different kinds. Small obstacles accumulate, as might be expected, very considerable banks and islands ; a large tree arrested in its progress downwards, or even a sunken boat, being sufficient for the purpose. As these islands are quickly formed, so are they easily swept away by any change in the mighty current.

At the junction of the Ganges and Burrampooter below Luckipoor, there is a large gulf in which the water is scarcely brackish, even at the extremity of the islands, some of which are described by Major Rennel as equalling the Isle of Wight in size and fertility. The sea is represented as perfectly fresh to the distance of several leagues from this place during the rainy season.

---

* Gleanings of Science, vol. iii. p. 185 ; Calcutta 1831. On the other hand, it appears that the average discharge of water from the Ganges is much greater than was estimated by Major Rennel, being about 500,000 cubic feet per second.—*Ibid.*

It will be seen that deltas not only occur in situations where there is neither tide nor considerable current to prevent a great accumulation of new land, as at the embouchures of the Nile and Po, but also where the tides are small (Mississippi), and even where they are considerable (Ganges). The deltas thus produced are no doubt large, and the amount of animal and vegetable matter which they may entomb very considerable; but we must not be led away by measurements and comparisohs with the length, breadth, or superficies of districts with which we may be familiar, and which we may, from habit, consider important. They should be regarded with reference to their relative importance as portions of dry land, when it will be seen that they do not expose so considerable a surface as might at first be supposed. The augmentation of deltas will correspond with the detritus carried forward to the embouchures of rivers ; and it will be obvious that the facility of the transport will depend, all other circumstances being the same, on the length and fall of the channel. Now the course will be shortest and the declivity greatest at the commence- ment of the delta, and therefore it might be concluded that deltas would accumulate heavier materials, and increase most rapidly at the first periods of their formation, and that this increase would gradually diminish as the fall of the river channel became less, and its length increased ; without reckon- ing on the innumerable checks given to the stream by the in- creasing divisions in the delta. It may also be supposed that the detritus from the high lands would become gradually less, from the equalization of levels, and the fewer asperities that meteoric agents have to act on. Should these remarks, made under the supposition of the non-interference of man, be correct, it will follow that the increase of deltas would gradu- ally diminish if these were the only circumstances which regulated them. But it must be admitted that heavy rains, more particularly in tropical countries, would tend to cut up and destroy the delta itself, (still accumulating at its highest parts,) and force the detritus into the sea. The dense aquatic vegetation, common at the extremities of deltas, would render this transport difficult, yet still some detritus would escape. The amount of such additions to the outskirts of the new land would not, perhaps, be considerable, but it would corre- spond with the size of the delta, and consequently the larger this was, the greater would be the increase thus derived.

Between those rivers, such as the Ganges, which obtrude deltas into tidal seas, and those which have large open em- bouchures, such as the Maranon, St. Lawrence, Tagus, and Thames, there are such variations, produced by local causes, that it would be exceedingly difficult, even if useful, to classify them. In the delivery of their detritus, therefore, such rivers will either produce deltas or estuaries at their embouchures, as they either partake of the characters of the Ganges or the St. Lawrence; if of the latter, the detritus will be dealt with according to the mode of deposit or transport in estuaries.

Action of the Sea on Coasts.

Breakers, or the waves falling on sea beaches or coasts, are continual and powerful agents of destruction in some situations; while in others they pile up barriers against themselves. Their destructive influence is principally felt when the rocks on which they are discharged are composed of soft materials, and rise somewhat abruptly above the level of the sea. Their protecting influence is most commonly experienced in front of low level lands, and across the mouths of valleys, on each side of which a hard rocky point supports the ends of a beach.

The destruction of coasts of equal hardness almost always bears a proportion to the extent of open sea to which such coasts are exposed, all other circumstances being the same. The configuration of most coasts will be seen to be determined by the hardness of the rocks composing them; the softer strata giving way before the battering power of the breakers, while the harder rocks preserve their places for a greater length of time. If the rocks forming a coast be stratified, much depends on the dip of the strata relatively to the breakers. Thus, in many situations on the southern coasts of Devon and Cornwall, the slaty rocks dip in such a manner towards the sea, that the waves have never effected more than the removal of some loose superficial matter, the same that covers all the hills in the vicinity. In fact, a skilful engineer could not have protected the coast better than has been accomplished by the dip of the strata. The destructive power in other situations is well known; and of this, the eastern coast of our island presents abundant proof, where very considerable encroachments of the sea have been recorded within the lapse of a few centuries. The substances so forced away by the action of the breakers will be acted on according to their weight, form, and solidity. The tides or currents will remove so much of them as they are able to transport, and the rest will remain on the shore within the immediate influence of the breakers, which constantly tend to grind them down into smaller portions, and finally into sand.

In the destruction of a cliff of unequal hardness, it not unfrequently happens, that the harder portions, when large, such

as many concretions in sandstones and marls, or blocks of indurated strata, remain at the base of the cliff, and in a great
measure protect it from the more powerful effects of the
breakers, as will be seen in the annexed figure.

a, a defence of blocks, derived
from the hard strata *b,* and the
concretions *c.*

Fig. 16.

Among the unstratified rocks,
great variety of hardness prevails,
so that they frequently present an
uneven front to the sea, resulting
from the quicker decomposition and
destruction of some parts than of others. Veins of one substance, or rock, traversing another are generally of different
textures and solidity from that which they cut, and consequently nothing is more frequent, on sea shores, than to observe them either standing out in relief or hollowed into coves.

When a shingle or sandy beach, but more particularly the
former, is partly torn up and held in temporary mechanical
suspension by the breakers during a heavy gale, the action of
the waves is very considerable, even on the hardest rocks, so
as to scoop them out near the ordinary level of the sea.   In
exposed situations, the hardest rocks are often drilled into
holes or caverns, from the force of the broken wave being
driven, by local circumstances, more in one direction than
another, or from the inferior hardness of different portions of
the rock.   The most beautiful of ocean caverns, Fingal's
Cave in Staffa, owes its existence to the circumstance of the
basaltic columns being jointed in that place, while the general
character is to be without divisions in the columns*.

After the sea has formed a cavern, the vault of which does
not rise above high water, it sometimes works its way upwards
at the inmost extremity, partly by means of the compressed
air held between each wave as it rolls into the cave.   Of this
kind of cavern Bosheston Mere in South Wales is an example
on the large scale.   It is formed through strata of carboniferous limestone, and the noise caused by the blast of compressed air and sea water upwards is heard at a considerable
distance.

The protecting influence of breakers is shown in long lines
of shingle and sandy beaches, which often defend low and
marshy land, particularly at the mouths of valleys, from the
destructive power of the sea.

* MacCulloch, Western Islands of Scotland.

Shingle Beaches.

In the case of shingle beaches, it will be observed, that during a heavy gale every breaker is more or less charged with the materials composing the beach; the shingles are forced forward as far as the broken wave can reach, and in their shock against the beach drive others before them that were not held in momentary mechanical suspension by the breaker. By these means, and particularly at the greatest height of the tide, the shingles are projected on the land beyond the reach of retiring waves. Heavy gales and high tides combined seem to produce the highest beaches; they do indeed sometimes cause breaches in the rampart they have raised against themselves, but they quickly repair them. The great accumulation of beach upon the land being effected at high water, the ebb tide, it is clear, cannot deprive the land of what it has gained. In moderate weather, and during neap tides, various little lines of beach are formed, which are swept away by a heavy gale; and when these little beaches are so obliterated, it might be supposed, by a casual observer, that the sea was diminishing the beach; but attention will show that the shingles of the lines, so apparently swept away, are but accumulated elsewhere. These remarks do not apply to situations where the sea, during gales, has access to cliffs or piers, from whence there might be a retiring wave carrying all before it; but to such situations—and they are abundant— where the breakers meet with no resistance of that kind, and strike nothing but the more or less inclined plane of a shingle beach. Even in cases where the waves in heavy gales and high tides do reach cliffs, and for the time remove shingle beaches, it is curious to see how soon these latter are restored, when the weather moderates, and when the breakers, in consequence of a diminished projecting force, cease to recoil from the cliff behind.

Shingle beaches travel in the direction of the prevalent winds, or those which produce the greatest breakers: of this there are abundant examples on our own southern coast, where the prevalent winds being W. or S.W., the beaches travel eastward until arrested by some projecting land, when the sea forms a barrier against itself, and not unfrequently leaves a space between it and the cliff which it formerly cut: this space, under favourable circumstances, is covered by vegetation, suited to such a situation, even the cliff being sometimes studded with sea-side plants, when they can find root. Works are sometimes constructed to arrest beaches, either to protect land behind or to prevent their passage round pier

heads into artificial harbours; and thus engineers are practically aware of their travelling power in the direction of certain winds. This progressive march of beaches is far from
rapid, and can only be in proportion to the greater power or
duration of one wind to another; moreover, the pebbles become comminuted in their passage, and thus the harder can
only travel to considerable distances.

The Chesil Bank, connecting the Isle of Portland with the
main land, is about sixteen miles long, and, as a general fact,
it may be stated that the pebbles increase in size from west to
east. It protects land which has, evidently, never been exposed to the destructive power of the Atlantic swell and seas,
which break with great fury against the bank; for the land
behind is composed of soft and easily disintegrated strata,
which would speedily give way before such a power. Perhaps
a gradual sinking of the land might produce the present appearances; for though the sea would have attacked the land
when the relative levels were different, the form of the bay,
and the projection of the Isle of Portland, would soon cause
a beach to be formed, which would rise as the land sunk, so
that, finally, no traces of a back cliff could be observed.
Under this hypothesis, Portland would not have formed an
island, but merely the projecting point of a bay, which, with
its exposure, would soon have accumulated the beach required.
It may be remarked, that this supposed gradual sinking of
the land is in accordance with appearances more westward
on the same coast, where the facts presented seem to require
this explanation. The sea separates the Chesil Bank from
the land for about half its length, so that, for about eight
miles, it forms a shingle ridge in the sea. The effects of the
waves, however, on either side are very unequal; on the
western side the propelling and piling influence is considerable, while on the eastern, or that part between the bank and
the main land, it is of trifling importance. The following is
a section.

Fig. 17.

a, the Chesil Bank: *b,* the water called the Fleet: *c,* small
cliffs formed by the waves of the Fleet and land springs: *d,*
various soft rocks of the oolite formation, protected from destruction by the Chesil bank *a* : *e,* the open sea.

Another curious example of land protected by a shingle
bank occurs on the southern coast of Devon, and is remarkable, as it shows that the sea, at its present relative level with

the land, has never reached the land behind the beach,—a fact that will admit of the same explanation as that previously given for the Chesil Bank. At the bottom of Start Bay, and for the distance of about five or six miles, a considerable bank, principally composed of small quartz pebbles, has been thrown up by the sea. The line of coast faces the east. Between Tor Cross and Beeson Cellar, a point of land comes within the reach of the breakers; but here, as well as elsewhere behind the bank, the land has evidently gained on the sea, or, in other words, the latter has piled up a barrier which prevents its reaching the cliff, as it once did, even during heavy gales. This bank, generally known as Slapton Sands, though composed wholly of small pebbles, protects and blocks up the mouths of five valleys. Between Slapton Sands (properly so called) there is a fresh-water lake, divided into two at Slapton Bridge, where the waters of the northern lake drain into the southern. The northern portion is nearly silted up by the detritus borne down by a river that drains a few miles of country, and is nearly covered by bulrushes and other aquatic plants. The southern and larger portion is open, and of many acres in extent. The waters are supplied by the rivers behind, and commonly percolate through the pebbles into the sea. When, however, the tides are high, and the waters kept up by heavy gales, it sometimes happens that, the relative levels being altered, the sea-water passes through the shingles into the lake, and renders it to a certain extent brackish. This usually happens in winter; but, generally speaking, the relative levels are such, that the lake drains into the sea and remains perfectly fresh. It contains a great abundance of trout, perch, pike, roach, and flounders. The presence of the latter, a marine or estuary fish, shows that it can be gradually accustomed to fresh water. The percolation of the sea through the pebbles, during heavy gales, does not seem to injure the fresh-water fish; but when a breach was made through this beach during the gale of November 1824, they were nearly all killed by the sudden influx of the sea. Those which escaped up the streams were sufficient, in five years, again to stock the lake abundantly.

The breach made through Slapton Sands continued open for nearly a year, becoming gradually smaller. The complete restoration of the sands was hastened by throwing a few bags, filled with shingles, into the gap, upon which two or three gales soon piled up a heavy beach.

The old bank must have remained undisturbed for a long period; for vegetation had become active upon it, as we see by those portions which remain uninjured, where turf and even furze-bushes have established themselves upon the shingles.

G

Fig. 18.

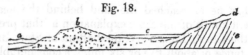

The above exhibits a section of the beach and lake.—*a*, the
sea which throws up the beach *b*: *c*, the fresh-water lake behind the beach: *d*, several feet in depth of pieces of slate and
sand derived from the slate-rocks *e*.

This diagram shows that the sea could not have acted upon
the hill *d e* since the accumulation of the loose substances *d*,
which it would have instantly removed.

The great size of rock fragments moved by the action of
the breakers attests their power. During heavy gales, blocks
of many tons in weight have been forced from their places;
and others, even squared and bolted together in the form of
piers and jetties, have been torn asunder by the battering
power of the waves. During the gale of November 1824,
which ravaged a considerable part of the southern coast of
England, a square block, from a ton and a half to two tons in
weight, strongly trenailed down, was torn away from a jetty
at Lyme Regis, and tossed upwards by the force of a breaker.
Mr. Harris, of Plymouth, informs me that, during the same
severe gales, and at the commencement of 1829, blocks of
limestone and granite, from 2 to 5 tons in weight were washed
about on the Breakwater like pebbles; about 300 tons, in
blocks of these dimensions, being carried a distance of 200
feet, and up the inclined plane of the Breakwater. These
blocks were thrown over on the other side, where they remained, after the gale, scattered in various directions. A
block of limestone, weighing 7 tons, was washed round the
western extremity of the Breakwater, and carried 150 feet.
Two or three blocks of this size were washed about. At the
Pier in Bovey Sand Bay, on the east side of Plymouth Sound,
a piece of masonry may be now seen, which was washed back
about 10 feet, being, at the time it was struck, 16 feet above
the level of an 18-feet spring tide. This piece of masonry
weighs about 7 tons, and consists of a few blocks of limestone
cemented together and covered by a large block of granite.
The mass was dovetailed into, and formed part of, a parapet
facing the sea.

At the Scilly Islands the blocks of granite that fall from the
cliffs are ground by attrition into great boulders, which become
the sport of the heavy Atlantic seas in tempestuous weather.

The effect produced by a heavy sea must depend considerably on the form of the block on which the sea acts. Thus,
a flat front would present the greatest resistance to the shock,
and the mass so struck would have a tendency to be more

easily moved than a rounded mass, if it were not that the resistance to removal offered at its base, is very considerably
greater than in a rounded mass.

The wedging power of the breakers is also very considerable where heavy blocks of difficult removal are mixed with
smaller stones easily transported. A beach of this nature
sometimes acquires much solidity, as the smaller pieces are
often forced among the larger so tightly as to require very
great force, and even fracture, before they can be taken out.

It would appear that, though shingle beaches, or those
composed partly of pebbles and partly of larger masses, may
be moved in the direction of the predominating and heaviest
breakers, we have no evidence of their being transported
outwards, or into the depths of the ocean, but that, on the
contrary, the waves of the sea strive to throw them upon the
land; and this, not only in the case of substances derived from
the land, but also in that of corals, shells, and marine plants
which have been produced in the sea itself. In tropical
countries it is found that many coral reefs and islands are
defended on their windward sides by beaches of coral shingles,
and even large fragments of coral. Lieut.-Col. Hamilton
Smith informs me, that, during a hurricane which he witnessed
at Curaçoa in September 1807, large pieces of coral were
torn up from a depth of ten fathoms, and thrown on the bank
uniting Punta Brava with the land. Beaches composed
wholly or entirely of comminuted marine shells are not uncommon, and will be noticed in the sequel.

The seaward front of most shingle beaches, particularly
when they defend tracts of flat country, is bounded by a line
along the edge of the beach; above this line the beach generally makes a considerable angle with the sands, in cases of
sandy flats. In cases where shingle beaches are not entirely
quitted by the tide, sandy, shelly, or very fine gravel soundings are commonly obtained at a short distance from the shore,
unless the bottom be rocky. It would appear that, if the
present continents or islands were elevated above, or depressed
beneath, the present ocean-level, shingle beaches would be
found to fringe the land, but not to extend far seaward *.

---

* We should be careful, when we obtain shingles in various soundings,
to consider that the probability is as great of finding pebbles at the bottom
of the sea as on the dry land; and that their presence there, is no proof that
they have been transported by existing currents, unless it can be shown that
the velocity of the existing current is sufficient to transport such detritus, and
that the direction of the current is that which would carry the fragments from
the known place of the parent rock. Without attention to this circumstance,
it might be supposed that the small shingles, covering the bottom of the
newly discovered bank off the north-west coast of Ireland, were carried there
by the present currents, when they are quite as likely to have been otherwise

Sandy Beaches.

The observations made respecting shingle beaches apply, in a great measure, to those composed of sand. The sand is derived either from the detritus borne down by rivers, from the attrition of sea-shore shingles against each other, or immediately from the sand and sandstones of the land. The breakers have the same tendency to force sand upon the land, as was observed in the case of shingles; but, being so much lighter than the latter, sand can be transported by coast tides or currents whose velocity would be insufficient to move shingles. On the other hand, however, smaller forces and bodies of water can throw sand on the shore. The spray that could not transport a pebble can carry sand, and thus this substance can be, and is, conveyed far beyond situations where the reflux of a wave can be felt. When the tide is low, or the sea less agitated, sand, dried by the sun or winds, is transported by the latter to great distances, so that whole districts of once fertile land have been overwhelmed by it.

Such transported sand, when sufficient to form hills, is known by the name of *dunes*, more or less common behind sandy shores or beaches over the globe. A striking example of the progress of such drifted sand inland, is to be found in the Bay of Biscay, on the eastern shore of which the sands have overwhelmed and are continuing to cover large tracts of country. Cuvier states the advance of these dunes as perfectly irresistible, forcing lakes of fresh water before them, derived from the rains which cannot find a passage into the sea. Forests, cultivated lands, and houses disappear beneath them. Many villages noticed in the middle ages have been covered, and in the department of the Landes alone, ten are now threatened with destruction. " One of these villages, named Mimisan, has been striving for twenty years against them; and one sand-hill, more than sixty feet high, may be said to be seen advancing. In 1802, the lakes invaded five fine farms belonging to Saint Julien; they have long since covered a Roman causeway which led from Bourdeaux to Bayonne, and which was seen, about forty years since, when the waters were low. The Adour, which was once known to flow by Vieux Boucaut, and fall into the sea at Cap Breton, is now turned aside more than a thousand toises[*]."

---

produced. That they are not now rolled about to any extent, is evident from the serpulæ and other marine productions attached to some of them brought up by Captain Vidal, during his survey, by the arming of the sounding lead.

[*] Cuvier, Dis. sur les Rév. du Globe.

M. Bremontier calculated that these dunes advance at the
rate of sixty, and even seventy-two, feet per annum.
Under favourable circumstances, sands, transported from a
beach into the interior, become consolidated : of this a good
example is found on the north coast of Cornwall, where the
matter thrown up is formed from comminuted sea-shells, and
the consolidation is principally effected by means of oxide of
iron. From the drift having taken place at different times,
this recent calcareous sandstone is stratified, with occasionally
interposed vegetable remains. Houses have been overwhelmed,
and human remains entombed where churchyards have existed. Mr. Carne describes a pot of old coins dug out of it.
The induration of this rock is so considerable, that holes are
drilled in it at New Kay, for the purpose of securing vessels
to the cliff. It is also used for architectural purposes, and
according to Dr. Paris the church of Crantock is built with
it. The same author states that the high cliffs of this recent
rock, which extend several miles in Fistrel Bay, are occasionally intersected with veins of breccia. " In the cavities,
calcareous stalactites of rude appearance, opaque, and of a
gray colour, hang suspended." " The beach is covered with
disjointed fragments, which have been detached from the cliff
above, many of which weigh two or three tons*."
Indurated dunes occur in various parts of the world: they
have been noticed by Peron in New Holland ; and the rock
in which the human remains of Guadaloupe have been found
would appear to be similar. These latter are discovered at
the Port du Moule, in an indurated beach composed of comminuted shells and corals. The specimen in the British
Museum is formed of coral and small pieces of compact limestone, and in it Mr. König has observed *Millepora miniacea*,
madrepores, and shells referred to *Helix acuta* and *Turbo
Pica.* According to Cuvier, the specimen in the Jardin du
Roi, at Paris, exhibits a gangue of travertin containing shells

* Paris, Geol. Trans. of Cornwall. Not only sands but shingle beaches
are sometimes indurated.—Captain Beaufort describes a plain several miles
in length, near Selinty, coast of Karamania, as bounded by a gravel beach,
which has become consolidated from the top of the crest to some distance
into the sea ; the consolidation extending to the depth of from one to two
feet, and being generally covered with loose sand and gravel, so that it is
not easily observed. The pebbles are cemented by a calcareous paste, and
the whole is so hard, that a blow " more frequently fractures even the
quartz pebbles than dislodges them from their bed." Other beaches of the
like kind, but on a smaller scale, were observed on other parts of the coasts
of Asia Minor and of Greece. Rocky ledges of a similar nature occur to
the westward of Sidé, partly above and partly under the water. They contain broken tiles, shells, bits of wood, and other rubbish. They are very
hard, and are cemented by calcareous matter, probably derived from some
calcareous slate in the vicinity.—Beaufort's Karamania, pp. 182 and 185.

of the neighbouring sea, and terrestrial shells, especially the *Bulimus guadaloupensis* of Férussac. Near Messina, loose sand becomes consolidated on the beach, and is used for building. It is stated that the cavities thus made are again filled up by sand, which becomes consolidated and used in its turn.

Von Buch notices a limestone and sandstone formation of a similar kind on the coast of the Great Canary. The violent north wind of summer raises the light fragments of broken shells, and little rounded grains of trachyte and basalt, (between the town and Isleta,) and drives them over the narrow tongue of land of Guanateme; thus forming dunes about 30 or 40 feet in height. The waters behind these dunes act on the sand, uniting it into a compact mass, which is broken away during the ebb. These waters are, during the greater part of the year, at a temperature of 77° F., which greatly promotes their action on the calcareous particles. The sandstone is produced on the Confidal shore of the Isleta, but not at Catalina, which is exposed to the N.E. wind. It is a real oolite, most of the grains being round and calcareous, surrounding a nucleus of basalt, trachyte, or the fragment of a shell. Similar rocks containing shells, resembling those still living on the coasts, occur at the height of 300 or 400 feet above the level of the sea, showing that a difference of level to that amount has taken place since the formation of these rocks *.

Dr. Clarke Abel describes a large bank, rising from the sea to the height of about a hundred feet, to the eastward of Simon's Town, Cape of Good Hope, formed of shell and sand, thrown up by the S.E. wind. In this he discovered singular cylindrical bodies, which resembled bones bleached by the air. " On a closer examination, many of them are found to be branched; and others are discovered rising through the soil, and ramifying from a stem beneath, thicker than themselves. Their vegetable origin immediately suggests itself, and is confirmed by a further inquiry. They are seldom solid, their centres being either hollow or filled with a blackish granular substance, which in many specimens, except in colour, resembles the substance called roestone by mineralogists. Their outer crust is chiefly composed of a large proportion of sand and a small proportion of calcareous matter, and in many specimens contains fragments of ironstone and quartz an inch square. That they are really incrustations formed on vegetables which have afterwards decayed, is proved by the different degrees of change which the internal parts of different specimens have undergone. In some the organiza-

* German Transl. of Manual.

tion of the plant sufficiently remains to leave its nature unequivocal; and near the sea the very commencement of the
process of incrustation may be witnessed on the large *Fuci*
which strew the shore*."

Peron's previous description of the change undergone by
vegetable substances in similar situations on the coast of Australia, is nearly the same. He considers that the shells undergo decomposition, and form a cement with the sand; and
that the vegetables become altered and finally replaced by this
sandstone, leaving nothing to show its origin but its general
form. On our coasts the sands thrown on shore by the action of the sea, and afterwards drifted by the winds, are often
comparatively considerable. Mr. Ritchie describes a district of ten square miles in Morayshire, once termed the Gra-
nary of Moray, as having been overwhelmed. " This barren
waste may be considered as hilly; the accumulation of sand
composing these hills frequently varying in their height, and
changing their situation †."

The following account by Mr. Macgillivray affords an additional example of the tendency of coast-seas to throw even
the substances formed in them upon the land. " The bottom
of the sea, along the whole west coast of the Outer Hebrides,
from Barray Head to the Butt of the Lewis, appears to consist
of sand. Along the shores of these islands this sand appears
here and there in patches of several miles, separated by intervals of rock of equal or greater extent. In some places the
sandy shores are flat, or very gently sloping, forming what
are here called Fords; in others, behind the beach, there is
an accumulation of sand to the height of from twenty to sixty
feet, formed into hillocks. This sand is constantly drifting;
and in some places islands have been formed by the removal
of isthmi. The parts immediately behind the beach are also
liable to be inundated by the sand; and in this manner most
of the islands have suffered very considerable damage......
The sand consists almost entirely of comminuted shells, apparently of the species which are found in the neighbouring
seas. It is rather coarse in the grain; but during high winds,
by the rubbing of its particles on each other, a sort of dust
is formed, which at a distance resembles smoke, and which,
in the island of Berneray, I have seen driven into the sea to
the distance of upwards of two miles, appearing like a thin
white fog ‡."

It would be useless to accumulate notices of these various

* Clarke Abel, Voyage to China, p. 308.
† Notes appended to Cuvier's Theory of the Earth, by Jameson.
‡ *Ibid.*

sand drifts, which often contain seams of vegetable matter that have been successively covered up, and of which sections are afforded *. The action of the waves round coasts tends to disturb the bottom at certain depths, and to move the shells, sands, and other substances, of which this bottom is composed, towards the land. The exact depth to which the moving action of waves extends, seems never to have been very accurately estimated ; indeed, when we consider that the power of the wave is continually varying, such an estimate becomes exceedingly difficult. Ninety feet, or fifteen fathoms, has been sometimes considered as the limit, in depth, to which this disturbing power extends; but this requires confirmation. Around coasts and on shores which do not much exceed ten or twelve fathoms, the action of the waves is very apparent in the discoloration of the water during heavy gales. This turbid character of the sea is due to the moving power of the waves on the bottom, and becomes more marked as the water becomes more shallow, either in approaching the land or over shoals. The transporting power of the waves will therefore be in proportion to the depth of water beneath them, the transport being greatest in the shallowest places. The waves will tend to throw substances on coasts, because the off-shore wind produces smaller waves than the wind blowing upon the land. On shoals distant from the land, the effect will be somewhat different, and the piling or propelling power will be greatest on the side of the prevalent or more violent winds. Shoals will be also liable to shift, as the turbid waters on the crown of a shoal will be forced over on the lee side.   Accordingly, we do find that shoals shift, more particularly when near the surface, unless there be an equal counteracting effect in a current or tide. We may, in some measure, learn the effects of waves at different depths, from the form of the outer talus of the *Digue*, or Breakwater, at Cherbourg, where they have, to a certain extent, arranged the stones, four-fifths of which

* Not only are sand-hills thrown up by the sea, but also by the waves of extensive fresh-water lakes. Dr. Bigsby (Journal of Science, vol. xviii.) and Capt. Bayfield (Trans. of Lit. and Hist. Soc. of Quebec, vol. i.) both remark the beaches thrown up in the bays of Lake Superior. The latter author notices some curious lines of ancient beaches rising one above the other, like the seats of an amphitheatre, in valleys at some distance from the shores of the present lake, and hence infers that the level of Lake Superior has fallen. Similar beaches are observable on other lakes of North America. Capt. Bayfield noticed seven ridges of shingle, rising above each other, near Cabot's Head, Lake Huron : the highest was overgrown with spruce firs ; the second had bushes or smaller trees of the same kind ; the third, shrubs and flowers ; the fourth, lichens and mosses ; the rest being bare of vegetation. Dr. Bigsby and Capt. Bayfield also notice the sand-hills thrown up on the shores of Lake Superior by the prevailing N.W. winds.

are small, in the manner best fitted to resist themselves. According to M. Cachin, there are four kinds of taluses, arranged one beneath the other. The upper line of talus, being only touched by the higher break of the waves, presents a height proportioned to its base, as 100 is to 185. The second line, comprising the whole distance between the line of high and low water at the equinoxes, and thus exposed to the battering power of the breakers during the whole flood and ebb, is consequently the most inclined, and its height is to its base as 100 to 540. The third line, being below the lowest water at the equinoxes, is only acted upon during the first of the flood or the last of the ebb. Its height is to its base as 100 to 302. The fourth line, or the base of all, not being acted on by the waves, maintains a talus, of which the height is to the base as 100 to 125 *.

The action of waves on coasts is not only exhibited by piling up detritus in the direction of their greatest force on the shore, by which embouchures of rivers are turned on one side, but also by heaping up bars, as they are termed, even at their mouths, rendering their navigation dangerous, and in many instances preventing it altogether; though, behind these barriers, the rivers may have considerable depth and breadth. In some situations these bars are partially dry at low water, at others they are never uncovered, though rendered visible by the breaking of a furious surf. To produce examples would be useless, as they are common in all parts of the world. In many cases, the bars are liable to shift, particularly after a gale of wind, so that vessels are frequently lost by keeping the direction of the old channels; and it requires the constant attention of pilots to be aware of the exact position of the new passages.

When the rivers are small, the force of the waves frequently blocks up their embouchures, and artificial means are necessary to permit the escape of the pent-up waters, that would otherwise form a lake in the low country behind. If the dam be a shingle beach, the water usually percolates through it; but if composed of sand, the water will accumulate until its level enables it to cut a passage through the barrier and escape. This done, the breach will be again repaired, and another accumulation of water take place behind, and so on. But, in the mean time, the level of the low land would rise, first, by deposition from the river waters; and, secondly, from the sand blown over the bank. In such an alluvial land there would probably be found remains of terrestrial, freshwater, and even marine shells, the latter worn or broken.

Rivers are deflected from their courses into the sea by beaches extending from one side, and produced by the winds and breakers; both forcing detritus before them, if it be composed of sand or comminuted shells, while the latter acts upon the shingles alone, except when light pebbles are caught up in the heavier spray, and are thus driven by the wind. Examples of this deflection may be seen in many situations, and the harbour of Shoreham, on our southern coast, is a marked one *.

Rivers, when thus deflected from their courses by beaches, generally escape into the sea by the sides of cliffs, which seem to give them such support that they can cut channels.

In tropical countries the breakers commonly throw up barriers against the advance of the mangrove trees, either from a deep bay or creek, or at the mouths of rivers, if they come within their influence. Capt. Tuckey remarks, that " the peninsula of Cape Padron and Shark Point, which forms the south side of the estuary (of the Zaire), has been evidently formed by the combined depositions of the sea and river, the external or sea shore being formed of quartzy sand constituting a steep beach; the internal or river side, by a deposit of mud overgrown by mangroves; and both sides of the river towards its mouth are of similar formation, intersected by numerous creeks (apparently forming islands), in which the water is perfectly torpid." This mangrove tract appears to extend inland, on both banks, about seven or eight miles, and is represented as impenetrable. Did not the sea pile up a barrier against it, and thus afford it protection from its own attacks, it would be destroyed †. Similar phænomena, though on a much smaller scale, are seen at the mouths of the Rio Minho, and other rivers in Jamaica. Beaches are accumulated in front of mangrove trees, under somewhat similar circumstances, in the same island, on the south side of which, particularly near Albion estate, lakes are formed on the inside of a shingle beach thrown up by the sea. The lake near Albion has a small opening in the protecting bank, permitting the surplus water to escape; this water being apparently derived from the drain of the mountains behind, and the splash of the sea during gales. The mountain drainage has carried much mud into the lake, upon which mangrove trees have

* See Geological Notes, pl. 1. fig. 2.; and Phil. Mag. and Annals of Philosophy, N. S. vol. vii. pl. 11. fig. 2.

† Expedition to the Zaire or Congo, p. 85. This author further remarks, that "small islands have in many places been formed by the current (of the river); and doubtless in the rainy season, when the stream is at its maximum, these islands may be entirely separated from its banks, and the entwined roots keeping the trees together, they will float down the river, and merit the name of floating islands."

established themselves. These by their roots entangle various substances, and form land, the accumulation being a compound of mineral, vegetable and animal substances*. A much larger lake of the same description is found under Yallah's Mountain, the most projecting part of the beach forming Yallah's Point †.

The bank called the Palisades, at the end of which stands Port Royal, Jamaica, seems thrown up by the action of the prevalent breakers, caused by the sea breezes, or winds from the east and south-east, which propel the materials of the beach from east to west. This bank is between eight and nine miles long, of little elevation above the sea, having a beach on the seaward front, with mangrove trees on many parts of the inward side. If the passage between the western end of this bank and the land opposite to it should be barred up by a continuation of the bank, a large lake would be inclosed, into which the Rio Cobre would discharge itself. The mangrove trees would assist in the formation of new land, in which a mixture of marine, fresh-water, and terrestrial remains might be entombed.

Mangrove trees afford support to beaches thrown up by the sea; and if such a beach originated from a shoal, there is always a tendency to increase land to leeward by their agency. Protection being once afforded, the mangrove trees establish themselves, and accumulate silt, mud, and drift-rubbish about their roots. Thus, support is afforded to the original bank, and new materials are piled upon it to windward by the action of the breakers, additional consolidation being produced by the tropical sea-side creepers. Meanwhile the advance to leeward continues, until the land immediately against the beach becoming too dry for the support of the mangrove trees, others, more suited to the new land, establish themselves; and, finally, a grove of cocoa-nut trees may gradually appear ‡.

Tides and Currents.

The principal motions in the waters of seas and oceans are

* For a section of this lake, see Sections and Views illustrative of Geological Phænomena, pl. 35. fig. 6.

† These waters contain a multitude of alligators.

‡ For a section of such an island near Jamaica, see Sections and Views illustrative of Geological Phænomena, pl. 36. fig. 2.

According to M. Gutsmuth, the great band of alluvial matter, deposited by the sea for a distance of 200 miles between the Maranon and Oronoco, is increased by the mangrove trees, which, when the deposits still continue submerged, advance into the shallow sea and soon form woods. (Hertha, vol. ix. 1827.) In this and similar cases we may consider that, from the shallowness of the sea, heavy breakers cannot reach the mangrove trees, and therefore a beach is not thrown up against them.

produced by tides and currents; the former due to the action of the sun and moon, the latter probably caused by the winds and the motion of the earth.

The streams of water caused by tides are chiefly felt on coasts, while the currents produced by winds are more or less experienced over the whole surface of the ocean. It must frequently happen that the direction of a tide and a current being the same, they add mutually to the velocity of each other, while the contrary arises with opposed courses.

The streams of water produced by tides and currents are geologically important, as they may be the means of distributing the detritus derived from the land over spaces at a greater or less distance from the shore; their power of affecting this being proportioned to their velocity and depth.

Tides.

The velocity of a stream of tide depends on the obstacles it encounters. These obstacles generally present themselves in the form of projecting headlands, a gradually diminishing channel, or a group of islands and shoals. In the former case the velocity of the tide is considerably increased round the opposing capes, gradually diminishing to its usual rate at a short distance on either side, or in the offing. The English Channel will present us with many examples, more or less striking, according to circumstances. Round the Start and the Bill of Portland the tides run exceedingly strong, causing dangerous Races when opposed to the winds. But these considerable streams of tide are merely local; for in the bays, and at a short distance out at sea, the velocity of the tides does not exceed a mile and a half or two miles; while at the headlands above noticed, it frequently flows at the rate of four or five miles *. Generally speaking, the increased velocity of the tidal stream round capes is in proportion to the body of water forced into the bays of which they form the extreme points.

The greatest obstacle opposed to the tidal wave flowing up the English Channel, is the great bight on the west of Cap la Hague, where we find innumerable islands and rocks, of which the principal are Guernsey, Jersey, and Alderney. The stream of flood being completely opposed to the line of coast, and pent-up by the islands and rocks, it rises to a very considerable height, and escapes through the Race of Alderney, between the island of the same name and the main land, with a velocity of seven miles an hour. It continues to run with great rapidity round Cap Barfleur, gradually decreasing in

* All the miles mentioned in the following notice of tides and currents are nautical, sixty being equal to one degree.

strength until the general level is restored. Some idea may
be formed of the variation in the Channel level, caused by
this obstacle, by the differences in the rise of tide observed between the mouth of the Channel and the Straits of Dover.
The perpendicular rise of tide on each side of the mouth of
the Channel is nearly the same, being twenty-one feet at
Ushant, and twenty feet at the Land's End. In the great
bight or bay west of Cap la Hague, the tide rises forty-five
feet between Jersey and St. Maloes, and thirty-five feet at
Guernsey. At Cherbourg this great elevation of the level is
diminished; the tide there rising about twenty-one feet. On
the opposite side of the Channel, on the English coast, the
perpendicular rise of the tidal wave is comparatively trifling,
being thirteen feet at Lyme Regis, seven feet in Portland
Road, fifteen feet at Cowes, and eighteen feet at Beachy Head.
Therefore, the elevated level of the Guernsey and Jersey
waters produces no perceptible effect on the English coast opposite. Between Beachy Head and Dover, there is a rise of
twenty-four feet on the west of Dungeness, and twenty feet at
Folkestone. On the opposite coast there is a rise of twenty
feet at Havre, nineteen feet at Dieppe, and nineteen feet at
Boulogne. The tides are twenty feet at Dover, and nineteen
feet at Calais.
The Bristol Channel is a familiar example of a high rise of
tide caused by a gradually contracted channel, at the end of
which there is no outlet. At St. Ives, Cornwall, the perpendicular rise of the spring tides is eighteen feet, of the neap
tides fourteen feet*. At Padstow the tide rises twenty-four
feet; at Lundy Island, thirty; at Minehead, thirty-six; at King
Road, near Bristol, from forty-six to fifty; and at Chepstow,
about the same.
The difference of level, produced by obstacles to the tide,
is remarkably exhibited on each side of the isthmus separating
Nova Scotia from the main land of North America. In the
Bay of Fundy, on the south side, the tides have a very considerable rise, amounting, according to Des Barres, to sixty and
seventy feet at the equinoxes; while on the northern side, in
Baie Verte, they rise and fall only eight feet. The tidal stream
is, as might be expected, very rapid in these gradually diminished channels, particularly where the rise and fall is most
considerable. This unusual rapidity ceases by degrees as we
approach the mouths of such channels, and arrive at the more
common levels.
From the great diversity in the line of coasts, innumerable
modifications are effected in tidal streams, causing them to

* The rise of tide at St. Ives is sometimes stated at twenty-two feet.

flow with augmented or diminished velocity. As such streams are only visible on coasts, it seems fair to infer that the effects produced by them do not extend to any considerable distance beyond the land.

The tide in the offing, and the tide along shore, do not exactly correspond, the flood tide continuing in the offing some time after the ebb has commenced on shore; the ebb tide the same. It has been stated that "the length of time between the changes of the tide on the shore and the stream in the offing, is in proportion to the strength of the current and the distance from the land ; that is, the stronger the current, and the greater the distance that the current is from the land, the longer it will run after the change on the shore*."

Among the small islands of the Pacific Ocean the tide rises about two feet, there being no great range of coast near them to produce a greater elevation. At the islands of the Atlantic. Ocean the rise is greater, being at the Azores from six to seven feet; at Madeira, eight or nine; among the Canaries, eight or ten ; at the Cape Verde Islands, from four to six; at the Bermudas, five or six; at St. Helena, three; at Fernando Noronho, six ; and at Tristan da Cunha, eight or ten feet.

The stream of tide along a coast is greatly increased at the time of full and new moon, so that at spring tides the current often runs at double the rate experienced at neap tides. The transporting power of tidal streams is therefore perpetually changing, independent of the variations produced by winds upon them.

From various circumstances the tides of flood and ebb are sometimes unequal. Thus, at the Land's End the flood runs nine hours to the north, and the ebb three to the south. In the expedition under Captains Parry and Lyon, it was found that in the higher part of Davis's Straits the flood tide set from the north at the rate of three miles an hour for nine hours, the tide of ebb making only three hours.

A current setting into the Straits of Malacca, during part of the year, causes the tide to run nine hours one way and three hours the other. The tides are irregular through the Straits of Banca, with an easterly wind. The ebb sets to the northward for sixteen hours, while the flood only lasts eight

---

* Purdy, Atlantic Memoir, 1829. In the same work it is stated that "the time which the flood-stream runs in the middle of the English Channel after the time of high water on shore, is, westward of the meridian of Portland, about three hours ; but to the eastward, off Beachy Head, only one hour and three quarters. In the offing, between the meridians of Dungeness and Folkestone, the North Sea and Channel tides seem to meet; and the ebb of the one uniting with the flood of the other, set in an easterly direction off the French coast, more than four hours after high water on the western shore of Dungeness." p. 88.

hours. In common tides there are two floods and two ebbs in twenty-eight hours in these straits, the duration of which is in some sort regulated by the winds: the flood lasts six hours, and the ebb eight hours; or there are five hours flood, and nine hours ebb.

The tides are very trifling and irregular in the West Indies, perhaps owing to the accumulation of water pent up by the equatorial current and trade winds. At Vera Cruz there is only one tide in twenty-four hours, and that irregular. Among these islands the tide varies in perpendicular rise from a few inches to two feet or two feet and a half. The stream or current produced by them must consequently be very trifling.

Theoretically, all bodies of water, even large fresh-water lakes, have tides; but they are so insignificant that inland seas, such as the Mediterranean and Black Seas, are generally termed tideless.

The current setting into the Mediterranean from the Atlantic is somewhat modified by the tides. In the middle of the Straits of Gibraltar the current sets eastward; on each side, however, the flood tide sets to the westward.

" On the European side, west of the island of Tarifa, it is high water at 11^h, but the stream without continues to run until 2^h. On the opposite shore of Africa, it is high water at 10^h, and the stream without continues to run until one o'clock; after which periods it changes on either side, and runs eastward with the general current. Near the shore are many changes, counter currents and whirlpools, caused by and varying with the winds. Near Malaga the stream runs along shore about eight hours each way. The flood sets to the westward[*]."

The strongest tides of which I can find mention, occur among the Orkney and Shetland Isles, and through the Pentland Frith, between the main land of Scotland and the former. The flood comes from the north-west, and is not of unusual strength until it encounters the obstacles of the islands and main land. The tides change near the shores sooner than at a distance from them. The difference of time varies according to situation, amounting in some places to two or three hours. The velocity of the tide through Stronsa Frith is about five miles an hour during spring tides, and a mile or a mile and a half at neaps. In North Ronaldsha Frith, the springs run at five miles an hour; the neap tides at one mile and a half. The flood divides near the shore at Fair Isle, forming a large eddy on the east side. The springs here run six miles an hour, the neaps two. These tides increase in velocity when supported by the winds. The most rapid stream of tide occurs

---

[*] Purdy, Atlantic Memoir, p. 90. The tide rises three feet at Malaga.

at the Pentland Frith, its velocity being nine miles an hour
during the springs, though it runs only three miles an hour at
neap tides.

Tides in Rivers and Estuaries.—These are necessarily much
modified by circumstances; but, generally speaking, the tide
of ebb is stronger than the flood, from the body of fresh water
being pent up by the flood, to which the rivers must always
present a certain resistance, proportioned to their velocity and
abundance of water;—the greatest resistance to the flood, and
increased velocity of the ebb, being during freshets, or when
the rivers have a surcharge of water produced by rains in the
interior.

When the flood tide takes place in rivers of sufficient depth,
the first operation of the tide appears to be that of a wedge,
elevating the fresh water from its inferior specific gravity to a
higher level. The flood gradually opposes greater resistance
to the outflow of the river, and in the end succeeds in damming
it up. I have found many fishermen aware of this " creeping," as they have termed it, of the salt water beneath the
fresh at the commencement of the flood, and have seen a rise
of five or six feet caused in water in the higher parts of tidal
rivers, while the water so raised has continued perfectly
fresh at the surface.

At the ebb, if the fresh or river waters be abundant, they
will, after the salt water has been discharged, flow over the
salt water to greater or less distances from the shore according to circumstances. After the rains, a strong freshet sets
do vn the Senegal, and a powerful current of fresh water runs
some distance out at sea. Masters of vessels crossing this
stream have been surprised by the sudden increased draught
of their ships, caused by their entrance into a fluid of inferior
specific gravity.

Captain Sabine states, that while proceeding in his voyage
faom Maranham to Trinidad, on September 10, 1822, the
general current running at the great rate of ninety-nine miles
in twenty-four hours (more than four miles per hour), they
crossed discoloured water in 5° 08′ N. lat., and 50° 28′ W.
long. He considers this water as that of the river Amazons
or Maranon, which had preserved its original impulse three
hundred miles from its embouchure, having flowed over the
waters of the ocean, from its less specific gravity. The line
between the ocean water and discoloured water was very
distinct, and great numbers of gelatinous marine animals were
floating on the edge of the river water. The temperature of
the ocean water is stated as $= 81°\cdot1$, and that of the supposed
river water $= 81°\cdot8$, both near the division line: " the specific
gravity of the former was 1 0262, and of the latter 1·0204."
From experiments made, the depth of the discoloured water

was superficial, and did not amount to 126 feet. There was no bottom at 105 fathoms. In this discoloured water the ship was set N. 38° W., sixty-eight miles in twenty-four hours, or rather less than three miles per hour. The western side of the fresh water was gradually lost in that of the sea. Captain Sabine attributes the unusual velocity of the ocean-current of ninety-nine miles per day, to the obstacle which this freshwater current opposes to it*.

In the river St. Lawrence we have a striking example of the superior velocity of the ebb tide to the flood. " At the Isle of Coudre, in spring tides, the ebb runs at the rate of two knots. The next strongest tide is between Apple and Basque Isles; the ebb of the river Saguenay uniting here, it runs full seven knots in spring tides; yet, although the ebb is so strong, the flood is scarcely perceptible; and below the Isle of Bic there is no appearance of a flood tide †."

The great difference in the ebb and flood of river tides must depend on many local causes, but be principally in proportion to the perpendicular rise of tide on the one side, and the mass of fresh water on the other. The flood tide sets up many rivers so suddenly, as to cause a wave of greater or less magnitude, according to circumstances, called the *bore*, appearing

* Experiments to determine the Figure of the Earth.

We have other accounts of discoloured waters in the Atlantic, which would render it necessary that the specific gravity and relative freshness of simply discoloured water should always be ascertained, as was done by Captain Sabine, before we can be certain that waters even flowing in the necessary direction were derived from rivers. Captain Cosmé de Churruca states, that 128 leagues to the eastward of St. Lucia, and 150 to the N.E. of the Orinoco, there is always discoloured water as if on soundings, but there is no bottom at 120 fathoms. The same appearances are observed about seventy or eighty leagues to the eastward of Barbadoes. Humboldt notices a place in the latitude of Dominica at about 55° W. longitude, where the sea is constantly milky, although it is very deep; and seems to think that there may possibly be a volcano beneath it. Captain Tuckey observed the same kind of milkiness upon entering the Gulf of Guinea; but considered it due to multitudes of crustacea which were caught, and which produced great luminosity at night.

Sir Gore Ouseley mentions that on February 12, 1811, when off the Arabian shore, a partial line of *green water*, such as generally indicates shallows, and perfectly different from the *blue* of a deep sea, was perceived extending considerably. It appeared eight or nine miles from the land. The change from the blue to the green waters was sudden, so that the ship was in green and blue waters at the same time. Having entered the green water they sounded, and found bottom at seventy-nine fathoms; proving that the change of colour was not due to a shoal; for previous to entering this water they sounded in the blue water, and found sixty-three fathoms, so that the blue was more shallow than the green water. This was observed not far from the Persian Gulf.—Sir Gore Ouseley, Travels, vol. i.—In this case there was no great river near to produce the difference of colour. " Green Sea " is the name given to the Persian Gulf by Eastern geographers.

† Purdy, Atlantic Memoir, p. 91.

as if the flood suddenly overcame the resistance of the ebb. The bore of the Ganges is very considerable. According to Major Rennel, it " commences at Hughly Point, below Fulta, the place where the river first contracts itself, and is perceptible above Hughly Town; and so quick is its motion, that it hardly employs four hours in travelling from one to the other, although the distance is near seventy miles. At Calcutta, it sometimes causes an instantaneous rise of five feet; and both here and in every other part of its track, the boats on its approach immediately quit the shore, and make for safety to the middle of the river *."

According to Romme, there is a considerable *bore* at the mouth of the Amazons or Maranon during three days at the equinoxes. It is observed between Maraca and the North Cape, and opposite the mouth of the Arouary. A wave of twelve or fifteen feet in height is suddenly formed, and is followed by three or four others. The advance of this *bore* is exceedingly rapid, and the noise caused by it is stated to be heard at the distance of two leagues. It occupies the whole breadth of the river, and in its progress carries all before it until it has passed the banks into deeper and wider water, where it ceases. M. De la Condamine has described this phænomenon, and has observed that there are two opposing currents during the flood, one superficial, the other deep. There are also two superficial currents, one setting by the shore on each side, while a central but retarded current descends. Tides are stated to be felt two hundred leagues up the Amazons, so that there are several in the river at the same time, and the surface of the water for that distance forms an undulating line.

The most curious *bore* which I find recorded, was observed by Monach, Port Commandant at Cayenne: he states, that " the sea rises forty feet in less than five minutes in the Turury Channel, river Arouary; that this suddenly elevated water constitutes the whole rise of tide, the ebb immediately taking place, and running with great velocity†."

In the Zaire or Congo we have an example of the comparatively small effect of the tide upon a large body of fresh water discharged with sufficient velocity. Notwithstanding the aid of Massey's machine, bottom was not found in Tuckey's expedition at 113 fathoms in mid-channel and at the mouth, and the stream ran at the rate of four and five miles an hour‡. This stream became checked but not overcome in mid-channel, and the tide only produced counter currents near the shore. The rise of water is felt between thirty and forty miles up the

---

* Phil. Trans.
† Romme: Vents, Marées et Courants du Globe, tom. ii. p. 302.
‡ It has been since supposed that this stream had greater velocity.

river. Alluvial land is continually forming into flat islands, which are covered by mangrove trees and papyrus, and are often partially or wholly carried by the river into the ocean*. Professor Smith describes a floating isle of this kind which he saw further north off the coast of Africa; it was " about 120 feet long, and consisted of reeds resembling the *Donax*, and a species of *Agrostis?* among which were still growing some branches of *Justicia*†."

Currents.

Currents are sometimes classed as constant, periodical, and temporary.

The great current which flows from the Indian Ocean round the Cape of Good Hope, up the coast of Africa to the equatorial regions, whence it strikes across the Atlantic to the West Indies, is considered a constant current, produced by the tropical or trade winds, assisted by the motion of the earth. The current having driven, by these means, a body of water to the continent of America, through which it cannot escape, passes up through the channel offered it at the Straits of Florida, flows considerably to the northward, and then bends to the eastward, and south-east, taking its course to the west coast of Europe and the upper part of Africa. It is considered that the latter division of the current again unites with the northern portion of the equatorial current, and again traverses the Atlantic.

Between Cape Bassas in Africa and the Laccadives or Lakdivas, there is a constant current to the westward, mostly to the S.W. or W.S.W. Its rate is supposed to be from eight to twelve miles per day. The current south of the equator, in the Indian Sea, runs to the west. During the N.E. monsoon the currents of the Mosambique Channel run to the south along the African coast, and even in the offing; their usual velocity being about seven or eight leagues in twenty-four hours. On the coast of Madagascar the currents take an opposite direction, and set towards the north. At the southern extremity of Africa, the currents set round the bank of Agulhas, or Lagullas as it is more commonly termed, a bank of considerable extent, the soundings in which are described as *mud* to the westward of Cape Lagullas, and *sand* to the eastward, the latter containing numerous small shells. Rennel informs us that this current is strongest during the winter, and that the outer verge of the stream runs into 39° S. before it turns to the northward, after which it proceeds slowly along the western coast of Africa to, and even beyond,

---

* Tuckey's Expedition to the Zaire or Congo.   † *Ibid.* p. 259.

the equator*. The general velocity of the current round the bank is not stated; but it appears that one vessel was carried by it one hundred and sixty miles in five days, or thirty-two miles per day†.

Beyond St. Helena, the current above noticed unites with the equatorial current of the Atlantic, and sets across from the Ethiopic sea to the West Indies. The velocity of this current has not been well ascertained, but is generally considered as about one mile and a half per hour, increasing as it proceeds westward, and setting off the coast of Guyana at the rate of two or three miles per hour. Captain Sabine states that, sailing from Maranham in 1822, and entering the current, he estimated it as running at the rate of ninety-nine miles in twenty-four hours, or a little more than four miles per hour. The central direction of this current is W.N.W.

" On the Colombian coast, from Trinidad to Cape la Vela, the currents sweep the frontier islands, inclining something to the south, according to the strait they come from, and running about a mile and a half an hour with little difference. Between the islands and the coast, and particularly in the proximity of the latter, it has been remarked, that the current at times runs to the west, and at others to the east. From Cape la Vela, the principal part of the current runs W.N.W.; and, as it spreads, its velocity diminishes: there is, however, a branch which runs with the velocity of a mile an hour, directing itself towards the coast about Cartagena. From this point, and in the space of sea comprehended between 14° of latitude

* Captain Tuckey in his expedition to the Zaire or Congo, found a current setting to the N.N.W. after making St. Thomas off the African coast. Its velocity was thirty-three miles in twenty-four hours.

† As the current round the Lagullas Bank evidently conforms to the bank, we may, perhaps, consider that it there has considerable depth, that is, a depth equal to about sixty or seventy fathoms. But of this we cannot be quite certain, for we do not know to what distance water thrown off by the bank at lesser depths may be carried round it.

There is an easterly or counter current which sets to the south of this main current. Capt. Horsburgh mentions having been carried by it at the rate of 20 to 30 miles in 24 hours ; and, in two instances, at the rate of 60 miles in the same time.

My friend, Mr. Becher of the Hydrographical Office, informs me that there is also an easterly or counter current close to the shore, so that the main or western current flows between two eastern or counter currents. The survey of the coast of Africa, to the east of the Cape of Good Hope, was made by Capt. Owen with the assistance of this current against the full strength of the trade wind. It also appears to exist more northward of the Cape, in the Atlantic.

Respecting the depths to which surface currents descend, we learn from Capt. Belcher, that he found a current running with nearly the same velocity at 40 fathoms as on the surface, off the west coast of Africa, in lat. 15° 27′ 9″ N., and long. 17° 31′ 50″ W., the rate being 0·75 per hour, and the direction south.—Jour. Geog. Soc., vol. ii.

and the coast, it has however been observed, that in the dry season the current runs to the westward, and in the season of rains to the eastward *."

It is asserted, that there is a constant stream entering the Mexican Gulf by the western side of the channel of Yucatan; and that there is commonly a re-flow on the eastern side of the same channel around Cape Antonio†.

On the northern coasts of St. Domingo and Cuba, in the windward passages, at Jamaica, and in the Bahama passages, the currents appear variable, their greatest observed velocity being about two miles per hour.

The accumulation of water in the Caribbean and Mexican seas does not raise the level of those seas so much as was, perhaps, once supposed. The difference of level observed by Mr. Lloyd, in his researches on the Isthmus of Panama, between the Mexican Sea and Pacific Ocean, was in favour of the greater height of the Pacific Ocean by 3·52 feet,—an unexpected result; but the measurements were conducted with such care, that we can scarcely doubt it. The high-water mark at Panama is 13·55 feet above high-water mark of the Atlantic at Chagres; but from the difference in the tides on each side the isthmus, the Pacific is lower than the Atlantic at low water by 6·51 feet‡. If we consider the body of water pent up by the effects of currents over so large a space as the Mexican Sea at eight feet, or even less, above the Atlantic Ocean, we need not be surprised at the velocity of the current produced by its escape through the Straits of Florida.

If the temperature of the waters, heated in the Gulf of Mexico and Caribbean Sea, be greater, as we know it is, than that of the waters north of the tropics through which the Gulf Stream flows, the specific gravity of the former waters will be less, and consequently they will flow onwards over the colder waters or those of greater specific gravity, precisely as river-water flows out to sea over that of the ocean, and will continue to do so until their progress be gradually checked and finally stopped.

From a mass of information that has been collected, it appears that the Gulf Stream varies considerably in breadth, length, and velocity. It has been found that winds much affect the current, diminishing its breadth and augmenting its velocity, or augmenting its breadth and diminishing its velocity.

In mid-channel, on the meridian of the Havanna, the direction is E.N.E., and the velocity about two miles and a half per hour. Off the most southern parts of Florida, and at about one third over from the Florida Reefs, it runs at the rate of about four miles per hour. Between Cape Florida and the Bemini Isles it runs to the N. by E., with a velocity of more than four miles an hour. The stream is weak on the Cuba side, and sets to the eastward.

A re-flow or counter current sets down by the Florida Reefs and Kays to the S.W., and W., and by its aid many small vessels have made their passages from the northward*. To the northward of Cape Canaveral there is no stream of tide, along the southern coast of the United States, further from the shore than in ten or twelve fathoms of water; from that depth to the edge of soundings, a current sets to the southward at the rate of a mile an hour; out of soundings, the Gulf Stream is found setting to the northward†. It is also stated that there is a re-flow or counter current on the eastward of the stream.

Capt. Sabine remarks, that in the latter part of 1822 the velocity of the current after passing Cape Hatteras was seventy-seven miles per day‡. Rennel, considering the force of the stream as determined at different points, calculates that the water requires about eleven weeks to run in the summer, when its rapidity is greatest, from the Gulf of Mexico to the Azores, a distance of about 3000 miles. Capt. Livingston, however, observes, that the calculations of the velocity of the Gulf Stream are not to be depended on. He found it setting at the rate of five knots and upwards on the 16th and 17th of August 1817. On the 19th and 20th of February 1819, it seemed to be almost imperceptible. In September 1819, it set at about the rate described in the charts§.

Lieut. Hare has found in the meridian of 57° W., that the stream ranges to $42\frac{3}{4}$° N. in the summer, and even to 42° N. in the winter.

It would appear, that the waters after issuing through the Straits of Florida, run off from the eastern edge of the stream to the eastward, as might be expected from their tendency to equalize their level, particularly in those parts not carried forward with considerable velocity.

A strong current sets from the Polar Seas, and through Hudson's Bay and Davis's Strait, commonly denominated the Polar or Greenland current. It sets southerly down the coast

---

* Purdy, Atlantic Memoir.   † *Ibid.*

‡ Capt. Livingston observes that the current set him, off Cape Hatteras, 1° 8' to the northward of his dead reckoning; this he ascertained by stellar and solar observations.—Atlantic Memoir.

§ Purdy, Atlantic Memoir.—These observations appear to have reference to the stream between Cape Florida and the Bemini Isles.

of America to Newfoundland, bringing down large icebergs beyond the Great Bank. Captains Ross and Parry found the velocity of the current from three to four miles per hour in Baffin's Bay and Davis's Strait.

A current from the polar regions sets into the North Atlantic between America and Europe : it produced such a drift of the ice to the south in Capt. Parry's attempt to reach the North Pole over the ice, that the expedition was finally abandoned in consequence of it.

The Polar current coming from Davis's Strait, may be said to unite with the Gulf Stream, and then to set eastward, directing its course to the coasts of Europe and Africa. Off the coast of Newfoundland, the current sometimes runs at the rate of two miles an hour, but is much modified by winds. About five degrees to the westward of Cape Finisterre the current has a velocity of thirty miles in twenty-four hours.

Between Cape Finisterre and the Azores there is a tendency of the surface waters to the S.E., being variable in winter. Lieutenant Hare, in September 1823, found a current setting E.S.E. with a velocity of a mile and a half per hour between N. latitude 45° 20′ and 43° 40′, and W. longitude 22° 30′ to 16°. Rennel remarks, respecting the currents between Cape Finisterre and the Canary Islands, that " it may be taken for granted, that the whole surface of that part of the Atlantic from the parallel of 30° to 45° at least, and to 100 or 130 leagues off shore, is in motion towards the Straits of Gibraltar."

" Near the coasts of Spain and Portugal, commonly called The Wall, the current is always very much southerly (as it is more easterly towards Cape Finisterre), and continues as far as the parallel of 25°, and is, moreover, felt beyond Madeira westward; that is, at least 130 leagues from the coast of Africa; beyond which a S.W. current takes place, owing, doubtless, to the operation of the N.E. trade wind." The same author observes, that the velocity of the current varies considerably, being from twelve to twenty, or more, miles in twenty-four hours. He considers sixteen as below the mean rate.

A current sets along the coast of Africa from the Canaries to the Gulf of Guinea, running westerly out of the Bight of Biafra. The rainy seasons, and Harmattan wind, interrupt this stream. From Cape Bojador and the Isles de Los, the velocity of the current has never been found to exceed a mile and a half per hour on the coast and on the outer edge of the bank. Its more common rate is less than a mile. At the distance of four leagues from the coast it becomes half a mile, and even less. In the meridian of 11° W. the current runs twenty-five miles to the E.S.E. in twenty-four hours. Off Cape Palmas it sets to the E. at forty miles; off Cape Three Points, and thence to the Bight of Benin, at from fifteen to thirty miles. It then decreases in strength, runs to the southward, turns to the S.W. between 6° and 8° S., and thence flows N.W. to the Cape Verde Islands. It is considered that the portion flowing eastward into the Gulf of Guinea, is not altogether continuous with that which comes from Cape Bojador to the south.

A current is described to pass round Cape Horn and Terra del Fuego, from the Pacific into the Atlantic, during the greater part of the year*. From the Straits of Magellan to the equator, a current sets northward along the western coast of South America. At eighty leagues from the coast, between 15° S. latitude and the equator, and even to 15° N. latitude, the currents generally run westward. Captain Hall found a constant current setting off the Galapagos, to the N.N.W. At Guayaquil a strong current sets out of the Gulf at the rate of forty miles in twenty-four hours. Between Panama and Acapulco, and at about 180 miles from the latter place, Captain Hall met with a steady current running E. by S. at rates varying from seven to thirty-seven miles per day. Great quantities of wood are drifted from the continent of America to Easter Island by the force of a current setting in that direction. Currents have been found at Juan Fernandez, and 300 leagues to the westward of it, running W.S.W. at sixteen miles per day. At the Marquesas they flow with a velocity of twenty-six miles in twenty-four hours. Between the Marquesas and the Sandwich Islands they have been found to run westward at the rate of thirty miles a day, in April and May. A southerly current has been observed at California; and a northerly current along the N.W. coast of America, from Cape Orford, the latter having a velocity of a mile and a half per hour.

A northerly current sets through Behring's Straits†, and is supposed to run along the north coast of America, and deliver itself, through Baffin's Bay and Hudson's Straits, into the Atlantic.

King found a current setting N.E. near the Japanese Is-

---

* Captain Hall states, that he did not meet with any current round Cape Horn. A naval officer, however, assures me that a current runs out of the Pacific into the Atlantic during nine months; and this is rendered probable from the prevalence of strong westerly winds during the greater part of the year, which would drive the waters before them. Kotzebue found a current which turned rapidly to E.N.E. near Staaten Land, having had another direction (S.W.) off Cape St. John.

† Kotzebue describes this current as setting through the Straits with a velocity of three miles per hour to the N.E. At Anchorage, near East Cape, the current was found to set at the rate of one mile per hour; but shortly afterwards, notwithstanding a brisk wind, the expedition under Kotzebue made but little way against it, though going, by the log, at the rate of seven miles per hour.

lands, the velocity five miles per hour; but he also found it to vary considerably in direction and strength.

Among the Philippine Islands a current comes from the N.E., and runs with considerable force among the passages dividing the islands; it has been found with a strength of twenty miles a day near these isles. This current varies.

Cook found a southerly current in August, flowing ten or fifteen miles a day, between Botany Bay and 24° S. On the same side of Australia a vessel was set forty miles to the southward in twenty-four hours, in the month of March; and in July another vessel was carried thirty miles in two days in the same direction.

A constant current sets eastward into the Mediterranean, with a velocity of about eleven miles in twenty-four hours. It has been considered that there is an under or counter current setting westward, and carrying out the dense water, rendered more than usually saline from evaporation within the Straits of Gibraltar; but this has lately been controverted. It was remarked by Dr. Wollaston, that the salt carried into the Mediterranean by the current from the Atlantic must remain there after the evaporation of the water which held it in solution, unless it could escape by some means. He inferred its escape to be by an under current, usually thought to exist, and this he considered proved by experiment; for water brought up from the depth of 670 fathoms about fifty miles within the Straits, by Captain Smyth, was found to contain about four times the usual quantity of saline matter. Water taken from depths of 450 and 400 fathoms, at 680 and 450 miles within the Straits, did not exceed in its saline contents many ordinary examples of sea-water. He further observed, that if the under current moved only with one fourth the velocity of the upper current, and was of the same depth and breadth as it, the former would convey out as much salt as the latter brought in *. Mr. Lyell infers that this dense water cannot pass out, because the bottom of the sea rises between Capes Spartel and Trafalgar, and has only 220 fathoms of water upon it; and therefore, if the under and more saline water be as deep as is supposed, it would be impossible for it to escape, and it would deposit great quantities of salt in the bed of the Mediterranean †. It is much to be regretted that we do not possess better information on this subject, and that direct experiments have not been made on this supposed under current. That this has not been done is the more remarkable, when we consider the numerous opportunities afforded by the

---

* Wollaston, Phil. Trans. 1829.
† Lyell, Principles of Geology, vol. i.

continual passage of ships, and the proximity of such esta-
blishments as those of Gibraltar. Mr. Lyell's theory of a
great deposit of salt at the bottom of the Mediterranean, though
very ingenious, can scarcely be true; for, supposing it to be
so, the sea would, as the depth increased, be more and more
charged with saline matter, until it finally became mere salt,
the density increasing at the same time. This being the case,
we should bring up salt with the sounding-lead, and little else.
But the fact is, that the deep soundings, as shown by Captain
Smyth, are mud, sand and shells. Sand and shells form the
bottom, beneath 980 fathoms of water, a little east of the meridian of Gibraltar; and the same bottom is found in the
Straits beneath 700 fathoms of water. Now these places are
near where the sea-water, so highly charged with saline matter, was brought up; and where, according to the theory,
there should be a bottom of salt. The same may be said of
other situations *.

The current entering the Mediterranean passes along the
southern shores of that sea, and is felt at Tripoli and the Island of Galitta. At Alexandria there is a stream flowing east,
as well as between the coast of Egypt and Candia : arrived on
the coast of Syria, it runs north, and then advances between
Cyprus and the coast of Karamania. A strong current flows
from the Black Sea into the Mediterranean, through the Dardanelles.

A constant current flows out of the Baltic, through the
Sound and Cattegat, into the German Ocean. Its velocity in
the narrowest part of the Sound is about three miles per hour;
but the ordinary rate, in fine weather, is about one mile and
a half or two miles. The currents out of the Sound and two
Belts are directed towards the Scau or Skagen, and flowing
thence, turn N.E. towards Marstrand, at the rate of about
two miles per hour. It is not impossible that a counter and
under current setting into the Baltic from the ocean may exist;
for Captain Patton observed, when at anchor a few miles from

---

* In all our remarks on the changes that may be supposed to occur at the
bottom of the Mediterranean, we should be careful to remember that this
bottom is divided into two great basins (See Smyth's Charts) by a winding
shoal, which connects Sicily with the coast of Africa. This shoal, known
as the Skerki, has the following line of soundings upon it, proceeding from
the African to the Sicilian coast; namely, 34, 48, 50, 38, 74, 20, 70, 52,
91, 16, 15, 32, 7, 32, 48, 34, 54, 70, 72, 38, 55, and 13 fathoms, from
whence an idea of its inequalities may be formed. There are soundings in
140, 157, and 260 fathoms, on either side, as also places where 190 and
230 fathoms of line have been run out, without finding bottom. It may be
here remarked, that, at the entrance of the Dardanelles into the Mediterranean, there are only thirty-seven fathoms of water ; so that the quantity
of solid matter requisite to bar the communication between the Black and
Mediterranean seas, would not be very considerable.

Elsinore, in an upper current setting at the rate of four miles an hour outwards, that in sounding in fourteen fathoms, he found the line continue perpendicular to his hand, when the lead was raised a little from the ground. Hence he concluded that there was an under current that prevented the line from being carried away.

In the Indian and Chinese seas we have good examples of periodical currents, evidently referable to the periodical winds or monsoons.

From St. John's Point to Cape Cormorin there is a nearly constant current in the direction of the coast from N.N.W. to S.S.E.; except that between Cape Cormorin and Cochin, it flows from S.E. to N.W. from October to the end of January.

The current sets from the ocean into the Red Sea from October to May, and runs out of that sea from May to October. A current commonly sets from the Gulf of Persia towards the ocean during the whole time that the current flows into the Red Sea, and runs into the Gulf from May to October.

In the Gulf of Manar, between Ceylon and Cape Cormorin, the current flows northward from May to October, setting the remaining six months to the S.W. and S.S.W. From Pedro Point on the north of Ceylon, to Pointe de Galle on the south, the current runs S.E, S.S.E., S., S.W., and W., according to the nature of the coast, uniting at the Pointe de Galle with the current that comes out of the Gulf of Manar. The ordinary velocity of the stream on the south coast of Ceylon is about a league an hour. The Ceylon currents are weak in June and November. In the Bay of Bengal the currents run with the wind towards the N.E. during the S.W. or W. monsoon, and slacken in September. On the coast of Orissa, about eight days before the equinox their direction is towards the south, and they become strong at the end of the month. During the N.E. and E. monsoon, the currents are, as before, with the wind, and strong in proportion to it.

In the S.W. monsoon the current between the coast of Malabar and the Lakdivas sets to the S.S.E. with a velocity of twenty, twenty-four, or twenty-six miles in twenty-four hours. Between the Lakdivas its direction is to the S.S.W. and S.W., its rate being from eighteen to twenty-two miles in twenty-four hours. The current, setting W. or W.S.W. to the westward of these islands, varies in velocity from eight to eleven miles per day. There is a strong current among the Maldives: among the southern isles the direction is generally to the E.N.E. in March and April; in May it sets to the eastward; in June and July the currents often run to the W.N.W., particularly to the south of the equator. Between these isles and Ceylon they frequently set strongly to the westward during the months of October, November, and December.

The currents in the China seas, at a distance from shore, generally flow, more or less, towards the N.E. from the middle of May to the middle of August, and have a contrary direction from the middle of October to March or April. The velocity of the currents from the N.E. is usually greater in October, November, and December, along the adjacent shores, than that of the opposite set in May, June, and July. Their strength is most felt among the islands and shoals near the coast.

The strongest currents of these seas are experienced along the coasts of Cambodia, during the end of November. They run with a velocity of from fifty to seventy miles to the southward, in twenty-four hours, between Avarella and Poolo Cecir da Terra. Some part of the stream setting into the Straits of Malacca, causes the tide to run nine hours one way and three hours the other. The currents to the northward commence running in April through the Straits of Banca, past the Straits of Malacca, and along the west coast of the Gulf of Siam, setting along the north-east side of the same Gulf to the E.S.E. until to the eastward of Point Ooby; they then bend to the N.E. running along the coasts of Cambodia, Cochin-China, and China, till September, when the opposite monsoon and currents prevail from the N.E., and continue to March or April.

Periodical currents occur, according to M. Lartigue, along the west coast of South America, from Cape Horn to latitude 19°. The S. and E.S.E. winds produce a current, setting to the N.W., off the coast of Peru, of which the maximum velocity is fifteen miles in twenty-four hours, and the mean velocity about nine or ten miles. Between this current and the coast, there is a counter current flowing to the S.E. During the prevalence of the wind from N. to W. the current flows S.E., but is only sensible near the land*.

Temporary currents are innumerable, every severe gale of any duration producing one. Nothing is more common than these partial currents, which are more particularly felt along coasts and through channels.

The direction and velocity of the currents above enumerated may be considered rather as approximations to the truth, than as the truth itself, for the determination of currents is liable to many errors. The usual manner of ascertaining them is by comparing the true place of a ship, determined by means of chronometers and astronomical observations, with the position

---

* Lartigue, Déscription de la Côte du Pérou.

of the same ship as deduced by dead reckoning. The latter is a calculation of the vessel's way through the water in a given direction. The rate of the vessel's way is estimated by means of a contrivance called a log-line, or a line at the end of which there is a float. According to the quantity of line run out in a given time, with allowances for the agitation of the sea, &c., is the rate of the vessel's way calculated. This operation is liable to numerous errors; and even with the line and glasses in the highest order, requires a nicety of execution seldom practised. The direction of the vessel's course is estimated by the compass, with allowances for magnetic variation. Here we have a most fruitful source of error, for until lately no allowance whatever was attempted for the local attraction of the ship. It is now well known that the disposition of iron in a vessel is such, that no two ships will be found to have the same local attraction; consequently no rules can be adopted for correcting the error of aberration by means of placing the magnets in any particular situation, though some situations have been found more favourable for true observations than others. It was not until Mr. Barlow invented his plate of iron for counteracting the effect of aberration, that the error arising from it could be fully known. Now nearly all the preceding observations, as to the direction and velocity of currents, were made before this great source of error was understood; consequently many of them are erroneous, and require that re-examination which the advance of science has rendered necessary. It is clear, that if a vessel is steering one course, and those on board consider they are taking another, the position deduced from dead reckoning must wander from the truth in proportion to the amount of aberration, even supposing the rate of way through the water and other necessary observations correct.

If, in the annexed diagram, a vessel, without any allowance for aberration, be supposed to hold her course from *a* to *b*, while in reality her course, with proper allowance for aberration, is from *a* to *c*, the distance from *b* to *c* will, according to the usual practice, be referred to current, after an observation shall have shown that her true place is at *c*. It will be clear that in this case no such current exists, and that the difference between the true and calculated situations of the ship arises solely from want of attention to local attraction.

Another great source of error in estimating the value of currents has been noticed by Captain Basil Hall. This author observes, that the usual method of laying down ships' tracks by two lines, one representing the course as estimated from the dead reckoning, and the other as deduced from chronometers and lunar observations, leads to no information as " to where the current began, or where it ceased, or what was its set, or its velocity." He proposes instead of this, that the position of the ship found at each good observation should form the point of departure, both for the line representing the distance and direction to the next observed true position, and for that representing the ship's course as estimated by dead reckoning*. A very superior plan, and one that should supersede the old method.

Although these causes of error render the exact velocity and course of currents heretofore observed vague and uncertain, so that many minor streams may be found imaginary, and that the navigator may be exposed to great danger from implicitly depending upon them ; yet to the geologist, perhaps, they may not be so formidable; as, probably, the general velocity of currents will not be found greatly altered ; and as it is with their velocity and consequent transporting power that he is principally concerned.

Transporting Power of Tides.

The stream caused by tides varies much in strength, but a common velocity appears to be one mile and a half per hour, when head-lands, shallow banks, and other obstacles are not opposed to it; and therefore, even supposing the superficial velocity to extend to the bottom, which would not be the case except in comparatively shallow seas, the general transporting power of such tides would appear, judging from the effects we witness near shores, to be but small. This the unchanged character of soundings for a great length of time, though principally composed of mud and sand, seems to attest.

Where obstacles are opposed to the tides, the transporting power will be increased, and the changes produced more rapid. The tide through the Pentland Firth having a velocity of nine miles per hour, would scour out pebbles of considerable size from its channel; but its power to do this would cease at each extremity, where the tides flow at the rate of two or three miles per hour, and the local cause would merely produce a local effect. The same with the Race of Alderney, and other similar places.

Changes in the shape of sand-banks frequently take place when they approach the surface; but as they then come within the influence of another cause,—the action of the waves, the transporting power of which is very considerable,—too much must not be attributed to the mere force of a tidal stream.

* Edinburgh Philosophical Journal, vol. ii.

The transporting power of tidal rivers outwards, or into the waters of the sea, is considerable, more particularly during the time of freshets or floods. As has been seen, the tide of ebb in rivers is always greater than the flood; therefore, although estuary waters are very turbid, and a great proportion of them merely carried backwards and forwards, detritus will escape into the open sea in proportion to the difference of velocity between the ebb and flood. It should be remarked, that all estuaries have a tendency to be filled up by deposition of the matters held mechanically in suspension by their waters. The heads of estuaries are very frequently alluvial plains, formed of the same kind of mud and silt as are at present brought down by the rivers; and it often appears as if the tides had flowed up to much greater distances than they now do, the higher parts having been gradually silted up*. These appearances are so common, that it is useless to insist upon them; but the extent of flat lands, evidently accumulated in this way on the sides and heads of estuaries, is often very remarkable, and would seem to have required a long lapse of ages for their formation; more particularly when the present deposits of the same estuaries are considered.

Notwithstanding this deposit in the estuary itself, and the bars and banks accumulated at the mouths of so many tidal rivers, above noticed, mud and silt escape into the sea, and are transported by the tides to greater or less distances from the rivers; as may often be seen at low water, on coasts where tidal rivers discharge themselves.

The transporting power of tides and currents being proportioned to their velocity, and this being greatest when obstacles are opposed to either, it is in these situations that we should look for the greatest transporting power.

The difference between the velocity of tides on the surface and at moderate depths must be very considerable, otherwise the previously noticed power of water to tear up different kinds of substances at given velocities must be incorrect; for if the velocities were nearly as great at moderate depths as on the surface, tidal streams would be little else than a mass of turbid waters.

The discoloration of the sea to greater or less distances from the shore, according to the depth, is well known to be effected during heavy gales, and is due to the action of the waves, and not to that of the tide merely passing over sand or mud with a certain strength, and therefore must not be confounded with it.

* If we could always give implicit confidence to old maps and charts, great deposits of this nature would seem to have taken place within historical times.

To take an example of tidal waters running over a certain bottom:—At the Shambles, a well-known bank near the island of Portland, the tides run at the rapid rate of about three nautical miles per hour, over soundings of gravel which do not alter. Now, if the calculations above noticed were correct, and the inferior velocity not very considerably different from that on the surface, stones, the size of eggs, could be torn up by water with a velocity of three feet in a second, or 3600 yards in an hour; consequently the pebbles on the bank would be carried away, and nothing but bare rock or masses of stone would be left; but the soundings on the Shambles are the same at present as they are represented to have been, by the charts, many years ago.

The preservation of the same kind of bottoms or soundings, over which tides or currents pass with considerable velocity without their being altered, is familiar to most mariners; and it would seem that we are far from being acquainted with the respective velocities required to tear up mud, sand, and pebbles at various depths in the sea. Tidal streams flow over mud banks in some estuaries at the rate of a mile and a half or two miles per hour, without removing them; though, if the above-noticed calculations were always applicable, the current would be sufficiently strong to remove pebbles of some size. The same remark applies to innumerable sand-banks *.

Transporting Power of Currents.

In estimating the transporting power of currents, we should consider the causes which produce them, and the nature of the fluid in which they are produced. The motion of the earth,

---

* While on the subject of soundings, it may be noticed that the British Islands are in reality united to the continent beneath the sea by banks of various kinds, at greater or less depths ; the principal soundings on which are mud or sand. The whole is more or less known by the name of soundings, because bottom can be easily obtained by a line of eighty or ninety fathoms in length. The boundary of these soundings is traced on all good charts, and is seen to commence at the bottom of the Bay of Biscay, then to run round the British Isles, and to communicate with the shallows of the German Ocean.

The bed of the sea in these soundings can only be considered as so much of the continent, which happens to be at no great depth beneath the ocean level. The upper part of the bottom, tenanted by various animals whose exuviæ are daily left in it, is probably in a great measure derived from the detritus of the British Islands and such parts of the continent of Europe as are either bathed by, or discharge their waters into, these seas. The depths being comparatively inconsiderable, the tides, currents, and waves are probably enabled to act, according to circumstances, in the distribution of the detritus.

The course of the tides round the British Islands is represented in Dr. Young's Natural Philosophy, vol. i. pl. 38. fig. 521; see also Lubbock on Tides, Phil. Trans. 1831.

although it would seem to give a certain general movement to the waters of our globe, does not appear capable, taken by itself, of producing currents of geological importance. The great cause of ocean-currents seems to be prevalent winds; and accordingly we find that in the equatorial regions of the world, over which the more or less easterly winds, commonly called the Trade Winds, prevail, there is a tendency of the waters to flow westward in the Pacific Ocean, in the Atlantic, and in those parts of the Indian seas free from the monsoons. That the winds are the great cause of ocean-currents, is a fact sufficiently proved by the velocity and direction of such currents in the Indian and Chinese seas, varying with the force and direction of the monsoons. On this subject Major Rennel observes, " It is well known how easily a current may be induced by the action of the wind, and how a strong S.W., a N.W., or even a N.E., wind on our own coasts raises the tide to an extraordinary height in the English Channel, the river Thames, the east coast of Britain, &c., as those winds respectively prevail. The late ingenious Mr. Smeaton ascertained, by experiment, that in a canal of four miles in length, the water was kept up four inches higher at one end than at the other, merely by the action of the wind along the canal. The Baltic is kept up two feet at least by a strong N.W. wind of any continuance; and the Caspian Sea is higher by several feet, at either end, as a strong northerly or southerly wind prevails. It is likewise known that a large piece of water, ten miles broad, and generally only three feet deep, has, by a strong wind, had its waters driven to one side, and sustained so as to become six feet deep, while the windward side was left dry. Therefore, as water pent up so that it cannot escape acquires a higher level, in a place where it can escape the same operation produces a current, and this current will extend to a greater or less distance according to the force by which it is produced or kept up*."

It is also considered that the moon exercises an influence on the waters of the tropical regions, increasing their velocity by drawing them from E. to W. The current setting six hours one way and six hours the other through the Straits of Messina, though there is no rise or fall of water with it, is attributed to the influence of the moon, and may be considered as a tide. Capt. Livingston observes, that "when the sun's declination is N., the N.E. trade wind blows fresher, and extends further to the northward than when the sun's declination is S., thus forcing a greater body of water into the Caribbean Sea†."

The current setting into the Mediterranean through the

Straits of Gibraltar is commonly attributed to the evaporation of that sea, which also receives a large supply of water from the Black Sea through the Dardanelles. The easterly in-draught from the Atlantic is stated to commence nearly one hundred leagues to the westward of the Straits of Gibraltar. It has been supposed that an under and counter current sets outwards; but this, as has been above noticed, has been lately controverted *. That under currents do, however, occur in the Mediterranean, Capt. Beaufort affords us sufficient proof. After remarking that from Syria to the Archipelago there is a constant current to the westward, slightly felt at sea, al-though very perceptible on shore, amounting to three miles per hour, between Adratchan Cape and the opposite island, he observes, "The counter currents, or those which return beneath the surface of the water, are also very remarkable: in some parts of the Archipelago they are sometimes so strong as to prevent the steering of the ship; and in one instance, on sinking the lead, when the sea was calm and clear, with shreds of bunting, of various colours, attached at every yard of the line, they pointed in different directions all round the compass†."

These observations of Capt. Beaufort are of the highest importance when we consider the transporting power of cur-rents, because they seem to show that we cannot judge of the direction of under currents from those known to flow on the surface.

The winds being, generally speaking, the cause of the great ocean-currents, and effects being only in proportion to their causes, the streams of water thus produced will not extend deeper than the propelling power of the winds can be felt. Now, as the ocean varies in density according to its depth, the cause sufficient to move waters on the surface, and to certain depths beneath it, will constantly meet with opposi-tion, at an increasing ratio; until finally, the moving power and the resistance being equal, no effect whatever is pro-duced; and all water beneath a certain depth would be, as far as respects surface causes, immoveable, and consequently would have no transporting power.

Hence it would appear that the transporting power of cur-rents will depend on the depth of the sea, all other things being equal, and that the smaller the depth the greater the transporting power. Consequently, coasts are the situations where we may look for this power.

If the current entering into the Mediterranean from the Atlantic be due to the evaporation of the former, this also is a superficial cause, and its effects will gradually become less, until, in deep water, it ceases altogether.

---

* Lyell's Principles of Geology.   † Beaufort's Karamania.

We have seen that tides as well as currents have their greatest velocity in shallow water, across headlands, or in contracted channels; consequently, their greatest transporting power exists in the same situations, and will be local. Tides commonly exert an equal transporting power in two directions, for the most part opposite to each other, except in the case of rivers, where this power is greater on the ebb than at the flood. Unless the rivers be very considerable, the detritus brought through their embouchures by the superior velocity of the ebb, enters into the power of the coast-tides, and is carried backwards and forwards by them until deposited. But in the case of great rivers, such as the Maranon, St. Lawrence, and Orinoco, the unchecked detritus is borne forward, until stopped and turned by the ocean-currents. Large additions are daily made to the coast of South America by the deposit from the waters of the Maranon, which are carried toward the shore by the prevailing current *.

Upon a review of what has been stated respecting the streams of water caused by tides and currents, it would appear that their geological importance will depend upon the relative depth of water which they traverse, and their proximity to land, by which their velocity is increased. Round coasts they have a transporting power, which varies according to circumstances; being greatest, all other things remaining the same, nearest the land. In great depths we have no reason to suppose that this transporting power exists; or if it does, the causes must be different from those which produce motion on the surface. It does not appear that we are acquainted with the velocities which could tear up mud, sand, or gravel; for currents pass over the bottom in shallow water, composed of mud and sand, without mixing them, with a considerable surface velocity. The changes produced on the bottom are scarcely perceptible, within the periods we should consider long, unless in shallow water, and near the mouths of great rivers, the deposits from which must gradually accumulate, and diminish the depth of the water. In the soundings round coasts, we do not generally find any great inequalities; but in the ocean these must exist to a very great extent, as is shown by the rocks, shoals, and small islands scattered over it, the tops of mountains emerging from the water, which is generally of great depth close to them.

Active Volcanos.

The surface of the earth is irregularly marked by orifices,

---

* The water upon this coast is so shallow, that the land is dangerous to approach without great care, the only harbours being the mouths of rivers.

through which various gases, cinders, ashes, stones, and streams of red-hot melted rocks are projected. From this continued propulsion of matter through a vent or vents, a conical mass is accumulated, to which the name of *volcano* is applied. Volcanos differ materially in the quantity of matter ejected, but agree in such a general resemblance to each other, that they seem all referable to the operations of the same causes.

Various theories have been formed for the explanation of volcanic phænomena; but it must be confessed, that they are all more or less defective, and that the real causes of such phænomena are mere subjects of conjecture. With some of the effects we are familiar; though with the districts most ravaged by erupted matter, we are far from being well acquainted; our principal knowledge of volcanos being derived from the two largest active vents of Europe, Etna and Vesuvius, but principally from the latter. Etna certainly covers a considerable surface, but Vesuvius sinks into insignificance before some of the great volcanos of the world.

From their general proximity to, or occurrence in, the sea, it has been supposed that the active state of volcanos is produced by the percolation of sea-water to certain metallic bases of the earths or alkalies at various depths beneath the surface; which metallic bases being thus inflamed, cause the phænomena observed in volcanic eruptions. The volcanos in the interior of Mexico, as also the volcanos of Tartary, have been accounted for by the advocates of this theory: the former, by supposing a connexion between the vents of Colima, Jorullo, Pococatepetl, and Orizaba, all situated on the same line;—the latter, by considering that the waters of salt lakes may percolate to their foci. Recent researches would, however, appear to render this hypothesis untenable; for according to MM. Klaproth, Abel Rémusat, and Humboldt, there is a volcanic region, with an area of about 2500 square geographical miles in central Asia, between 300 and 400 leagues distant from the sea. The principal seat of volcanic action is the range of the Thian chan, in which are the two volcanos of Pé chan and Ho tcheou, distant about 105 miles in an east and west direction from each other, the former being about 225 leagues from Lake Aral *. Recent observations also, in the central chain of the Andes, by MM. Roulin and Boussingault, show that volcanic eruptions take place at a considerable distance from the sea. The Peak of Tolima (according to Humboldt in lat. 4° 46′ N. and long. 77° 56′ W. from Paris,) was seen to be in eruption by M. Roulin in 1826; and M. Boussin-

---

* Humboldt, Fragmens Asiatiques.

gault observed a volcano of the same district to be in activity in 1829. M. Roulin discovered from an ancient document that there had been a great eruption of Tolima in March 1595 *.

As the first chemical operation, if the theory of a percolation of sea-water to the metallic bases of earths or alkalies were true, would be the union of the oxygen with the metallic base, and the escape of an immense quantity of hydrogen, M. Gay Lussac has objected to it, that pure hydrogen gas is not evolved from volcanos; and as a proof of it, observes, that if it were present, it would be inflamed by the red-hot matter ejected from the craters. Dr. Daubeny endeavours to meet this objection, by supposing the hydrogen " to have combined in its nascent state with sulphur, and the two bodies to have been evolved in the form of sulphuretted hydrogen gas." He also considers that the presence of large quantities of muriatic acid would destroy the inflammability of the hydrogen†.

According to the same author, the gases evolved from volcanos consist of muriatic acid gas, sulphur combined with oxygen or hydrogen, carbonic acid gas, and nitrogen; to which must be added a great quantity of aqueous vapour‡.

Sir Humphry Davy found the sublimations of Vesuvius to consist of a common salt (one specimen containing a minute quantity of muriate of cobalt), chloride of iron, sulphate of soda, muriate and sulphate of potassa, and a small quantity of oxide of copper§.

Volcanic eruptions are usually preceded by detonations in the mountain, and agitations of the earth, or earthquakes in the vicinity, after which the mountain vomits forth an abundance of ashes, cinders, and stones; and streams of melted lava flow from apertures made in the side of the cone, the resistance of which becomes unequal to the pressure of the melted mass within. The lava very rarely seems to proceed from the lip of the crater.

The following is a summary, from various authorities, of the heat and appearances of a lava-current. " Lava, when observed as near as possible to the point from whence it issues, is for the most part a semifluid mass of the consistence of honey, but sometimes so liquid as to penetrate the fibre of wood. It soon cools externally, and therefore exhibits a rough unequal surface; but as it is a bad conductor of heat, the internal mass remains liquid long after the portion exposed to the air has become solidified. The temperature at which it continues fluid is considerable enough to melt glass and silver, and has been found to render a mass of lead fluid in four

* Humboldt, Fragmens Asiatiques. † Description of Volcanos, p. 377.
‡ *Ibid.* p. 376. § Phil. Trans., 1828.

minutes; when the same mass, placed on red-hot iron, required double that time to enter into fusion." The heat does not, however, appear to be always equal; for it is stated, that when bell-metal was thrown into lava (of 1794), the zinc was melted and the copper remained unfused *.

The volcanic eruption which produced the greatest quantity of lava known to have been thrown out at one time, is that recorded as having proceeded in 1783 from the low country near Shaptar Jokul, in Iceland. The lava burst out, according to Sir G. Mackenzie, at three different points, about eight or nine miles from each other, and spread in some places to the breadth of several miles†.

The whole of Iceland may be considered as little else than a volcanic mass, in which there are many apertures through which lava, ashes, and other products have been ejected. The igneous matter struggles to escape in various places, and, consequently, many single eruptions from different points have been recorded since historical times; nevertheless, volcanic discharges have taken place at various times through the same apertures. Thus, there have been twenty-two eruptions from Hecla since the year 1004; seven from Kattlagiau Jokul since 900; and four from Krabla since 1724.

As might be expected in such a region as that of Iceland, the eruptions are not confined to the immediate dry land, but have pierced through the sea in the vicinity. In January 1783, a volcanic eruption, described as flame, rose through the sea, about thirty miles from Cape Reikianes; several islands were observed, as if raised from beneath, and a reef of rocks exists where these appearances occurred. "The flames lasted several months, during which, vast quantities of pumice and light slags were washed on shore. In the beginning of June, earthquakes shook the whole of Iceland; the flames in the sea disappeared; and the dreadful eruption commenced from the Shaptar Jokul, which is nearly two hundred miles distant from the spot where the marine eruption took place‡."

Another submarine eruption occurred near the same island, on June 13th, 1830. An island was produced, and consequent eruptions were feared in the interior, as in the case above cited ‖.

An example of a volcano forcing its way from beneath the sea into the atmosphere was observed off St. Michael's, Azores, in 1811. It was first seen above the sea on June 13th. On the 17th it was observed by Captain Tillard and some other gentlemen from the nearest cliff of St. Michael's. The appearances were exceedingly beautiful, the volcano

---

* Daubeny, Description of Volcanos, p. 381.
† Sir George Mackenzie, Travels in Iceland, 2nd edit.   ‡ *Ibid.*
‖ Journal de Géologie, tom. i.

shooting up columns of the blackest cinders to the height of between 700 and 800 feet above the surface of the water. When not ejecting ashes, an immense body of vapour or smoke revolved almost horizontally on the sea. The bursts are described as accompanied by explosions resembling a mixed discharge of cannon and musketry, and by a great abundance of lightning *. By the 4th of July a complete island was formed, described by Capt. Tillard (who landed upon it,) as nearly a mile in circumference, almost circular, and about 300 feet in height. In the centre there was a crater, then full of hot water, which discharged itself through an opening facing St. Michael's. To this island, which afterwards disappeared, Capt. Tillard gave the name of Sabrina, from that of the frigate which he commanded.

By reference to the manuscript journals of the Royal Society of London, I find that a volcanic island was thrown up among the Western Islands about the middle of the seventeenth century. Sir H. Sheres is described, in the account of the meeting of the Royal Society, on January 7th, 1690--91, as having informed those assembled, " that his father passing by the Western Islands went on shore on an island that had then been newly thrown up by a volcano, but that in a month or less it dissolved, and sunk into the sea, and is now no more to be found†."

The volcano which rose through the sea, between the island of Pantellaria and the coast of Sicily, in July 1831, affords us a recent example of the propulsion of igneous and other rocks through the sea into the atmosphere, forming an island. The water was observed, by Neapolitan vessels, to be heaved up and agitated, and smoke to be evolved over the spot in the early part of July. Intelligence of the circumstance having been received at Malta, vessels were dispatched to ascertain the exact position of the new volcano, and to warn other ships of the danger. On the 18th and 19th of July, Capt. Swinburne estimated the crater, then above the sea, at seventy or eighty yards in external diameter, and twenty feet above the water in the highest place, the agitated and heated water in the crater escaping by an outlet on one side.

Dr. Davy visited this volcano on the 5th of August, and has presented us with a detailed description of the various

* For a view of this scene, and a plan and elevation of the island, see Sections and Views illustrative of Geological Phænomena, pl. 34 & 35.

† These manuscript and unpublished journals of the Royal Society contain a fund of curious information, highly illustrative of the science of the time, the heads of the conversations at each meeting being entered. They moreover afford a valuable insight into the progress of science since the first establishment of the Royal Society.

phænomena he observed. The most common appearance is represented to have been that of a dense white vapour, probably the vapour of water, thrown up to a great height. In violent erupticns, columns of black matter rose to the elevation of three or four thousand feet, spreading out widely even to windward. Vivid lightning, accompanied by thunder, darted amid the atmosphere of eruption, and the sounds thus produced were far more intense than those resulting from the eruptive explosions. Dr. Davy watched in vain for any appearance indicative of the presence of inflammable gas. When in the midst of a dark shower of ashes to the leeward of the volcano, this author remarks, that excepting once or twice, when a slight smell of sulphur was perceived, " no unusual odour, not the slightest bituminous smell, or smell of sulphuretted hydrogen, or of sulphurous acid, or of any other acid fume, was observable *." The various solid matters examined were impregnated with saline matter, resembling that contained in the sea, and afforded slight traces of sulphur. Among the substances ejected were fibres, resembling vegetable fibres, and which emitted a smell like that of burning sea-weed, and it is conjectured that they were portions of weed sucked into the crater.

M. C. Prévost, charged by the Academy of Sciences of Paris with the examination of this volcanic island, visited it on the 28th of September. The island was then about 2300 feet in circumference, and from 100 to 230 feet high, having two principal elevations, on different sides of the crater. The crater filled with boiling water.

From the various accounts which have been received, it does not appear that any lava-current was ejected. The island seemed merely the result of an accumulation of ashes, cinders, and small fragments, among which were occasionally some of larger size. Among the ejected substances were pieces of dolomite and limestone; one fragment, of some pounds weight, is said to resemble grauwacke. As was to be expected, the island, when the eruptions ceased, was unable to resist the action of the breakers; accordingly it has given way before this force, and is now reduced to a shoal.

The bottom whence this island was thrown up is represented to have been at little depth beneath the surface of the sea. Capt. Smyth has, however, clearly shown that this was not the case, but that on the contrary it rose in deep water, unconnected with the Adventure Bank, to the westward of it †. Some miles to the S.E. of this island, which has been named Sciacca, Julia, Hotham, Graham, and Corrao, is the

* Dr. Davy, Phil. Trans., 1832. Smyth, Phil. Trans., 1832, pl. 7.

Nerita bank, which may perhaps be the subaqueous summit of another volcanic mass.

If we consider that the heat of the volcano would destroy the various marine animals, which either lived in or on the sand, mud, or gravel of the previous bottom, such animals would probably be buried beneath the mud and cinders derived from the explosions.

The volcano would seem to have been in activity some little time beneath the sea before it reared itself above the surface; for Capt. Swinburne, while passing over the same spot on the 28th of the previous month (June), experienced shocks which were then attributed to an earthquake*.

It can only have been since historical times, and by mere accident, that instances of volcanos so forcing themselves from beneath the sea could have been recorded. Now, the power of man to do this is so recent, that we may conclude such occurrences to have been far from rare; and that, even in the present day, they may happen in remote regions, into which civilized man rarely, if ever, enters, and therefore they remain unknown.

There are numerous islands in the ocean, composed almost entirely of volcanic matter, and in which active volcanos still exist, that may have been thus formed; the dome or cone not giving way before the pressure of the water, but gradually accumulating a mass of lava, cinders and ashes, so that the islands have become firm, and even of considerable size. Owhyhee, or Hawaii, is perhaps a magnificent example of such an island. The whole mass, estimated as exposing a surface of 4000 square miles, is composed of lava, or other volcanic matter, which rises in the peaks of Mouna Roa and Mouna Kaah, to the height of between 15,000 and 16,000 feet above the level of the sea. Mr. Ellis describes the crater of Kirauea as situated in a lofty elevated plain, bounded by a precipice fifteen or sixteen miles in circumference, apparently sunk from two hundred to four hundred feet below its original level. " The surface of this plain was uneven, and strewed over with loose stones and volcanic rock; and in the centre of

---

* It is impossible not to be struck, in the drawings and plans of the islands of Sabrina and Sciacca, with the resemblance they bear to those volcanic islands which have basins in them, into which there is a narrow passage communicating with the sea. Deception Island, New South Shetland, (of which there is a description and a plan in the Journal of the Geographical Society,) affords a good idea of such islands. The interior basin is there five miles in diameter and ninety-seven fathoms deep. Many other examples will readily present themselves to the geographer. The communication between the interior basin and the sea would seem produced, in the cases of Sabrina and Sciacca, by the rush of the waters out of the crater during the explosions.

it was the great crater, at a distance of a mile and a half from the place where we were standing. We walked on to the north end of the ridge, where, the precipice being less steep, a descent to the plain below seemed practicable. After walking some distance over the sunken plain, which in several places sounded hollow under our feet, we at length came to the edge of the great crater, where a spectacle sublime, and even appalling, presented itself before us. Immediately before us yawned an immense gulf, in the form of a crescent, about two miles in length, from N.E. to S.W., nearly a mile in width, and apparently 800 feet deep. The bottom was covered with lava, and the S.W. and northern parts of it were one vast flood of burning matter in a state of terrific ebullition, rolling to and fro its ' fiery surge' and flaming billows. Fifty-one conical islands of varied form and size, containing so many craters, rose either round the edge, or from the surface of the burning lake; twenty-two constantly emitted columns of gray smoke, or pyramids of brilliant flame; and several of these at the same time vomited from their ignited mouths streams of lava, which rolled in blazing torrents down their black indented sides into the boiling mass below." Mr. Ellis concluded, from the existence of these cones, that the mass of boiling lava resulted from the streams poured from the craters into this upper reservoir, which appeared to vary in its level; for there were marks on the rocks bounding it, which showed that the great crater had been recently filled up 300 or 400 feet higher to a black ledge, from whence there was a slope to the hot fluid mass*.

It will be obvious that this crater by no means resembles those with which we are more familiar. Instead of the more or less rounded orifice usually found, we have a semicircular crack in a level of considerable extent, and, by the description, this level does not appear to have been ravaged by lava streams flowing from the crater over it. The depth of water round Owhyhee, and indeed round the Sandwich Islands generally, is so great, that they are somewhat dangerous to approach in stormy weather, as anchorage cannot be obtained except close to the land ; seeming to show that these volcanic masses rise from considerable depths, and are only partly out of the water.

The number of volcanos which fringe the Pacific Ocean, or occur in it, or in that part of the Indian Seas which contains Java and the neighbouring islands, far exceeds that of any

* Ellis, Tour through the Sandwich Islands. An interesting account of the state of Kirauea, in 1829, will be found in Stewart's Visit to the South Seas. The general description is not materially different, the changes being principally in the crater.

other part of the world. From Terra del Fuego they occur
northerly through the range of the Andes, often attaining very
considerable elevations. In Mexico the northerly line is met
by an east and west line, connecting it with the volcanos in
the West Indian Islands. In California there are three volcanos, of which one, Mount St. Elia, is variously estimated
from 13,000 to 17,000 feet in height. America is connected
with Asia by means of the volcanic vents of the Aleutian Isles.
From Kamtschatka southwards we observe volcanos in the
Kurule Islands, Japan, the Loo Choo Isles, Formosa, and the
Philippines. From the latter islands a range of volcanic vents
proceeds to nearly lat. 10° S., ranges westward along this parallel for about twenty-five degrees of longitude, and then turns
up N.W. diagonally through about twenty degrees of latitude.
This line, which when represented in maps * resembles an
enormous fish-hook, passes from the Philippines, by the N.E.
point of Celebes, Gilolo, the volcanic isles between New Guinea and Timor, Floris, Sumbawa, Java, and Sumatra, to
Barren Island.

Active volcanos are by no means relatively so abundant in,
or on the shores of, the Atlantic. Indeed the shores of this
ocean in Europe, Africa, and America, appear free from them,
if we except Mexico and the land connecting the main body
of North America with the Southern continent, and which
may be considered as common both to the Atlantic and Pacific Oceans †.

Teneriffe affords the greatest volcanic elevation in the Atlantic, the Peak rising 12,216 feet above its surface. Iceland,
though its volcanos do not attain any considerable elevation,
presents the largest accumulation of volcanic matter above the
level of the same mass of waters.

We have seen that in Iceland high cones or elevations of
land do not always accompany volcanic eruptions, for the lava
of 1783 seems to have flowed from comparatively low apertures. Elevations seem more especially formed when the
erupted matter consists of cinders, ashes, or stones, which
being ejected, arrange themselves in a conical manner around
the central aperture, where the amount of melted rock or lava
may vary. The escape of this melted rock will, in a great
measure, depend on its relative proportion to the cinders,
ashes, or stones thrown out. If these be in comparatively

* See Von Buch's Canary Islands, pl. 13; and a corrected reduction of
this in Lyell's principles of Geology, pl. 1.
† Mr. Scoresby notices a volcano off the main land of Greenland. This
volcano is situated in the island of Jan Mayen, presented marks of recent
eruption, and had a crater about 500 feet deep, and 2000 feet in diameter.
—Edin. Phil. Journal.

small quantity, the lava will have the less difficulty to escape, and may easily break down its barrier and rush forth. But when the proportions are inverted, a large cone may be raised without the escape of any lava-current. Between the two extremes there will be every kind of variation, and lava-currents will flow from various apertures and at various heights. By repeated action a volcano acquires considerable solidity at its base, for the loose erupted matter is, independently of the consolidation produced by other causes, bound together by lava radii proceeding from the central aperture. Rents are often produced in the base, particularly when the great vent has accumulated matter to a considerable height, and through these, lava is protruded; the streams so thrown out serving to brace the lower parts of the mountain more firmly together. The occurrence of such apertures is precisely what we should expect in a volcano, which had accumulated materials upon it nearly equal to the average force of the elastic vapours propelling igneous matter upwards; for the pressure of the elevated column being very considerable, and in proportion to its height, it will always struggle to free itself in the direction of the least resistance. Now the sides of a volcanic mountain are not likely to be homogeneous, but to vary much in their resisting powers, being most solid where crossed by lava-currents, and weakest where merely formed of ashes or substances of the like nature. If to these causes of unequal resistance to pressure, we add the fractures and rents produced by shocks in the mountain itself, we should always expect to find lateral discharges of lava common, while similar streams from the mouth would be rare.

M. von Buch is the author of a theory respecting the elevations of volcanos, which has been adopted by many geologists, while it has been combated by others. He observes, that the appearances of many craters are such, that we can scarcely consider them as erupted in the ordinary way; because they do not seem to present either lava-currents, or such an arrangement in the deposit of other volcanic substances as to justify such a conclusion. To these craters he has given the name of Craters of Elevation (*Erhebungskrater*). It has been opposed to this theory, that it presupposes a horizontal accumulation of lava or other volcanic matters, previously to the propulsion of elastic vapours through it, which should elevate the flat mass in a dome or cone, and burst through the highest part, presenting the appearance of a crater of eruption. How far this objection may be valid would seem to depend on the possibility of forming sheets of volcanic matter, which heat might soften and elastic vapours force up, so that the necessary forms should be produced. It may be questionable whether, under a great pressure of the sea, there is the same tendency to produce cinders and ashes as in the atmosphere; and whether the superincumbent weight would not so act upon the solid matter ejected, that it would be forced into fusion, and sheets of melted matter be the result, if the elastic vapours beneath a column of melted rock were sufficiently powerful to overcome the resistance both of the column of lava and the superincumbent water.

It being by no means probable that the density of sea-water, beneath any depth which we can reasonably assign to the ocean, would be such as to render it of greater specific gravity than liquid lava ejected from a volcanic rent, situated beneath the sea; it would follow, that so long as the lava continued in a state of fusion, it would arrange itself horizontally beneath the fluid of inferior specific gravity. The question then arises, how long a body of lava in fusion would remain fluid beneath the waters of the sea. The particles of water in contact with the incandescent lava would become greatly heated, and consequently, from their decreased specific gravity, would immediately rise, their places being supplied from above by particles of greater density and less temperature. Thus a cooling process would be established on the upper surface of the lava, rendering it solid. Now as the particles of fluid lava would be prevented from moving upwards by the solid matter above, pressed downwards by its own gravity and the superincumbent water, they would escape laterally, where not only the cooling process would be less rapid, from the well-known difficulty of heated water moving otherwise than perpendicularly upwards, but where also the power of the fluid lava to escape resistance would be greatest. Let *a*

Fig. 20.

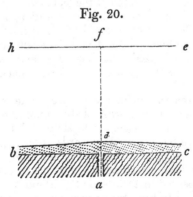

(Fig. 20) be a volcanic rent, through which liquid lava is propelled upwards in the direction *d f*. The lava being of greater specific gravity than the water *b h e c*, it would tend to arrange itself horizontally in the directions $d\,b$, $d\,c$. The surface $b\,d\,c$ having become solid, the lava would escape from the sides b and c, spreading in a sheet or tabular mass around ; and this effect would continue so long as the propelling power at a was sufficient to overcome the resistance opposed to the progress of the lava, or until the termination of the eruption, if that should first happen.

If such would be the state of things beneath a considerable depth of water, the tendency to produce ashes and cinders in a volcanic vent would increase with its approach to the surface of the water; and therefore all the phænomena of eruptions from beneath the surface of the sea would differ but little from those observed in the atmosphere. Another objection to the theory of craters of elevation is, that the stratification of such supposed craters is precisely that of craters of eruption; and that therefore the inference from this circumstance would be in favour of the latter, because we now have daily examples of such modes of formation, while of the other we have none. Data on this subject are so few, that it seems difficult to estimate the value of this objection. The fact, however, that solid rocks can be raised by elastic vapours, is shown in the case of the Little and New Kameni, (Island of Santorino,) where brown trachyte, of a resinous lustre and full of crystals of glassy felspar, was upraised ; the former in 1573, and the latter in 1707 and 1709. The elevation of the Little Kameni was " accompanied by the discharge of large quantities of pumice, and a great disengagement of vapour [*]." By terming this rise an earthquake, we merely seem to be using two names for the same thing. That there were elastic vapours it is clear, and that these vapours were the propelling power may be fairly inferred ; therefore the fact is the same, whether we call it an earthquake or a volcanic elevation, and it would be somewhat difficult to draw fine lines of distinction between the two. The trachyte of New Kameni was observed to have shells upon it when raised, and limestone and marine shells are described as composing a part of these otherwise igneous islands [†].

These occurrences at Santorino are quite sufficient to show, that volcanic rocks, with shells upon them, may be raised bodily to the surface. Clayslate and limestone appear also to have been forced upwards at some previous period, as they are seen at Mount Elias, dipping from the interior outwards. Langsdorff notices a trachyte rock 3000 feet high, which appeared in 1795 near the island of Unalaschka, and which seemed to have been thrown up as a mass from the

* Lyell, Principles of Geology, vol. i. p. 386.   † *Ibid.*

bottom of the sea*. M. Omalius d'Halloy cites M. Rein-
wardt as stating, that on the western side of the Isle of Banda,
a bay was, in 1820, replaced by a promontory formed of
huge blocks of basalt. The rise of land is described as having
been so gradual, that the inhabitants were not aware of the
change until it was nearly completed. It was accompanied
by a bubbling and great heat of the sea †. Considering this
account as correct, it is a remarkable example of the quiet
rise of land above the level of the ocean.

Ingenious explanations have been given to account for the
large orifices which have been termed craters of elevation.
Mr. Lyell considers that the crater resulting from the destruction of the summit of Etna in 1444, was as large as those
noticed in other places and named craters of elevation; and
supposes that a series of great explosions might so reduce the
cone, that finally there would be a circular bay, forty or fifty
miles round, in an island seventy or eighty miles in circumference, wholly composed of volcanic rocks which should dip
outwards. But supposing such appearances to have been
produced, the whole base of Etna, a kind of circular island,
would still show its lava-currents, sections of which would be
observed in the interior bay, or might be exposed outside, and
no doubt would remain that it was a crater of eruption. How
far the so-called " craters of elevation" may resemble the supposed case of Etna remains to be seen ; yet if they should not,
as is considered they do not, present traces of lava-currents,
radiating from a centre or centres, but large envelopes of trachyte or other fused volcanic rock, they can scarcely be referred to the same origin. There does seem a possibility of
producing craters of elevation by the action of heat and elastic
vapours on a sheet of lava, therefore the subject should be
fairly investigated, without bias, with proper caution, and in
the necessary detail.

It is supposed that after the craters of elevation were formed,
the eruptive action poured forth the usual volcanic substances,
which, when it was continued sufficiently long, produced a
cone like the Peak of Teneriffe; but when such eruptive action
was small, or the crater comparatively recent, the appearances
were such as we now observe at Barren Island in the Bay of
Bengal, where a central cone, in activity, in the midst of a
basin of water, is surrounded by a circular range of volcanic
ground, which, according to the figure given by Mr. Lyell‡,
rises at an angle of about 45° from the sea. The height of

---

* Daubeny, Description of Volcanos, p. 310.
† Omalius d'Halloy, Elémeuts de Géologie, p. 405.
‡ Principles of Geology, vol. i. p. 390.

the central cone is about 1800 feet above the water, and the
elevation of the surrounding volcanic circle being nearly the
same, the interior is only viewed through a break in it. It
would appear that the rocks of this island are extremely hot;
for Capt. Webster, landing upon it in March 1822 or 1823,
found the water almost boiling at one hundred yards from the
shore; the stones upon the beach, and the rocks exposed by
the ebb tide, hissing and steaming, and the water bubbling
around them*.

Von Buch adduces the Caldera in the Isle of Palma, Canaries, as a good example of the craters of elevation. A large
precipitous cavity or crater exists in a lofty range sloping
outwards, which incloses it on all sides but one, where a gorge
forms the only communication from the exterior to it. The
sides of this great cavity expose a section of beds of basalt, and
conglomerates composed of basaltic fragments, dipping regularly outwards. White trachyte, and a rock composed of
hornblende and white felspar, are also noticed. Now if the
beds be so regular, and not composed of scoriaceous matter
or ashes, as it is stated they are not, their formation would
seem not to have taken place in the air or beneath a small
pressure of water, but under different circumstances, which
would permit the basalt to be flattened into tabular masses,
not presenting the appearance of lava-currents which have
flowed in the atmosphere.

Jorullo affords a striking example of the outburst of volcanic
action in the interior of dry land, where no active volcanos
then existed, though the rocks in the vicinity would seem to
indicate their previous presence. Judging from the direction
of the vents, a cleft seems to extend east and west across
Mexico to the Revillagigedo Isles in the Pacific. Previous
to June 1759, the space where the volcano of Jorullo now
stands was covered by indigo and sugar-canes, bounded by
two brooks, the Cuitimba and San Pedro. In June, hollow
subterranean noises were heard, accompanied by earthquakes,
which lasted from fifty to sixty days. Tranquillity seemed
re-established at the commencement of September, but on the
28th and 29th of this month the subterraneous noises again
commenced, and, according to Humboldt, the ground, with a
superficies of three or four square miles, rose up like a bladder.
The extent of this movement is considered to be now marked
by an elevation round its edges of 39 feet, gradually acquiring
a height of 524 feet towards the centre of the present volcanic district. The eruption appears to have been very
violent, fragments of rock were hurled to great heights, ashes

* Edin. Phil. Journal, vol. viii.

were thrown up in clouds, and the light emitted was seen at considerable distances. The Cuitimba and San Pedro are described as having precipitated themselves into the volcanic vent, and to have assisted, by the decomposition of their waters, the fury of the eruption. "Eruptions of mud, and especially of strata of clay, enveloping balls of decomposed basalt in concentric layers, appear to indicate that subterraneous water had no small share in producing this extraordinary revolution. Thousands of small cones, from 6 to 10 feet in height, called by the natives Hornitos (ovens), issued forth from the Malpays. Each small cone is a fumirole, from which a thick vapour ascends to the height of from 22 to 32 feet. In many of them a subterraneous noise is heard, which appears to announce the proximity of a fluid in ebullition." From amid these cones, six volcanic masses, varying from 300 to 1600 feet in height above the old plain, were ejected from a chasm having a direction N.N.E. and S.S.W. The most elevated mass is named Jorullo, and from its north side a considerable quantity of lava, containing fragments of other rocks, has been thrown out. The great eruptions ceased in February 1760, and afterwards became gradually less frequent.— The opponents to the theory of craters of elevation consider the raising of the ground in the form of a bladder as not altogether proved, resting on Indian accounts of appearances, which have been considered with reference to a particular theory.

The well-known Monte Nuovo near Naples was thrown up in a day and a night in 1538. This is also described as ejected from a fissure. The present height of this volcanic elevation is 440 feet above the sea, and its circumference about a mile and a half.

Various descriptions of volcanic eruptions will be found in works dedicated to the subject, and could not be admitted within the necessary limits of this volume. The following account, however, obtained by the exertions of Sir Stamford Raffles, of a great eruption from Tomboro, in the island of Sumbawa, is too important to be omitted.—The first explosions were heard at various distant places, where they were very generally mistaken for discharges of artillery. They commenced on the 5th of April 1815, and continued more or less until the 10th, when the eruptions became more violent; and such a great discharge of ashes took place that the sky was obscured, and darkness prevailed over considerable distances. It appears that a Malay prow, while at sea on the 11th, far from Sumbawa, was enveloped in utter darkness, and that, afterwards passing the Tomboro mountain at the distance of about five miles, the commander observed that the lower part

appeared in flames, while the upper portion was concealed in
clouds. Upon landing, for the purpose of procuring water,
he found the ground covered to the depth of three feet by
ashes, and " several large prows thrown on shore by the con-
cussion of the sea." Quitting Sumbawa, he with difficulty
sailed through a quantity of these ashes floating on the sea,
which he described as two feet thick, and several miles in extent. This person also stated that the volcano of Carang
Assam, in Bali, was convulsed at the same time. The most
interesting account is that presented us by the commander of
the East India Company's cruiser Benares, which is nearly
as follows:—At the commencement of the explosions this vessel
was at Macasar, and the reports so closely resembled those of
cannon, that it was supposed there was an engagement of pirates somewhere in the neighbourhood. Troops were consequently embarked on board the Benares, and the vessel stood
out to sea in search of the supposed pirates. On the 8th of
April she returned, without having found any cause for alarm.
On the 11th the apparent discharges of cannon were again
heard, sometimes shaking the ship and Fort Rotterdam. The
vessel proceeded southward to ascertain the cause of these
explosions. At eight o'clock on the morning of the 12th,
" the face of the heavens to the southward and westward had
assumed a dark aspect, and it was much darker than when
the sun rose; as it came nearer it assumed a dusky red appearance, and spread over every part of the heavens; by ten it was
so dark that a ship could hardly be seen a mile distant; by
eleven the whole of the heavens was obscured, except a small
space towards the horizon to the eastward, the quarter from
which the wind came. The ashes now began to fall in show-
ers, and the appearance was altogether truly awful and alarm-
ing. By noon the light that remained in the eastern part of
the horizon disappeared, and complete darkness covered the
face of day. This continued so profound during the remainder
of the day, that I (the commander of the Benares) never saw
anything to equal it in the darkest night; it was impossible to
see the hand when held close to the eyes. The ashes fell
without intermission throughout the night, and were so light
and subtile that, notwithstanding the precaution of spreading
awnings fore and aft as much as possible, they pervaded every
part of the ship."

"At six o'clock the next morning it continued as dark as
ever, but began to clear about half-past seven, and about eight
o'clock objects could be faintly observed on deck. From this
time it began to clear very fast . . . . . The appearance of the
ship when day-light returned was most singular; every part
being covered with the falling matter. It had the appearance

of calcined pumice-stone, nearly the colour of wood-ashes; it lay in heaps of a foot in depth on many parts of the deck, and several tons weight of it must have been thrown overboard; for though an impalpable powder or dust when it fell, it was, when compressed, of considerable weight. A pint measure of it weighed twelve ounces and three quarters; it was perfectly tasteless, and did not affect the eyes with a painful sensation; had a faint smell, but nothing like sulphur; when mixed with water it formed a tenacious mud difficult to be washed off."

The same vessel left Macasar on the 13th, and made Sumbawa on the 18th. Approaching the coast she encountered an immense quantity of pumice-stone, mixed with numerous trees and logs with a burnt and shivered appearance. When arrived at Bima Bay, the anchorage was found to be altered, as the vessel grounded on a bank where a few months previously there had been six fathoms of water. The shores of the bay were entirely covered with the ashes ejected from Tomboro, which is distant about forty miles. The explosions heard at Bima were described as terrific, and the fall of ashes so heavy as to break in the Resident's house in many places. There was no wind at Bima, but the sea was greatly agitated, the waves rolling on shore, and filling the lower parts of the houses a foot deep. When off the Tomboro mountain, about six miles distant, on the 23rd, the commander of the Benares observed the summit to be enveloped in smoke and ashes, while the sides showed lava-currents, some of which had reached the sea.

The explosions were heard at very considerable distances. Not only were they noticed at Macasar, which is 217 nautical miles from Tomboro, but also throughout the Molucca Islands; at a port in Sumatra, distant about 970 nautical miles from Sumbawa; and at Ternate, distant 720 miles.

Lieut. Phillips being dispatched to relieve the wants of the inhabitants, who were perishing from famine and disease, learned from the Rajah of Saugar, that about seven o'clock in the morning of the 10th of April there was an appearance of three distinct columns of flame, all within the crater, which united at a great height upwards; and that, subsequently, the whole mountain appeared like a mass of liquid fire. How far the appearance of flame may be correct, it would be difficult to say, as nothing is so common as deceptive appearances of this kind; its character, however, would seem remarkable.

The Rajah's account proceeds:—" The fire and columns of flame continued to rage with unabated fury, until the darkness caused by the quantity of falling matter about eight P.M. Stones at this time fell very thick at Saugar, some of them as large as two fists, but generally not larger than walnuts."

Soon after 10 P.M. a violent whirlwind arose, "which blew down nearly every house in the village of Saugar, carrying the tops and light parts along with it. In the part of Saugar adjoining Tomboro its effects were much more violent, tearing up by the roots the largest trees, and carrying them into the air, together with men, houses, cattle, and whatever else came within its influence." The sea was agitated, rising twelve feet higher than it was ever known to do before. The water rushed upon the land, sweeping away houses and all within its influence, and destroying the few rice-grounds which previously existed at Saugar. As might have been expected amid such a convulsion, a great destruction of life was effected, and many thousand inhabitants were killed. The vegetation on the north and west sides of the peninsula was completely destroyed, with the exception of a high point of land where the village of Tomboro previously stood, and where a few trees still remained *.

The changes produced by such eruptions as that here recorded, would, independently of the alteration in the shape of the volcano itself, and of the streams of lava which flowed from it, extend to very considerable distances. On the dry land, vegetables and animals would be entombed beneath stones and ashes, the quantity of the covering matter probably increasing with the proximity to the volcano. And if it should chance, as sometimes happens, that the aqueous vapours discharged from the volcanic vent were suddenly condensed, the torrents produced would sweep away not only the looser parts of the volcano, but also the plants and animals which they might encounter, embedding them in a thick mass of alluvial matter.

The vegetable and animal substances enveloped by the discharged ashes, cinders, and stones falling into the sea, would be both marine and terrestrial; and a very curious mixture, as far as regarded its organic contents, would be observed: trees, men, cattle, fish, corals, and a great variety of marine remains, would be encased, and it might so happen that both on the land and in the sea a bed of lava might cover such accumulations.

In the case of the great discharge of lava in Iceland, in the year 1783, many terrestrial remains may have been covered by the igneous matter, possibly some in such situations as to preserve their form. Should a similar eruption take place in the sea, where, as before observed, the conditions are favourable for the production of a sheet of lava, sands and clays, perhaps full of marine remains, would be covered over, and very considerable changes might be produced by such a

* Life of Sir Stamford Raffles.

superincumbent mass of heated matter. Upon these, after a certain time, sands and clays, charged with organic remains, might be accumulated, and again covered up by a new eruption. Thus producing an alternation of igneous and aqueous rocks.

Mr. Henderson notices an alternation of fossil wood, clay, and sandstone in Iceland, surmounted by basalt, tuff, and lava. When this accumulation of vegetable matter was so covered is not so clear; but if Mr. Henderson be right in considering many of the fossil leaves as those of the poplar, it is not, probably, very recent, for it supposes a change of climate, as poplars do not now grow in Iceland*.

During great explosions, volcanos cannot be approached sufficiently near for the purposes of very minute observation; therefore we can only judge of some of the probable effects from appearances at their calmer periods, and consequently a minor state of activity is very favourable for such examinations. After ineffectual attempts to observe the workings of the fluid mass within the crater of Vesuvius at the commencement of 1829, when that mountain was somewhat active, I was fortunate enough on the 15th of February to have ascended on a calm day, when the vapours darted majestically upwards as they were propelled from the small cone in the middle of the grand crater†. The incandescent matter in the vent was at times distinctly visible,—a rare circumstance; for should there be the slightest movement in the air, every object is obscured by the vapours. After the more continued detonations there was a lull or calm, succeeded by a violent explosion, throwing up stones to a considerable height, mixed with pieces of red-hot lava, the latter falling like lumps of soft paste on the sides of the small cone. When the vapour cleared away, the red-hot mass appeared as if in ebullition from the passage of the gaseous matters through it. The light emitted varied exceedingly in intensity, being brightest at the moment of the great explosion, when a great volume of vapour suddenly forced its way through the fiery mass, darting up with great velocity, and carrying all before it. Wishing to profit by my good fortune, I continued many hours on the mountain, until night closed in, hoping that objects might be perceived within the crater not previously observed. In this I was disappointed, appearances being the same, though more distinctly visible. The picturesque effect, however, was greatly heightened; the solid ejected substances darted upwards like a grand discharge

* Henderson's Iceland, vol. ii. p. 115. According to this author, the lignite deposit occurs extensively in the N.W. peninsula of Iceland.

† For a sketch of the crater at this time, see Sections and Views illustrative of Geological Phænomena, pl. 22.

of red-hot balls, while the reflection of the incandescent matter within, on the vapour above, was at times extremely
brilliant, producing, at a distance, those false appearances of
flames so frequently noticed. Flame, that is, the combustion
of some inflammable gas, does, however, seem to issue, though
very rarely, from volcanic vents. Sir Humphry Davy notices
a jet of flame as proceeding from an aperture on the side of
Vesuvius facing Torre del Greco, in May 1814 : it rose to
the height of sixty yards, accompanied by a violent hissing
noise. It continued for three weeks, but Sir Humphry Davy
was unable to collect any inflammable gas *.

The products of active volcanos, though man seems to ex-
haust his language in finding terms to express his horror and
dismay at their mode of ejection, do not constitute such an
addition to dry land as at first sight would appear probable,—
for their mass must be regarded relatively to the mass of dry
land generally, and not with reference to particular districts.
Moreover, cavities corresponding with the quantity of matter
thrown out will sometimes occur not far beneath the surface ;
and when the weight above shall overcome the resistance below, either suddenly from a violent convulsion, or slowly from
gradual change, the mass above will fall into the abyss beneath,
and matter be, in some measure, restored to its place. Among
volcanic changes it is by no means uncommon to hear of hills
disappearing, and being converted into lakes. The most me-
morable example, perhaps, of the disappearance of a volcano,
is that which took place in Java in 1772. The Papandayang,
on the south-western part of the island, reputed one of its
largest volcanos, was observed at night, between the 11th and
12th of August, to be enveloped by a luminous cloud. The
inhabitants being alarmed, betook themselves to flight ; but
before they could all escape, the mountain fell in, accompanied
by a sound resembling the discharge of cannon. Great quantities of volcanic substances were thrown out, and carried over
many miles. The extent of ground thus swallowed up was
estimated at fifteen miles by six. Forty villages were engulfed
or covered by the substances thrown out, and 2957 persons
were reported to have been destroyed †.

Extinct Volcanos.

From a similarity of appearances, rocks existing under certain circumstances where there are at present no active vents,
have been attributed to a volcanic origin. To draw fine lines
of distinction between volcanos now in activity and those

---

* Davy, Phil. Trans. 1828.   † Horsfield, as quoted by Daubeny.

which appear extinct, would be almost impossible, for there is
no certainty that the one may not soon be converted into the
other. Of this we probably have a good example in Vesuvius,
which after being, as far as we can judge from historical re-
cords, for a long period extinct, became convulsed in the year
79, destroyed the higher part of its old cone, part of which
now remaining is named Monte Somma, and overwhelmed
Herculaneum, Pompeii, and Stabiæ, entombing not only men,
but theatres, temples, palaces, and innumerable works of art,
which have afforded by their disinterment more real knowledge of the manners and customs of the ancient inhabitants of
these beautiful regions of Italy, than all the writings which
have escaped destruction.

Solfataras, as they are termed, are usually considered as
semi-extinct volcanos, emitting only gaseous exhalations and
aqueous vapour; but there can be no certainty that they also
may not again enter into activity. According to Dr. Daubeny,
sulphuretted hydrogen and a small portion of muriatic acid
are contained in the steam which rushes out of the fuma-
roles at the Solfatara near Naples. The rocks of the crater
and vicinity are greatly decomposed by the action of these
gaseous exhalations; and, among other salts thus formed, the
muriate of ammonia is the most abundant. Solfataras, variously modified, are by no means rare in volcanic countries.

Not only do extinct volcanic vents occur in regions where
active volcanos now exist,—so that we may imagine a mere
change of fiery orifice,—but they are also found in districts
where all trace of activity has been lost since the earliest historical times, if we except the presence of mineral and thermal
springs. In Central France and in Germany such appearances are particularly remarkable, and it has been attempted
to draw a line of distinction between those volcanos which
have existed in a state of activity since the establishment of the
present order of things, and those whose activity was previous
to this state. The subject is full of difficulty, more especially
as respects Central France, where volcanic ejections have taken
place at different periods; so that there is no ready mode of
making geological distinctions between the ejections, which
would seem little else than productions from new orifices
opened for the discharge of volcanic matter in the same region.
We may be able to observe the extremes, but to mark
striking and easily distinguishable points intermediate between
them would be exceedingly difficult. Volcanic ejections were
probably continued through nearly the same orifices for a long
period of time, during which many and great geological
changes were taking place around them, and on the surface of
the earth generally.

It has been attempted to determine the relative ages of volcanos by the absence or presence of craters; as also on the supposition that some have existed prior to the excavation of valleys, while others have been produced after their formation,
their lava-currents having been discharged into them. Such
distinctions can scarcely be made; for craters may be easily
obliterated, and relative age, from the excavation of valleys,
cannot be very satisfactorily established amid circumstances
which could so easily produce changes in this respect. A
more direct mode has been to try their relative antiquity by
means of the mineral structure of their lavas; and if this should
hold good, it would be the safest guide; but it may be doubted
how far our knowledge of volcanic products authorizes so general a conclusion. That there is a great difference in the
mineral character, generally, between the igneous rocks of the
older periods of the world and those at present formed, few
will doubt. We know of no granite or serpentine streams
thrown out from modern volcanos; but when igneous rocks so
closely allied in geological dates as those produced by active
and extinct volcanos are under consideration, such distinctions
should not be too hastily adopted.

Dr. Daubeny considers that the more modern volcanic products of Auvergne are more cellular, have in general a harsher
feel, and possess a more vitreous aspect than the more ancient *.

In Auvergne and the Vivarais there are numerous examples
of the more modern extinct volcanos, the craters of which are
frequently very perfect, or merely broken down by the discharge of large lava-currents from them. Details respecting
them will be found in works written expressly on the subject,
and pictorial representations among the views contained in
Mr. Scrope's work on Central France†.

In the district of the Eifel, near the Rhine, there are also
extinct volcanos which have been considered as comparatively
recent, from the situations which they occupy; having been
apparently produced after the formation of the valleys in the
neighbouring country. In the volcanic district of Central
France the lava-currents have in some places traversed valleys,
and dammed up the waters that pass through them. The
waters so dammed up accumulated into a lake, which was
subsequently drained through a gorge cut in the rocky barrier
by means of the surplus water; which not only effected this,
but also cut, by continual erosion, into the rock beneath,
forming a part of the original valley.

Many other examples of extinct volcanos have been noticed

---

* Description of Volcanos.

† Part of one of the most striking of these views is copied in "Sections
and Views illustrative of Geological Phænomena," pl. 24.

in districts where active volcanos do not now exist. Their relative antiquity is however so little understood, that a general classification of them cannot be attempted.

Mineral Volcanic Products.

Various classifications of volcanic substances have been proposed, among which the division into Trachytic and Basaltic seems to be that most commonly adopted ; trachyte being considered as essentially composed of felspar, and containing crystals of glassy felspar; while basalt is supposed to be essentially composed of felspar, augite, and titaniferous iron. Lavas, however, present such various mixtures of different minerals, that exact classifications of them would appear exceedingly difficult; and when we consider that these different compounds may be infinitely modified by circumstances, such classifications cannot be of much value. These products are of such a compound nature, consisting of felspar, augite, leucite, hornblende, mica, olivine, and other minerals, that definite names can scarcely be attached to them. Mr. Poulett Scrope has distinguished the rocks termed trachyte, basalt, and graystone (the latter a name proposed by himself,) under the following heads:—1. *Compound trachyte,* with mica, hornblende, or augite, sometimes both, and grains of titaniferous iron. 2. *Simple trachyte,* without any visible ingredient but felspar. 3. *Quartziferous trachyte,* when containing numerous crystals of quartz. 4. *Siliceous trachyte,* when apparently much silex has been introduced into its composition. 1. *Common graystone,* consisting of felspar, augite, or hornblende, and iron. 2. *Leucitic graystone,* when leucite supplants the felspar. 3. *Melilitic graystone,* when melilite supplants the felspar, &c. 1. *Common basalt,* composed of felspar, augite, and iron. 2. *Leucitic basalt,* when leucite replaces the felspar. 3. *Olivine basalt,* when olivine replaces the felspar. 4. *Hauyine basalt,* when hauyine replaces the felspar. 5. *Ferruginous basalt,* when iron is a predominant ingredient. 6. *Augite basalt,* when augite composes nearly the whole rock*.

As all fused substances will tend to crystallize, or arrange their component parts more compactly, where their liquidity continues the longest, and their loss of temperature is the slowest, we find that lava-currents are always more crystalline or compact in their interior parts, and that dykes cutting volcanic cones are generally more compact and crystalline than the lavas which flow from them ; such dykes being also more crystalline towards their interior parts than towards their walls or sides. It has been inferred from the appearance and distribution of the ejected matters, that many volcanic rocks have not been formed in the atmosphere, but beneath seas, and that they have been subsequently elevated. The ashes and pumice ejected from volcanos seem merely, if I may so express myself, the frothy part of the great fused and incandescent matter within, produced by the action of elastic vapours, or by the intumescence of that matter under diminished pressure. The force required to eject such light substances is evidently far inferior to that necessary for the propulsion of the more solid lava, and consequently the one is in general more common than the other. As might be expected from the nature of such mineral productions, volcanic substances vary, from the lightest ash to a highly crystalline rock, the intermediate states being vitreous, and of the character of obsidian. The quantity of minerals detected in volcanic products is exceedingly great, a circumstance by no means surprising when we consider the various elementary substances acted on by heat in the bowels of a volcano, and striving to combine with each other in various ways*.

Not only are fused substances ejected, but also various portions of rocks traversed by the volcanic vent; and as this is very variously situated, so are the rocks various which are thrown out. Vesuvius having been under observation for so long a period, its products have received greater attention than falls to the lot of most volcanos; and it has been observed, though no doubt volcanos vary most materially in this respect, that such ejected substances are far from being either rare or of one kind. The Chevalier Monticelli's invaluable collection of Vesuvian products at Naples contains a great variety of these substances, among which may be seen fragments of the compact limestones of the district, with organic remains in them, seeming to show that the vent traverses the limestones, and that the fiery mass rends off portions of them, as indeed might be expected from the nature of the country. The limestones so ejected are often impregnated with magnesia, supposed to have been acquired in this great natural crucible.

Volcanic Dykes, &c.

Dykes or fissures in the sides of volcanos, subsequently filled by melted lava, are sufficiently common. M. Necker de Saussure mentions numerous dykes which traverse the beds

---

* Sulphur is exceedingly common, and is often sublimed in such quantities as to be carried away for economical purposes.

of Monte Somma. These veins are nearly all of the same composition, differing somewhat from the lava-beds they cut; augite being more abundant, while leucite, so common in the beds, occurs rarely in the dykes, with the exception of one vein of Monte Otajano, and another near the foot of the Punte del Nasone, which contain large crystals of leucite. The lava of the dykes also contains minute crystals of felspar (?), with a considerable abundance of a yellow substance, which may be olivine. The rock composing the veins is fine-grained on the sides, and more crystalline in the middle. These veins vary from one to twelve feet in width.

One remarkable dyke, different from the rest, occurs at Otajano. It is about ten feet and a half wide, and rises perpendicularly to the crest of the mountain, having apparently turned up the alternating beds of porous and compact lava which it traverses. Another singular dyke cuts the rocks of the Primo Monte. It rises perpendicularly, and is formed of a slightly greenish gray and homogeneous rock. At its base (that of the mountain) it is only eleven inches wide, and for twelve feet of its height is bordered by a line of vitreous lava, half an inch thick, separating it from the porous volcanic breccia which it cuts. Above the twelve feet, the vitreous lava ceases entirely, the solid rock occupying the whole vein*.

Dr. Daubeny notices tuff traversed by dykes of a cellular trachytic lava at Stromboli, and at Vulcanello in the island of Lipari†. Dykes, described as resembling greenstone, were noticed by Sir George Mackenzie traversing alternate beds of tuff and scoriaceous lava in Iceland.

Dykes of porphyry traverse the older lavas of Etna. Their formation is by no means difficult of explanation, by supposing fissures, which sometimes have, and sometimes have not, penetrated to the surface, injected with incandescent lava. Of fissures extending to the surface, the cleft twelve miles long and six feet broad, which opened on the flank of Etna, between the plain of St. Lio and a mile from the summit, at the commencement of the great eruption of 1669, is an example‡. This fissure gave out a vivid light; from which Mr. Lyell with great probability concludes that it was filled to a certain height with incandescent lava. After the formation of this, five other fissures were produced, and emitted sounds heard at the distance of forty miles§.

While on this subject it may be as well to notice the pro-

---

* Necker, Mémoire sur le Mont Somma, Mém. de la Soc. de Phys. et d'Hist. Nat. de Genève, 1828.

† Daubeny's Description of Volcanos, p. 185—187, where there are views of these appearances.

‡ Lyell, Principles of Geology. § *Ibid.* vol. i. p. 364.

bable effects of a column of lava passing through stratified
rocks, insinuating melted matter among the strata, or through
fissures formed in them.

Let *a b* in the annexed diagram represent a column of liquid lava, traversing horizontal strata. It
is obvious that it will strive to overcome
the resistance of the sides, and such resistance will always be less between the
strata than elsewhere. If it obtain an
aperture in that direction, it will endeavour to separate one stratum from another;
and it will the more readily accomplish
this, as to the pressure of the column of
lava will be added the mechanical action of
the wedge; and eventually an injection of

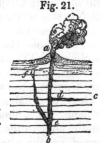

Fig. 21.

liquid lava may be made, and carried laterally, so far as the
pressure will permit. Thus, if a separation of the strata can
be commenced at *d*, it will be carried on in the direction *d c*
as far as the pressure of the column *a d* will permit. If, instead of this kind of injection, we consider the strata to have
been fractured, as is very likely to be the case near volcanic
action, the fissure will be filled, and forced asunder as far as
resistances will permit. Thus, if a fracture *e f* be made, it
will be filled by liquid lava as far as can be effected by the
pressure of the column *a e.* The strata have here been supposed horizontal, for the sake of illustration, but as they
might occur in all modes, the effects would be varied accordingly, the principle remaining the same.

Earthquakes.

The connexion between volcanos and earthquakes is now
so generally admitted that it would be useless to enumerate the
various circumstances that point to this conclusion. They
both seem the effects of some cause as yet unknown to us. The
motion of the ground produced by earthquakes is not always
the same; sometimes resembling the undulatory movement of
a heavy swell at sea, though much quicker, and being at
others tremulous, as if some force shook the ground violently
in one spot. The former of these is far the most dangerous,
as it forces walls and buildings off their centres of gravity,
crushing whatever may be beneath them.

It has been considered that earthquakes are presaged by
certain atmospheric appearances, but it may be questionable
to what extent this supposition is correct. Historians of earthquakes seem to have been generally desirous of producing effect in their descriptions, adding all that could tend to heighten

the horror of the picture. They have not always, moreover, been anxious or able to separate accidental from essential circumstances. As far as my own experience goes, which is however merely limited to four earthquakes, the atmosphere seemed little affected by the movement of the earth; though I would be far from denying that it may be so; for we can scarcely imagine such movements to arise in the earth, without some modification or change of its usual state of electricity which would affect the atmosphere. If animals be generally sensible of an approaching shock, it might arise as well from electrical changes as from the sounds which they may be supposed capable of distinguishing.

Earthquakes very frequently precede violent volcanic explosions, even though they may be felt far from a fiery vent. Thus, the great earthquake which ravaged the Caraccas, March 26th, 1812, was followed by the great eruption of the Souffrier in St. Vincents, on April 30th of the same year; when, according to Humboldt, subterranean noises were heard the same day at the Caraccas and on the banks of the Apure.

Earthquakes are felt over very considerable spaces; and of this no better example has yet been recorded than the celebrated earthquake of Lisbon in 1755, the shock of which was felt over nearly the whole of Europe, and even in the West Indies. The force capable of causing such extensive vibrations must have been very considerable; and, with every allowance for the easy transmission of motion and sound laterally through rocks, must have required considerable depth for its production. Motion seems always to be communicated to water during earthquakes, the vibratory movement being very frequently felt by vessels at sea, and waves of greater or less magnitude, according to the force of the shock, being commonly driven on shore. The wave produced during the great Lisbon earthquake rose sixty feet high at Cadiz, and eighteen feet at Madeira, causing various movements of the water on the coasts of Great Britain and Ireland. Similar waves, though of proportionally less size, are common during volcanic eruptions; motion being produced in the surrounding water, which being unable to rend and crack like the land, communicates the impulse it has received to the waters around, and thus a wave is propagated which will diminish in height in proportion as it recedes from the disturbing cause. In almost all ports irregularities in the motion of the sea are at times observable, which cannot be reconciled with the tides or motions communicated to water by temporary currents or winds in the offing. The movement is generally a quick flow

or reflow of the water, and is often so trifling as to escape the
attention of all but seamen or fishermen, who, constantly
engaged with their vessels or boats in harbours, are surprised
to find them suddenly floated or left dry, and this sometimes
several times repeated. May not these movements be caused
by earthquakes beneath the depths of the sea, or be so trifling
as to escape observation on land? If, as it seems reasonable to conclude, earthquakes are propagated laterally through
considerable distances, in the same manner as sound is conveyed through the air, the intensity of the shock will depend
on the medium through which it is conveyed; and if this view
should be correct, earthquakes will not be equally felt on
every description of rock. I once observed a fact, which,
though it struck me much at the time, cannot in itself form
the basis of any reasonable hypothesis, but as it may be the
means of exciting inquiry it is as well to mention it. While
sitting in a house in Jamaica, situated on a hill, near the verge
of the white limestone of that island, where the large gravelly,
sandy, and clay plain of Vere and Lower Clarendon meets it,
I experienced a slight shock of an earthquake. Having occasion about half an hour afterwards to descend to some houses
at the foot of the hill, and on the gravel plain, I inquired if
the inhabitants had felt the earthquake; they, however, ri-
diculed the idea, stating that if any such had occurred they
must have known it, as they also had been sitting quietly,
and were too much accustomed to shocks not to have observed
the earthquake if it had really occurred. I then considered
that I had deceived myself, and thought no more of the subject until the evening; when some negroes, who had been
employed on their own account, a few miles distant in some
mountains composed of the white limestone, reported that they
had felt the shock of an earthquake; and it subsequently appeared, that a much more considerable shock had been felt in
the vicinity of Kingston, about forty miles distant. The importance of this fact certainly rests on the little apparent sensation produced at the lower house; and therefore, as the
shock may have escaped the attention of those then present,
this circumstance is in itself of no great value, and is merely
stated to promote inquiry. It may, however, be remarked,
that gravel would transmit a vibration less readily than compact limestone, though it might more easily give way before
a vertical movement. Humboldt has remarked that during
the Caraccas earthquake of 1812, the Cordilleras were more
shaken than the plains. This may have arisen from the more
easy transmission of the vibration through the gneiss and
mica slate, than through the rocks in the plains; or, as

might be also the case in Jamaica, the inferior rocks might be more shaken through their continuity than the superior rocks, being nearer the disturbing cause.

It may also be remarked that rocks would transmit sounds unequally from variations in their texture and continuity, and that subterranean noises might be audible, while the shock which produced them could not be distinctly felt. Various sounds are recorded as accompanying earthquakes, but the most general seems a low rumbling noise like that of a waggon passing rapidly along. The first shock I ever experienced was, during a beautiful night, on the north side of Jamaica, when it appeared as if a waggon, rolling rapidly to the house, gave it a smart rap and then passed on.

It has been considered, and with much probability, that the very great distances at which volcanic explosions from surface-vents have been heard, arises from the transmission of the sound through the rocks. The great explosion at Sumbawa above noticed is described as having been heard in Sumatra, a distance of 970 geographical miles, and at Ternate, 720 miles in another direction *. It is also stated that the eruption from the Aringuay, in the island of Luçon, Philippines, in 1641, was heard in Cochin-China †.

Earthquakes produce changes in the level of the land, raising and depressing ground, and causing clefts, slips or faults, and various other modifications of surface. The raising of the surface implies either an expansion of the solid matter beneath, or a separation of parts, which should form a cavity, filled either by gaseous or liquid substances. We are not aware of anything that could produce the expansion required but heat, so that if the temperature were again diminished, contraction would ensue. If a separation of parts were effected, and the upper portion raised, the gaseous or liquid support could scarcely be considered permanent, unless the injected matter became solid, as might happen with liquid lava, and the hollow produced by such injection be far removed from the surface.

The best example of the bodily elevation of land with considerable surface appears to be that recorded by Mrs. Maria Graham, as having taken place during the Chili earthquake of 1822. The shock extended along the coast for more than a thousand miles, and the land was raised for a length of one hundred miles, with an unknown breadth, but certainly extending to the mountains. The beach was raised about three or four feet, as was also the bottom near the shore; on the former, shell-fish were still adhering to the rocks on which

* Life of Sir S. Raffles.   † Chamisso, Kotzebue's Voyage.

they grew. It was also observed that there were other lines
of beach, with shells intermixed, above that newly elevated,
attaining in parallel lines a height of about fifty feet above the
sea; seeming to show that other elevations of the same land
had been effected by previous earthquakes. During this earthquake the sea flowed and ebbed several times. No visible
change in the atmosphere was produced previous to the shocks,
but it is supposed that some effect, perhaps electrical, may
have been caused by the earthquake, for the country was subsequently deluged by storms of rain *.

Mr. Lyell has accumulated a considerable mass of evidence
to show that such elevations have been the consequence of
earthquakes in other places, and that considerable depressions
have also occurred †. Thus, during the Cutch earthquake of
1819, the eastern channel of the Indus was altered, the bed
of which was in one place deepened about seventeen feet, so
that a spot once fordable became impassable. It further appears, from the observations of Lieutenant Barnes, that not
only was there considerable subsidence in this case, but also
a remarkable elevation of upwards of fifty miles in length,
running parallel to the subsidence, across the delta of the
Indus from east to west. The greatest observed height of this
elevated land, in some places sixteen miles broad, was ten
feet‡.

Various surface-changes were effected during the great
earthquake in Calabria in 1783. Of these a summary has
been given from various authorities by Mr. Lyell, whose account will be perused with interest, however little we may feel
inclined to adopt the theoretical conclusions that have been
deduced from it. The earth had a waving motion; numerous and deep rents were formed; faults were produced, even
through buildings; large land-slips took place; lakes were
formed,—one about two miles long by one broad, from the
obstruction of two streams; the usual agitation of the neighbouring sea was produced, and heavy waves broke upon the
land, sweeping all before them.

The great earthquake in Jamaica of 1692, generally described
as having swallowed up Port Royal, has been adduced as an
example of great derangement. It is a common tradition in
that island, that many of the accounts which have appeared
respecting this earthquake have been much exaggerated; nor
need this surprise us, when we reflect how difficult it is to
obtain a clear account of unusual natural phænomena from
those who have been dreadfully alarmed by them. In order

---

* Journ. of Science; Geol. Trans. vol. i.   † Principles of Geology.
‡ Lyell's Principles of Geology, vol. ii. p. 269.

to estimate the changes that may be supposed to have taken place during this earthquake, it becomes necessary to notice the condition of Port Royal and of the neighbouring coast previous to its occurrence. The site of the present town of Port Royal is the same as that of the old town, being at the western extremity of a sand-bank, known as the Palisades, about eight miles long, thrown up apparently by the sea. Immediately seaward are numerous shoals and coral reefs, known by the name of the Keys.

Now it appears from the evidence of Captain Hals, who went to Jamaica with Penn and Venables in 1655, that the land upon which Port Royal then stood, was joined to the Palisades, distant about a quarter of a mile, by a narrow ridge of sand just appearing above the water; and it further appears, that when Jackson invaded St. Jago de la Vega, about seventeen years before this time, the same land formed an island; the narrow ridge resulting from the drift of sand by the prevalent E. or S.E. winds, and the action of the breakers. By a continuance of the same forces the whole space between the Palisades and Port Royal was eventually filled up, aided by the contrivances of the inhabitants, who drove in piles and formed wharfs, close to which the water was so deep that vessels of 700 tons came alongside and unloaded *. Upon this newly formed land the greater part of the town was built, and consisted of heavy brick houses. Now the part of Port Royal, described as having been swallowed up or sunk, was situated upon this new formed land. "The ground gave way as far as the houses stood, and no further, part of the fort and the Palisades at the other end of the houses standing †."

Sir Hans Sloane says: "the whole neck of land being sandy (excepting the fort, which was built on a rock and stood) on which the town was built, and the sand kept up by palisades and wharfs, under which was deep water, when the sand tumbled, on the shaking of the earth, into the sea, it covered the anchors of ships riding by the wharfs, and the foundations yielding, the greatest part of the town fell, great numbers of the people were lost, and a good part of the neck of land, where the town stood, was three fathoms covered with water." We have next to consider the state of the sea

---

* The variation in the depth of this water would be trifling, for the tides only rise or fall eleven inches or a foot at Port Royal.

† Phil. Trans. for 1694. Long, who from his office was so well qualified to obtain the best information, says: " the weight of so many large brick houses was justly imagined to contribute, in a great measure, to their downfall; for the land gave way as far as the houses erected on this foundation stood, and no further."

during the shocks. The harbour is described as having had
"all the appearance of agitation as in a storm; and the huge
waves rolled in with such violence, as to snap the cables of
the ships, and drive some from their anchors." Again, we
find that the houses near the water sunk at once: "and a
heavy rolling sea followed, closing immediately over them."

The Swan frigate, which was by the wharf careening, was
carried over the tops of the houses by the sea, and some hundreds of persons escaped by clinging to her. Some houses
sunk or settled perpendicularly, so that they remained from
the balcony upwards above water; but the greater part were
rendered a mass of ruins. Finally, the land with the fort on
it is reported to have formed an island, as at the time of Jackson's expedition *. This state of things has not however continued; for the same causes, which once joined the Palisades
with the fort, continuing, the whole now forms a continuous
piece of land.

A plan of Port Royal formed from authentic and existing
documents, and representing the extent of that town previous
to, and immediately after, the earthquake of 1692, as also
about a century afterwards, was published in the Jamaica Al-
manac for 1806, having first appeared about ten or twelve
years previously. The annexed wood-cut (Fig. 22) is a reduction of this plan (the lines of streets being omitted), with
the addition of the present extent of this point of land, according to Dr. Miller, of Jamaica.   *a, a, a, a, a,* and *L,* the

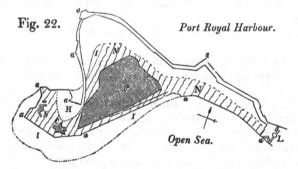

Fig. 22.   *Port Royal Harbour.*

Open Sea.

boundary of the town and point of Port Royal previous to
the great earthquake. The darkly shaded parts *P* and *C,* the
portion that remained after the earthquake; *C* being Fort
Charles. The lightly shaded part *N, N, N,* the extent of
the town and point at the end of the eighteenth century; the
accumulation of sand having been principally produced by

* Phil. Trans. 1694; Sloane, Nat. Hist. of Jamaica; Long, Hist. of Jamaica; and Bryan Edwards, Hist. of the West Indies.

the natural drift of the sand. The spaces *I, I, I,* and *H,* the additional land, formed by the drifting of the sand since that time, and constituting with the lightly shaded portions *N, N, N,* the present extent of the town and point of Port Royal. The space *H,* formerly known as Chocolate Hole, now filled up, forms part of the garrison parade.

It also appears that the portion of Port Royal which remained above water, after the shock, is generally considered to be based upon white limestone, as is the case with Fort Charles. This rock is now known to constitute the base of part, if not a large proportion, of the ridge of land, named the Palisades, which commences close to Port Royal (*L,* Fig. 22.), and very probably also forms the base of the various coral reefs known as the Port Royal Keys.

Upon a review of what has been adduced respecting this earthquake, it does not appear that there is evidence of subsidence, that is, the bodily subsidence of a mass of land of great depth; though I would be far from denying that there may have been something of the kind. The whole may be explained by the settlement of loose sand, charged with the weight of heavy houses, during the violent shocks of an earthquake, and by the inroad of the sea; for had there been a general subsidence, the rocks would have disappeared with the rest *. The evidence of the ruins of houses commonly stated to be seen beneath the sea, in calm weather, close to the present town, will do for either hypothesis; for they would be similarly situated either from the settlement of the sand, or by the subsidence of land, in the usual acceptation of this term.

While in Jamaica, in 1824, I endeavoured in vain to see what are commonly termed the Ruins of Old Port Royal, and it now appears that they are covered up with sand, forming considerable inequalities beneath the sea, to the W. and N.W. of the present naval hospital. When vessels touch on these inequalities they are said to "ring the bells of Old Port Royal." It however by no means follows that the ruins were not more distinctly visible in 1780, as is stated to have been the case by Sir Charles Hamilton †, and others: on the contrary we must expect them to have been so, for we have seen that the transport of sand, principally produced by the breakers, driven forward by a prevalent wind (the trade wind), is

---

* Dr. Miller informs me, that when he resided in Port Royal, prior to the great fire of 1815, there were many old people then living, descendants of the early settlers, and it was the traditional opinion among them, that the great damage was produced by the slipping of the sand, an opinion in accordance with that previously noticed.

† Lyell, Principles of Geology, vol. ii. p. 269.

very considerable at Port Royal, so that the ruins would be
gradually covered up.

The earthquake was generally destructive of buildings in
Jamaica, and masses of rocks were detached from the heights;
—no great difficulty in a country abounding with precipices
and steep mountains. According to one account, two mountains met in the Sixteen Mile Walk; if they did so, they
have since been so complaisant as to separate, for there is nothing at present existing there to warrant a conclusion that
they ever did meet. That heavy fragments of rock, and considerable masses of earth, blocked up the passage for the time,
is exceedingly probable; but there is a great difference between such an event and the meeting of mountains.

Funnel-shaped, or inverted conical cavities are by no means
unfrequent on plains after earthquakes; and are so much
alike wherever they occur, that they must have some common
cause for their production. Circular apertures were produced
in the plains of Calabria by the earthquake of 1783: they are
described as commonly of the size of carriage-wheels, but often
larger and smaller; they were sometimes filled by water, but
more frequently by sand. Water seems to have spouted through
them *. During the earthquake in Mercia in 1829, numerous small circular apertures were produced in a plain near
the sea, which threw out black mud, salt water, and marine
shells †. After the earthquake at the Cape of Good Hope, in
December 1809, the sandy surface of Blauweberg's Valley is
described as studded with circular cavities, varying from six
inches to three feet in diameter, and from four inches to a foot
and a half in depth. Jets of coloured water are stated, by the
inhabitants of the valley, to have been thrown out of these
holes to the height of six feet during the earthquake‡. It
seems somewhat difficult to account for these appearances,
though the common aqueous discharges through rents or
chasms can be more readily understood. During the Chili
earthquake, previously noticed, sands were forced up in cones,
many of which were truncated with hollows in their centres§.

The courses of springs are, as would be anticipated, often
deranged amid such motions of the ground; and flashes of
light, or bright meteors, are so frequently mentioned that we
can scarcely doubt their occurrence, and they may, perhaps,
be considered as electrical.

If we now withdraw ourselves from the turmoil of volcanos

* Lyell, Principles of Geology; where a view and section of these cu-
rious cavities are given: pp. 428, 429.
† *Ibid.*; and Férussac's Bulletin, 1829.
‡ Phil. Mag. and Annals, January 1830.
§ Journal of Science.

and earthquakes, and cease to measure them by the effects
which they have produced upon our imaginations, we shall
find that the real changes they cause on the earth's surface are
comparatively small, and quite irreconcileable with those
theories which propose to account for the elevations of vast
mountain-ranges, and for enormous and sudden dislocations
of strata, by repeated earthquakes acting invariably in the
same line, thus raising the mountains by successive starts of
five or ten feet at a time, or by catastrophes of no greater importance than a modern earthquake. It is useless to appeal
to time: time can effect no more than its powers are capable
of performing: if a mouse be harnessed to a large piece of
ordnance, it will never move it, even if centuries on centuries
could be allowed; but attach the necessary force, and the resistance is overcome in a minute.

Hurricanes.

These are of geological importance, as by the sudden application, if I may so express myself, of a furious wind and de-
luges of rain to the surface of land, very considerable changes
are in a short time produced on that surface. It has been
considered that the wind, during hurricanes, travels with a
velocity of from eighty to one hundred miles per hour; but it
must be confessed that we possess no very satisfactory information on this head. Be, however, the velocity of the wind
what it may, its force is sufficient to level forests, throw down
buildings, and destroy a large amount of animal life; in a few
hours converting a beautiful and luxuriant country, studded
with villages and towns, into a scene of desolation and mourning. Furious torrents are suddenly formed, which not only
sweep away a large proportion of the uprooted trees, and the
bodies of numerous terrestrial animals destroyed by the effects
of the wind; but also act most powerfully on all the drainage
depressions, producing the maximum effects of running water
in such situations. In mountainous regions the land-slips are
then also frequently considerable; and if these fall into the bed
of a torrent, they add to its destructive effects, by damming
up the waters for a time, which, when they have forced their
passage through the obstacle, rush onwards with increased
velocity and power.

In the hurricane in the West Indies of August 1831, we
have a melancholy example of the destruction of animal and
vegetable life caused by these scourges of that portion of the
world. Not only were buildings of various kinds levelled with
the earth, and numbers of persons buried beneath their ruins,
but a large amount of animal life was also destroyed; and

those trees which were not uprooted by the fury of the wind, were deprived of their foliage, many even of their branches, so that the unfortunate island of Barbadoes presented that strange phænomenon, a mass of leafless trees on a tropical island. This hurricane also ravaged the islands of St. Vincent and St. Lucie, and was even felt at the eastern end of Jamaica.

The sea is, as might be expected, violently agitated during hurricanes, and causes great destruction, particularly on low coasts. Thus, in the great Jamaica hurricane of 1780, the sea suddenly burst in upon the small town of Savanna la Mar, and swept it, and every thing in it, entirely away. The hurricane of August 1831, was sufficiently powerful at Hayti to raise the sea at Aux Cayes to a considerable height, and the swell consequent on it was so great on the coast of Cuba as to throw every vessel on shore at St. Iago de Cuba.

Hurricanes are often more partial, but they are not the less destructive to the land they traverse on that account. The hurricane of 1815, which traversed Jamaica from North to South, was one of this description; it took its way across the western portion of the Blue Mountains, and was exceedingly destructive. Not only was the wind furious, but the quantity of rain which fell in a given time was considered quite unexampled even in the tropics. The flood which descended the Yallahs river, swept away all the fish in it, and ten years afterwards it was considered that there were no fresh-water fish in that river. The land-slips in the Port Royal, St. Andrews, and Blue Mountains were very considerable; and when I visited these mountains several years afterwards, many a bare cliff bore evidence to the changes that had been thus produced. When these land-slips descended to the bottom of the ravines, they dammed up the waters for the time, and then giving way, were partially swept onwards. The loss of life was considerable, and many buildings were either washed away or buried beneath detritus. The land communications between Kingston and the Eastern coast were stopped; and Mr. Barclay relates, that being thus compelled to pass by sea to Morant, the vessel was obliged to make "a considerable offing to keep clear of the enormous quantity of trees, which literally covered the water to a considerable distance." Though so destructive about the centre of its course, this hurricane was neither felt at St. Jago de la Vega (Spanish Town), forty miles to the westward, nor at Morant Keys, fifty miles to the eastward.

The force of a hurricane is well shown by the facts observed at Guadaloupe, on the 25th of July 1825. Strongly-built houses were blown down, and many tiles driven through the doors of warehouses. A deal plank, 39 inches long, 9·8 inches wide, and ·9 inch thick, was transported with such rapidity that it traversed the trunk of a palm-tree 17·7 inches in diameter. A piece of wood, 7·8 inches square, and from four to five yards long, was forced by the wind into a hard, frequented, and stoned road, to the depth of about a yard. The exactitude of these and other facts illustrative of the great power of this hurricane, was verified on the spot by General Baudrand, of the French Engineers*.

It will be obvious that during hurricanes a comparatively large amount of terrestrial animals and vegetables may, in addition to the land-detritus, be carried outwards into the seas which bathe the shores of tropical islands, such as those of the West Indies, more particularly when such islands are mountainous, as is the case with Cuba, Hayti, Jamaica, and others. Not only men, quadrupeds, birds, and land reptiles, but also fresh-water tortoises and crocodiles, may be surprised and carried out to sea, where they would have a poor chance of escape amid the turmoil of the waves at such times. A large proportion of the creatures thus borne by torrents outwards would, most probably, be devoured by sharks and other voracious inhabitants of the sea; but there is still a possibility that the river-detritus, and the sands and mud, stirred up by the action of the waves in shallow seas, would, when tranquillity was restored, envelope various terrestrial, fluviatile, and marine remains: such a deposit would thus, to a certain extent, resemble one formed in an estuary, but would so far differ from it as probably, in the case here supposed, the remains would exhibit the marks of violent transport. In the immediate vicinity of the coast, the breakers would throw a considerable quantity of these remains on shore.

Gaseous Exhalations.

In several situations removed from any volcanic action, so far as is visible on the surface, natural jets of inflammable gases are seen to issue, affording decisive evidence of chemical changes that are taking place at various depths beneath. Of these, some have served the purpose of the priest to delude mankind, while part of the others have been more usefully employed.

Carburetted hydrogen gas is well known to be the "firedamp" of the coal districts, and to issue from the coal strata; collecting in the ill-ventilated galleries of collieries, and, when sufficiently mixed with atmospheric air, exploding with great violence if approached incautiously with an unprotected flame,

* Pouillet, E'lémens de Phys. Expérimentale, t. ii. p. 718, 2nde E'dit

spreading mourning and misery among the families of the miners. If the genius of Davy had merely produced his safety-lamp, it would alone have entitled him to the applause and thanks of mankind.

As carburetted hydrogen is so freely liberated in coal-mines, it would be expected that it should occasionally be detected on the surface, and accordingly it has been so discovered *. Inflammable gas also occurs in other situations, where there is no reason to suspect the presence of coal strata. Of this, the well-known jets of gas in the limestone and serpentine district of the Pietra Mala, between Bologna and Florence, afford an example.

Captain Beaufort describes an ignited jet of inflammable gas, named the Yanar, near Deliktash, on the coast of Karamania, which perhaps once figured in some religious rites. He states that, "in the inner corner of a ruined building, the wall is undermined, so as to leave an aperture of about three feet in diameter, and shaped like the mouth of an oven : from thence the flame issues, giving out an intense heat, yet producing no smoke on the wall." Though the wall was scarcely discoloured, small lumps of caked soot were found in the neck of the opening. The hill is composed of crumbly serpentine and loose blocks of limestone. A short distance down the hill there is another aperture, which from its appearance seems once to have given out a similar discharge of gas. The Yanar is supposed to be very ancient, and is possibly the jet described by Pliny †.

Colonel Rooke informed Captain Beaufort, that high up on the western mountain at Samos there was an intermittent flame of the same kind ; and Major Rennell stated that a natural jet of inflammable gas, inclosed in a temple at Chittagong, in Bengal, is made use of by the priests, who also cooked with it.

The village of Fredonia, in the State of New York, is lighted by a natural discharge of gas, which is collected by means of a pipe into a gasometer. The quantity obtained is about eighty cubic feet in twelve hours. It is carburetted hydrogen, and is supposed to be derived from beds of bituminous coal. The same gas is discharged in much larger quantities in the bed of a stream about a mile from the village.

According to M. Imbert, gaseous exhalations are employed at Thsee-Lieou-Tsing, in China, to distil saline water obtained from wells in the neighbourhood. " Bamboo pipes carry the

---

* It appears very remarkable that in the coal districts of the British Isles, where such a large amount of carburetted hydrogen is annually produced, means have not been adopted for making an economical use of this gas, both as respects light and heat.

† Beaufort's Karamania.

gas from the spring to the place where it is to be consumed. These tubes are terminated by a tube of pipe-clay, to prevent their being burnt. A single well (of gas) heats more than three hundred kettles. The fire thus produced is exceedingly brisk, and the caldrons are rendered useless in a few months. Other bamboos conduct the gas intended for lighting the streets and great rooms or kitchens *." These wells of inflammable gas were, according to M. Imbert, formed for the purpose of obtaining salt water, which they in fact first gave out. The water failing, the wells were sunk to a considerable depth in order to find the water; instead, however, of finding salt water, inflammable gas suddenly rushed forth with considerable noise†.

M. Klaproth notices other jets of inflammable gas in China; one, now extinguished, is stated to have burnt from the second to the thirteenth century of our era. This *Ho tsing*, or fiery well, was situated 80 li to the S.W. of Khioung tcheou, and like those above mentioned produced salt water‡.

This connexion of inflammable gas with saline springs or salt is not confined to China, but has also been observed in America and in Europe. While boring for salt at Rocky Hill, in Ohio, and near Lake Erie, the borer suddenly fell, after they had pierced to a depth of 197 feet. Salt water immediately spouted out, and continued to flow for several hours; after which a considerable quantity of inflammable gas burst forth through the same aperture, and, being ignited by a fire in the vicinity, consumed all within its reach§.

It also appears that M. Rœders, inspector of the salt-mines of Gottesgabe, at Reine in the county of Tecklenberg, has for two or three years used an inflammable gas which issues from these mines, not only as a light, but for all the purposes of cookery. He obtains it from the pits that have been abandoned, and conveys it by pipes to his house. From one pit alone a continuous stream of this gas has issued for sixty years.

* Bibl. Universelle; and Edin. New Phil. Journal, 1830.

† Humboldt, Fragmens Asiatiques.

‡ *Ibid.* In the same work will be found an interesting account of the mode in which the Chinese sink these wells, in search of salt water, to considerable depths. This is effected by the constant striking of a piece of steel, of about 300 or 400 pounds weight, against the rock, upon the same principle that a hole is made in the solid rock by an iron or steel bar for the purpose of blasting with gunpowder. In the Chinese work, however, the steel weight is suspended by a cord to one end of a piece of wood, placed over a support in such a manner that a workman by dancing or jumping on the other end, raises the weight about two feet at each motion, and suddenly lets it fall again. By these slow, but somewhat sure means, a round perpendicular hole is formed, about five or six inches in diameter, very smooth, and, according to M. Imbert, from 1500 to 1800 French feet in depth.

§ Trans. New York Phil. Soc.

It is supposed to consist of carburetted hydrogen and olefiant gas*.

Inflammable gases are also found to proceed from ground charged with petroleum and naphtha. The inhabitants of Baku, a port on the Caspian Sea, are supplied with no other fuel than that derived from the petroleum and naphtha with which the earth in the neighbourhood is strongly impregnated. About ten miles to the N.E. of this town there are many old temples of Guebres, in each of which there is a jet of inflammable gas, rising from apertures in the earth. The flame is pale and clear, and smells strongly of sulphur. Another and a larger jet issues from the side of a hill. The ground is generally flat, and slopes to the sea. If in the circumference of two miles, holes be made in the earth, gas immediately issues, and inflames when a torch is applied. The inhabitants place hollow canes into the ground, to convey the gas upwards, when it is employed for the purposes of cookery as well as for a light†. M. Lenz, describing an eruption of mud and flame near the village of Iokmali, fourteen wersts to the west of Baku, would seem to attribute the gaseous exhalations of this district to a volcanic origin, but the facts adduced will scarcely admit of this interpretation. He notices this eruption as having taken place on November 27th, 1827. A column of flame burst out, where no flame had been previously seen, and rose for three hours to a considerable height, then lowered itself to the height of three feet, and burnt for twenty-four hours. After this the mud rushed forth and covered the country over an area of 200 toises by 150, to the depth of two or three feet. There is sufficient evidence that other eruptions of mud or clay had previously taken place from the same, or nearly the same, place. This and other " salses" noticed in the same territory cannot be termed volcanic, in the usual acceptation of the word. Moreover we learn from the observations of the same author, that at the *Atech-gah,* or the great fires of Baku, the principal jet rises through a calcareous rock, with a dip of 25° to the S.E., the fissures or cracks being rendered blue by it‡.

Carbonic acid gas is evolved abundantly in coal-pits and volcanic regions. Its occurrence in the Grotto del Cane, of which such overcharged descriptions have been given, is well known. MM. Bischof and Nöggerath notice a pit, on the side of the lake of Laach, in which they found dead birds, squirrels, bats, frogs, toads and insects, killed by the evolution of carbonic acid gas. M. Bischof estimates that the ex-

---

* Journal of Science. † Edin. Phil. Journal, vol. vi.
‡ Humboldt, Fragmens Asiatiques.

halations of carbonic acid gas, in the vicinity of the lake of Laach, amount to 600,000 pounds daily, or 219,000,000 pounds in a year.

In the Brohlthal on the Rhine, an old volcanic country, there is a considerable evolution of carbonic acid gas, which is employed by M. Bischof in the manufacture of chemical preparations on the large scale. Six hundred pounds of this gas are calculated to be discharged from only one of the jets in twenty-four hours, being at the rate of 219,000 pounds in the year *

Carbonic acid gas is so abundantly evolved in a district of mineral springs in Armenia, near Fort Diadine, on the Euphrates, that it produces a noise while issuing through the cracks in the limestones of the country, killing the birds which come within its influence †.

The Guevo Upas, or Valley of Poison, in Java, would appear to be a cavity filled to a certain height with carbonic acid gas. Mr. Loudon describes it as about half a mile in circumference, and of an oval form. The depth is from thirty to thirty-five feet, the bottom being flat, without vegetation, and covered with skeletons of men and of various animals, such as tigers, hogs, deer, &c., which have perished by their entrance into the gas. The destructive gas did not rise, so as to be dangerous, above eighteen feet from the bottom ‡.

A very copious discharge of carbonic acid gas occurs on 'the Kyll, nearly opposite Birresborn. The gas rises through fissures of the rock, and traverses a pool of rain-water, resting on it, with such violence that the noise is stated to be heard at the distance of 400 yards. Birds are killed when they approach too close, and persons wishing to drink are driven away by the gas, a stratum of which covers the surrounding turf§.

In many situations gaseous vapours come to the surface mixed with water or petroleum, with sufficient force to produce "salses" or mud volcanos. Dr. Daubeny considers those of Maculaba in Sicily as independent of volcanic action, but due to the combustion of the sulphur existing among the rocks. Mud eruptions from the discharge of gaseous vapours and water are known in many other places‖.

* German Trans. of Manual.
† Voskoboinikov, Goinoi Journal, 1829; and Boué's Mém. Geol. et Paléontologiques, t. i. 1832.
‡ Journal of Geographical Society, vol. ii.
§ Bischof and Nöggerath, Edin. Phil. Journal.
‖ Those near Modena have long been celebrated.

Deposits from Springs.

Springs are seldom or ever quite pure, owing to the solvent property of water, which percolating through the earth, always becomes more or less charged with foreign matter. Carbonate, sulphate, and muriate of lime, muriate of soda, and iron, are frequently present in spring waters. Some are more highly charged with these and other substances, such as carbonate of magnesia and even silica, than others, and have hence obtained the name of mineral springs. Many are thermal, as before noticed, and seem not immediately derived from the waters of the atmosphere; as may also be the case with many that are cold, their more elevated temperature having been lost in their passage upwards through colder strata.

Many thermal springs contain silica, though this substance is of exceedingly difficult solution. The siliceous deposits from the Geysers in Iceland are well known. Sir George Mackenzie describes the leaves of birch and willow converted into stone, every fibre being discernible. Grasses, rushes, and peat are in every state of petrifaction. There are also deposits of clay containing iron pyrites, which decompose and communicate very rich tints to it. The deposits from the Geysers extend to about half a mile in various directions, and their thickness must be more than twelve feet, for that depth is seen in a cleft near the Great Geyser.

The finest exhibition of such deposits as yet noticed, occurs in the volcanic district of St. Michael, Azores. Dr. Webster describes the hot springs of Furnas as respectively varying in temperature from 73° to 207° Fahr., and depositing large quantities of clay and siliceous matter, which envelope the grass, leaves, and other vegetable substances that fall within their reach. These they render more or less fossil. The vegetables may be observed in all stages of petrifaction. He found " branches of the ferns which now flourish in the island completely petrified, preserving the same appearance as when vegetating, excepting the colour, which is now ash-grey. Fragments of wood occur, more or less changed; and one entire bed, from three to five feet in depth, is composed of the reeds so common in the island, completely mineralized, the centre of each joint being filled with delicate crystals of sulphur *."

The siliceous deposits are both abundant and various: the most abundant occur in layers from a quarter to half an inch in thickness, accumulated to the depth of a foot and upwards.

* Edin. Phil. Journal, vol. vi.

The strata are nearly always parallel and horizontal, though sometimes slightly undulating. The silex forms stalactites, often two inches in length, in the cavities of the siliceous deposits, and these are frequently covered with small brilliant quartz crystals. Compact masses of siliceous deposits, broken by various causes, have been re-cemented by silica, and the compound is represented as very beautiful. Some of the elevations of this breccia Dr. Webster considers upwards of thirty feet in height. The general deposit appears to be considerable, and to form low hills. The colours of the clay and siliceous substances are very various, and even brilliant,—white, red, brown, yellow, and purple being the principal tints. Where the acid vapours reach the rocks, they deprive them of their colours. Sulphur is abundant, and the springs occur in a district of lava and trachyte *.

According to James†, the thermal springs of the Washita deposit a very copious sediment, composed of silex, lime, and iron. This shows that hot springs, when propelled through a non-volcanic district, may yet contain silica. The same may be said of some of the springs in India. Dr. Turner found that the thermal springs of Pinnarkoon and Loorgootha, in that country, which produced 24 grains of solid matter in a gallon, contained 21·5 per cent of silica, 19 of chloride of sodium, 19 of sulphate of soda, 19 of carbonate of soda, 5 of pure soda, and 15·5 of water‡. The following is an analysis of the Geyser waters and hot springs of Reikum, Iceland, by Dr. Black. A gallon of each produced:—

	Geyser.	Reikum.
Soda	5·56	3·00
Alumina	2·80	0·29
Silica	31·50	21·83
Muriate of soda	14·42	16·96
Sulphate of soda	8·57	7·53

These analyses do not show the presence of lime, but Sir G. Mackenzie mentions a calcareous deposit from boiling springs (temp. 212°) in the valley of Reikholt, in Iceland, charged with carbonic acid gas. Many thermal and other springs contain this gas, which seems very abundant in volcanic regions. To its power of dissolving lime, when passing through calcareous rocks, those deposits are due, that are so common in some countries, particularly when volcanic, which are known under the general name of Travertin or calcareous tufa. Probably, also, many hot springs may contain carbonic acid gas, which, not meeting with calcareous or magnesian strata, is thrown off when in contact with the atmosphere.

* Edin. Phil. Journal, vol. vi.   † Expedition to the Rocky Mountains.
‡ Elements of Chemistry.

Travertins are of greater geological importance than the siliceous deposits from modern springs, at least so far as their
extent of surface and depth are concerned; though both these
have been greatly exaggerated, from the usual mode of comparing such deposits, not with the superficies of the land generally, but with their magnitude relatively to the valleys or
plains in which they may occur, and not unfrequently with
that of man himself.

The deposit from the fountain of St. Allire, near Clermont,
formed a bridge which was, in 1754, one hundred paces long,
eight or nine feet thick at its base, and twenty or twenty-four
inches in its upper part *.

Mr. Lyell notices the calcareous deposits from the baths of
San Vignone, and states that one stratum, composed of several layers, is fifteen feet thick, and that large masses are cut
out of it for architectural purposes †. According to Dr. Gosse,
the thermal waters which deposit this travertin are sufficiently
hot to boil eggs.

The thermal waters at the baths of San Filippo, not far from
the above, have a temperature of 122° Fahr., one spring being
about a degree or two higher. They contain silica, sulphate
of lime, carbonate of lime, sulphate of magnesia, and sulphur;
and, notwithstanding their elevated temperature, *Confervæ*
flourish in them. The ground around is formed of travertin
deposited by the springs. There are many fissures; one 30
feet deep, and 150 to 200 feet long. In it the water is whitish,
and in a state of ebullition, whence its name, Il Bollore. It
emits copious discharges of steam and sulphurous vapour.
There are other fissures in which sulphur is sublimed in the
same manner as at the Solfatara near Naples, and the produce
was sufficient to constitute a branch of industry, now however
abandoned. The surfaces of these fissures are penetrated by
sulphuric acid. Dr. Gosse observed the siliceous stalagmites
mentioned by Professor Santi, and describes them as covering
the surface of the travertin to the depth of one eighth of an
inch ‡. Mr. Lyell notices the spheroidal structure of the travertin deposited, and compares it with the magnesian limestone
of Sunderland. What the amount of magnesia may be in the
San Filippo travertin is not stated, but according to Dr.
Gosse it is combined with sulphuric acid. Sulphate of lime
exists in great abundance in these springs; so much so, that
before the water is conducted to the places where the well-
known medallions are formed, it is allowed to stagnate for the
purpose of depositing the sulphate of lime. That the sulphates

* Daubuisson, t. i. p. 142.   † Principles of Geology, p. 202.
‡ Gosse, Edin. Phil. Journal, vol. ii.

should be common, would be expected where so much sulphurous vapour is evolved; and it is even stated that sulphur exists
in the travertin, though it is principally composed of carbonate of lime.

Deposits of travertin are by no means uncommon from cold
springs in the Apennines, particularly near the volcanic region
of southern Italy. The celebrated Falls of Terni are, as is well
known, artificial, and have been formed by cutting through a
previous calcareous deposit, to form a channel for the Velino,
which now rushes over a precipice into the Nera beneath.
Upon the flat land above, a considerable deposit of lime has
taken place;—when, it does not so clearly appear, but probably since the establishment of the present order of things.
Notwithstanding the velocity of the water, its cutting powers
are trifling, and the upper channel preserves all the appearance of art. The Velino contains much carbonate of lime,
which it deposits after the great leap, even in the bed of the
Nera, which does not cut it off, but is obstructed to a certain
degree by it, as may be seen at a place called the Bridge, over
which I crossed the Nera, by taking one or two leaps at the
chasms cut by the latter torrent. At this place there must be
a constant struggle between the destructive power of the Nera,
and the lapidifying power of the Velino. The country around
exhibits abundant examples of calcareous deposits from springs
charged with carbonate of lime. The usual explanation of this
phænomenon seems very probable. It supposes the carbonic
acid to be derived from the volcanic regions beneath, (and they
appear not far distant on the surface,) which, passing with the
water through the calcareous strata, dissolves as much lime as
it can take up, giving off the excess of carbonic acid under diminished pressure in the atmosphere, and causing the carbonate
of lime to be deposited. The carbonic acid found so abundantly
in acidulous springs is ascribed by Von Buch, Brongniart, Boué,
Von Hoff, and other geologists, to volcanic or igneous action at
various depths beneath the surface. M. Hoffman has further
shown that, in certain valleys of elevation, mineral springs are
frequent, and cites the valley of Pyrmont as a good example,
where the waters are charged with carbonic acid gas*. In the
marshy meadows of the valley of Istrup (one of elevation),
mounds of mud, from fifteen to twenty feet high, and 100 feet
in circumference, are produced by currents of carbonic acid
gas, and on their surface many small reservoirs of water are

---

* The following are the contents of these waters, according to Bergman,
in a wine pint: Carbonic acid, 26 cubic inches; carbonate of magnesia, 10
grains; carbonate of lime, 4·5; sulphate of magnesia, 5·5; sulphate of lime,
8·5; chloride of sodium, 1·5; and oxide of iron, 0·6.—Henry's Elements,
and Turner's Elements.

kept in a state of ebullition by bubbles of gas of the size of the fist *. After producing other examples of this evolution of carbonic acid gas, either combined with water, or nearly if not altogether free, M. Hoffman observes, that " the country situated on the left bank of the Weser in the direction from Carlshafen to Vlotho, up to the foot of the Teutoburg-Wald, may be compared to a sieve, whose apertures, as yet unclosed, permit the escape of gas, disengaged from volcanic depths by means unknown †.

The travertin of Tivoli, and the famous Lago di Zolfo, near Rome, have been much appealed to by those who ascribe all geological appearances to such causes only as are now in operation; but the former is a mere incrustation, considerable it is true in some situations, if measured by our own magnitude, but insignificant if compared with the country in which it occurs; and the latter is but a pond of water, dignified somewhat strangely by the name of a lake, and containing, according to Sir H. Davy, a saturated solution of carbonic acid, with a very small quantity of sulphuretted hydrogen. The spring is thermal, being about 80° Fahr.; plants thrive in and about it, and they are encased in stone beneath, while they vegetate above, and thus they may become fossil, their most delicate structure preserved, and their ramifications uncompressed.

All the examples hitherto produced of deposits, that can fairly be traced to existing springs, are relatively unimportant; and though they may lead us to understand how great geological deposits may, chemically, have taken place, as the cabinet experiments of the chemist teach us the laws which govern nature on a large scale, they no more could have produced the great limestone or siliceous deposits observed on the earth's surface, than the experiments above alluded to could produce the great chemical phænomena they illustrate, however long continued.

Mr. Lyell has presented us with an account of calcareous deposits in Scotland, which are remarkable, not for their extent, but for the circumstances which attend them. It appears that the Bakie Loch, Forfarshire, has produced a marl used in the agriculture of the country. The following is a section of the beds: 1. Peat, containing trees, one to two feet; 2. Shell-marl, containing in parts tufaceous limestone, provincially termed "*rock-marl*," one to sixteen feet; 3. Quick-sand, without pebbles, cemented together in some places by carbonate of lime, two feet; 4. Shell-marl of good quality for agriculture, (almost every trace of shell is often obliterated,) one to two

---

* Hoffman, Journal de Géologie, t. i.; and Poggendorf's Annalen, 1829.
† *Ibid.*

feet; 5. Fine sand, without pebbles, resting on transported detritus, at least nine feet. The rock-marl is limited to the vicinity of the springs, irregularly distributed over the lake. The Bakie shell-marl is white, with a yellow tint. The rock-marl has the same yellow tint, and consists almost wholly of carbonate of lime, compact, and even crystalline.

Organic remains of the marl. Horns of stags and bulls; wild boar tusks. *Cypris ornata,* Lam. *Limnæa peregra, Valvata fontinalis, Cyclas lacustris, Planobris contortus, Ancylus lacustris,* all of Lamarck. Mr. Lyell considers this calcareous rock as not immediately due to the springs, but to have been produced through the agency of the testaceous inhabitants of the lake; for though the springs do contain lime, it is in such small quantities, that they could not directly produce the marl. He considers that the testaceous animals obtained the lime either from the water or from the *Charæ* which they fed upon, and that, dying, they left their calcareous exuviæ to form, by accumulation, the shell-marl, which was converted into calcareous rock by the action of the water upon it; the water containing carbonic acid, and forming a solution of carbonate of lime, which might produce a crystalline limestone. Seeds of *Charæ,* or *Gyrogonites,* are converted into carbonate of lime, in which the nut is sometimes found within; but commonly that space is empty, and the integument alone preserved. The *Chara* here found mineralized is the *Chara hispida,* a plant, which now abounds in the Bakie Loch, and in the other lakes in Forfarshire. It contains such a proportion of carbonate of lime, as strongly to effervesce with acids when dried.

Mr. Lyell, noticing the deposits of marl in the Loch of Kinnordy, states that it is thickest at that end of the lake where the springs are most common. The shells are the same here as at the Bakie Loch, and are, like them, nearly all young, scarcely one in ten being full-sized. A large skeleton of a stag (*Cervus Elaphus*) was dug out of the marl, and was remarkable as being found in a vertical position, the points of the horns being nearly at the surface of the marl, while the feet were about two yards below it. The marl is covered by peat, and in this peat were discovered other skeletons of stags, and (in 1820) the remains of an ancient canoe, hollowed out of the solid trunk of an oak *.

There is something in the formation of these lakes which reminds us strongly of the epoch of the submarine forests and of the lacustrine deposits of East Yorkshire, which will be noticed in the sequel; like them they seem to have succeeded a considerable transport of detritus, and to have been gradually filled up, being surmounted by peat; previous to the formation

---

* Lyell, Geol. Trans. 2nd series, vol. ii.

M

of which latter production man certainly was an inhabitant of
these islands, as his works are entombed in it: the lakes being
then, probably, more or less open spaces of water, or else his
boat would have been of little service to him.

Naphtha and Asphaltum Springs.

These are distributed over various parts of the world, and
cannot be considered as rare. According to Dr. Holland, the
petroleum springs of Zante are much in the same state as in
the time of Herodotus. They are situated on a small marshy
flat, bounded by the sea on one side, and by limestone and bituminous shale-hills on the others. The principal pool is about
50 feet in circumference, and a few feet deep: the sides and bottom of this and the others are thickly covered with petroleum,
which by agitation is brought to the surface of the water, and
collected. The amount obtained is estimated at 100 barrels
annually *.

James states that about 100 miles above Pittsburgh, and near
the Alleghany river, there is a spring, on the surface of which
float such quantities of petroleum, that a person may collect
several gallons in a day. He considers that it may probably
be connected with coal strata, as is the case with similar springs
in Ohio, Kentucky†.

The pitch lake of Trinidad, estimated at about three miles
in circumference, has long been celebrated. According to Dr.
Nugent, the asphaltum is sufficiently hard in wet weather to
support heavy weights, but during the heats it approaches
fluidity. It is intersected with numerous cracks filled with
water; and it appears that these cracks sometimes close up
again, leaving marks on the surface of the pitch lake. When
slightly covered with soil, as it is in some situations, good
crops of tropical productions are obtained. From this covering of soil it is difficult to estimate the exact boundaries of the
lake ‡.

Captain Alexander states, that at Pointe la Braye masses
of pitch advance into the sea, and have the appearance of
black rocks among the trees. At the hamlet of la Braye, the
coast is covered with pitch, which runs out to sea and forms
a bank under water§.

* Holland's Travels in the Ionian Isles, Albania, &c.
† Expedition to the Rocky Mountains.
‡ Nugent, Geol. Trans. vol. i.
§ Alexander, Jameson's Edin. Phil. Journal, Jan. 1833. The same author
notices an assemblage of salses or mud volcanos at Pointe du Cac, forty miles
southward from the pitch lake. The largest of the salses is about 150 feet
in diameter.

Large quantities of naphtha are obtained on the shores of the Caspian. The inhabitants of the town of Baku, a port on that sea, are supplied with no other fuel than that obtained from the naphtha and petroleum, with which the neighbouring country is highly impregnated. In the island of Wetoy and on the peninsula of Apcheron, this substance is very abundant, supplying immense quantities which are taken away. Thermal springs are found near those of naphtha *.

The naphtha springs at Rangoon, Pegu, appear to be exceedingly abundant. Mr. Coxe estimates their produce at 92,781 tons per annum. In the Indian Islands there are also similar springs. Marsden notices them in Sumatra, at Ipu, and elsewhere.

Coral Reefs and Islands.

In consequence of the numerous situations where these are observable in the Pacific Ocean and Indian Seas, very exaggerated ideas have generally been entertained of their relative importance. Large masses, supposed to be the work of myriads of polypifers, were considered to have been raised by the labour of these animals from great depths, while immense sheets of coral rock were supposed to cover the bottom of the seas. During Kotzebue's voyage, M. Chamisso enjoyed opportunities of visiting some remarkable groups of islands, arranged in a circular or oval manner, with openings among them which permitted the passage of a vessel from the outer ocean into the central basin. These islands seemed merely higher portions of a circular or oval ridge of coral reefs of unequal heights. M. Chamisso presented a description of what he considered the stages which the coral reef passed through before it became an island habitable for man. This description has been so often quoted that it must be familiar to most readers.

Subsequently to Kotzebue's voyage, MM. Quoy and Gaimard, who sailed with the expedition of M. Freycinet, paid particular attention to the coral islands and reefs which they had opportunities of examining ; and the result of their observations was, that the geological importance of these islands and reefs had been greatly exaggerated. Far from supposing that the polypifers raise masses from great depths, they consider that they merely produce incrustations of a few fathoms in thickness. In those situations where the heat is constantly intense, and where the land is cut into bays, with shallow and quiet water, the saxigenous polypi increase most considerably, incrusting the rocks beneath. The same authors observe,

* Edin. Phil. Journal, vol. v.

that the species which constantly formed the most extensive
banks belong to the genera *Meandrina, Caryophyllia,* and
Astrea, but especially to the latter ; and that these genera are
not found at depths exceeding a few fathoms.   It is therefore
concluded, that unless we are to suppose these animals en-
joying the prerogative of inhabiting all depths, under various
pressures of water, and different temperatures, they cannot
have produced the masses attributed to them.   From these
and other considerations they infer, that the appearance of
coral reefs and islands depends on the inequalities of the
mineral masses beneath, the circular character of some being
due to the crests of submarine craters*.   This conclusion
seems far from improbable, for we know that volcanic vents
are common in the same seas ; and that in the West Indies,
and the tropical parts of the Atlantic, where corals are sufficiently numerous, we do not observe these circular groups of
islands, the volcanic vents, though existing, not attaining the
importance of those in the Pacific Ocean or Indian Seas.

MM. Quoy and Gaimard observe, that, neither with the
anchor nor the lead, have they ever brought up fragments of
Astreæ, alone capable of covering large spaces, except where
the water was shallow, about twenty-five or thirty feet in
depth, though they found that the branched corals, which do
not form solid masses, lived at great depths†.   They agree with
Forster, that the polypifers may form small isles, when masses
of land shelter them, by raising their habitations to the level
of the sea : thus exposing a surface on which sands and other
matters are heaped and consolidated : a mode of formation
in accordance with what I have observed on the coasts of
Jamaica.

With regard to the great depth of water frequently observed
close to the coral reefs, the same authors consider, that they
may be accounted for on the supposition that the polypifers
have erected their dwellings upon the verge of a steep cliff, such
as is commonly observed on the sides of mountains and coasts.
In support of this opinion they cite the isle of Rota ; where
corals, resembling those now found in the neighbouring seas,
occur on cliffs.   There are, however, certain situations where
coral reefs run, as it were, in a line with a coast, but separated
from it by deep water, which would seem to require a different
explanation.

---

* Quoy et Gaimard, Sur l'Accroissement des Polypes Lithophytes con-
sidéré géologiquement, Ann. des Sci. Nat. tom. vi.

† Sounding off Cape Horn at about 56° S., and in about fifty fathoms of
water, they brought up small live branched corals; and sounding in one hundred fathoms on the bank of Laghullas (off the southern point of Africa,)
they obtained *Reteporæ.*

In situations such as those in which these coral isles and reefs abound, where recent, and comparatively recent, volcanic action is so apparent, we should expect to find evidences of the rise of such reefs above the level of the sea; and, accordingly, navigators have presented us with them. MM. Quoy and Gaimard state, that the shores of Coupang and Timor are formed of coral beds, which induced Peron to consider that the whole island was the work of polypifers. But it appears, that, proceeding towards the heights, vertical beds of slate, traversed by quartz, are met with at about five hundred yards from the town: and upon these and other rocks do the coral beds rest, which MM. Quoy and Gaimard estimate as not exceeding twenty-five or thirty feet in thickness. At the Isle of France a similar bed, more than ten feet thick, occurs between two lava-currents; and at Wahou, one of the Sandwich Isles, coral beds extend some little distance into the interior. To this we may add, that round the east coast, and on the northern side of Jamaica, there is an extensive bed that merely fringes the land, about twenty feet thick, which has every appearance of a coral bank raised above the waters, and brought within the destructive action of the breakers.

In situations like those in the Pacific, where volcanos and coral reefs are both abundant, we should expect to find some curious combinations of volcanic matter with coral banks, and even alternations: even admitting, for the argument, that the principal rock-forming polypifers do not build beneath twenty-five or thirty feet of water; still with the movements of land which may accompany volcanic action, such banks may be depressed, and covered by lava-currents, and again raised and brought to view. The example adduced in the Isle of France is sufficient to show, that at least one coral bed may be inclosed between lava-currents.

We cannot conclude this sketch without noticing a singular fact, observed, as we have been informed, by Mr. Lloyd while engaged in his survey of the Isthmus of Panama. Seeing some beautiful polypifers on the coast, he detached specimens of them; and, it being inconvenient to take them away at the time, he placed them on some rocks, or other corals, in a sheltered and shallow pool of water. Returning to remove them a few days afterwards, it was found that they had secreted stony matter, and fixed themselves firmly to the bottom. Now this property must greatly assist in the formation of solid coral banks; for if pieces of live corals be struck off by the breakers, and thrown over into calm water or holes, they would affix themselves, and add to the solidity of the mass.

Submarine Forests.

At various points round the shores of Great Britain, of the northern parts of France, and of the Baltic, accumulations of wood and plants, which do not appear to differ from those now existing, but on the contrary to be identical with them, occur at levels beneath those of high-water. To these ligneous and other vegetable remains, which are commonly seen at the retreat of the tide, a temporary removal of the beach, or an encroachment of the sea on tracts of land but slightly raised above it, the name of *Submarine Forests* has been given.

Correa de Serra describes the submarine forest on the coast of Lincolnshire as composed of the roots, trunks, branches, and leaves of trees and shrubs, intermixed with aquatic plants; many of the roots still standing in the position in which they grew, while the trunks were laid prostrate. Birch, fir, and oak were distinguishable, while other trees could not be determined. In general, the wood was decayed and compressed, but sound pieces were occasionally found, and employed for œconomical purposes by the people of the country. The subsoil is clay, above which were several inches of compressed leaves, and among them some considered to be those of the *Ilex aquifolium*, as also the roots of *Arundo Phragmites*.

These appearances are not confined to the coast, but extend considerable distances into the interior, so that the former merely presents a natural section of that which occupies a large area inland. A well sunk at Sutton afforded the following section:—

1 Clay.	16 feet.
2. Substances similar to the submarine forest	3 to 4 feet.
3, Substances resembling the scouring of a ditch-bottom, mixed with shells and silt .	20 feet.
4. Marly clay	1 foot.
5. Chalk rock*	1 foot to 2 feet.
6. Clay	31 feet.
7. Gravel and water	Not known.

Another boring made inland by Sir Joseph Banks afforded a similar section. This "moor" as Correa de Serra terms it, is considered to extend to Peterborough, more than sixty miles south from Sutton†.

Mr. Phillips presents us with very interesting details respecting some lacustrine deposits in Yorkshire, which are apparently of the age of these submarine forests, and which have

---

* This would seem not to be chalk, properly so called, but merely a chalky substance.   † Correa de Serra, Phil. Trans. 1799.

become in some places submerged. He remarks that the following may be considered as their general section. 1. Clay, generally of a blue colour and fine texture. 2. Peat, with various roots and plants; and in large deposits, containing abundance of trees, nuts, horns of deer, bones of oxen, &c. 3. Clay of different colours, with fresh-water *Limnææ.* 4. Peat, as above. 5. Clay, with fresh-water *Cyclades,* &c. and blue phosphate of iron. 6. Shaly curled bituminous clay. 7. Sandy coarse laminated clay, filling hollows in the diluvial formation. Mr. Phillips considers the accumulations of peat along the banks of the Humber and its tributaries as of the same epoch as these deposits, of which, he observes, the most constant beds are Nos. 1, 2, and 5. The species of deer enumerated as found in the peat, are,—the great Irish elk, (*Cervus giganteus*), the red deer (*C. Elaphus*), and the fallow deer (*C. Dama*). The peat deposit of the marsh-lands is covered by silt and clay, sometimes thirty feet thick, such as is now deposited by the Humber*. The peat is represented as beneath low-water mark.

Dr. Fleming describes a submarine forest on the shores of the Frith of Tay, extending in detached portions on each side of Flisk beach, three miles to the westward and seven miles to the eastward. It rests on clay of unknown depth. The clay is similar to the carse ground on the opposite side of the Frith, and to the banks in the channel. "The upper portion of this clay has been penetrated by numerous roots, which are now changed into peat, and some of them even into iron pyrites. The surface of this bed is horizontal, and situated nearly on a level with low-water mark. In this respect, however, it varies a little in different places. The peat bed occurs immediately above this clay. It consists of the remains of leaves, stems, and roots of many common plants of the natural orders *Equisetaceæ*, *Gramineæ*, and *Cyperaceæ*, mixed with roots, leaves, and branches of birch, hazel, and probably also alder. Hazel-nuts destitute of kernel are of frequent occurrence. All these vegetable remains are much depressed or flattened where they occur in a horizontal position, but when vertical, they retain their original rounded form. The peat may be easily separated into thin layers, the surface of each covered with leaves. The lower portion of this peat is of a browner colour than the superior layers; the texture is likewise more compact, and the vegetable remains more obliterated†."

The same author further observes, that stumps of trees,

* Phillips, Illustrations of the Geology of Yorkshire, 1829.
† Trans. Royal Soc. of Edinburgh, vol. ix.

with the roots attached, are observed on the surface of the peat, and no doubt can exist that they are in the positions in which they grew. No alluvial soil stratum was observed above the peat, the surface of which does not occur at a higher level than from four to five feet *below high-water mark.*

Dr. Fleming also describes another submarine forest in the Frith of Forth, at Largo Bay. It rests on a brown clay, into which the roots of the trees have penetrated. The author considers it as lacustrine silt. Over this there is an irregularly distributed covering of sand and fine gravel. The peat is composed of land and fresh-water plants, among which are the remains of birch-, hazel-, and alder-trees; hazel-nuts are also seen. Dr. Fleming traced the root of one tree, apparently an alder, more than six feet from the trunk[*].

If we pass from the main land of Scotland to its isles, we shall observe that the same appearances present themselves. Mr. Watt notices a submarine forest in the bay of Skaill, on the west coast of the mainland of Orkney. Stems of small fir-trees, ten feet long and five or six inches in diameter, are found partly imbedded in, and partly resting on, the surface of an accumulation of vegetable matter principally composed of leaves. The stems were still attached to their roots, and the whole was greatly decayed, so as to be easily cut by the spade. Many seeds of the size of a turnip-seed were discovered among the vegetable matter[†].

The Rev. C. Smith describes a submarine forest on the coast of Tiree, one of the Hebrides. Beneath a plain of 1500 acres in extent there would appear to be moss-land, similar to that previously noticed, under twelve or sixteen feet of alluvial covering. The moss-land is seen to bound the plain on the east, and the bay in which it appears is open to the whole force of the Atlantic. The general depth of the peat or moss-land amounts to several feet, but at its appearance on the shore it does not exceed four or five inches. This is firm, and adheres strongly to a sandy clay, on which it is based. Besides the remains of trees, which are obvious, there are other and smaller plants, and numerous seeds, which at first looked quite fresh, but afterwards became darker from exposure. " The seeds have the appearance of belonging to some plant of the natural order of *Leguminosæ*; and Mr. Drummond suggests that they may probably be those of *Genista anglica*[‡]."

According to the same author, submarine forests are by no means uncommon on the shores of Coll. He also cites the Rev. H. Maclean as having noticed similar appearances, not observed by himself, in the island of Tiree.

[*] Journal of Science.   [†] Edin. Phil. Journal, vol. iii. p. 100.
[‡] Smith, Edin. New Phil. Journal, 1829.

Returning again to the main land, we find similar appearances described by Mr. Stephenson, on the shores of the flat
lands between the Mersey and the Dee, on the coast of Cheshire. Stumps of trees, ramifying in all directions, are stated
to appear as if cut off about two feet from the ground. The vegetable matter rests on blueish marl, and is covered by sand *.
  Mr. Yates notices a submarine forest on the coast of Meri-
onethshire and Cardiganshire, divided into two by the estuary of the river Dovey. A sandy beach and wall of shingles
bound it on the land side, beyond which is marsh-land, partially drained by the oozing of the water through the sand
and shingle. The *Pinus sylvestris*, or Scotch Fir, occurs
among the other trees†.
  Mr. Horner describes a submarine forest on the coast of the
S.W. part of Somersetshire. It is well seen between Stolford
and the mouth of the Parret, where the shore is low; a high
shingle beach, principally composed of lias (the rock of the
vicinity), protects the level land behind from the sea. The
vegetable remains present themselves here, as in the other
places, as a stratum of peat or decayed leaves, containing the
trunks, stems, and branches of trees. Among these are twigs,
nuts, and a plant, (commonly found entire,) which Mr. Brown
considered might be the *Zostera oceanica* of Linnæus. Some
of the stems of trees were twenty feet long, and the woods
were considered to be oak and yew, not generally decayed,
but sufficiently hard and tough to be used as timber, and for
fuel. Even those trees which were soft when taken out, became hard when dried. The brown vegetable matter was
generally a foot or eighteen inches thick, and rested on blue
clay‡.
  From this coast there is an extensive tract of flat land, which
extends a considerable distance inland, and from it the hills
rise in promontories, islands, and other forms, precisely as
they would rise from a level sea. Mr. Horner cites De Luc
as stating, that while new channels were digging between the
Brue and the Axe, a bed of peat was found beneath the surface.
This stratum, if it may be so called, has been noticed in other
parts of the same flats, and even trees have been reported as
found in it; seeming to show that the forest noticed on the
shore may be only a section of a large deposit beneath the
Bridgewater levels.

---

* Edin. Phil. Journal, vol. xviii. Mr. Smith cites the Liverpool Courier of
December 1827, to show that after a heavy gale, trunks and roots of trees
were found under the sand below high-water mark, which had all the ap-
pearance of having grown where then found.
  † Yates, Proceedings of the Geol. Soc., November 7, 1832.
  ‡ Horner, Geol. Trans. vol. iii. p. 380, &c.

A very important addition to our knowledge of submarine forests has been made by Dr. Boase in his description of that in Mount's Bay, Cornwall. The vegetable bed consists of a brown mass, composed of the bark, twigs, and leaves of trees, which appear to be almost entirely hazel. In this there are numerous branches and trunks of trees. The greater part of this wood is hazel, mixed with alder, elm, and oak. "About a foot below the surface of this bed, the chief part of the mass is composed of leaves, amongst which hazel-nuts are very abundant. In this layer may also be found filaments of mosses, and portions of the stems and seed-vessels of small plants, many of them evidently belonging to the order of Grasses; together with the fragments of insects, particularly of the elytra and mandibles of the beetle tribe, which still display the most beautiful shining colours when first dug up, but on exposure to the air all these minute objects soon crumble into dust." Beneath this, the vegetable matter becomes closer, and finally earthy and of a lamellar structure. It rests on granitic sand, and this again on clay slate. The vegetable stratum slopes from the interior to the sea at about an angle of two degrees. It is covered by a bed of smoothly polished shingles, composed of hornblende rock, about two or three inches in diameter. The bed is sixteen feet thick, and is crowned by a granitic sand about ten feet thick. The vegetable bed, by its rise, appears beneath a marsh inland, having passed under its covering of pebbles and sand *.

M. De la Fruglaye observed that after a heavy gale in 1811, a beach near Morlaix, which previously seemed to consist of sand, presented, from the sand being washed away, an appearance of a large mass of vegetable matter and trees united together and extending along shore for a considerable distance. The leaves were well preserved, but the trunks and branches of trees were rotten. Oak was observed among the wood, and insects with their colours preserved were discovered in the mass. A few days after this event this accumulation of vegetable matter was again covered up by sand +.

The peat moors which occur on different points of the shores of the Baltic, near Greifswald, near Gnageland, on the S.E. side of the Haff, at its confluence with the Oder near Swinemünde, in the island of Usedom, and in the vicinity of Colberg, agree in many respects with the submarine forests above noticed. They are in many places from ten to fourteen feet below the level of the Baltic, above which they only rise a few feet. They are separated from the sea by a slip of coast of various breadths, by dunes and sandbanks, under which they

---

* Boase, Trans. Geol. Soc. Cornwall.   † Journ. des Mines, t. xxx.

do not extend, but gradually disappear. In these peat moors, the greatest depths of which are in their central portions, land, marsh, and fresh-water plants, with their seeds, are alone discovered, to the exclusion of marine vegetation. Trunks of trees with their roots, among which are oaks and pines (*Pinus sylvestris*), are found in them, not at the bottom, but at some height above it. The roots occur in their natural positions, often several times above each other, still however beneath the present level of the Baltic to the depth of five feet.

In some places the *Arundo Phragmites* so abounds that the peaty moss seems entirely composed of it. The lower layers contain *Ceratophyllum demersum, Potamogeton pusillum, Najas major, Nymphæa lutea. Scirpus palustris,* and *Hippuris vulgaris,* are also discovered with the *Arundo.* Seeds, especially of the *Menyanthes trifoliata,* are also frequent in the lower layers. The ground beneath the peat contains fresh-water shells: *Paludina impura,* Lam., *Planorbis imbricatus, Cyclostoma acutum,* and *Limneus vulgaris**.

Having cited so many examples to show their general similarity, I shall merely notice that I have observed submarine forests on the coasts of Normandy, one to the east of the Vaches Noires cliffs, and the other near St. Honorine, both at the mouths of valleys; and that at the mouth of the Char, coast of Dorset, there are traces of another.

To account for the occurrence of these submarine forests two very different explanations have been proposed; one supposing a change in the relative level of sea and land on certain coasts of Western Europe; the other referring their origin to an accumulation of vegetable matter in stagnant pools or lakes of either fresh or brackish water, defended from the sea by sandbanks or dunes: the bottom of such pools or lakes being beneath the level of the sea. The latter explanation would appear sufficient, if, in all cases, that portion of the peat which shows the growth of such woods as certain firs and oak, was above *low-water mark.* This, however, does not appear to be the fact. We can readily imagine an accumulation of beach or sandy dunes in front of a level tract of country, or at the mouths of valleys; and the formation of lakes or pools of fresh or brackish water behind such beaches or dunes, because this we daily witness. Therefore if these lakes or pools were gradually filled up, and a change took place in the action of the sea on the same coast, so that it proceeded to cut away the beach or dunes which it once accumulated, we should have peaty matter protruding through the beach, and even extending beneath the level of the sea. If, however, we consider

* German Transl. of Manual.

the relative level of the sea and land to have remained constant, there is a difficulty in conceiving the growth of certain
firs and other trees, not being marsh plants, in situations
which could have received no drainage at any time of tide; the
levels being in fact so much against such a drainage, that even
the level of low-water was higher than that of the situations
where the firs and oaks grew. Of a similar growth we have
now no examples; indeed it supposes a change in the nature of
some of the plants discovered in the submarine forests.

We may probably approximate nearer to the truth by partly
admitting both explanations, the one for the accumulation of
the vegetable matter, the other for its occurrence at the present levels of sea and land. The vegetable matter might
easily have been produced by the gradual filling up of pools
or lakes, if we suppose the process to have been so far advanced when the firs and oaks appeared, that the ground
was no longer beneath a shallow lake or marsh, but was such
as usually supports the growth of these and other trees of similar habitats. This, however, supposes a sufficient drainage,
and by no means a state of levels, such as is now often seen,
in which the planes of the stumps and roots of firs and oaks
are several feet beneath the present level of *low-water.* To
account for this in such situations a change in the relative level
of sea and land seems requisite, bringing either the lower
ends of extensive and inclined woody plains, or accumulations
of vegetable matter behind banks and dunes, into the action of
the breakers, and even beneath low-water mark. Such a
change is in accordance with general geological phænomena;
it is one now frequently taking place in various parts of the
globe, and has been exceedingly common during the formation
of a large portion of the earth's crust, as numerous facts sufficiently attest.

The above details are perhaps too copious for the plan of
this volume; but it seemed important to show that changes in
the relative levels of the ocean and land had taken place round
our own shores at such geologically recent times, more particularly as it will be attempted to prove beneath, that at least
a partial difference of levels on our southern shores, of quite a
contrary kind, has preceded it.

Raised Beaches and Masses of Shells.

At Plymouth and the neighbouring coast there are the remains of a beach, of which the maximum elevation is about
thirty feet above high-water mark, sloping gradually to the
sea *. The following is a section at the Hoe.

* Professor Sedgwick informs me that the Rev. R. Hennah pointed out

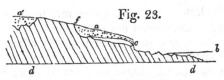

Fig. 23.

Upon the grauwacke limestone beds *d d,* which dip at a considerable angle southwards, rests an accumulation, *c,* of rounded pebbles and sand, with here and there a larger and angular piece of limestone intermixed. The accumulation has every appearance of a sea-beach raised above the present level of the sea *b,* and the shingles and sand are so arranged that the resemblance is quite perfect, more particularly when shells are found in it[*]. The shingles consist of limestone, slate, red sandstone, reddish porphyry (which occurs, in place, in another part of Plymouth Sound), and of various rocks that form part of the grauwacke series of the neighbourhood. The section annexed is exposed by blasting the rock, the limestone being taken away in great quantities. It will be observed that the beach, *c,* did not extend to *f,* which seems formerly to have been a cliff, in the same manner as the present beach is backed by a low cliff. The beach and part of the limestone hill are covered by a gravel or loose breccia of angular limestone fragments, *a a',* which clearly have not received attrition from the action of water upon them. This circumstance seems to afford us a relative date for the beach, as the reader will recollect that under the head of degradation of land it was observed that the whole of this part of Devon afforded a superficial detritus of the rocks beneath. Now the angular pieces of limestone, *a,* are derived from the hill above, and have slipped by the force of gravity, assisted by meteoric causes, over the beach *c,* as they have also fallen into the cavity *a',* which being above the old beach *c,* does not contain either pebbles or sand, but is precisely similar to those clefts in the Oreston quarries near Plymouth, where the remains of elephants, rhinoceroses and other animals, occur beneath fragments of the same kind. It therefore seems fair to infer that the beach was raised during the existence of these animals, and previous to that long period of time, during which the action of the atmosphere slowly, though considerably, destroyed the

this beach to him several years since; and Mr. Hennah has noticed it in his account of the Plymouth limestones.

[*] I was only fortunate enough to see fragments, and these apparently consisted of pieces of *Patellæ* and small *Neritæ,* the latter with their colours preserved, and resembling those now found on the coast; but many hundreds were found in a cavity of the limestone filled with sand and thrown away by the quarry-men. Beneath the citadel the sand is composed of fragments of shells.

surface of the hills. It seems, moreover, to show a configuration of land in the vicinity not very different from that which
now exists. This view is strengthened by a minute examination of the coast from hence towards Tor Bay, along which,
similar appearances may here and there be seen ; but it must
be evident that this will depend upon the quantity of cliff cut
back by the sea at its present level, as will be seen by the annexed diagram.

Fig. 24.

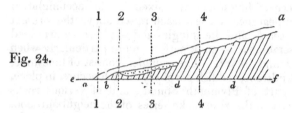

Thus, if *a a* represent the angular detritus derived from the
slate rock *d d*, which rises into a high hill behind, and *b* an
ancient beach now raised above the level of the sea *e f*, and
covered by the detritus *a a*, a section made at 1 1 would only
show the detritus ; another made at 2 2 would only expose detritus or slate, the bottom rising, as is very commonly the
case on sea-beaches where rocks project among the shingles,
particularly on coasts such as are now under consideration.
If the sea cut back the cliff to 3 3, we should have such a section as that at the Hoe ; but if the cliff be cut to 4 4, the
whole beach is removed, and no traces of it left. This is precisely what happens on this coast, where all the above varieties are observed. Under Mount Edgecumbe, near Plymouth,
the rolled shingles are covered by fragments of slate and red
sandstone. At Staddon Point, sand is covered by compact
red sandstone fragments. Further south, on the eastern side
of the Sound, and nearly opposite the Shag Rock, there is
the following section, which may or may not be ancient beach
covered with detritus: *c*, the main
rock of argillaceous and arenaceous
schist ; *b*, detritus, some of which approaches a sandy earth, mixed with
small pieces of slate rarely exceeding
the size of a shilling or sixpence ; *a*, detritus, composed of
angular pieces of schist and sandstone of the size of an egg and
upwards, mixed with others of smaller dimensions.

Fig. 25.

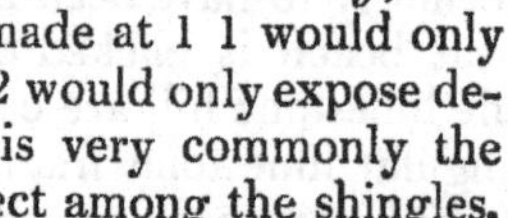

This change in the relative level of sea and land is not confined to the coast above mentioned, for Mr. Fox notices an
horizontal bed of rolled quartz pebbles, gravel, and sand, precisely similar to that of the neighbouring coast, as occurring

between Falmouth and Helford. This bed does not extend far into the cliff, is generally nine or twelve feet above high-water mark, and extends for a distance of about four miles, its continuity being interrupted in those places where solid rock cliffs alone constitute the coast. It would appear that this beach is also covered with detritus, whether angular or rounded is not stated*. From the observations of Dr. Boase, traces of elevated beaches are observable in several other situations on the coasts of Cornwall, as well on the north as the south. He particularly notices the granitic pebbles and boulders on the shores of the Land's End district (described by Mr. Carne), and the beds of gravel and sand at Fistral and Gerran's Bay. The maximum height of the former beach is estimated at fifty feet. The lower parts of these beaches sometimes touch the sea, while at others they are seen on the cliffs, high above its present level, resting on the rocks of the coast†.

From the apparent date of the elevated Plymouth beach, this notice might perhaps have been more in its place in the next section, but it is so intimately connected with the subject of the alternate rise and fall of land, that it seemed better to let it follow the notice of submarine forests. The conclusions from both phænomena, which should by no means be hastily generalized, but confined for the present to the places noticed, would seem to be:—1. A configuration of land not greatly differing from the present, when elephants and rhinoceroses, *perhaps*, existed in this climate. 2. The beach elevated. 3. A considerable but quiet destruction of the surface of the hills, covering over the ancient beach, the general shape of the hills and valleys being not very different from those we now see. 4. A depression of the land, submerging woods and forests, and bringing the detritus of epoch 3 into destructive contact with the sea, from which it was in a great measure previously protected by the usual slopes and beaches: and, 5. The changes effected since the establishment of the present relative levels of sea and land.

Captain Vetch describes six or seven terraces or lines of beach on the Isle of Jura in the Hebrides, which appear to have been successively raised above the present level of the ocean. The lowest is on a level with high-water, the most elevated about forty feet above it. The terraces or beaches rest partly on the bare rock, and partly on a thick compound of clay, sand, and angular pieces of quartz. Their continuity is here and there interrupted by mountain torrents, or the ac-

* Fox, Phil. Mag. and Journ. of Science, Dec. 1832.
† Boase, on the Geology of Cornwall, Trans. Geol. Soc. Cornwall, vol. lv.

tion of the sea on the supporting compound. They are well
seen at Loch Tarbert. Their aggregate breadth varies " according to the disposition of the ground : where the slope is
precipitous, it may be a hundred yards ; where gentle, as on
the north side of the loch, three quarters of a mile from the
shore." These terraces or beaches are formed of round
smooth white pieces of quartz, of the size of cocoa-nuts. They
are precisely similar to those which constitute the present
beach of the Atlantic on this side of the island, and from their
forms they must have been produced by the united action of
tides and waves. Captain Vetch mentions, in confirmation of
this opinion, that a series of caves is to be found on the same
level along the north side of Loch Tarbert, at a considerable
height above the sea; and as he never observed any caverns
formed in the quartz rock of Isla, Jura, and Fair Island, except those on the sea shore, he considers these to have been
thus produced*.

M. Brongniart describes a singular accumulation of shells,
precisely similar to those which exist in the neighbouring sea,
at Uddevalla, in Sweden. Their abundance is very considerable, for they have been long employed on the roads ; they
are nearly free from any earthy mixture, and though many
are broken, there are numbers entire. The largest mass rises
among gneiss rocks, to the height of sixty-six metres above
the level of the sea. This author, considering that he might
find traces of the former residence of the sea upon the fundamental rock, gneiss, searched around with considerable attention, and was rewarded by the discovery of *Balani* still adhering to the rocks on which they grew, now become the summit of a hill. MM. Berzelius, Wöhler, and Ad. Brongniart,
were present at this discovery†.

At Nice the sub-fossils of St. Hospice have long attracted
attention ; they correspond with the present inhabitants of the
Mediterranean, and often retain their colours, though they
are generally blanched. Of these shells M. Risso has given a
long list‡. From personal observation, I have little doubt
that the whole has been raised, in comparatively recent times,
above the present level of the Mediterranean. Beneath Baussi
Raussi, a neighbouring cliff, and from thence to the principal
deposit of sub-fossil shells, there is apparent evidence of a
raised beach, the pebbles being rounded, and intermixed with
sand, in which shells similar to those now existing in the

---

* Vetch, Geol. Trans. 2nd series, vol. i.
† Brongniart, Tableau des Terrains qui composent l'Ecorce du Globe,
p. 89. For a list of the shells found, consult Hisinger, Esquisse d'un Tableau
des Petrifications de la Svède, Ed. 2me, Stockholm, 1831.
‡ Hist. Nat. de l'Europe Meridionale.

neighbouring sea are discovered. Indeed, between the peninsula of St. Hospice and the cliff above mentioned, the old beach much resembles that near Plymouth, with the exception that the latter has been higher raised*. The elevation near Nice must have taken place after the land had, in a great measure, received its present configuration.

M. de la Marmora presents us with a very interesting account of a bed containing sub-fossil shells and the remains of coarse pottery in Sardinia, which would appear not only to come under the head of a raised beach, but also of the bottom of a shallow sea connected with it. Where most distant from the present sea-coast, and consequently where it most probably constituted the shores of the ancient coast, before the elevation of the land, or the depression of the sea-level, the bed is earthy and ferruginous, and contains the remains of terrestrial, fluviatile, and marine shells, mixed with fragments of coarse pottery;—a state of things we should expect on an inhabited coast, particularly of a nearly tideless sea, such as the Mediterranean. Where nearest the sea, and where we may consequently consider it to have been formerly beneath the sea, (for the bed rises gradually inland), this bed is formed of a calcareous sandstone, the pottery disappears, and *Cerithia* and *Lucinæ* become more rare. On the N.W. of Cagliari, where the bed rises about 150 feet above the present level of the Mediterranean, and is there about a mile and a quarter distant from the sea, oysters (*Ostrea edulis*) are found adhering to the rock upon which they evidently grew. The sub-fossil shells are of the same species with the shells now living on the same shores, and are described as in a good state of preservation. Among the pottery, M. de la Marmora discovered, on the N.W. of Cagliari, a round ball of baked earth, about the size of an apple, with a hole in the centre, as if to pass a cord through. This ball M. de la Marmora considers may have belonged to fishermen to whom the use of lead was unknown, fishermen who followed their calling before the change of level was effected which converted the bottom of a shallow sea into dry land†. We here appear to have an example of an elevation of land, or a depression of the sea-level, in this part of the Mediterranean, after the island of Sardinia was inhabited by man. If M. de la Marmora be correct in considering that he can identify similar beds on the shores of

---

* For a more detailed description of these localities, with a view and a section of Baussi Raussi cliff, see my paper in the Geological Transactions, vol. iii. 2nd series.

† De la Marmora, Journal de Géologie, tom. iii.

Tuscany, of the Roman States, and of Sicily*, the change of level would appear not to have been altogether local.

M. Boblaye notices various lines of worn rock on the limestones of Greece (similar to the lines now produced by the action of the waves on the coasts of the same country), raised at various heights above the present level of the Mediterranean. He also points out the existence of small horizontal terraces, and lines of holes drilled by perforating shells;—circumstances which M. Boblaye attributes to successive elevations of the land above the level of the sea. A littoral cavern, near Napoli di Romania, contains a breccia referrible to the present epoch, for it contains fragments of antique pottery. This cavern has apparently been raised five or six yards above the present level of the Mediterranean†.

It has been previously observed, that on the west coast of South America a beach was raised during the earthquake of 1822, and there were evidences of former beaches having been so elevated. M. Lesson also observed at Conception, more southerly on the same coast, banks of shells, corresponding with those of the neighbouring sea, now dry and raised above it‡.

It is almost impossible not to remark in these raised beaches and sea beds, the action of the same forces which have been noticed under the head of Earthquakes. The land has been liable to rise and fall at various epochs, as will be seen in the sequel; the intensity of the force, producing these changes, varying materially. It is exceedingly difficult to assign dates to the Plymouth raised beach, to the shells at Uddevalla, and to the other similar appearances above noticed; but we learn from them, that since the establishment of animal life, such as we now observe it, the relative levels of the sea and land have been liable to change, as they have been previously to this period, and, referring to the Temple of Serapis, near Naples, as they have been even since man has erected his temples and other works of art§.

---

* M. de la Marmora, carefully distinguishes the sandstone from the rock daily forming at Messina.

† Boblaye, Journal de Géologie, tom. iii.

‡ Brongniart, Tab. des Ter. qui composent l'Ecorce du Globe, p. 92.

§ For a detailed account of the geological appearances connected with the celebrated Temple of Serapis, at Puzzuoli, near Naples, consult Lyell's Principles of Geology, vol. i. The rise and fall of land seem to have been as follows: 1. After the original building of the temple, a sinking of the land, and a covering of the lower part of the columns, so that the boring shell (*Lithodomus*) only attacked them about twelve feet above their pedestals. The height to which the shells have bored is also about twelve feet; therefore the columns, without being overthrown, were certainly lowered to

Organic Remains of the Modern Group.

These will necessarily consist of existing animals, but may also include some no longer found in a living state. Man not only greatly modifies the present surface of the land, by destroying tracts of forests, preventing the inundations of low countries, turning torrents, and directing the surface-water through innumerable channels to satisfy his own wants and conveniences, but he also drives all animals before him which do not suit his purposes; thus circumscribing the domain of those which are not useful to him, while he covers the country with those that are, and which never could exist in such numbers but for his care and protection. Consequently all terrestrial remains would correspond with the increasing power of man, and therefore a very different suite of such remains would be now entombed, than when his power was more limited. Over the inhabitants of the waters he would exercise little control, excepting in rivers, small lakes, and round some coasts.

One very material difference would be effected in the quantity of trees and shrubs transported to the sea, more particularly in the temperate and colder regions, where man requires wood, not only for the purposes of various constructions, but also for fuel. We see in the delta of the Mississippi that an abundance of wood is now transported there by the river, but this will daily diminish as man converts the forests, whence it is derived, into pastures and corn-fields.

The gigantic animal *Cervus giganteus*, commonly known as the Irish Elk, was once imagined to have existed only at an epoch anterior to man, but it is now considered that he was co-existent with him; although this by no means proves that it did not live upon the earth previous also to him, as seems to have been the case. We have no great certainty when the Mastodons of North America ceased to exist; it is commonly supposed that they became extinct previous to the commencement of the modern group, but of this we have no good proof. The same may be said of some other animals.

The Dodo seems to afford us an example of the extinction of an animal in comparatively recent times; for it is now almost certain that this curious bird existed on the isle of Mauritius, during the voyages of the early navigators to the East

---

the depth of twenty-four feet above their pedestals in water. 2. Elevation of the temple, still standing, above the level of the sea, or nearly so, for the pavement is not flooded to any considerable depth, not more than about one foot.

Indies. The relative antiquity, therefore, of animals whose remains are only now found entombed, must not be too hastily inferred. The bone of the wolf is that of an extinct animal, as far as the British islands are concerned. In the darkness of ages many animals may have perished, not a tradition of whose existence remains, not only from the advance of man, and the power which civilization affords him, but also from the destruction caused by predaceous animals,—though the latter is not so probable as the former.

Section III.

ERRATIC BLOCK GROUP.

Erratic Blocks and Gravel.

We must impress upon the geological student the necessity of considering this group as simply one of convenience, formed provisionally for the purpose of presenting certain phænomena to his attention, which in the present state of science could not so easily be done under any other head.   The origin of the various transported gravels, sands, blocks of rocks, and other mineral substances scattered over hills, plains, and on the bottoms of valleys, often referred to one epoch, may belong to several.   In a word, all that transported matter commonly termed *Diluvium*, requires severe and detailed examination.   At the present time, there would appear to be three principal opinions connected with the subject.   One, supposing the transport to have been effected at one and the same period; —another, that several catastrophes have produced these superficial gravels;—while a third would seem to refer them to a long continuance of the same intensity of natural forces as that which we now witness.   Perhaps these various opinions may arise from our present inadequate knowledge of the phænomena on which we attempt to reason, and probably also from premature generalizations of local facts.   These different opinions, though they cannot each be correct in explanation of all the observed facts, may each be so in part; and it were to be wished that the phænomena here arranged under one head solely, as above stated, for convenience, were examined without the control of a preconceived theory.

At the close of the last section, a local elevation of land was noticed, of somewhat difficult arrangement in our systems. In order to illustrate the changes which have taken place in the same district, without, however, attempting to consider such appearances as general, I shall continue the description of it.   At Oreston quarries, Plymouth, clefts and caverns in limestone rocks have afforded numerous remains of the elephant, rhinoceros, bear, ox, horse, deer, &c. buried, more particularly in the case of clefts, beneath considerable angular masses and smaller fragments of limestone.   In one instance which I noticed, the animal remains occurred beneath ninety feet of such accumulations, the bones and teeth being confined to a black clay under the fragments.   The remains of bears,

rhinoceroses, hyænas, and other animals contained in the celebrated Kent's Hole, near Torquay, belong to the same district. In the superficial gravel of this part of the country,
the remains of animals, of the same kind as those detected in
the caverns, have not yet been discovered; but if we continue
our researches eastward, we shall find them in the valleys of
Charmouth and Lyme*, where they occur in situations which
would appear anterior to the great weathering, if I may so express myself, of the circumjacent hills: thus apparently giving
these remains of elephants and rhinoceroses the same relative
antiquity as those beneath fragments in the clefts of rocks near
Plymouth, and probably also as those contained in the caverns
at the same place, and in Kent's Hole. Now the raised beach
in Plymouth Sound seems to afford evidence of a configuration
of land not widely different, in that place, from the present,
and therefore we may perhaps infer the existence of inequalities in the land, or hill and dale, in this district generally, not
widely different from those we now observe. It will be remarked that the animal remains which seem to imply a warmer
climate existing at that time than at present, occur in low
grounds, fissures, and caves. Upon the former they may have
lived, and into the two latter they may have either fallen or
been dragged by beasts of prey. The elephants probably
browsing on branches and herbage, rhinoceroses preferring
low grounds, the bears and hyænas inhabiting caves, and the
deer, the ox, and the horse, ranging through the forest and
the plain; all which supposes land fitted for them, and therefore
hill and dale, level plains and rocky escarpments with open caverns. Consequently valleys were scooped out previous to the
existence of the elephants; and if a mass of waters acted on the
land, destroying these animals, it must have been influenced in
its direction by the previously existing inequalities of surface.

The next question may be, does this district present evidences of the exertion of a greater intensity of natural force
than that which we now observe? The answer may be, that
it does. The whole district is fractured, or, to use geological
terms, so broken into faults, that the spaces in which, with
careful examination, they may not be detected, are very inconsiderable. Such dislocations may, or may not, have been
contemporaneous with the raised beach. Perhaps they were
previous to it, for there has evidently been a very considerable
dispersion of rock fragments, and this apparently by water,
which would have scattered such a beach as that noticed at
Plymouth. The following section at the Warren Point, near

---

* The line of coast has been preferred in this description, because the
sections are there more clear and less equivocal.

Dawlish, is not only a good example of a compound fault, but also of transported gravel upon it.

b b, conglomerates, and *c c*, sandstones of the red sandstone formation, fractured or broken into faults at *f f*, so that continuous strata are displaced. Upon these fractured strata rests a gravel, *a a*, composed of chalk flints, and green sand chert, mixed with a few pebbles similar to those in the conglomerates *b b*. It has evidently been deposited subsequent to the fracture, for it rests quietly upon it and is unfractured.

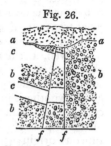

Fig. 26.

The chalk and green sand of this district have once covered very considerable spaces, though the latter is now only seen on Haldon Hills; near this section, it is true, but separated from it by an intervening valley. There are many other dislocations so covered on the same coast, where these appearances can be observed with the greatest ease, particularly at low water.

It might be supposed that these flints and pieces of chert were merely the remains of superincumbent masses of chalk and green sand, which have been destroyed by meteoric agents, the harder parts falling down on the top of the fracture. We can scarcely consider this physically probable, if even possible; for it supposes the removal of more than 600 feet of sandstone and conglomerate (for not until that height above this section would the green sand and chalk come on), without scarcely leaving any of the pebbles, or large masses of the red conglomerate, while the flints and cherts, which belonged to upper, and consequently first destroyed rocks, remain.

Let us now consider another class of appearances. Over the whole district, wherever transported gravel occurs, the surface of the rocks (it being of no importance what they happen to be,) is drilled into cavities and holes, similar to those well known on the chalk of the east of England. The following sections will illustrate this:

Fig. 27.

(Fig. 27.) *a a*, gravel, principally of flint and chert, resting in a hollow of the red sandstone *b b*, between Teignmouth and Dawlish, the lines in the gravel following the outline of the cavity.

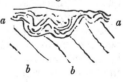

Fig. 28.

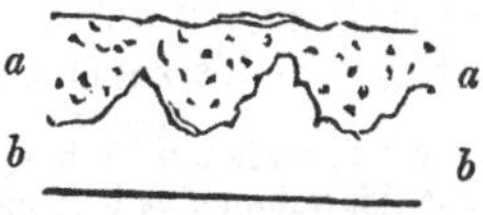

(Fig. 28.) *a a*, gravel composed in a great measure of flints, among which are some large rounded pieces of siliceous breccia (the same as that which occurs in blocks on the top of the chalk hills near Sidmouth), resting in cavities in the pipe-clay, near Teign Bridge, which constitutes a part of the Bovey coal formation, and which is not, as has been supposed, contemporaneous with the superficial transported gravel.

Other examples might easily be adduced; but these are here given, because the geological student can very readily observe them*. Not, however, to deceive the reader, or oneself, by any attempt to overlook difficulties, or to render any given hypothesis plausible, it must not be concealed, that, from certain appearances in the vicinity of Dawlish, it is by no means impossible some of these gravels may be of contemporaneous origin with the Bovey coal deposit, to be noticed in the sequel. The phænomena may therefore be explained by the action of water moving with no small velocity, (for the transported fragments are often large,) at different periods of time, not, geologically, far removed from each other. In addition to the mere transport of detritus, we have, still in the same country, evidences of a washing of the rock beneath, by which portions of it are mixed with the transported substances, and even, in fortunate sections, have the false appearance of surmounting the transported matter, as the annexed section of the cliff near Dawlish well illustrates.

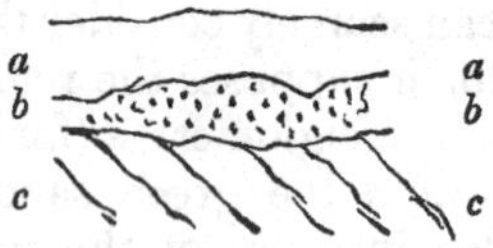

a a, regenerated red sandstone; *b b*, gravel composed of chalk flints, chert, and pebbles derived from the conglomerate interstratified with the red sandstone *c c*, upon which it rests. To a person unaccustomed to geological investigations it would easily be imagined, from this section alone, that the flints were included in the red sandstone; but the true arrangement is very apparent, even if the stratification of *a a* and *c c* did not show it; for the section is entirely fortuitous, every variety being observable in the vicinity, and this merely selected as an extreme case.

Fig. 29.

Our limits will not permit greater detail, which would require the necessary maps, but it would go far to support the supposition that masses of waters had passed over this land. The question might now arise, may there be any connexion between the masses of water supposed to have passed over the land, and the fractures or faults so common? In answer to this it may be replied, that such a supposition is not inconsistent with possibilities or probabilities. We have seen that during such vibrations and comparatively small dislocations of the earth's surface as those which we now witness, the water

---

* The same motive has governed me in the selection of sections throughout this volume, as it cannot be expected that the student should so readily observe difficult facts as the accomplished geologist.

is thrown into movement, and breaks with greater or less fury on the land. Still confining our attention to one district, it should be observed, that the dislocations are far greater, and the faults, evidently produced at a single fracture, far more considerable, than any we can conceive possible from modern earthquakes. It is not, therefore, unphilosophical to infer, that a greater force causing vibrations and fractures of the rocks would throw a greater body of water into more violent movement, and that the wave or waves bursting upon the land, or acting on the bottom at comparatively small depths, would have an elevation, and a destructive sweeping power, proportioned to the disturbing force employed.

The next question that will arise is, are there any other marks of masses of water passing over the land? To this it may be replied, that the forms of the valleys are gentle and rounded, and such as no complication of meteoric causes, that ingenuity can imagine, seems capable of producing; that numerous valleys occur on the lines of faults; and that the detritus is dispersed in a way that cannot be accounted for by the present action of mere atmospheric waters. I will more particularly remark, that on Great Haldon Hill, about 800 feet above the sea, pieces of rock, which must have been derived from lower levels, occur in the superficial gravel. They are certainly rare, but may be discovered by diligent search. I there found pieces of red quartziferous porphyry, compact red sandstone, and a compact siliceous rock, not uncommon in the grauwacke of the vicinity, where all these rocks occur at lower levels than the summit of Haldon, and where certainly they could not have been carried by rains or rivers, unless the latter be supposed to delight in running up hill.

It may be stated, before we quit this local description, that the faults do not all range in one direction, though east and west are not uncommon; and that as we approach the Weymouth district, this direction predominates. Near Weymouth there is one east and west fault, fifteen miles of which can be traced, but it probably extends further, for it enters the chalk on the east, and therefore cannot be easily observed, while it plunges into the sea on the west. There seems also every probability that these Weymouth faults are connected, as has already been remarked by Prof. Buckland and myself in another place, with the east and west dislocations through the Isle of Wight, and probably also with the east and west upraised, and afterwards denuded country of the Wealds of Sussex *. It should also be remarked, that the accumulations

* Many faults in the Blackdown Hills range N. and S., and are therefore at right angles to the Weymouth faults, from which they are not far distant. These also have been produced since the deposit of the chalk, and probably also of the plastic clay.

of gravel are often most considerable on the eastern sides of the valleys, in the vicinity of Sidmouth and Lyme.

Let us now proceed to consider to what extent these local facts may be more or less general. To begin with England. Lowland valleys, often very considerably broader than those before noticed, and therefore more favourable to the supposition of moving masses of water, occur very generally; for the surface composed of lowland valleys is very considerably greater than that exhibiting mountain valleys, though both have been modified by rivers and other agents now in operation. Over these valleys, foreign matter, not detritus derived from the weathering of the rocks beneath, is variously distributed. It may sometimes be possible, with the aid of ingenuity, to produce a case of transport by a long continuance of such natural effects as are now seen, but in other situations such explanations seem altogether valueless and unphilosophical. In like manner also faults covered only by gravel are common, the lines of faults being frequently lines of valley. I would by no means infer that all faults, only covered by gravel, have been contemporaneous; on the contrary, it seems only reasonable to conclude, that faults or fractures have accompanied every great convulsion, and that as these have been frequent, so faults may also have been frequent.

Not only are gravels brought from various distances, but even huge blocks, the transport of which by actual causes into their present situations seems physically impossible. Mr. Conybeare has remarked on the great accumulation of transported gravel in midland England, more particularly at the foot of the inferior oolite escarpments on the borders of Gloucestershire, Northamptonshire, and Warwickshire, and observes that it is composed of such various materials that a nearly complete suite of English geological specimens may there be obtained. " Portions of the same gravel have been swept onwards through transverse valleys affording openings across the chains of the oolite and chalk hills, as far as the plains surrounding the metropolis; but the principal mass of diluvial gravel in this latter quarter is derived from the partial destruction of the neighbouring chalk hills, consisting of flints washed out from thence, and subsequently rounded by attrition [*]." Mr. Conybeare also notices the occurrence of great blocks among the transported rocks of Bagley Wood, Oxfordshire, as also the presence of flints on the summits of the Bath Downs. Prof. Buckland mentions that he found, among the transported gravel of Durham, twenty varieties of slate and greenstone, which do not occur, in place, nearer than the lake district of Cumberland. He also notices a large block of

---

[*] Conybeare and Phillips, Outlines of the Geology of England and Wales.

granite at Darlington, composed of the same granite as that
of Shap, near Penrith. Blocks of the same granite occur in
the valley of Stokesley, and in the bed of the Tees, near Bar-
nard Castle. Similar blocks are also found on the elevated
plain of Sedgefield, near Durham. In many of these cases
blocks are mixed with rolled pieces of various kinds of greenstone and porphyry, probably derived from Cumberland *.
    Prof. Sedgwick notices large transported boulders on parts
of the Derbyshire Chain, which overhang the great plain of
Cheshire. He also remarks on the boulders accompanying
the transported detritus at the base of the Cumberland mountains from Stainmoor to Solway Firth, the plain bordering the
hilly region on the north presenting boulders and pebbles that
have been transported across the Firth from Dumfriesshire.
In the transported rubbish capping a hill near Hayton Castle,
about four miles N.E. of Maryport, there are large granitic
boulders resembling the rocks of the Criffel. "Among them
was one spheroidal mass, the greatest diameter of which was
ten feet and a half, and the part which appeared above the
ground was more than four feet high." From St. Bees Head
to the southern extremity of Cumberland, the coast region is
covered by transported detritus, among which are boulders of
granite, porphyry, and greenstone, some of large size. In
low Furness similar phænomena are observable. Prof. Sedgwick further remarks, that large blocks derived from the green-
slate district are found on the granitic hills between Bootle
and Eskdale. Millions of large blocks are scattered over the
hills on the N.W. boundary of the mountainous region. The
syenitic blocks of Carrock-fell can be traced " through the
valleys and over the hills of the mid region, to the very foot of
the parent rock." Numerous boulders of the Carrock syenite
rest on the side of High Pike; the largest, termed "The
Golden Rock," being twenty-one feet long, ten feet high, and
nine feet wide. Rolled masses of St. John's Vale porphyry
abound near Penruddock, and descend the valleys thence into
the Eamont. Rounded boulders of Shap granite are numerous on the calcareous hills south of Appleby; some being
twelve feet in diameter. Rounded blocks, apparently derived
from the green-slate at the head of Kentmere and Long Sled-
dale, are found on the flat-topped calcareous hills W. of
Kendal. Prof. Sedgwick remarks that the blocks of Shap
granite, which cannot be confounded with other rocks in the
North of England, are not only drifted over the hills near
Appleby, but have been scattered over the plain of the new
red sandstone; rolled over the great central chain of England

* Buckland, Reliquiæ Diluvianæ.

into the plains of Yorkshire; imbedded in the transported detritus of the Tees; and even carried to the eastern coast*.

By comparing these statements with the little district first noticed, we find that the evidences of a transporting power by water are far greater in midland and northern England than in Devon and Dorset, the gravel having been carried far greater distances, and huge blocks added to the transported mass. How far these gravels may be contemporaneous can only be determined by future and exact observation. We shall, therefore, merely confine ourselves to a detail of facts, which must be taken into account in all generalizations on this subject. Between the Thames and the Tweed, pebbles and even blocks of rock are discovered, of such a mineralogical character, that they are considered as derived from Norway, where similar rocks are known to exist. Mr. Phillips states, that the accumulation, at present termed *diluvium*, in Holderness, on the coast of Yorkshire, is composed of a base of clay, containing fragments of pre-existent rocks, varying in roundness and size. "The rocks from which the fragments appear to have been transported are found, some in Norway, in the Highlands of Scotland, and in the mountains of Cumberland; others, in the north-western and western parts of Yorkshire; and no inconsiderable portion appears to have come from the sea-coast of Durham, and the neighbourhood of Whitby. In proportion to the distance which they have travelled, is the degree of roundness which they have acquired†."

Patches of gravel and sand are stated to occur in the great mass of clay, sometimes amounting to considerable accumulations. In one of these, at Brandesburton, the remains of the fossil elephant were detected.

If, quitting England, we proceed northwards to Scotland, there are evidences of a similar force having acted in that country; and Sir James Hall even considers that a rush over the land has left traces of its course in the shape of furrows, which the transported mineral substances, moving with great velocity, have cut in the solid rocks beneath. From the direction of these marks Sir James Hall infers that the current had a western course in the vicinity of Edinburgh‡. Continuing our course still northwards, the evidence of a transport continues; for Dr. Hibbert found fragments of rocks at Papa Stour, Shetland Isles, which must have travelled twelve miles from Hillswick Ness, the latter bearing from the former, N. 47° E. He also remarks on the large blocks near the mansion of Lunna, on the east of Shetland, named the stones of

---

* Sedgwick, Ann. of Phil. 1825.
† Phillips, Illustrations of the Geology of Yorkshire.
‡ Sir James Hall, Trans. Royal Soc. Edinburgh.

Stefis, which appear to have been removed a mile or more by a shock from the N.E. The same author mentions many other interesting circumstances : among others, that at Soulam Voe, open to the Northern Ocean, there are boulders about three or four feet high, which do not correspond with any known rock in the country, and were probably derived from the northward *. It is also probable, from Landt's notice, cited by Dr. Hibbert, that similar phænomena are observable in the Feroe Islands.

The probability therefore, as far as the above facts seem to warrant, is, that masses of water have proceeded from north to south over the British Isles, moving with sufficient velocity to transport fragments of rock from Norway to the Shetland Isles and the eastern coast of England ; the course of such masses of water having been modified and obstructed among the valleys, hills, and mountains which they encountered ; so that various minor and low currents having been produced, the distribution of detritus has been in various directions.

If the supposition of masses of waters having passed over Britain be founded on probability, the evidences of such a passage or passages should be found in the neighbouring continent of Europe, and the general direction of the transported substances should be the same. Now this is precisely what we do find. In Sweden and Russia large blocks of rock occur in great numbers, and no doubt can be entertained that they have been transported southward from the north. In Sweden, the transported materials were observed by M. Brongniart to run in lines, sometimes inosculating, but having a general direction north and south †. Similar observations had been previously (1819) made by Count Rasoumovski on the transported blocks of Russia and Germany, which, having been unknown to M. Brongniart, render his account of the Swedish blocks the more valuable. Count Rasoumovski observes, that, where many blocks are accumulated they form parallel lines, with a direction from N.E. to S.W. He states that the erratic blocks are very numerous, and composed of Scandinavian rocks between St. Petersburgh and Moscow; and remarks, that in some places, especially in Esthonia, the blocks appear and disappear at greater or less intervals, apparently owing to the form of the land at the time of their transport; for these masses are discovered where escarpments presented themselves, while, where the land sloped away, or became more or less horizontal, they disappear ; thus seeming to show that the steep escarpments caught them in their passage on-

---

* Hibbert, Edin. Journal of Science, vol. vii.
† Ann. des Sci. Nat. 1828.

190   *Erratic Blocks and Gravel.*

wards. Count Rasoumovski also remarks that the blocks occur abundantly on the heights, and but rarely, or thinly scattered, over the lowlands *. Prof. Pusch observes, that the
erratic blocks from the Duna to the Niemen are composed of
granite resembling that of Wiborg in Finland; of another
granite, with Labrador felspar from Ingria; of a red quartzose
sandstone from the shores of Lake Onega; and of a transition
limestone from Esthonia and Ingria. In Eastern Prussia, and
in that part of Poland situated between the Vistula and the
Niemen, the granitic blocks are abundant: three varieties of
granite are the same as those found in Finland, at Abo and
Holsinfors; another coarse-grained granite and a sienite are
also from the north. The hornblende blocks of the same countries are from southern and central Finland; the quartzoze
blocks are exactly the same as the rocks named *Fjall Sandstein,*
between Sweden and Norway; and the porphyry blocks are
of the same mineralogical character as the porphyries of Elf-
dalen in Sweden. " From Warsaw to the west, towards Ka-
lisch and Posen, the blocks of the red granite of Finland diminish in number, but those composed of hornblende rocks
and gneiss become more abundant, as is also the case with
those of porphyry. Few Finland rocks are in general there
found, while those of Sweden are common †." A great quantity of erratic blocks covers the plains of Upper Silesia, occurring chiefly on elevated land, even attaining the height of
1000 feet above the level of the sea. The blocks reach the
foot of the Carpathian mountains between Nicolai and Birdul-
tau, and penetrate, in masses, to the declivity of the Moravo-
Silesian mountains. Granite blocks of extraordinary magnitude occur in the island of Gristow, near Camin. The
country of Oderberg and Liepe, between Lewenberg, Dan-
nenberg, Straussberg, and Bukow abound in large granite
blocks. The ranges of hills are often so thickly covered with
them that cultivation of any kind is impossible. The well-
known great block of granite (the Markgrafenstein) from the
Rauen mountains at Fürstenwald, has been cut into a magnificent basin 22 feet in diameter. Red sandstones are so common in some tracts between Fürstenwald and Trebus, south
of Berlin, that it might be supposed the red sandstone was in
place in the vicinity. Limestone blocks, with Trilobites and
Orthoceratites, are found from Mecklenberg and Pomerania,
through the Marches, to Sorau, where they are in such abundance that they have supplied the lime-kilns for centuries past.
The limestone blocks extend on to Münsterberg, but they are

* Ann. des Sci. Nat., t. xviii.
† Pusch, Journal de Géologie, t. ii. p. 253.

not discovered in Saxony and Silesia. The quantity of flints, evidently derived from the chalk, which are found in the Marches, between the Oder and the Elbe, and beyond the latter river to the S. and S.W., likewise show a northern origin, for in the island of Rügen in Denmark and Scania, there are the remains of very extensive beds of chalk with flints. They are discovered as far south as Leipzig, and westward as far as the S.E. border of the Harz, in the county of Mansfeld.

The northern border of the Harz opposes a great obstacle to the advance of the erratic blocks towards the south, yet they occur at some points in the valleys between the Harz and the Thuringerwald; they are found in the deep ravine which begins at Tonna, and extends, in the range of hills running from the Haynich, to the village of Ballstädt. Westward of the Harz, as far as the Weser, the base of the hilly country generally marks the southern limit of the blocks. In most of the valleys, especially those which open towards the north, blocks are discovered high up. Thus they advance into the Innerstethal and its adjacent valleys. They spread in the Leinethal to beyond Wirpenstein, through the narrow defile at Brunkensen to the Reuberg. They may be traced in the valley of the Weser to Holzminden, and on the right to Coppenbrügge. A large collection of erratic blocks lies close to the Porta Westphalica, 150 feet above the level of the Weser. They rest singly in the country of Horn at Schwalenberg, and are more frequent, but smaller, near Pyrmont. The valleys of the Werra and the Bega are particularly remarkable, for the blocks become gradually more numerous in the upper narrow parts of these valleys, at Detmold and E. of Lemgo. At Lage, on the north-western projection of the chain of hills between the two valleys, where the plain of Ravensberg ends, rests a block of granite (named the Johannisstein) 24 feet in diameter, with four smaller portions, that have been broken off, lying around it. On the inner slope, near Ravensberg, their number is very great: on the Sieker, near Bielefeld, they are not much inferior in size.

Erratic blocks occur in the long valleys of the Teutoburgerwald, having been enabled to pass over the mountain ridges. On the declivities of the valleys of the Werra and Bega, the blocks observe a defined upper limit, which they do not pass. In the great basin, in the centre of which stands Münster, between the Teutoburgerwald and the Rhenish-Westphalian slate mountains, erratic blocks are not found, but they occur again, forming a strictly defined south border, to the southward of Paderborn, where the land again rises on the northern frontier of the Westphalian slate mountains. At Alfeln and

Tudorf, S. of Salzkotten, they are innumerable, but towards
the west they rapidly decrease in number. At Unna and
Dortmund none are known, and but few in the neighbourhood
of Bochum, between Emsecke and Ruhr. Quartziferous porphyries predominate; next to these follow coarse-grained
greenstones; the smaller portion consisting of granite and
gneiss. Near Aix-la-Chapelle and in Belgium, the blocks do
not reach the base of the high lands, and they remain in the
levels of Holland *. Therefore proceeding south, the course
of the waters seems to have continued in that direction over
the low districts of Germany, to the Netherlands, depositing
huge blocks in their passage; these blocks proved by their
mineralogical composition to have been derived from rocks
known to exist in the northern regions.

Such a movement as this over part of Europe would, if the
supposition of a mass of waters were correct, be observed in
other northern regions, for the waters thrown into agitation
would cause waves around the centre of disturbance. In America, therefore, we should expect to find marks of such a deluge, the evidences pointing to a northern origin †. Now in
the northern regions of that country we do find marks of an
aqueous torrent bearing blocks and other detritus before it,
the lines of their transport pointing northward, according to
Dr. Bigsby, and reminding us of the same appearances observed in Sweden and Germany. The quantity of transported
matter covering various large tracts in North America seems
quite equal to that scattered over Northern Europe; and as
they both point one way, we can scarcely refuse to admit that
the course of the disturbance or disturbances was towards the
north, the undulations of the waters having been caused by
some violent agitation, perhaps beneath the sea in those regions, for it is by no means necessary that it should be above
its level.

A convulsion or convulsions of this magnitude, reasoning
from the analogy of those minor agitations which we term
earthquakes, would be felt over a considerable portion of the
globe, and the waters over a large surface would be thrown
into agitation. A part of the earth would be greatly disturbed,
and we should expect fractures and faults produced in strata
where the convulsion was most felt, as similar minor effects
are produced at present from the exertion of a less intense
force.

Ice would seem to afford a possible explanation of the transport of many masses; for the glaciers which descend the valleys of high northern regions are, like those of the Alps,

* German Transl. of Manual.   † Journal of Science, vol. xviii.

charged with blocks and smaller rock fragments, which have fallen from the heights. Waters rushing up or down such valleys would float off the glaciers, more particularly as northern navigators have shown that they project into the sea. It is considered that the huge masses of ice known as icebergs, are the projecting portions of these boreal glaciers, which having been detached from the parent mass, are borne into more temperate climes, in some cases transporting blocks and smaller fragments of rock. This debris will, as Mr. Lyell has observed, be deposited at the bottom of the seas over which they pass; and therefore, if such bottoms were raised so as to become dry land, blocks might be discovered scattered over various levels of that land, presenting appearances that might be mistaken for the action of diluvial currents. If the present continents bore evident marks of long submergence beneath an ocean immediately previous to their present appearance, and if the blocks were merely scattered here and there, this explanation would by no means be without its weight : but there are too many circumstances tending to other conclusions, to render it probable. The supposition of masses of ice, covered by blocks and smaller rock fragments, borne southwards with violence, though it may account for some appearances, does not, it must be confessed, seem applicable to all, more particularly where blocks can be traced to their sources at comparatively small distances. Supposing a wave, or waves, discharged over Europe and America from the northwards, many phænomena would depend on the time of year at which the catastrophe, or catastrophes, took place ; for if in the winter, waters rushing from that quarter would transport a greater quantity of ice, and many superficial blocks and gravels, bound by ice together, might be torn up and carried considerable distances, from the possible small specific gravity of the mass ; for even in the case of rivers, it has been found that large masses of rocks have occasionally been transported from places, when encased in ice and acted on by the stream. In Sweden and Russia it is more than probable that many blocks would be thus encased during winter, and therefore a flood of waters passing over them would cause them to rise, to float, and to be borne onwards, until the ice melting, the blocks would sink and be finally brought to rest.

Upon the hypothesis of a convulsion, or convulsions, in the North, the effects would become less as we receded from the centre of disturbance, and, finally, all traces of them would be lost.

We now arrive at another question,—how far the distribution of blocks from the Alps may have been contemporaneous with the supposed transport of erratic fragments from Scandi-

navia ? To answer this question, without more direct information than we possess, would be difficult; and we should be
particularly cautious in applying preconceived theories before
we have the requisite data. All that we can safely remark on
this subject seems to be, that the blocks in both cases appear
to a certain extent superficial and uncovered by deposits which
would afford us information respecting their difference of age;
and that it is possible a great elevation of the Alps, and distribution of blocks on both sides of the chain, may have been
contemporaneous, or nearly so, with a convulsion in the
North.

An immense quantity of debris has, at a comparatively recent epoch, been driven from the central chain of the Alps
outwards; the consequence, according to M. Elie de Beaumont, of a great elevation in those mountains, extending from
the Valais to Austria. MM. Von Buch, De Luc, Escher,
and Elie de Beaumont, have presented us with a detail of numerous and well observed facts, which all tend to one conclusion; namely, that the great valleys existed previously to the
catastrophe which tore blocks and other fragments from the
Alps, and scattered them on either side of the chain. M. Elie
de Beaumont observes * that the valleys of the Durance, of
the Drac, of la Romanche, of the Arc, and of the Isere, present the same appearances as those of the Arve, the Rhone,
the Aar, the Reuss, the Limmat, the Rhine, and the valleys
which descend into the plains of Bavaria, noticed by different
geologists. On the Italian side of the chain, appearances are
also similar, and no doubt can exist that the blocks and debris have passed down the respective valleys, where they have
left unequivocal marks of their transit. M. Elie de Beaumont
has presented us with very detailed accounts of these appearances in the valleys of the Durance, of the Drac, and others,
where they are precisely what would have been expected from
the passage of a rock-charged mass of waters down the respective channels, the largest fragments having been transported the shortest distances, being most angular, while the
smaller and most rounded have been carried the furthest.
Thus, in the valley of the Durance, the transported substances
become more angular and of greater volume, as we proceed
from the great mass of pebbles, called the Crau, to the mountains beyond Gap, whence the debris, judging from its mineralogical characters, have very clearly been derived. Similar
phænomena will be observed up the valley of the Drac, which
proceeds by another course to the neighbourhood of the same

* Recherches sur les Rév. de la Surface du Globe; Ann des Sci. Nat.,
1829 et 1830.

mountains, the two streams of debris not mingling until they join in the Crau *.

From my own observations, I can fully confirm the remarks of various authors respecting the situations of the Alpine blocks, and their probable derivation from the respective valleys, which they, as it were, appear to face. But I have nowhere observed such striking masses of erratic blocks as those which occur in the vicinity of the lakes of Como and Lecco. They are particularly remarkable on the northern face of the Monte San Primo, a lofty mountain ridge presenting one of its sides to the more open and northern part of the lake of Como, where the latter stretches towards the high Alps; thus presenting a bold front to any shock which should come from the north, leaving open passages to the right and left of it, one down the southern part of the lake of Como, the other down that of Lecco. Not only in front, facing the high Alps, but also round the flanks and shoulders of this mountain, and even behind it, where the eddy-current would have transported them, blocks of granite, gneiss, mica-slate, and others from the central chain, of various sizes, and often accompanied by smaller fragments and gravel, are seen in hundreds, nay thousands, scattered over the dolomite, limestone, and slate of the mountain, and nearly filling up a previously existing valley which faced the north, the direction whence the rock-charged fluid descended. Proceeding down the side valleys, partly occupied by the lower lake of Como, and the lake of Lecco, we find the evidences of such a current in the presence of blocks occurring, as they should do, where direct obstacles were opposed to its course, or in situations where eddies would be produced behind the shoulders of the mountains. One very remarkable instance of such occurrence is behind, or on the southern side of, the Monte San Maurizio, above the town of Como; where numerous blocks are accumulated on the steep flank of the mountain, precisely where a body of water, rushing down the great valley, would produce an eddy at its discharge into the open plains of Italy†. The blocks, though no doubt many have descended from their first positions in consequence of the long-continued action of atmospheric agents, occupy an elevated line, as also on other but lower heights in the vicinity, which opposed more direct obstacles to the debacle: seeming to show that the blocks occurred near the surface of the fluid mass, and were whirled by the eddy, at nearly the same level, against the steep sides of this calcareous moun-

* Elie de Beaumont, Recherches sur les Rév. du Globe.
† For illustrations of these appearances, see Sections and Views illustrative of Geological Phænomena, plates 31, 32.

tain, as well as thrown against the more direct obstacle of a range of conglomerate hills.

The following is a section of the Monte San Primo, exhibiting the manner in which the erratic blocks rest on its surface.

Fig. 30.

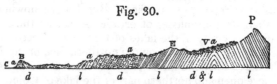

P, Monte San Primo: B, bluff point of Bellaggio rising out of the lake of Como C: *a a a a*, blocks of granite, gneiss, &c., scattered over the surface of the limestone rocks *l l l l*, and the dolomite *d d d*. V, the Commune di Villa, where a previously existing depression or valley is nearly filled with transported matter. E, the Alpi di Pravolta, on the northern side of which is the large granite block figured beneath, remarkable not so much for its size as for its angular character.

Fig. 31.

The accumulation of erratic blocks of the Alps in groups has been particularly remarked by M. de Luc (nephew), who has very carefully examined them round the lake of Geneva and neighbouring country *. The levels which the blocks keep on the Jura and other places have been often observed by various authors. Such a common mode of occurrence must, we should suppose, have some common cause, and can scarcely be accidental.

Solutions of the problem of erratic blocks seem not very practicable at present, and our attempts at general explanations can be considered little else than conjectures that may appear more or less probable. The student, therefore, should be careful not to consider such explanations as well ascertained truths, but merely as hypotheses, which future and extensive observations may, or may not, prove to be correct.

It has been above remarked, that the Alpine erratic blocks

---

* De Luc, Mém. de la Soc. de Phys. et d'Hist. Nat. de Genève, vol. iii.

frequently occur in groups.   To present a general explanation of this phænomenon would, at present, be somewhat difficult; but it may be asked, as a mere conjecture, whether masses of floating ice charged with blocks and other detritus, rushing down the great valleys into the more open country of lower Switzerland, might not be whirled about by the eddies, and the icy masses be destroyed by collision against each other, so that groups of blocks would afterwards be found beneath the places where the whirlpools had existed.   Masses of ice, charged with blocks and pent up for the moment within such basins as might be formed between the Alps and Jura, might also be carried at certain levels against the sides of the opposing mountains, such as the Jura, and be there deposited in groups and in lines of level.

Such passages of bodies of water over land, as have been above noticed, whether contemporaneous or not, could scarcely have failed to destroy the larger portion of the animals previously existing on that land.   At the time when the remains of extinct elephants, mastodons, and rhinoceroses, were considered to characterize one set of gravels or transported matter, it was natural to conclude that all such debris were contemporaneous : but as these animals are now found to have existed earlier, if not also later, than was imagined, this supposed guide has failed us ; and we gain no very definite ideas relative to the age of the transported matter in which they may occur, further than that they probably come within a certain range of the more recent geological deposits.

The following is a list of those animals which are commonly enumerated as found in deposits referrible to the group under consideration and which, whether exactly contemporaneous or not, are found in superficial gravels, sands, and clays *.

1. Elephas primigenius, *Blumenbach*.   Scattered over various parts of Europe.   Very common in the northern parts of Asia, where the ivory of the fossil tusk, or defence, is so far uninjured as to be used for ornamental purposes.   Found also on the northern coast of the American continent.   United States of North America.   Mexico, Quito, *Humboldt*. (Highest transported gravel near Lyons. *Beaum*.)
1. Mastodon maximus, *Cuv.*   North America.   Various Authors †.

---

* The student should be careful, if he be so fortunate as to discover any of these remains, to remark, whether they occur in detritus evidently moved from a distance, or in that great mass of weathered fragments which often covers hills and valleys, and which seems principally due to the action of the atmosphere upon them.

† The relative age of the deposit, in which the remains of the *Mastodon maximus* are found, cannot be considered as very satisfactorily ascertained. Some geologists, indeed, suspect that these animals have disappeared more

198   *Organic Remains in the Gravel, &c.*

2. Mastodon angustidens, *Cuv.* Simorre; Italy; France, *Cuv.* Damstadt, *Sömmering.* Austria, *Stutz.* Peru; Columbia, *Humboldt.*
3. ———— Andium, *Cuv.* Cordilleras; Santa Fé de Bogota. *Humboldt.*
4. ———— Humboldtii, *Cuv.* South America. *Humboldt.*
5. ———— minutus, *Cuv.* Europe, *Al. Brong.*
6. ———— tapiroides, *Cuv.* Europe, *Al. Brong.*
1. Hippopotamus major, *Cuv.* Walton in Essex; Oxford; Brentford. *Buckl.* Bavaria, *Holl.* Italy; France, *Cuv.*
2. ———— minutus, *Cuv.* Landes of Bourdeaux. *Cuv.*
1. Rhinoceros tichorhinus, *Cuv.* Very common in Europe.
2. ———— leptorhinus, *Cuv.* Common in Europe.
3. ———— incisivus, *Cuv.* Germany; Appelsheim. *Al. Brong.*
4. ———— minutus, *Cuv.* Moissac. *Al. Brong.* Magdeburg. *Holl.*
Elasmotherium. Siberia, *Fischer.*
1. Tapirus giganteus, *Cuv.* Allan; Vienne in Dauphiné; Chevilly; and other parts of France. *Cuv.* Furth, Bavaria; Feldsberg, Austria. *Holl.*
1. Cervus giganteus, *Blum.* Ireland; Silesia; Banks of the Rhine; Sevran, near Paris.
2. ———— Americanus, *Harlan.* Big Bone Lick, Kentucky.
Cervus. Several different species, common in various parts of Europe.
1. Bos bombifrons, *Harlan.* Big Bone Lick, Kentucky.
2. ———— Urus. Eschscholtz Bay, North America, *Buckl.*
Bos. Remains of, common.
Auroch (fossil), *Cuv.* Siberia, Germany, Italy, &c.
1. Trogontherium Cuvieri, *Fischer.* Sea coast near Taganrock, Sea of Azof, *Fischer.*
1. Megalonyx laqueatus, *Harlan.* Big Bone Lick, Kentucky, *Harlan* *.
Megotherium, *Cuv.* Buenos Ayres; Lima.
Hyæna (fossil), *Cuv.* Lawford, near Rugby; Herzberg and Osterode; Canstadt, near Stutgart; Eichstadt, in Bavaria, *Buckl.*
Ursus. Krems-Münster, Higher Austria, *Buckl.*
Equus. Common in many places in Europe. Big Bone Lick, Kentucky. Eschscholtz Bay.

recently than is commonly supposed. Among some of these remains discovered at Withe, Virginia, there was found a mass of small branches and leaves, among which it was considered that there was a species of reed still common in Virginia. The whole appeared enveloped in a kind of sack, considered to be the stomach of the animal.—(Cuvier, Oss. Foss. t. i. p. 219.)

According to Mr. Cooper, the fundamental rock of the valley and country round the celebrated Big Bone Lick, is a limestone. At the Lick the valley is filled up by various unconsolidated beds, generally about thirty feet deep. The upper bed is a light yellow clay, produced by the soil washed away from the higher grounds by the rains. This contains the bones of buffaloes and other recent animals. Beneath this is another bed of a different character, being more gravelly, and darker coloured. It contains the remains of reedy plants, smaller than the abundant Kentucky cane, fresh-water shells, and the remains of the mastodon, elephant, &c. Mr. Cooper remarks the broken character of even the smallest bone found in this bed, and considers that this effect must have been produced by violent action.—(Featherstonhaugh's American Journal, vol. i.)

* Dr. Harlan describes the bones of the same species as having been found on the *surface* of White Cave, Kentucky. With these were received bones of the *Bos, Cervus,* and *Ursus,* as also the metacarpal bone of the human species. The remains of the Bear alone appeared of equal antiquity with the *Megalonyx.* Harlan, Jour. Am. Nat. Soc. 1831. The remains of

We cannot quit the subject of the large mammalia entombed in superficial gravel, sands, and clays, without adverting to the elephant found encased in ice near the embouchure of the river Lena in Siberia. It had been preserved entire, having undergone no decomposition since death; on the contrary, when detached from the ice, it afforded food to various animals, and parts of its skin and hair were collected, and are now preserved with its skeleton in the Museum at St. Petersburgh. Mr. Adams, to whom the scientific public are indebted for the preservation of what remained of the animal, and for the account of its original discovery, relates that Schumachof, a Tungusian chief and owner of the peninsula of Tamset, where the elephant was discovered, first observed a shapeless mass among the ice in 1799; but it was not until 1804 that this mass fell on the sand, and disclosed the ice-preserved elephant, whose tusks were cut off and sold by the Tungusian chief. Two years afterwards Mr. Adams visited the spot, and collected the remains as above stated. According to this observer, the escarpment of ice in which the elephant had been preserved, extended two miles, and rose perpendicularly about 200 or 250 feet. On this ice, which is described as pure and clear, there was a layer of friable earth and moss, about fourteen inches thick*.

M. Cuvier mentions that in 1805 M. Tilesius had received, and had sent to M. Blumenbach, some hair torn from the carcase of a mammoth, or elephant, by a person named Patapof, near the shores of the Icy Sea. He further observes, that some of the hair and skin of this individual was presented to the Jardin du Roi at Paris, by M. Targe, who had received it from his nephew at Moscow †.

Pallas mentions the discovery (in 1770) of an entire rhinoceros with its skin and hair, enveloped in sand on the banks of the Wiluji, which falls into the Lena below Jakoutsk. The animal is described as being very hairy, particularly on the feet. It was an individual of the *Rhinoceros Tichorhinus,* Cuvier ‡.

Considerable light has recently been thrown on the remains of the elephant and rhinoceros of Northern Asia, by the observations made at Eschscholtz Bay, within the Arctic Circle,

---

Megalonyx Jeffersonii were found two or three feet *beneath* the surface of a cavern, in Green Briar County, Virginia.

The remains of a *Megalonyx* have been brought from the Brazils. According to Dr. Wagner, they were discovered in a cavern.

* From the account of the elephant found in the ice of Siberia, London 1819;—taken from the Mem. of the Imp. Acad. of Sciences of St. Petersburgh, vol. v.

† Cuvier, Oss. Fossiles, t. i. éd. 1822.   ‡ *Ibid.* t. ii.

North America, during the expedition of Captain Beechey to those regions. These observations have been arranged and commented on by Prof. Buckland *; and it now appears that, instead of the remains of elephants being found in the ice at this place, as was considered to be the case during the expedition of Kotzebue, they are enveloped in frozen mud and sand, emitting a strong odour of burnt bones †. The remains thus entombed were referrible to the elephant, *Bos Urus*, deer, and horse, with the cervical vertebra of an animal not known. Prof. Buckland supposes that the frozen elephant of Siberia, above noticed, was also encased in frozen mud, the front of the mud or sand cliff being only a facing of ice, as was found to be the case in Eschscholtz Bay; and this supposition is rendered the more probable, as we know that the rhinoceros of the Wiluji was thus enveloped.

The causes, whatever they were, which destroyed the elephant at the mouth of the Lena, have, as Prof. Buckland observes, been common to all the shores of the two continents within the Arctic Circle; and this is further proved by the researches of M. Hedenstrom, who visited the shores of the Icy Sea, under the direction of the Russian Government, between the Lena and the Colyma, and who states that there are hundreds of elephants, rhinoceroses, oxen, and other animals, in the ice or frozen ground of those regions ‡.

It seems probable, therefore, that there has been a great change of climate on the northern coasts of Asia and America since these animals existed there; for, with every allowance for the adaptation of the particular species of elephant, so commonly found fossil, to much colder climates than the existing species now inhabit, (and that they were so adapted seems exceedingly probable, from the woolly hair discovered on the individual encased in ice at the mouth of the Lena,) we must grant them something to live upon, food fitted to their powers of mastication and digestion; and this they could scarcely find, if the climates were such as they now are, permitting only the existence of a comparatively miserable vegetation, and that only during part of the year §.

* Appendix to Beechey's Voyage to the Pacific and Behring's Straits.

† Mr. Brayley, commenting on the evidence adduced of this odour, observes that many circumstances render it more probable that it should always arise, in the places noticed, from the decomposition of animal matter, than from any other cause, though Prof. Buckland was inclined to consider the odour produced by other circumstances.—Phil. Mag. and Annals, vol. ix. p. 411.

‡ Journal de Géologie, tom. ii. p. 315.

§ The tigers, apparently in every respect the same with those of Bengal, which are now ascertained to roam into Siberia, up to the parallels of Berlin and Hamburgh, by no means render it more probable that elephants

Ossiferous Caverns, and Osseous Breccia.

It is to Prof. Buckland that we are indebted for our more intimate acquaintance with the various circumstances under which organic remains are found in caverns; for though bones of bears and other animals, occurring in caves, had long attracted attention, more particularly in Germany, it was not until after the discovery of the celebrated Kirkdale cavern in Yorkshire, that the subject acquired a new interest, and became as much a part of general geological investigation, as the fossil contents of any well established rock had previously been. It is gratifying to observe, that even those who are opposed to the theoretical conclusions that have been deduced from cavern bones, are still willing to pay their tribute of praise to the zeal and activity with which Prof. Buckland conducted his researches.

Prof. Buckland pointed out that the general arrangements in caverns, are: 1. The original sides of the cave, which may or may not be covered with stalagmite. 2. A deposition of animal remains, mixed with mud, silt, rolled stones, or broken fragments; many circumstances sometimes attending this deposition seeming to attest the long-continued residence of animals in certain caverns for successive generations: some, the hyænas for instance, having there dragged their prey, often consisting of parts of the elephant and rhinoceros. 3. The deposition of stalagmite covering up the animal remains, the mud, silt, &c., with a greater or less depth of carbonate of lime; so that to all appearance, in a newly discovered cavern, the bottom is a mere mass of stalagmite, beneath which the organic riches would for ever remain unknown, unless the concealing crust should be fractured by accident, or broken through by the geologist, now aware that animal remains may be found beneath it.

Since the discovery and description of Kirkdale cavern, notices of other ossiferous caves have become so numerous that a mere list of them would be somewhat long; and they multiply so fast, that we may anticipate, at no distant period, a very singular body of evidence on this subject alone.

The remains of animals, similar to those contained in caverns, are frequently found in fissures of the rock; in some situations, the whole mass of bones, fragments of rock, and cementing matter being so hard and compact, that it frequently

---

once existed in climates similar to that of the present Arctic Circle ; for, the former subsisting on flesh and the latter on vegetables, it is obvious that the tigers could live, as far as respects food, further north than the elephants.

equals, and sometimes exceeds, in durability, the rock within which it is inclosed. Of this the *osseous breccias* of Nice and many other places in the Mediterranean are examples.

It becomes daily more necessary to ascertain, as far as may be, the relative ages of these various accumulations of animal remains, investigating the subject with proper attention, and as much as possible without preconceived theory. It also becomes important to examine with attention, in those cases where the mouths of ossiferous caverns are covered up with detritus, whether such detritus be composed of angular fragments of the rock in the vicinity, which might have been gradually accumulated over the external aperture during the long lapse of ages, by causes and effects similar to those in daily operation; or whether it is composed of transported fragments more or less rounded, and which must have travelled from a distance: in the latter case, endeavouring to ascertain whether such transported matter could have been carried to its present situation by actual causes, or whether we must seek a greater intensity of force to account for its presence, physical obstacles opposing its carriage by any other means. If angular fragments, derived from the immediate vicinity, alone cover the cavern's mouth, we have no certainty when it was finally closed; and therefore, even supposing that one set of animals may have been overwhelmed by a rush of waters into the cavern, there is nothing to prevent another race of animals from frequenting the same place, whose bones might become, to a certain degree, mixed with the others, and entombed beneath fragments of rock and stalagmite, from the constant change operating in the interior of caves. Thus the bones of man, and his early rude manufactures, such as unbaked pottery, may become, to a certain extent, mingled, in a mass of stalagmite and rock fragments, with the remains of elephants, rhinoceroses, cavern bears, and hyænas; and the whole might, after the cave became deserted, and the accumulation at the mouth considerable, be covered with a crust of stalagmite: so that upon the discovery of such a cavern, it might be described, if attention had not been paid to the kind of detritus which blocked up the mouth, as being closed externally, and open to a certain height inside, beneath which there was a crust of stalagmite, covering an accumulation of rock fragments and bones, among which those of man were found mingled with those of the elephant and other animals. It might hence be concluded, that all these remains were of contemporaneous origin, and, consequently, that man existed at the time when the elephants roamed the forests, and hyænas and bears lurked in the caverns of Europe. If the mouths of ossiferous caverns be closed by fragments of rock transported from a distance, such transport being clearly not due to the operation of actual causes, but to the exertion of a greater intensity of force; and if we then find the remains of man entombed with those usually contained in caverns, there would seem little reason to doubt, unless other communications from the surface could be traced, that man was a contemporary with the extinct species of elephants, rhinoceroses, hyænas, and bears, found not only in the caves, but also in masses of transported gravel, and that he existed previous to the catastrophe, or catastrophes, which overwhelmed him and them.

When the bones of man were discovered by MM. Tournal, De Christol, Marcel de Serres, and other geologists, in caverns of the southern parts of France, they inferred that the mode in which the human remains occurred among those of extinct animals, was such, that man must have been contemporaneous with these animals. We are indebted to M. Tessier for a description of the cavern of Miallet, near Anduze, department of the Gard, which throws much light on this subject. This cavern occurs on a steep slope, thirty yards above the valley, in a dolomitic rock, subordinate to the lias. The lowest bed deposited on the bottom of the cavern, consists of a dolomitic sand, irregularly covered with thin stalagmite, and here and there by an argillo-ferruginous clay, more than a yard thick. This bed contains the abundant remains of bears. Beneath stalagmite and a bed of clayey sand, from eight to sixteen inches thick, human remains were discovered in different parts of the cavern. At the inmost end they were decidedly mixed with those of bears, which predominated; but at the entrance the human bones prevailed. On the ossiferous clay, and beneath a small rocky projection, a nearly entire human skeleton was discovered, and close to it a lamp and a baked clay figurine, copper bracelets being found at a short distance. In other places were the remains of coarse pottery, worked bones, and small flint tools, exhibiting a ruder state of the arts than the preceding.

M. Tessier infers from the facts which he observed: 1. An epoch when the cavern was inhabited by bears. 2. A time when man, little advanced in civilization, inhabited, and probably was buried, in the cave. 3. The Roman epoch, shown by the remains of more advanced art. As to the mixed bones of man and the bears, it is inferred that this is accidental, as men and bears could not have lived together in this cavern *.

It is a singular circumstance, and one which demands attention, notwithstanding the ingenious remarks that have been made on the subject, that the remains of the monkey tribe

---

* Bulletin de la Société Géologique de France.

should not yet have been discovered among the undisturbed bones and other substances in caves, or in the old transported gravel, or *diluvium* of Prof. Buckland. It has been objected to a remark that man and the monkey tribe were perhaps created about the same period, and were of comparatively modern appearance on the earth's surface, that the countries have not been geologically well examined where the monkey race now exist. This is perfectly true. But is there any reason why monkeys should not have lived in climates and in situations where elephants, rhinoceroses, tigers, and hyænas were common? for the climates and regions in which existing elephants, rhinoceroses, tigers, and hyænas abound, are precisely those where monkeys are now found. To the objection, that if they did then exist, their bones would not be discovered, as their activity would secure them from falling a prey to hyænas and other predaceous animals; it may be opposed, that they must have died like other animals, and that their dead carcases must have fallen to the ground, and that they were quite as likely to have become the food of less nimble creatures, as the birds found in the cavern of Kirkdale.

Kirkdale cavern was discovered by cutting back a quarry, in the summer of 1821, and was visited by Prof. Buckland in December of the same year. Its greatest length is stated at 245 feet, and its height generally to be so inconsiderable, that there are only two or three situations where a man can stand upright. The following is a section*.

a a, a a, horizontal beds of limestone, in which the cave is situated; *b*, stalagmite incrusting some of the bones, and formed before the mud was introduced; *c*, stratum of mud containing the bones; *d d*, stalagmite formed since the introduction of the mud, and spreading over its surface; *e*, insulated stalagmite on the mud; *f f*, stalactites depending from the roof.

Fig. 32.

" The surface of the sediment when the cave was first opened was nearly smooth and level, except in those parts where its regularity had been broken by the accumulation of stalagmite above it, or ruffled by the dripping of water : its substance is an argillaceous and slightly micaceous loam, composed of such minute particles as would easily be suspended in muddy water, and mixed with much calcareous matter, that seems to have been derived in part from the dripping of the roof, and in part from comminuted bones. . . . . . At about 100 feet within the cave's mouth the sediment became more coarse and sandy†."

* From Buckland's Reliquiæ Diluvianæ.   † *Ibid.*

According to Dr. Buckland, the following are the animals, the remains of which were found in the Kirkdale cavern: *Carnivora ;*—Hyæna, Tiger, Bear, Wolf, Fox, Weasel. *Pachydermata;*—Elephant, Rhinoceros, Hippopotamus, Horse. *Ruminantia ;*—Ox, and three species of Deer. *Rodentia ;*— Hare, Rabbit, Water-rat, and Mouse. *Birds ;*—Raven, Pigeon, Lark, a small species of Duck, and a bird about the size of a Thrush.

From the mode in which these remains were strewed over the bottom of the cavern when the mud was removed, the great proportion of hyæna teeth over those of other animals, and the manner in which many of the bones were gnawed and fractured, Prof. Buckland inferred that this cavern was the den of hyænas during a succession of years; that they brought in, as prey, the animals whose remains are now mixed with their own; and that this state of things was suddenly terminated by an irruption of muddy water into the cave, which buried the whole in an envelope of mud. The inference of the hyænas having been long resident in the cave, was strengthened by the occurrence of their fæces, precisely as would happen in a den of hyænas at the present day. In addition to which it was further observed, that many bones were rubbed smooth and polished on one side, while the opposite side was not. This, Prof. Buckland considered, was produced by the friction of the animals walking or rubbing themselves upon the bones.

The German caverns of Gailenreuth, Küloch, Bauman, &c. contain an abundance of bones, nearly identical, according to Cuvier, over 200 leagues : by far the greatest proportion being referrible to two extinct species of Bear, *Ursus spelæus* and *U. arctoideus.* The remainder consisted of the extinct Hyæna (the same as at Kirkdale), a Felis, a Glutton, a Wolf, a Fox, and Polecat *. These caverns so far resemble Kirkdale cave, that there is more or less of a stalagmitic crust, beneath which the bones are discovered, the stalagmitic matter being frequently transfused through the previously deposited sediment†. There is, however, one fact connected with these caverns, wherein they differ very considerably from the Yorkshire cave. In the latter, no rolled pebbles were observed ; while in the former they have been noticed in some places. Thus, in Bau-

---

* Sections of some of these caves will be found in Prof. Buckland's Reliquiæ Diluvianæ.

† Buckland, Reliquiæ Diluvianæ. According to M. Wagner, the Muggendorf caverns contain the remains of the *Ursus spelæus, Ursus arctoideus* (Cuv.), *Ursus priscus* (Goldf.), *Hyæna spelæa* (Goldf.), *Felis spelæa* (Goldf.), *Canis spelæus* (Goldf.), *Canis minor, Gulo spelæus* (Goldf.), a *Cervus,* and a *Bos.*—Wagner, Leonhard and Bronn's Jahrbuch für Geologie, &c. 1830.

man's Höhle, pebbles of various sizes are stated to occur among crushed and pounded bones; leading to the presumption that the pebbles broke the bones, for the sand and mud of the same chamber contain them nearly entire. It would therefore appear that water had rushed into the cave, bringing with it rolled pebbles of the surrounding country, crushing and distributing the previously accumulated bones. By reference to Prof. Buckland's section of this cave *, we find the gorge of Bode exposes the entrance of the cavern, from whence there is a descent into the chamber where the crushed bones and pebbles occur: so that the same phænomena may here be explained by two different hypotheses; the one supposing a fracture of strata produced during a great convulsion permitting the sudden inroad of waters from above; the other, the gradual cutting of the gorge by the river Bode, which, so long as it cut across the mouth of the cavern, would throw rounded pebbles into it, very considerable rushes of water and pebbles taking place during floods. We thus obtain little information on the subject. The same remarks apply to the caves of Rabenstein, and others in Franconia. The Zahnloch may, perhaps, admit of only one explanation; for it is described as being on a hill 600 feet above the valley of Muggendorf. The ossiferous mass is stated to be composed of "brown loam, mixed with *numerous pebbles* and angular fragments of limestone†."

Be the origin of the pebbles, sand and mud, what it may, it seems clear that the remains of various animals were enveloped by them; since which, there has been a long continuance of repose, permitting, in most cases, the deposit of stalagmite upon the ossiferous mass.

Dr. Buckland informs me that Mr. McEnery found rounded pebbles of granite, of the size of an apple, mixed with the bones under the stalagmite in Kent's Hole, Torquay; and he states that he has found pebbles of greenstone, completely rounded, in the same place; and that in some parts of Kent's Hole, particularly the lowest, the bone breccia is full of fragments of grauwacke and slate, some of them rolled, some angular. The cave itself is situated in a limestone resting on shale, and the grauwacke and slate are rocks of the country; but the granite is at some distance, not nearer than Dartmoor.

M. Thirria describes the Grotte d'Echenoz, on the south of Vesoul, near the summit of a high plateau, between the villages of Echinoz, Andelarre, and Chariez (Haute Saone), as formed in the lower system of the Jura limestone, or oolitic

* Reliquiæ Diluvianæ, pl. 15.   † *Ibid.* p. 131.

group. The upper part of this cave is very irregular, and in
one place (the Grand Clocher) rises so high, that there must
be little space remaining between it and the surface of the
plateau. The bottom is not far removed from a level, here
and there interrupted by stalagmites. These stalagmites are
not numerous; but there are some which rise high, and cover
a considerable surface. No researches had been attempted in
this cavern previous to those of M. Thirria, in August 1827.
He broke up the ground "at different points of the four
chambers of the cavern, and all afforded bones in greater or
less abundance. The researches carried on in the fourth
chamber were the most productive, for each blow of the pick-
axe brought up a bone. The depth at which the bones were
discovered, varied from four to thirty-nine inches: they occurred in the midst of a red clay, mixed with a great number
of rounded pebbles with a smooth surface, the size of which
often attained that of a man's head. They are all composed of a
gray lamellar limestone, resembling that which forms the sides
of the cavern and many rocks of the vicinity. Independently of
these pebbles, which have evidently been rolled by waters, and
could not have penetrated into the cave except through some
fissures in its roof no longer visible, pieces of stalagtites and
stalagmites are discovered with their angles worn down, showing that they have been moved. The clay deposit, the thickness
of which does not appear to exceed four feet three inches, is
nearly everywhere covered by stalagmite two or three inches
deep; and upon this crust, which is mammillated, there rests
a bed, from four to ten inches thick, composed of a clay more
unctuous but less red than that situated beneath, and frequently
blackish from the remains of vegetables, of which it still contains some debris. No rounded pebbles are found above the
stalagmitic crust, and they are only seen on the surface when
the stalagmite does not exist. Hence it appears evident, that
the ossiferous clay containing the rounded pebbles has been
carried by the waters and deposited in the cavern, anterior to
the formation of the stalagmitic crust, produced by droppings
from the roof, before the deposit of the clay bed by which this
crust is covered *." M. Thirria further infers, from the resemblance of these pebbles to those of the transported matter
(termed diluvium) in the vicinity, that the introduction of the
pebbles and clay mixed with the bones in the Grotte d'Echenoz was contemporaneous with the transport of the diluvium.
The bones were most commonly discovered beneath a certain
thickness of clay; but in many situations they occurred imme-

---

* Thirria, Mém. de la Soc. d'Hist. Nat. de Strasbourg, t. i., where good
sections of the cave will be found.

diately beneath the stalagmitic crust, and sometimes even entirely in it. " In general the bones constituted a thickness of
about eight to sixteen centimetres in the middle of the clay:
they crossed in various directions, and covered each other
with small intermediate spaces, without having preserved their
relative position. They have not, however, suffered complete
dislocation ; for the dorsal vertebræ were nearly always discovered near the skull and jaws ; the humerus and cubitus
near the pelvis ; and the os calcis, the metatarsal and metacarpal bones or phalanges, near the femurs, the tibias and the
cubitus." The bones, examined by Cuvier, were found to
belong to the *Ursus spelæus*, *Hyæna*, *Felis*, *Deer*, *Elephant*,
and *Boar*; by far the largest proportion belonging to the
Ursus spelæus*.

M. Thirria also describes the Grotte de Fouvent, situated
at Fouvent near Champlitte (Haute Saone). This cavern was
accidentally discovered by quarrying the rock in such a manner as to strike into a natural cleft, through which the matters
contained in the cave are supposed to have entered, there
being apparently no other aperture. The cave is considered
too small for the habitation of beasts of prey; its upper part
is only about two yards beneath the surface of the plateau ;
and it was completely filled with bones, a yellow marl, and
angular pieces of the surrounding rock and of those in the
vicinity; the whole mixed pell-mell, and resembling the detritus, termed *diluvium*, covering many plains and valleys in the
neighbourhood. A thin red clay bed covers the bottom of
the cave, and a small thickness at the top did not contain
animal remains. According to M. Cuvier, these remains belong to the Elephant, Rhinoceros, Hyæna, *Ursus spelæus*,
Horse, Ox, and Lion. M. Thirria remarks that this ossiferous mass merely requires a compact cement to become an
osseous breccia.

A very common condition of cavern bones is their being
found mixed with angular fragments of the rock in which the
caverns occur. Banwell Cave, in the Mendip Hills, is a good
example of a large accumulation of the remains of *Ursus*,
Cervus, *Bos*, and other animals, with fragments of carboniferous or mountain limestone, the rock in which the cavern is
formed. The contents of this cave merely require, as M.
Thirria has observed respecting that at Fouvent, a calcareous
cementing matter, to become an osseous breccia, such as is
found at Nice and other places on the shores of the Mediterranean. The osseous breccia of the Chateau Hill at Nice
appears indeed to have been partly a cavern, which has been

---

* Thirria, Mém. de la Soc. d'Hist. Nat. de Strasbourg, t. i.

quarried away by the works constantly carried on there. The following is a section, fresh when I observed it in the winter of 1827.

q, quarry; *a a*, hard brecciated dolomite; *l l l*, holes bored in the dolomite by some lithodomous shell; *c*, rounded pebbles, composed principally of rock fragments transported from a distance, cemented by a compact calcareous paste; *o*, osseous breccia, united by a reddish calcareous cement.

Fig. 33.

This section seems to point to the following conclusions:—1. An open fissure beneath water, the sides pierced by some boring shell. The lithodomous shells being of all ages, the time does not appear to have been short. 2. The lower part of the fissure filled by gravel transported from a distance. 3. The remainder of the fissure filled by the broken bones of animals, shells (marine and terrestrial), and fragments of rocks, mostly, but not solely, those of the vicinity. 4. The rise of land, or the fall of the sea, to their present relative positions.

Other osseous breccias are common in the vicinity, some being at least 500 feet above the level of the present Mediterranean: the cement reddish, and often vesicular; the vesicles being lined with carbonate of lime. A portion at least of this osseous breccia would seem to have been formed beneath the sea, for it contains marine remains; and among other things those of a *Caryophyllia* at Villafranca. Independent of the fissures containing the remains of terrestrial animals, there are others, merely affording marine remains, which remains do not seem to differ from the actual inhabitants of the Mediterranean, and the breccia appears to have been contemporaneous with the osseous breccias; the mineral compound in all cases taking its character from the rock in which it occurs.

The osseous breccia of Cagliari, Sardinia, occurs in clefts and small caverns of a supracretaceous rock, about 150 feet above the sea. The remains of a *Mytilus* are discovered mingled with the other organic exuviæ*. Dr. Cristie describes the osseous breccia at San Ciro, near Palermo, as not confined to the cave itself, but as forming part of the external talus, resting upon the upper supracretaceous (tertiary) beds, with a thickness of about 20 feet. The same author considers this deposit to have been effected in water, and to have been subsequently raised above the sea, for parts of the cavern are

* De la Marmora, Journal de Géologie, t. iii. p. 310.

perforated by lithodomous shells, reminding us of the osseous breccia of Nice. Dr. Cristie also notices the osseous breccia, 70 feet above the sea, near the bay of Syracuse, as containing an admixture of sea shells. He infers, as the osseous breccia of the Beliemi Caves, near Palermo, does not present any marks of having been formed in the sea, and as it rises 100 feet above the San Ciro cave, itself about 200 feet above the sea, that the breccia at Beliemi was above the surtace of the sea at the time that the breccia of San Ciro was beneath it; and that their present heights mark the extent to which the tertiary formation has been raised at that part by the great convulsion which elevated a large part of Sicily*.

Similar osseous breccias occur at Gibraltar, Cette, Antibes, Corsica, and various other places on the shores of the Mediterranean. The bones found at these places consist, according to Cuvier, (besides those referrible to Horses, Oxen, and large Deer,) of Deer of the size of the Fallow Deer (Gibraltar, Cette, Antibes); Deer resembling, in their teeth, some in the Indian Archipelago (Nice); a smaller species (Nice); a species of Antelope or Sheep (Nice); two species of Rabbit (Gibraltar, Cette, Pisa, &c.), one resembling the common Rabbit, the other smaller; *Lagomys* (Corsica, Sardinia); species of *Mus*; *Felis* (Nice); *Canis* (Sardinia); Lizard (Sardinia); Land Tortoise (Nice).

M. Brongniart considers that many of the pisiform iron-ores which occur in the clefts of some rocks, particularly in the Jura, are of contemporaneous origin with the osseous breccias. In support of this opinion, M. Necker de Saussure observes, that at Kropp, in Carniola, clefts of rocks containing iron-ore worked for profitable purposes, contain the remains of the *Ursus spelæus.* It also appears that the remains of mammalia have been discovered under similar circumstances in the district of Wochein†. According to MM. Thirria and Walchner, there are two deposits of pisiform iron-ore in the north-west part of the Jura (Haute Saone) and in the environs of Bâle, one probably derived in a great measure from the partial destruction of the other, which occurs between the oolitic group and the supracretaceous rocks. The most recent deposit sometimes contains the remains of the rhinoceros and bear, and is considered of the same geological date as the osseous breccias‡.

There would appear to be much analogy between many

---

* Cristie, Phil. Mag. and Annals, Dec. 1831. The bones from the San Ciro cave were ascertained by Cuvier to be those of the *Elephant, Hippopotamus, Deer,* and of animals of the genus *Canis.*

† Ann. des Sci. Nat. Jan. 1829.

‡ Mém. de la Soc. d'Hist. Nat. de Strasbourg.

ossiferous caverns, the osseous breccias, and some clefts containing iron-ore, leading to the presumption that the animal remains contained in them have been introduced under certain general circumstances. The great cleft before noticed, at Oreston near Plymouth, seems to have been quite open when the elephant and rhinoceros remains were introduced into it; the accumulation of angular fragments, many of them very large, and ninety feet deep, having taken place since the remains were deposited; marking no transport from a distance, but a simple falling in of fragments, of the same nature as that of the rock on each side (grauwacke limestone).

Osseous breccias, occurring under similar circumstances, are not confined to Europe, for it now appears that they are discovered in Australia. According to Major Mitchel, the principal ossiferous cavity is situated near a large cave in Wellington Valley, about 170 miles from Newcastle, through which valley flows the river Bell, one of the principal sources of the Macquarrie. This cavity is described as a wide and irregular kind of well or fissure, accessible only by ladders or ropes, and the breccia is a mixture of limestone fragments of various sizes, and bones enveloped in an earthy red calcareous stone. Such of the bones, forwarded to Europe, as were inspected by Mr. Clift, were referred by that anatomist to the *Kangaroo, Wombat, Dasyurus, Koala,* and *Phalangista,* all animals at present existing in Australia. With these were found two others; one of which, considered to be that of an elephant, was obtained in a singular manner by Mr. Kankin, who first visited this fissure; for, supposing it to be a projecting portion of the rock, he fastened the rope by which he descended to it, and was only undeceived by the support breaking, and showing itself to be a large bone.

Mr. Pentland considers that the bones from the Australian breccia, forwarded to Paris, and examined by Baron Cuvier and himself, belong to fourteen species of animals, referrible to the following genera: *Dasyurus,* or Devil of the colonists, three species, one of which does not seem to differ from the *D. Macrourus* of Geoffroy; *Perameles,* one species; *Hypsiprymnus,* or Kangaroo Rat, one species; *Macropus,* or Kangaroo proper, three or four species; *Halmaturus,* three species; *Phascolomys,* or Wombat, one species; a small animal of a new genus, and of the order *Rodentia;* Elephant, one species; and a saurian reptile allied to the genus Gecko. Of these, three or four only are known to zoologists as recent animals. Mr. Pentland found many fragments of bones evidently gnawed, and the epiphysis of the long bones was always absent[*].

---

[*] Pentland, Jameson's Edin. Phil. Journal, April 1832, and Jan. 1833.

Major Mitchel notices other and similar breccias on the Macquarrie, eight miles N.E. from the Wellington cavity; as also at Boree, fifty miles to the S.E., and at Molony, thirty-six miles to the E.; the latter, according to this author, containing bones apparently larger than those of the animals now existing in the country*.

Before we conclude this subject, we should notice an ossiferous cavern on the banks of the Meuse, at Chockier, about two leagues from Liége, which exhibits some curious circumstances. Fragments of limestone, of the same kind as that in which the cavern occurs, are mixed with some quartz pebbles, and with bones, mostly broken; the whole being united by a calcareous cement. The bones and teeth occur equally in the solid breccia and mud, which, with three beds of stalagmite, nearly fill the cavern. It is stated that bones were discovered beneath each of these three distinct beds of stalagmite. The remains belong to at least fifteen species of animals,—elephant, rhinoceros, cavern bear, hyæna, wolf, deer, ox, horse, &c. The most abundant being those of bears, hyænas, and horses†.

---

* Jameson's Edin. Phil. Journal, 1831; and Phil. Mag. and Annals, June 1831.

† Journal de Géologie, t. i. 1830.

SECTION IV.

SUPRACRETACEOUS GROUP.

Syn. Superior Order, *Conyb.*; Tertiary Rocks, *Engl. Authors*; Terrains Tertiares, *Fr. Authors*; Tertiärgebilde, *Germ. Authors*; Terrains Izémiens Thalassiques, *Al. Brong.*

Prior to the labours of MM. Cuvier and Brongniart on the country round Paris, the various rocks comprised within this group were geologically unknown, or were considered as mere superficial gravels, sands or clays. Subsequently to the publication of their memoir (1811), it has been found that the geological importance of these rocks is very considerable, and that they occupy a large part of the superficies of the present dry land, entombing a great variety of terrestrial, fresh-water, and marine remains. It was observed that in the vicinity of Paris, and for certain distances around, the organic remains detected in the different beds were not all marine, but that fresh-water shells and terrestrial animals of genera now unknown were not uncommon; and by prosecuting the discovery, it was found that these remains were deposited in beds, each holding a certain place in a certain series*.

---

* While these discoveries were proceeding in France, Mr. William Smith, —a name that must always be remembered with respect by the geologists of Britain,—was working on more ancient rocks, and, amid a thousand difficulties, identified strata in various parts of England by means of organic remains. It is true that he did not publish regular works until 1815; but it is equally true and well known, that fossils constituted his mode of tracing equivalent beds long previous to this period.

According to M. Keferstein, Fuchsel (a German geologist) had observed that certain beds between the Hartz and Thuringerwald, and around Rudelstadt, were characterized not only by their mineralogical structure, but by their organic contents, as early as 1762 and 1775. This is proved by two works of Fuchsel, one in 1762, entitled *Historia Terræ et Maris, ex Historia Thuringiæ per Montium Descriptionem erecta;* the other in 1775, entitled, *Entwurf zu der ältesten Erd-und-Menschengeschichte.* Fuchsel seems to have determined the relative position of the rocks now known as the muschelkalk, red or variegated sandstone, the zechstein, the copper slate, and the rothe todte liegende. His theoretical geology is remarkable, and far superior to that of Werner, which afterwards became so prevalent. " He states that the continents were formerly covered by the sea until after the formation of the muschelkalk: but as certain beds only contained vegetables or terrestrial animals, this sea must have been surrounded by a continent more elevated than it, and which occupied the place of the present ocean. This land has *by degrees* been swallowed up by the sea. Debacles have often carried masses of vegetables into the sea, which have been covered by ma-

As might have been expected from their labours and those of Mr. Smith on the older rocks of England, the presence of fossils in particular strata was instantly generalized ; and it became a well received theory for a considerable time, that every formation or particular set of beds contained the same organic remains, not to be discovered in those above or beneath. This opinion has gradually given way before facts; and the present theory seems to be, that though certain shells may not be precisely peculiar to certain beds, they are more abundant in them than in others, and that the uniformity of organic contents is greater as we descend in the series of fossiliferous rocks : so that the older the beds, the greater will be the uniformity over considerable spaces; and the newer the series, the less the uniformity. How far this opinion may be correct, can only be determined by an accurate examination of rocks in distant parts of the world ; and most probably we shall be indebted to the American geologists for the first great advance on this subject. But while we thus wait for information, it may be remarked, that such an opinion is not inconsistent with that which supposes the world to have once been a heated mass, which has gradually cooled at the surface. These observations have been rendered necessary, as, in the group of rocks under consideration, a great variety of organic remains, in many cases of a different character, is found in deposits not far distant from each other.

During the deposit of the different rocks comprehended

---

rine mud. Similar changes may now take place ; *for the earth has always presented phænomena similar to those of the present day."* " He (Fuchsel) found that in the formation of deposits Nature must have followed existing laws ; every deposit forms a stratum, and a suite of strata of the same composition constitutes a formation, or an epoch in the history of the world : the currents of the ancient sea may be determined by the direction of the formations. There are many chemical deposits the formation of which remains inexplicable. All the sedimentary deposits have been formed horizontally, and have accommodated themselves to the inferior surface. The inclined beds occur in that position in consequence of earthquakes or oscillations of the ground, catastrophes which have produced a considerable quantity of mud, which distinguishes the deposits which pass from one into the other." (Keferstein, Journal de Géologie, t. ii.)—The above and other observations are mixed with remarks characteristic of an infant science, but such remarks are comparatively few in number. Altogether, Fuchsel seems to have been a very remarkable man ; and, as M. Keferstein observes, it was little creditable in Werner, that while he adopted his ideas as to strata and formations, he should have followed them so much less logically.

It may be here noticed that the celebrated Dr. Hooke also considered highly inclined and vertical strata as so placed in consequence of earthquakes ; for I find by reference to those curious documents, the MS. journals of the Royal Society, that he stated this opinion to the meeting of that Society on June 27, 1667 ; and he further inferred that shells which he had observed in a cliff in the Isle of Wight, were raised above the level of the sea by the same forces.

within this group, the various operations of Nature would seem to have proceeded, uninterrupted by a catastrophe so violent, or by any condition so common to a large surface, as to produce a deposit of similar substances, characterized by great depth and by similar organic remains, over Europe; for to this comparatively limited area it would yet seem prudent to confine our generalizations. Under this state of things, springs would deposit the different substances which they are capable of holding in solution: and if the theory of internal heat and of a great decrease of surface temperature be well founded, they would generally be hotter than at present; *i. e.* the number of thermal springs might be greater;—so far an important consideration, as perhaps more silica would be dissolved and deposited then, as indeed might be the case with many other substances *. It may be here remarked, that this consideration would have weight throughout the deposits of an older date; so that the older the class of rocks, the greater would be the probability of an increased number of thermal springs, and consequently, the greater the abundance of the siliceous and some other deposits.

Whether this hypothesis be correct or not, it is geologically certain that the superficial temperature has decreased, and, as Mr. Lyell has observed, shows itself in the rocks under consideration, even when the organic remains they contain are of the same species of animals as those which now exist; for they are found, as is to be seen in Italy, larger than those which live in the neighbouring seas, thereby pointing out their probable growth beneath the influence of a warmer climate.

A difference in climate would also produce other variations visible in the supracretaceous rocks, as also in those which were previously formed. The warmer and more tropical the climate, the greater, perhaps, might be the evaporation and the fall of rain, as also the power of many meteoric agents. Consequently, under this hypothesis, the earlier the deposits, the more they would present evidences of having felt the influence of such climates. Tropical rains bursting upon high mountains like the Alps, even supposing a portion of them not to have been so lofty as at present, would produce very

---

* The manner in which some solutions of silica are effected seems as yet unexplained. It is well known that the Grasses, Canes, and other plants of the same natural family, have an external coating of silica,—a wise provision of Nature for their protection. But the most remarkable siliceous secretion with which we are acquainted, seems to be that which takes place in the cavities of the Bamboo, and is known by the name of *tabasheer*. Dr. Turnbull Cristie informs me, that the *tabasheer* found in the *green* bamboo of India is perfectly translucent, soft, and moist; but that after its exposure to the atmosphere its moisture evaporates, and it becomes opaque, hard, and of a white or gray colour, such as it appears when brought to Europe.

different effects from those we now witness in the same regions. Torrents of water would be suddenly produced, of which the present inhabitants of those mountains have no conception; and the body of detritus borne down by them would be vastly greater than that carried forward by the present Alpine torrents, though these are by no means inconsiderable. So that the differences produced on land by the greater power of meteoric agents in warm regions should always, supposing this hypothesis correct, be taken into account; particularly when it is apparent, from a succession of beds observed in the same district, that the temperature under which the deposits have taken place has gradually diminished.

Let us now inquire how far vegetation could counteract the superior decomposing and transporting power of atmospheric agents in tropical or warm climates. It appears that, all other circumstances being equal, the warmer the climate, the greater the body of vegetation produced in it. The question then is, does vegetation protect land from the destructive agency of the atmosphere? We can scarcely reply, except in the affirmative. Indeed, if we wanted evidences of it, we might find them in the artificial mounds of earth, or barrows, so common in many parts of England, which have been exposed to the action of the atmosphere in this climate for about two thousand years, and yet have not suffered any marked alteration of form, though only covered with a short turf for at least a considerable portion of that time. Now if it be admitted that vegetation, to a certain extent, protects land beneath it, it will follow that the greater the vegetation, the greater the protection; and consequently, that land is always defended from the destructive agency of the atmosphere in proportion to the protection required. Without this provident law of nature, the softer rocks in tropical regions would speedily be washed away, and the soil would be unable to support animal and vegetable life; for though in many tropical countries large tracts of apparently barren wastes suddenly seem to spring into life, and are covered with a brilliant green herbage, as if by enchantment, after two or three days of rain; the roots, which when wetted send up such vigorous shoots, and those of the bygone annuals, whose seeds now develope green leaves, are matted together in such a manner as to produce considerable resistance to the destructive power of the rains *.

It is by no means intended to infer that the degradation of land is not greater in the tropics generally than in milder climates, but merely to state that there is a relative proportion of

---

* In the savannahs of the western world, there is frequently very little vegetation, and the consequent loss of surface is considerable.

vegetable protection in both.   Suppose a rainy season, such
as is common in the tropics, to fall on England ; who would
doubt that large tracts of land would be bared, and that the
barrows before noticed would speedily disappear : and that if
the rains of the English climate were to fall in the tropics, there
would be scarcely such a thing as vegetation in the lowlands, the
water thus produced being insufficient to support the tropical
plants.   The water, though it might tend to degrade the land,
would be so speedily evaporated that little would be effected
in that manner.   The rains and the vegetation are proportioned to each other : but the destruction of the land still remains in proportion to the quantity of rain and the superior
force of many meteoric agents ; so that, all other circumstances
being the same, the heavier the rains, the greater the destruction of land ; and consequently the warmer the climate, the
greater the degradation of the hills *.

It must also be borne in mind, that during the epoch in
which the supracretaceous rocks were formed, subterraneous
forces would probably be not less active than they were previously or have been since.   We should expect to find igneous
rocks of various kinds intermixed with the aqueous deposits ;
and, under favourable circumstances, interstratified with them ;
approaching through a succession of ages so nearly to the character of modern volcanos, more particularly as their exposure to ordinary destructive causes would be gradually less,
that it would be exceedingly difficult to say where the modern
volcano commenced and the ancient volcano ceased.   There
is also no reason why the same vent should not have continued
to vomit forth various substances for a long succession of ages
and during various changes on the earth's surface, as has been
previously noticed ; so that our endeavours to classify their
products may not be very successful.   Great movements in the
land may have been effected, altering the general levels of various districts ; and even ranges of mountains may have been
thrown up, producing consequent effects that may have greatly
influenced certain deposits.

It has been observed that the supracretaceous rocks present
numerous instances of fresh-water deposits, scattered over a

---

* In tropical countries the parasitical and creeping plants entwine in every
possible direction, so as to render the forests nearly impervious, and the
trees possess forms and leaves best calculated to shoot off the heavy rains,—
thus affording protection to innumerable creatures which seek shelter, at
such seasons, beneath them.   The pattering of the tropical rains on such
forests is heard at distances which an inhabitant of the temperate regions
would little suspect, and is particularly striking to a stranger.   The rain,
thus broken in its fall, is quickly absorbed by the ground beneath, or thrown
into the drainage depressions, where, it must be confessed, the torrents thus
produced are sufficiently furious, and cause great destruction.

considerable surface,—a fact which seems to point to a large
continuous body of land; in other words, to the presence of
considerable continents or large islands. And this opinion
seems strengthened by finding the remains of large mammiferous animals entombed in the same rocks, which are termed
fresh-water, because marine remains are not detected in them,
their organic contents being either the exuviæ of animals of
which the analogous kinds inhabit lakes or rivers at the present day, or else of animals or vegetables whose analogues are
found only on the dry land. It is also inferred that these remains could only have been entombed beneath deposits in
rivers or in lakes, whence also they are often named lacustrine
rocks. Independent of these lacustrine or fresh-water formations, there are others of a mixed character, wherein the organic remains are terrestrial, fresh-water, and marine; and
these are considered as deposited in estuaries, from analogous
assemblages of this kind now supposed to be forming in such
situations. The rocks containing only marine remains speak
for themselves: but it by no means seems to follow, that because a rock may contain terrestrial or fresh-water remains,
the origin of the deposit is necessarily an estuary; for if analogies be always sought in the present state of things, we know
that such remains are frequently carried far beyond the mouths
of rivers.

It is a common practice to describe the supracretaceous
rocks as occurring in basins, such as the London, Paris, Vienna,
Swiss, and Italian basins: but this term seems often exceedingly misapplied; for great marine deposits were, one would
suppose, no more liable to have been formed in basins formerly
than now, when certainly, unless we often term the great bed
of the ocean a basin, we should by no means characterize the
deposit as taking place in such a cavity. Thus we should ill
characterize the delta deposit of the Ganges by terming it
basin-shaped. It is a common thing to speak of the London
basin, when the supracretaceous rocks which occur in this
supposed basin, seem little else than the continuation of a
great belt of these rocks which extends through Europe by the
north of Germany towards the Black Sea. We also hear of
the Isle of Wight basin, as if there had existed a separate cavity or depression in that particular place; while there is very
good reason for supposing, (as has been stated by Prof. Buckland,) that the supracretaceous deposits of London and the
Isle of Wight have once been continuous, and that this continuity has been destroyed by the upheaving of the chalk beneath, subsequently to the deposit of these rocks; the intervening upraised portion having been removed by denudation,
as has happened to much thicker and harder rocks. The

same with the Paris basin, which may have easily been connected with those above mentioned, and as easily separated from
them, by movements of the earth, and by denudation. It may
therefore have happened, that these so-called basins were formerly continuous portions of one whole, which various circumstances have disunited, perhaps even during the deposit of the
rocks in question; their commencement having been in a sea
which washed the older strata, and extended from the west of
Europe, between Scandinavia and northern Germany, towards
the Black Sea. Outliers of these rocks, similar to those of
other deposits, are seen on the hills in the West of England,
attesting their elevation, and the denudation which has destroyed the continuity of their mass, and left detached portions
like islands fringing a continent. In consequence of the various movements of lands, and the denudation either consequent on them or some other cause, the barriers of many a
fresh-water deposit are removed ; and though, from analogy,
we consider them as formed beneath the waters of lakes, we
are totally unable to point out the shores of such pieces of
water. The student should be careful to keep this great denudation in mind, not only as applicable to the rocks under
consideration, but also to the various changes and deposits
that have previously occurred : indeed he may consider that
no considerable portion of the earth's surface has ever remained
long, geologically speaking, in a state of rest; but that the rise
and depression of land, and the removal of a large proportion
of it, have been frequent. Even in the rocks now treated of,
he will be called upon to consider that there has been an alternate rise and depression of land, to account for an alternation of marine and fresh-water deposits ; and this he will perhaps be the more ready to do, as he has already seen that such
movements of the land have happened at a more recent period.

Amid so great a variety of deposits, attesting such different
modes of formation, it is no easy task to know where to begin
in the descending series, or what may be precisely contemporaneous. In this difficulty, perhaps the safer course is to consider those deposits the most modern which contain organic
remains bearing the closer resemblance to the animals and
vegetables now existing. Now all the terrestrial animals
found in caves and superficial gravels, marls, and sands, whatever may be the theory formed to account for their disappearance, must have lived upon lands existing at the period under
consideration : and even supposing them in a great measure
destroyed by a catastrophe, there is nothing to prevent their
having been abundantly entombed during their residence on
the earth. For while the extinct bears and hyænas were the
inhabitants of caverns, generation succeeding generation in

their possession, the great work of nature was proceeding; and the elephants, rhinoceroses, hippopotami and other animals, some of which were dragged into the hyænas' dens, were perishing from old age or accident, and their remains included in the various deposits then forming. The same with land and fresh-water animals, marine remains, and vegetables.

The nearer also, judging from organic remains, that the climates can be considered like those now existing, the greater would appear the probability, that the rocks containing them occupied the higher part of the supracretaceous series. Thus in the tropics we should expect to find, among the most recent of these beds, remains analogous to those now existing in similar regions; while as we approached each pole, we should be prepared to discover organic remains corresponding with the various latitudes. As far as facts have yet gone, this would seem to be the case; for the fossil vegetables found in the more recent strata in the tropics are tropical, while those discovered in contemporaneous deposits in Europe are not so, but more suited to the climate; as, for instance, the vegetable remains of Œningen *.

Should it eventually be found, that the organic remains discovered in tropical countries are always characteristic of such climates, or of one which may be termed ultra-tropical, it will go far to prove that the present equatorial regions have always been under the influence of considerable heat, which, though it may have decreased with that of the surface of the world generally, still produces a far more vigorous vegetation than is to be found in the north or south. Should attentive examination also show that at a certain term in the series of rocks, the nature of the vegetable and animal remains found entombed in the tropics, does not point to a comparatively more elevated temperature than a similar term in a general series in Europe, or in any more northern or southern latitude, it would seem to show that the cause of this equal temperature has not been external but internal; for, with any arrangement that may be made in the relative positions of the earth and sun, we cannot conceive one which should produce an equal, or nearly equal

---

* The most recent observations on the Œningen deposit, which has at various times engaged the attention of naturalists, are by Mr. Murchison, who has shown that it is purely lacustrine, and formed in a depression previously existing in the Molasse of the district; the deposit being afterwards re-excavated to a considerable depth, thus affording a passage to the Rhine. Of the many instructive remains detected in this rock, those of a fossil fox discovered by Mr. Murchison are among the most remarkable, as, according to Mr. Mantell, they belonged to an animal which approached, if it was not identical with, the *Vulpes communis.* The various organic remains are described as belonging to species undistinguishable from those now existing, and to others decidedly extinct. Geol. Trans., 2nd series, vol. iii.

temperature over our spheroid ; while we might conceive such a state of things possible, if an internal heat be capable of producing an equable surface temperature, independent, in a great degree, of solar heat.

In the supracretaceous rocks of Italy and the south of France, and probably also of other Mediterranean countries, there seems better evidence of the nearer approach of organic life to that now existing, than has yet been pointed out elsewhere, though other evidence is not wanting. Indeed, it may be exceedingly difficult to separate the actual state of animal and vegetable life from that which preceded it in the more recent deposits of Italy, or precisely to say when marine remains similar to those now existing in the Mediterranean were raised to various heights above it.

In the more modern supracretaceous deposits of the Apennines, commonly termed Sub-Apennine rocks, it is well known that there is a mixture of species such as now exist in the Mediterranean, and of those found in warmer climates. The deposit noticed by Mr. Vernon in Yorkshire may not be far removed from this date, as land and fresh-water shells were found precisely similar to those now existing, though mixed with the bones of elephants, &c.*

M. Deshayes having most attentively studied the conchology of the supracretaceous rocks of Europe, considers that they can be zoologically divided into three groups. 1. The more modern group, including the Sub-Apennine deposits, with many rocks in Sicily, the Morea, and at Perpignan. Of 700 species of shells found in this group, more than one half are analogous to existing species. The English *Crag* is included in this group. 2. A group consisting of the Fahluns of Touraine, and many beds in Austria, Hungary, Poland, of the Gironde, and in the environs of Turin. Of 900 species found in this group, 161 have living analogues. 3. The most ancient group, consisting of rocks deposited in the basins of London, Paris, the Isle of Wight, and Valognes, and in certain cantons of Bourdeaux and Italy. Of 1400 species found in these beds, thirty-eight only are analogous to living species.

The analogues of the first group live in the neighbouring seas. Of the 161 species in the second group, the greatest part of the analogues are discovered in Senegal, Madagascar, and the Indian Archipelago; a less number in the Mediterranean, and a few only in other European seas. Of the thirty-eight species in the more ancient group, some are found in various latitudes, but the greater portion within the tropics†.

* Phil. Mag. and Annals of Philosophy, 1829—1830.
† Deshayes, Annales des Sci. Nat. 1831.

We should be careful to recollect, when estimating the value of the remains of any particular species or genus of animals found, not only in these deposits, but in the fossiliferous rocks generally, that great variations are produced in the kind of animals inhabiting the present seas, by depth of water, the strength of tidal streams or currents, the greater or less exposure to heavy seas, the kind of bottom in particular situations, and the nature of the climate. We therefore cannot, if we reason from the existing state of things, expect to find the same, and only the same, organic remains entombed in a contemporaneous deposit over a considerable area; for such a supposition would infer precisely the same conditions over the whole area, a state of things that cannot be considered probable.

According to M. Elie de Beaumont, there exists in the valleys of the Isère, Rhone, Saone, and Durance, a large deposit of rolled pebbles and sands, clearly distinguishable from that which accompanies the transported blocks, and more ancient than it. It is not in general distinctly stratified, but seems rather to constitute a deep mass, sometimes several hundred yards thick. The rolled pebbles can all be traced to the Alps, and are unmixed with the fragments of distant rocks. Lignite occurs in it, and apparently bears the marks of slow deposit. At one place, (Vallon de Roize, near Pommiers,) the lignite is covered and supported by rolled pebbles, and is itself inclosed in a fine-grained and earthy bed: the carbonaceous mass is divided into even strata, between which numerous shells of *Planorbes* are discovered. M. Elie de Beaumont remarks, that in places where the parts are slightly agglutinated, the sands, mixed with mica, strongly remind us of those now brought down by the Rhone, the Isère, and the Durance. This sand sometimes becomes marly and schistose, containing fragments of lignite, which often accumulate into sufficient masses to be profitably worked, the lignite being included between strata of clay, marl, or fine sand, alternating with the rolled pebbles. The lignites of St. Didier are composed of the flattened trunks of trees, in which the woody fibre can still be traced. It is considered that these lignites are contemporaneous with those in Savoy, at Novalése, Barberaz, Bisses, Motte-Serrolex, and Sonnaz, near Chambery. This deposit of pebbles and sands is traced through the plain of Bresse; it is observable in the escarpments of the Rhone between the embouchure of the Ain and Lyon, with the same characters as are observable in the department of the Isère. It may be well studied near Lyon, and is seen at the foot of the Jura near Ambronay and Ambrutrix. Near Ajou there is a deposit of bituminous wood, described by M. Héricart de

Thury, who notices beneath a mass of rolled pebbles and argillaceous marls: 1. Blue Clay; 2. Lignite; 3. A bed of pebbles; 4. Blue clay; 5. Lignite; 6. Blue clay, containing the branches, trunks, and roots of trees, more or less well preserved; 7. Red and blue clays; 8. A bed of bituminous wood, very thick and compact. In the first bed of lignite there was sometimes an admixture of pebbles; and numerous terrestrial and fluviatile shells were discovered in the mass.

M. Elie de Beaumont traces the deposit in other directions, and considers it may have been one formed in the waters of a shallow lake, which existed subsequent to the elevation of the Alps of Savoy and Dauphiné, but prior to that of the main chain from the Valais into Austria. The various pebbles seem clearly to be derived from the Alps, and the different lignite deposits appear to show that they were not suddenly transported in a mass. It may not therefore be unreasonable to infer that they were carried forward by the action of rivers from the Alps into the situations where we now find them. The time required for this would be very considerable; but with the lignite deposit, as a part of the mass, we can scarcely refuse it a gradual formation*.

The same author points out that this mass of pebbles should not be confounded with those collections of Alpine pebbles and sands which constitute a very considerable deposit on either side of the Alps, known commonly by the name of Nagelfluhe and Molasse; and which had not only been previously formed and consolidated, but also upheaved before the pebbles and sands under consideration were transported. These observations in the same district are highly important; for it must rarely happen that the Nagelfluhe and Molasse, the pebbles and sand now treated of, and the transported substances of the erratic block group, can be distinctly seen, as it were, together, under circumstances which mark their difference.

We should expect that, previous to the supposed convulsion of the Alps at the erratic block period, such marks of degradation should be everywhere apparent; and that the occurrence of river-borne pebbles, sands and clays, would be sufficiently common; and would, when not removed by subsequent debacles, be often found beneath deposits formed by such debacles.

The precise age of the celebrated Bovey coal cannot at present be well determined, but may conveniently find a place here. A body of water has evidently passed over it, working

* Elie de Beaumont, Recherches sur les Rév. du Globe; Ann. des Sci. Nat. 1829 et 1830.

hollows in the clay, and leaving a large deposit of transported substances in some situations. It also appears to have been tranquilly deposited in a previously existing depression at this spot. The area comprising the surface of the deposit is far more considerable than is usually given, and it has certainly once occupied a greater elevation, as a mass, than it now does, the upper portion having been removed by denudation. The principal deposit of lignite occurs near Bovey Tracy, Devon, at the north-western end of the deposit. The upper part is composed of quartzose sand, probably derived from the granite country near, of portions of rocks of the immediate neighbourhood, and of rounded pieces of clay, which appear portions of the clay that accompanies the Bovey coal formation.

Beneath about twenty feet of this *head,* as the workmen term it, there is an alternation of compressed lignites; shales, or clays; the whole mass dipping at about 20° to the S.E. or S.S.E. The lignite is evidently composed of dicotyledonous trees, many of which are knotted; and among them a curious seed is occasionally discovered.

Other similar parts of the deposit are worked for profitable purposes: but its most useful product is a clay used in the potteries, in some cases so fine as to constitute what is termed pipe-clay. Large quantities of both these varieties of clay are annually shipped at Teignmouth. Lignite more or less accompanies the clay throughout, occurring, when not in beds, as small detached pieces. Animal remains must be exceedingly rare; for I could not, after diligent search, obtain any traces of them, though I was given to understand some shells had been seen near Teignbridge. This deposit has been considered as part of the transported gravels named Diluvium, as also a representative of the plastic clay. It will have been seen that it existed previous to a great transport of pebbles in this district; and it seems more recent than the plastic clay, as there is good reason to suppose that deposits of that age once covered the chalk and green sand, now so extensively denuded in Devonshire, as will be noticed hereafter. And it does not seem improbable that various undulations of this district had been formed subsequent to the deposit of the plastic clay series; which undulations did not very materially differ in character from those we now see, though they may have been greatly modified since. Now the Bovey coal deposit seems to have taken place in a kind of basin, after a general arrangement of hill and dale in the vicinity; for it is exceedingly conformable to their windings, even seeming to run up some valleys, as at Aller Mills, not far distant from Newton Bushel, where there has evidently been an old valley excavated in red sandstone conglomerate, grauwacke, limestone, and grauwacke slate; and in this the alternate beds of lignite and clay, now worked, have been deposited.   The deposit has evidently been at one time more considerable in this valley, and has been denuded; for on Milber Down on the one side of it, and on some hills on the other, there are large accumulations of sands and rolled flints; and although it is possible some portion of them may be the remains of the green sand, and even of the plastic clay series, the remainder seems to have formed part of the Bovey coal deposit.   The following is a section of these rolled flints and coarse sand, apparently composed of triturated quartz and flints, and possibly also chert, on that part of Milber Down facing Ford.

a a, rolled flints; *b b*, coarse sand.   Fig. 34.

The disposition of the two is strongly characteristic of the unequal wash of water, the velocities of which have not been constantly the same over the same spot.   A similar mixture of the clay and sand may be seen near Aller Mills.
On an inspection of the whole formation, there would, apparently, be little doubt that it was, as before stated, deposited in a pre-existing depression in a variety of rocks.   The only question is,—when was this depression formed?   For my own part, I should answer,—after the plastic clay on the chalk to the westward had been upheaved.   Without, however, the more direct testimony of characteristic organic remains, I should give this answer with much hesitation, it being one for the confirmation or rejection of which future observations are very necessary *.   Considering that the relative age of the valleys in this part of England is geologically very important, I have been induced to offer the above notice, as it may lead to further inquiry; though the detail here given somewhat exceeds the limits that should be assigned it.

No doubt, future and delicate observations will detect numerous passages or transitions, in various countries, from a different state of animal and vegetable life to that which now exists, more particularly in marine remains, not so liable to

* According to Mr. Whiteway and Mr. Kingston, who have possessed the great advantage of continued local observation, the Bovey deposit consists chiefly of five clay beds, and as many of gravel, the latter varying from 50 to 100 feet in width.   The clay beds are described as undulating like the waves of the sea; and it is stated that beneath the four more western beds the Bovey coal is found; while below the more eastern or pipe-clay bed (frequently worked to the depth of 80 feet,) there is sand and white quartz. Near the S.E. corner of Bovey Heathfield, (the name given to this low district,) the deposit has been bored to the depth of 200 feet without traversing it.—Nat. Hist. of Teignmouth, Tor Quay, Dawlish, &c.; by Turton and Kingston.

Q

destruction as the inhabitants of dry land. It was long considered that the remains of elephants, rhinoceroses, and mastodons were confined to superficial gravels; but we now know
that they are entombed deeper in the series of rocks, and were
inhabitants of the globe before the *Palæotherium* and some
other mammiferous genera became extinct.

It was also once considered that the supracretaceous rocks
of England and Paris presented us with all the deposits which
were formed between the chalk and the present times; and this
theoretical opinion being strongly impressed on the minds of
geologists, it was very natural that all supracretaceous or tertiary deposits should be considered as the equivalents of some
one or other of those detected in the Paris basin. Such generalizations of local circumstances are common in the history of
geology, and are such as would be expected in the progress
of any science; for until our knowledge of facts becomes extensive, there is nothing to check such opinions. We must therefore be exceedingly careful not to consider our power of
checking such generalizations, as evidence of a clearer sight
than those of our predecessors; while, in point of fact, we are
merely in possession of a greater mass of facts, and are therefore enabled to turn them to a different account. Neither
should we be unthankful for these generalizations, for they
have promoted inquiry, and have probably contributed, far
more than we are often inclined to admit, to that knowledge
which we now possess, and which permits us to see that such
generalizations are untenable.

The Italian deposits, commonly termed Sub-Apennine from
occurring at the lower part of the Apennines, have been appealed to as good examples of a transition or passage from the
present state of things to one wherein animals were somewhat
different: and this appeal seems well founded; for among the
shells discovered in them, there are some which closely correspond with those now existing in the Mediterranean; while
there are others whose analogues seem to live in warmer climates, and many are wholly unknown.

In 1829, M. Desnoyers endeavoured to show, 1st, That all
tertiary or supracretaceous basins were not contemporaneous,
but successively formed and filled. 2nd, That this succession
of basins may have resulted from frequent oscillations of the
soil, produced during the long series of supracretaceous deposits, by the influence of volcanic agents, then very considerable. 3rd, That this difference in the epoch of the formation
of basins may allow us to distinguish many great periods in
the supracretaceous or tertiary deposits, some stable, others
transitory. 4th, That each of these periods would comprehend
deposits formed in the sea, either by the sea waters, or by the

rivers; and deposits formed at the same time out of the sea, by lakes, thermal springs, and rivers; both the one and the other offering, according to the basins, every possible variety of sediment. 5th, That the basins of Paris, London, and the Isle of Wight only contain the ancient and middle supracretaceous deposits. 6th, That the last lacustrine rocks of the Seine basin did not therefore terminate the series of these rocks, but that many formations, both marine and fresh-water, have succeeded it in other and more modern basins. 7th, That these more recent formations appear to indicate at least two periods; to which we may add that with which we are contemporaneous. 8th, That all these periods presented, in their deposits and in their fossils, a progressive and insensible passage from one to the other, from the ancient state of nature to the present, from the more ancient supracretaceous basins to the actual basins of our seas. This author also endeavours to establish other opinions, which may however be more questionable; but it seemed necessary to state the above, because there would appear to be much truth in them, and because he was one of the first to point out the probable zoological passage of the ancient supracretaceous deposits to the present state of things, though he was not the first, as he himself remarks, to attribute the variations observed in tertiary or supracretaceous basins to the differences produced by the local action of such causes as we now witness, this having already been done by MM. Prevost, Boué, and other geologists. He also remarks that the continental waters would carry terrestrial and fresh-water shells into the sea, together with the remains of the large mammalia, such as the Elephant, the Rhinoceros, Mastodon, and Hippopotamus, with fluviatile and terrestrial reptiles, which would thus become mixed up with Cetacea and other marine remains[*].

Without following M. Desnoyers through many cases, of which the relative dates may be questionable, we will proceed to a striking example, where there would appear little doubt as to the occurrence of the remains of the large mammiferous animals buried in more ancient strata than those noticed in the erratic block group. There is a mixture of the remains of *Mastodon* and *Palæotherium* in the basin of the Loire, in the Touraine faluns. According to M. Desnoyers the bones are broken and worn, their substances black and hard, often siliceous, and altogether resembling, in these respects, the marine mammalia which accompany them. The bones are stated to be found in many points of the great faluns to the

---

[*] Desnoyers, Obs. sur un Ensemble de Depôts marins plus recents que les Terrains tertiaires du Bassin de la Seine;—Ann. des Sci. Nat. 1829.

east of Saint Maure. Some are covered with *Serpulæ* and *Flustræ*, showing that they have remained as bare bones for some time in the sea. The remains are stated to be those of the *Mastodon angustidens, Hippopotamus major? H. minutus, Rhinoceros minutus,* and also one of the larger species of the *Tapirus giganteus,* of a small *Anthracotherium, Palæotherium magnum,* the Horse, of one of the *Rodentia* of the size of a Hare; and of one or two Deer. The same author states that a mixture of bones of the *Lophiodon* and *Palæotherium,* with those of the *Mastodon tapiroides* and middle-sized Rhinoceros, are found accompanied by terrestrial and fluviatile shells, at Montabuzard *.

It has long been known that at Mont de la Molière, near Estavayer, Switzerland, the remains of the Elephant, Rhinoceros, Hog, Hyæna, and Antelope occurred in the molasse of that hill †; and I remember having had the remains of the Mastodon and Castor pointed out to me by Professor Meisner of Berne in 1820, as having been obtained from the lignite of the Swiss molasse ‡, so that the probable antiquity of these large mammalia has been for some time remarked.

Mr. Murchison states that at Georgesgemünd, near Roth, beds of sandy marl and whitish concretionary limestone occur in isolated patches on heights about 150 feet above the present drainage of the district. In these beds are subordinate layers of calcareous, ferruginous, and bony breccia; portions of which, collected by this author, and examined by Mr. Pentland and Mr. Clift, were found to contain the remains of *Palæotherium magnum; Anoplotherium* (new species); a new genus allied to *Anthracotherium* or *Lophiodon; Hippopotamus;* Ox; Bear, &c. According to Mr. Murchison, Count Munster had previously collected nearly similar remains from the same place, with the addition of those of *Palæotherium Orleani; Mastodon minutus; Rhinoceros pygmæus,* Munst.; *Ursus spelæus,* and a small species of Fox §.

It would appear that this curious mixture of existing and extinct genera is also found at Friedrichsgemünd; for M. Meyer states, that a calcareous rock there contains the remains

---

* Desnoyers, Ann. des Sciences Naturelles, 1829. M. Desnoyers has inserted a tabular view in the Bulletin de la Société Géologique de France, t. ii. p. 336, entitled, The Geological and Geographical Relations, in the Basin of the Loire, of the bone deposits of terrestrial and marine Mammalia, and fluviatile Reptiles, contained in the marine tertiary rocks more recent than those of the Seine, &c.

† Bourdet de la Nièvre, Soc. Lin. de Paris, 1825.

‡ Professor Meisner had printed a notice of them, with a plate, in a work then in the course of publication at Berne, but of which the exact title has escaped my recollection.

§ Murchison, Proceedings of Geol. Soc. May 1831.

of *Mastodon Arvernensis; Mast. angustidens; Palæotherium Aurelianense; Rhinoceros incisivus; Chæroptamus Sœmmeringii; Lophiodon;* a small carnivorous animal; *Cervus;* Tortoise, &c. The calcareous rock also contained the remains of a *Helix**. The same author also notices a mixture of the remains of the *Mastodon angustidens; Mast. Arvernensis; Rhinoceros incisivus; Lophiodon; Tapirus giganteus;* three species of pig-like animals; *Cervus;* Gigantic Pangolin; carnivorous and other animals, as discovered at Eppelsheim, near Alzey, Hesse†.

How far the various deposits, to which the English crag has been added, and which have been referred to one epoch, may really be contemporaneous, it will probably require much time to determine; but at all events the facts stated are important, as they show that the Mastodons, Rhinoceroses, and Hippopotami existed as genera at the same time with the *Lophiodon* and *Palæotherium*, and that the former continued to inhabit certain parts of ' Europe when many molluscous animals existed, similar or analogous to some of those contemporaneous with ourselves.

Great mammalia are stated to be found in the blue marl of Italy, at Peruggia, Parma, and the Val di Metauro, as also in the sandy deposits of other places of the same country.

The English crag occupies a surface with a variable outline in Norfolk and Suffolk, as will be seen by Mr. Taylor's map, and moreover appears to be somewhat changeable in its character. The same author has given sections of it in his " Geology of East Norfolk," where it will be seen to rest indifferently on chalk and London clay. The following is a list of some of its organic remains, as appears in Mr. Woodward's " British Organic Remains," including the same author's MS. notes on the Norfolk crag. POLYPIFER: *Turbinolia sepulta.* To this may be added a great variety in the possession of Mr. Taylor. RADIARIA: *Fibularia Suffolciensis.* CIRRIPEDA: *Balanus crassus, B. tessellatus, B. balanoides?* (Woodward.) CONCHIFERA: *Solen siliqua? Panopæa Faujasii, Mya arenaria, M. Pullus, M. lata, M. subovata, M. truncata? Mactra arcuata, M. dubia, M. ovalis, M. cuneata, M. magna, M. Listeri? Corbula complanata, C. rotundata, Saxicava rugosa, Petricola laminosa, Tellina obliqua, T. ovata, T. obtusa, T. prætenuis, Lucina antiquata, L. divaricata, Astarte plana, A. antiquata, A. obliquata, A. planata, A. oblonga, A. imbricata, A. nitida, A. bipartita, Venus æqualis, V. rustica, V. lentiformis, V. gibbosa, V. turgida, Venericardia senilis, Ven. chamæformis, Ven. orbicularis, Ven. scalaris, Cardium Parkinsoni, C. an-*

---

gustatum, C. edulinum, Isocardia Cor? Pectunculus variabilis, Nucula lævigata, N. Cobboldiæ, N. oblonga, Pecten compla-natus, P. sulcatus, P. gracilis, P. striatus, P. obsoletus (3 var.), *P. Princeps, P. grandis, P. reconditus, Ostrea Spectrum, Tere-bratula variabilis.* Mollusca: *Chiton octovalvis? Dentalium costatum, Patella æqualis, P. unguis, P. ferruginea,* jun., *Emarginula crassa, E. reticulata, Infundibulum rectum, I. tenerum, Bulla convoluta, B. minuta, Auricula pyramidalis, A. ventricosa, A. buccinea, Paludina subaperta, Natica de-pressa, N. hemiclausa, N. cirriformis, N. patula, N. glauci-noides* (var.), *Acteon Noæ, A. striatus, Scalaria frondosa, S. subulata, S. foliacea, S. minuta, S. similis, S. multicostata, Trochus lævigatus, T. similis, T. concavus* (var.), *Turbo rudis, T. littoreus, Turritella incrassata, Tur. punctata, Tur. stri-ata, Fusus alveolatus, F. cancellatus, Murex contrarius, M. striatus* (2 var.), *M. rugosus* (2 var.), *M. costellifer, M. echinatus, M. Peruvianus, M. tortuosus, M. alveolatus, M. corneus, M. elongatus, M. Pullus, M. bulbiformis, M. lapilli-formis, M. gibbosus, M. angulatus, Cassis bicatenata, Buccinum granulatum, B. rugosum, B. reticosum, B. tetragonum, B. pro-pinquum, B. labiosum, B. sulcatum* (2 var.), *B. incrassatum, B. elongatum, B. elegans, B. Mitrula, B. Dalei, B. crispatum, B. tenerum, Voluta Lamberti, Ovula Leathsi, Cypræa coccinel-loides, C. retusa, C. avellana.*

It has been stated that the remains of the great mammalia are mixed with these fossils in the crag, but it does not so clearly appear that this has been the case. According to Smith, the remains of a Mastodon have been there found; and although the bones of Elephants and other animals dis-covered in the transported rocks above it may, without great care, be easily confounded with the fossils of the crag, there does not appear to be any good reason why such remains should not be discovered in this rock as well as in similar, or nearly similar, strata in other parts of Europe.

The following is, according to Mr. Taylor, a section of the crag strata at Bramerton, near Norwich, whence a large pro-portion of the organic remains noticed in this rock have been derived. 1. Sand, without organic remains, five feet. 2. Gravel, one foot. 3. Loamy earth, four feet. 4. Red ferru-ginous sand, containing occasionally hollow ochreous nodules, one foot and a half. 5. Coarse white sand, with a vast number of crag shells, one foot and a half. 6. Gravel, with fragments of shells, one foot and a half. 7. Brown sand, in which is a seam of minute fragments of shells, six inches thick; fifteen feet. 8. Coarse white sand with crag shells, similar to No. 5.; the *Tellinæ* and *Murices* are the most abundant; three feet and a half. 9. Red sand without organic remains, fifteen feet.

10. Loamy earth, with large stones and crag shells, one foot.
11. Large irregular black flints crowded together, one foot.
12. Chalk, excavated to the level of the river*.

It will be observed from this section that the transporting power of water has been sufficient to carry coarse sand, and even gravel, and that at one time (No. 7.) there has been a drift of broken shells. Mr. Taylor has shown me other sections of the crag strata which present those diagonal lines so frequent in mechanical rocks of all ages, where there have been irregular currents of water. From this circumstance, and from the variations in the component parts of the sections, there would appear reason to believe that the crag strata were deposits from irregular currents of water, varying in their velocities and consequent transporting powers. With regard to the unrolled chalk flints upon which the crag strata rest, they remind us of the apparent dissolution of a portion of the chalk in place, so common over a large part of England and France, previous to the deposit of the supracretaceous rocks.

If we look to the Alps, we find on all sides of that chain beds of various depths of sandstones and conglomerates, forming a whole of very considerable thickness. If we also attentively examine the component parts of the sandstones and conglomerates, we find that the former are generally mere comminuted portions of the latter, and that both have been derived from the Alps. The whole is evidently a detritus of the Alpine rocks, and in it organic remains are by no means common, though they occur in certain situations. Such general appearances would seem to indicate a common origin, and that origin to be the Alps themselves. Rolled and comminuted detritus of the kind found may either be derived by the continued action of what are termed actual causes, or some more violent exertion of forces, which, producing rapid motions in water and greater destruction of the land, should accomplish a far greater quantity of work in a given time.

It is quite evident that in certain parts of the Alps, whatever may be the case in others, these detritus beds rest unconformably on many limestone and other rocks, of which some may be referred to the cretaceous and others to the oolitic series. It also clearly appears that subsequent to their deposit they have been thrown up by some force, which, from the evidence of position of strata, must have proceeded from the interior of the Alps, as the strata are tilted up from it on either side ; it thus appearing as if a force had endeavoured to thrust the main body of the Alps higher upwards, and had consequently upheaved the lateral deposits of conglomerates and

* Taylor, Geol. Trans. 2nd series, vol. i.

sandstones with it. The two following sections, one on the
north side of the main chain of the Righi near Lucerne, the
other on the south side of the same chain near Como, show
the disturbed appearance of the conglomerates. Fig. 35. is
from the observations of Dr. Lusser; Fig. 36. is a sketch by
myself.

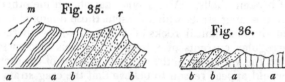

Fig. 35. *m*, Murteberg; *r*, Righi; *a a*, limestone and shales,
containing nummulites and other fossils; *b b*, conglomerate of
rolled pebbles, composed of pieces of pre-existing Alpine rocks.
Fig. 36. *a a*, vertical or nearly vertical beds of gray limestone
(containing much silica) covered by the conglomerates and.
sandstones *b b*, also composed of pre-existing Alpine rocks
There will be little doubt in the mind of the reader, that the
conglomerates have been upraised since their deposit, and
even have been thrown over at the Righi, if the appearances
between that mountain and the Murteberg may not be caused
by a fault*.   There is also another curious fact, which is,
that limestone strata near Como have been upheaved before
the deposit of the conglomerate.

If we transport ourselves from Como to the Maritime Alps,
we find that these also have been upheaved before the deposit
of the rolled fragments, which are clearly derived from the
high adjacent country.   The rocks upheaved in the vicinity of
Nice are compact white limestones, with gypsum, or arenaceous limestones and beds charged with green grains; which
latter may, perhaps, be referred to the cretaceous group: but
there are other rocks more eastward charged with Nummulites
and other fossils, which may belong to some deposits that will
be noticed in the sequel.

While on the subject of Nice, it may be as well to notice
the supracretaceous rocks of that place generally.   After the
more regular strata, before noticed, were upheaved, the relative
level of the sea and the Maritime Alps must have been very
different from what it is at present, for at the height of 1017
feet on the western side of Mont Cao (or Calvo), blocks of
the same rock of which the mountain is composed, namely,

* M. Ebel assured me (while at Zurich in 1829,) that this overthrown
character was more considerable in other situations in the line of the Righi:
it is exceedingly desirable that this should be distinctly determined to be an
overthrow of the strata, and not a great longitudinal fault, which might
easily accompany a great longitudinal uprise of strata.

white compact limestone and dolomite, bear marks of having
been pierced by lithodomous shells; and this has been accomplished during a period of comparative tranquillity; for the
fragments of rock are angular, and have evidently not been
transported from any considerable distance. The same kind
of breccia covers the side of the mountain, separating a great
mass of rolled pebbles and sandstones from the mass of disturbed white limestone of which the mountain is composed,
the blocks still drilled with holes, as may here and there be
observed. At the base of Mont Cao, and at a place named
the Fontaine du Temple, there is an excellent section of this
breccia, where the limestone blocks are angular, and sometimes of large size, weighing many hundred pounds. They
are still drilled with holes, such as are formed by boring shells,
and are encased in a cement composed of siliceous grains agglutinated by calcareous matter. The whole therefore appears
like a state of repose; but if further proof were requisite, it
would be found in certain shells which have much the appearance of *Spondyli*, the lower valves of which are attached to the
blocks, not only near the Fontaine du Temple but higher up
in the mountain, and the cement has gently covered up the
finest of their edges. Now if there had been any considerable ·
motion of the water, more than is common in moderate currents, this could not have happened; for the fine edges of the
shells (and they are very fine,) must have been destroyed.

If we proceed to the shores of the present sea, we also find
evidences of a tranquil residence of water over the disturbed
strata; for beneath the Chateau de Nice we observe an open
crack pierced by lithodomous shells, and these shells still remaining in the holes; and we know that this happened before
the epoch when such an abundance of rolled Alpine pebbles
was carried over this district, because we find part of the cleft
filled up by them, burying many of the holes and their inmates.
That this residence of the sea was not momentary is shown,
as before observed under the head of Osseous Breccia, by the
different size of the lithodomous shells and holes, great and
small being mixed with each other, affording evidence of a
difference of their ages.

In this deposit I have only detected a very large Pecten,
found also in Piedmont, the lithodomous and other shells
before noticed, the tooth of (perhaps) a Saurian, and some
smaller species of Pecten; but no doubt a more extended
search would amply repay the geologist.

Near the Fontaine du Temple are some gray marls, resting
on the above, which probably constitute the base of the blue
or gray marly clay, which next succeeds in the order of superposition. This clay contains a great abundance of marine

remains, which have been enumerated by M. Risso*, and of
which many are identical with those noticed by Brocchi in the
Sub-Apennines. With them vegetable remains are discovered,
but these are rare. There is nothing in this deposit which
does not mark a continuance of a comparative state of repose.
The most delicate shells are well preserved, and all their fine
edges are uninjured. Next follows a very different state of
things, one in which pebbles of the Alps have been rounded
by attrition and conveyed by the force of water over the deposits that have been proceeding so quietly. This force has
often torn up the superficies of the clay beds, as must necessarily happen, whether the currents of water thus produced
be considered as the currents of rivers or those of the sea;
for the force or velocity of water capable of transporting pebbles must necessarily be too great to permit clay or marl to
remain at rest; it consequently must cut it up, and leave the
surface uneven, producing an irregular mixture of clay, gravel,
and sand at the line of junction. Now this is precisely what
it has done, as may be seen by the following section, which is
not uncommon in the valleys formed in the supracretaceous
deposit near Nice, and which only exhibits the unconformable
character of the two rocks, it being almost superfluous to adduce examples of the mixture.

Section of the Valley of La Madelaine.   Fig. 37.
c c, bed of the torrent; *a,* blue marly clay;
b b, beds of rolled Alpine pebbles. This *b*
gravel and sand deposit is of very considerable thickness, and dips gently seaward, *a*
sloping up to the hills. It spreads out like
a fan, the point or centre of the radii being *c*
inwards towards the mountains. This form will not help us
in determining whether the deposit was successive during a
long series of years, by means of a river, or was more sudden,
and caused by more violent rushes of water. Be this as it
may, it is quite clear that the causes which have operated in
this district have not been always the same. A period of
comparative repose has been succeeded by one of somewhat
considerable motion; and if the whole were considered as
derived from river detritus, we must suppose that this river
was at first by no means rapid, and afterwards acquired considerable velocity; that it continued a quiet river for a considerable period, after which it became a rapid current, no
longer transporting mere argillaceous and calcareous particles,
but sand and pebbles. The only mode of reconciling these
appearances with the river hypothesis seems to be the suppo-

* Hist. Nat. de l'Europe Méridionale.

sition, that originally, up to and including the period of the clay with its shells, the river was one with a small current, and that the silt was deposited at a distance from the shore;—that the relative levels of sea and land, owing to the elevation of the latter, were somewhat suddenly changed, and that the river-course was lengthened, and the velocity of the current, from the increased declivity of the bed, became sufficient to transport pebbles over the clay*.

Whether we admit this hypothesis, or that of a more sudden rush of waters, a considerable rise of land would seem requisite, as also that the force was exerted between the deposit of the clay and that of the pebbles. If we suppose a sudden rise of land, causing a difference of levels, to the height required, probably a thousand feet and more, the body of waters in the vicinity would be thrown into motion; the waves being in proportion to the disturbing force, and the upraised and fractured land being exposed to all its violence, rounded pebbles would be formed in abundance, and the superficies of the clay washed into inequalities.

It might be considered from a glance at the Maritime Alps, that the clay and the pebbles alternated, and that these alternations merely showed a deposit of one kind at one time, and of another deposit at another; and certainly there are places where they do seem to alternate to a certain extent, particularly at the line of junction. This occurs at Vintimiglia, where the alternating clays contain organic remains; but, nevertheless, the base of the deposit at that place is clay, many hundred feet deep (beneath the Castel d' Appio), and the top is a mass of pebbles. So that, under either hypothesis, we are compelled to admit a great change in the velocities of water passing over the same situation, one from slow to rapid; and it seems difficult to explain this on any other principle than a change, more or less sudden, in the relative levels of the sea and land.

This superposition of gravel, in which the rolled fragments are sometimes by no means small, showing a considerable change in the velocity with which water has passed over the same country, is not confined to the environs of Nice and Vintimiglia, but is to be noticed in other situations between these places and Genoa, and extends on the other side of the gulf into other parts of Italy. The clay is not always present, the causes that produced it not having acted; but I have here and there observed fragments of rock beneath the mass of

* It should be observed, that in certain situations the marl becomes arenaceous at top, changing into a sand; seeming to show that the transporting power had increased more gradually in some situations than in others.

sand and pebbles, which, by their angularity, position, and occasional mixture with unbroken fossils, seem to show that they have not participated in the transport of rolled pebbles. If we enter the body of Italy, and continue towards Florence and Rome, we find a series of sands, marls or clays, which contain many of the organic remains of the Nice rocks, and were probably contemporaneous with them; and we may here also observe a change in the velocity of the water which has deposited the different substances. Thus between Sienna and Florence we shall observe a succession of clay or marl, sand and pebbles, the latter particularly abundant on the approach to Florence, and apparently constituting the upper beds. It would therefore appear that the phænomena noticed near Nice are not altogether local, though they may be modified by local causes, but somewhat general. Indeed the structure of many rocks on the other, or Adriatic side of the Apennines, shows that they merely form a part of some great whole, if we look at their mode of deposit, even independent of organic remains, which are found closely to agree. It would no doubt be easy to state generally certain facts that may be observed in the great gulf of supracretaceous rocks which extends into the northern part of Italy, between the Apennines and Alps, and thus to present an appearance of knowledge, and an intimate acquaintance with the whole mass. The more, however, I have looked into parts of this mass, the more I am convinced that our knowledge of those data, that we ought to possess before we generalize, is imperfect. Certainly the Sub-Apennine marls and sands preserve a general character down the whole range of the Apennines into the Adriatic, and from the abundance and nature of their fossils have attracted considerable attention; but their various connexions with other rocks, more particularly with those beneath them, and these again with others, yet requires much attention, judging at least from published documents. If the geologist would make a careful section from Rimini to Foligno, on the road to Rome, over the Apennines, he would find much to reward his labours; or if, instead of pursuing the high road, he were to keep the coast from Ancona, and observe the various rocks as they successively plunge into the Adriatic, and thus avail himself of coast sections, he would be rendering good service to science. He would find the white limestone of the main chain contorted and twisted in every direction, and many of those rocks which rest upon it not quite so quietly arranged upon it as, theoretically, they ought to be. He would also observe some curious instances of denudation in the more modern rocks, producing numerous isolated and steep hills, crowned by towns and villages, the picturesque arrangement of which, if he be

also an admirer of beautiful scenery, will add not a little to
the pleasures of his journey.

In the basin between the Jura and Alps, in Switzerland and
thence into Austria, there are immense accumulations of rolled
pebbles and sands, known, generally, by the names of Nagelfluhe and Molasse, the whole composed of Alpine detritus,
and entombing terrestrial, fresh-water, and marine remains.
Various artificial divisions have been made in this mass, and
parts of it have been considered equivalent to deposits in the
Paris basin, *i. e.* of contemporaneous formation with them.
M. Studer, who has examined this mass in Switzerland with
considerable attention*, agrees with M. Brongniart in referring the molasse to an epoch posterior to the gypseous deposit
of the Paris basin. To whatever relative age parts of this
mass may be referred, the mineralogical character of the accumulation would seem to show that it was, as a mass, produced by nearly similar causes, such as effected a degradation
of, and a transport from, the Alps.

The pebbles are generally of that magnitude which it would
require water moving with considerable velocity to transport.
We therefore should inquire what current or currents would
be able to produce the effects required. If we can obtain a
probable explanation of the maximum effects, we may perhaps
search for the minor effects in a less intense exertion of the
same forces. M. Studer considers that there is evidence of
the more recent beds being furthest from the Alps, and nearest
to the Jura; this is precisely what would be expected either
by the hypothesis of the continued action of meteoric causes,
or by that of a series of debacles from the Alps. If rivers
have effected the transport of the pebbles, they must, from
the size of the pebbles, have had considerable velocity. We
should expect the rivers to push forward their detritus into
the great basin between the Alps and the Jura: but being
once freed from the high mountains and their rocky channels,
they would endeavour, as all rivers do when not cut off by
rapid tides and currents, to produce deltas, and these might
at first cause mixtures of gravel, sand and clay; but the more
they were advanced, the greater would be their tendency to a
horizontal arrangement, and, consequently, the less the velocity of the current, and the smaller the transporting power.
Therefore the same river which could once carry large pebbles to the sea would after a lapse of time be unable to do so,
unless an elevation of the mountains, whence it flowed, should
cause a new system of levels, and the river thus acquire increased velocity, transporting pebbles over the ground, which
it formerly only covered with silt or sand. The question then

* Monographie der Molasse: Berne, 1825.

arises, Will the height of the Alps, compared with the distance from these mountains at which large pebbles are found, permit us to consider the transport of these pebbles possible by rivers? In answering this, we must be careful to exclude those superficial gravels scattered over the lower lands, and down the great valley of the Rhine, the transport of which it seems difficult to conceive, except by means of water moving with a greater velocity, and in a greater body, than any river flowing from the Alps could possess. We should only consider those sand- and pebble-beds which constituted the hills on the outskirts of the Alps before they were denuded as we now see them. To do this fairly would require some exceedingly delicate calculations; and we should remember that the warmer the climate the higher the line of perpetual snow, and consequently the greater would be the fall of the running waters. On the river theory, we shall also have to account for the extraordinary equalization of the Alpine pebble-beds, and their general resemblance throughout so long a line of country,—a somewhat difficult task; for if rivers formed the mass, each river would transport its own detritus and push this forward; and though their various deltas might ultimately meet, there would be no stratification common to the whole mass, but one peculiar to each delta. The older or first transported Alpine detritus, marking the commencement of this great degradation of the Alps, rests remarkably even, over considerable spaces, on the rocks beneath them, which is scarcely consistent with their delta or river formation. That these latter now form as much a part of the great transverse valleys as any rock beneath, rising to the height of several thousand feet, is no objection to the river hypothesis ; for the causes which upheaved the Alps would upheave these beds with the rest, and they would be traversed by the transverse cracks equally with the lower rocks.

Upon the hypothesis that the pebbles and sands have, in a great measure, been transported from the Alps by debacles, caused by movements in the Alps themselves, which produced corresponding agitation in the seas that bathed their sides, it is not required that these mountains should have been so lofty as seems necessary under the river hypothesis; and the whirling of the waters and currents produced, might equalize the detritus in beds,—not only that detritus which might be broken away during a convulsion, but all that previously formed by the rivers, and on the beaches and deltas, which would give way before the force employed.

While on this subject, let us for a moment consider the Swiss lakes, which occur precisely where they should not, if rivers are to be considered the only excavating forces. The

lake of Constance is contained in the rocks under consideration; the lake of Geneva partly in them, and partly out of
them in older rocks; the lake of Lucerne the same; and the
lake of Neufchatel, with one of its sides bounded by the Jura,
and the other by the molasse and nagelfluhe. No supposition
of river excavation can meet these cases; for the moment the
velocity ceases, then will the excavating power cease with it;
and we cannot conceive a river cutting out a deep basin
bounded on all sides by equal levels, the drainage of which is
nearly on a level with the entrance of the river into the basin:
but under the hypothesis of a mass of waters thrown into
agitation, the difficulty does not appear to be great; for amid
the various whirls and great eddies of water, inequalities of
all kinds must be formed; and although the depressions may
appear to us considerable, they are, when compared with the
general superficies of land, trifling. If we suppose a body of
waters suddenly poured out of the great transverse valleys of
the Alps, it would have a tendency to cut up the ground
where first discharged upon the low lands, before it had lost
its great velocity. I admit that this supposition does not account for all the difficulties; indeed the present remarks are
merely made to call attention to the subject, for the lake of
Constance is not close to the valley. The position of the lake
of Neufchatel is, however, not inconsistent with the idea of a
mass of water striking the sides of the Jura. The lake is unequally excavated; and during some soundings which I once
made upon it, I found a hill in the middle, but a few fathoms
beneath the surface, and with a steep escarpment on one side*.
These remarks on the lakes amid the nagelfluhe and molasse
have been introduced merely to show that other excavating
forces than those of rivers would seem necessary to explain
some phænomena now observable in this district; and that if
such forces have once acted, there does not appear any reason,
from the nature of the country generally, that they may not
have acted at other times.

In many parts of the mass there would appear evidence of
a quiet deposit, as, for instance, the deposits of lignite, such
as those of Kæpfnach, which contain the remains of the *Mastodon angustidens*, a Rhinoceros, and a *Castor*. One of the
plants is noticed under the name of *Endogenites bacillaris*.
Other lignites occur at Lausanne, Vevay, Ugg, &c., and occur
in the lower part of the molasse; *Flabellaria Schlotheimii* being,
according to Brongniart, found in that of Lausanne. The remains of the *Palæotherium* have also been discovered in the

---

* This may be a portion of the more solid rock of the Jura, close to it,
which, being harder, better resisted the excavating action than the more
easily removed sands and pebbles.

molasse used for building, near the lake of Zurich. These remains would appear to point to a period when a part of this deposit was forming quietly, and, if fresh-water remains be alone mixed with them, as is stated, by means of fresh water.

The upper parts of these rocks appear, however, more decidedly to mark the presence of the sea; for they contain marine remains, such as *Turritella imbricataria,* Lam., *T. Terebra,* Broc., *T. triplicata,* Broc., *T. subangulata,* Broc., *Natica glaucina,* Lam., *Mitra mitræformis,* Broc., *Cancellaria cassidea,* Broc., *Buccinum corrugatum,* Broc., *Cerithium Lima,* Brug., *C. quadrisulcatum,* Lam., *Murex rugosus,* Sow., *M. minax, Pyrula ficoides (Bulla ficoides,* Broc.), *Ostrea virginica,* Lam., *O. edulina,* Sow., *Pecten latissimus,* Broc., *P. medius,* Studer, *Meleagrina margaritacea,* Studer, *Arca antiquata,* Lam., *Cardium edulinum,* Sow., *C. oblongum,* Broc., *C. semigranulatum,* Sow., *C. hians,* Broc., *C. clodiense,* Broc., *C. multicostatum,* Broc., *Tellina tumida,* Broc., *Venus Islandica,* Lam., *Venus rustica,* Sow., *Astarte excavata,* Sow., *Cytherea convexa,* Brong., *Corbula Gallica,* Lam., *Panopæa Faujasii, Solen Vagina,* Lam., *S. strigilatus,* Lam. (analogue now living), *S. Legumen,* Linnæus, *Balanus perforatus,* Studer*.

Prof. Sedgwick and Mr. Murchison, in describing the continuation of these rocks on the flanks of the Salzburgh and Bavarian Alps, mention great alternating masses of conglomerate, sandstone and marl, north of Gmunden; and still further north, in the higher part of the series, beds of lignite. Detailing the section of the Nesselwang, they state that the lowest supracretaceous or tertiary strata are of great thickness, and are applied vertically against the Alps. The conglomerates are mentioned as extremely abundant, the molasse and marl being entirely subordinate to them. According to these authors, there are three or four distinct lines of lignite, separated from each other by thick sedimentary deposits. Hence they infer that the presence of lignites alone is unimportant, as these occur in very different situations. In a section taken through the hills at the east end of the lake of Constance, the lower part of the supracretaceous or tertiary system is described as composed of green micaceous sandstone, in which beds of conglomerate are subordinate, and it is considered identical with the molasse of Switzerland. The conglomerates alternating with greenish sandstone and variously coloured marls are noticed as forming the upper supracretaceous group, and composing the mass of the mountain ridge extending northwards from Bregenz. Supracretaceous rocks are noticed in the valley of the Inn, containing coal, worked for profitable purposes, thirty-four feet thick, near Haring. The

---

* Brongniart, Tableau des Terrains qui composent l'Ecorce du Globe.

coal is described as accompanied by fetid marls variously indurated. In the coal and overlying beds there are many terrestrial and fluviatile shells, and also in the latter beds numerous impressions of dicotyledonous and other plants. Several
marine shells are discovered in these strata. The authors consider that the various sections which they observed, prove the
comparatively recent elevation of the neighbouring Alpine
chain; and the more recent supracretaceous deposits noticed
by them, bear the same relation to the neighbouring Alps as
the Sub-Alpine rocks in Northern Italy do to the high mountains near them; whence they infer that the northern and
western basins of the Danube, and the supracretaceous basin
of the Sub-Alpine and Sub-Apennine regions, have been left
dry at the same period *.

According to Prof. Sedgwick and Mr. Murchison, the supracretaceous rocks of Lower Styria consist, in the ascending
series of a section from Eibeswald to Radkersburg,—1. Of
micaceous sandstones, grits and conglomerates, derived from
the slaty rocks on which they now rest at a highly inclined
angle. 2. Of shale and sandstone with coal. At Scheineck,
where the coal is extensively worked, it contains bones of
Anthracotheria, and in the shale *Gyrogonites* (*Chara tuberculata* of the Isle of Wight), flattened stems of arundinaceous
plants, *Cypris, Paludinæ*, fish-scales, &c. 3. Of blue marly
shale and sand. 4. Of conglomerate, with micaceo-calcareous
sand and millstone conglomerate, occupying the whole hilly
region of the Sausal. 5. Of coralline limestone and marl.
The organic contents of this rock are stated to be,—many
corals of the genera *Astrea* and *Flustra ; Crustacea ; Balanus
crassus, Conus Aldrovandi, Pecten infumatus, Pholas, Fistulana*,
&c. The authors refer this rock to the epoch of the Sub-
Apennine formations and English crag. 6. Of white and
blue marl, calcareous grit, white marlstone, and concretionary
white limestone. At Santa Egida, concretionary white limestone, alternating with marls, contains *Pecten pleuronectes,
Ostrea bellovicina, Scalaria, Cypræa*, &c. 7. Of calcareous
sands and pebble beds, calcareous grits and oolitic limestone.
At Radkersburg, where the hills sink into the plains of Hungary, the strata are charged with shells, some being identical
with living species (*Mactra carinata* and *Cerithium vulgatum*).
The authors consider this group as similar to the more recent
rocks of the Vienna basin.

In describing another section, Prof. Sedgwick and Mr.
Murchison notice that, at the Poppendorf, the marls, sands

* Sedgwick and Murchison, Proceedings of the Geol. Soc. of London,
Dec. 4, 1829.

and conglomerates, are crowned by a micaceo-calcareous sand, containing concretionary masses of a perfect oolite, affording a good example, if any were wanting, of the trifling value of mineralogical character in determining rocks far distant from each other *.

Let us now proceed to those parts of the South of France which border the Mediterranean, observing that M. Elie de Beaumont, when remarking on the period at which he considers the Alps to have been thrown up in a direction between Marseille and Zurich, notices numerous situations where the newer supracretaceous strata are characterized by the remains of Oysters, Polypifers, *Patellæ*, the *Balanus crassus* (fig. 38), (which M. Deshayes considers may only be a variety of *Balanus Tulipa*), *Patella conica*, and other shells. He also identifies these rocks in Provence, Dauphiné, and Switzerland. In the molasse of Pont du Beauvoisin, M. Elie de Beaumont discovered shells which M. Deshayes recognised to be *Balanus crassus*, *Patella conica*, and a *Pecten* partaking of the characters of *P. Beudanti*, *P. Jacobæus*, and *P. flabelliformis†*.

Fig. 38.

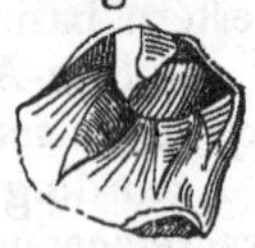

According to M. Marcel de Serres, the marine supracretaceous rocks of the South of France rest on each other in the following descending order:—1. Sands, generally yellow or white, and more or less argillaceous, calcareous, or siliceous, according to circumstances. These sands abound in the remains of terrestrial and marine mammalia, reptiles, and fish, mixed with the remains of birds, and some wood. Shells are not common, with the exception of *Ostreæ* and *Balani*. 2. Yellow and calcareous marls, of no great thickness, sometimes alternating with stony beds. 3. Beds of limestone, to which the same author has given the name of *calcaire moellon*, usually worked as a building-stone in the South of France. The upper beds generally contain the greater quantity of shells; these and the middle strata also contain the remains of mammalia, fish, crustacea, annulata, and zoophytes. Terrestrial mammalia are very rare, consisting principally of a few bones and isolated teeth, which mostly approach those of the *Palæotherium* and *Lophiodon*. The lower beds contain but few shells. 4. Argillaceous blue marls, well known as the blue Sub-Apennine marls. These marls vary much in their mineralogical character, being more or less calcareous, argillaceous, or sandy, according to circumstances. They have nearly the

* Sedgwick and Murchison, Proceedings of the Geol. Soc. of London, March 5, 1830.

† Elie de Beaumont, Rév. de la Surf. du Globe;—Ann. des Sci. Nat. 1829 et 1830.

same colour, passing from a greenish or blueish gray into a
blue of greater or less intensity. Their thickness seems to
depend on the inequalities of the surface on which they rest,
their depth being sometimes very considerable, while at others
it is trifling. They contain a large collection of marine remains, principally shells. Terrestrial mammalia and reptiles
are exceedingly rare. M. Marcel de Serres only mentions
one stag's horn, the bones of a land tortoise, and the vertebræ
of a crocodile. Marine mammalia and fish are scarce, as are
also the remains of zoophytes*.

The following section, by M. Marcel de Serres, of the strata
of Banyuls, through which the Tech has cut its bed, will remind the geologist of sections to be seen at Nice, and in various
parts of Italy; 1. (upper bed.) Transported substances, named
by the author *diluvium of the plains*, rolled pebbles of primary
rocks, cemented by a brownish red gravelly clay; thickness
from one to three yards. 2. Another deposit of transported
detritus, named *mountain diluvium* by the author, stated to be
distinctly separated from the above, composed of rolled pieces
of granite, mica-slate, gneiss, and quartz, cemented by a
slightly red clay, more gravelly than the first. The size of
the rolled fragments is considerable, the smallest being equal
to that of the head; thickness, two to three yards. 3. Yellowish siliceous sands, indurated in parts, the beds thick,
varying from four to six yards. Lower portion contains
shells and lignites. 4. Argillo-arenaceous marls, blueish gray,
and micaceous; sometimes alternating with the upper yellow
sands. Shells very abundant; thickness, six to eight yards.
5. Blueish argillaceous and tenacious marls. They contain
few shells, and even these become less abundant as the section
increases in depth; thickness not known. These marls are
supposed to rest upon micaceous clay-slates, from the structure of the Albères chain, at the foot of which these beds of
Banyuls dels Aspre are found. Nos. 3. and 4. are stated to
contain the remains of mastodons, deer, lamantins, land-tor-
toises, and sharks, disseminated among the marine shells, but
they are represented to be scarce†.

There are many lignite deposits in this part of France, of
which the relative ages have not been determined so accurately
as could be wished. M. Marcel de Serres, however, shows
that some of them are inferior to the calcaire moellon, and
probably occur at the lower part of the blue marls. The following is a section at Saint Paulet, about a league and a half

---

* The organic exuviæ discovered in these marls are enumerated in the
lists at the end of the volume.

† Marcel de Serres, Géognosie des Terrains Tertiaires du Midi de la
France. Montpellier, 1829.

from Saint Esprit (order descending): 1. Yellowish calcareo-
siliceous sands, containing the remains of marine shells. 2.
Thick beds of the calcaire moellon, containing numerous casts
of *Cytherea, Venus,* and *Cerithia.* 3. Sands with marine shells
resembling No. 1. 4. Alternation of fresh-water limestone
(containing *Gyrogonites*), earthy lignite, and sandy marls.
5. Compact limestone, with *Cerithia* or *Potamides* and *Paludinæ.* 6. Thin argillaceous marls, with small oysters. 7. Thin
earthy lignite. 8. Argillo-arenaceous marls, with traces of lignite. 9. Compact fresh-water limestone, with *Limnææ* and
Cyrenæ. 10. Thin yellowish and calcareous marls. 11. Argillaceous blue marls, with traces of more or less fibrous lignite. 12. Argillo-bituminous marls, containing numerous marine and fluviatile shells. These marls, as well as the lignite
which succeeds them, contain small pieces of amber. 13. Lignite in beds of two or three yards in thickness, preserving the
woody structure, even resembling charcoal : contains amber.
14. Argillo-bituminous marls, with marine and fluviatile shells,
the same as No. 12. 15. Lignite with the same characters as
No. 13.—All these beds rest parallel on each other with great
regularity, and show that they have been deposited tranquilly
and successively*.

Many species contained in the rocks above noticed are analogous with those now existing in the Mediterranean, pointing
to some kind of connexion between the ancient state of that
sea and the present. We therefore seem to arrive at something like a probability that the blue marls were deposited in
a sea, perhaps somewhat similar to the Mediterranean, but
presenting more surface than it.

From the recent observations of Colonel Silvertop it would
appear, that the Sub-Apennine deposits are also discovered at
Malaga, and other parts of the South of Spain, the blue marls
occupying the same relative position†.

M. de la Marmora shows us that the supracretaceous deposits of Sardinia correspond with those of the South of France,
of the South of Spain, and of a large part of Italy. The following is his account of their superposition (in the descending
order):—1. A fine-grained white, or yellowish white, limestone; 2. A yellow and very earthy *calcaire moellon,* mixed
with sand; 3. Calcareous, sandy, and siliceous strata; 4. Blue
marls, sometimes whitish; 5. Some very rare strata of calcareous conglomerates, with traces of lignite, or else trachytic
tuffa cemented by carbonate of lime. No. 5. is rare. The
characteristic shells of the blue marls are stated to be *Pecten*

* According to M. Dufrénoy, these beds rest unconformably on strata
equivalent to the green sand ; Annales des Mines, 1830, pl. v.
† Silvertop, MSS.

pleuronectes and *Venus rugosa.* They likewise contain numerous remains of crabs, but univalves are described as rarely
found*.

The remains of large mammalia, which have rendered the
Upper Val d'Arno so celebrated, would appear to be discovered in beds of somewhat contemporaneous origin; a difference in the circumstances attending the deposit of the superior
rocks having produced a difference of the remains detected in
them, inasmuch as marine exuviæ are absent.

M. Bertrand-Geslin distinguishes three basins between
the source of the Arno, and Florence; namely, the basins
of Casentino, Arrezzo, and Figline; the whole valley of the
Arno, for that distance, being bounded by a sandstone named
macigno, or by dark-coloured limestone. According to the
same author, the following section (in the descending order)
may be observed between Arrezzo and Incisa:—1. Thick bed
of yellow argillaceous sand. 2. Thick beds of rolled quartzose
pebbles, intermixed with coarse sand. 3. Fine gray and micaceous sands, many fathoms thick, containing thin beds of
blue sandy marl; these sands being, in the middle and lower
parts, exceedingly rich in the bones of mammiferous animals.
4. Very thick argillaceous blue marl, constituting the lowest
deposit in the basin, and containing many fossils in its upper
part.

From his various observations on the Val d'Arno, M. Bertrand-Geslin concludes;—1. That the rolled pebbles are
larger and more abundant in proportion as they approach the
mountain chain on the north, whence they appear to have
been derived: 2. That the coarse sands occupy the central
part of the valley, while the finest sands skirt the foot of the
mountain range on the south: 3. That the lower sands and
blue marls are deposited in horizontal beds: 4. That the
bones of mammalia are very abundant towards the central
part of the Val, on the right bank of the Arno, and are rare
on the left bank: 5. That these bones, in good condition, and
sometimes disseminated, are generally deposited in different
planes, as if not all at one time: 6. That the yellow sands
contain fluviatile shells at Monte Carlo; and 7. That this
transported mass contains neither the remains of marine shells,
solid stony beds, nor lignites†.

The animals whose remains are stated to have been discovered in the Upper Val d'Arno are:—*Elephas primigenius,*
Hippopotamus major, Rhinoceros, Tapir, Deer, Horse, Ox,
Hyæna, *Felis,* Bear, Cavern Fox, and Porcupine. The presence of these remains would appear to indicate that the de-

* De la Marmora, Journal de Géologie, t. iii. p. 319.
† Ann. des Sci. Nat. t. xiv. 1828.

posit containing them was not far removed, as to date, from
the transported gravels and sands, mingled with volcanic substances, in Auvergne, and which will be noticed in the sequel.

During this state of comparative repose, in which similar
mineralogical substances enveloped similar animal remains
over a considerable surface, there were some situations in
which vegetable matter was more abundantly collected than
in others, as might now happen at the embouchures of rivers
when the streams possessed no great velocity. After the production of the blue marl, circumstances became somewhat
altered, and this over a considerable surface,—for the deposit
no longer continued the same; sands, showing a greater velocity
or transporting power of water, commonly covering these blue
marls in the South of France and Italy. There were, however, modifying circumstances; for sheets of calcareous matter,
frequently producing limestones, occur mixed with these sands,
enveloping terrestrial, fresh-water, or marine remains, as these
came within their influence.

M. Elie de Beaumont notices the following section near the
Pertuis de Mirabeau; which, while it shows that the rocks
belonging to the cretaceous and oolitic groups of that neighbourhood were disturbed and contorted, previous to the deposit of the supracretaceous rocks which rest upon them, also
exhibits the superposition of certain supracretaceous strata of
that part of France with which we have been occupied, and
which, in the neighbourhood of Aix, presents such a curious
approach, in their organic contents, to some of the terrestrial
inhabitants of the present country.

Fig. 39.

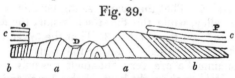

a a, rocks of the oolitic group: *b b*, rocks of the cretaceous
group, containing Ammonites and *Belemnites mucronatus.*
D, bed of the Durance at the Pertuis de Mirabeau, on both
sides of which rest nearly horizontal beds of supracretaceous
rocks, *c c*, on the upturned edges of the older strata.

On the side P, that of Peyrolles, the supracretaceous rocks
constitute a thick fresh-water deposit, " principally composed
of gray compact limestone, penetrated by numerous irregular
tubular cavities, and of sandstone, analogous to that which
near Aix alternates with the variegated marls of the fresh-
water series *." On the other side of the Durance, and near

* Elie de Beaumont, Rév. de la Surf. du Globe: Ann. des Sci. Nat.
1829 et 1830.

the chapel of La Magdelaine, o, the supracretaceous rocks are
seen resting on the edges of the older strata, and the following
beds are observed, in the ascending order:—1. A calcareous
sandstone, without shells, in some strata containing calcareous
pebbles, and passing into a conglomerate. 2. The above beds,
with the remains of marine shells. In these beds M. Elie de
Beaumont observed dolomite. 3. A bed containing some
limestone pebbles, and a great number of oysters, their hinges
elongated, among which are probably the *Ostrea virginica* of
the shelly molasse of Piolene and Narbonne; also other shells,
among which M. Deshayes recognised *Anomia ephippium,
Balanus crassus,* and an undescribed Pecten, resembling the
P. Jacobæus, P. Beudanti, and *P. flabelliformis.* 4. A considerable thickness of molasse, not very shelly, in one bed of
which there are vegetable remains. 5. An oyster-bed, analogous to No. 3, covered by a certain thickness of shelly molasse. 6. A thickness of three yards of a yellow sand, covering
an alternation of calcareous sandstone, and a compact blueish
gray limestone, with irregular tubular cavities, containing terrestrial and fresh-water shells. M. Elie de Beaumont does
not consider this limestone as the same as that noticed on the
other side of the Durance, but as forming the upper part of
the supracretaceous series at this place; while the beds near
Peyrolles constitute the lower part of the same series.

The exact relations of these rocks with the fresh-water deposit at Aix, remarkable for the insects found entombed in
part of it, do not appear to have been yet well determined.
According to Messrs. Lyell and Murchison, the following is
a section of the beds rising above the level of the town of Aix
(in the descending order):—1. White calcareous marls and
marlstone, passing gradually into a calcareo-siliceous grit,
containing *Cyclas gibbosa,* Sow.; *Potamides Lamarckii, Bulimus pygmæus,* and an undescribed species of *Cypris;* thickness
about 150 feet. 2. Marls, with plants and shells. 3. Marls,
with fish and plants. 4. Bed with insects, with occasionally
Potamides and plants. This bed is described as a brownish
green, or light gray calcareous marl, composed of very thin
laminæ. 5. Gypsum, with plants. 6. Marls. 7. Gypsum, with
fish and plants. 8. Marls, with traces of gypsum. 9. Pink
limestone, containing *Potamides, Cyclas gibbosa,* Sow., and
Cyclas Aquæ Sextiæ, Sow. This limestone is often highly
contorted, and passes either into a calcareous grit or red
sandstone, and, still lower, into compact calcareous breccia;
the whole is based on a coarse conglomerate. The lower beds
dip N.N.E. at about 25° or 30°. From the section accompanying the memoir of Messrs. Lyell and Murchison, it would
appear that these conglomerates rest, beyond Aix, on red marl,

fibrous gypsum, and gray limestone, with *Limnææ* and *Pla-norbes*; and these again on the compact limestone, sand, and shale, containing coal at Fuveau, accompanied by the remains of an *Unio, Melania scalaris,* Sow., *Cyclas concinna,* Sow., *C. cuneata,* Sow., and *Gyrogonites**.

The preservation of the insects is very great, permitting the determination of genera and species. According to M. Marcel de Serres, *Arachnides* accompany the insects, properly so called; the latter, however, being far more abundant than the former, two or three genera only of *Arachnides* having been determined, while sixty-two genera of insects have been observed. The most curious circumstance attending these remains is, that some are considered identical with those now existing in the country; *Brachycercus undatus, Acheta campestris, Forficula parallela,* and *Pentotoma grisea,* being, according to M. Marcel de Serres, the more remarkable. It is also worthy of observation, that the greater part of the insects are of those kinds which generally inhabit arid and dry places. Although they occur in various positions, they are sometimes spread out, as if by an entomologist for the purpose of displaying their wings. Their colour is generally an uniform tint of brown or black. Some of the fish discovered in the same marls are so small that they do not exceed ten or eleven millimetres in length†.

The place in the series of the supracretaceous rocks to which the brown coal formation of Germany should be referred, does not appear to be as yet well determined. This deposit is characterized by an immense quantity of vegetable remains, and is probably of different ages. The brown coal may be traced from the environs of Aix-la-Chapelle to the Rhine. It there occurs in a narrow plateau between the latter and the Erft, and acquires a thickness of above 100 feet between Bonn and Cologne, without any extraneous bed. The brown coal deposit rests on the declivity of the grauwacke mountains on the right bank of the Rhine, and is connected with the trachytic conglomerates and basaltic formations of the Siebengebirge. It rises on the plateau of the grauwacke mountains further south in the vicinity of Linz, (Orsberg, Mendenberg,) and spreads in detached portions to the Westerwald, where it is greatly extended, and is interrupted and covered by basalt. The deposit extends to the south side of the grauwacke mountains into the Wetterau.

---

* Lyell and Murchison, Edin. New Phil. Journal, 1829.

† Marcel de Serres, Géog. des Ter. Tertiaires du Midi de la France, in which some of the insects are figured; as also in the Memoir of Messrs. Lyell and Murchison above noticed, in illustration of the remarks of Curtis on the specimens brought to England.

Further separated, but still with similar characters, it occurs in the Habichtswalde near Cassel, at the Meissner with basalt, the latter being more recent than it. It again occurs, not far distant, in the basin of Thuringia, the beds being of considerable thickness at Artern in the Unstrutthal. At Dürrenberg and Halle, where the brown coal (near Langenbogen) occurs in beds nearly fifty feet thick, it is broken into separate basins.

The brown coal deposit extends to the country around Leipsic, to the Elbe as far as Torgau, and occurs frequently in the low tracts between Magdeburg and the Hartz. It is found in the level country between the Elbe and the Oder. It is discovered at Bockup near Domitz (Mecklenburg), and near Buckow, Freienwalde on the Oder, and to the east at Zielenzig and Gleissin.

The brown coal generally rests upon a compact tenacious clay, and is covered by large masses of sand. The organic remains which abound in the superincumbent rocks in Hesse and Magdeburg, have not been accurately determined; those which occur in the inferior strata of the Rhine are better known. They consist of—Pisces; *Cyprinus papyraceus*, Bronn (Geistingerbusch near Siegburg); an undetermined species from Mendenberg and Friesdorf near Bonn.—Reptilia; *Rana diluviana*, Goldf.; *Salamandra ogygia; Triton Noachinus; Ophis dubius*, from Orsberg.—Crustacea; a small crab from Geistingerbusch.—Insecta; species of the genera *Lucanus, Meloe, Dytiscus, Buprestris, Cantharis, Cerambyx, Parandra, Belostoma, Cercopis, Locusta, Anthrax*, and *Tabanus*, all from Orsberg.—Plantæ; seeds of *Ervum hirsutum*, and *E. tetraspermum*, from Geistingerbusch; reed and dicotyledonous leaves, resembling those still growing in the country, but not specifically the same. Large trunks of trees are very common *.

A large part of the South of France, bounded by the ocean, or rather by the sandy dunes it has thrown up, between the districts of Bordeaux and Bayonne, and extending far into the interior, particularly at the foot of the Pyrenees, is composed of supracretaceous rocks; an exact and detailed account of whose varied relations to each other may still perhaps be considered as wanting, though much has been done respecting them. This superficies comprises, among other districts, that extensive and monotonous region named the Landes, where the traveller finds little to relieve the sameness which surrounds him, except the peasants stalking over the country mounted on stilts, for the greater convenience of seeing objects afar off.

* German Transl. of Manual.

M. de Basterot has presented us with a very valuable detail of the fossil shells obtained by him from the districts of Bordeaux and Dax, which is inserted in the Lists of Organic Remains, considering that such lists are of the greatest utility to the geological student; referring him, however, to M. de Basterot's memoir for the detailed description of each shell. This author remarks, that out of the 330 species of shells noticed by him in the great sandy deposits of the Landes, forty-five only have existing analogues in the neighbouring seas, comprising the Mediterranean; and he further observes, that if the basin of the Gironde be taken as a centre, the shells in similar supracretaceous basins will the more resemble each other as the distances are less. Thus, out of the 330 species collected in the vicinity of Bordeaux, ninety-one are found in the deposits of Italy, sixty-six in those of the environs of Paris, eighteen in those of Vienna*, and twenty-four in the supracretaceous rocks of England†.

If reference has been made to M. de Basterot's list, it will have been observed that, though many shells found in this part of France are also discovered at Paris, there is likewise a very considerable correspondence between them and those of Italy. It would appear, from the mention of the freshwater limestone at Saucats, that there was a change of the relative level of sea and land in that situation, which permitted the envelopment of fresh-water shells in carbonate of lime; and that after this deposit, a change of level was effected, which enabled marine lithodomous shells to bore extensively into the fresh-water rock, and permitted an accumulation of mineral matter and marine shells above it. The analogues of existing species are forty-five; the living species being remarkable for the diversity of their habitats,—some being found in the Atlantic and Pacific Oceans, and the Indian and Mediterranean Seas, while not a few inhabit the coasts of the Channel and the Bay of Biscay, to which, from the fall of the land, the Bordeaux and Dax deposits seem naturally to belong. When the ocean covered this part of France, it seems necessary to suppose that the mean temperature of the situation was above that which it now is, in order to suit the animals, many of whose analogues exist in warm climates.

We now proceed to give a short notice of the supracretaceous rocks of the Paris basin, as they long constituted the type to which all deposits of this epoch, wherever found, were

---

* M. de Basterot observes, that this number will probably become increased as the Vienna basin shall become better known; which we may expect it soon will be, from the labours of M. Parsch.

† De Basterot, Description Géologique du Basin Tertiaire du Sud-Ouest de la France, 1ère partie; Mém. de la Soc. d Hist. Nat. de Paris, t. ii.

referred. However the rocks of this group may be eventually discovered to differ from this type, the labours of MM. Cuvier and Brongniart on the rocks of the Paris basin will not the less retain that place in the annals of Geology, which by common consent has been assigned them. Nor will the zoological discoveries of Cuvier, constituting as they did such a brilliant epoch in the history of geological science, the less claim the gratitude of geologists in succeeding ages.

The following is the classification of the Paris rocks, according to MM. Cuvier and Brongniart (order ascending):

1. First fresh-water formation . $\begin{cases} \text{Plastic clay.} \\ \text{Lignite.} \\ \text{First sandstone.} \end{cases}$

2. First marine formation. . . . Calcaire grossier.

3. Second fresh-water formation $\begin{cases} \text{Siliceous limestone.} \\ \text{Gypsum, with bones of animals.} \\ \text{Fresh-water marls.} \end{cases}$

4. Second marine formation . . $\begin{cases} \text{Gypseous marine marls.} \\ \text{Upper marine sands and sandstones.} \\ \text{Upper marine marls and limestone.} \end{cases}$

5. Third fresh-water formation . $\begin{cases} \text{Millstone, without shells.} \\ \text{Shelly millstone.} \\ \text{Upper fresh-water marls.} \end{cases}$

Plastic Clay.—So named because it easily receives and preserves the forms given to it, and is used in the potteries. It rests on an unequal surface of chalk beneath, which is hollowed and furrowed in various ways, so as to present hills, valleys, and outstanding knolls, which sometimes have not been covered by the newer and superincumbent rocks; at least, if they have covered them, the strata which did so have been removed by denudation *. This clay is variously coloured, being white, gray, yellow, slate-gray, and red. It differs considerably in thickness, as might be expected from the nature of the surface on which it reposes. Above these beds, to which, strictly speaking, the term "plastic clay" is alone applicable, there is often another clay, separated from the former by a bed of sand ; the latter clay being black, sandy, and sometimes containing organic remains. In it occur lignites, amber, and shells (both fresh-water and marine). It is stated, that in this deposit, considered as a mass, the lower parts do not contain organic remains; that in the central portion the remains are *commonly* those of fresh-water animals; and that in the upper part there is a mixture and even an alternation of marine and fresh-water remains, the latter gradually becoming more scarce, and the former finally prevail-

---

* A breccia of chalk fragments cemented by clay is found at Meudon, separating the chalk and plastic clay.

ing.—The following is a list of the organic remains most commonly found in the plastic clay.

Fresh-water Remains.— PLANORBIS *rotundatus,* Al. Brong.;
P. incertus, Defr.; *P. Punctum,* Defr.; *P. Prevostinus,* Defr.
PHYSA *antiqua,* Defr. LIMNEUS *longiscatus,* Al. Brong.
PALUDINA *virgula,* Defr.; *P.* indistincta, Defr.; *P. uni-
color,* Olivier; *P. Desmarestii,* Prevost; *P. conica,* Prev.;
P. ambigua, Prev.

MELANIA *triticea,* Defr.

MELANOPSIS *buccinoidea,* Poiret; *M. costata,* Olivier.

NERITA *globula,* Defr.; *N. Pisiformis,* Defr.; *N. sobrina,*
Defr.

CYRENA *antiqua,* Defr.; *C. tellinoides,* Defr.; *C. cuneiformis,* Sow.

Marine Shells contained in the mixture of the upper part.—
CERITHIUM *funatum,* Sow.; *C. melanoides,* Sow.; another
Cerithium not determined.

AMPULLARIA *depressa,* Lam.? (var. *minor*); *Ostrea bello-
vaca,* Lam.; *O. incerta,* Defr.

Fossil Vegetables.—*Exogenites ; Phyllites multinervis ; Endogenites echinatus.*

Calcaire grossier.—This, as its name implies, is composed
of a coarse limestone, and is more or less hard, so as to be
employed for architectural purposes. It alternates with argillaceous beds, and is remarkable for the constancy of its character throughout a considerable extent of country. It is often
separated from the plastic clay beneath by a bed of sand. The
organic remains are stated to be generally the same in the corresponding beds, presenting rather marked differences when
the beds are not identical. The inferior beds are very sandy,
often more sandy than calcareous, and almost always contain
green earth, disseminated either in powder or grains, which,
according to the analysis of M. Berthier, appears to be a silicate of iron. These beds are remarkable for the abundance of
their organic contents.—The following is a list of those fossils
which are considered to characterize the different parts of this
deposit.

In the lower beds.—MADREPORA, at least three species.
ASTREA, three species at least.

TURBINOLIA *elliptica,* Al. Brong.; *T. crispa,* Lam.; *T.
sulcata,* Lam.

RETEPORITES *digitalia,* Lam.

LUNULITES *radiata,* Lam.; *L. urceolata,* Lam.

FUNGIA *Guettardi.*

NUMMULITES *lævigata ; N. scabra ; N. numismalis ; N.
rotundata.*

CERITHIUM *giganteum.* LUCINA *lamellosa.*

CARDIUM *porulosum.* VOLUTA *cithara.*
CRASSITELLA *lamellosa.* TURRITELLA *multisulcata.*
OSTREA *flabellula ; O. cymbula.*
In the central beds *.—OVULITES *elongata,* Lam. ; *O. mar-garitula,* Deroissy.
ALVEOLITES *milium,* BOSC. ORBITOLITES *plana.*
TURRITELLA *imbricata.* TEREBELLUM *convolutum.*
CALYPTRÆA *trochiformis.* CARDITA *avicularia.*
PECTUNCULUS *pulvinatus.*
CITHEREA *nitidula; C. elegans.* MILIOLITES. CERITHIUM?
In the upper beds.—MILIOLITES. AMPULLARIA *spirata.*
CERITHIUM *tuberculatum; C. mutabile; C. lapidum; C. petricolum.*
LUCINA *Saxorum.* CARDIUM *Lima.*
CORBULA *anatina ? C. striata* †.
Vegetable Remains, according to M. Ad. Brongniart, in the Calcaire Grossier of Paris :—
NAYADÆ—*Caulinites parisiensis.*
EQUISETACEÆ—*Equisetum brachyodon.*
CONIFERÆ—*Pinus Defrancii.* PALMÆ—*Flabellaria parisiensis.*
MONOCOTYLEDONS, OF UNCERTAIN FAMILY—*Culmites nodosus ; C. ambiguus.*
DICOTYLEDONS, OF UNCERTAIN FAMILY—*Exogenites; Phyllites linearis, Ph. nerioides, Ph. mucronata, Ph. remiformis, Ph. retusa, Ph. spathulata, Ph. lancea* ‡.
Siliceous. Limestone.—A limestone, sometimes white and soft, sometimes gray and compact, penetrated by silica, infiltrated in every direction and at all points. It is often cellular, the cells sometimes large and communicating with each other in all directions, the silica lining their sides with mammillary concretions, or with small transparent quartz crystals.
Osseous Gypsum (Fresh-water), and Marine Marls.—The gypseous rocks consist of an alternation of gypsum and calcareous and argillaceous marls. Above this alternation there are thick marl beds, sometimes calcareous, at others argillaceous. In these latter strata are found abundant remains of *Limnææ* and *Planorbes,* and in their lower parts, palms of considerable size are discovered prostrate. The gypseous strata contain the remarkable remains of extinct mammalia and other animals, which the genius of Cuvier may almost be said to have restored to life. Above these beds, which, from the nature of their organic remains, are considered to have

* Nearly all the well-known fossils from Grignon are found in these beds.
† MM. Cuvier and Brongniart, Desc. Géol. des Envir. de Paris. éd. 1822.
‡ Ad. Brongniart, Prod. d'une Hist. des Veg. Fossiles, 1828.

been deposited in fresh water, there is a succession of marls, considered as deposited in the sea, because they contain marine remains ; the marine and fresh-water systems being separated by calcareous or argillaceous marls, often thick.   The upper marl beds contain numerous remains of oysters, considered to have certainly lived in the places where now entombed, more particularly, as M. Defrance discovered them at Roquencourt attached to rounded pieces of marly limestone, which latter are sometimes pierced by *Pholades.*

Organic Remains in the Gypseous Beds.—MAMMALIA: *Palæotherium magnum,* Cuv. (fig. 40, *a.*) *; *P. medium,* Cuv.;

Fig. 40.

P. crassum, Cuv.; *P. latum,* Cuv.; *P. curtum,* Cuv.; *P. minus,* Cuv. (fig. 40. *b.*); *P. minimum,* Cuv ; *Anoplotherium commune,* Cuv. (fig. 40. *c.*); *A. secundarium,* Cuv.; *A. gracile,* Cuv.; *A. murinum,* Cuv.; *A. obliquum,* Cuv.; *Chœroptamus parisiensis,* Cuv.; *Canis parisiensis,* Cuv.; *Coati ; Didelphis parisiensis,* Cuv.; *Sciurus ;* &c.

BIRDS. REPTILES: *Crocodile ; Trionyx ; Emys.* FISH.

Organic Remains of the Fresh-water Marls.—MAMMALIA: *Palæotherium aurelianense,* Cuv. (Orleans); *Lophiodon major,* Cuv. (Soissons, &c.); *L. minor,* Cuv. (Paris); *L. pygmæus,* Cuv. (Paris).

BIRDS. FISH. SHELLS: *Cyclostoma mumia,* Lam.; *Limnæa longiscata,* Al. Brong.; *L. elongata,* Al. Brong.; *L. acuminata,* Al. Brong.; *L. Ovum,* Al. Brong.; *Planorbis Lens,* Al. Brong.; *Bulimus pusillus,* Brard.

---

* The forms of the animals above represented are such as they are considered to have been by Cuvier, Oss. Foss. t. iii. pl. 66.

In the Marine Marls (Yellow).—Fish bones; Cytherea? convexa; Cytherea? plana; Spirorbes; Cerithium plicatum. *Yellow Marls separated from the above by Green Marls.*— Spears and palates of the Ray; Ampullaria patula? Cerithium plicatum; C. cinctum; Cytherea elegans; C. semisulcata?? Cardium obliquum; Nucula margaritacea. *Calc. Marls, with large Oysters.*—Ostrea hippopus; O. Pseudochama; O. longirostris; O. canalis. *Calc. Marls, with small Oysters.*—Ostrea cochlearia; O. cyathula; O. spatulata; O. linguatula; Balani; Crabs' feet.

Upper Marine Sands and Sandstones.—These are composed of irregular beds of siliceous sandstone and sand, the lower portion without organic remains that can be supposed to have existed in the places where now found, these being broken and very rare. In some situations, where the broken shells are more common, millions of small bodies are discovered, to which M. Lamarck has given the name of *Discorbites.*

These non-fossiliferous sands are in many places covered by a limestone, sandstone, or calcareo-siliceous rock filled with marine shells, of which the following is a list: Oliva mitreola; Fusus? approaching F. longævus; Cerithium cristatum; C. lamellosum; C. mutabile? Solarium; Melania costellata? Melania? another species; Pectunculus pulvinatus; Crassatella compressa? Donax retusa? Cytherea nitidula; C. lævigata; C. elegans? Corbula rugosa; Ostrea flabellula.

Upper Fresh-water Formation.—This rock varies very considerably in its mineralogical character, being sometimes composed of white friable and calcareous marls, at others of different siliceous compounds; among which are the well-known millstones, sometimes without shells, at others charged with Limnææ, Planorbes, Potamides, Helices, Gyrogonites (seeds of the Charæ), and silicified wood.

Organic Remains.—ANIMAL. Cyclostoma elegans antiqua; Potamides Lamarckii; Planorbis rotundatus; P. Cornu; P. Prevostinus; Limneus corneus; L. Fabulum; L. ventricosus; L. inflatus; Bulimus pygmæus; B. Terebra; Pupa Defrancii; Helix Lemani; Helix Demarestina *.

VEGETABLE. Muscites? squamatus; Chara medicaginula; C. helicteres; Nymphæa Arethusæ; Culmites anomalus; Carpolithes thalictroides †.

As has been often remarked, there is evidence in the various organic remains entombed in the strata above noticed, that the space comprised within what is commonly termed the Paris basin, has not always been exposed to the influence of

---

* Cuvier and Brongniart, Desc. Géol. des Env. de Paris.
† Ad. Brongniart, Prod. d'une Hist. des Veg. Fossiles.

the same circumstances since the deposit of the chalk, but that there has been an alternation of three lacustrine or fresh-water deposits, with two which are marine; the former constituting the lower and the upper part of the series. It remains to inquire the probable cause of these variations.

By employing the term *basin* for this collection of supracretaceous rocks, we, as before observed, seem to assume that of which we have no great evidence; the fresh-water deposits may have been, and probably were, effected in basins, but the marine do not require this form. It would seem reasonable to infer that there may have been here, as has been shown to have happened elsewhere, movements in the land, changing its level relatively with the sea. When we regard the mode in which the various deposits are now arranged, we find that, as a mass, they do not repose horizontally on each other; but that, according to MM. Cuvier and Brongniart, there were various inequalities at different times, commencing with those of the chalk, presenting hills and valleys. In various parts of this unequal soil the lignite and plastic clay were deposited, thus to a certain extent filling up some of the inequalities. Upon this the calcaire grossier was formed, following more or less the inequalities of the surface beneath. To the calcaire grossier succeeded a gypseous deposit, showing an absence of the sea, and the presence of fresh water, of unequal depth. Then followed a large deposit of sand covering up the pre-existing inequalities, in the upper part of which sand are numerous marine remains; the whole presenting a vast plain. A new state of things followed; the sea disappeared; and fresh-water remains became entombed[*].

The mechanical and chemical circumstances attending these deposits have also curiously varied. We will not stop to inquire whether the inequalities of the chalk were produced suddenly or slowly, for on this head we possess no very decided evidence; but the deposit of the plastic clay (properly so called) would appear to have been slow, even if the detritus, mechanically suspended, may have resulted from a somewhat violent wash of the inferior rocks. In the sands above this, we have the evidence of a transport by water moving with sufficient velocity to carry sand onwards. This is followed by a deposit, to a certain extent quiet, composed of vegetables and amber derived from them. The nature of the other organic remains mingled with them, at first indicates the presence of fresh-water animals; but finally, some variation in the relative level of the land and sea, apparently occurring gradually rather than suddenly, (for there is no evidence of a rush of waters,)

[*] Cuvier and Brongniart, Env. de Paris.

introduces marine animals, which existed at the same time with many fresh-water animals that have gradually become accustomed to live in the same medium with them. This state of things was destined to disappear, and we have a movement of water sufficient to transport sand. This was succeeded by a calcareous deposition, when carbonate of lime, probably in a great measure derived from the ruin of older rocks, was washed away by water, and deposited over a considerable space. It is obvious, from the structure of these rocks, that the materials of which they consist must have been in a state of fine mechanical division, such as to have required no violent rush of waters for their removal: they probably subsided during a period of tranquillity. After the deposit of the calcaire grossier, the production of calcareous rocks, remarkable for their cellular structure, took place. The origin of these cells is unknown; but they probably arose from the calcareous matter, during the act of subsidence, enveloping foreign matter more soluble or perishable than itself, which has subsequently been removed by the agency of water. It is remarkable that the cavities are now lined by silica in such a manner as scarcely to admit of any other supposition, than that the silica was deposited within the cells from a liquid in which it had been previously dissolved.

The osseous gypsum presents us with a decidedly new state of things. Singular animals, of which the very genera are now extinct, must have existed somewhere in the district, the remains of which became in some manner entangled in sulphate of lime, considerable deposits of which were then in progress. The question will arise, Whence did such a quantity of sulphate of lime proceed? Certainly it is a new ingredient, at least in any abundance, in this district; and there is no evidence that it was deposited in a sea, as was the case with the carbonate of lime of the calcaire grossier; on the contrary, as it only contains terrestrial and fresh-water remains, it would seem to have been formed through the medium of fresh water. If so, the previous level of the land and sea had been altered, and the springs of the district, if the gypsum was derived from them, must, instead of carbonate of lime, have produced an abundance of sulphate of lime. This state of things changed; the sulphate of lime ceased to be produced or deposited in abundance, the relative level of sea and land again became altered, the result was a formation of marls with marine shells in them; during which, there were at least some places where rolled pebbles were produced, to which oysters became attached, some of the pebbles being pierced by boring shells. These deposits are described as conforming more or less to the surface beneath each, and there is no evidence of any par-

ticular movement of water; but to them succeeds a vast quantity of sand, the organic remains in which are broken, and the
mass fills up inequalities and forms a plane surface. This appears to show a long continued action of water, with a velocity
equal to the transport of sand over a considerable space. At
the close of this period the causes, whatever they were, that
prevented the envelopment of organic remains, ceased, and
marine exuviæ became entombed in great abundance. Finally,
to crown this curious series, we have a deposit of a very various mineralogical character, containing the remains of such
animals and vegetables as are only known to exist on dry
land, marshy places, or in fresh water. This variety of mineralogical structure is what we should consider probable in
a shallow lake, into which springs, holding various substances
in solution, entered at various parts. That the water was
shallow, at least in part, has been considered probable by
MM. Cuvier and Brongniart, from the remains of *Charæ*, so
commonly found in this deposit; an opinion exceedingly
strengthened by the observations of Mr. Lyell on the *Charæ*
of the Bakie Loch, Scotland. To produce the friable calcareous marls, it is not necessary that the waters should be thermal; but judging from the phænomena of existing springs,
this condition would seem requisite for the siliceous deposit;
for we do not know of any such formation now in progress,
except in such springs. If the millstone and other siliceous
substances were thus produced (and it seems difficult to obtain their formation in any other manner consistent with existing causes), these thermal waters have disappeared, and silica is no longer deposited in this district; seeming to show
that very great changes in the solvent powers of water, and in
the temperature of springs, may take place in the same district
at different epochs. Thus we have a great deposit of carbonate of lime at the epoch of the calcaire grossier; another of
sulphate of lime at the period of the osseous marls, and, finally,
one of silica at the time of the millstone formation.

Supracretaceous Rocks of England.—Let us now compare
the supracretaceous rocks of England with those of the Paris
basin. Those of the former country are commonly known by
the names of Plastic Clay, London Clay, Bagshot Sands, the
Fresh-water formations of the Isle of Wight, and the Crag
formerly noticed.

Plastic Clay.—Unlike the deposit to which the same name
is applied in the environs of Paris, this rock, though occasionally containing a considerable abundance of clay, employed for
various useful purposes, presents us with pebble beds, irregularly alternating with sands and clay; but, like the strata of
the same name at Paris, they rest upon an unequal surface of

chalk beneath. The organic remains also are not principally terrestrial and fresh-water, but for the most part marine, though the others are intermingled with them. These remains are, according to Mr. Conybeare: UNIVALVES—*Infundibulum echinatum ; Murex latus, M. gradatus, M. rugosus, Cerithium funiculatum, C. intermedium, C. melanoides ; Turritella ; Planorbis hemistoma.* BIVALVES—*Ostrea pulchra, O. tener; Pectunculus Plumstediensis ; Cardium Plumstedianum ; Mya plana ; Cytherea ; Cyclas cuneiformis, C. deperdita, C. obovata.* In addition to this, traces of lignite and vegetables are observed in several places. The three following sections will convey an idea of this deposit in the neighbourhood of London, according to Prof. Buckland; and in the Isle of Wight according to Mr. Webster.

Section near Woolwich (series ascending).—Chalk with flints, above which: 1. Green-sand of the Reading oyster-bed, containing green coated chalk flints, but no organic remains ; 1 foot. 2. Light ash-coloured sand, without shells or pebbles; 35 feet. 3. Greenish sand, with flint pebbles; 1 foot. 4. Greenish sand, without shells or pebbles; 8 feet. 5. Iron-shot coarse sand, without shells or pebbles, and containing ochreous concretions disposed in concentric laminæ; 9 feet. 6. Blue and brown clay, striped, full of shells, chiefly *Cerithia* and *Cythereæ*; 9 feet. 7. Clay striped with brown and red, and containing a few shells of the above species; 6 feet. 8. Rolled flints, mixed with a little sand, occasionally containing shells like those of Bromley; e. g. *Ostrea, Cerithium,* and *Cytherea,* disseminated in irregular patches; 12 feet. 9. Alluvium*.

Section at Loam-Pit Hill, three miles S.W. of Woolwich (order ascending).—Chalk with flints, above which: 1. Green sand, identical with the Reading bed, and in every respect resembling No. 1. at Woolwich; 1 foot. 2. Ash-coloured sand, slightly micaceous, without pebbles or shells; 35 feet. 3. Coarse green sand, containing pebbles ; 5 feet. 4. Thick bed of ferruginous sand, containing flint pebbles; 12 feet. 5. Loam and sand, in its upper part cream-coloured, and containing nodules of friable marl; in its lower part sandy and iron-shot; 4 feet. 6. Three thin beds of clay, of which the upper and lower contain *Cythereæ,* and the middle, oysters; 3 feet. 7. Brownish clay, containing *Cythereæ*; 6 feet. 8. Lead-coloured clay, containing impressions of leaves; 2 feet. 9. Yellow sand; 3 feet. 10. Striped loam and plastic clay, containing a few pyritical casts of shells, and some thin leaves of coaly matter; 10 feet. 11. Striped sand, yellow, fine and iron-shot; 10 feet. At a higher level than No. 11. on the same hill, the line of the London clay commences†.

* Buckland, Geol. Trans. 1st series, vol. iv.   † *Ibid.*

*Section of the vertical beds in Alum Bay, Isle of Wight (order
ascending).*—Above, or rather next to, the chalk: 1. Green,
red, and yellow sand; 60 feet. 2. Dark blue clay, containing
green earth and nodules of dark limestone, in the latter of
which *Cythereæ, Turritellæ,* and other shells are found; 200
feet. 3. A succession of variously coloured sands; 321 feet.
4. Beautifully coloured sands, alternating with pipe-clay, coloured white, yellow, gray, and blackish; 543 feet. In the
central parts of these latter deposits are three beds of lignite,
and above them, at some distance, five other lignite beds; each
1 foot thick. 5. Strata of rolled black flint, contained in a
yellow sand. 6. Blackish clay, containing much green earth
and septaria; analogous to London clay*.

It will be observed, from these sections, that the transporting powers of water have not been precisely similar near London and at the Isle of Wight. At the former place, there would
appear to have been a greater movement than at the latter;
the mass of the strata near London containing more pebbles
in proportion to its depth than the beds of the Isle of Wight,
where there would appear to have been a more calm, as well as
a more abundant, deposit. This may perhaps in some measure be accounted for, by supposing the Isle of Wight strata,
now thrown into a vertical position, to have been gradually
accumulated in a hollow or cavity, more remote from the disturbing power of currents or motions in the water, than in
shallower depths. At all events, the transporting power of the
waters appears to have been irregular; their velocities varying
in such a manner that pebbles are carried forward at one time,
while fine particles of detritus are alone moved at another. In
the Isle of Wight beds we also see that circumstances have
been favourable to the accumulation of vegetable matter, which
is not irregularly disseminated, but occurs in beds; the circumstances which attended this deposit being continued at irregular intervals, such as might be expected at the mouths of
rivers.

London Clay.—This name has been applied to the great
argillaceous deposit which underlies the London district. The
clay is mostly blueish or blackish, and composed of argillaceous and calcareous matter in variable proportions, the latter
rarely attaining a sufficient quantity to constitute marl or imperfect limestone. Layers of calcareous concretions, known
by the name of Septaria, are by no means unfrequent; and it
is stated that beds of sandstone are occasionally observed in it.

It has been often remarked, that if the description of the
Paris rocks had not preceded that of the country round Lon-

---

* Webster, Geol. Trans. 1st series, vol. iv.

don and of the Isle of Wight, it never would have been considered that the, so called, Plastic Clay was separated from the
London Clay, but rather that they constituted different terms
of the same series. It will have been observed that in the
above-noticed section at Alum Bay, in the Isle of Wight,
there was nothing to warrant such a separation; neither does
there appear to be any good reason why in the London district they should not be regarded as upper and lower portions
of a deposit formed under nearly similar general circumstances.
The deposit of the London Clay would appear to mark a
comparatively quiet state of things; and the clay named Plastic
marks a similar state, although it occurs among sands and pebbles. The whole seems merely to show that the velocities of
the transporting waters varied, and that they continued for a
longer period of little importance during the deposit of the
London clay.

This clay varies very considerably in thickness. Thus, one
mile east of London it is only 77 feet deep; at a well in St.
James's-street, 235 feet; at Wimbledon it was not pierced
through at 530 feet; and at High Beech, 700 feet*.

Organic Remains.—A Crocodile; a Turtle. Fish. Crustacea, a great variety, few of which have been noticed; among
these few, Cancer tuberculatus, *König ;* C. Leachii, *Desmarest;*
Inachus Lamarckii, *Desm.* CONCHIFERA—Clavagella coronata, *Desh.*, cal. gros., Paris; Fistulana personata, *Lam.*, cal.
gros., Paris; Gastrochæna contorta; Pholadomya margaritacea, *Sow.;* Solen affinis, *Sow. ;* Panopæa intermedia, *Sow. ;*
Mya subangulata, *Sow. ;* Lutraria oblata, *Sow. ;* Crassatella
sulcata, *Lam.*, cal. gros., Paris; C. plicata, *Sow;* C. compressa;
Corbula globosa, *Sow. ;* C. Pisum, *Sow. ;* C. revoluta, *Sow. ;*
Sanguinolaria Hollowaysii, *Sow. ;* S. compressa, *Sow. ;* Tellina
Branderi, *Sow. ;* T. filosa, *Sow. ;* T. ambigua, *Sow. ;* Lucina
mitis, *Sow. ;* Astarte rugata, *Sow. ;* Cytherea nitidula, *Lam.*,
cal. gros., Paris, Bourdeaux; Venus incrassata, *Sow. ;* V.
transversa, *Sow. ;* V. elegans, *Sow. ;* V. pectinifera, *Sow. ;*
Venericardia Brongniarti, *Sow. ;* Ven. planicosta, *Lam.*, cal.
gros., Paris, Ghent; Ven. carinata, *Sow.;* Ven. deltoidea, *Sow.;*
Ven. oblonga, *Sow. ;* Ven. globosa, *Sow. ;* Ven. acuticostata,
Lam., cal. gros., Paris; Cardium nitens, *Sow. ;* C. semigranulatum, *Sow.*, molasse, Switzerland; C. turgidum, *Sow. ;* C.
porulosum, *Lam.*, cal. gros., Paris; C. edule, *Brander*, Bordeaux, analogous to the existing species; Cardita margaritacea,
Sow. ; Isocardia sulcata, *Sow. ;* Arca duplicata, *Sow. ;* A. Branderi, *Sow. ;* A. appendiculata, *Sow. ;* Pectunculus decussatus,

* Conybeare and Phillips's Outlines of the Geology of England and Wales:
art. *London Clay.*

Sow. ; P. costatus, *Sow.* ; P. scalaris, *Sow.* ; P. brevirostris, *Sow.;*
P. pulvinatus, *Lam.*, cal. gros., Paris, Bourdeaux, Turin,
Traunstein; Nucula similis, *Sow.* ; N. trigona, *Sow.* ; N. mi-
nima, *Sow.* ; N. inflata, *Sow.* ; N. amygdaloides, *Sow.* ; Axinus
angulatus, *Sow.* ; Chama squamosa, *Sow.;* Pinna affinis, *Sow.*;
P. arcuata, *Sow.* ; Avicula media, *Sow.* ; Pecten corneus, *Sow.*;
P. carinatus, *Sow.* ; P. duplicatus, *Sow.* ; Ostrea gigantea, *Sow.*,
Traunstein; O. flabellula, *Lam.*, cal. gros., Paris, Bourdeaux;
O. dorsata, *Sow.* ; O. cymbula, *Lam.*, cal. gros., Paris, Bourdeaux ; O. oblonga, *Brander;* Lingula tenuis, *Sow.* Mollusca—Patella striata, *Sow.* ; Calyptræa trochiformis, *Lam.*,
cal. gros., Paris; Infundibulum obliquum, *Sow.* ; I. tuberculatum, *Sow.* ; I. spinulosum, *Sow.* ; Bulla constricta, *Sow.* ; B.
elliptica, *Sow.* ; B. attenuata, *Sow.* ; B. filosa, *Sow.* ; B. acuminata, *Sow.* ; Auricula turgida, *Sow.* ; Au. simulata, *Sow.* ;
Melania sulcata, *Sow.* ; M. costata, *Sow.* (Qu. M. costellata,
Brander and *Lam.*, cal. gros., Paris?); M. minima, *Sow.* ; M.
truncata, *Sow.;* Paludina lenta, *Sow.* ; P. concinna, *Sow.;* Ampullaria ambulacrum, *Sow.* ; Am. acuta, *Lam.*, cal. gros., Paris;
Am. patula, *Lam.*, cal. gros., Paris; Am. sigaretina, *Lam.*, cal.
gros., Paris; Neritina concava, *Sow.* ; Nerita globosa, *Sow.* ; N.
aperta, *Sow.* ; Natica Hantoniensis ; N. similis, *Sow.* ; N. glauci-
noides, *Sow.* ; N. striata, *Sow.;* Sigaretus canaliculatus, *Sow.*, cal.
gros., Paris, Bourdeaux ; Acteon crenatus, *Sow.* ; A. elongatus,
Sow. ; Scalaria acuta, *Sow.* ; S. semicostata, *Sow.* ; S. interrupta,
Sow. ; S. undosa, *Sow.* ; S. reticulata, *Sow.* ; Solarium patulum,
Lam., cal. gros., Paris, Bourdeaux ; Sol. discoideum, *Sow.* ;
Sol. canaliculatum, *Sow.* ; Sol. plicatum, *Lam.*, cal. gros.,
Paris; Trochus Benettiæ, *Sow.* ; Piacenza, Turin, Bourdeaux;
T. extensus, *Sow.* ; T. monilifer, *Lam.*, cal. gros., Paris; Tur-
ritella conoidea, *Sow.* *; Tur. elongata, *Sow.;* Tur. brevis, *Sow.;*
Tur. edita, *Sow.* ; Tur. multisulcata, *Lam.*, cal. gros., Paris;
Cerithium dubium, *Sow.* ; C. Cornucopiæ, *Sow.* ; C. gigan-
teum, *Lam.*, cal. gros., Paris; C. pyramidale, *Sow.* ; C. gemi-
natum, *Sow.*; C. funatum, *Sow.* †; Pleurotoma attenuata, *Sow.;*
P. comma, *Sow.* ; P. semicolon, *Sow.* ; P. colon, *Sow.* ; P. exerta, *Sow.* ; P. rostrata, *Sow.* ; P. acuminata, *Sow.* ; P. fusiformis, *Sow.* ; P. lævigata, *Sow.* ; P. brevirostra, *Sow.* ; P. prisca,
Sow. ; Cancellaria quadrata, *Sow.* ; C. læviuscula, *Sow.* ; C.
evulsa, *Sow.* ; Fusus deformis, *König* ; F. longævus, *Lam.*, cal.
gros., Paris; Fusus rogosus, *Lam.*, cal. gros., Paris, Bourdeaux ; F. acuminatus, *Sow.* ; F. asper, *Sow.* ; F. bulbiformis,

---

* According to M. Deshayes, *Turritella conoidea, T. elongata* and *T. edita,*
of Sowerby, are the same shells, referrible to *T. imbricataria* of Lamarck.

† It is remarkable that, out of the numerous species of *Cerithium* found in
the calcaire grossier of Paris, the *C. giganteum* should be the only one yet
noticed in the London clay.

Lam. (4 var.), cal. gros., Paris; F. ficulneus, *Sow. ;* F. errans,
Sow. ; F. regularis, *Sow. ;* F. Lima, *Sow.;* F. carinella, *Sow.;*
F. conifer, *Sow. ;* F. bifasciatus, *Sow.;* F. complanatus, *Sow.;*
Pyrula nexilis, *Sow. ;* P. Greenwoodii, *Sow. ;* P. lævigata,
Lam., cal. gros., Paris, Traunstein; Murex Bartonensis, *Sow. ;*
M. fistulosus, *Sow. ;* M. interruptus, *Sow. ;* M. argutus, *Sow. ;*
M. tricarinatus, *Lam.,* cal. gros., Paris, Vicentin; M. bispi-
nosus, *Sow. ;* M. frondosus, *Lam.,* cal. gros., Paris; M. de-
fossus, *Sow.;* M. Smithii, *Sow.* (2 var.); M. trilineatus *Sow. ;*
M. curtus, *Sow.;* M. tuberosus, *Sow.;* M. minax, *Sow. ;*
Switzerland; M. cristatus, *Sow. ;* M. coronatus, *Sow. ;* Rostellaria Parkinsoni, *Sow.* (var.); R. lucida, *Sow. ;* R. rimosa,
Sow. ; R. macroptera, *Sow.* (2 var.); R. Pes-Pelicani (Strombus Pes-Pelicani, *Linn.*), Piacenza, &c., analogous to the existing species; Cassis striata, *Sow.;* C. carinata, *Lam.,* cal.
gros., Paris; Harpa Trimmeri, *Parkinson;* Buccinum jun-
ceum, *Sow. ;* B. lavatum, *Sow. ;* B. desertum, *Sow. ;* B. canaliculatum, *Sow. ;* B. labiatum, *Sow. ;* Mitra scabra, *Sow. ;* M.
parva, *Sow.;* M. pumila, *Sow. ;* Voluta Luctator, *Sow. ;* V.
spinosa, *Lam.,* cal. gros., Paris; V. suspensa, *Sow. ;* V. mon-
strosa, *Sow. ;* V. costata, *Sow. ;* V. Magorum, *Sow. ;* V. Athleta, *Sow. ;* V. depauperata, *Sow. ;* V. ambigua, *Sow. ;* V. nodosa, *Sow. ;* V. Lima, *Sow. ;* V. geminata, *Sow. ;* V. bicorona,
Lam., cal. gros., Paris; Volvaria acutiuscula, *Sow. ;* Cypræa
oviformis, *Sow. ;* Terebellum fusiforme, *Sow. ;* T. convolutum,
Al. Brong., cal. gros., Paris; Ancellaria canalifera, *Lam.,* cal.
gros., Paris, Bourdeaux; A. aveniformis, *Sow. ;* A. Turritella,
Sow.; A. subulata, *Sow. ;* Oliva Branderi, *Sow. ;* O. Salisbu-
riana, *Sow. ;* Conus Dormitor, *Sow. ;* C. concinnus (2 var.),
Sow. ; C. scabriusculus (2 var.), *Sow. ;* C. lineatus, *Brander ;*
Nummulites lævigata, *Lam.,* cal. gros., Paris, Bordeaux,
Traunstein; Num variolaria, *Sow.;* Num. elegans, *Sow. ;* Nau-
tilus imperialis *Sow.,* cal. gros., Paris; N. centralis, *Sow. ;* N.
ziczac, *Sow. ;* N. regalis, *Sow.**

Vegetable Remains.—The Isle of Sheppy has long been
known as affording a great variety of fruits and seeds; and
small portions and masses of wood are found in the London
clay elsewhere, the argillo-calcareous concretions frequently
enveloping pieces of it. Some fragments are pierced by a
boring shell analogous to the *Teredo navalis,* which shows that
the wood must have floated in the sea†.

Bagshot Sands.—These rest on the London clay, and consist, according to Mr. Warburton, of ochreous meagre sand,

* Sowerby's Mineral Conchology; Woodward's British Organic Remains;
Al. Brongniart, Tableau des Terrains qui composent l'Ecorce du Globe.
† Outlines of Geol. of Engl. and Wales.

foliated green clay alternating with a green sand, and alternations of white, sulphur-yellow, and pinkish foliated marls, containing abundant grains of green sand, and fossil shells of the genera *Trochus ? Crassatella, Pecten **.

Fresh-water Formations, Isle of Wight and Hampshire.—We are indebted to Mr. Webster for the discovery of these beds, not long after the labours of MM. Cuvier and Brongniart on the supracretaceous rocks round Paris so strongly excited the attention of geologists. The fresh-water strata of the Isle of Wight are divided into two deposits by a rock characterized by the presence of marine remains, and named the Upper Marine Formation, from being a supposed equivalent to the sands which intervene between the two fresh-water deposits of Paris. The lower fresh-water deposit of Binstead, near Ryde, consists of a limestone formed of fragments of fresh-water shells, white shell marl, siliceous limestone and sand ; at Headen the equivalent rock is composed of sandy, calcareous, and argillaceous marls. According to Mr. Pratt, one tooth of an *Anoplotherium* and two teeth of a *Palæotherium* have been discovered in the lower and marly beds of the Binstead quarries ; and he further states, that these remains were " accompanied, not only by several other fragments of bones of Pachydermata (chiefly in a rolled and injured state), but also by the jaw of a new species of Ruminant, apparently closely allied to the genus *Moschus*†."

Prof. Sedgwick observes, that in the upper part of this deposit there is a mixture of fresh-water and marine species, especially in Colwell Bay, where a single specimen of rock contained the following genera : *Ostrea, Venus, Cerithium, Planorbis, Lymnæa.* The common fossils in the lower fresh-water deposit would appear to be : *Paludina, Potamides, Melania,* (more than one species), *Cyclas* (2 species), *Unio, Planorbis, Lymnæa* (both the last more than one species), *Mya, Melanopsis*‡.

The Upper Marine Formation, first noticed by Mr. Webster, was called in question by Mr. G. B. Sowerby, who showed that all the shells detected in it were not marine ; and he hence inferred that there was no real separation between the fresh-water formations of the Isle of Wight§. Subsequently to Mr. Sowerby's remarks, Prof. Sedgwick has presented us with an account of these strata, in which he remarks that "the lower calcareous beds appear to have been tranquilly deposited in fresh water. But if we ascend to the argillaceous marl which

* Warburton, Geol. Trans., vol. i. 2nd series.
† Pratt, Proceedings of the Geol. Soc. 1831.
‡ Sedgwick, On the Geology of the Isle of Wight; Annals of Philos. 1822.
§ G. B. Sowerby, Annals of Philos. 1821.

rests immediately upon them, we not only find a complete
change in the physical circumstances of the deposit, but a new
suite of organic remains; some of which are of a marine origin,
others of a doubtful character, and a few are identical with
those in the lower beds*." With regard to the organic remains contained in this rock, Mr. Webster points out a thick
oyster-bed in Colwell Bay; and Prof. Sedgwick gives the
following list of shells: *Murex* (at least two species), *Buccinum, Ancilla subulata, Voluta* (resembling *V. spinosa), Rostellaria rimosa* (two last species rare), *Murex effossus,* BRANDER,
M. innexus, BRANDER, *Fusus* (fragments), *Natica, Venus,
Nucula, Corbula, Corbis? Mytilus, Cyclas, Potamides, Melanopsis, Nerita* (2 species, one approaching *N. fluviatilis),* together with other fresh-water shells. These beds would therefore appear to have been deposited, as Prof. Sedgwick observes, in an estuary. But to have produced this estuary, and
the circumstances requisite for the presence of marine shells,
some physical change, some alteration of the relative levels or of
the geographical features of the sea and land, seems necessary,
for the previous deposit does not contain marine remains.

Upper Fresh-water Formation.—This, according to Mr.
Webster, principally consists of yellowish white marls, in
which there are more indurated, and apparently more calcareous portions. The organic remains are either fresh-water
or terrestrial; and therefore the circumstances, whatever they
were, which permitted a mixture of marine shells in the beds
beneath, no longer existed; and a tranquil deposit in some
lake was, probably, the mode in which these beds, about 100
feet thick, were formed.

The fresh-water formation of Hordwell Cliff, Hampshire,
was first described by Mr. Webster in 1821. The cliff is
noticed as composed of alternations of clays and marls, some of
a fine blueish green colour, in which there were also beds of hard
calcareous marls, apparently derived from shells of the genera
Lymnæa and *Planorbis.* The whole is surmounted by a mass of
transported gravel, which covers the various rocks of the vicinity.
Mr. Webster observed that these beds seemed the equivalent
of the lower fresh-water deposit of the Isle of Wight. Subsequently to these observations of Mr. Webster, Mr. Lyell published a more detailed account of the Hordwell beds; whence
it would appear that the upper strata do not show a passage
into a marine deposit, as was first supposed, but that all the
fossil contents of the beds point to a fresh-water origin, equivalent to the lower fresh-water rocks of the Isle of Wight.
The following are the organic remains discovered at Hordwell,

according to Mr. Lyell: Tortoise scales, (a Tortoise found at Thorness Bay, Isle of Wight); *Gyrogonites*, or seed-vessels of *Charæ* (*C. medicaginula*); seed-vessel named *Carpolithes thalictroides*, AD. BRONG.; teeth of crocodile, and scales of fish? *Helix lenta*, BRANDER, abundant; *Melania conica*; *Melanopsis carinata*; *M. brevis*; *Planorbis lens*; *P. rotundatus*; *Lymnæa fusiformis*; *L. longiscata*; *L. columellaris*; *Potamides*; *P. margaritaceus?* *Neritina*; *Ancylus elegans*; *Unio Solandri*; *Mya gregarea*; *M. plana*; *M. subangulata*, perhaps the young of *M. plana*; *Cyclas* (2 species). Mr. Lyell observes, that though the species are few, the individuals are numerous,—a common characteristic of fresh-water deposits *.

Both in the Isle of Wight and on the opposite coast of Hampshire, these fresh-water deposits rest upon a considerable thickness of sand. As a similar sand occurs in the fresh-water rocks of Hordwell, Mr. Lyell considers that there is as much probability of its fresh-water, as of its marine origin. Be this as it may, there must have been a difference in the transporting power of water carrying the sands, from that which permitted the deposit of the marls, which seems to have been very quiet. The sands certainly do not require any considerable velocity of water; still there must have been a difference in the circumstances attending the deposit of the one mass and of the other, though those, which give rise to the mass of sand, partially returned during the formation of the marls.

A very material difference, it will be observed, must have attended the deposit of the supracretaceous rocks in the Parisian and English districts (London and Isle of Wight), as far as respects their mineralogical nature. In the former we have deposits of carbonate of lime (calc. grossier), sulphate of lime, (gypseous deposits), and silica (millstones); formations only in part mechanical; while in the latter we have little that may not be considered altogether mechanical, with the exception, perhaps, of the fresh-water marls and the calcareous concretions in the London clay, which latter may have been chemical separations, after deposition, from the argillo-calcareous mass. There is, nevertheless, such an analogy between the organic character of the calcaire grossier of Paris and the London clay, that though not strictly identical, they may have been nearly contemporaneous; so that however the mineralogical character of these deposits may vary, we may suppose them to have been formed at the same or nearly the same epoch, local circumstances and accidents having determined the character of each.

* Lyell, Geol. Trans. 2nd series, vol. ii.

Our limits prevent a proper notice of the labours of Prévost, Boué, Voltz, Parsch, Lill Von Lillienbach, Pusch *,
Du Bois, and many other geologists, on the rocks of this age
in various parts of Europe; but the following section seems
so important that it requires a place here.

Prof. Pusch, describing the rocks of Podolia and southern
Russia, states, that near Krzeminiec, in Volhynia, (where
mountains rise above a plain covered with chalk flints and
sand,) upper supracretaceous sandstone, occupying a thickness of 396 feet above the river Ikwa and sixty feet beneath
it, is composed of: 1. Twenty feet of sand, cemented by a
little carbonate of lime, containing many small shells and madrepores, the latter approaching *M. cervicornis.* 2. Forty
feet of calcareous sandstone, containing many shells of the
genera *Cardium, Venericardia,* and *Arca.* 3. Sixty feet of a
compact quartzose and porous sandstone, the cavities filled
with sand; contains many *Venericardiæ;* lowest part most calcareous. 4. Eighty feet of a marly limestone, containing many
striated *Modiolæ,* Pectens, and other shells. 5. At sixty feet
beneath the surface, a quartzose and slightly calcareous white
sandstone, containing numerous *Venericardiæ, Trochi,* and
Paludinæ or *Phasianellæ.* " According to M. Jarocki,—while
sinking a well in June 1829, the tusk and molar tooth of an
elephant were found in the last-mentioned bed (No. 5), which
are now preserved in the museum of Krzeminiec. Many
other bones were also observed, but they were too firmly fixed
in the rock to be extracted †." M. Pusch further remarks,
that this rock is the same, both mineralogically and zoologi-

* Amid a great variety of supracretaceous deposits in Russia and Poland,
this author remarks some with an oolitic character, especially near Tiraspol,
Latyczew, and Kaluez, on the Dniester, and in the Cecin hills at Czerno-
witz. The pisolitic structure of some supracretaceous limestones is particularly remarkable in parts of Poland. The grains are either reniform or
rounded, and generally of the size of a pea or a bean, though they here and
there become two or three inches in diameter. Good examples of this rock
are seen at Rakow. M. Pusch states that repeated observations have convinced him that these concretions are derived from corals, especially *Nulli-
poræ.* He observes that the large reniform concretions of Rakow are only
the *Nullipora byssoides,* Lam., or the *N. racemosa,* Goldf. In some places,
particularly at Skotniki, near Busko, a rock of this kind appears as if composed of bullets and cannon-balls.

It should be stated that Prof. Pusch, from a careful comparison of the
shells contained in the supracretaceous limestone of Poland with those figured
by various authors, considers that the tertiary shells of Poland bear a much
greater resemblance to those found at the foot of the Italian Alps and in the
Sub-Apennine hills, than those discovered in England or the North of
France; moreover, that the species which at first sight do appear identical
with those of France and Italy, are found to be varieties of them when examined with attention.

† Pusch, Journal de Géologie, t. 2.

cally, as the tertiary sandstone of Szydtow and Chmielnik,
in Poland; and that this fact is analogous to the occurrence
of an elephant's molar tooth and tusk in the tertiary sandstone
of Rzaka, Wieliczka, which contains *Pecten polonicus*, *Saxi-
cavæ*, and many other marine shells. The reader will also
observe that it corresponds with the occurrence of the remains
of the great Pachydermata, previously noticed as found mingled with marine exuviæ in other parts of Europe.

It will have been remarked, that throughout this detail of
supracretaceous rocks, (perhaps too long for a work of this
nature,) the observations have been confined to certain parts
of Europe. Rocks of the same nature no doubt abound in
other parts of the world; indeed we are well assured that very
extensive districts are composed of them,—as for instance in
India; but our knowledge of them is as yet so imperfect, that
we cannot with safety compare them with known European
deposits. Dr. Buckland, from the information which he obtained from Mr. Crawfurd, who collected an abundance of
organic remains on the banks of the Irawadi, considered that
supracretaceous rocks probably existed in the kingdom of Ava,
containing shells of the genera *Ancillaria*, *Murex*, *Cerithium*,
Oliva, *Astarte*, *Nucula*, *Erycina*, *Tellina*, *Teredo*; mixed with
sharks' teeth and fish scales : these remains are contained in
a coarse shelly and sandy limestone. A great abundance of
mammiferous and other remains were discovered in the vicinity of some petroleum wells, between Prome and Ava, apparently mixed with much silicified wood in a sandy and gravelly deposit. The bones or teeth of vertebrated animals consist of those of the *Mastodon latidens*, Clift; *M. elephantoides*,
Clift; *Hippopotamus ; Sus ; Rhinoceros ; Tapir ; Ox ; Deer ;
Antelope ; Trionyx ; Emys ;* and *Crocodiles* (2 species) [*].
Mr. Scott met with beds, probably of the supracretaceous
epoch, in the Caribári hills, left bank of the Brahmputra.
The following section (order ascending) was observed:—
1. Slate clay. 2. Ferruginous concretions and indurated
sand. 3. Yellow or green sand. 4. Slate-clay. 5. Sand
and small gravel. Fossil wood is found on the indurated clay;
and in a small isolated hill in the vicinity the following remains : Teeth and bones of sharks, fish palates and fin bones,
teeth and bones of crocodiles, remains of quadrupeds, *Ostreæ*,
Cerithia, *Turritellæ*, *Balani*, *Patellæ*, &c. [†]. These exuviæ
have subsequently been examined by Mr. Pentland, who found
that the mammiferous remains were referrible to the genus
Anthracotherium, Cuv., to a species allied to the genus *Mos-*

---

[*] Buckland and Clift, Geol. Trans. 2nd series, vol. ii.
[†] Colebrooke, Geol. Trans. 2nd series, vol. i.

chus, to a small species of the order *Pachydermata,* and to a carnivorous animal of the genus *Viverra.* The *Anthracotherium* he proposes to name *A. Silistrense* *.

These observations are sufficient to show that rocks, probably supracretaceous, exist extensively in India. According to Prof. Vanuxem and Dr. Morton, the supracretaceous or tertiary rocks are extensively distributed over parts of the United States, occurring in Nantucket, Long Island, Manhattan Island, the adjacent coasts of New York and New England; sparingly in New Jersey and Delaware, but extensively in Maryland and to the southward. The deposit is stated to be composed of limestone, buhr-stone, sands, gravels, and clays; and contains the remains of the genera *Ostrea, Pecten, Arca, Pectunculus, Turritella, Buccinum, Venus, Mactra, Natica, Tellina, Nucula, Venericardia, Chama, Calyptræa, Fusus, Panopæa, Serpula, Dentalium, Cerithium, Cardium, Crassatella, Oliva, Lucina, Corbula, Pyrula, Crepidula, Perna,* &c. Of 150 species of these shells, found in a single locality in St. Mary's county, Maryland, Mr. Say has described and figured more than forty as new †. According to Dr. Morton, the upper supracretaceous beds of Maryland and the more southern states contain the following species of shells, still found in a recent state on the coasts of the United States :—*Natica duplicata,* Say ; *Fusus cinereus,* Say ; *Pyrula carica,* Lam.; *P. canaliculata,* Lam.; *Ostrea virginica,* Linn.; *O. flabellula,* Lam.; *Plicatula ramosa,* Lam.; *Arca arata,* Say ; *Lucina divaricata,* Lam.; *Venus mercenaria,* Linn.; *V. paphia?* Lam.; *Cytherea concentrica,* Lam.; *Mactra grandis,* Linn.; *Pholas costata,* Linn.; *Balanus tintinnabulum?* Lam.; *Turbo littoreus?* Linn.; and a *Buccinum* ‡. That deposits of a similar age are not wanting in South America seems also certain; but as yet they have not been examined in sufficient detail to enable us to institute any useful comparison with rocks of the same antiquity in Europe. Neither can we, for the same reason, judge of the relative antiquity of innumerable igneous formations scattered over various parts of the world. As the science of geology advances, great insight must be obtained into the superficial appearance of the world at this period, leading to the most important conclusions; but we must anticipate very serious obstacles to this advancing knowledge, arising from hasty generalizations of local facts, and the too common endeavour to force conclu-

* Pentland, Geol. Trans. 2nd series, vol. ii.
† Vanuxen and Morton, Journal of the Academy of Natural Sciences of Philadelphia, vol. vi.   ‡ Morton, *Ibid.*

sions, more particularly as to the identity or parallelism of
deposits.

It is impossible to close this sketch of the supracretaceous
rocks without noticing the important observations of Dr. Boué
on those of Gallicia, wherein he establishes the fact, that the
celebrated salt deposit of Wieliczka constitutes a portion of
the supracretaceous series. Dr. Boué describes this deposit
as 2560 yards long, 1066 yards broad, and 281 yards deep.
The salt is termed green salt in the upper part of the mine,
where it occurs in nodules with gypsum in marl. The salt
sometimes contains lignite, bituminous wood, sand, and small
broken shells. In the lower part the marl becomes more arenaceous, and there are even beds of sandstone in the salt.
Beneath this is a gray sandstone, rather coarse, containing
lignite, and impressions of plants, with veins and beds of salt.
In the lower part of this stratum an indurated calcareous marl
is observed, containing sulphur, salt, and gypsum. Beneath
this is an aluminous and marno-argillaceous schist. From the
fossils and various other circumstances, Dr. Boué concludes
that this great salt deposit forms part of a muriatiferous and
supracretaceous clay, subordinate to sandstone (molasse).
Most frequently the marly clays are merely muriatiferous ; an
abundance of salt, such as at Wieliczka, Bochnia, Parayd in
Transylvania, and other places, being more rare *.

Volcanic Action during the Supracretaceous Period.—We
have already seen that there was much difficulty in stating at
what periods certain products of extinct volcanos had been
thrown out. This difficulty is by no means lessened as we
descend in the series; for the seat of volcanic action seems to
have continued nearly, or very nearly, in the same place for
long periods; and the mere circumstance of the interstratification of volcanic matter with aqueous rocks, whose relative
age may to a certain extent be known, will not always give
that of the igneous rocks so circumstanced, because we cannot
be sure that they have not been injected among the aqueous
deposits; and when this may have happened it would be difficult to say. Thus Etna would appear to have been the seat
of volcanic action through a long series of ages, commencing
with the supracretaceous rocks, on which much of the igneous
mass is now based.

In Central France, amid the extinct volcanos which there
constitute such a remarkable feature in the physical geography of the country, we certainly approach relative dates in
some instances. Thus the volcanic mass of the Plomb du

---

* Boué, Journal de Géologie, t. i. 1830.

Cantal appears to have burst through, to have upset, and to have fractured the fresh-water limestones of the Cantal, which, according to Messrs. Lyell and Murchison, may be equivalent to the fresh-water deposits of the Paris basin, and to those of Hampshire and the Isle of Wight. The following is a list of organic remains obtained by them in the fresh-water rocks of the Cantal :—The rib of an animal resembling that of an *Anoplotherium* or a *Palæotherium ;* scales of a tortoise; fish teeth ; *Potamides Lamarckii ; Limnæa acuminata ; L. columellaris ; L. fusiformis ; L. longiscata ; L. inflata ; L. cornea ; L. Fabulum ? L. strigosa ? L. palustris antiqua ; Bulimus Terebra ; B. pygmeus ? B. conicus ; Planorbis rotundatus ; P. Cornu ; P. rotundus ; Ancylus elegans.* Plants: *Chara medicaginula,* the seeds (*gyrogonites*), and stems; carbonized wood. It is remarked, that out of this short list there are eight or nine species identical with those found in the upper fresh-water rocks, and five or six with those in the lower fresh-water deposits of the Paris basin *. Here we seem to obtain a relative date for the upburst of the igneous products of the Plomb du Cantal; one posterior to the deposit of the fresh-water rocks of Paris and the Isle of Wight.

With regard to the relative date of the igneous rocks of Auvergne, it would appear from the labours of MM. Croizet and Jobert, that the Montagne de Perrier, N.W. from the town of Issoire (Puy de Dome), is divided into two stages or terraces, the first about twenty-five yards above the valley of the Allier, the second occupying a height of about 200 yards. The mountain may be considered as based on granite, above which there is a considerable thickness of fresh-water limestone, surmounted by numerous beds of rolled pebbles and sand, of which one in particular is remarkable for the abundant remains of mammalia found in it ; the whole crowned by a mass of volcanic matter.

MM. Croizet and Jobert consider that in this locality and in the neighbouring country there are about thirty beds above the fresh-water limestone, which may be divided into four alternations of alluvial detritus and basaltic deposits. Among the beds there are four which contain organic remains: three belonging to the third of the ancient alluvions, that which succeeded the second epoch of volcanic eruptions ; the third fossiliferous deposit being referrible to the last epoch of ancient alluvion. The whole of these beds are not seen in the Montagne de Perrier, but are determined from the general structure of the country.

* Lyell and Murchison, Sur les Dépôts Lacustres Tertiaires du Cantal, &c. Ann. des Sci. Nat. 1829.

The principal ossiferous bed is about nine or ten feet thick, and can be traced a considerable distance at the foot of the Montagne de Perrier, and in the Vallée de la Couse on the opposite side. The fossil species, according to MM. Croizet and Jobert, are very numerous, consisting of:—Elephant, one species; Mastodon, one or two; Hippopotamus, one; Rhinoceros, one; Tapir, one; Horse, one; Boar, one; Felis, four or five; Hyæna, two; Bear, three; Canis, one; Castor, one; Otter, one; Hare, one; Water-Rat, one; Deer, fifteen; and Ox, two. The animals were of all ages, and the various remains mixed pell-mell with each other. The bones are never rolled, though often broken, and sometimes gnawed. Mingled with these exuviæ are the abundant fæcal remains of the Carnivora, appearing to occupy the place where they have been dropped. Hence the authors conclude that the remains have not been far removed from the places where the animals existed, and that the lignites found among these beds are the exuviæ of the vegetation upon which many of them subsisted.

MM. Croizet and Jobert notice the following remains in the fresh-water sands, clays, and limestone of the country, over which they consider that the first basaltic currents flowed:— *Anoplotherium?* two species; *Lophiodon,* one; *Anthracotherium,* one; *Hippopotamus,* one; a Ruminant; *Canis,* one; Marten, one; *Lagomys,* one; a Rat; Tortoise, one or two; Crocodile, one; Serpent or Lizard, one; Birds, three or four (among the latter remains are their eggs, perfectly preserved); *Cypris faba; Helix; Lymnæa; Planorbis; Cyrena; Gyrogonites,* and other vegetable exuviæ. It should be observed that M. Bertrand-Roux * had some time previously observed the remains of a *Palæotherium* in a similar rock in the Puy en Velay, and that the fresh-water rocks at Volvic contain birds' bones †.

M. Bertrand de Doue describes the occurrence of bones entombed in and beneath volcanic matter near St. Privat-d'Allier (Velay). After stating that the discovery was due to Dr. Hibbert, who communicated it to him, and that he proceeded to the spot pointed out, accompanied by M. Deribier, he notices the following descending section:—*a*, third and last flow of basaltic lava; *b*, second flow, four yards thick; *c*, grayish volcanic cinders, two to four decimetres thick; *d*, agglutinated scoriæ and tuff, one or more yards thick, in the upper part of which the bones were discovered; *e*, oldest plateau of

---

* Now M. Bertrand de Doue.

† Croizet and Jobert, Recherches sur les Oss. Foss. du Depart. du Puy de Dome; and Ann. des Sci. Nat. t. xv. 1828.

basaltic lava; *f,* gneiss. The osseous remains were those of
the *Rhinoceros leptorhinus, Hyæna spelæa,* and a large proportion of bones, referrible to at least four undetermined species of *Cervi.*

The same author considers, from the fractured character
and irregular distribution of the bones over a horizontal and
limited space, that this place was the retreat of hyænas, affording them, from the nature of the country, the best shelter
they could find. Into this it is considered they dragged their
prey, as appears to have been done in the case of Kirkdale.
It is observed that the lava-current which passed over the
cinders containing these remains has very little altered the
bones.

M. Bertrand de Doue does not consider the detrital deposits
of the country as produced by transport in a body of waters
from a distance, but by a succession of local causes, the substances being all derived from the vicinity. He supposes the
distribution of the lateral valleys connected with the Allier,
(among which is that where the bones were discovered,) the
same now as when the neighbouring volcanos were in activity;
and remarks on the " incertitude in establishing the chrono-
logical relations between the epoch when the volcanos of the
Velay became extinct, and that in which these animals disappeared from our climates*."

M. Robert, describing the position in which numerous
bones have been discovered at Cussac (Haute Loire), mentions that marls, without fossils, rest on the granitic rocks of
the country. At Solilhac these marls are surmounted by
clayey marls about two or three feet thick, containing plates
of mica, grains of quartz, volcanic ashes, basaltic gravel, and
impressions of gramineous plants; they also contain the entire
skeletons of unknown Deer and Aurochs, with other bones.
Above these are beds of volcanic sand two or three yards
thick, with small basaltic and granitic pebbles, containing the
remains of Ruminants and *Pachydermata,* the bones being
more or less broken. On these rest alluvions of greater solidity, composed of the same volcanic sand, large granitic and
basaltic blocks (of which the angles are not rounded), geodes
of hydrate of iron, and bones, which appear to have been exposed to the air before they were enveloped. All these substances are cemented by oxide of iron, and beds of ferruginous sands either alternate with, or repose on, the alluvions.
M. Robert extracted from these ferruginous beds at Cussac
the remains of the *Elephas primigenius;* the *Rhinoceros leptorhinus;* the *Tapir Arvernensis;* the Horse, two species;

* Bertrand de Doue, Edin. Journal of Sci. vol. ii. new series, 1830.

T

Deer, seven species (to two of which he assigns the names of *Cervus Solilhacus*, and *C. Dama Polignacus*), the *Bos Urus* and *Bos Velaunus*, and the Antelope. The same author refers the entombment of these remains to a more ancient date than the accumulation of bones at St. Prevat and Perrier, considering it due to some particular cataclysm, which surprised the animals: thus explaining the occurrence of entire skeletons of young and old individuals found mingled at Solilhac; a state of things differing from the accumulations at St. Prevat and Perrier, where the bones seem to have been dragged into their present position by carnivorous animals, whose bones are also mixed with those of their prey *.

Dr. Hibbert considers that the lowest supracretaceous rocks of the Velay were deposited in fresh-water lakes, entombing the remains of the *Palæotherium* and *Anthracotherium*, of terrestrial and fresh-water shells, and of the vegetation which then existed; such deposit being of long continuance, as shown by its depth, which amounts to 450 feet. This deposit ceased, and the land became covered with forests and animals; the forests being of a marshy growth. The common degradation of land taking place, parts of this vegetation were variously entombed, as were also the remains of animals which then existed; such as various species of *Cervi*, some of large size, animals of the *Bos* kind, the *Rhinoceros leptorhinus*, and the *Hyæna spelæa*. Volcanic explosions now took place through various vents, ejecting trachyte and. basalt, the latter predominating, piercing the fresh-water deposit in some places, and covering it with lavas in others. Notwithstanding these convulsions, vegetation still flourished in certain situations, and became entombed amid volcanic products, as is seen at Collet, Ronzal, and other places, where vegetable matter contained in black carboniferous clays, " accompanied with ferruginous sands, alternate with rolled masses of trachyte, phonolite, basalt, or volcanic cinders." During the progress of these eruptions, the water-courses became much deranged, lava-currents crossing these channels, damming up the passage, and forming lakes, in which various singular compounds and rock-mixtures were produced. It would appear from the large size and rounded angles of many of the fragments of basalt, that great currents of water had acted upon them in certain situations. After a time this great confusion seems to have ceased, and the large fragments became covered by a deposit of sand and clay, formed into regular strata, as may be observed near Cussac. During this state of things near Cussac, animals of the *Bos* kind, and gigantic stags, became

---

* Robert, Férussac's Bulletin de Sci. Nat. et de Géologie, Oct. 1830.

entombed. After this, the district seems to have become the haunt of hyænas, which, issuing from their dens in search of food, dragged their prey into their retreats*, in the manner of the Kirkdale hyænas.

In these various localities in central France, the evidence seems generally in favour of the great outburst of volcanos after the deposit of very extensive fresh-water rocks, the volcanic action continuing more or less from that period up to a comparatively recent date.

Quitting central France and proceeding either in the direction of Aix or Montpellier, we find remains of volcanos, which probably were more or less contemporaneous with those of Auvergne. Beaulieu near Aix has been known since the time of De Saussure.

Spain, Italy, and Germany, present us with various igneous rocks, which appear referrible to the epoch in which the supracretaceous rocks were in the course of formation. As yet, the volcanic rocks of Spain are little known; but those of Germany and Italy, and especially those of the latter, have long engaged the attention of geologists.

The Euganean Hills, south of Padua, present a mass of trachytic and other volcanic products, which belong to the supracretaceous epoch; as they rest in certain situations on *scaglia*, the equivalent of chalk. Dr. Daubeny mentions that the trachyte is associated with basalt at Monte Venda. The same author informs us, that at the hill of Belmonte in the Vicentine, a rivulet section exposes five basaltic dykes, which from their mode of occurrence might be mistaken for an interstratification of chalk and basalt. " Dykes of basalt are also frequently seen traversing this formation at Chiampo, Valdagno, and Magre, but without altering the adjacent rock†." An extensive formation of porphyritic augite rock covers the whole district, resting in some places on chalk, in others on older rocks, filling up the preexisting inequalities in each; the upper part is amygdaloidal: this is surmounted by various alternations of calcareous beds, with others composed of fragments, basalts, volcanic sand, and scoriform lava; the aggregate or mixture of volcanic substances containing fossil remains, as well as the calcareous deposits, and being often as fully charged with them‡. The long celebrated fossil fish from Monte Bolca are derived from the calcareous beds of this deposit. At Ronca there are six alternations of volcanic substances with the calcareous beds, the lowest volcanic product being a cellular basalt.

* Hibbert, On the Fossil Remains of the Velay; Edin. Journ. of Sci. vol. iii. 1830.
† Daubeny, Description of Volcanos. ‡ *Ibid.*

M. Al. Brongniart presents us with the following list of the shells and zoophytes in these beds of the Vicentine, the locality of each being marked (R. for Ronca; C. G. Castel-Gomberto; V. S. Val-Sangonini; M. M. Monteccio-Maggiore;): —*Nummulites nummiformis,* Defr., R.; *Bulla Fortisii,* Al. Brong., R.; *Helix damnata,* Al. Brong., R.; *Turbo Scobina,* Al. Brong., C. G.; *T. Asmodei,* Al. Brong., R.; *Monodonta Cerberi,* Al. Brong., V. S.; *Turritella incisa,* Al. Brong., R.; *T. asperula,* Al. Brong, R.; *T. Archimedis,* Al. Brong., R.; *T. imbricataria,* Lam., R.; *Trochus cumulans,* Al. Brong., C. G.; *T. Lucasianus,* Al. Brong., C. G.: *Solarium umbrosum,* Al. Brong., R.; *Ampullaria Vulcani,* Al. Brong., R.; *A. perusta,* Defr., R.; *A. obesa,* Al. Brong., M. M. and C. G.; *A. depressa,* Lam., R.; *A. spirata,* Lam., V. S.; *A. cochlearia,* Al. Brong., C. G.; *Melania costellata,* Lam., (var. *roncana,* Al. Brong.,) R., and V. S.; *M. elongata,* Al. Brong., C. G.; *M. Stygii,* Al. Brong., R.; *Nerita conoidea,* Lam., R.; *N. Acherontis,* Al. Brong., R.; *N. Caronis,* C. G.; *Natica cepacea,* Lam., Val de Chiampo; *N. epiglottina,* Lam., R.; *Conus deperditus,* Broc. (var. *roncanus,* Al. Brong.), R.; *C. alsiosus,* Al. Brong., R.; *Cypræa Amygdalum,* Broc., R.; *Cyp. inflata,* Lam., R.; *Terebellum obvolutum,* Al. Brong.; *Voluta subspinosa,* Al. Brong., R.; *V. crenulata,* Lam., V. S.; *V. affinis,* Broc., R.; *Marginella Phaseolus,* Al. Brong., R.; *M. eburnea,* Lam., R., and V. S.; *Nassa Caronis,* Al. Brong., R.; *Cassis striata,* Sow., R.; *C. Thesei,* Al. Brong., R.; *C. Æneæ,* Al. Brong., R.; *Murex angulosus,* Broc., various parts of the Vicentine; *M. tricarinatus,* Lam., Vicentine; *Terebra Vulcani,* Al. Brong., R.; *Cerithium sulcatum,* Lam. (var. *roncanum,* Al. Brong.), R.; *C. multisulcatum,* Al. Brong., R.; *C. undosum,* Al. Brong., R.; *C. combustum,* Defr., R.; *C. calcaratum,* Al. Brong., R.; *C. bicalcaratum,* Al. Brong., R. &c.; *C. Castellini,* Al. Brong., R.; *C. Maraschini,* Al. Brong., R.; *C. corrugatum,* Al. Brong., R.; *C. saccatum,* Defr., R.; *C. ampullosum,* Al. Brong., C. G.; *C. plicatum,* Lam., R.; *C. lemniscatum,* Al. Brong., R.; *C. Stropus,* Al. Brong., C. G.; *Fusus intortus,* Lam. (var. *roncanus,* Al. Brong.), R.; *F. Noæ,* Lam., R.; *F. subcarinatus,* Lam. (var.) R.; *F. polygonus,* Lam., R.; *F. polygonatus,* Al. Brong., R.; *Pleurotoma clavicularis,* Lam., M. M.; *Pteroceras Radix,* Al. Brong., C. G.; *Strombus Fortisii,* Al. Brong., R.; *Rostellaria corvina,* Al. Brong., R.; *Ros. Pes-carbonis,* Al. Brong., R.; *Hipponyx Cornucopiæ,* Defr., R.; *Chama calcarata,* Lam., C. G.; *Spondylus cisalpinus,* Al. Brong., C. G.; *Ostrea,* R.; *Pecten lepidolaris?* Lam., R.; *P. plebeius?* Lam., R.; *Arca Pandoræ,* Al. Brong., C. G.; *Mytilus corrugatus,* Al. Brong., R.; *M. edulis?* Linn., R.; *M. Antiquorum,* Sow., R.; *Lu-*

cina Scopulorum, Al. Brong., R.; *L. gibbosula*, Lam., R.;
Cardita Arduini, Al. Brong., C. G.; *Cardium asperulum*,
Lam., C. G.; *Corbis Aglauræ*, Al. Brong., C. G.; *Cor. la-
mellosa*, Lam., R.; *Venus? Proserpina*, Al. Brong., R.; *V.?
Maura*, Al. Brong., R.; *Venericardia imbricata*, Lam., C. G.;
Ven. Lauræ, Al. Brong., C. G.; *Mactra? erebea*, Al. Brong.,
R.; *M.? Sirena*, Al. Brong., R.; *Cypricardia cyclopæa*, Al.
Brong., R.; *Psammobia pudica*, Al. Brong., V. S.; *Cassidu-
lus testudinarius*, Al. Brong., R.; *Nucleolites Ovulum?* Lam.,
R.; *Astrea funesta*, Al. Brong., R.; *Turbinolia appendiculata*,
Al. Brong., R.; *T. sinuosa*, Al. Brong., Vicentine *.

It has been concluded, and with great probability, that
these rocks were produced by the alternate eruptions of volcanos in the vicinity, and the deposit of calcareous matter in
shallow seas. M. Brongniart mentions that parasitical shells
and certain corals are seen adhering to fragments of igneous
rocks, which shows that these rocks have had abundant time
to cool and form the bottom of the sea previous to the deposits
above them. And as in some places igneous products and
calcareous deposits often alternate, we may infer that a long
period elapsed during the formation of the whole.

On the north and south of Rome there is abundant proof
of extinct volcanic action. At Viterbo basaltic rocks rest on
a compound of pumice and volcanic tuff, in which the bones
of mammalia have been discovered; reminding us of Auvergne.
Rome itself is founded on rocks of volcanic origin, mixed with
others which are aqueous, and mostly of contemporaneous
formation. Proceeding hence to Sicily, we find it very difficult to conceive when the volcanic action commenced which
now finds a vent at Etna; as volcanic products are found
mixed with supracretaceous rocks. Dr. Daubeny observes,
that the supracretaceous blue marl which occupies a considerable portion of Sicily, contains sulphur, various sulphuric
salts, and muriate of soda; all substances sublimed from modern volcanos, and which may have been produced by exhalations from beneath.

Among the variety of volcanic products in the vicinity of
the Rhine and neighbouring parts of Germany, are many
which seem clearly to belong to the supracretaceous epoch.
Among these may be mentioned the Siebengebirge, the Westerwald, the Habichtswald near Cassel, and the Meisner near
Eschwege. The Siebengebirge are composed of trachyte,
basalt, and volcanic conglomerates, traversed by dykes. The
Westerwald is composed of the like substances. Basaltic
knolls are scattered over the country between the Westerwald

---

* Brongniart, Terrains Calcaréo-Trappéens du Vicentin, 1823.

and the Vogelsgebirge. The Kaiserstuhl and the igneous rocks on the north of the lake of Constance would appear to be examples of volcanic rocks which may have been ejected at the supracretaceous epoch.

According to M. Beudant there are five principal volcanic groups in Hungary, referrible to the age with which we are now occupied:—1. That in the district of Schemnitz and Kremnitz. 2. That constituting Dregeley mountains, near Gran on the Danube. 3. That of the Matra, in the centre of Hungary. 4. The chain commencing at Tokai, and extending north about twenty-five leagues. 5. That of Vihorlet, connected with the volcanic mountains of Marmorosch (borders of Transylvania). The whole composed of different varieties of trachytic rocks.

According to Dr. Boué, volcanic rocks of undoubted supracretaceous origin occur in Transylvania. They constitute a range of hills separating Transylvania from Szecklerland, and extending from the hill of Kelemany, north of Remebyel, to the hill Budoshegy, on the north of Vascharhely. They are principally composed of varieties of trachyte, and trachytic conglomerate*.

From the observations of Von Buch and Dr. Daubeny, it appears that Gleichenburg, not far from Gratz, Styria, is composed of trachyte, round which are mantle-shaped strata of volcanic products and supracretaceous beds, alternating with each other.

If we turn from these igneous products on the continent of Europe to our own islands, we find that great igneous eruptions have taken place in the north-eastern parts of Ireland, after the deposit of the chalk, and consequently in the supracretaceous period. The basaltic ranges of the celebrated Giant's Causeway, Fairhead, &c. belong to this eruption, which in its upburst has torn and rent all which it encountered, entangling enormous masses of chalk, as may be seen at Kenbaan. We find the mass of this erupted igneous rock to be basaltic,— sometimes columnar, at others not; the two varieties being so arranged on the coast between Dunseverie Castle and the Giant's Causeway, that they have the appearance of being interstratified. At Murloch Bay, Fairhead, and Cross Hill, the basalt rests on coal measures; at Knocklead and other places, on chalk†. As an intermixture with supracretaceous rocks has not yet been observed, the relative date of this eruption cannot be well determined. Both the basaltic mass and the rocks on which it rests have been traversed, at a period

---

* Daubeny's Volcanos.

† Buckland and Conybeare, Geol. Trans. vol. iii.; and Sections and Views illustrative of Geological Phænomena, pl. 19.

posterior to the first overflow of the former, by dykes of
igneous matter: one of these has pro-
duced a singular change in the chalk,
which it cuts, together with superincumbent basalt, in the Isle of Raghlin,
as will be best explained by the annexed
section. *a a a,* trap dykes cutting
through chalk *b b,* which it has converted into granular limestone *c c c c.*

Fig. 41.

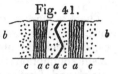

It now only remains to consider those recent observations
on the Alps, Pyrenees, and the vicinity of Maestricht, which
seem to point to at least a zoological passage of this group
into the next; appearing to show, that from the progress of
science, the clear line of separation once supposed to exist
between the secondary and tertiary classes, as they are termed,
cannot be drawn, but that the zoological character of the upper
part of the one and the lower portion of the other would approach each other, as indeed might be expected; for we cannot conceive a natural destruction of life so general as to cause
the complete annihilation of animals, particularly those which
are marine, existing at any given time, so that a totally new
creation should be necessary. Such a supposition would not
appear to accord with what is observable in other rocks, as
will be noticed in the sequel. It is not contended that there
may not be great specific distinctions in the remains entombed
in this and the next group in many parts of Europe, but merely
that it does not necessarily follow,—because Europe may present us with two classes of rocks, one of which may be named
tertiary and the other secondary, from the general nature of
their organic contents,—that in many parts of the world the
whole may not constitute a series in which lines of distinction
cannot be drawn. Suppose some violent cause should produce a great debacle which should rush over Europe, the
land and fresh-water animals and plants would probably be
destroyed; and we will even consider, for the sake of the
argument, that the marine inhabitants of our seas perished
also,—does it necessarily follow that the marine, fresh-water,
and terrestrial inhabitants would also be annihilated in Australia? Should we not rather consider that these would be
entombed, if rocks were there forming, as well after and during
the destruction of European life, as previous to it? and that
the rocks formed in those regions, about this supposed period,
would by no means show any alteration in their zoological
character? That very great changes have taken place in the
organic character of deposits in the same districts, and that
somewhat suddenly, does not admit of a doubt; but it is

a subject on which we are, as yet, far from seeing our way clearly.

There is always great difficulty in comprehending why the marine remains should be so suddenly changed in certain deposits, which do not exhibit marks of being the results of violent commotion; for although we can understand why terrestrial and fresh-water animals should be destroyed by a change in the relative levels of sea and land, or an inroad of the sea, it is difficult to comprehend why, from these causes alone, the general character of the marine animals should be changed. The depth of a sea may, indeed, be so changed, from a movement in the bottom, that deep water may be rendered shallow, and shallow water deep; but it does not seem to follow that the inhabitants should necessarily all perish, the species being so completely destroyed as never to be found afterwards. On the contrary, we should be led to infer that, though such changes would by no means suit the habits of the respective inhabitants of such waters, they would merely induce them to seek situations better calculated for their various modes of live.

From an examination of portions of the Austrian and Bavarian Alps, in 1829, Professor Sedgwick and Mr. Murchison concluded that they had discovered a series of beds intermediate between the chalk and commonly known supracretaceous rocks, affording as it were a passage of the, so called, tertiary class into the secondary; yet, as they were above the true chalk, being considered as tertiary. The correctness of this determination is questioned, more particularly by Dr. Boué, who contends that the disputed rocks belong to the cretaceous series. According to the former authors, the valley of Gosau, in the Salzburg Alps, presents a good example of the correctness of their views. This valley is described as about 2600 feet above the level of the sea, exhibiting these newer strata brought suddenly into contact with more ancient rocks on one side. The following is stated to be a section of them, in the descending order. "1. Red and green slaty micaceous sandstone, several hundred feet thick (cap of the Horn). 2. Green micaceous gritty sandstone, extensively quarried as whetstone, succeeded by yellowish sandy marls (Ressenberg). 3. A vast shelly series consisting of blue marls alternating with strong beds of compact limestone and calcareous grit, the upper beds of which are marked by obscure traces of vegetables, and the middle and inferior strata by a prodigious quantity of well preserved organic remains*." The fossils found in the lowest

* Proceedings of the Geol. Soc., Nov. 1829.

strata at Gosau bear the impress, according to these authors, of the cretaceous period; while those of the overlying blue marls approach so nearly to many species of the lower supracretaceous or tertiary formations, that they refer the whole deposit to an age intermediate between the chalk and those formations hitherto considered as tertiary *.

Dr. Boué is by no means willing to admit this deposit of Gosau as a tertiary or supracretaceous rock, but as constituting a part of the cretaceous series which extends along the Alps, as will be seen in the next section, from Austria into Savoy†.

It may here be remarked that M. Brongniart long since (1823) considered that certain rocks constituting the upper part of the Diablerets (Valais), were referrible to the supracretaceous or tertiary series. From the section of this mountain, made by M. Elie de Beaumont, and produced by M. Brongniart, it appears that the strata are singularly contorted, so that the newer beds have been twisted between the older strata in such a manner that the latter not only occur beneath the former, but also above them‡. The beds considered supracretaceous are described as composed of calcareous sandstone, anthracite, and a black, compact, and carbonaceous limestone, containing *Nummulites; Ampullaria* (two species); *Melania costellata,* Lam.; *Cerithium Diaboli,* Al. Brong. (very abundant); *Turbinella? Hemicardium; Cardium ciliare,* Broc.; *Caryophyllia; Madrepora.*

The nummulites found so abundantly in the Alps by no means mark a distinct geological epoch, as they would appear to do in Northern France and in England; for instead of being confined to the supracretaceous group, they pervade the cretaceous, and possibly also some older rocks.

The observations of Dr. Fitton on the Maestricht beds would appear to throw light on these Alpine deposits, as far at least as their zoological character is concerned; it being understood that the celebrated deposit of the Mont St. Pierre contains a mixture, to a certain extent, of the, so called, secondary and tertiary remains; that "it is throughout superior to the white chalk, into which it passes gradually below, but the top bears marks of devastation, and there is no passage

* The various labours of Prof. Sedgwick and Mr. Murchison on the Alps will be found in the second part of vol. iii. of the Geol. Transactions, 2nd series, where there are also figures of some of the fossils discovered by them at Gosau. For a list of the Gosau fossils, see Lists of organic remains at the end of this volume.

† Boué, various memoirs, Edinburgh Phil. Journal, 1831; Journal de Géologie, 1830; and Proceedings of the Geol. Soc. of London, 1830.

‡ Brongniart, Sur les Terrains Calcaréo Trappéens du Vicentin, p. 47; and Sections and Views illustrative of Geological Phænomena, pl. 38, fig. 5.

from it to the sands above. The siliceous masses which it includes are much more rare than those of the chalk, of greater
bulk, and not composed of black flint, but of a stone approaching to chert, and in some cases to chalcedony; and of about
fifty species in the author's (Dr. Fitton's) collection, about
forty are not found in Mr. Mantell's catalogue of the chalk
fossils of Sussex*."

According to M. Dufrénoy, a similar mixture of the organic
remains, usually considered as characterizing the cretaceous
and supracretaceous rocks respectively, is discovered in the
upper part of the chalk series of the Pyrenees. This author
observes, that out of numerous species obtained from this deposit, many are such as are commonly referred to the supracretaceous epoch†.

From these data, it would appear that at Maestricht, in
the Pyrenees, and in the Alps, there do exist deposits containing organic remains common to the supposed great classes
of secondary and tertiary rocks ; therefore it seems established
that no line can, zoologically, be drawn between them. How
far other characters may distinguish them, remains to be seen ;
and probably minute researches will eventually afford the necessary information.

* Fitton, Proceedings of the Geol. Soc. 1830.
† Dufrénoy, Annales des Mines, 1831 and 1832.

Section V.

CRETACEOUS GROUP.

Syn.—Chalk, (*Craie*, Fr.,—*Kreide*, Germ.,—*Scaglia*, It.,). Chalk Marl, (*Crai Tufau*, Fr.,—*Kreidemergel*, Germ.) Upper Green Sand, (*Glauconie, Crayeuse*, Fr.,—*Chloritische Kreide, Planerkalk*, Germ.). Gault. Lower Green Sand, (*Glauconie Sableuse*, Al. Brong.,—*Grünsand*, Germ. —Part of the German *Quadersandstein*.)

The upper portion of the cretaceous group partakes of a common character throughout a considerable portion of Western Europe, generally presenting itself under the well-known form of chalk. The upper part of the chalk throughout a large portion of England is characterized by the presence of numerous flints, more or less arranged in parallel lines: seams of this substance not only occur in a line with the flints, but also traverse the beds diagonally. The white chalk, when freed from the flints or siliceous grains mixed with it, is found to be a nearly pure carbonate of lime. According to M. Berthier, the chalk of Meudon, when the sand disseminated in it was separated by washing, contained in 100 parts,—carbonate of lime 98, magnesia and a little iron 1, alumine 1. In the lower parts of the English chalk deposit, the flints disappear, becoming gradually more rare in the passage from the upper to the lower parts. From this circumstance, the white chalk has not unfrequently been divided into upper, or chalk with flints, and lower, or chalk without flints. This supposed characteristic is not available to any great distances; for at Havre the lower chalk contains an abundance of flint and chert nodules, where it passes into the upper green sand. There is, however, along the line of coast from Cap la Hêve to the eastward, a considerable accumulation of chalk, in which flints are rare, apparently interposed between the Havre beds and the chalk with numerous flints. In the cliffs of Lyme Regis (Dorset), and Beer (Devon), we observe how little dependence can be placed on minute divisions of rocks, even within the distance of a few miles; for considerable differences in the development of the cretaceous series will be observed between the two places, as I had formerly occasion to remark *. There are,

* Geol. Trans. 2nd series, vol. ii.—An example of a notable change in a much shorter distance is observable at Cap la Hêve (Havre). Towards the town, and even under the light-houses, a marl or clay may be observed separating the arenaceous beds into two masses, which marl gradually disap-

however, a few beds which are remarkably persistent throughout the district, extending to Weymouth; they are characterized by the presence of small and irregularly rounded grains
of quartz, probably of mechanical origin, occasionally disseminated through the mass in great abundance. These beds are also
remarkable for a great variety of organic remains. Notwithstanding the very general presence of these beds, they sometimes
become almost suddenly replaced by others, wherein the grains
of quartz are not seen. Thus at Beer, the Beer stone, worked
during centuries for architectural purposes, seems the equivalent of them, though composed of a white rock, principally
carbonate of lime, with some argillaceous and siliceous matter.
Probably the Beer stone may be the equivalent of the Malm
rock of Hants and Surrey described by Mr. Murchison, and
the Merstham firestone noticed by Mr. Webster, and considered as the upper green sand. It may be here observed that
the lower part of the chalk, or its passage into the green sand
beneath, is extensively used as a building stone in Normandy,
and that some of the inferior chalk beds of that country are
considerably indurated, even approaching a whitish compact
limestone, as may be well seen on the high road, bordering
the Seine, between Havre and Rouen. M. Passy remarks,
that many beds of the Norman chalk, divided from each other
by flints, are so compact as even to approach a crystalline
fracture (Elbeuf, Gouy, Duclair, &c.). The lower portion
of the cretaceous group has, in England more particularly,
received various names, though the mass is very commonly
known as green sand. These subdivisions, for the accurate
determination of which, and their separation from the Weald-
en rocks, we are indebted to Dr. Fitton *, should be borne
in mind, more particularly in the study of English geology;
as by tracing them as far as possible, we may obtain an insight
into the causes which have produced them. These divisions
are, Upper Green Sand, Gault, and Lower Green Sand; and
can be best studied in the south-eastern parts of England†.

The upper green sand generally appears to graduate into
the cretaceous mass above, and is charged with a large quan-

---

pears, so that at a short distance eastward from the light-houses the bed
(about 18 feet thick) becomes a slightly argillaceous green sand. M. Passy
gives a tabular view of these changes in his Description Géologique de la
Seine Inférieure : Rouen, 1828, p. 235.

* Fitton, On the Beds between the Chalk and Purbeck Limestone: Annals
of Philosophy, 1824;—a memoir in which the general relations of all these
beds were first pointed out.

† The student should consult Dr. Fitton's Memoir (above cited); Mr.
Murchison's Memoir on North-western Sussex, Geol. Trans. 2nd series,
vol. ii.; Mr. Mantell's Geology of Sussex; Mr. Martin on West Sussex; and
Dr. Fitton's Geological Sketch of the Vicinity of Hastings, 1833.

tity of green grains, which, according to the analysis of M. Berthier, made on those of the equivalent deposit at Havre, contain:—Silica 0·50, protoxide of iron 0·21, alumine 0·07, potash 0·10, water 0·11. The same author found that the green or reddish nodules disseminated through the same rock, also at Havre, contained:—Phospate of lime 0·57, carbonate of lime 0·07, carbonate of magnesia 0·02, silicate of iron and alumine 0·25, water and bituminous matter 0·07. The reader will at once observe the different composition of the nodules and grains. Respecting the former, M. Al. Brongniart observes, that the phosphate of lime sometimes so abounds as nearly to constitute the whole substance *.

The gault (or galt) is an argillaceous deposit of a blueish gray colour, frequently composed of clay in the upper, and marls in the lower part, containing disseminated specks of mica ; it effervesces strongly with acids.

The lower green sand is formed of sands and sandstones of various degrees of induration, but principally of ferruginous and green colours, the former usually constituting the upper part, and the latter being most prevalent in the lower portions, which are not unfrequently argillo-arenaceous, particularly at bottom.

Without entering further into the smaller divisions of the cretaceous group, it may be remarked that the whole, taken as a mass, may in England, and over a considerable portion of France and Northern Germany, be considered as cretaceous in its upper part, and arenaceous and argillaceous in its lower part. The divisions established in south-eastern England have been observed by Mr. Lonsdale in Wiltshire; and M. Dumont considers that the inferior portion of the cretaceous group, which occurs between the Meuse and the Roer, and is rather thick near Aix la Chapelle, may be well divided into Upper Green Sand, Gault, and Lower Green Sand †. In northern England the arenaceous deposit is scarcely observable, the white chalk resting on red chalk, the latter based on an argillaceous rock, named Speeton clay by Mr. Phillips. In south-western England the chalk rests on a great arenaceous deposit somewhat variable in its composition, sometimes containing thick regular seams of chert, at others being nearly without them ; the lower portion being very generally an argillo-arenaceous deposit, characterized by the presence of a great abundance of green particles, and a great variety of organic remains. The central part is formed of yellowish-brown and loosely aggregated sand, in which organic remains

---

* Cuvier and Brongniart, Desc. Géol. des Env. de Paris, 1822, p. 13.
† Omalius d'Halloy, Eléments de Géologie.

are rare; the superior, of a mixture of brownish-yellow and green sands, with and without chert seams, the organic remains being frequently fractured.

In Normandy the sands beneath the chalk assume a great variety of characters. Advancing into the interior of France, amid the sands which emerge from under the chalk, and extend from the coasts of Normandy by Mortagne to the banks of the Loire at Tours, and thence by the vicinities of Auxerre and Troyes to the northward, we soon become sensible of the utility of abandoning the smaller divisions, so valuable in England, and of adopting two great divisions,—Chalk and Green Sand.

This group is extensively distributed over Europe. The chalk and *mulatto*, or green sand of northern Ireland, is the most western portion known in the British Islands. It occurs on the Spanish side of the Pyrenees, and would appear, from the observations of Colonel Silvertop, to extend much further westward into the kingdom of Spain. It is probably found in the provinces of Sevilla and Murcia. Of the geological structure of Portugal so little is yet known, that we are not aware of the existence of cretaceous rocks in that country.

According to M. Nilsson, the chalk of Sweden (the continuation of that in Denmark,) is generally incumbent on gneiss, more rarely on rocks of the grauwacke group, and has only been observed resting on beds of the oolitic group, at one place near Limhamn, in Scania. In one locality, near Hammer and Käseberga, it has a large capping of sand with bituminous wood, which M. Nilsson refers to the cretaceous group, as the vegetable remains are associated with cretaceous fossils. The chalk deposit of Sweden is occasionally of considerable thickness, and abounds in organic remains. The northern portion of the deposit is white or grayish white, more or less abundantly mixed with siliceous substances. The southern portion is stated to present the various modifications from green sand to white chalk *.

From the observations of Professor Pusch it appears that the cretaceous group occurs extensively in Podolia and southern Russia, being a continuation of that of Lemberg and Poland. It occupies the country in the shape of marly chalk between the Bog and the Dniester round Janow, Lubin, Mikolajew, Uniow, and Rohetyn. Concealed beneath the supracretaceous rocks it is prolonged from Halicz to Zalezczyki on the Dniester. On the west of this river it occupies the environs of Tlumacz, Otynia, and other places to the foot of the

---

* Nilsson, Petrificata Suecana Formationis Cretaceæ descripta, et iconibus illustrata: 1827.

Carpathians. On the north of the Dniester it exists beneath the supracretaceous rocks between that river and Brzezan ; it extends to Brody and into the plains of Volhynia. " In many places, and especially around Krzeminiec, it is covered by more recent deposits, but its presence is indicated by an abundance of flints and chalk fossils scattered through the sands." The chalk forms considerable eminences round Grodno in Lithuania. According to M. Eichwald, the chalk of the latter country abounds in belemnites, which are wanting in Volhynia, where they are replaced by *Echinites*, *Terebratulæ*, *Ostreæ*, *Placunæ*, *Inoceramus (Catillus)*, &c. The flints of the two countries contain *Reteporæ*, *Escharæ*, *Ananchytes*, *Encrinites*, &c. *.

According to M. Eichwald, chalk without flints, with shells of the genera *Plagiostoma*, *Pecten*, *Ostrea*, &c., rests on argillaceous slate at Ladowa, on the Dniester. At about seven wersts from thence, near Bronnitza, it alternately rests on a coarse sandstone, grauwacke, and argillaceous slate †. Further south, and in the plains of Moldavia, Podolia, and Bessarabia, it only appears' in detached portions, as between Jaroszow and Mohilew on the Dniester, from Raszkow to Jaorlik on the Pruth, near Kolomea, Sniatyn, Sadagora, Seret, Roswan, Illina, and Jassy. "The chalk is found on the south side of the granitic steppe, in the Crimea, and on the borders of the Sea of Azof, between the Berda and the Don ; it also occurs on the west of the Don, across the south-east and middle of Russia. In the country of the Don Cossacks, in the governments of Worenech, Koursk, and Toula, it here and there appears in hills, and on the banks of the rivers beneath the vegetable soil, and probably constitutes the base of that great and fertile plain. The marly clay of eastern Gallicia and of Podolia is connected, as in Poland, with gypsum, at Mikulnice, Seret of Podolia, to the east of Trembowla, but more particularly at Zbrycz near Czarnokozienice. The graphic chalk is there more abundant than in the centre of Poland, and more abounds in flints ‡."

It further appears from the interesting details of M. Pusch, that there is a deposit of lignite upon the upper part of the chalk, reminding us of the lignite sand noticed by M. Nilsson in Sweden, which would thus appear to be similarly situated at various distant points. It seems to be wanting in central Poland, but is found in many situations in eastern Gallicia, and abundantly along the Carpathians, in Pocutia and Bukowine, from Otynia towards Maydan, Lanczyn, Kniazdwor, and mounting the Pruth, from Miszyn to Seret, and near

---

* Journ. de Géol., t. ii. p. 62.   † *Ibid.* p. 61.   ‡ Pusch, Journ. de Géol. t. ii.

Czorthow and Ulaszkowce, and on the Dniester near Cho-
chim and Mohilew. This lignite deposit is described as a
blueish or greenish gray calcareous sandstone, alternating
with sand and clay, more or less calcareous, and with laminated marl: it sometimes contains amber, but more frequently
pieces of bituminous wood, thin beds of lignite, and trunks of
fossil trees. It contains many shells, among which are *Pectunculus pulvinatus*, *P. insubricus*, *Pecten* (smooth species),
and more rarely *Nummulites discorbinus*, *Dentalium eburnium*,
and small *Cerithia*. This sandstone is considered distinguishable from the well-known lignite deposits of western and
northern Poland by its fossil shells; but it may perhaps admit of a question, how far local circumstances may not have
caused a great difference in this respect.

Prof. Pusch describes the cretaceous rocks as extensively
deposited in Poland, and as divisible into marly chalk and
white chalk: the marly chalk is a soft calcareous marl, either
white or light gray, becoming sandy in some districts (Mie-
chow, Kazimirz); while other beds are coloured green by silicate of iron (Czarkow, Szczerbakow); it alternates with
more compact white limestone. A shaft sunk through this deposit at Szczerbakow showed that it was 697 English feet thick
at that place. M. Pusch considers that certain gypseous deposits of Poland are connected with the marly chalk. The
white chalk is described as identical with that of England,
containing a much larger proportion of flints than the marly
chalk *.

Rocks of the cretaceous group occur in the great plain of
Münster, skirting the northern edge of the Westphalian slate
mountains to the south-western border of the Teutoburgerwald.
Though the rocks of this district undoubtedly form a continuation of the English chalk series, they differ materially from
it in mineralogical structure. There is not throughout the
whole tract a trace of white chalk with flints. On the southern borders of the district, whitish, yellow, blueish, and green
marls predominate, containing beds of indurated marl full of
green grains, which become sandy. These beds sometimes
rest immediately on rocks of much greater antiquity, such as
the coal measures of Frohnhausen near Essen, and sometimes
occur, twenty feet thick, in the yellowish marl, as may be observed from Aplerbeck, near Werl, to the vicinity of Paderborn. At the salt-works of Koenigsborn, where the marl
deposit is 470 feet thick, two of these beds, in contact with
each other, furnish excellent building stone.

In the central portion of the district the light-coloured marls

* Pusch, Journal de Géologie, t. ii. p. 253.

become more calcareous, and pass into limestone. The following section is observable in the chain of the Teutoburgerwald. At the top, a light gray calcareous marl, which readily crumbles in the air. To this succeeds a compact white limestone, often of a splintery fracture, which becomes green in the lower beds. Next follows, always in the descending order, a dark gray and friable calcareous marl, with globular detached masses of compact limestone (Horn, and on the Lausberge at Bielefeld). Then sandy or gravelly yellowish gray clayey marls, striped black, and containing detached pieces of chert and chalcedony. The lowest portions mostly consist of a light-coloured, and sometimes ferruginous, thick sandstone, rarely associated with beds of conglomerate.

The cretaceous rocks which occur on the borders of the hill country of northern Germany, join the northern side of the Hartz, and extend towards the north-west into the plain between Brunswick and Hildesheim. Cretaceous marl sometimes comes into immediate contact with the grauwacke rocks of the Hartz, and extends, under the sand, into the plain of Peina. On the other hand, the sandstone is much developed in the country about Halberstadt and Blankenburg. It contains a thick bed of sandy marl, full of green grains, which passes into a gray white earthy limestone with flints, resembling that which occurs above the sandstone. Coal occurs in this sandstone near Quedlinburg. The great mass of the cretaceous rock, named *Quadersandstein*, of Saxony and Bohemia, is separated from the border of the north German basin; but Prof. Weiss * has shown that the crystalline rocks which surround it on the N.E. side, from Weinbohla to Hohnstein, have been forced up since it was deposited, and thus caused the separation. The sandstone forming the well-known rocks of Schandau and Adersbach is very uniform, large-grained, and white, containing but little cementing matter.

There are many patches of quadersandstein in Silesia, the most considerable of which rests on the northern edge of the Riesengebirge from Goldberg to Löwenberg, and thence from the Queiss to the Neisse. It extends N.W. towards the low country, and is bounded on the E. by older rocks. The lowest beds often become conglomerates, as in the mountains of Goldberg and Prausnitz, and on the ridge between Löwenberg and Neuland, where millstones are quarried (Wehrau on the Queiss, Wartha). Similar beds of conglomerate are found in the Moiser quarry near Löwenberg, where numerous organic remains are seen in it. The sandstone is clayey at

---

* Weiss, Ueber einige geogn. Punkte bei Meissen und Hohnstein, Karsten's Archiv für Bergbau, &c. B. xvi. p. 3.

U

Töllendorf (Buntzlau), and contains beds, a foot and a half thick, of red and white potters' clay, out of which much pottery is manufactured. In the neighbourhood of Wehrau the sandstone passes into compact quartz rock, containing beds of clay and the well-known clay ironstone. Several coal beds, from six to twenty-four inches thick, are found in it near Wenig Rackwitz, Ottendorf, and Newen.

The cretaceous group occurs extensively on the southern side of the great level to the eastward of the Oder. It fills the basin, forty-five miles broad, between the oolitic high range from Wielun to Cracow, and the Sandomirer grauwacke range. It extends down the Vistula as far as Pulavy, and thence further eastward through the southern part of the province of Lublin to Lemberg and the Dniester. It is here connected with the great chalk plain of Volhynia and the plateau of Podolia.

In the upper division of the cretaceous series on the Nida (west of the Vistula) in the province of Cracow, and in the basin of Lemberg and Podolia, at Mikulnice and Zbrycz as far as the river Podhorec, there is a gypsum deposit, 100 feet thick, consisting of large yellow and gray crystals. The isolated patches of gypsum in Upper Silesia probably belong to this bed *. The sulphur bed at Czarkow on the Nida occurs between this gypsum and the chalk marl. Sulphurous springs, and occasionally weak saline springs, accompany this line from Busko to the Vistula, from Lübien near Lemberg to Jassy †. The fossils of these strata in general resemble those of the English and French chalk, though there are some small local differences ‡.

The cretaceous rocks of France have been already noticed; but it may be remarked that they rest on the coal measures of Mons and Valenciennes, and that the rocks of the Isle d'Aix and the embouchure of the Charente, are considered referrible to this group. They are well known as contained in some of the valleys of the Jura, and as ranging along a considerable portion of the French side of the Pyrenees. They occur on both sides of the Alps, and range down a large portion of the Apennines.

This group occurs extensively in the maritime Alps, containing among its fossils an abundance of *Nummulites*, remains once considered as wholly supracretaceous. Its usual appearance in that district is that of a marno-arenaceous limestone, the arenaceous matter sometimes predominating, and forming a sandstone. Beds of light-coloured limestone charged with green

* German Transl. of Manual.

† Pusch, Ueber die geogn. Konstitution der Karpathen und der Nord-Karpathen-Länder.—Karsten's Arch. für Min. &c. Bd. i. p. 29.

‡ German Transl. of Manual.

grains, and full of *Belemnites, Ammonites, Nautili,* and *Pectines,* constitute its lower part, and even appear intimately connected with the upper part of a light-coloured limestone deposit, among which crystalline dolomite abounds. The latter rocks are very difficult to classify, and may either belong to the lower part of the cretaceous, or upper part of the oolitic, group. Be the age of these beds what it may, they seem, according to M. Elie de Beaumont, intimately connected with a large proportion of the Alpine nummulitic rocks, the light-coloured limestones of Provence, of Mont Ventoux, of the departments of the Drome, Isère, &c.; the nummulitic rocks being connected with the cretaceous series of Briançonnet (Basses Alpes), of Villard le Lans (Isère), of the mountains of the Grande Chartreuse, of the Mont du Chat, of the high longitudinal valleys of the Jura, of the Perte du Rhone, of Thonne, and of la Montagne des Fis.

Having premised thus much respecting the geographical distribution of the cretaceous group, we will take a slight sketch of the variations in its mineralogical character. Throughout the British Islands, a large part of France, the northern parts of Germany, in Poland, Sweden, and in various parts of Russia, there would appear to have been certain causes in operation, at a given period, which produced nearly, or very nearly, the same effects. The variation in the lower portion of the deposit seems merely to consist in the absence or presence of a greater or less abundance of clays or sands, substances which we may consider as produced by the destruction of previously existing land, and as deposited from waters which held such detritus in mechanical suspension. The unequal deposit of the two kinds of matter in different situations would be in accordance with such a supposition. But when we turn to the higher part of the group, into which the lower portion graduates, the theory of mere transport appears opposed to the phænomena observed, which seem rather to have been produced by deposition from a chemical solution of carbonate of lime and silica, covering a considerable area*. For the reader will have observed, that white chalk, very frequently containing flints, extends from Russia, by Poland, Sweden, Denmark, Northern Germany, and the British Islands, into France. The great European sheet of chalk and green sand, produced at the cretaceous epoch, has since been so covered up, shattered, upheaved and destroyed by various

---

* If we regard present appearances, we find that silica is held in solution by thermal waters, which also, as in the case of those of St. Michael in the Azores, may contain carbonate of lime. No springs or set of springs that we can imagine are likely to have produced this great deposit of chalk, so uniform over a large surface.

causes, that we have mere remnants presented to our examination. Still, however, we have enough to show that it overlapped a great variety of pre-existing rocks, from the gneiss of Sweden to the Wealden deposits of south-eastern England inclusive.

Thus far no very material difference in the arrangement and mineralogical character of the mass has been observed, of course disregarding small local variations: but arrived at the Alps we meet with rocks, which certainly, from their mineralogical characters alone, would never have been referred to the cretaceous group: yet, unless we disregard the evidence of organic remains, they have been formed at the same epoch. Instead of the soft and white chalk, and the abundance of loosely aggregated sands, which constitute so large a proportion of the group in England and northern France, we have compact limestones and sandstones vying in hardness with the oldest rocks, so as, in the earlier days of geology, to have been considered only referrible to them. Such is the hard black limestone (containing an abundance of *Scaphites, Hamites, Turrilites,* and other fossils,) which crowns the summits of the Fis, the Sales, and other mountains of Savoy, that range up to the Buet.

The rocks referrible to this group, on the southern side of the Alps, and facing the great Lombardo-Venetian plains, are not so far removed from the mineralogical character of the chalk of western Europe, being often composed of white, greenish, and reddish beds, occasionally very argillaceous. In some parts of the Apennine range, in which a large mass of rocks would seem referrible to this epoch, the character is quite cretaceous.

How far the Alpine rocks of this age have been altered since their deposit, in consequence of the disturbances they have experienced, or how far their present condition can be attributed to original formation, which must always have been influenced by local causes, yet remains a problem to be solved: but it may be remarked, that we can scarcely imagine them to have been exposed to the various circumstances attending great disturbances, without having suffered from such circumstances.

According to M. Dufrénoy, the cretaceous series of southern France not only contains a curious mixture of organic remains, but also presents mineralogical characters different from those of the contemporaneous deposit of the northern part of the same country. That portion which reposes on the central elevations of France, is composed, in its lowest parts, of marls and sandstones, more or less charged with oxide of iron, and containing lignite in some situations. M. Dufrénoy refers these beds, such as they are seen at Rochefort, Angoulême,

Sarlat, Pont St. Esprit, and other places, to the inferior arenaceous rocks of the cretaceous series. At Angoulême, and
some other localities, these deposits are surmounted by regular beds of a nearly saccharine limestone,—a fact which shows
that a slow chemical deposit here took place; so that if we consider the white chalk of northern Europe as chemically formed,
it would appear that there was a slower deposit in some localities than in others. The same author also states, respecting that portion which either constitutes a part of the Pyrenees, or is continuous with it, that although the limestones
which rest on the arenaceous deposits (containing lignites and
vegetable impressions) are commonly compact, there are some
which are crystalline. It should however be observed, that
there are evidences of mechanical action in the upper portion
of the Pyrenean chalk, for it is stated that thick beds of calcareous conglomerates alternate with the limestones in the
upper part of the series *.

The same author states that the cretaceous rocks of the
Spanish Pyrenees closely resemble those on the French side
of the same mountains. The celebrated salt mine of Cardona
is contained in the upper part of the series, and the rock salt
of Mon Real is also included in it. Saline springs occur near
Orthez, between Jaca and Pampeluna, and other places, accompanied by gypsum, trap rocks, and dolomite,—always, it
is stated, in lines of fractured country. Coal is discovered in
this series at Pereilles near Bellesta, Ernani near Irun, at
Saint-Lon in the Landes, &c., and sulphur and bitumen at
Saint-Boes near Orthez †.

M. Partsch describes a series of calcareous and arenaceous
rocks containing nummulites in Dalmatia and the neighbouring provinces, which appears to belong to the cretaceous
group. These rocks form high mountains, particularly in
Croatia. From the direction of the mountain chains, M. Elie
de Beaumont infers that these rocks may extend into Livadia
and the Morea. Facts can alone determine how far this inference is correct; but in the mean time it may be remarked,
that rocks of the Dalmatian character seem to prevail extensively in parts of Greece, and even along the coast of Karamania.

From the various memoirs of MM. Keferstein and Boué,
Prof. Sedgwick, Mr. Murchison, and M. Lill von Lillienbach,
it seems clear that the cretaceous group exists extensively in
the Alps of Austria and Bavaria, and in the Carpathians.
There may be certain differences of opinion as to where the
series commences, or where it ends, but the main fact of the

* Dufrénoy, Annales des Mines, 1831. † *Ibid.* 1832.

presence of the group itself would appear to be undisputed: it would also appear that the deposit was in a great measure arenaceous.

After remarking on the stability of the cretaceous rocks of the Carpathians since their deposit, contrasted with their dislocation in the main chain of the Alps, (a fact subsequently fully confirmed within a certain distance from Vienna by Mr. Murchison,) M. Elie de Beaumont proceeds to observe that "nearly in the prolongation of the Carpathians, to the environs of Dresden, the right and northern side of the Elbe valley is bordered by a continuation of granite and sienite mountains, which extend from Hinterherms, on the frontier of Bohemia, to Weinböhla, about a league and a half east from Meissen, rising suddenly above the plain of quadersandstein and planerkalk (cretaceous rocks). When the contact of the granitic and cretaceous rocks is examined, it is observed that the former cut, and even horizontally cover, the latter in many places; clearly proving that the granitic and sienitic rocks were elevated to the surface since the deposit of the green sand and chalk: and it is not the less remarkable, that the little chain formed of them runs in the direction of the valley of the Elbe, and exactly parallel to that which reigns in the Pyreneo-Apennine system *."

The most remarkable point is at the quarry of Weinböhla, where, according to M. Weiss, the chalk there worked contains the *Plagiostoma spinosum, Podopsis, Spatangus,* &c. This rock is in horizontal beds; but near the sienite they gradually dip until they plunge beneath it, so that the sienite conformably covers the chalk. A marly and clay bed, partly bituminous, covers the chalk, occurring between it and the granitic rock. M. Klipstein remarking on these appearances, observes, that mounting the valley of Polenz, from the foot of the Hochstein, the green sand beds on the right, which are generally horizontal, begin gradually to dip, the angle increasing with their approach to the granite; near the latter, dipping at 46° or 48° beneath it; and he states that of this fact there can be no doubt. "Coming from Brand, the height of the green sand diminishes in such a manner in the descent of the valley, that a few feet of it are alone visible. In a valley extending into the mountains towards the Rothenwald, the chalk with its marls and clays appears between the green sand and granite; and there are places where galleries have been driven through the granite and chalk into the green sand."

---

* M. Elie de Beaumont cites these curious appearances of the superposition of granitic rock, as obtained from the descriptions of Prof. Weiss, inserted in Karsten's Archiv für Mineralogie, &c. t. xvi., and new series, t. i.

From these works it would appear that "the chalk with its clays and marls gradually diminishes, so that the granite at first resting on chalk, comes into contact with green sand. The superposition of the granite is quite evident at some distance from this point, when suddenly there is a change, and the granite cuts the arenaceous beds without at all deranging or altering them : it is even stated, that beneath it commences taking a position under the green sand *." Prof. Naumann remarks, that the fact of the increased dip of the cretaceous rocks as they approach the granite, so that they finally are covered by it, is also seen near Oberau ; and that near Zscheila and Niederfehre, the cretaceous rocks rest horizontally on the granite. The same author remarks that the connection of the two rocks is sufficiently evident at both these localities, for the limestone and granite are, as it were, entangled in each other, and irregular portions and veins of hard limestone with green grains and cretaceous fossils are here and there imbedded in the granite. The gorge of Niederwarta, on the left bank of the Elbe, is pointed out as a very interesting point. "The chalk is horizontal in the village, but at about the third of a league beyond it, the beds rise and dip at about 25° or 30°; a hundred paces further on, the dip is from 70° to 80°, and the rocks, fractured near the granite, rise in steep mountains above the chalk country." At Lichtenhain and Ottendorf the limits of the sandstone and granite, are exposed, and at twenty paces from the granite the sandstone is seen to be horizontal; but on approaching the granite, the beds, or fragments of beds, rise, and some dip at an angle of 60° †.

For the following section, representing the contact of the sienite or granite with the chalk at Weinböhla, I am indebted to Mr. Killaly.

Fig. 42.   Fig. 43. 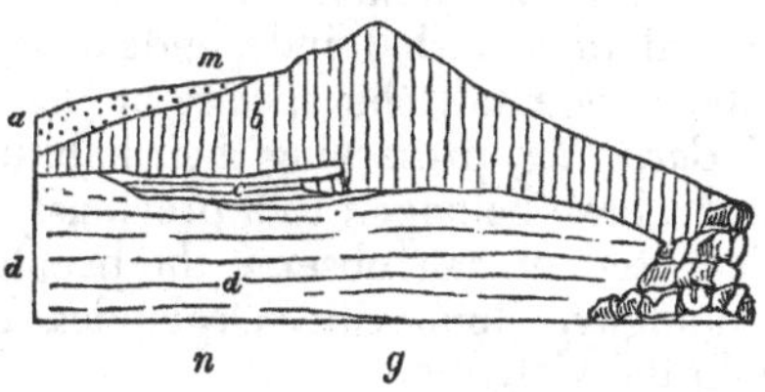   Fig. 44.

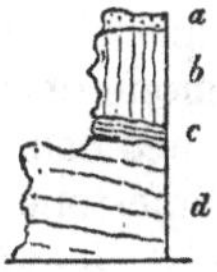

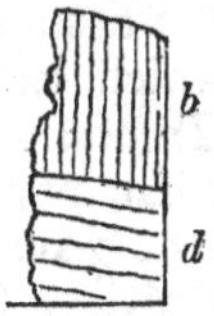

Fig. 43. is a front view of the quarry from the west. *a*, gravel; *b*, granite or sienite ; *c*, gray clay containing nodules of iron

* Journal de Géologie, t. ii. p. 182.
† Naumann, Poggendorf's Annalen ; and Journal de Géologie, t. iii. 1831.

pyrites, varying in size from that of a nut to an egg; *d*, the cretaceous rock. Fig. 42. is a section of the quarry at *m n* (fig. 43):
a, gravel; *b*, granitic rock; *c*, clay; *d*, chalk. Fig. 44., a section at *f g*, where the granitic rock rests immediately on the
chalk, the clay bed being absent. The granitic rock on the
northern side of the quarry, resting on the clay and chalk in
that direction, contains numerous cubes of iron pyrites *.

Before we terminate this sketch, we should notice certain
beds found in the Cotentin (Normandy) and in Seeland; which,
if they do not show a passage of the chalk into the supracretaceous rocks, exhibit an interesting juxtaposition of strata,
containing chalk fossils, and those with organic remains commonly referred to a more recent date. The baculite limestones,
as they are termed, of the Cotentin had often been visited,
and more or less noticed; but their real position in the series
was not pointed out until they were described by M. Desnoyers
in 1825. The baculite limestone is white or yellow, and for
the most part compact, varying, however, in its mineralogical
character, being sometimes cretaceous and even arenaceous.
It contains the organic remains of the cretaceous group, several being also found at Maestricht; such as *Bacculites vertebralis, Thecidea radians, T. recurvirostra, Terebratula*, four
or five particular species not named, &c. These beds are surmounted by others (the whole collectively being of no considerable thickness), composed principally of calcareous matter,
not presenting any marked difference in appearance, though
they do not precisely resemble those beneath. They contain
organic remains, such as are found in the calcaire grossier;
and M. Desnoyers considers that a well defined zoological
line can be drawn between the two deposits; observing, however, that at the contact of the lower portion of the one with
the upper part of the other, when the rocks were without
much coherence, there was sometimes an apparent mixture of
the fossils. "But at the same time," observes M. Desnoyers,
"it has appeared to me, that independent of this confusion,
which may be accidental, the species of the compact chalk,
Trochus and *Bacculites*, preserving their habitual mode of petrifaction, might have belonged to a previously formed bed †."
The geological student can observe the baculite limestone at
Fréville, Cauquigny, Bonneville, Orglandes, Hauteville, and
other places in the Cotentin.

M. Desnoyers remarks on the absence of *Turrilites, Gryphœa
Columba, G. striata, Ostrea carinata, O. pectinata, Pecten spi-*

---

* Killaly, MSS.
† Desnoyers, Sur la Craie et les Terrains Tertiaires du Cotentin; Mém.
de la Soc. d'Hist. Nat. de Paris, t. ii.

nosus, Chenendopora, Hallirhoa, Ventriculites, Spongus, &c.
amid a mass of fossils found in the chalk and green sand.
MM. Passy and Graves have noticed some beds at Saint
Germain-de-Laversines (Normandy), which are referred to the
same age as the baculite limestone. The rocks of Saint-
Germain-de-Laversines form two distinct beds, the highest of
which consists of yellow limestone, containing numerous casts
of shells and polypifers, and is about six or seven yards thick.
The lowest bed, which rests immediately on white chalk, is very
hard, and contains the same organic remains as the upper bed.
These remains are stated to differ, with the exception of some
cretaceous echinites, both from those usually found in the
chalk and in the calcaire grossier*.

The base of the cliff at Stevensklint (Seeland) is formed of
chalk with beds of nodular flints. Upon the chalk, which is represented as having an undulated surface, rests a thin bed (about
six inches thick) of a bituminous clay, containing a *Zoophyte*,
Sharks' teeth, a *Pecten*, impressions of a bivalve, and traces of
vegetable remains. Incumbent on this is a hard yellowish white
limestone, containing the remains of the genera—*Patella*,
1 species; *Cypræa,* 2; *Fusus*, 1; *Cerithium,* 2; *Ampullaria,* 1;
Trochus, 1; *Dentalium,* 1; *Arca,* 1; *Mytilus,* 1; *Serpula,* 1;
Spatangus, 1; *Favosites,* 1; and *Turbinolia,* 1; with *Fishes'*
teeth, and undeterminable univalves, bivalves, and corals. This
limestone contains green grains, seldom exceeds three feet,
and is sometimes only a few inches in thickness, but is nowhere
entirely wanting. It is covered by another limestone, from
thirty to forty feet thick, almost entirely composed of fragments
of corals, and forming the upper part of the cliff. This is divided by chert into many beds, the chert being bent and curved.
It is remarkable that the organic remains of this superior deposit are such as are considered characteristic of the chalk,
consisting of *Ananchytes ovata, Ostrea vesicularis, Belemnites
mucronatus,* &c. Dr. Forchhammer observes that the remains
of *Ananchytes ovata* are occasionally so abundant that the limestone consists almost entirely of them†. In this case it might
be inferred that there was an alternation of the fossils, commonly considered as cretaceous, with others of a supracretaceous character.

From an inspection of the List of organic remains found in
the cretaceous group‡, it would appear that the remains of
mammalia have not yet been detected in the cretaceous group;
while reptiles, one of them of considerable size, the *Mososaurus
Hoffmanni*, have been observed in Yorkshire, Sussex, Maes-

* Passy, Description Géologique de la Seine Inférieure: Rouen, 1832.
† Forchhammer, Edin. Journ. of Science, vol. ix. 1828.
‡ See List of organic remains at the end of the volume.

tricht and Meudon. Fish have been observed in France, and
in various parts of England. Sharks' teeth and the tritores of
some fish are far from uncommon. Crustacea have been noticed in Denmark, Yorkshire, Sussex, the Isle of Wight, Dorsetshire, and Maestricht. Among the polypifers the most
abundant would appear to be different species of the genera
Spongia and *Alcyonium* of some authors;—genera, many species of which have been classed by Goldfuss under the heads
of *Achilleum, Manon, Scyphia,* and *Tragos,* so that there is
much difficulty in presenting a list which should give the different species under any one arrangement. *Manon pulvinarium,* and *M. Peziza,* Goldf., are found at Maestricht, and at
Essen in Westphalia; *Spongia ramosa,* Mant., is discovered in
the chalk of Yorkshire, Sussex, and Noirmoutier; *Alcyonium
globosum,* Defr., at Amiens, Beauvais, Meudon, Tours, Gien,
and in the baculite limestone of Normandy; *Hallirhoa costata,*
Lam., in the green sand of Normandy, and the upper green
sand of Wiltshire; *Ceriopora stellata,* Goldf., Maestricht and
Westphalia; *Lunulites cretacea,* Defr., at Maestricht, Tours,
and in the baculite limestone of Normandy; *Orbitulites lenticulata,* Lam., in Sussex, and at the Perte du Rhone. According to Goldfuss, numerous polypifers are discovered at Maestricht; consisting of *Achilleum,* 2 species; *Manon,* 4; *Tragos,*
1; *Gorgonia,* 1, *Nullipora,* 1; *Millepora,* 2; *Eschara,* 9; *Cellepora,* 6; *Retepora,* 5; *Ceriopora,* 13; *Fungia,* 1; *Diploctenium,* 2; *Meandrina,* 1; *Astrea,* 13; to which should be added,
according to M. Desnoyers, *Lunulites,* 1. Among the Radiaria, the *Apiocrinites ellipticus,* Miller, is found in the chalk of
Yorkshire, Sussex, Westphalia, Maestricht, Normandy and
Touraine; the *Cidaris variolaris,* Al. Brong., in Sussex, and
Normandy, at the Perte du Rhone, in Westphalia, and Saxony; the *C. granulosus,* Goldf., at Maestricht, Aix-la-Chapelle,
and Westphalia; the *C. saxatilis,* in Sussex and Normandy;
the *Galerites albogalerus,* Lam. (Fig. 46.), in Yorkshire, Sus-
sex, Dorset, Normandy, Quedlinburg, Aix-la-Chapelle, and
Poland; the *G. vulgaris,* Lam., in Sussex and France, at
Quedlinburg, and Aix-la-Chapelle; the *Ananchytes ovata,* in
Yorkshire, Sussex, Normandy, at Meudon, in Westphalia,
Poland and Sweden; the *A. hemisphærica,* in Yorkshire and
Normandy; the *Spatangus Cor-anguinum,* Lam. (Fig. 45.) in
Yorkshire, Sussex, Dorsetshire, various parts of France, the
Savoy Alps, various parts of Germany, Poland, and Sweden;
the *Sp. ornatus,* at Aix-la-Chapelle, Normandy, and Bayonne;
Sp. Bufo, Al. Brong., Sussex, Normandy, Maestricht, and
Aix-la-Chapelle; the *Sp. Cor-testudinarium,* at Maestricht and
Quedlinburg. Among the shells, the most widely distributed
would appear to be *Lutraria Gurgitis,* found at the Perte du

Rhone, and in Sweden; *Mya mandibula,* Sussex, Isle of Wight, Normandy, and in the South of France; *Trigonia alæformis,* Sussex, Isle of Wight, West of England, South of France, Aix-la-Chapelle; *Inoceramus* (or *Catillus*) *Cuvieri* (Fig. 47 and 48.), discovered in the chalk of Yorkshire, Sussex, Normandy,

Fig. 45.　　　Fig. 46.　　　Fig. 47.

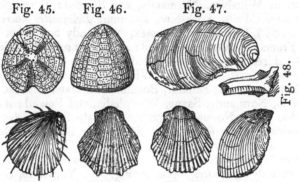

Fig. 48.

Fig. 49.　　　Fig. 50.　　　Fig. 51.　　Fig. 52.

Meudon, the South of France, and Sweden; *Inoceramus* (or *Catillus*) *Brongniarti,* in the chalk of England, Poland, and Sweden; *Ino. concentricus,* in Sussex and in Wiltshire, West-phalia, at the Perte du Rhone, and in the Savoy Alps; *Ino. sulcatus,* in Sussex, at the Perte du Rhone, in the Savoy Alps, and in Sweden; *Plagiostoma spinosum* (Fig. 49.), in the chalk of Sussex, Dorsetshire, Normandy, Meudon, the South of France, Saxony, Poland, and Sweden; *Gervillia solenoides,* Maestricht, Sussex, Wilts, Dorset, and Normandy; *Pecten quinquecostatus* (Fig. 50.), in Sussex, the West of England, Normandy, at Meudon, the Perte du Rhone, Sweden, &c.; *P. quadricostatus* (Fig. 51.), in Sussex, the West of England, Normandy, at Maestricht, and in the Alps of Dauphiné; *P. asper,* Wilts, Normandy, Germany, and Poland; *Podopsis truncata* (Fig. 52.), in Normandy, Dorset, Touraine, and Swe-den; *Pod. striata,* in Yorkshire, Westphalia, and Normandy; *Ostrea vesicularis** (Fig. 53.), in Sussex, Normandy and other places in France, at Maestricht, and in Sweden; *O. carinata,* in Germany, Sussex, Normandy, and the South of France; *O. serrata,* Sweden, Maestricht, and in the South of France; *Gryphæa auricularis,* at Périgueux, South of France, in the Alps of Dauphiné, and Poland; *G. Columba* (Fig. 54.), North-amptonshire, Normandy, South of France, Maritime Alps, Germany, and Poland; *G. sinuata,* Yorkshire, Isle of Wight, Normandy, Dauphiné, South of France, and the Pyrenees; *Terebratula plicatilis,* Moen, in Sussex, at Meudon, South of France, and the Alps of Savoy and Dauphiné; *T. subplicata,*

---

* *Gryphæa globosa,* Sowerby.

in Yorkshire, Sussex, Maestricht, Normandy, and at Tours
and Beauvais; *T. Defrancii,* in Yorkshire, Sussex, at Meudon,
Maestricht, and in Sweden; *T. alata,* Normandy, South of
France, at Meudon, and in Sweden; *T. octoplicata,* in Nor-
mandy, South of France, Quedlinburg, and Sweden; *T. pec-
tita,* in Wiltshire, Normandy, and Sweden; *T. semiglobosa,*
Sweden, Moen, Yorkshire, Bochum; *Belemnites mucronatus*
(Fig. 56.), in Yorkshire, Sussex, Normandy and other parts
of France, Sweden, and Poland; *Ammonites varians,* in Sussex,
Wiltshire, Germany, Normandy, and the Savoy Alps; *Am.
Rhotomagensis,* in Sussex, Wiltshire, and Normandy; *Am.
Mantelli,* Sussex, Saumur, Bochum, and Hanover; *Am. Selli-
guinus,* Normandy, Savoy, Westphalia, and Poland; *Am. in-
flatus,* Wilts, Normandy, and the Perte du Rhone; *Am. Hip-
pocastanum,* Dorsetshire, and Normandy; *Bacculites Faujasii*
(Fig. 55.), Sussex, Norfolk, Maestricht, Bochum, Aix-la-
Chapelle, and Sweden; *Bac. obliquatus,* Sweden, Sussex, and
Normandy; *Hamites rotundus* (Fig. 58.), Yorkshire, Sussex,
Normandy, the Perte du Rhone, and Aix-la-Chapelle.

Fig. 53.   Fig. 54.   Fig. 55. Fig. 56.

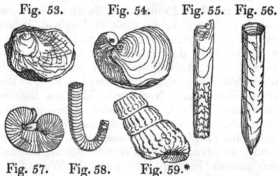

Fig. 57.   Fig. 58.   Fig. 59.*

It will be observed that this list is far from large, when we
consider the number of species enumerated in the catalogue,
and that, perhaps, some of those considered identical may be
different species. No doubt when we reduce our view to smaller
distances and more minute divisions of the cretaceous group,
other species than those above enumerated will be found occurring under similar circumstances in different situations;
but even then, certain species do not seem to be so constant
to particular beds as has been supposed, though some certainly
are found over considerable distances in similar parts of the
group.

The following summary of the organic remains stated by
various authors to have been discovered in the cretaceous

* Fig. 57. *Scaphites obliquus,* Sow. (Sc. striatus, Mant.); Fig. 59. *Turrilites tuberculatus.* Figured to show the forms of these genera, common in
the cretaceous group.

group, though not pretending to perfect accuracy, may yet be useful, as presenting a general view of the subject, and as being an approximation towards the truth.

Plantæ.—Confervites, 2 species; Fucoides, 9; Zosterites, 4; Cycadites, 1; Thuites, 1; and various vegetable remains not yet determined.

Zoophyta.—Achilleum, 3; Manon, 7; Scyphia, 12; Spongia, 12; Spongus, 2; Tragos, 5; Alcyonium, 2; Choanites, 3; Ventriculites, 3; Siphonia, 4; Halirrhoa, 1; Serea, 1; Gorgonia, 1; Nullipora, 1; Millepora, 5; Eschara, 10; Cellepora, 7; Coscinopora, 2; Retepora, 5; Flustra, 3; Cœloptychium, 3; Ceriopora, 21; Lunulites, 1; Orbitolites, 1; Lithodendron, 2; Caryophyllia, 2; Anthophyllum, 1; Turbinolia, 2; Fungia, 3; Chenendopora, 1; Hippalimus, 1; Diploctenium, 2; Meandrina, 1; Astrea, 15; Pagrus, 1.

Radiaria.—Apiocrinites, 1; Pentacrinites, 1; Marsupites, 1; Glenotremites, 1; Asterias, 1; Cidaris, 9; Echinus, 5; Galerites, 9; Clypeus, 1; Clypeaster, 3; Echinoneus, 4; Nucleolites, 12; Ananchytes, 8; Spatangus, 29.

Annulata.—Serpula, 30.

Cirripeda.—Pollicipes, 2.

Conchifera.—Magas, 1; Thecidea, 3; Terebratula, 54; Crania, 8; Orbicula, 1; Hippurites, 8; Sphærulites, 15; Ostrea, 22; Exogyra, 4; Gryphæa, 8; Sphæra, 1; Podopsis, 5; Spondylus? 1; Plicatula, 4; Pecten, 28; Lima, 3; Plagiostoma, 15; Avicula, 2; Inoceramus, 19; Pachymya, 1; Meleagrina, 1; Gervillia, 3; Pinna, 4; Mytilus, 5; Modiola, 2; Chama, 2; Trigonia, 11; Nucula, 12; Pectunculus, 3; Arca, 6; Cucullæa, 6; Cardita, 4; Cardium, 3; Venerecardia, 1; Astarte, 1; Thetis, 2; Venus, 9; Lucina, 1; Tellina, 3; Corbula, 6; Crassatella, 2; Cytherea, 2; Lutraria, 2; Panopæa, 1; Mya, 2; Pholas? 1; Teredo, 1; Fistulana, 1.

Mollusca.—Dentalium, 4; Patella, 1; Emarginula, 2; Pileopsis, 1; Helix, 1; Auricula, 3; Paludina, 1; Ampullaria, 2; Nerita, 1; Natica, 2; Vermetus, 4; Delphinula, 1; Solarium, 1; Cirrus, 4; Pleurotomaria, 3; Trochus, 8; Turbo, 4; Turritella, 1; Cerithium, 1; Pyrula, 2; Fusus, 1; Murex, 1; Pteroceras, 1; Rostellaria, 5; Strombus, 1; Cassis, 1; Dolium, 1; Eburna, 1; Nummulites, 2; Lenticulites, 2; Lituolites, 2; Planularia, 2; Nodosaria, 2; Belemnites, 7; Nautilus, 7; Scaphites, 2; Ammonites, 50; Turrilites, 6; Baculites, 5; Hamites, 21.

Crustacea.—Astacus, 4; Pagurus, 1; Scyllarus, 1; Eryon, 1; Arcania, 1; Elyæa, 1; Coryster, 1; Orythia, 1.

Pisces.—Squalus, 3; Muræna, 1; Zeus, 1; Esox, 1; Salmo? 1; Amia? 1.

Reptilia.—Mososaurus, 1; Crocodile, 1.

Thus making, *Plantæ*, 5 genera, 17 species. *Zoophyta*, 35 genera, 146 species. *Radiaria*, 14 genera, 85 species. *Annulata*, 1 genus, 30 species. *Cirripeda*, 1 genus, 2 species, *Conchifera*, 48 genera, 300 species. *Mollusca*, 40 genera, 167 species. *Crustacea*, 8 genera, 11 species. *Pisces*, 6 genera, 8 species. *Reptilia*, 2 genera, 2 species.—Total, 160 genera, 768 species.

Fossil vegetables are by no means common in the mass of the true or white chalk, and those that are found are stated to be principally marine. The distribution of vegetable remains would appear to be very unequal in the lower parts of the group; for while vegetable matter has been so abundant in some places as to constitute coal beds, at others traces of vegetables are exceedingly rare. Dicotyledonous wood, pierced by some boring shell, seeming to show that it had been drifted about, is not rare in the green sands of Dorsetshire.

The reader will observe that the genera *Ammonites, Scaphites, Hamites, Turrilites, Baculites,* and *Belemnites,* are now first introduced into the lists of organic remains; these genera not having as yet been noticed in the supracretaceous rocks. It was once considered that the genera *Scaphites, Hamites, Turrilites,* and *Baculites,* were confined to the series under consideration; but though their species may be more abundant in the cretaceous group, they are not confined to it; for, as will be seen in the sequel, *Hamites* and *Scaphites* are found in the oolitic group. Moreover, a *Turrilite* has been mentioned, though with doubt, as occurring in the Coral Rag of the North of France. The presence therefore of these genera in distant places may not be alone sufficient to identify the rocks containing them with the cretaceous group; yet if the species are in any abundance, our present knowledge would lead us to suspect that such deposits might be contemporaneous with the cretaceous series. If we reason from the analogy of the existing state of things, there is nothing to oppose the inference that the same genera may equally characterize contemporaneous deposits in North America and in Europe; for according to Dr. Morton, several *species* are now common to the shores of Europe and the United States.

Dr. Morton considers that rocks equivalent to the cretaceous group do exist somewhat extensively in North America. He has named it the *Ferruginous Sand Formation* of the United States, and describes it as occupying " a great part of the triangular peninsula of New Jersey, formed by the Atlantic, and the Delaware and Raritan rivers, and extending across the state of Delaware from near Delaware city to the Chesapeak: appearing again near Annapolis, in Maryland; at Lynch's Creek, in South Carolina; at Cockspur Island, in

Georgia; and several places in Alabama, Florida, &c." In
New Jersey there is a very extensive development of marl.
Taken as a mass, the deposit varies considerably in its mine-·
ralogical character; most frequently presenting itself in minute
friable grains, with a dull blueish or greenish colour, often
with a grey tint. The predominant constituent parts of this
marl, as it is termed, are described as silica and iron. There
are subordinate beds of clay, of siliceous gravel, (the pebbles
varying in size from coarse sand to one or two inches in diameter,) and calcareous marl. The marl is sometimes yellowish brown and filled with green specks of silicate of iron,
and sometimes contains a considerable quantity of mica.—The
following is a list, according to Dr. Morton, of the organic
remains found in this deposit, and described by Mr. Say,
Dr. Dekay, and himself*.

Ammonites Placenta, Dekay; *A. Delawarensis,* Morton;
A. Vanuxemi, Morton; *A. Hippocrepis,* Dekay; *Baculites
ovatus,* Say; *Scaphites Cuvieri,* Morton; *Belemnites Americanus,* Morton, abundant, (allied to *B. mucronatus*); *B. ambiguus,*
Morton; *Turritella; Scalaria annulata,* Morton; *Rostellaria;
Natica; Bulla? Trochus; Cyprœa* (cast); *Terebratula Har-
lani,* Morton; *T. fragilis,* Morton; *T. Sayi,* Morton; *Gry-
phœa convexa,* Morton; *G. mutabilis,* Morton, (some varieties
of this species closely approach *Ostrea vesicularis,* Lam.);
G. Vomer, Morton; *Exogyra costata,* Say; *Ostrea falcata,*
Morton; *O. Crista-Galli; Ostrea,* two other species; *Anomia
Ephippium?* Lam.; *Pecten quinquecostatus,* Sow.; *Pecten,*
another species; *Plagiostoma; Cardium; Cucullœa vulgaris,*
Morton; *Cucullœa,* another species; *Mya; Trigonia? Tellina; Avicula; Pectunculus; Pinna,* resembling *P. tetragona,*
Sow.; *Venus; Vermetus rotula,* Morton; *Dentalium Serpula;
Spatangus Cor-anguinum?* Park.; *Sp. Stella,* Morton; *Ananchytes cinctus,* Morton; *An. fimbriatus,* Morton; *An.? cru-
cifer,* Morton; *Cidaris? Clypeaster.* Crustaceous remains:
Anthophyllum atlanticum, Morton. *Eschara; Flustra; Retepora,* resembling *R. clathrata,* Goldf.; *Caryophyllia; Alcyonium; Alveolites.* Teeth and vertebræ of the shark. *Sauro-
don Leanus,* Say. Remains of the *Crocodile* (frequent); of
the *Geosaurus;* of the *Mososaurus* (Sandy Hook and Wood-
bury, New Jersey); of the *Plesiosaurus;* of a Tortoise; and
of some gigantic animal. Lignite pierced by the *Teredo,*
abundant.

It is almost impossible not to be struck, in the foregoing
list, with the great zoological resemblance of this ferruginous

* Say, American Journal of Science, vol. i. and ii.; Dekay, Annals of
the New York Lyceum; and Morton, Journal of the Acad. of Nat. Sciences
of Philadelphia, vol. vi. ; and American Jour. of Sci. vol. xvii. and xviii.

sand deposit with the cretaceous rocks of Europe. The *Pecten quinquecostatus* is a well known and widely distributed cretaceous fossil. But it is not so much by individual parts as by the general character of the whole, that Dr. Morton's inference seems in a great measure established. How far the cretaceous group of the United States may be separated beneath and above from other deposits more or less contemporaneous with those in Europe, remains an interesting problem, which it is hoped that Dr. Morton and other American geologists will endeavour to solve. From some notices scattered through the memoirs of Dr. Morton and other authors, it would seem far from improbable that the cretaceous rocks may pass into the supracretaceous group.

Assuming that the American ferruginous sand formation belongs to the group under consideration, of which there seems great probability, it would appear that the great white carbonate of lime deposit, or chalk, did not extend there, but that a series of sands, marls, clays and gravels, constituted the whole group. How far the marls or clays may be altogether mechanical is perhaps uncertain; but the gravel would seem to attest the former presence of water, moving with some velocity, for the pebbles even attain one or two inches in diameter.

WEALDEN ROCKS.

SYN. Weald Clay, (*Argile Veldienne*, Al. Brong.; *Wealdthon*, Germ.)
Hastings Sands, (*Iron Sand; Sable Ferrugineux; Kurzawka* of Poland.)
Purbeck Beds, (*Calcaire Lumachelle Purbeckien*, Al. Brong.)

These rocks, characterized in England by the presence of abundant terrestrial and fresh-water remains, occur beneath the lower green sand of the English series. The Weald clay, which constitutes the upper part of the rocks under consideration, does not present a clear line of separation from the marine deposits above it; the lower part of the one and upper portion of the other alternating, according to Mr. Murchison[*] and Mr. Martin[†], in the western part of Sussex;—an important fact, as it shows that the change of circumstances, which permitted the residence of marine animals over a surface previously only covered by fresh-water animals, was not sudden but gradual[‡].

Weald Clay.—According to Dr. Fitton, (to whom we are

[*] Murchison, Geol. Trans. 2nd series, vol. ii.
[†] Martin, Geol. Mem. on Western Sussex, 1828.
[‡] For particular descriptions of the Wealden rocks of Sussex, and their organic contents, the reader should consult the various works of Mr. Mantell:—Illustrations of the Geology of Sussex; Illustrations of Tilgate Forest, &c.

indebted for our knowledge of the nature of the Wealden rocks of England, which were previously confounded with the marine argillaceous and arenaceous beds beneath the chalk,) this clay is composed in the Isle of Wight, where there are fine sections of it, of slaty clay and limestone, with beds of iron-stone; the laminæ of the clay frequently coated with the remains of *Cypris Faba*, Desm.* Mr. Martin defines the clay of the Weald of Sussex (whence the name,) as " a stiff clay, brown on the surface, and blue and slaty beneath, containing concretional iron-stone†." It appears that the iron-stone was once worked, and slags from the ancient furnaces are found in different situations. The thickness of the clay is estimated at 150 or 200 feet in western Sussex. Beneath this there is an alternation of clays and sands, including the limestones full of the *Paludina vivipara*, and known as the Petworth marble.

Hastings Sands.—The following is the order, according to Dr. Fitton, of the beds of Hastings sands, in Sussex, &c. 1. Ferruginous and fawn-coloured sands, and sandstone, including small portions of lignite, with stiff gray loam. 2. Sandstone. 3. Sandstone containing concretional courses of calciferous grit. 4. Dark-coloured shale. 5. White sandstone of Hastings cliffs, 100 feet. 6. Clay, shale, and thin beds of sandstone, containing lignite and silicified wood (*Endogenites erosa*). 7. Sandstone, without concretions; dividing into rhomboidal masses; numerous veins of argillaceous iron ore, and of clay, approaching to pipe-clay at the lower part. 8. Dark-coloured shale, with roundish masses of sandstone, and several layers of rich iron-stone, thin layers of lignite, and innumerable fragments of carbonized vegetables‡. The same author observes that the equivalent beds in the Isle of Wight are composed of sands and sandstones, "frequently ferruginous, with numerous alternations of reddish and variegated sandy clays, and concretions of calcareous grit§."

There are certain local variations, which will be found described in the works treating of particular districts. The Hastings beds, however, would appear, as a mass, to be principally arenaceous. According to Mr. Mantell, the lower part of the Hastings deposits (the Ashburnham beds,) are composed of argillaceous limestone alternating with schistose marls, which are probably connected with the following.

* Fitton, Ann. of Phil. 1824.
† Martin, Geol. Mem. Western Sussex.
‡ Fitton, Geological Sketch of the Vicinity of Hastings, 1833. Dr. Fitton considers the Tilgate beds, so well known from the remarkable fossil animals discovered by Mr. Mantell, as belonging to the upper part of the Hastings Sands.—*Ibid.* p. 44.
§ Fitton, Ann. of Phil. 1824.

Purbeck Beds.—These are composed of various limestone strata, alternating with marls, many of the former being extensively used for the pavement of London. Mr. Webster observes, that at Warbarrow Bay, Lulworth Cove, and other places on the coast of Dorsetshire, the upper bed of the Purbeck strata, supporting the Hastings Sands, contains a large proportion of green earth, the calcareous matter being apparently derived from the fragments of a bivalve shell.

From the lists of organic remains found in the Wealden rocks, it will appear that this deposit of limestones, sands, sandstones, and clays, was formed in water which permitted the existence of shells analogous to those which now live in fresh water. With these are discovered the remains of estuary animals, and a few shells (*Corbula, Tellina, Bulla*), which may be considered as marine, the species of the analogous genera of the present day inhabiting sea coasts. The presence of the latter will not, however, invalidate the general evidence in favour of a lake, river, or estuary; for not only may these shells have been introduced accidentally, but the animals inhabiting them may also have been gradually accustomed to live in fresh or estuary waters, as is the case in the present day with the species of some genera usually considered marine.

It would appear that the *dirt-bed*, first noticed by Mr. Webster in the Isle of Portland, and which has since been observed in the vicinity of Weymouth and elsewhere, commences the phænomena which attest dry land, succeeded by submersion of the same land beneath fresh or estuary waters, in which the whole of the Wealden rocks of south-eastern England were formed; not suddenly, for there are no conglomerates to mark a possible state of violence; but quietly, the shells being tranquilly enveloped by the calcareous, argillaceous, or arenaceous matter which now entombs them. It will be seen that the oolitic group, immediately preceding this state of things, was, judging from the nature of the organic remains, formed beneath a sea. Therefore we must suppose a rise of the land, or depression of the sea, to such an amount as to permit the sea-formed rocks to become dry land, upon which *Cycadeoideæ* and dicotyledonous plants of a tropical nature flourished. This land was then depressed; but so tranquilly, that the vegetable soil, mixed with a few pebbles of the subjacent rock, was not washed away; neither were the trees considerably displaced, but they were left much as we have seen other trees in the submarine forests which surround Great Britain in various places, and occur on the coasts of France. Like them, also, the trees of the *dirt-bed* are found, some prostrate, others inclined, and others nearly in the position in which they grew;

the upright portions being partly included in the limestone strata above. The only difference in the trees in the *dirt-bed*, and those in the submarine forests, would appear to consist in the tropical nature of those in the *dirt-bed*, and the near approach, if not the identity, of the submarine forest vegetation with that now existing in Great Britain and France. There is, therefore, nothing singular in the gradual depression of the land, so quietly as not to cause the removal of the trees and other vegetable matter, as this has happened at various periods.

Instead of the depression having been effected, in the first instance, beneath the waters of the sea, circumstances have so existed that it took place beneath fresh water, which gradually acquired sufficient depth to permit a deposit of various mineral substances several hundred feet thick. The circumstances attending this deposit have not been constant. At first calcareous matter was thrown down, with somewhat regular interruptions, which introduced a sufficient quantity of argillaceous matter to produce marl. Although fresh-water and terrestrial animals were now imbedded, there would also appear to have been at least one time when the water near Weymouth and in the Isle of Wight was capable of supporting the life of oysters and cockles, and therefore at least brackish. After this first period, sands were accumulated in great abundance, and in them were entombed a great variety of land and fresh-water Tortoises, Crocodiles, Plesiosauri, Megalosauri, Hylæosauri, and huge Iguanodons*, those monstrous terrestrial reptiles. These must have sported in the waters, or roamed along the banks of this lake or estuary, into which trees and different vegetables were drifted. A clay deposit crowns this succession of rocks, still however not showing any other than a fresh-water origin. How far we may consider the change of the relative level of sea to have produced a constant depression of the land, is uncertain; but be this as it may, the sea was destined again to cover the land and resume its empire, for above the last-noticed clay reposes the whole mass of the cretaceous rocks of south-eastern England, of marine origin. This change, like that which preceded it, was not sudden; there are no marks of violence between the Weald clay and the green sand; on the contrary, there is a passage of one into the other, an alternation of the two at their junction. There is every probability that the sea did not make a furious inroad over the land, but that there was a quiet and gradual change of level, as in the case of the *dirt-bed*. I shall not trace the subsequent changes that have taken place over this spot on

---

* For descriptions of the remains of this creature, consult Mantell, Phil. Trans. 1825, and Illustrations of Tilgate Forest, 1827.

the earth's surface, further than to remark, that the sea again disappeared (Isle of Wight), and fresh-water or estuary deposits succeeded *.

These conclusions can scarcely be termed hypothetical, for they appear such, however remarkable, as may be considered honest deductions from the phænomena observed.

The extent of the area over which the *dirt-bed*, or a contemporaneous ancient soil, may be traced, is very remarkable, when we reflect upon the various circumstances which must have combined to preserve such a surface of ancient dry land. Dr. Fitton notices an earthy bed in precisely the same geological position in the cliffs of the Boulonnois, and also in Buckinghamshire and in the Vale of Wardour. It further appears that silicified wood is found in the bituminous bed from Boulogne to Cap Gris-nez†.

To form such a deposit as that we have been noticing would be a work of time, and therefore we may infer that equivalent formations were taking place elsewhere, the great operations of nature proceeding in their usual course. The fresh-water character of the deposit can only be considered accidental or local; precisely as formations at the present day, though contemporaneous, may be either marine or lacustrine. Therefore, even supposing various perpendicular movements in the land to have taken place extensively over certain portions of Europe, it does not follow that they should have produced a constant rise of that land above the surface of the sea. On the contrary, we may consider that such movements very frequently caused a mere change in the relative depth beneath the surface-water, and that all deposits in the course of formation, and so circumstanced, partook of the marine character of the surrounding aqueous medium.

The observations of MM. Graves‡ and Passy§ leave little doubt that beds of the same relative age with the Wealden rocks occur in the departments of the Oise and Seine Inférieure. The country usually known as the Pays de Bray, which runs N.W. from between Auneuil and Beauvais to and beyond Neufchâtel, is a denudation in the midst of the great chalk district of that part of France, extending down to the beds of Kimmeridge clay. There are various sandstones and clays above the Kimmeridge clay, and beneath the chalk and a mass of the green sand series, containing a considerable

* For further observations on these curious facts, accompanied by sections of Portland, &c., consult a memoir on the Weymouth district by Dr. Buckland and myself, Geol. Trans. 2nd series, vol. iv.
† Fitton, Geological Sketch of the Vicinity of Hastings, 1833, p. 76.
‡ Graves, Précis statistique du Canton d'Auneuil (Oise).
§ Passy, Descr. Géol. de la Seine Inférieure, 1832.

number of vegetable remains, among which is the *Lonchopteris Mantelli*, well known as found in the Hastings Sands of Sussex. There would also appear to be a bed or beds of a limestone abounding in the remains of *Paludinæ*, reminding us of similar beds in the Wealden rocks of southern England. We may gather from the observations of MM. Graves and Passy, and from those of various members of the Geological Society of France who assembled at Beauvais in September, 1831 *, that though these Wealden rocks of the Pays de Bray contain abundant terrestrial, and some fresh-water remains, there are also numerous marine remains, characteristic of the green sand series. Dr. Fitton, who also notices the occurrence of contemporaneous rocks in the Boulonnois, suggests that the Wealden rocks of Sussex, the Boulonnois, and of the Pays de Bray may have been formed in a single estuary, the area in that case not being greater than that now occupied by some deltas. He, however, at the same time remarks that this, though a plausible explanation, should only be considered as provisional†. It certainly by no means follows that because these deposits should contain fluviatile shells, and are of the same age, they should necessarily have been produced in the same estuary, even when the shells have been observed within distances which might admit of this explanation. The estuaries of the Thames and Seine are now in all probability the depositaries of mineral substances and organic remains which do not widely differ, and consequently in some future state of the world, when these deposits shall have been heaved above the ocean level and partially covered with other rocks, they would exhibit similar geological characters.

M. Thirria describes a considerable superficial deposit of clay with pisiform iron-ore in the department of the Haute Saone, part of which he considers referrible to the green sand, and may be equivalent to the Wealden rocks. Above rocks which seem equivalent to the Portland beds of England, there are strata of sand and clay, apparently the denuded remains of a deposit, once more extensive, which has suffered aqueous destruction, the water mixing up portions of the removed strata with the bones of Bears and Rhinoceroses; so that the mass upon reconsolidation much resembles the mineralogical composition of the original beds. The following is a section of beds, which M. Thirria considers as in place, the list of fossils being increased by those which he discovered, also in place, in the department of the Haute Saone: 1. Unctuous green clay; 2. Fine and slightly argillaceous yellow sand; 3. Nodules of yellow limestone contained in greenish clay;

---

* Bulletin de la Soc. Géologique de France, t. ii.
† Fitton, Geological Sketch of Hastings.

4. Yellow and slightly argillaceous sand; 5. Greenish-yellow and unctuous clay; 6. Greenish clay, with nodules of marly limestone and grains of iron ore; 7. Pisiform iron-ore, contained in an ochreous clay, with *Ammonites binus, A. planicostata,* Sow., *A. coronatus,* Schlot., and other species; *Hamites* (new species); *Nerinæa; Cirrus; Terebratula coarctata,* Sow., and other species; and *Pentacrinites;* 8. White marl, with nodules of greenish clay and concretions of marly limestone. The whole forming a thickness of about forty feet, and resting on beds considered equivalent to those of Portland *.

The extraordinary mixture of fossils contained in the pisiform iron ore is commented on by M. Thirria, who further remarks that the reniform pieces of ore sometimes contain the empty casts of Jura limestone fossils.

In support of the opinion that some of these pisiform and reniform iron-ore beds are of contemporaneous formation with either the Wealden rocks or green sand and chalk of England, we may cite the observations of Professor Walchner on similar beds near Candern in the Brisgau. He remarks, "that the reniform and pisiform iron-ore deposits in the vicinity of Candern belong to two formations of very different ages; one of which rests on a compact Jura limestone, apparently corresponding with either the coral rag or Portland stone of the English. It is composed of a mass of sandy clay, containing reniform iron-ore in the lower, and pisiform iron in the upper part; and at the same time spheroids of flint (silex) and jasper. The reniform ores, and the flints which accompany them, contain organic remains; the former of *Astreas* and *Ammonites,* the latter of *Pectines* and *spines of Cidaris.* The whole is covered with the solid beds of conglomerate, more ancient than the molasse, or by the molasse itself. This iron-ore formation may be considered as one of the last of the Jura limestone (oolitic group), and it, without doubt, closely approaches the chalk; perhaps it may be like the green sand, intermediate between the Jura limestone and the chalk †."

In further support of this conclusion, Professor Walchner quotes the remarks of MM. Merian and Escher, on parts of the Jura, both of whom describe a clay with pisiform or reniform iron-ore, intermediate between the upper beds of the Jura limestone and the molasse (one of the supracretaceous rocks of Switzerland); but being sometimes wanting, so that the molasse rests directly on the Jura limestone. M. Merian states that, near Aarau, the ferriferous bed sometimes con-

---

* Thirria, Notice sur le Terrain Jurassique du Département de la Haute Saone; Mém. de la Soc. d'Hist. Nat. de Strasbourg, tom. i. 1830.

† Walchner, Sur les Minérais de Fer pisiforme et réniforme de Candern en Brisgau; Mém. de la Soc. d'Hist. Nat. de Strasbourg, tom. i.

tains large angular fragments of the limestone on which it rests, as also nodules of flint and jasper; angular fragments of the former containing organic remains, which are the same as those detected in the iron-ore itself. The same author observes, that " the pisiform ore of Aarau is immediately covered by a sandstone and bituminous schist, passing into lignite, which sometimes clearly exhibits a woody texture." The schist, and its accompanying clays, contain an abundance of fossils, among which *Planorbes* and other fresh-water shells could be distinguished.

M. Brongniart notices among the cretaceous rocks of the Isle d'Aix and the embouchure of the Charente, a marl, which he refers to the Wealden clay, containing nodules of amber, pieces of lignite and silicified wood, in which holes, formed by some perforating animal, are replaced by agates *. The latter fact agrees with the presence of pieces of silicified wood, occasionally of large size, found on the green sand of Lyme Regis, where the holes, formed by some perforating animal, are filled with chalcedony or agate;—both examples appearing to show that the wood had drifted, and remained some time in the sea.

According to Professor Pusch there is a ferriferous deposit in Poland, situated between the Jura limestone and the cretaceous rocks, which may be considered as the equivalent to the Weald clay and iron sand (Hastings Sands) of England. The following is Prof. Pusch's account of these beds, which is too valuable to be abridged : " It fills the valleys (in Poland) of Czarna Przemsa as far as Siewirz, that of Mastonica, that of the Wartha from its origin at Kromolow towards Czenstochau, and of the Liziwarta; extending across Higher Silesia to the Oder, and running up this river to the country of Ribnyk. It is composed of horizontal beds, often alternating and of little continuity, of a slightly calcareous and schistose clay, either blue or variegated, named *kurzawka*; of a siliceous, quartzose, and compact conglomerate; of a brown ferriferous sandstone; of beds of loose sand, and of thin beds of white or variegated marly limestone. In the country of Kromolow, Poremba, and Siewirzce, this formation contains horizontal beds from six inches to fourteen feet in thickness, of a coarse coaly substance (*moorkohl*), often accompanied with bituminous wood and much pyrites. This combustible is little worked, as the deposit occurs in marshy valleys, but the want of wood may render it useful in the country between Pelica and Czenstochau. From Siewirz, the carbonaceous beds lose themselves on the north. Faint traces of them are

* Tab. des Terrains, p. 218.

found round Czenstochau, Krzepice, and Klobucho; while the unctuous and blue schistose clays are largely developed in these countries, with, as on the top of the carbonaceous deposits, numerous beds of iron-ore, consisting of ranges of spheroidal nodules of compact argillaceous iron-ore, containing numerous Ammonites, (especially *Ammonites bifurcatus,*) and bivalves, of the genera *Cardium, Venus, Trigonia, Sanguinolaria,* &c., fossils which partly correspond with those of the Jura limestone. This ferriferous deposit abounds near Panki, near Krzepice, between this point and Wielun, and on the north of Upper Silesia. It furnishes iron for the foundries of Poremba, Miaczow, Panki, Zarki, and various places in Silesia, producing 50 per cent. of iron. A brown ferruginous sandstone, agglutinated by hydrate of iron, covers the blue schistose clays, especially round Kozieglow, Panki, and Prauska *."

The reader will at once perceive the great resemblance of this ferriferous deposit to that above noticed in the Jura; such resemblance being heightened by the occurrence of organic remains, of which Ammonites constitute a portion, in the iron-stone nodules of both situations. There would appear to be little difficulty in considering this deposit, with M. Pusch, as the equivalent of the Wealden rocks of England, showing that where local circumstances did not interfere, and the deposit continued to be effected beneath the sea, its zoological character marked a certain connexion with the oolitic group; the species of animals existing during the formation of at least a portion of the latter rocks not being suddenly cut off: thus exhibiting a zoological passage of the oolitic into the cretaceous groups, when local circumstances did not interfere, as they have done on the south-east of England. It is remarkable that, notwithstanding the different character of the organic remains, apparently entombed in beds of the same age, which would seem to point out deposits in different waters, iron-ore should be so common in the Wealden rocks of England, the Jura, and Poland.

When the upper beds of the oolitic series formed dry land, and sustained vegetation in southern England, it seems reasonable to conclude that many parts of the land now constituting Europe were similarly circumstanced; and therefore contemporaneous deposits of various characters may have been produced in different situations; some, by the nature of their organic remains, marking the presence of large lakes, or the embouchures of considerable rivers :—in fact, a state of things, during which there was a mixture of dry land, fresh waters, and sea in this part of the globe. Some cause, with which as yet we are imperfectly acquainted, subsequently produced a great change in the relative levels of sea and land, and the cretaceous rocks (chalk and green sand) became deposited over a very considerable area, one apparently extending over a much larger superficies than that in which the last-formed rocks of the oolitic series were deposited.

Section VI.

OOLITIC GROUP.

Syn.—*Oolite formation,* Engl. authors; *Calcaire de Jura, Calcaire Jurassique,* Fr. authors; *Oolithenbildung, Jurakalk,* Germ. authors.

This group is, in the southern parts of England, composed
of various alternations of clays, sandstones, marls, and limestones; many of the latter being oolitic, whence the name
oolitic series. At a very early period in the history of English
geology, Mr. William Smith affixed names to various portions
of this series, many of which are still employed by the geologists of Europe. Several of the divisions and subdivisions are,
undoubtedly, very arbitrary, and perhaps separate those things
theoretically which nature has united; but their convenience
seems proved by their very general adoption. In consequence
of three great clay or marl deposits appearing to divide the
series in the south of England into three natural groups, Mr.
Conybeare has separated it into three systems, as follows, (the
Purbeck beds only, for reasons before assigned, being omitted): 1. Upper system, containing, in the descending order,
a. Portland oolite; *b.* calcareous sand and concretions; *c.* an
argillo-calcareous deposit, named Kimmeridge clay. 2. Middle system, *a.* coral rag, and its accompanying oolites; *b.* calcareous sand and grit; *c.* Oxford clay. 3. *a.* Calcareous
strata, (sometimes divided by clays or marls,) named cornbrash, forest marble, great or Bath oolite, and inferior oolite;
b. calcareo-siliceous sands, usually termed sands of the inferior
oolite; *c.* an argillo-calcareous deposit named lias.

These three principal divisions, marked by argillaceous deposits, have been traced to various distances, though their
subdivisions have not been so readily identified. The extent
to which a few fossil shells of each division can be observed, is
also deserving of attention.

Mr. Phillips distinguishes this group in Yorkshire into,
a. Kimmeridge clay; *b.* upper calcareous grit; *c.* coralline
oolite; *d.* lower calcareous grit; *e.* Oxford clay; *f.* Kelloway
rock (a name given to stony portions of the Oxford clay, near
Kelloway Bridge in Wiltshire); *g.* cornbrash limestone;
h. upper sandstone, shale, and coal; *i.* impure limestone (Bath
oolite); *k.* lower sandstone, shale, and coal; *l.* ferruginous
beds (inferior oolite); *m.* upper lias shale; *n.* marlstone series;

and *o.* lower lias shale. It will be observed that these divisions do not very materially differ from those of the southern parts of England, except in the presence of certain shales and sandstones containing coal, above and beneath a bed considered equivalent to the Bath oolite. These carbonaceous beds are stated to have a collective thickness of 700 feet, the supposed representative of the Bath oolite being abstracted.

We are indebted to Mr. Lonsdale for a detailed and highly valuable account of the oolitic district of Bath, a district which, independently of other considerations, must always be interesting to British geologists from having been the scene of Mr. William Smith's early labours, and as having long constituted the type to which geologists directed their attention, when describing the oolitic series of Western Europe. M r Lonsdale divides the group into : *a,* Kimmeridge Clay (thickness unknown); *b,* Coral Rag, subdivided into Upper Calcareous Grit, Coral Rag, and Lower Calcareous Grit (in all 190 to 130 feet) ; *c,* Oxford Clay, based on the calcareous sandstone named Kelloway Rock (thickness unknown); *d,* Cornbrash (a thin bed); *e,* Forest Marble (100 feet); *f,* Bradford Clay, which the author remarks should be united with the Forest Marble (40 to 60 feet); *g,* Great Oolite (40 to 125 feet); *h,* Fuller's Earth (about 140 feet); *i,* Inferior Oolite (130 feet); *k,* Marlstone; *l,* Lias (280 to 290 feet)*.

By a rigorous examination of the beds, from the marlstone to the cornbrash inclusive, in Gloucestershire, the same author was enabled to point out several important modifications of the oolitic rocks of the Bath district, even within that distance. It is remarked that in the South of Gloucestershire the inferior oolite " consists of nearly equal divisions of soft oolite and slightly calcareous sand; but in the northern part of the county, the latter, for the greater part, is replaced by a yellow sandy limestone. The freestone beds, which are not to be lithologically distinguished from those of the Great Oolite, gradually increase in number and thickness, from the neighbourhood of Bath to the Cotteswolds, east of Cheltenham, where they constitute the whole of the escarpment. This vertical importance is retained through the north of the country examined ; but to the eastward of the valley, ranging from Stow-on-the-Wold to Barrington, near Burford, a change takes place, both in the structure and thickness of the formation. The freestone beds are there replaced by strata of nodular coarse oolite, containing numerous specimens of *Clypeus sinuatus ;* the sandy portion consists of only a thin bed, and the thickness of the whole formation is diminished from 150 to 50 feet†." Other impor-

* Lonsdale, Geol. Trans., 2nd series, vol. iii.
† Lonsdale, Proceedings of the Geol. Soc., Dec. 1832.

tant changes are remarked in the Fuller's Earth, Great Oolite, and Cornbrash, tending to show the variable nature of the oolitic subdivisions, even in short distances. In his memoir on the Bath district, Mr. Lonsdale also points out the thinning off of the Bath oolite in the vicinity of Norton, by which the Bradford Clay and Fuller's Earth are brought into contact *. To this it may be added that the Bath Oolite is no more seen to the southward in England; a thick bed of clay, probably the continuation of the two clays above noticed, separating the Forest Marble from the Inferior Oolite in Dorsetshire.

The same author, in his memoir on the Gloucestershire oolites, establishes, by a close comparison of the lower part of the Great Oolite, as it exists at Burford, with the Stonesfield slate, so celebrated for its organic contents, that the latter, instead of being subordinate to the Forest Marble, is referrible to the lower part of the Great Oolite. Now this is a very important correction of an error, inasmuch as all conclusions at which we might previously have arrived, in our endeavours to trace the circumstances under which a particular part of the deposit might have been formed, would have been vitiated, by supposing the Stonesfield slate to occupy one portion of the series, when it really occurs in another.

The oolitic series of Normandy presents a close analogy in its general, and even in some of its minor divisions, with those of southern England. Commencing with the vicinity of Havre, and extending our observations to the Cotentin, we find the following series : *a.* Kimmeridge clay, in which certain sandstones named *Glos* sandstones are subordinate; *b.* limestone and oolitic beds, referrible, from their geological and zoological characters, to the coral rag; *c.* a ferruginous and calcareous sandstone; *d.* Oxford clay; *e.* a series of beds, including the well-known Caen stone, and representing the forest marble and great oolite ; *f.* inferior oolite; *g.* lias†. M. Boblaye divides the oolitic series of the north of France as follows‡ : *a.* beds referrible to the coral rag, (the highest of the oolitic series in the district); *b.* a sandy and ferruginous oolite; *c.* a series of beds representing the cornbrash, forest marble, and great oolite; *d.* ferruginous limestone, micaceous marls, and sandy limestones, equivalent to the inferior oolite and its sands; *e.* lias.

---

* Lonsdale, Geol. Trans. 2nd series, vol. iii. p. 254, where there is also a section representing the manner in which the Bath oolite fines off.

† De la Beche, Geol. Trans. vol. i. 1822 ; De Caumont, Essai sur la Topographie Géog. du Calvados, 1828 ; Herault, Tableau des Terrains du Calvados, Caen, 1832.

‡ Boblaye, Sur la Form. Jurassique dans le Nord de la France; Ann. des Sci. Nat. 1829.

In Burgundy, M. Elie de Beaumont, who has remarked on the constancy of the geological facts observable in the oolitic belt of the great geological basin which contains London and Paris, has found beds which he considers referrible to those of Portland, beneath which is a marly limestone with the *Gryphæa Virgula*, a remarkable shell of the Kimmeridge clay, particularly in France. These beds are succeeded by compact earthy or oolitic limestones, beneath which is gray marly limestone, supposed equivalent to the Oxford Clay. This is followed, in the descending order, by a series of oolite and other beds, beneath which there is a limestone remarkable for containing an abundance of *Entrochi*, and considered equivalent to the inferior oolite, under which are rocks corresponding with the lias *.

M. Thirria describing the oolitic series of the department of the Haute Saone, where it constitutes the north-western limits of the Jura, notices the following beds (the lias being excluded from the list according to the views of some of the continental geologists):—*a.* inferior oolite, composed of various limestones, oolitic, sublamellar, lamellar, and compact, reddish, gray, and yellow; some of the beds being studded with *Entrochi*, or joints of *Crinoidea*. One bed is remarkable for oolitic hydrate of iron, so abundant as to be worked for profitable purposes at Calmontiers, Oppenans, Jussey, and other places; *b.* a yellow marl, considered equivalent to the Fuller's earth of the English (two yards thick); *c.* great oolite, composed of oolitic beds, containing among other shells *Ostrea acuminata* and *Avicula echinata ; d.* limestones with much red oxide of iron, schistose, suboolitic, or compact, considered equivalent to forest marble; *e.* marly limestone, gray or yellowish, full of oolitic grains, supposed equivalent to the cornbrash of England ; *f.* schistose blackish gray marls with marly limestone, resting on gray schistose marls containing oolite grains of hydroxide of iron, worked for profitable purposes in the districts of Orrain and Saguenay. The whole of this subdivision, *f*, is based on dark gray and schistose argillaceous limestone, and contains many fossils, particularly in the ferruginous oolite, among which is *Gryphæa dilatata*, a very characteristic shell of the Oxford clay, to which, and to the Kelloway rock, the whole is referred; *g.* a series of clay and limestone beds, the latter mostly oolitic; the upper part containing Corals, and the lower portion numbers of *Nerinæœ*, the whole considered equivalent to the coral rag; *h.* gray marls and marly limestone, based on compact gray limestone, the latter containing abundant remains of

* Elie de Beaumont, Note sur l'uniformité qui regne dans la constitution de la Ceinture Jurassique qui comprend Londres et Paris;—Ann. des Sci. Nat. 1829.

Astarte, while the other parts present the *Gryphæa Virgula ;*
these marls are consequently referred to the Kimmeridge clay;
i. various limestone beds, principally of a gray colour, sometimes whitish and yellowish, at others of a deeper tint, considered equivalent to the Portland stone*.

M. Thurman divides the oolitic series of the central part of
the Jura, named the Porrentry (the ancient Eveché de Basle,
and the present Bernese Jura) into groups similar to those
which have been formed in Normandy and southern England.
a. fine oolites and various compact limestones (considered
equivalent to Portland stone), 65 feet; *b.* yellowish marls and
marly limestones with *Gryphæa Virgula* (Kimmeridge clay), 50
feet; *c.* compact limestone, with *Astarte minima* (Phil.), 100
feet; *d.* compact or cretaceous white limestone with *Nerenææ,*
65 feet; *e.* oolitic and pisolitic limestone, 65 feet; *f.* compact
gray polypiferous limestone, 18 feet (*c, d, e,* and *f,* are regarded
as equivalent to coral rag); *g.* marly and sandy limestones,
with the concretions named *chailles* (calcareous grit), 75 feet;
h. blue marls, smoke gray compact limestones, and ferruginous
oolite (Oxford clay), 50 feet; *i.* oolitic lumachella limestone,
20 feet; *k.* reddish sandy limestones and marls, 30 feet; *l.* fine-
grained oolite (considered equivalent to the great oolite), 18
feet; *m.* marls and suboolitic limestone, with *Ostrea acuminata*
(Fuller's earth), 13 feet; *n.* subcompact oolite, 120 feet; *o.* ferruginous oolite, 20 feet (*n,* and *o,* considered as equivalent to
the inferior oolite); *p.* reddish green and micaceous sandstones
and marls ('marly sandstone), 18 feet.   The whole based on
lias.   M. Thurman observes that the thickness of the rocks
here enumerated is often more considerable, and points out the
Mont Terrible as exhibiting an excellent section of nearly the
whole series†.

M. Dufrénoy, in his remarks on the rocks of this age which
occur in the south-western parts of France, divides the oolitic
group into three distinct systems ; admitting, however, at the
same time, that these divisions are not well pronounced, the
beds which apparently correspond with the Oxford and Kimmeridge clays being replaced by marly limestone.   He further
observes, that "the numerous subdivisions noticed by the
English geologists are but very imperfectly seen in the secondary basin under consideration; some, nevertheless, being
sufficiently constant."   The lower portion rests on lias, and is
composed of micaceous marls, with *Gryphæa Cymbium, Belemnites,* and other shells, which, as he observes, may be referred to the sands of the inferior oolite.   There are beds of

* Thirria, Notice sur le Terrain Jurassique du Département de la Haute-
Saone; Mém. de la Soc. d'Hist. Nat. de Strasbourg, 1830.
† Thurman, Essai sur les Soulèvemens Jurassiques du Porrentry, 1832.

limestone with oolitic iron, and oolites, considered equivalent to the Bath oolites, the latter only well developed at Mauriac, Aveyron. This lower division is represented as of considerable thickness. Above this there is a system of marly limestone beds, in some places associated with considerable masses of polypifers and thick beds of irregular and earthy oolite (Marthon, forest of la Braconne, and other places). M. Dufrénoy infers, from the great abundance of the corals, the presence of the oolite and many fossils, that these beds are equivalent to the coral rag and Oxford oolite. Upon this system, rests another, composed of marls and marly limestone, abounding in the *Gryphæa Virgula,* supporting an oolite (from the environs of Angoulême to the ocean), in which this gryphite is also found. These rocks are referred to the Kimmeridge clay and Portland oolite respectively, and are stated to be surmounted by rocks of the cretaceous group *.

It would thus appear, that throughout a considerable portion of France and England, and in the Jura, the causes which have produced the deposit of the oolitic group have not varied materially. Before, however, we attempt any remarks on this apparent uniformity of mineralogical structure over a considerable area, it will be necessary to present a sketch of this deposit in Scotland, Germany, and Sweden.

Our knowledge of the oolitic group of Scotland is more particularly due to Mr. Murchison. The coal deposit of Brora, in Sutherlandshire, has been shown to correspond with the carbonaceous series of Yorkshire, described by Mr. Phillips as occurring between the inferior oolite and cornbrash, and including in its central part a rock considered equivalent to the Bath or great oolite. In the vicinity of Brora there would appear to be various sandstones and shales, containing coal and vegetable impressions. The freestone of Braambury and Hare hills is described as covered by a rubbly limestone, "an aggregate of shells, leaves, stems of plants, lignite, &c." Mr. Murchison considers the organic remains of this bed, and the casts in the freestone, as referrible to such as occur in the lower part of the coral rag. At Dunrobin Castle calcareous sandstones are succeeded by beds of "pebbly calciferous grit," covered by shale and limestone with fossils. Other varieties of this oolitic deposit occur on this coast, which consists, in the descending order, of rubbly limestone, white sandstone and shale, shelly limestone, sandstone, shale, and limestone, with plants and coal, considered the same with the Yorkshire carbonaceous deposit.

This oolitic deposit is not confined to the main land of Scot-

* Dufrénoy, Annales des Mines, tom. v. 1829.

land, but is found in the Hebrides. According to Mr. Murchison, it occurs at Beal near Portree, Sky, the higher part presenting a calcareous agglomerate of fossils, resembling many portions of the English cornbrash and forest marble: it is identical with the shelly limestone of Sutherland, above noticed. At Holm the sandstone rises to a considerable height from beneath the limestone. Impressions of plants are found in the sandstone on the north-east of Holm. Near Tobermory in Mull, sandstone, considered as equivalent to that of the inferior oolite, rests on lias, containing the *Gryphæa incurva.* It also appears that rocks of the oolitic series, including lias, occur in other parts of Mull, the opposite coast of Ross-shire, and in the islands of Rasay and Pabbla, often cut and covered by trap rocks*.

M. von Decken observes, that the oolitic group of northern Germany, which occurs extensively from Bramsche on the Haase to Minden on the Weser, and thence to the country near Hildesheim and Eimbeck, as also northwards of the Hartz between Wolfenbüttel and Helmstadt, approaches in its characters to the same series of Yorkshire and some parts of Scotland. Marls and sandstones predominate, and the oolitic limestones are confined to subordinate beds. Beds of coal accompany the sandstone, and are worked for economical purposes, more particularly at Obernkirchen (Bückeburg).

The connexion of the upper part of this group with the superincumbent cretaceous series is not clearly seen, and fixed points are wanting to compare it with the English divisions. The inferior division, the lias, is on the contrary well developed. The black bituminous marls contain layers of bituminous limestone, and the marl itself is in some places (Essen, Osnaburg; Ostercappeln) used for slate pencils. A thick sandstone bed, of a dark brown colour, and traversed by stripes of brown iron ore, rests upon the lias in the countries on the Weser. It contains beds of dark gray slate clay, in which are nodules of oolitic brown ironstone, and may be considered as the lowest member of the inferior oolite. Above this there is an oolitic limestone (in the Weser chain, eastwards of Hildersheim, Ith, and Lauensteinberge), which is at first sandy, and then contains veins of chalcedony and chert. There are also beds of dark slate clay, marl, and yellow brown sandstone, which sometimes resemble grauwacke and the quartz rock associated with it. The latter predominates from Lübbecke to Bramsche. M. von Decken remarks, that it is by no means decided to which part of the English series this mass of rock, 700 feet thick, should be referred, and that probably the various opinions on

* Murchison, Geol. Trans. 2nd series, vol. ii.

this head can only be settled by an accurate examination of the fossils contained in it. If, however, it be assumed that the sandstone and coal of Obernkirchen and Böhlhorst, correspond with the oolitic coal of Yorkshire, this mass of rock would answer to that part of the English series comprised between the inferior oolite and the upper part of the great oolite.

Above this limestone formation there rests, at Böhlhorst and the Bückeberg, a mass of black marl 400 feet thick; then sandstone, 200 feet thick, containing beds of coal, surmounted by black slaty marl, covered by sand and loose gravel. On the Deister and Osterwald, the schistose marl above noticed is wanting, and the sandstone rests immediately on the oolitic limestone. The series of beds between these last rocks of the oolitic group and the lias seem nowhere interrupted in northern Germany. Pursuing the comparison with the oolite coal of Yorkshire, the schistose marl above noticed (the highest member of the oolitic group of the Weser chain,) should be considered equivalent to the Oxford Clay*.

The oolitic series of southern Germany forms the immediate continuation of the Swiss Jura towards the N.E., cut through by the Rhine at Schaffhausen. These rocks extend to Siegmaringen on the Danube, from whence they follow the left side of the valley of that river. They constitute the plateau known by the name of the Swabian Alps. The oolitic group extends northwards from Ratisbon to the Maine, to Banz, Lichtenfels, Staffelstein, and in detached portions to Coburg. On the east of this range the older strata rise on the slopes of the Böhmerwald and Fichtelgebirge. Some difficulty has attended the comparison of these rocks with the English and French divisions of the oolitic group. The lias is so completely developed, and so similar to that in England, that its identity has long been placed beyond all doubt. Several divisions may be observed in it, characterized by their organic remains. The lower portion contains numerous beds of limestone, and among other organic remains the *Gryphæa incurva*, Sow. Above this reposes aluminous marl, and dark smoky marl, remarkable for the prodigious number of *Belemnites* discovered in it, as also for containing the *Gryphæa Cymbium*, Lam. These are surmounted by black shales, with *Posidoniæ*, *Fishes*, and *Sau-*

* M. Hoffman (Uebersicht der orog. und geogn. Verhaltniss evom N.W. Deutschland,) compares this marl to the Weald Clay, the sandstones with coal of the Bückeberg to the Hastings Sands, and terminates the oolitic group with the marl beneath. The Kahlenberg near Eichte (on the N.W. flank of the Hartz,) affords, however, sufficient proof of the age of these beds. Many fossils correspond with those of the coral rag, while some are referrible to the Kimmeridge clay and Portland stone. Von Dechen, German Transl. of Manual.

Y

rians. The sandstone with clay iron-stone at Aalen and Was-seralfingen, has been sometimes considered as forming part of the lias; it has, however, been shown by Count Münster to represent the inferior oolite of England*. The mass of the compact light-coloured Jura limestone succeeds, but from the want of marked differences in the compactness of the strata, sharp escarpments do not occur. Calcareous sandstone, with a bed of blue clay, rests on the ferruginous oolite of Wasse-ralfingen. This may comprise the rocks of the English series up to the Kelloway rock, represented by a repetition of ferru-ginous oolite in the succeeding clay, equivalent to the Oxford Clay, and containing the *Gryphæa dilatata.* Upon this rests white marl and white compact limestones, the representative of the coral rag of England, more developed, but containing numerous characteristic fossils. A large portion of country is formed of these beds, and there is no clay stratum above them, which might be considered analogous to the Kimmeridge clay. The lithographic slates of Bavaria take their place above them, and contain such an extraordinary mixture of organic remains that they may be considered as local, not constituting an ex-tended bed which can be identified in distant places. These lithographic slates of Pappenheim, Solenhofen, and Monheim near Eichstädt, form the upper part of the oolitic group of Southern Germany; therefore their position in the series is doubtful. As, however, many fossils of the white limestone beneath occur in them also, they may be considered as not far removed from it †.

In the whole range from the Danube to Coburg there are thick and extensive masses of dolomite, which take the place of a part of the white limestone (coral rag). These masses are for the most part non-fossiliferous, and it is only in a few places that organic remains can be detected. They are immediately covered by the lithographic slates ‡.

Von Buch was the first to point out that the coral rag con-stituted the elevated plateau between the Maine and Switzer-land, and that it was found in the mountains of Streitberg, at Donzdorf in Swabia, at Rathshausen near Bahlingen, and at Mont Randen near Schaffhausen. He observes, that at the latter place there are several beds of polypifers, in which *Cne-midium lamellosum, Cn. striatum,* and *Cn. rimulosum,* are the most characteristic fossils. Beneath these are beds full of

---

* Münster, Über den Oolithischen Thoneisenstein in Süd-Deutschland.

† It is understood that M. von Buch is preparing a detailed account of the oolitic series of Southern Germany, which will no doubt afford us a mass of valuable information and points of comparison with the English portion of the same series.

‡ German Transl. of Manual.

Ammonites, such as *A. placatilis, A. triplicatus,* large and very abundant, *A. perarmatus, A. biplex, A. flexuosus, A. bifurcatus,* and *A. canaliculatus.* These coral-rag beds rest on clays and marls, containing the *gryphæa dilatata* and *Ammonites sublævis*.* The List of Organic Remains will show that polypifers are abundant on this rock at Streitberg, Muggendorf, &c. M. Merian has afforded us very valuable details respecting the structure of the Jura near Bâle, and of its continuation into Germany in the same vicinity; whence it appears that the inferior oolite (*Eisen Rogenstein*) and the lias (*Gryphiten Kalk*) constitute clearly marked rocks of the series. The beds which rest on the *Eisen Rogenstein* are divided into older and newer Jura limestone (*Alterer Rogenstein* and *Jüngerer Jurakalk*), the former being considered in a great measure equivalent to the Great or Bath oolite, and separated from the latter by beds of clay†.

So far, if we except the dolomite in Germany, we have found no great change in the oolitic group, taken as a mass: there is nothing which shows that in the particular parts of Europe above noticed any forces were called violently into action during its deposit. On the contrary, a greater or less degree of repose seems characteristic of it, as also the presence of a large proportion of calcareous matter. The lowest portion, or the lias, preserves certain general characters over a considerable area; and why some geologists have separated it from the oolitic series is not easily understood; for if an apparent passage into the rocks beneath in some situations be the reason, such a reason would hold equally good for not separating it from those above, into which it also passes: if its zoological character be brought forward, there can be little doubt that throughout Western Europe this would place it in the group under consideration.

The lias of Western Europe may be considered, taken in the mass, as an argillaceous and calcareous deposit, in which sometimes one substance predominates, sometimes the other; sometimes presenting a great abundance of marls or clays, at others of limestones: the latter are however generally most common in the lower portions of the rock. In the Vosges district the lower part of the lias is formed of a sandstone, described by M. Elie de Beaumont as yellow and quartzose, containing mica, a few flattened argillaceous nodules, and small

* Von Buch, Recueil de Planches de Pétrifications Rémarquables, Berlin, 1831.

† Merian, Geognostischer Durchschnitt durch das Jura-Gebirge von Basel bis Kestenholz bey Aarwangen; Denkschriften der allgemeinen Schweizerischen Gesellschaft für die gesammten Naturwissenschaften. Zurich, 1829.

white or black quartz pebbles*. The presence more particularly of the pebbles seems to point to a transport by water. This sandstone extends into the neighbouring parts of Germany, and is one of those to which the name of *Quadersandstein* has been applied. Beneath the oolitic group which comes into contact with the granitic rocks of central France, M. de Bonnard has described an arenaceous rock, which he has named *Arkose*, and which may represent the arenaceous beds constituting the lowest part of the same rocks in the district of the Vosges. M. Dufrénoy describes an arenaceous deposit corresponding in geological position and external characters with the arkose of M. de Bonnard in the south-western part of France. He also states, that from Châtre, where the coalmeasures terminate, to beyond Brives, the separation of the oolitic series and the granitic rocks is marked by the presence of this sandstone, composed of quartz grains and felspathic portions, cemented by matter generally marly, but sometimes siliceous; the silica in the latter case becoming sometimes so abundant as to obliterate its character of a sandstone, so that it passes into a jasper. This sandstone seems to pass into the lias limestone, presenting an arenaceous limestone between the two. M. Dufrénoy considers it as the inferior sand of the lias. The same author describes the lias of the south-west of France; and states that it contains masses of gypsum. Although sulphate of lime, in the shape of crystals of selenite, is by no means uncommon in the lias marls of other countries, its presence, in that form, does not appear to mark a chemical deposit so much as in the gypsum above noticed. Taken as a whole, the lias seems very persistent in its characters throughout a considerable part of France, England, and Germany, pointing to a somewhat common origin. In the lias of Lyme Regis, Dorset, there would appear evidences of slow deposit in some parts, while in others the animals entombed seem to have been suddenly killed and preserved, so that the animal substances had not time to decay. The ink-bags of fossil *Sepiæ*, noticed by Prof. Buckland, afford perhaps the best evidence we can adduce of this fact; for had the animal substances which contained ink been exposed but for a short time to decomposition or the attacks of other animals, the ink must have flowed out of the bags. Now the actual forms of this fossil ink are precisely those of the ink-bags found in the *Sepiæ* and other animals possessing organs of a similar description at the present day; and therefore they appear to have been preserved entire and suddenly in a soft deposit.

* Elie de Beaumont, Mém. pour servir à une Description Géologique de la France, tom. i.

In the lias of southern England and many parts of France and Germany, the calcareous matter has been more abundant in the lower parts; and limestone beds have been the consequence, interstratified with marl, the latter sometimes schistose. Above the lias we have an arenaceous deposit, into which the marls graduate; and these sandy beds would seem to have been formed over a considerable area, embracing a large portion of France and England, and parts of Scotland and Germany. These are surmounted by limestones, one of which, characterized by the presence of oolitic iron-ore, though not precisely continuous, is remarkable for its occurrence in a similar part of the series, whether it be in the southern parts of England, in the north of France, in the Jura, or in some parts of Germany. Above these beds, termed the Inferior oolite, there is a series which varies much in its mineralogical character, presenting modifications of clays, marls, and limestones; the latter, which are often oolitic, affording beautiful materials for architectural purposes, as is seen in the towns of Bath, Caen, Nancy, and other places. This variety is commonly known by the name of the Bath or Great oolite, while other portions have received the names of Fuller's earth, Bradford clay, Forest marble, and Cornbrash. There can be little doubt that in tracing these supposed minor divisions over many parts of Europe, too much attention has been given to them as they exist in southern England and in Normandy, and that conclusions respecting their complete identity elsewhere have been somewhat forced. This is not the case with the next division,—one like the lias composed of argillaceous and calcareous matter, known as the Oxford clay, which, with certain modifications, seems to extend through England, and over a considerable portion of France, including the Jura, into Germany. The next superior rock, termed Coral rag, (from containing in certain situations a great abundance of polypifers,) separating an argillaceous deposit termed Kimmeridge clay from the Oxford clay, seems also to have a wide range, and presents a mixture principally calcareous, and often oolitic, the grains being not unfrequently so large that the rock is named Pisolite. The Kimmeridge clay is also an argillaceous and calcareous mixture, which has a considerable range, particularly over England and France. Its covering, or the beds termed Portland beds, seems very irregularly dispersed, the causes that produced the beds not being so constant as those which formed the clay beneath: it will however have been seen that rocks considered equivalent occur in the south-west of France, and in the Jura.

As yet we have seen the oolitic group composed of nearly similar mineral substances, and abounding in organic remains.

In Poland, however, there would appear, according to Prof.
Pusch, to be a change in the general mineral structure, preparing us for other greater changes, which will be noticed in
the sequel. M. Pusch describes the lower member of the
group under consideration in that country as more or less
white and marly. On this rests dolomite, generally of a
dazzling whiteness, affording the forms so remarkable in the
rocks of this nature, and composing the picturesque country
between Oldkusz and Cracow, and near Kromolow, Niego-
womie, and other places, rising to the height of 1200 or 1400
feet above the sea. The upper part of the dolomitic limestone
from Oldkusz towards Zarki, and especially near Wladowice,
contains pisiform iron-ore; it there becomes mixed with a
coarse sandstone, and constitutes a problematical agglomerate
and red sandstones. The upper portion of the group is formed
of gray and oolitic limestones and calcareous agglomerates,
and is represented as passing into the beds considered equivalent to the Wealden rocks. The rocks of the oolitic group
are seen to rest unconformably on the coal-measures and
muschelkalk of Poland; and it is necessary to use some caution not to confound them with the latter rock, when they are
in contact, as at Oldkusz and Nowagora. Taken on the large
scale, the Polish rocks of this age are stated to have a general
direction N.N.W. and S.S.E. From Wielun they plunge
beneath the great plain of Poland, here and there appearing
in islands above it, and are considered to be its support, being
met with in sinking through it. The organic remains contained in this deposit are stated to be such as to establish its
identity with the oolitic series of other parts of Europe[*].

We have now to consider a series of equivalent deposits,
with little or no mineralogical resemblance to those noticed
above, occurring in the Alps, the Carpathians, and in Italy.
Numerous memoirs have been written by different geologists,
and some have even considered that certain minor divisions
might be established; but it must be confessed,—though the
evidence is greatly in favour of a considerable development of
the oolitic group, with altered mineralogical characters, in the
situations above noticed,—that the termination of the group
either above or beneath is far from possessing that clear and
certain character which could be desired. The mineralogical
character being so different, recourse has generally been had
to organic remains; there are, however, such singular mixtures
of these, in the Alps more especially, that the determination
of particular deposits is far from certain. Instead of tender,
soft marls, clays, sands and light-coloured limestones, we have

* Pusch, Journal de Géologie, t. ii.

dark-coloured marbles, masses of crystalline dolomite, gypsum, and schists approaching talcose and micaceous slates. The Alps are also particularly difficult of examination, as from the convulsions by which they have been upraised or otherwise visited, whole mountain masses are thrown over, and the rocks really deposited the latest occur beneath the older strata; and this not in limited spaces, but over considerable distances. These dark-coloured rocks were during the prevalence of the Wernerian theory referred, as was natural, to the transition class; and we are indebted to Dr. Buckland for first pointing out that they were of more recent origin: since that time, other geologists have shown the probable relative antiquity of different portions; and among these, M. Elie de Beaumont holds a distinguished place, particularly as respects Savoy, Dauphiné, Provence, and the Maritime Alps. In a note on the geological position of the fossil plants and *Belemnites* found at Petit Cœur near Moutiers in the Tarentaise, published in 1828[*], this author observes that the system of beds described by M. Brochant in his memoir on the Tarentaise, and which in many places contains considerable masses of granular lime-stone and micaceous quartz rock, as well as large masses of gypsum, belongs to the oolitic group. He is of this opinion, as he considers that the most ancient secondary rocks of that country, in which no fossil shells have been found that have not been discovered in the lower part of the oolitic series, can be traced to the environs of Digne and Sisteron (Basses Alpes), where they afford a great abundance of those remains supposed to be characteristic of the lias.

In a notice on the geological position of the fossil plants and graphite found at the Col du Chardonnet (Hautes Alpes), M. Elie de Beaumont observes, that as the traveller quits the Bourg d'Oisans (Piedmont) and approaches the continuous range of masses, termed primitive, that extend from the Monte Rosa towards the mountains on the west of Coni, he will per-ceive that the secondary rocks gradually lose their original character, though certain distinguishing marks may still be seen,—thus resembling a half-burnt piece of wood, in which the ligneous fibres may be traced far beyond the part that re-mains wood[†]. He has also remarked on the original differ-ences that may have existed between these secondary rocks of the interior of the Alps, and those in the same series of other countries; and thence concludes, that very little importance should be attached to the difference of mineralogical structure observed in the beds above mentioned, and in the lower part of the oolitic group, occurring undisturbed in other parts of

[*] Annales des Sciences Naturelles, t. xiv. p. 113.
[†] Ibid. 1828, t. xv. p. 353.

Europe, and of which these Alpine rocks appear to him the enlarged prolongation. The vegetables found by M. Elie de Beaumont in the situations above noticed, were examined by M. Ad. Brongniart, and many were found by him to be generally the same with those discovered in the coal-measures. The following is a list of those which he obtained from the Alps, apparently all similarly situated as to geological position: *Calamites Suckowii,* Ad. Brong., at Pey-Ricard, near Briançon (also in the coal-measures of Newcastle and other places) ; *C. Cistii,* Ad. Brong., the same locality (also at Wilkesbarre in Pennsylvania); *Lepidodendron,* 2 sp., Pey-Ricard and Pey-Chagnard, near Lamure ; *Sigillaria,* the above localities, and La Motte near Lamure; *Stigmaria,* Pey-Chagnard; *Neuropteris gigantea,* Ad. Brong., Servoz, Savoy (also in the coal-measures of Bohemia); *N. tenuifolia,* Ad. Brong., Petit-Cœur, and Col de Balme (also in coal-measures of Liége and Newcastle); *N. flexuosa,* Stern., La Roche Macot, Tarentaise (also coal-measures of Liége and Bath); *N. Soretii,* Ad. Brong., same locality ; *N. rotundifolia,* Ad. Brong., La Roche Macot, and Col de Balme (also in the coal-mines of Plessis, Calvados); *Odontopteris Brardii,* Ad. Brong., Petit-Cœur (also coal-mines of Terrasson, Dordogne); *Od. obtusa,* Ad. Brong., Col de l'Ecuelle, near Chamonix; Petit-Cœur (also at Terrasson); *Pecopteris polymorpha* *, Petit-Cœur (also in the coal-measures of St. Etienne, Alais, Litry, Wilkesbarre); *Pe. pteroides,* Ad. Brong., Pey-Chagnard (also in coal-measures at Liége, Mannebach, St. Etienne, and Wilkesbarre); *Pe. arborescens,* Ad. Brong., Val Bonnais, near Lamure; Petit-Cœur (also at Mannebach and Aubin, Aveyron); *Pe. platyrachis,* Ad. Brong., Val Bonnais (also at St. Etienne); *Pe. Beaumontii,* Ad. Brong., Petit-Cœur; this new species is described as resembling the *Pe. nervosa, Pe. bifurcata,* Stern., and *Pe. muricata,* Schlot., found in the coal-measures, and *Pe. tenuis,* found in the oolitic series of Whitby and Bornholm; *Pe. Plukenetii?* Petit-Cœur; Col de l'Ecuelle (also at Alais); *Pe. obtusa,* Ad. Brong, Petit-Cœur (also in coal-measures near Bath); *Asterophyllites equisetiformis,* Tarentaise (also at Alais and Mannebach); *Annularia brevifolia,* Col de Balme (also at Alais and Geislautern)†.

These vegetable remains are so far associated with *Belemnites,* that the latter occur both above and beneath them; so that there can be no doubt as to the *Belemnites* having existed previous to and after the vegetable deposit; and therefore these localities would involve the question of the preference that should be given to the *Belemnites* or to the vegetables, if

* This species is common in the coal-measures of France according to M. Ad. Brongniart.
† Ad. Brongniart, Ann. des Sci. Nat. vol. xiv. pp. 129, 130.

M. Elie de Beaumont did not appear certain that the same series of beds was continued to Digne and Sisteron, and there contained characteristic lias remains.

M. Necker de Saussure has described a series of beds that composes the upper part of the Buet (Savoy), and which constitutes the lowest calcareous deposit of that portion of the Alps, resting, like those above noticed at Petit-Cœur and the Col de Chardonet, on older and non-fossiliferous rocks. The following is a section, in the ascending order :—1. Mica slate, which may form part of the protogine rocks of this district. 2. A sandstone, formed of numerous grains of quartz, mixed with a few crystalline grains of felspar, and sometimes with a little talc or chlorite. 3. Red and green argillo-ferruginous schist. This rock is sometimes wanting in the section; but on the east of the Vallée de Vallorsine it alternates with the well-known Vallorsine conglomerate, which is but a similar schist, filled with rounded pebbles of gneiss, mica slate, protogine, &c., among which we neither observe true granite nor limestone;—an important fact, as is observed by M. Necker, for it appears to show that the Vallorsine granite, which cuts through the gneiss, did not exist before the formation of the conglomerate. 4. A black schist, with impressions of ferns, the vegetable remains being converted into thin talc*. 5. Black or dark bluish-gray limestone, filled with grains of quartz. 6. A black argillaceous schist, containing nodules of Lydian stone. Ammonites are found in this rock, as also in an argillo-talcose schist which alternates with it. 7. A gray calcareous and arenaceous schist, containing *Belemnites*†. The last bed constitutes the summit of the Buet, 10,099 English feet above the sea.

It has been observed by M. Elie de Beaumont, that the calcareous portions of these regions of the Alps are separated from the older and non-fossiliferous rocks by a sandstone more or less coarse, which passes into a conglomerate, seen not only at the Vallée de Vallorsine above noticed, but also at Trient, Ugine, Allevard, Ferrière, and Petit-Cœur. The same circumstance is observable to the east of the Bourg d'Oisans and

---

* When crossing and wandering over the Col de Balme in 1819, I picked up specimens of sandstone with impressions of plants upon them; these plants I then considered, from their general character, to be such as are usually found in the coal-measures (Geol. Trans. 2nd series, vol. i. p. 162) ; an opinion which has since been confirmed by M. Ad. Brongniart, though it now appears that they may belong to a more modern deposit.

† Necker, Mém. sur la Vallée de Vallorsine, Mém. de la Soc. de Phys. et d'Hist. Nat. de Genève, 1828.—For a section of the Buet, see the same Memoir; and Sections and Views illustrative of Geological Phænomena, pl. 27. fig. 5.

Huez, and in other places *. This evidence of the action of
water possessing sufficient velocity to transport coarse sands
and pebbles should be borne in mind, as, however such sands
and pebbles may have been since altered in appearance, it
shows that the deposits were not produced quietly; though
subsequently, from a change of circumstances, and the establishment of comparative tranquillity, limestones were formed.
These appearances are not confined to the Savoy and French
Alps, but are seen on the shores of the Lake of Como and of
the Gulf of La Spezia. The calcareous beds, of which such
fine sections are afforded in the Lakes of Como and Lecco,
are separated from the gneiss and mica-slate of the higher
Alps, by a conglomerate composed of rounded pieces of quartz,
red porphyry, and other rocks, associated with sandstone
beds.

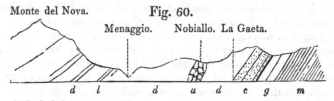

Monte del Nova.   Fig. 60.

Menaggio.   Nobiallo. La Gaeta.

d, *d*, *d*, dolomite. *l*, limestone. *a*, gypsum included in dolomite. *c*, red conglomerate and sandstone separating the
calcareous and dolomitic rocks from the gneiss and mica-slate.
g, gneiss. *m*, mica slate†.

The limestone series incumbent on the conglomerate is in
some situations strangely mixed with dolomite more or less
crystalline, as will be noticed in the sequel.   Taken as a mass,
the limestones occupy a thickness of many thousand feet, and
are more or less gray.   They are siliceous, and contain seams
of chert in the upper part (near Como), become slaty, with
apparently little siliceous matter in their central parts, and are
finally compact and more thickly bedded in their lowest situations.   Ammonites greatly resembling *A. Bucklandi* and *A.
heterophyllus* are discovered in it, as are also *Turritellæ*, and
other shells.   Anthracite is here and there found.   I have
little doubt that the oolitic group is represented by at least a
part of this calcareous mass ; but how much, and what other
equivalents there may be, my present information will not
permit me to hazard an opinion.   The general circumstances
are however so similar, that it does not seem unreasonable to
conclude that the causes, whatever they were, which produced

* Elie de Beaumont, Ann. des Sci. Nat. t. xv. p. 354.
† For a map, other sections, and a description of this district, see Sections
and Views illustrative of Geological Phænomena, pl. 31, 32.

the Vallorsine conglomerates and the sandstone associated with them in that part of the Alps, were contemporaneous with those which formed the conglomerates and associated sandstones of the lakes of Como and Lugano.

To present a detail of the various observations on those Alpine rocks which are considered as referrible to the oolitic group, would far exceed our limits; the student will consult with advantage the various labours of Studer, Boué, Sedgwick, Murchison, Lill von Lillienbach, Lusser, and others. There may be occasionally some difference of opinion among authors, as to where the series may commence, or where it may end; but the main fact, the existence of the group itself, seems established beyond all doubt. When we consider the disturbed nature of the country to be examined, and the difficulty of attaining certain situations perfectly necessary to a right understanding of the subject, except under very favourable circumstances, we should be more surprised that so much has been accomplished in so short a time, than at finding discordant opinions on certain minor points.

Mr. Murchison observes that, accompanied by M. Lill von Lillienbach, he found in the dark-coloured limestone and shale, at the gorge of the Mertelbach, below Crispel (Austrian Alps), —*Ammonites* 2 species (one approaching *A. Conybeari*), *Pecten* 3 species, small *Gryphæa*, *Mya*, *Perna* 2 species, *Ostrea*, *Corallines*, &c. This group is referred to the lias. An overlying red encrinite limestone contains several species of *Ammonites*, and some *Belemnites*. According to Professor Sedgwick and Mr. Murchison, most of the salt-mines of the Austrian Alps are contained in the oolitic group (Halstadt, Aussee, &c.). The upper part of the oolitic series of this part of the Alps contains semi-crystalline, brecciated, compact, and dolomitic limestones *.

I cannot conclude this sketch of the oolitic group, without adverting to certain limestones of La Spezia which may be referrible to it. On the west side of the celebrated Gulf of La Spezia, there is a range of mountains extending along the coast nearly to Levanto, their breadth augmenting as they advance N. W. The sections of these mountains expose the following rocks, easily observed up any of the cross valleys. The annexed wood-cut exhibits a section over Coregna.

Coregna.   Fig. 61.

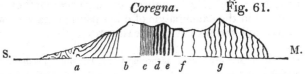

S. _____ M.

a   b   c d e f   g

* Proceedings of the Geological Society, 1831. Phil. Mag. and Annals, vol. ix. 1831.

S. Gulf of La Spezia. M. Mediterranean. *a.* Limestone
series:—Upper beds compact and gray, varying in intensity
of tint; more or less traversed by calcareous spar; here and
there interstratified with schistose beds, and even argillaceous
slate. The beds most commonly thick. The limestone with
light-brown veins, so long known by the name of Porto Venere
marble, forms part of these beds. *b.* Dolomite:—varying in
appearance; not unfrequently crystalline; when most so nearly
white; in some places beds may be distinguished, in others
stratification cannot be traced. *c.* Numerous thin beds of
dark-gray limestone. *d.* The same kind of beds alternating
with light-brown schist, containing an abundance of small nodules of iron pyrites, *Belemnites, Orthoceratites,* and *Ammonites,* enumerated beneath. The limestones which alternate
with the schist become occasionally light-coloured as they approach the next rock, from which however they are separated
by a repetition of the dark-coloured limestone and brown
schist. *e.* Brown shale which does not effervesce with acids.
f. Variegated beds:—greenish-blue and argillo-calcareous
rocks; more or less schistose, the calcareous matter being
often in very small quantity. *g.* Brown sandstone;—principally siliceous, though some of it does contain calcareous matter. It is sometimes micaceous, and occurs either in thick,
thin, or schistose beds. It has sometimes been called grauwacke, and it is one of the *macignos* of the Italians.

The organic remains from Coregna were first discovered
by M. Guidoni, of Massa; a few indications only of the presence of such bodies in the limestone under consideration
having been noticed by M. Cordier some years previously.
The strata being perpendicular, the weather acts on the edges
of the shale beds, in which the remains are found, and they
are thus brought to light. At my request Mr. Sowerby examined the remains that I brought from thence, and he considers that out of fifteen different species of *Ammonites,* one
seemed the same with the *A. erugatus,* Phil., discovered in the
lias of Yorkshire, while two resembled *A. Listeri* * and *A. bi-
jormis,* shells discovered in the coal-measures of the same part
of England. The remainder he considers undescribed. From
the great scarcity of organic remains of these limestones in
Italy, I have inserted Mr. Sowerby's descriptions of the various species, together with figures, considering that they may
be of service in the examination of other parts of Italy, as well
as Greece, and various countries eastward.

* This shell is also discovered, according to M. Hœninghaus, in the coal-
measures at Werden.

Fig. 62.   Fig. 63.   Fig. 64.   Fig. 65.

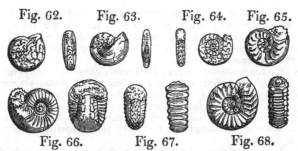

Fig. 66.   Fig. 67.   Fig. 68.

Fig. 62. *Ammonites cylindricus.* Inner whorls perfectly concealed; sides slightly concave about their centres, flat towards the margin; surface smooth; aperture oblong, deeply indented by the preceding whorl; the front square, which distinguishes it from *A. heterophyllus*, Sow.

Fig. 63. *A. Stella.* A small portion of the inner whorls exposed; the sides rather convex, largely umbilicated; of the inner whorls plain; of the outer, two thirds covered by large convex rays; aperture elongated, its front elliptical, its inner angles truncated.

Fig. 64. *A. Phillipsii.* Inner whorls almost wholly exposed; whorls slowly increasing, about four, their sides flat, irregularly and obscurely undulated; aperture four-sided, rather longer than wide, the sides nearly straight. The cast is contracted at distant intervals by the periodical thickening of the edge of the aperture. Named in honour of Mr. Phillips[*].

Figs. 65 and 67. *A. biformis.* Inner whorls partly visible; whorls three or four, rapidly increasing, crossed by many prominent sharp ribs; each rib suddenly becomes obscure, and spreads into two as it passes over the broad convex front; aperture transversely oblong, twice as wide as long, slightly arched.

Upon the inner whorls, which have the front plain, the ribs are contracted into round tubercles. The extremities of the longer ribs almost form spines. This species is found in the coal-measure near Leeds.

Fig. 66. *A. Listeri.* See Min. Conch. tab. 501. Also discovered in the coal-measures of Yorkshire.

Fig. 68. *A. Coregnensis.* Inner whorls much exposed; whorls three or four, crossed by many straight, prominent, sharp ribs, which bend forward, and suddenly terminate upon the nearly plain front; aperture transversely obovate.

This shell is intermediate between *A. biformis* and *A. planicostata*, Sow.: it is, however, nearer the former, as it has tubercles upon the inner whorls, where *A. planicostata* is quite smooth.

Fig. 69.   Fig. 70.   Fig. 71.

Fig. 72.   Fig. 73.   Fig. 74.   Fig. 75.

[*] Author of Illustrations of the Geology of Yorkshire.

Fig. 69. *A. Guidoni.* Inner whorls much exposed; whorls few, their sides flat and crossed by distant flattened ribs; each rib split, the posterior branch most prominent, and raised into a low tubercle before it passes over the narrow convex margin. Named in honour of Sig. Guidoni, the discoverer of these remains at Coregna.

Fig. 70. *A. articulatus.* Inner whorls nearly exposed; whorls few, each divided by eight or ten furrows into as many imbricating joints; the anterior edge of each joint elevated, and crossed by the edges of the septa.

Fig. 71. *A. discretus.* Inner whorls partly exposed in a large umbilicus; globose; whorls three or four, crossed by many prominent ribs, which split as they cross over the convex front; keel sharp, entire; aperture transversely oval, slightly arched.

Fig. 72. *A. ventricosus.* Inner whorls slightly exposed; whorls about three; half the fourth whorl much inflated; sides ornamented with arched ribs, that are often flattened and united in pairs as they pass over the front, which in the last whorl has a furrow along it; aperture circular, large.

Fig. 73. *A. comptus.* Inner whorls almost wholly exposed, rapidly increasing in size; sides flat; whorls crossed by very numerous, sharp, straight radii, which terminate in obscure spines near the narrow concave front; aperture oblong, narrowest towards the front.

Fig. 74. *A. catenatus.* Inner whorls much exposed; whorls rapidly increasing, crossed by strong curved ribs, which enlarge as they approach the margin; front ornamented with a chain of hollow squares; apertures rather square, notched by the preceding whorl; the hollow squares around the margin united by two of their angles to the extremities of corresponding radii.

Fig. 75. *A. trapezoidalis.* Inner whorls exposed; whorls three or four, rapidly increasing in size, crossed by many prominent nearly equal ribs reaching to the narrow front; aperture trapezoidal, indented by the preceding whorl; the acute angle truncated by the front.

The above figures are all of the natural size of the *Ammonites.* The remains of *Orthoceratites*, which abundantly accompany the *Ammonites*, resemble the *O. Steinhaueri*, found in the coal-measures of Yorkshire; they also approach the *O.? elongatus* of the Dorsetshire lias. The remains of *Belemnites* consist only of their alveoles, and are somewhat common.

As far therefore as the evidence of the *Ammonites* and *Orthoceratites* extends, we may refer the limestone of La Spezia either to the lias or the coal-measures. There will be observed a curious correspondence in the organic character of the rocks of the Savoy and French Alps above noticed, and considered as lias by M. Elie de Beaumont, with that of the limestones of La Spezia. In the former, coal-measure plants are found with *Belemnites;* in the latter, coal-measure *Ammonites* also occur with *Belemnites.* The organic character of the oolitic group in the Alps is far from being well ascertained, and the undescribed organic remains found in the same series of the South of France are exceedingly numerous, so that it may be possible to discover some of the La Spezia *Ammonites* in both situations; and the organic remains of the south-east of France, the Alps, and La Spezia, may hereafter mutually assist in determining the relative ages of the rocks in which they are discovered *.

The dolomite found among the limestones of La Spezia rises so perpendicularly, that it might be considered as a dyke elevating the strata; while at the same time it has the appearance of an included bed, or series of beds. It preserves a very constant position, and extends in a line across the mountains of La Castellana, Coregna, Santa Croce, Parodi, and Bergamo, towards Pignone. M. Laugier, at the request of M. Cordier, very obligingly made for me an analysis of some crystalline dolomite of La Castellana. One hundred parts were found to contain,—carbonate of lime, 55·36; carbonate of magnesia, 41·30; peroxide of iron and alumine, 2; silica, 0·50; loss, 0·84.

These limestones occur on the other or eastern side of the Gulf of La Spezia, and dolomitic rocks are also found among them. The mode on which they repose on the older rocks is particularly instructive, and is well seen at Capo Corvo, of which the annexed wood-cut is a section, laid bare by the sea.

Fig. 76.

G. Gulf of La Spezia. M. Embouchure of the Magra. *a.* Gray compact limestones mixed with schist. *b.* Thick beds of gray compact limestone. *c.* Schist with mica. *d.* Thick beds of hard conglomerate, containing pieces of quartz, varying from the size of a pea to that of a walnut, and even larger, agglutinated by a siliceous cement. Two or three beds of coarse sands are associated with this. *e.* The same, mixed with chlorite schist, often in the same bed. The quartzose beds contain veins of specular iron-ore. *f.* Brown micaceous and schistose beds, with a small proportion of limestone. *g.* A mixture of brown and white crystalline limestone. *h.* Compact chloritic rock. *i.* White saccharine limestone. *k.* Brown micaceous beds. *l.* White saccharine limestone, rendered schistose by mica. *m.* Brown semi-crystalline limestone, mixed with white. *n.* Micaceous schist, curving round to the eastward.

The crystalline limestones and micaceous schist of this section would seem to form part of the system of rocks, which in

---

* It should be observed, that M. Passini states he has discovered red ammonitiferous limestones in the midst of sandstones in Tuscany, which he considers may be referred to the same age as the limestones of La Spezia. Journal de Géologie, t. ii. p. 98.

the neighbouring mountains of Massa Carrara, now again known by the name of Alpi Apuani, furnishes the long celebrated Carrara marbles. The gray limestones appear the same as those on the western side of the Gulf of Spezia; but instead, like them, of resting upon a mass of sandstone, they repose upon a conglomerate, seen, between the mouth of the Magra and Ameglia, to become far more developed than at the Capo Corvo section, where it is in some manner squeezed between the crystalline limestones and the compact gray limestones. Amid this greater development, which appears to mark an unconformable superposition, a conglomerate will be observed (particularly on the shore of the Magra), closely resembling that commonly known as the Vallorsine conglomerate, and noticed above.

I cannot avoid connecting this conglomerate, and that of the Lake of Como, with the conglomerates and sandstones of the Vallorsine and other parts of the Western Alps, and referring them to the same epoch of formation;—one in which water, with a certain velocity, ground down portions of pre-existing rocks, and which was succeeded by a state of things when a great abundance of carbonate of lime was deposited. This deposit appears to have been extensive, not only in the Alps, but in Italy; and in both situations, where it occurs close to the rocks of an older date, such as protogine, gneiss, micaceous slates, associated saccharine marble, and talcose rocks of that age, it seems to be separated from them by strata which mark a mechanical origin. As we may suppose great inequalities to have existed during this deposit, and others immediately preceding it, we may perhaps in this way account for the almost close contact of the gray compact limestones with the saccharine limestone and other associated rocks at Capo Corvo, while on the western side of the gulf they rest on arenaceous rocks of considerable thickness, which again repose on gray siliceo-calcareous schists and sandstones, that extend over a considerable part of Liguria. How far these beds, which separate the limestones of the Alps, Liguria, and Tuscany, may be equivalent to the sandstone found beneath the lias in Southern Germany and various parts of France, may perhaps be now difficult to determine, but there is a certain general resemblance which seems to point to that conclusion.

Supposing that these Italian and Alpine limestones do represent the oolitic series of Western Europe, (and it seems very possible that they may do so,) it remains to account for the very great abundance of organic remains in the one, and their very great scarcity in the other. It has often struck geologists, that some deposits may have taken place in shallow seas, and others in deep water. This mode of viewing the subject has, if I mistake not, induced M. Elie de Beaumont to consider that the oolitic series of the Western Alps was deposited in a deep sea, at the same time that the same series was in the course of formation in shallow seas in other places. This observation may be extended into Italy and Greece, where the absence or very great scarcity of organic remains at this epoch seems to afford it support. That great inequalities existed at all periods on the earth's surface it seems fair to infer, as well beneath the sea as on land. It would be unphilosophical to conclude that marine animals were ever more capable of supporting very considerable differences of pressure than at the present day. Now we know that certain kinds of marine animals, particularly some *Mollusca* and *Conchifera*, are only found on coasts where they can find support beneath a moderate pressure of water; while others, such as the *Nautilidæ*, are so provided with floating apparatus, that they are discovered in parts of the ocean where there may be considerable depth. We have only to consider that in those parts of Western Europe where organic remains are abundant, shallow seas existed, while the same ocean was deep, with some exceptions, over that part of the globe's surface where we find Italy and Greece, and an explanation would seem to be afforded, not only of the abundance of shells in one place, and their scarcity in another, but also of the kind of shells found; for, as yet, camerated shells, such as *Belemnites, Orthoceratites,* and *Ammonites,* have been principally discovered in the oolitic rocks of central Italy; in other words, animals capable of swimming in deep seas*. Organic remains are not only scarce in the limestones in Italy, but also in the sandstones or macignos, which occur in great thickness above and beneath them. The organic remains as yet noticed in these sandstones are *Fucoides,* marine plants which may easily be drifted to considerable distances, as the *Sargasso Weed* now is. The differences of depth may also in some measure account for the different mineralogical structure of the rocks composing the oolitic group in different situations. Still, however, the question whence all this great mass of carbonate of lime was derived, remains unanswered. To attempt to account for it by means of springs neither more numerous nor abundant than those we now see, seems quite unphilosophical; and to con-

* M. Guidoni states in a memoir published in the Nuovo Giornale de letterati de Pisa, 1830; and the Journal de Géologie, 1831, that he has discovered in the limestone of La Spezia, not only a variety of *Ammonites* referrible to the oolitic group, but also many other univalves and bivalves; among the rest, the *Gryphæa arcuata,* Lam. (*G. incurva,* Sow.), which would appear to show a state of things at that place more resembling the oolite of Western Europe.

sider it entirely due to animals which have separated lime from
the water, leaving their shells produced through millions of
ages to be gradually converted into limestone, appears also a
cause inadequate to the effect required, though it cannot be
denied that the mass of many limestones is nearly made up of
organic remains. With every allowance for the limestone deposits of the oolitic series formed by springs and organic bodies,
there remains a mass of calcareous matter to be accounted for,
distributed generally over a large surface, which requires
a very general production, or rather deposit, of carbonate of
lime, contemporaneously, or nearly so, over a great area.

It appears from the lists of fossils discovered in the rocks of
the oolitic group*, that our knowledge of the vegetable remains is too limited to enable us to form any general conclusions respecting them. Mammalia have been found in one
locality only, Stonesfield ; where there are the remains of more
than one species of Didelphis. Pterodactyles have been discovered at Solenhofen, where there would appear to be many
species; and at Lyme Regis, where there is another species
found also at Banz, in Bavaria. The remains of this strange
genus probably also occur at Stonesfield. The Macrospondyli,
nearly allied to Crocodiles, are found in Northern France and
Germany. The Teleosaurus is discovered near Caen, Normandy. The Megalosaurus is found in Oxfordshire, in Normandy, and near Besançon. The Geosaurus has as yet been
noticed only in the lias of Wurtemberg, and in the Solenhofen
beds. Two species of Lacerta are discovered in the Solenhofen beds, which also contain the remains of the genera
Ælodon, Rhacheosaurus, and Pleurosaurus. Ichthyosauri
and Plesiosauri would appear to have been somewhat widely
distributed, and to have existed during the formation of the
whole oolitic series. Neither Pterodactyles, Crocodiles, nor
any of the above-noticed reptiles have as yet been detected in
the oolitic deposits of Southern France, of the Alps, or of
Italy. Tortoises have been noticed in England and Germany.
Fish would appear to be by no means rare ; those of Germany,
however, have only been examined with attention. Insects
have been detected in the oolite of Stonesfield and at Solenhofen. Polypifers occur in considerable abundance in particular places, more especially in the beds which have been
named Coral Rag, and in the upper part of the great oolite,
which has thus obtained, in Normandy, the name of Calcaire à
Polypiers. Of Radiaria, the genera detected in the oolitic
series are numerous, consisting of *Cidaris*, *Echinus*, *Galerites*,
Clypeaster, *Nucleolites*, *Ananchytes*, *Spatangus*, *Clypeus*, *En-*

---

* See lists at the end of the volume.

*crinites, Eugeniacrinites, Apiocrinites, Pentacrinites, Solano-
crinites, Rhodocrinites, Comatula, Ophiura,* and *Asterias.*

Respecting the shells, the following summary will show some
of those that have been discovered in the same division of the
oolite series*, in more than one moderately distant locality:
and the places where they have been observed will be found
by reference to the list of oolite fossils.

Fig. 77.

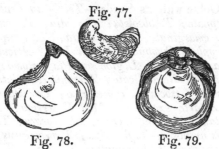

Fig. 78.   Fig. 79.

Kimmeridge Clay.—*Ostrea deltoidea* (Fig. 78.), a very characteristic shell in England; *Gryphæa virgula* (Fig. 77.), a
characteristic shell of this part of the oolitic series in France;
*Pinna granulata; Trigonia clavellata; T. costata; Mya de-
pressa; Pholadomya acuticostata; Pteroceras Ponti.*

Coral Rag.—*Ostrea gregarea; Pecten Lens; P. inæquicostatus; P. vimineus; P. vagans; Lima rudis; Plagiostoma
rusticum; P. læviusculum; P. rigidum; Modiola bipartita;
Gervillia aviculoides; Trigonia costata; T. clavellata; Turbo
muricatus; Trochus Tiara; Melania Heddingtonensis; M. striata; Ammonites plicatilis; A. vertebralis; A. Sutherlandiæ.*

Oxford Clay.—*Terebratula ornithocephala; Ostrea palmetta; O. Marshii; O. gregarea; Gryphæa dilatata* (Fig. 79.),
a very characteristic shell in England and France; *Pecten fibrosus; P. Lens; Gervillia aviculoides; Trigonia clavellata;
T. costata; Ammonites armatus; A. Kænigi; A. Calloviensis;
A. Duncani; A. sublævis; A. plicatilis; Patella latissima.*

Compound Great Oolite, including Fuller's Earth, Great
Oolite, Bradford Clay, Forest Marble, and Cornbrash.—*Terebratula subrotunda; T. intermedia; T. digona; T. obsoleta;*

---

* The student will have noticed, that in the list of oolite fossils, the same
shell is stated to have been discovered in places distant from each other, but
in various beds. Such shells are not here enumerated; and it may be
questionable how far some of those stated to be found in remote situations
in equivalent strata may really be so; for conclusions respecting the smaller
divisions of the oolite frequently appear much forced.

T. reticulata; T. globata; T. coarctata; T. media; Ostrea Marshii; O. costata; O. acuminata; Pecten fibrosus; Plagiostoma cardiiforme; Avicula echinata; Av. costata; Lima gibbosa; Modiola imbricata; Perna quadrata; Trigonia clavellata; T. costata; Nucula variabilis; Isocardia concentrica; Patella rugosa.

Inferior Oolite with its Sands.—*Terebratula sphæroidalis; T. ornithocephala; T. obsoleta; T. media; T. concinna; T. bullata; T. emarginata; T. punctata; T. resupinata; T. ovoides; Gryphæa Cymbium; Pecten Lens; Avicula inæquivalvis; Lima proboscidea; L. gibbosa; Plagiostoma giganteum; P. punctatum; Modiola plicata; Trigonia clavellata; T. striata; T. costata; Isocardia concentrica; Cardita similis; C. lunulata; Astarte excavata; Mya V scripta; Myoconcha crassa; Melania Heddingtonensis; M. lineata; Turbo ornatus; Trochus arenosus; T. fasciatus; T. promineus; T. punctatus; T. elongatus; T. abbreviatus; T. Tiara; T. angulatus; T. duplicatus; Pleurotomaria ornata; Ammonites læviusculus; A. discus; A. contractus; A. Blagdeni; A. Brocchii; A. acutus; A. Stokcsii; A. Murchisonæ; A. Braikenridgii; A. elegans; A. annulatus; A. Parkinsoni; Nautilus lineatus; N. obesus; Belemnites compressus.*

Fig. 80.   Fig. 81.   Fig. 82.   Fig. 83.

Fig. 84.   Fig. 85.

Lias.—*Spirifer Walcotii* (Fig. 85.), a very characteristic shell; *Terebratula ornithocephala; T. acuta; T. tetraedra; T. punctata; T. triplicata; T. bidens; T. serrata; Gryphæa incurva* (Fig. 81.), a very characteristic shell; *G. obliquata; G. gigantea; G. Maccullochii; Plicatula spinosa; Pecten æquivalvis; P. barbatus; Plagiostoma giganteum* (Fig. 82.); *P. punctatum; P. Hermanni; Lima antiqua; Avicula inæquivalvis* (Fig. 84.); *A. cygnipes; Inoceramus dubius; Modiola Scalprum; M. Hillana; Unio crassissimus; Amphidesma rotundatum; Pholadomya ambigua; Trochus Anglicus; T. imbricatus; Belemnites sulcatus; B. elongatus; B. apicicurvatus; B. pistilliformis; Ammonites Walcotii* (Fig. 80.), characteristic; *A. fimbriatus; A. Henleii; A. communis; A. planicostatus; A. falcifer; A. heterophyllus; A. brevispina; A. Jamesoni; A. Turneri; A. stellaris; A. Bucklandi* (Fig. 83.), characteristic; *A. obtusus; A. Stokesii* (A. Amaltheus); *A. sig-mifer; A. Conybeari; A. concavus; A. Humphresianus; A. Birchii; A. Bechii; Nautilus lineatus.*

Although this list may assist the student, so far as to show the shells stated to be found in the same rock in various situations, he must be cautious in referring any particular beds, wherein he may detect any of the above remains, to the rock under the head of which such remains are here noticed; but rather look at the general character of all the shells he may find in such beds, and thence infer their probable similarity, yet with much reserve, when the type and the rock considered equivalent to it are far distant from each other.

The following summary will convey an idea of the genera, with their respective number of species, stated by various authors to have been discovered in the beds of the group under consideration.

Plantæ.—Fucoides, 3 species; Equisetum, 1; Pachypteris, 2; Pecopteris, 6; Sphænopteris, 5; Tæniopteris, 2; Cyclopteris, 2; Glossopteris, 1; Neuropteris, 2; Lycopodites, 1; Pterophyllum, 4; Zamia, 11; Zamites, 4; Thuytes, 4; Taxites, 1; Bucklandia, 1; Mamillaria, 1.

Zoophyta.—Achilleum, 6; Manon, 3; Scyphia, 41; Tragos, 9; Spongia,·2; Alcyonium, 1; Cnemidium, 9; Limnorea, 1; Siphonia, 1; Myrmecium, 1; Gorgonia, 1; Millepora, 6; Madrepora, 1; Cellepora, 2; Retepora? 1; Flustra, 1; Ceriopora, 9; Agaricia, 3; Lithodendron, 3; Caryophyllia, 7; Anthophyllum, 3; Fungia, 2; Cyclolites, 1; Turbinolia, 2; Turbinolopsis, 1; Cyathophyllum, 6; Meandrina, 5; Astrea, 25; Thamnasteria, 1; Aulopora, 3; Entalopora, 1; Favosites, 1; Spiropora, 4; Eunomia, 1; Crysaora, 2; Theonoa, 1; Idmonea, 1; Alecto, 1; Berenicea, 1; Terebellaria, 2; Cellaria, 1; Sarcinula, 1; Intricaria, 1.

Radiaria.—Cidaris, 18; Echinus, 6; Galerites, 3; Clypeaster, 1; Nucleolites, 6; Ananchytes, 1; Spatangus, 4; Clypeus, 6; Encrinites, 2; Eugeniacrinites, 6; Apiocrinites, 8; Pentacrinites, 14; Solanocrinites, 3; Rhodocrinites, 1; Comatula, 4; Ophiura, 3; Asterias, 8.

Annulata.—Lumbricaria, 6; Serpula, 53.

Conchifera.—Spirifer or Delthyris, 3; Terebratula, 59; Orbicula, 3; Lingula, 1; Ostrea, 28; Exogyra, 3; Gryphæa, 15; Plicatula, 4; Pecten, 28; Monotis, 4; Plagiostoma, 18; Posidonia, 1; Lima, 5; Avicula, 12; Inoceramus, 1; Gervillia, 7; Perna, 3; Crenatula, 1; Trigonellites (*Phil.*), 2; Pinna, 7; Mytilus, 6; Modiola, 22; Lithodomus, 1; Chama, 3; Unio, 6; Trigonia, 15; Nucula, 18; Pectunculus, 2; Arca, 7; Cucullæa, 14; Hippodium, 1; Isocardia, 11; Cardita, 3; Cardium, 11; Myoconcha, 1; Astarte, 9; Cras-

sina, 7; Venus, 1; Cytherea, 4; Pullastra, 2; Donax, 2;
Corbis, 3; Tellina, 2; Psammobia, 1; Lucina, 4; Sanguinolaria, 2; Corbula, 4; Mactra, 1; Amphidesma, 5; Lutraria, 1; Gastrochæna, 1; Mya, 8; Pholadomya, 20; Panopæa, 1; Pholas, 2.

Mollusca.—Dentalium, 2; Patella, 8; Emarginula, 1; Pileolus, 1; Bulla, 1; Helicina, 4; Auricula, 1; Melanea, 5;
Paludina, 1; Ampullaria, 1; Nerita, 4; Natica, 5; Vermetus, 2; Delphinula, 1; Solarium, 2; Cirrus, 5; Pleurotomaria, 3; Trochus, 21; Rissoa, 4; Turbo, 8; Phasianella,
2; Turritella, 5; Nerinæa, 6; Cerithium, 3; Murex, 2;
Rostellaria, 3; Pteroceras, 3; Actæon, 5; Buccinum, 1;
Terebra, 4; Belemnites, 65; Orthoceratites? 1; Nautilus,
10; Hamites, 1; Scaphites, 2; Ammonites, 173; Aptychus,
4; Onychoteuthis, 1; Sepia, 1.

Crustacea.—Pagurus, 1; Eryon, 4; Scyllarus, 1; Palæmon, 3; Astacus, 6.

Insecta.—Libellula, 1; Æschna, 1; Agrion, 1; Myrmeleon? 1; Sirex? 1; Solpaga? 1.

Pisces.—Dapedium, 1; Clupea, 5; Esox, 2; Uræus, 1;
Sauropsis, 1; Ptycholepis, 1; Semionotus, 1; Lepidotes, 3;
Leptolepis, 3; Tetragonolepis, 4.

Reptilia.—Pterodactylus, 7; Macrospondylus, 1; Croco-
dilus, 3; Teleosaurus, 1; Megalosaurus, 1; Geosaurus, 2;
Lacerta, 2; Rhacheosaurus, 1; Ælodon, 1; Pleurosaurus, 1;
Plesiosaurus, 6; Ichthyosaurus, 4; Tortoise, 1.

Mammalia.—Didelphis, 2 *.

Thus making; *Plantæ,* 17 genera, 51 species. *Zoophyta,*
43 genera, 175 species. *Radiaria,* 17 genera, 94 species.
Annulata, 2 genera, 59 species. *Conchifera,* 55 genera, 406
species. *Mollusca,* 39 genera, 372 species. *Crustacea,* 5 genera, 15 species. *Insecta,* 6 genera, 6 species. *Pisces,* 10 genera, 22 species. *Reptilia,* 13 genera, 31 species. *Mammalia,* 1 genus, 2 species.—Total 208 genera, 1233 species.

Although this summary cannot be considered as strictly accurate, the present state of our knowledge respecting organic
remains rendering the catalogues of those contained in any
series of beds imperfect, it may still be useful as an approximation to the truth, and as affording a general view of the
fossils stated, for the most part by those now considered as
good authorities, to have been discovered in the group under
consideration.

It has been above remarked, that the surface on which the

* When the genus has been noticed, and the species is represented as
undetermined in the lists at the end of the volume, the genus has been considered to have only one species in this summary. Two species have been
assigned to Didelphis, as there seems little doubt of that fact.

oolitic group was deposited, was probably at very various depths beneath that of the sea; and that even during the deposit itself, the sea varied in depth over the same point, in consequence of movements in the land.   The nature of the organic remains also points to the proximity of dry land in some places, while it may have been comparatively remote in others.   It does not seem unphilosophical to infer that the bays, creeks, estuaries, rivers, and dry land, were tenanted by animals, each fitted to the situations where it could feed, breed, and defend itself from the attacks of its enemies.   That strange reptile the Ichthyosaurus * (one species of which, *I. platyodon,* was of a large size, the jaws being strong, and occasionally eight feet in length,) may, from its form, have braved the waves of the sea, dashing through them as the Porpess now does; but the Plesiosaurus, at least the species with the long neck (*P. dolichodeirus,* fig. 87.)†, would be better suited to

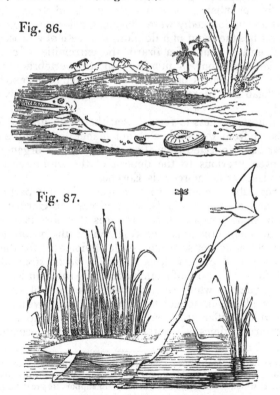

Fig. 86.

Fig. 87.

* It is attempted in the annexed wood-cut (fig. 86.), to convey an idea of the probable form of *I. communis*, and of the head of *I. tenuirostris.* The former is represented on dry land, where probably it never reposed, for the purpose of exhibiting its form.

† The animal is represented in the act of catching a *Pterodactyle.* It is

have fished in shallow creeks and bays, defended from heavy breakers. The Crocodiles were probably, as their congeners of the present day are, lovers of rivers and estuaries, and like them destructive and voracious. Of the various reptiles of this period, the Ichthyosaurus, particularly the *I. platyodon*, seems to have been best suited to rule in the waters, its powerful and capacious jaws being an overmatch for those of the Crocodiles and Plesiosauri. Thanks to Professor Buckland, we are now acquainted with some of the food upon which these creatures lived: their fossil fæces, named *Coprolites*, having afforded evidence, not only that they devoured fish, but each other; the smaller becoming the prey of the larger, as is abundantly testified by the undigested remains of vertebræ and other bones contained in the coprolites *. Amid such voracity, it seems wonderful that so many escaped to be imbedded in rocks, and after the lapse of ages on ages to tell the tale of their existence as former inhabitants of our planet. And strange inhabitants they undoubtedly were: for, as Cuvier says, the Ichthyosaurus has the snout of a dolphin, the teeth of a crocodile, the head and sternum of a lizard, the extremities of cetacea (being, however, four in number), and the vertebræ of fish; while the Plesiosauras has, with the same cetaceous extremities, the head of a lizard, and a neck resembling the body of a serpent†.

It is almost needless to remark that these two genera have disappeared from the surface of our planet; and, as the student may have collected from the various lists of organic remains, even previous to the deposit of the supracretaceous rocks, at least as far as regards Europe.

The vegetable remains have been accumulated in particular places, such as Yorkshire, Brora, and parts of Germany, at about the same period. Circumstances therefore must have existed at such situations not common to the whole area. These deposits do not seem the result of violence; for the vegetables are well preserved, as if, like the *hortus siccus* of the botanist, for the purpose of examination. By their aid we learn that the vegetation which then clothed some parts of this portion of our planet, no longer resembles that which we now see, but one widely different. Perhaps we may, in anticipation, look forward to times when the geologist may speculate on

---

figured as swimming high above the water for the purpose of showing its general form. It more probably swam beneath the surface, in the manner of crocodiles, which would enable it the better to support its great length of neck.

* For an interesting account of Coprolites and their contents, see Buckland's Memoir, Geol. Trans. 2nd series, vol. iii.

† Cuvier, Oss. Fossiles, t. v. This notice of the Plesiosaurus applies more particularly to *P. dolichodeirus*.

the proximity of certain lands near places where the abundant remains of vegetables and certain animals would seem to point to such conclusions, even though, from the various movements in the land, no part of such ancient continents or islands may now appear on the surface. We of the present day, however desirous we may be to elucidate this subject, seem to possess too few data to proceed in the inquiry. One thing, however, the student should bear in mind : he must not consider that all older rocks in the vicinity of others of more recent origin, though now rising in mountains high above them, necessarily formed the dry land previously to the deposit of the newer rocks: for amid the various surface-changes that have been effected, such older rocks have frequently been upheaved after the formation of the more recent, as is shown by the mode of stratification near the junction of the two, the one being tilted up with the other. We may, perhaps, more closely approach the truth, when we find, as in Normandy, the oolitic group resting quietly on, and surrounding, disturbed older strata; so that in that country (and the same observation applies to other situations) we may conceive the sea in which the oolitic rocks were formed, to have bathed the slaty and granitic districts of Normandy and Brittany.

Fig. 88.

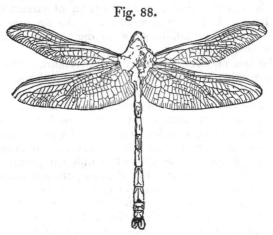

Scale = ⅓ of nature.

Those strange flying creatures, the Pterodactyles, must have sported on dry land, probably subsisting on insects, such, among others, as that figured above, which was obtained by Mr. Murchison from the quarries at Solenhofen, where the remains of Pterodactyles are also discovered.

That the Pterodactyles should be scarce fossils, is what we might expect, for the circumstances favourable to their preservation must have been extremely rare. Even supposing that they dashed out to sea in pursuit of their insect prey, there must have been a combination of fortunate accidents to have prevented the Pterodactyles and their intended prey from being devoured by the fish and other inhabitants of the sea, among the exuviæ of which their remains are now detected.

It is curious, and seems to establish a connexion between the insects and the Pterodactyles, that in the spot where the remains of the latter are most abundant (Solenhofen), the greatest quantity of fossil insects yet noticed in the oolitic group has been detected. At Stonesfield also, where the remains of insects are stated to have been discovered, the exuviæ of Pterodactyles, according to Prof. Buckland, are also observed. Not so, however, with the Pterodactyle of Lyme Regis, whose remains are mixed with those of Ichthyosauri and other marine animals, and where insects have not yet been detected. But when we consider the abundant exuviæ of Plesiosauri, perhaps we may not err greatly, in considering dry land not very far distant from the spot where we now find their bones entombed. Be the case as it may, a Pterodactyle in a sea, amid Ichthyosauri and other voracious creatures, must have had but a slight chance of escape; and geologists should be grateful that any combination of circumstances should have so far prevailed, as to permit the preservation of even a single individual, to show us the strange terrestrial creatures that then existed.

In the lias of Lyme Regis, the Ichthyosauri, Plesiosauri, and many other animals, seem to have suffered a somewhat sudden death; for in general the bones are not scattered about, and in a detached state, as would happen if the dead animal had descended to the bottom of the sea, to be decomposed, or devoured piecemeal, as indeed might also happen if the creature floated for a time on the surface, one animal devouring one part, and another carrying off a different portion;—on the contrary, the bones of the skeleton, though frequently compressed, as must arise from the enormous weight to which they have so long been subjected, are tolerably connected, frequently in perfect, or nearly perfect order, as if prepared by the anatomist. The skin, moreover, may sometimes be traced, and the compressed contents of the intestines may at times be also observed,—all tending to show that the animals were suddenly destroyed, and as suddenly preserved. Not only has this apparently happened to these reptiles which, breathing air, might under favourable circumstances be drowned simultaneously in great numbers, but also to the mollusca, to which constant, or nearly constant, immersion in water is absolutely necessary. Among the multitude of Ammonites discovered in the lias, I have often observed individuals, of which the large terminating chamber of the last whorl, where the body of the animal seems to have been placed, was hollow for half its distance upwards towards the aperture or mouth, as if the animal, when overwhelmed, had retreated as far as possible into this part of the shell, so that the muddy matter was prevented from completely filling it. This idea is rendered more probable from the condition of the calcareous matter filling the remaining part of the great cavity, which is exceedingly bituminous, as would happen from the decomposition of the animal within the remainder of the chamber.

The student should not, from what has been above remarked respecting the lias at a particular point, Lyme Regis, consider that such observations are applicable to the same rock generally; or even that the lias of Lyme Regis has suddenly been produced in its whole thickness at once: on the contrary, the lias varies materially at different points, as we should expect it to do, from different local causes; and the lias of Lyme Regis bears evidence of successive deposition, in part during a state of comparative tranquillity, and partly in consequence of a series of small catastrophes, suddenly destroying the animals then existing in particular spots. One observation is, however, necessary, and it will be often applicable to other parts of the oolitic rocks in various situations,—that during the formation of the lias in this part of England, there has been a certain change in the animal life of the same place. Thus the animals and shells in the upper part of this rock differ in the mass from those in the lower portion. Very frequently, also, particular strata afford certain organic remains, while all others are exceedingly rare.

Notwithstanding the temptation to treat of the probable circumstances that have accompanied the deposit of a particular rock, even within the distance of a few miles, we must abstain, as it would lead us into detail not compatible with this work. It may, however, be remarked, that the destruction of the animals, whose remains are known to us by the name of Belemnites, was exceedingly great at this place. When the upper part of the lias was deposited, multitudes seem to have perished simultaneously, as is attested by a bed composed of little else, beneath Golden Cap, a cliff between Lyme Regis and Bridport Harbour. Not only are millions entombed in this bed, but in the upper part of the lias generally. The production of such a bed would seem by no means difficult; for we have only to consider the occurrence of some circumstance destructive to molluscous creatures in the fluid containing, or otherwise carrying, the belemnites,—such as might happen to those swarms of mollusca which sometimes surround the navigator in warm latitudes,—and the floating animal mass, if not immediately, would eventually descend to the bottom; at least the part that escaped the predaceous animals, which indeed might be driven away by the circumstance, whatever it was, that destroyed the molluscous animals. Suppose a multitude of the common cuttle-fish to be suddenly killed by the irruption of, or their entrance into, water charged with carbonic acid; their internal bones, as they are commonly termed, would be distributed over a common surface after the decomposition of the animals, which were not likely to fall a prey to other creatures; for those which were not destroyed with the cuttle-fish, would avoid the water so charged with carbonic acid.

The vegetables of the lias of this place occur in two different states: the one showing that they have been scarcely injured before they were imbedded; the other seeming to point to the fracture of wood into junks, the small branches truncated, as if they had been broken either during, or previous to, their drift. These latter most frequently occur in argillo-calcareous nodules, often of large size: but the nodules are not concentric concretions; on the contrary, both these nodules, and those that frequently envelope the Ammonites and Nautili in the argillaceous beds, are fissile, the line of the laminæ being parallel to that of the general stratification; so that though the nodules, particularly those containing Ammonites and Nautili, are spheroidal, their fracture is lamellar; and a successful blow in the line of the laminæ, through the centre, discloses a fossil, not unfrequently a fish.

It being a very interesting inquiry to ascertain the chemical condition of organic remains entombed in various rocks, Dr. Turner was kind enough to analyse certain fossils from the lias of Lyme Regis. He found that a vertebra, a rib, and a tooth of an Ichthyosaurus examined by him, had all a highly crystalline texture, owing to the deposit of carbonate of lime, of which they chiefly consist. The colour is nearly, and in parts quite black, in consequence of bituminous matter, which in general amounts to not more than $\frac{1}{2}$, and he has not found it exceed $\frac{3}{4}$ per cent. The phosphate of lime in the vertebra amounted to about 29 per cent., while in the rib and tooth it was about 50 per cent. In fact, as might have been expected, the phosphate of lime remains in greater or less quantity in different specimens, probably depending on the situation where it was preserved, and on the compactness of the original bone.

Dr. Turner also ascertained that the scales of *Dapedium politum*, cleared as much as possible from adhering limestone, consisted of the same ingredients as the ichthyosaurian bones; but the phosphate of lime amounted only to 19 per cent.

Of course care was taken to select such specimens as were
not impregnated with sulphuret of iron, as sometimes happens;
and those examined were found to be remarkably free from
iron, manganese, alumina, and silica.

When we view the oolitic group as a whole, we cannot but
remark a certain general uniformity of its structure over a
considerable portion of Western Europe; showing that at the
time of its production some similar general causes were in action over a particular portion of the European area. While,
however, this general uniformity is sufficiently obvious over
such area, it is equally obvious that the various attempts which
have been made to detect certain minor divisions of the oolitic
group in the Alps, in Italy, and other places, have been by no
means successful. There can be little doubt that a very large
portion of the oolitic series has been mechanically produced.
Granting this, we can scarcely expect that perfect uniformity
to exist which was once considered probable. In point of fact,
minor changes in the nature of the beds are constantly taking
place; and from a multiplication of these minor changes, very
considerable differences in the subdivisions of the group are
produced.

The annexed proportional sections (Fig. 89. and 90.) will
exhibit the different development of the oolitic series in the
northern and southern portions of England; the superincumbent cretaceous rocks being also represented, to render the
general changes still more apparent.

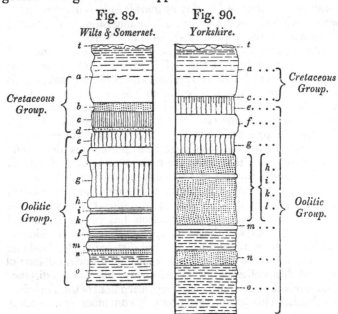

Fig. 89.   Fig. 90.

Wilts & Somerset.   *Yorkshire.*

As the same proportional scale has been adopted for both
sections, the eye will readily seize the different depths of the
cretaceous and oolitic groups in each. The same letters also
have been used for the minor divisions, so that the rocks in one
section can easily be compared with those in the other. *Cretaceous Group: a,* chalk; *b,* upper green sand; *c,* gault; *d,* lower
green sand. (*b.* and *d.* appear to be absent in the Yorkshire
section, being represented by *c.*)—*Oolitic Group: e,* Kimmeridge
clay; *f,* coral rag, and its calcareous grits; *g,* Oxford clay, and
the Kelloway rock in its lower part; *h,* cornbrash, and forest
marble; *i,* Bradford clay; *k,* great oolite; *l,* Fuller's earth; *m,*
inferior oolite; *n,* marlstone; *o,* lias; (*t,* gravel, &c.) It will be
observed that so far as regards the divisions *e, f,* and *g,* the two
sections do not much vary; but that a very considerable difference exists between the beds *h, i, k, l,* is developed in Southern and in Northern England. Not only is their mineralogical character different, but their organic contents are for the
most part distinct. The clays and limestones of the South are
full of marine remains, while the shales and sandstones of the
North abound with terrestrial plants, which have been so thickly
accumulated in certain beds as to form coal strata. The marlstone (*n*) and the lias (*o*) are also far more developed in Yorkshire than in Somersetshire; and the former, which in the South
may be considered as the passage of the inferior oolite sands
into the lias, has in the North become of more importance, and
is separated from inferior oolite by what is termed the upper
lias shale. A thin bed of clay or marl is indeed interposed between the inferior oolite sands and the marlstone in the South,
which may be the upper lias shale of the North.

The terrestrial character of the organic remains contained
in a certain portion of the oolitic group is not, as has been
seen, confined to Northern England and Scotland, but extends
into Germany. Now if it be fair to infer, as it seems to be,
that when we find accumulated vegetable matter sufficiently
abundant to constitute coal strata, accompanied by beds full
of delicately preserved plants, dry land could not have been
far distant at the time such vegetable matter was deposited;
we may conclude, that when the beds of Northern England,
Scotland, and a portion of Germany corresponding with the
compound great oolite of Southern England, were deposited,
dry land existed in that part of our planet now known as
Northern Europe. Mr. Murchison has observed vertical
stems of *Equisetum columnare,* apparently in the position in
which they grew, reminding us of the *dirt bed* of Portland and
other places, in the lower carboniferous shale and sandstone of
the Yorkshire oolite, not only on the coast, but also at a distance
of forty miles, on the north-western escarpment of the Yorkshire
moorlands. This author very properly concludes, from finding

the same plants in the same bed, and in the same position, over so large an area, that it is almost demonstrable that the vertical position of the *Equiseta* could not have been the effect of chance, but must have been the result of some general cause acting over the area. This cause, he further observes, has very probably been a great submergence of the area, permitting the plants to retain their original positions, and their gradual envelopment by mud, silt, and sand*. Now this fact shows, as in the case of the Portland *dirt bed* for another period, that the conversion of dry into subaqueous land has been exceedingly gradual in that part of England during the deposit of the oolites. How far this fact may be found more general remains to be seen; but it is, as far as it goes, extremely important.

If we continue our view of the course of the beds, contemporaneous with those above noticed, from Northern to Southern Europe, we observe that the character of the organic contents becomes marine, and that many of the beds, particularly the clays, can be traced over extensive areas. The multitude and character of their general organic contents may lead us to suppose that over a considerable area,—one extending from Southern England to the Jura, and embracing a large portion of France, and a part of Southern Germany,—the sea was by no means deep in which the beds deposited. At Stonesfield indeed, the slates of which place have recently been shown by Mr. Lonsdale to be a lower portion of the great oolite, we have evidence, in the remains of the Didelphis and other organic exuviæ, tending to show that land may not have been far distant. Now if we connect this with the vertical stems of the Yorkshire oolite, they would, both taken together, lead us to infer that the dry land was somewhat extensive at this period over what now constitutes a considerable portion of England.

It has been above remarked that the Alpine and Italian deposits, equivalent to the oolitic group generally, may have been formed in a deep sea, while other portions of the same group may have been deposited in shallow waters. If we now regard that particular part of the series corresponding to the compound great oolite, it appears very probable, that dry land was then abundant over what now constitutes Northern Europe, that over Southern Europe there was deep water, while an extensive area between the two was shallow water, with here and there, perhaps, dry land.

It would not accord with the plan of this volume further to investigate the probable condition of the European area during the deposit of various portions of the oolitic series; but we may

---

* Murchison, Proceedings of the Geol. Soc., and MSS.

remark that the clay beds, which can be traced with nearly similar characters over considerable areas, while the changes in
the calcareous and arenaceous masses are much more complicated, is precisely what we should expect from the transport
of comminuted matter mechanically suspended in water. For
the more comminuted the matter, the further will moving water
transport it; and consequently the uniformity of the deposit of
such a mixture will be greater than from less comminuted
matter,—supposing always the transporting force to be equal,
or nearly equal. While on this subject it should be remarked
that the sands of the inferior oolite cover a very considerable
area, extending from Northern far into Southern Europe.
Of all the divisions of the oolitic series the lias is most uniform
over the whole space occupied by it, for its variations are comparatively inconsiderable in Great Britain, France, and Germany. It is not until we enter the oolitic districts of the Alps,
and other portions of Southern Europe, that we find even the
mineralogical structure very materially altered.

SECTION VII.

RED SANDSTONE GROUP.

SYN.—Red or Variegated Marls (*Marnes Irisées*, Fr.; *Keuper, Bunte Mergel*, Ger.). Muschelkalk (*Calcaire Conchylien*, Al. Brong.). Red or Variegated Sandstone (*New Red Sandstone*, Eng. Auth.; *Grès Bigarré*, Fr.; *Bunter Sandstein*, Ger.). Zechstein (*Magnesian Limestone*, Eng. Auth.; *Calcaire alpin*, Fr.; *Alpenkalkstein*, Ger.). Rothliegendes (*New Red Conglomerate, Lower New Red Sandstone, Exeter Red Conglomerate*, Eng. Auth.; *Todtliegendes, Rothe Todtliegende*, Ger.; *Grès Rouge*, Fr.; *Pséphite Rougeâtre*, Al. Brong.).

THIS group, which is often one of very considerable thickness, succeeds, in the descending order, that previously noticed. Perhaps very fine lines of distinction should not be drawn between the two; for when the lower part of the one and the upper part of the other have been consideraly developed, they seem in some measure to pass into each other. This led M. Charbaut, who first observed the circumstance in the vicinity of Lons le Saulnier, to class the lias with the variegated marls which constitute the upper portion of the group under consideration. The rocks composing the red sandstone group occur in the following descending order:—1. Variegated Marls; 2. Muschelkalk; 3. Red or Variegated Sandstone; 4. Zechstein; and 5. Rothliegendes.

Variegated Marls.—In the district of the Vosges and in the neighbouring countries, these commence beneath the sandstone named lias sandstone, into which they gradually pass; the upper part of the variegated marls, which are green, presenting thin beds of black schistose clay, and of quartzose sandstone, nearly without cement, which latter gradually becomes the lias sandstone,—a rock that passes into the lias, and contains the same organic remains *. M. Elie de Beaumont observes, that in many countries the variegated marls can scarcely be separated from the lias sandstone, even artificially, as is done in the Vosges; for they appear to become one deposit, as in the environs of St. Leger-sur-Dheune, and Autun, and in the *arkose* of Burgundy. The variegated marls of the Vosges generally are, as their name implies, marked by different colours, among which the principal are wine-red and greenish or blueish gray; they break into fragments, which have no trace of a schistose structure. In the central portion of these marls there are beds

---

* Elie de Beaumont, Mém. pour servir à une Désc. Géol. de la France, t. i.

2 A

of black schistose clay, blueish gray sandstone, and grayish or yellowish magnesian limestone. The sandstone and clay contain vegetable impressions, and even coal. Masses of rock-salt occur in the lower part of the marls at Vic, Dieuze, and other parts of that district; and masses of gypsum are found in the upper and lower portions, but principally in the latter[*]. According to M. Charbaut, limestone beds, almost entirely composed of shells, are found in the upper part of this deposit.

M. von Dechen remarks, that the superior white sandstones are, in the neighbourhood of Stuttgard and Tübingen, covered by variegated clayey marls, by which they are separated from the lias. They are coarse grained; and the fragments of quartz, limestone, &c. are sometimes a foot in diameter, partly rounded, partly angular. Dark gray rolled pieces of limestone sometimes predominate in the lower beds, and are cemented by calcareous matter, containing grains of quartz and felspar. The cement of the other beds is quartzose. The sandstone, especially the upper portion, often contains so much carbonaceous matter that it becomes dark gray and even black. Fibrous anthracite and pitchcoal occur in it near Spiegelberg. At Erlaheim, near Balingen, similar coal, with large masses of iron pyrites, are found close to the lias boundary. These carbonaceous deposits are separated from the sandstones beneath by variegated marls, inclosing thin sandstone beds and traces of coal[†].

The variegated marls, not differing considerably in their mineralogical characters, occur in various parts of the north of France and Germany, and according to M. Dufrénoy they crown the red sandstone rocks of the South of France. How far the variegated marls may be traced in England remains questionable; but it would appear far from improbable, that the upper part of the red sandstone deposit of this country would answer sufficiently well in its mineralogical structure to the rocks above noticed in the Vosges. There is with us no apparent passage of the lias into the red sandstone series; on the contrary, we sometimes have, as at the Old Passage near Bristol, a kind of conglomerate of pieces of limestone, bones, teeth, and other remains of saurians and fish, with their fossil fæces or *coprolites*, which would seem to mark a period when comminuted deposits ceased, and currents of water sufficient to transport pebbles were in action, accumulating bones and other substances, as at the bottom of some seas. Where seen on the southern coast of England, between Lyme Regis and Sidmouth, the upper part of the red sandstone series is so like

---

[*] Elie de Beaumont, Memoir above cited.
[†] Von Dechen, German Transl. of Manual.

the variegated marls of the Vosges and parts of Germany, that I have little hesitation in considering them contemporaneous deposits. In this part of England these marls contain vegetable remains, and, though rarely, scales of fish and bones of pterodactyles(?).

As the lower lias sandstone passes into the variegated marls, and even seems in some measure equivalent to them, a deposit of sands having possibly taken place in one situation, while marls were produced in another, we should not, when considering the general subject, force our conclusions too far, nor carry those divisions which may be locally useful beyond the countries where they may be advantageously employed. Prof. Pusch, in his very interesting account of the Polish rocks, states that between the oolitic series of Poland and the muschelkalk there is an extensive and important deposit of sandstone, usually termed *white sandstone*, from its colour. The deposit is divisible into two portions; the upper being formed of the white sandstone, while the lower part is composed of alternations of fine white marly sandstone, schistose sandstone, shale, and other schistose and dark-coloured rocks, the whole inclosing beds of coal from three to twenty-five inches thick. The white sandstone of the upper part alternates with thick beds of gray blue marls, partly red, and more rarely variegated. Beds of limestone are also found in it; but the most valuable product is iron ore, which furnishes the largest amount of iron of any rock in Poland, twenty-seven furnaces affording annually 560,000 quintals of metal. Fossils are rare in this deposit, with the exception of vegetable remains. M. Pusch refers this rock to the lias sandstone, the same as it occurs in Suabia, in Scania, and in the Isle of Bornholm, in all which places it is rich both in iron and coal[*].

M. von Dechen considers it very doubtful whether we should refer this *white sandstone* to the lias sandstones and variegated marls, or to the inferior oolite, and rather inclines to the latter opinion, as it occurs between what is considered a middle member of the oolitic series and the muschelkalk. He moreover remarks, that the true geological position of the sandstone of Hör and the north of Lund in Scania, of which it is thought to be the continuation, is by no means settled.

Muschelkalk.—A limestone varying in texture, but being most frequently gray and compact. It is occasionally dolomitic, and passes into marls above and beneath. When very compact, with numerous remains of the *Encrinites moniliformis*, Miller (a very characteristic fossil of at least a considerable

---

[*] Pusch, Esquisse Géognostique du Milieu de la Pologne: Journal de Géologie, t. ii.

portion of the deposit), it has much the appearance of some varieties of the carboniferous limestone of England. The muschelkalk is sometimes, though rarely, oolitic (between Stühlingen and Bonndorf), and contains beds of chert (Wurtemberg, and some places in Germany). Gypsum and marl are not unfrequently mixed with it. Copper ore is found in a bituminous marl-slate in the neighbourhood of Horgen, on the eastern border of the Swarzwald, which has led to the erroneous impression that these beds belonged to the zechstein. Disseminated copper ore is also found in the lower portion of the muschelkalk in other parts of Wurtemberg *. It is sometimes, as at Epinal (Vosges), sufficiently hard to be employed as marble. In some situations organic remains would appear to be very abundant, while in others they are somewhat rare. According to M. Alberti, salt is contained in the muschelkalk of Wurtemberg†; and M. von Dechen states that salt is also found in it at Buffleben, between the Thüringerwald and the Hartz. This rock would appear to be unknown in England and in the North of France; but on the east and south of the latter country, and in parts of Germany, it is found interposed, in its place, between the variegated marls and the red or variegated sandstone. According to Prof. Pusch it occurs in Poland, and is described as being gray and yellow.

Red or Variegated Sandstone.—This rock is, as its name implies, of different tints,—these being red, white, blue, and green; the former, however, greatly predominating. It is principally siliceous and argillaceous, occasionally containing mica, masses of gypsum, and rock-salt. In the Vosges, the upper part of the variegated sandstone often presents, according to M. Elie de Beaumont, thin beds of marly limestone and dolomite, which gradually become more abundant; so that, finally, they constitute the lower part of the muschelkalk‡. An oolitic and calcareo-magnesian rock§ is found in this deposit in some parts of Germany, and conglomerates are also included in it.

A very extensive deposit, varying but little in its character, occurs in the Vosges, and has thence obtained the name of the *Grès de Vosges.* A difference of opinion seems to exist between M. Elie de Beaumont and M. Voltz respecting the exact member of the red sandstone series to which this rock should be referred; the former considering it the equivalent of

---

* Von Dechen, German Transl. of Manual.

† Alberti, Die Gebirge des Konigreichs Wurtemberg, 1826. M. Bronn notices salt in the Muschelkalk of Hasmerheim, and other places in the vicinity of Heidelberg.—Gæa Heidelbergensis, 1830.

‡ Elie de Beaumont, Terrains Secondaires du Système des Vosges.

§ The grains forming this oolitic rock are radiated from the centre to the circumference.

the rothe todte liegende, which occurs beneath the zechstein; the latter, that it is the lower portion of the red or variegated sandstone, which rests on the zechstein: as the zechstein is wanting in the district, there is perhaps but little essential difference in these opinions.

The *Grès de Vosges* is essentially composed of amorphous grains of quartz, commonly covered by a thin coating of red peroxide of iron; among which are discovered others which appear fragments of felspar crystals. It is often marked by cross and diagonal laminæ so common in arenaceous rocks, the result, probably, of deposit by cross currents of water. The rock contains quartz pebbles, sometimes so abundantly as to present a conglomerate with an arenaceous cement. From the mineral character of these pebbles, M. Elie de Beaumont considers that they are derived from the destruction of the older rocks, and are merely larger portions which have better resisted trituration than the smaller grains composing the body of the sandstone.

The variegated or red sandstone of some countries affords a good building-stone, and when nearly free from colour, as at Epinal, (Vosges,) one of handsome appearance. In situations where it becomes schistose from mica, it is often employed, like some varieties of the old red sandstone of the English, for flag-stones, and even tiles for houses.

According to Prof. Sedgwick, the red sandstone occurring above the magnesian limestone, in the North of England, represents the Bunter sandstein of Germany, the variegated marls surmounting it being the equivalent of the keuper of the same country. This sandstone is represented as of a complex character, from the variable mixtures of sand, sandstone, and marl. In its range from Nottinghamshire into Yorkshire it is generally coarse, often nearly incoherent, and here and there passes into a fine conglomerate. The superincumbent marls are red and gypseous*.

Fig. 91.

At Wasselonne, Marmoutier, and Sulz-les-Bains, more particularly at the latter place, numerous vegetable remains have been discovered in the red sandstone. These have been described by M. Adolphe Brongniart. The annexed figure (Fig. 91.) represents a spe-

* Sedgwick, Geol. Trans. 2nd series, vol. iii.

cimen of *Voltzia brevifolia,* from Sulz-les-Bains, remarkable
as exhibiting the fructification of the plant*.

Zechstein.—This name has, fortunately, been applied by
Humboldt to distinguish a limestone series of a very variable
character, to which different names were given, the term *zechstein* having been previously applied to only one of the varieties. The various beds were known to the German miners
by the names of Asche (friable marl), Stinkstein (fetid limestone), Rauchwacke, Zechstein, and Kupferschiefer (copper-
slate). According to Daubuisson, the mean thickness of the
copper-slate in these countries is about one foot. The zechstein is represented as sometimes from twenty to thirty yards
thick ; the rauchwacke, when pure and compact, one yard
thick, when cellular sometimes attaining fifteen to sixteen
yards ; the stinkstein, from one to thirty yards thick ; and the
asche, very variable. Notwithstanding these minor divisions,
to which an extraordinary value has been attached, it does
not appear that they can always be observed in the countries
where they have been established ; for Daubuisson observes,
that the upper portions pass into each other, and even sometimes into the zechstein.

The zechstein of Germany is composed at its lowest part of
a thin bed of marl-slate, known as the copper-slate (*Kupferschiefer*), from containing finely-disseminated ores of yellow
copper, purple copper, and vitreous copper, worked for centuries. Upon this rests the *Zechstein,* properly so called,
which is a compact dark-coloured limestone. At Thalitter
and Stadtbergen, fine seams of marl-slate, containing disseminated malachite, alternate from ten to thirty times with the
beds of zechstein, and must be considered as forming part of
it. On the Upper Saale, at Camsdorff, the marly slate lies
entirely in the zechstein. The upper series of beds is more
irregular, and appears to have been more influenced by local
causes. Masses of dolomite and gypsum sometimes attain
such thickness that the other rocks disappear. The dolomite
is known in Mansfeld by the name of *Rauchwacke.* Rock-
salt is found in the gypsum. Beds of bituminous limestone
(*Stinkstein*) commonly rest on the gypsum and the accompanying marls and clays. A pulverulent loose-grained mass, consisting of limestone fragments, frequently occurs in the upper
division, and is named *Asche.* In this division there are also
masses of carbonate of iron. They are irregularly combined
with limestone (named *Eisenkalkstein*). At Camsdorff a marly
slate accompanies the iron-stone. On the south border of the

* Taken from a figure by M. Ad. Brongniart, in the Ann. des Sci. Nat.
t. xv. pl. 16.

Hartz, where the largest masses of gypsum occur, limestones are found among them, but are so interrupted that no regular order of stratification can be observed. Beds of an oolitic character are discovered in the upper division of the zechstein series of Germany *.

According to Professor Sedgwick, the magnesian limestone of the North of England, which is the equivalent of this deposit in Germany, is divisible into, 1. Marl-slate and compact limestone, or compact and shelly limestone, and variegated marls. 2. Yellow magnesian limestone. 3. Red marl and gypsum. 4. Thin-bedded limestone. The same author considers No. 1. as equivalent to the kupferschiefer and zechstein, and Nos. 2. 3. and 4. to the rauchwacke, asche, stinkstein, &c. of Thuringia †.

Rothliegendes.—This name is given to a series of red conglomerates and sandstones which occurs between the zechstein or magnesian limestone and the rocks of the next group. The term was originally applied to those beds of Thuringia and other adjacent countries upon which the copper-slate reposes, with the intervention only of portions which are white. M. von Dechen remarks that the rothliegendes appears to be more developed on the east side of the Hartz than at any other point at which it has been observed. Under the *Weisliegendes* (the upper white portion of the rothliegendes), which disappears with the marl-slate, there is a red slate clay, and a fine-grained argillaceous sandstone. Beneath this is a porphyry conglomerate. The porphyry pebbles are remarkable, and do not always resemble the nearest quartziferous porphyry of the Saale, often becoming of greater volume as the distance from the latter increases. They vary in size from that of a walnut to that of the fist. This conglomerate is widely extended throughout the country, and it is worthy of remark that porphyry pebbles are not discovered in the remaining portion of this rothliegendes. Red, greenish, yellowish, and white sandstone, with an uniform grain, are sometimes associated with it; the quartz grains retaining their crystalline structure, and being cemented by argillaceous matter. On account of their roughness the sandstones are well adapted for millstones, as at Siebigkerode. These beds, with common red sandstones, schistose sandstones, and slate clays, constitute the upper division of the rothliegendes.

In the middle division, several thin seams of red and dark blue gray limestone alternate with argillaceous beds, sand-

* Von Dechen, German Transl. of Manual.
† Sedgwick, on the Geological Relations and Internal Structure of the Magnesian Limestone, &c.: Trans. Geol. Soc. 2nd Series, vol. iii.

stone, and breccia. They occur regularly on both sides of a ridge at Rothenburg on the Saale.

The lower division is characterized by a conglomerate, in which large pebbles of splintery gray quartz occur. The cement is a red friable clay. Rocks, which by their partial destruction may have furnished these quartz pebbles, are not known in Northern Germany. They occur only in the higher part of the lower division of the rothliegendes. No regularity is observable in this lower division, the inferior portions of which consist of schistose sandstone, schistose clay, and breccias. In the immediate vicinity of the Hartz grauwacke, pebbles of this rock occur in it. The general colour of the rothliegendes is cherry and violet red. The connexion of this rock with the porphyry at Wettin and Loebejun, at Ihlefeld, on the southern flank of the Hartz, and in the Thuringerwald, render the relations of the whole deposit difficult to determine. While in the Hartz, the pebbles and rock fragments correspond but little with the nearest rocks, in the Thuringerwald they invariably do so *.

Such are the characters of the rocks, known as Rothliegendes, in the countries whence the name has been derived. Taken as a whole, they are for the most part conglomerates, formed from the partial destruction of those rocks on which they rest, the fragments being sometimes angular, as well as rounded, and of considerable size.

The researches of Professor Sedgwick have shown that an arenaceous deposit, of a somewhat variable character, and known as the Pontefract rock of Smith, is in all probability the equivalent of the rothliegendes of Germany. It may be traced between the coal measures and zechstein (magnesian limestone), with a few interruptions, from the mouth of the Tyne to the confines of Derbyshire. Though of a very variable structure and thickness, it possesses a certain uniformity of character when viewed on the large scale. Conglomerates are rare, and are sometimes seen at the junction of this rothliegendes with the magnesian limestone above it. A coarse siliceous sandstone, usually of a red or purple tint, but sometimes gray or yellowish brown, seems most common. It contains pebbles of quartz more than an inch in diameter, generally ranged in lines parallel to the stratification, though sometimes irregularly disseminated. It is often a mere sand, with little or no cohesion; and this character is attributed to an abundance of earthy felspar. Though the want of cohesion would appear common, it occasionally becomes sufficiently hard to

---

* Von Dechen, German Transl. of Manual.

afford building-stone (Wetherby, Knaresborough, and Hart Hill, Yorkshire). The sandstones are associated with variegated micaceous sandy shale, and variegated marls. This rothliegendes sometimes becomes a brown or gray micaceous sandstone, not to be distinguished from some of the sandstones of the coal measures *.

Mr. Hutton, though he fully admits the variable thickness and structure of this rock, considers that it may, in the county of Durham, be satisfactorily divided into two parts; the upper consisting of an incoherent sand, generally of a buff colour, while the lower portion is more consolidated, even furnishing building materials. Though the colour of the latter varies materially, red or purple is by far the most prevalent tint. The division line between these two portions is generally well defined, though they sometimes pass insensibly into each other †.

Prof. Sedgwick remarks not only that this rothliegendes, or lower (new) red sandstone, rests unconformably on the coal measures beneath,—an opinion confirmed by the observations of Mr. Hutton,—but also that its upper surface, that on which the magnesian limestone rests, is uneven. In some places (Branham Moor, North Deighton, Knaresborough,) considerable degradation of the upper surface of the lower red sandstone has taken place prior to the deposit of the limestone; showing, as Prof. Sedgwick remarks, that the continuity of the two deposits was partially interrupted by disturbing forces ‡.

It seemed necessary to premise the above notices of the more remarkable mineralogical structures of the various rocks of this group, known as Variegated or Red Marl, Muschelkalk, Red or Variegated Sandstone, Zechstein, and Rothliegendes, in order that the student might be acquainted with the whole when fully developed. Taken as a mass, the group may be considered as a deposit of conglomerate, sandstone, and marl, in which limestones occasionally appear in certain terms of the series; sometimes one calcareous deposit being absent, as the muschelkalk is in England; sometimes the zechstein, as in the East and South of France; and sometimes both being wanting, as in Devonshire. The conglomerates, or rothliegendes, commonly occupy the lowest position, though conglomerates are occasionally noticed higher

---

* Sedgwick, On the Geological Relations and Internal Structure of the Magnesian Limestone, &c. : Trans. Geol. Soc. 2nd series, vol. iii.

† Hutton, On the New Red Sandstone of the County of Durham Trans. Nat. Hist. Soc. Newcastle, vol. i.

‡ Sedgwick, Geol. Trans. 2nd series, vol. iii. p. 74.

in the series ; the sandstones form the central part, and the
marls occur in the highest place.

When we look for the causes which have produced this
mass, we may, perhaps, in some measure approach them, by
observing the state of the rocks on which it rests. These are
found in the greater number of instances highly inclined, contorted, or fractured;—evidences of disturbance which the inferior and older rocks have suffered previous to the deposit of
the red sandstone group upon them. These appearances are
not confined to one particular district, but are, with a few exceptions, more or less general in Western Europe. From
an examination of the lower beds, no doubt can exist that the
fragments of rock contained in them have, for the greater
part, been broken off from the older rocks of the more immediate neighbourhood. It therefore does not appear unphilosophical to conclude, that, as far at least as regards these
lower conglomerate beds, we have approached to something
like cause and effect,—the cause being the disruption of the
strata, the effect being the dispersion of fragments, consequent
on this violence, over greater or less spaces by means of water, probably thrown into agitation by the disturbing forces.
That these forces have, in some places at least, not been
small, is attested by the large size of the fragments driven off,
and the rounded condition of some of them, as may be well
seen in the vicinity of Bristol, where the rolled masses of carboniferous limestone are sometimes considerable. Of the evidence of the great force employed, I know of no better or
more easily observed example, than that at the cliff named
Petit Tor, in Babbacombe Bay, Devon, whence so large a
portion of Devonshire marble is obtained. Of this the following is a section :

Fig. 92.

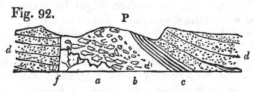

P. Petit Tor Cliff. *a.* Fractured limestone, the rents filled,
when sufficiently open, with the finer matter of the conglomerate above; when small, with carbonate of lime. *b.* A breccia composed of large blocks (some many tons in weight) of
the same marble limestone as that on which it rests, mixed
with others which are smaller. The cementing matter is
sometimes a red sandstone, at others a reddish clay. The
marble (known as Babbacombe marble) is wholly derived

from these blocks, which are detached from their situations,
and either partially worked on the spot or removed elsewhere.
Upon this rest beds of fine conglomerate, sandstone, and red
marl, at *c*, which are surmounted by a considerable thickness of
red conglomerate *d*, extending many miles eastward, and composed of angular pieces of limestone, numerous pieces of slate,
such as is of common occurrence in the surrounding country, as also of pebbles of flinty slate, grauwacke, &c. Among
these are rounded pieces of various red quartziferous porphyries. *f.* A fault or dislocation of the strata, bringing down
the conglomerates on the left hand against the fractured limestones on the right. Such faults or dislocations are common
in the district.

The annexed figure (93.) represents one of   Fig. 93.
the fissures in the fractured limestone at Petit
Tor, filled with the matter of the superincumbent conglomerate. *b, b.* Limestone. *a.* Fissure
filled with the smaller matter of the red conglomerate above.

It will, I think, be scarcely doubted that the
angular blocks of the conglomerate *b* (Fig. 92.)
have been detached by violence from the limestone *a*, and that during the commotion they were thrown
upwards, in such a manner that other and smaller detrital
substances were insinuated between them; the watery mass
being highly charged with sand, mud, and other substances
held in mechanical suspension. It may be right, while on the
subject of these Devonshire conglomerates, to adduce evidence of the unequal action of currents of water, in this vicinity, at the same period. There is perhaps no situation where
better examples of this can be observed than on the line of
cliffs between Babbacombe and Exmouth. The alternations
of conglomerates and sandstones at the upper part of the conglomerate series are very frequent, more particularly in the
vicinity of Dawlish; showing that the water had sometimes
the power of carrying forward rounded fragments of the size
of the head and even larger, while at others it merely accomplished a transport of sand. Not only do the alternations exhibit this difference in the velocity of water, but the structure
of the beds themselves shows that the directions of the currents have continually varied, as will be seen by the annexed
wood-cuts.

Fig. 94.   Fig. 95.

Fig. 94. in the cliff west of Dawlish. Fig. 95. on the east of the same place. *a.* Conglomerate. *b, b, b, b.* Sandstones deposited by changing currents. *c.* Wavy sandstone. The velocity of the currents must have varied considerably in the immediate neighbourhood of these sections; for amidst sandstones and moderately sized conglomerates on the west side of Little Haldon Hill, there are blocks of quartziferous porphyry, generally rounded, of a ton or more in weight. Being scattered on the side of the hill, they might be mistaken for superficial erratic blocks, did we not find them in their proper situations on the sea cliffs, imbedded in the mass of rock. The transport of these must have required water moving with considerable velocity, so great, possibly, as to grind down, by attrition against each other, the rock fragments of inferior hardness, while the pieces of quartziferous porphyry being exceedingly hard, and of very difficult fracture, have better resisted attrition.

The presence of these porphyries in the red conglomerate of South Devon is remarkable, inasmuch as, though rolled, masses of the same kind are not observed unconnected with the red conglomerate of the same country. The absence of such rocks on the exposed surface is certainly no proof that they may not be near; for when we consider the area covered by the red sandstone series in that district, there is ample space for the abundant occurrence of such rocks beneath the sandstone; and there are also many unexplored situations, where they may yet be detected among the rocks now uncovered by the sandstone series. The student must be careful not too hastily to generalize on such facts as have been above noticed in Devonshire, for the appearances may be more or less local. When however we extend our observations, we find that conglomerates are very characteristic of deposits of the same age in other parts of Britain, France, and Germany, and they most frequently, though not always, rest on disturbed strata. As we can scarcely conceive such a general and simultaneous movement in the inferior strata, immediately preceding the first deposits of the red sandstone series,—that every point on which it reposes was convulsed and threw off fragments of rocks at the same moment,—we should rather look to certain foci of disturbance for the dispersion of fragments or the sudden elevation of lines of strata, sometimes, perhaps, producing lines of mountains, in accordance with the views of M. Elie de Beaumont. The accumulation of the larger fragments, and the relative amount of conglomerate, would, under this hypothesis, be greatest nearest to the disturbing cause; and amid such turmoil we might anticipate the occurrence of igneous rocks thrown up at the same period. If we return for the moment to that part of Devonshire with which we commenced these remarks, we shall observe facts which seem to afford support to this view; for where the conglomerates are abundant, there is no want of trappean rocks in the vicinity, such as various greenstones and porphyries, which have cut and broken through the slates, limestones, and other older rocks, in various directions: and I had recently an opportunity of observing, that red quartziferous porphyry, precisely resembling some of that which occurs so abundantly in rolled fragments in the red conglomerate of the district, is found in mass among the lower portion of the latter, and even (at Ideston, near Exeter,) surmounts a portion of it. But notwithstanding the abundance of the greenstones and dark-coloured porphyries, fragments of them have not yet been discovered among the conglomerates, though rolled portions of the red porphyries are so abundant: and it should be observed, that good sections are by no means rare, particularly on the coasts. We have every reason to consider that the eruption of trap rocks did accompany, if partly not produce, the disruption of strata, whence the fragments in the conglomerate were derived : for we have seen that red quartziferous porphyry, in mass, surmounts a portion of the red conglomerate; and the occurrence of trappean rocks (principally of a red or brown tint, and containing much siliceous matter,) so blended with the conglomerates that lines of separation cannot be drawn between them, is by no means rare in the district (Western Town, Ideston, and other places in the vicinity of Exeter). Now if igneous rocks were ejected—a conclusion which the facts appear to justify—at the time of the production of the conglomerate, there would seem no reason why, under favourable circumstances, the two should not be in some measure blended with each other. Another circumstance also lends probability to this view, and that is the occurrence of pebbles cemented in certain inferior beds, (well observed on the coast and in-land between Babbacombe Bay and Teignmouth, at the Corbons, Torbay, in the vicinity of Exeter, and other situations,) by a kind of semi-trappean paste, containing crystals of that variety of felspar named Murchisonite by Mr. Levi. Such a cement might possibly have resulted from the upburst of igneous rocks, accompanied by various gases beneath a mass of water, when some of the erupted matter may have so combined as to form a cement, in which crystals of Murchisonite became developed : without some such hypothesis this cement seems of very difficult explanation.

We must now turn from this scene of disturbance, which may be one of the extreme cases, (though many analogous facts might be adduced,) to that state of things where no violent disrupting cause is to be surmised, but where, on the contrary, the causes which produced the arenaceous rocks that constitute the upper portion of the next, and inferior group, have not been interrupted by any sudden violence, one series of rocks passing into the other so that the exact lines of demarcation are imaginary. Such a state of things is perfectly consistent with local and violent disturbances; for the consequences of a violent disruption of the inferior rocks would extend no further than to distances proportioned to the agitating cause; and the effects would gradually become less, until finally the deposits at remote places would not be interrupted, though the disturbing causes may have produced such a general state of things in the fluid mass, and in the relative positions of land and water, that future deposits would have an altered character ;—one more common over a large area.

This supposed passage of certain lower parts of the red sandstone group into the upper part of the coal measures, seems also supported by facts; for such is stated to be the case in certain parts of the continent of Europe ; so that some geologists, and among them Humboldt, Daubuisson, and others, consider the two rocks as one.

Between such extremes there would be every variety of deposit, produced either by difference in the intensity of the disturbing forces, or by local circumstances. Thus, sands and little or no conglomerate might be found resting unconformably upon older rocks, even in the vicinity of greatly disturbed situations, as may be occasionally observed in the district first noticed.

After the causes, whatever they were, which produced the conglomerates and sandstones known by the name of *Rothliegendes*, had in some measure been modified, a considerable deposit of carbonate of lime, often charged with carbonate of magnesia, took place over certain parts of Europe. This is the *Zechstein*, which, though somewhat extensively developed in certain parts of Germany and England, seems little known in France. The causes which produced this limestone have therefore not been so general as those which have furnished the limestones formerly noticed under the head of the Oolitic Group, which are distributed over a far larger area. A deposit of bituminous or marly slate appears to have been contemporaneous at distant places, in parts of Germany and in the North of England, containing the remains of a marked genus of fishes, *Palæothrissum*. There is nothing in itself remarkable that the same fish should be discovered in rocks formed within the same geological epoch, at such distances as Mansfeld and Durham ; for if these districts were now beneath a common sea, no naturalist would be surprised that cod-fish, turbots, and many other fish, should be caught at the two places, being aware that cod-fish are found on the shores of North America and Europe, and that salmon ascend the rivers of both continents. The geologist, therefore, should expect to find the remains of similar fish entombed in contemporaneous deposits within certain reasonable limits of latitude and longitude.

As yet, these fish seem only to have been observed in the copper-slate, or its equivalent marly slate, and they have apparently perished by some common cause; what that cause was, is by no means clear; but certainly waters which held the component parts of the copper-slate of Thuringia either in chemical solution or mechanical suspension, would be far from favourable to their existence; and if the fish should by any chance be enveloped by, or enter into, such a medium, they are little likely to escape from it alive. When we consider the numerous marine animals always ready to prey upon fish either dead or alive, and the small chance that any part of them will remain undevoured, their occurrence in a fossil state would seem to show that the fossil individuals have been so circumstanced, that the creatures which preyed on them were either destroyed with them, avoided those situations which had been fatal to the fish, or were otherwise unable to get at them.

By reference to the lists of organic remains, it will be observed that marine vegetables occur with the fish in the copper-slate. Now certainly these could no more exist in a medium impregnated with copper, than the fish; and therefore one would suppose they existed prior to the presence of such medium: but as we cannot be certain that these grew near the spot where now entombed, (for marine plants may, like the Sargasso Weed in the Atlantic, be drifted considerable distances,) they do not afford direct proof that the copper-slate was of sudden formation. The remains of the Monitor and some plants seem to indicate a certain proximity of land.

The remainder of the zechstein deposit is of a very mixed character; part being such as we may consider mechanical, while much seems a deposit from a solution of carbonate of lime, carbonate of magnesia, and sulphate of lime. The very frequent occurrence of the two latter in rocks that have apparently originated from some common causes, is very remarkable, and has not yet received any satisfactory explanation.

In Somersetshire and the neighbouring districts, as will be found detailed in the valuable memoir of Prof. Buckland and Mr. Conybeare, the lower part of the red sandstone group is very frequently a conglomerate, composed of the broken fragments of inferior and older rocks, united by a cement containing much magnesia,—whence the term Magnesian or Dolomitic conglomerate. This rock sometimes graduates into a limestone of a more homogeneous character, apparently containing also much magnesia. This conglomerate seems the result of violent action on the carboniferous rocks of the district, detaching various portions of them; in fact, producing effects similar to those noticed under the head of Rothliegendes. How far it may be the exact equivalent of the latter deposit, that is, how far the epoch of disturbance may be precisely contemporaneous, may admit of doubt; for the disturbance which caused the deposit of the rothliegendes in Germany may have preceded that in Somersetshire, so that the latter may have been brought more within the influence of a spread of calcareous and magnesian matter. Still, however, the production of the magnesian conglomerates of Somersetshire and the rothliegendes of Thuringia would not appear to be widely separated from each other as to time; they both constitute the lower part of the red sandstone group in their respective situations, and both contain fragments of rocks commonly derived from their more immediate vicinities.

The organic character of the zechstein approaches, as far as researches have yet gone, that of the next, or carboniferous group; *Productæ*, which abound in the carboniferous limestone, being not only discovered in the zechstein, (and the student will observe that these shells are now introduced for the first time to his attention,) but also *Spirifers*, shells which likewise abound in the carboniferous limestone.

This resemblance in organic character will at all times render the determination of the two rocks difficult, when their geological position cannot be ascertained with certainty, as it may be in Germany and England; and this difficulty may in some cases be considered as insurmountable, should the deposit of the two groups have been continuous, without a violent break, the limestones of the carboniferous group being dispersed through the coal measures (the upper part of the next group,) in such a manner that they should approach the upper terms of the series on the one hand, while the zechstein should descend towards the lowest parts of the red sandstone group on the other. We might under these circumstances have a series of arenaceous and limestone rocks representing the carboniferous group and the lower part of the red sandstone group, with one common, or nearly common, organic character.

The zechstein is surmounted by a mass of rocks for the most part arenaceous, though occasionally argillaceous, gypseous and saliferous. The predominant colour is red, though it is not unfrequently variegated, whence the names *Bunter Sandstein, Grès bigarré.* Where the zechstein is wanting, this sandstone graduates into the inferior conglomerates; and when the muschelkalk is absent, as is commonly the case in England, into the red or variegated marls: therefore where both calcareous deposits are wanting, as in Devonshire, the whole group is composed of conglomerates in the lower part, sandstones in the central, and marls in the upper part; an arrangement which suggests the possibility of the whole, in that district, being the result of some violent commotion, which, as the disturbing causes ceased, deposited the various matters held by water in mechanical suspension, somewhat in the order of their specific gravities; always considering the deposit in the mass: for not only are there alternations of conglomerates and sandstones, sandstones and marls, where these pass into each other on the large scale, but frequent mixtures of them also occur on the small scale,—a circumstance easily accounted for by the various directions and velocities of the currents produced.

Viewed in the mass, circumstances appear to have been unfavourable in those parts of Europe which have been best examined, if not to the existence of animal and vegetable life, at least to their envelopment and preservation; for, with the exception of Alsace and Lorraine, few or no organic remains have been detected in it. The vegetables have been enumerated in the lists of organic remains, from the descriptions of M. Adolphe Brongniart, as also the remains of shells noticed by M. Voltz and others. It will be observed that these shells are not analogous to those found in the zechstein, but to those discovered in the muschelkalk, a rock well developed in the same district; and it is further important to observe, that the remains discovered by M. Elie de Beaumont in the sandstone under consideration in the district of the Vosges, were not far beneath the muschelkalk. I obtained numerous fragments of vegetables from the sandstone near Epinal, Vosges, and the quarrymen informed me that they very commonly discovered them.

We next arrive, in the ascending order, to the Muschelkalk, a limestone the general characters and known extent of which have been above noticed. We here have evidence, that probably at the same epoch a deposit of calcareous matter, mixed sometimes with carbonate of magnesia, sulphate of lime and salt, took place, if not continuously, at least at various places, from Poland to the South of France inclusive, and that the marine animal life distributed over this surface was nearly of the same kind. But it is a remarkable circumstance, that this

life was not of the same kind as that which existed at the time
when the zechstein was formed; the organic character of the
two rocks is distinct, and therefore those who found their divisions of strata solely on this character, do well in drawing
a line between the zechstein and muschelkalk. If, however,
we look at these rocks on the large scale, and see the mineralogical passages which exist between the muschelkalk and the
rocks above and beneath it, and observe that the latter is far
from being a constant rock in the series, and that when it is
absent, those beds between which it is interposed graduate
into one another,—there seems, theoretically, a difficulty in separating them, and practically, a very great inconvenience in
doing so. In whatever manner this may be considered, the
fact appears certain, that circumstances had arisen, changing
the character of marine life over certain portions of Europe;
that certain animals abounding previously, and apparently for
a great length of time, (for, as will be seen in the sequel, they
are enveloped in various thick and older deposits,) have disappeared never to reappear, at least as far as we can judge from
our knowledge of organic remains.

While on the subject of the muschelkalk, it may be useful
to show the manner in which this rock gradually disappears
in Germany, so that the Red or Variegated Marls (*Keuper*)
above, and the Red or Variegated Sandstone (*Bunter Sandstein*) beneath, approximate and finally constitute one mass,
such as is seen in England. In some places thick beds of a
red or variegated marl occur in the muschelkalk. These
marls gradually become more developed at the expense of the
muschelkalk, so that the latter appears merely as a few subordinate beds in a mass of red marl, which may be considered
as common both to the keuper above or the bunter sandstein
beneath. This fact is observable at the termination of the
north-western range of hills of Germany. It is very clearly
seen on the southern declivity of the coal range of Ibbenbühren, in the district of Oster Ledde. The bunter sandstein
here chiefly consists of red marls, with a few beds of sandstone.
Above this follows limestone, about fifty feet thick. It is of a
light gray colour, thinly laminated and friable, intermixed with
some thicker beds of dark-blue, gray, and yellowish brown.
Encrinites and other characteristic fossils are not wanting in it.
Above this rests red marl, exactly resembling that beneath, in
which there are two or three limestone beds, which can only
be considered as belonging to the muschelkalk. The upper
part of this marl is not to be distinguished from the keuper.
The limestone beds finally vanish between the red marls by
passing into rocks, which resemble the calcareous marls of the

keuper*.   Thus a rock, which has formed a very marked feature in the red sandstone group, disappears, even in a country where it is largely developed, and where its geological importance is considerable.

Among the organic remains of the muschelkalk, two of the most characteristic of which are considered to be the *Ammonites nodosus* (Fig. 96.) and *Encrinites moniliformis* (*E. liliiformis* Schl.), are reptiles of various forms.   That extraordinary genus the Plesiosaurus, and perhaps also his common fossil companion the Ichthyosaurus, then ex-

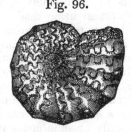

Fig. 96.

isted near what now constitutes the eastern part of France, and the adjoining portion of Germany.   How far these singular Saurians now first appeared in any numbers on this part of the globe, it would be premature to say; for it must always be recollected, that the preservation of such remains would seem in some measure to depend on local circumstances, possibly also, in some cases, on the proximity of land, and the chance that if drifted about at sea when dead, they escaped the other predaceous animals of the deep, all, great and small, ready to devour them.

The Red or Variegated Marls, which surmount the muschelkalk, possess a common mineralogical character over very considerable surfaces, such as would lead us to suppose some cause or causes exerting an influence of a similar kind over a large area.   At least some of the deposit would appear chemical, more particularly the masses of gypsum and rock-salt which exist in certain situations.   How far the mass of marls may be partly chemical or wholly mechanical, may in the present state of science admit of a doubt; but the sandstones with which they are in some countries connected, and which even seem to replace them in others, (as has been above noticed,) are of mechanical origin, inclosing beds of coal, the result probably of accumulated vegetable matter.   Some of the vegetable remains are still sufficiently preserved to be determined, as in the red or variegated marls of the Vosges.

If we now abstract our attention from these divisions, and regard the group as a mass, it would seem to constitute the base of a great system of rocks, which when not deranged by local accidents has filled numerous hollows and inequalities of land over considerable parts of Europe.   Such a hollow is well

---

* Hoffman, Über die grognostischen Verhältnisse der Gegend von Ibbenbühren und Osnabruck, Karsten's Archiv für Bergb. B. 12.

seen in our own island, where the central counties are occupied
by the red sandstone series, apparently filling up a previously
existing depression in that situation; but it is here without that
great capping of the oolitic group, which for the most part
rests so conformably upon it; so that taken as a whole, and
abstraction being made of minor derangements, they would
both seem to fill up great depressions in Europe; sometimes,
as is the case in Normandy, the oolitic rocks overlapping and
coming in contact with strata older than the red sandstone
group, upon which latter they nevertheless rest so conformably
that the one seems a tranquil deposit on the other. We must
of course consider that numerous local disturbances would produce a marked difference in the deposits, even amounting to a
perfectly unconformable position; yet the conformable nature
of the two groups taken in the mass is somewhat striking.
During their deposit, great and remarkable changes were effected in animal and, perhaps, vegetable life; and it seems somewhat necessary to admit, that considerable differences in the
relative levels of sea and land were produced at various times,
causing changes in the character of the inhabitants of the sea,
from variations of pressure and other circumstances, while no
small difference might be effected from the filling up and rise
of the bottom.

The following summary of the fossils stated to be found in
the Red Sandstone Group of Europe, will exhibit our present
knowledge of its organic contents.

In the Variegated Marls.—*Plantæ,* Equisetum, 3 species;
Pecopteris, 1; Tæniopteris, 1; Filicites, 2; Marantoidea, 1;
Pterophyllum, 3. *Radiaria,* Ophiura, 1. *Conchifera,* Plagiostoma, 1; Cardium, 1; Trigonia, 3; Mya, 2; Avicula, 3;
Posodonia, 2; Modiola, 1; Venericardia, 1; Lingula, 1; Saxicava, 1. *Mollusca,* Buccinum, 1. *Pisces,* genera not determined. *Reptilia,* Phytosaurus, 2; Mastodonsaurus, 1; Ichthyosaurus, 1; Plesiosaurus, 1.

Thus making, *Plantæ,* 6 genera, 11 species. *Radiaria,* 1
genus, 1 species. *Conchifera,* 10 genera, 16 species. *Mollusca,*
1 genus, 1 species. *Pisces,* number of genera and species not
determined. *Reptilia,* 4 genera, 5 species.—Total, 22 genera,
34 species.

In the Muschelkalk.—*Plantæ,* Neuropteris, 1; Mantellia, 1.
Zoophyte, Astrea, 1. *Radiaria,* Cidaris, 1; Ophiura, 2; As-
terias, 1; Encrinites, 1; Pentacrinites, 1. *Annulata,* Serpula, 2.
Conchifera, Terebratula, 4; Delthyris, or Spirifer, 1; Lin-
gula, 1; Ostrea, 9; Gryphæa, 1; Pecten, 4; Plagiostoma, 5;
Avicula, 4; Mytilus, 1; Trigonia, 6; Arca, 1; Cardium, 2;
Mya, 5; Venus, 1; Mactra? 1; Cucullæa, 1. *Mollusca,*
Calyptræa, 1; Capulus, 1; Dentalium, 2; Trochus, 1; Turri-

tella, 5; Buccinum, 2; Strombus, 1; Natica, 2; Turbo, 2;
Nautilus, 2; Ammonites, 2. *Crustacea,* Palinurus, 1. *Pisces,*
genera not determined. *Reptilia,* Plesiosaurus, 1; Ichthyosaurus, 1; Crocodilus, 1; Great Saurian, genus not named;
Chelonia, 1.
Thus making: *Plantæ,* 2 genera, 2 species. *Zoophyte,* 1
genus, 1 species. *Radiaria,* 5 genera, 6 species. *Annulata,*
1 genus, 2 species. *Conchifera,* 16 genera, 47 species. *Mol-*
lusca, 11 genera, 21 species. *Crustacea,* 1 genus, 1 species.
Pisces, number of genera and species not known. *Reptilia,*
5 genera, 5 species.—Total, 42 genera, 85 species.
In the Red or Variegated Sandstone.—*Plantæ,* Equisetum,
1 species; Calamites, 2; Anomopteris, 1; Neuropteris, 2;
Sphenopteris, 2; Filicites, 1; Voltzia, 5; Convallarites, 2;
Paleoxyris, 1; Echinostachys, 1; Ethophyllum, 1. *Conchifera,*
Plagiostoma, 2; Avicula, 2; Mytilus, 1; Trigonia, 1; Mya, 2.
Mollusca, Natica, 1; Turritella, 2; Buccinum, 1.
Thus making: *Plantæ,* 11 genera, 19 species. *Conchifera,*
5 genera, 8 species. *Mollusca,* 3 genera, 4 species.—Total,
19 genera, 31 species.
In the Zechstein.—*Plantæ,* Fucoides, 6 species; Pecopteris,
2; Lycopodites, 1; Asterophyllites, 1. *Zoophyta,* Gorgonia,
3; Calamopora, 1; Retepora, 2. *Radiaria,* Encrinus, 1;
Cyathocrinites, 1. *Conchifera,* Delthyris, or Spirifer, 4; Terebratula, 9; Producta, or Leptæna, 7; Orbicula, 1; Axinus,
1; Ostrea, 1; Pecten, 1; Plagiostoma? 1; Avicula, 1; Mytilus, 3; Modiola, 1; Arca, 1; Cucullæa, 1; Astarte? 1; Venus?
1. *Mollusca,* Turbo? 1; Pleurotomaria? 1; Melania? 1;
Ammonites, 1. *Pisces,* Palæothrissum, 8; Stromateus, 2;
Clupea, 1. *Reptile,* Monitor, 1.
Thus making: *Plantæ,* 4 genera, 10 species. *Zoophyta,* 3
genera, 6 species. *Radiaria,* 2 genera, 2 species. *Conchifera,*
15 genera, 34 species. *Mollusca,* 4 genera, 4 species. *Pisces,*
3 genera, 11 species. *Reptile,* 1 genus, 1 species.—Total, 32
genera, 68 species.
It would appear, more particularly from the descriptions of
Humboldt, that very extensive tracts of red sandstones and
conglomerates exist in Mexico and South America; how far
these may be of contemporaneous production with the red
sandstone series of Europe, the state of science does not permit us very satisfactorily to determine. The porphyries and
slates of New Spain are surmounted by red conglomerates and
sandstones, forming the plains of Celaya, Salamanca, and
Burras, and supporting a limestone which mineralogically resembles that of the Jura. The conglomerates contain fragments
of pre-existing rocks, cemented by an argillo-ferruginous, and
yellowish brown or brick red, paste. In Venezuela, the vast

plains are in a great measure covered by red sandstones and conglomerates, with limestones and gypsum ; the former being deposited in a concave manner between the coast mountains of the Caracas and the mountains of Parima, resting on slates, termed transition, on the north, while on the south they repose upon granite. This arenaceous deposit is covered, at Tisnao, by compact whitish gray limestone. An immense extent of red sandstone is described as " not only covering, nearly without interruption, the southern plains of New Grenada, between Mompox, Mahates, and the mountains of Tolu and Maria, but also the basin of the Rio de la Magdalena, between Teneriffe and Melgar, and that of the Rio Cauca, between Carthago and Cali." The conglomerates of this country are composed of angular fragments of lydian stone, clay-slate, gneiss and quartz, cemented by argillaceous and ferruginous matter. These conglomerates alternate with schistose and quartzose sandstones *. According to Humboldt, the Cordilleras of Quito presented him with the greatest extent of red sandstone which he had observed, covering the whole plateau of Tarqui and Cuença for twenty-five leagues. The sandstone is generally very argillaceous, with small grains of slightly rounded quartz; but it is sometimes schistose, and alternates with a conglomerate containing fragments of porphyry from three to nine inches in diameter. The same author considers that the red sandstone of Cuença also occurs in High Peru, and remarks on the resemblance of these rocks of New Grenada, Peru and Quito, to the red sandstone or *rothliegendes* of Germany†.

A series of red sandstones, intermixed with conglomerates, occurs extensively in Jamaica, particularly in the Port Royal and St. Andrew's mountains, stretching thence north-west towards the north side of the island. The sandstone is generally siliceous and compact, intermixed with marly red sandstone and marl, and, though rarely, with gypsum (Hope Valley). The conglomerate is formed of pebbles (from an inch to four inches in diameter) of granite, large-grained greenstone, sienite, quartz, hornstone, &c. Beds of a gray colour are intermixed with these rocks; and subordinate to them are strata of compact gray limestone and of shale, and schistose sandstone intermixed with coal. The higher portion of the mass is formed of a conglomerate in a great measure composed of pieces of trap rocks, principally porphyry, the cementing matter being most frequently reddish brown and argillaceous, varying in induration, sometimes so obscure that the pebbles seem joined by a trappean cement. Mixed more particularly with this superior portion there is a great variety of trappean rocks, such as

---

* Humboldt, Gisement des Roches dans les deux Hémisphères.   † *Ibid.*

sienite, greenstone, porphyries, &c. appearing as if an upburst of igneous matter had accompanied the production of the conglomerates. The red conglomerates and sandstones pass beneath into a rock, which at first differs from them only in colour, and finally presents the mineralogical character of grauwacke. The aggregate thickness of the whole is considerable, amounting to several thousand feet. These rocks appear to me the equivalent of those named red sandstone in the neigh-bouring continent of America*.

The mere mineralogical resemblance of this deposit, in America and Jamaica, with the sandstones and conglomerates of the red sandstone group of Europe, is in itself of no great value, and therefore we can only at present conclude that considerable forces have been exerted in both parts of the world (whether contemporaneous or not remains to be determined), which have dispersed fragments of pre-existing rocks, scattering them, most probably by the medium of agitated water, in various directions, the transporting powers being unequal, so that sandstones and marls alternate with conglomerates. These sandstones and conglomerates would appear, from the descriptions of geologists and intelligent travellers, to extend from Mexico far into the heart of North America; so that if different deposits have not been confounded under one head, as might easily happen in England if it were an uncultivated country and rapidly examined, (the old red sandstone of English geologists being confounded with their new red sandstone,) these sandstones and conglomerates of America would appear not the result of a limited disturbance, but of one common to a considerable surface.

* For a more detailed account of these and other Jamaica rocks, with sections, consult my Remarks on the Geology of Jamaica, Geol. Trans, 2nd series, vol. ii.

SECTION VIII.

CARBONIFEROUS GROUP.

SYN.—Coal-measures, Engl. Auth. (*Terrain Houiller*, Fr. Auth. *Steinkohlen-gebirge*, Germ. Auth.). Carboniferous limestone, *Conyb.* (*Mountain lime-stone*, Engl. Auth. *Calcaire carbonifère, Calcaire anthraxifère, Calcaire de Transition*, Fr. Auth. *Bergkalk, Kohlenkalk, Uebergangskalk*, and *Neuere Uebergangskalk*, Germ. Auth.). Old red sandstone, Engl. Auth. (*Grès rouge intermediaire*, Fr. Auth. *Jüngeres Grauwackengebirge*, Germ. Auth. *Alter rother sandstein*, Von Dechen.)

Coal Measures.

THESE are composed of various beds of sandstone, shale, and coal, irregularly interstratified, and in some countries inter-mixed with conglomerates; the whole showing a mechanical origin. The coal-measures abound in vegetable remains, and the coal itself is now, by very general consent, referred to a vegetable origin, being considered the accumulation of an im-mense mass of plants.

The sandstone, shale, and conglomerate beds vary much, as might be expected, in different situations. Some being, even in the same districts, more continuous than others; a sandstone bed, for example, becoming gradually thinner and finally disap-pearing, so that the beds above and beneath, should their conti-nuity continue, come into immediate contact with each other. The like happens with the shale, conglomerate, and even coal-beds. Undoubtedly some beds are persistent over a great area, but nothing can be more various than the areas occupied by any given beds. It thus becomes highly important to trace given beds with accuracy, noting whether they terminate by gradually fining off, or by an almost insensible change in their mineralogical character; such, for instance, as a sandstone bed acquiring argillaceous matter by degrees, and thus passing through a sandy into an argillaceous shale.

It is stated by Mr. Buddle, that as the strata of the New-castle coal-field rise or crop upwards, the sandstone beds in-crease in number and thickness; whereas the argillaceous shales increase in the opposite direction [*]. The same author re-marks, that the quality of the coal in the same district greatly depends on the kind of bed which immediately covers the coal stratum, and that the coal is always deteriorated when covered by sandstone, becoming in that case more or less mixed with

[*] Buddle, Trans. Nat. Hist. Soc. Newcastle, vol. i. p. 238.

iron pyrites; whereas it is of comparatively good quality beneath argillaceous shale*. This difference in the comparative value of the coal itself beneath sandstone or shale, would seem readily accounted for by the difference in the character of the two rocks, the first being generally pervious to gases and liquids, while the shale would by no means so readily permit the passage of either. Hence it would follow, not only that the vegetable matter would be better preserved beneath the shale from the comparative difficult escape of gaseous matter upwards, but that it would also be preserved from injury by the difficulty with which water, not only pure, but charged with foreign matter, would have in percolating downwards.

In tracing coal-beds, it has been observed, that the same bed frequently varies much in thickness, sometimes being much thinner than at others. Moreover shale and sandstone often get interstratified, as it were, with the coal-beds, thus parting them into minor beds. The area and depth occupied by these interstratified portions of sandstone and shale, though of little importance in some beds, become not unfrequently in other coal strata so considerable as seriously to injure the value of the coal.

It is by no means uncommon to describe the coal deposits as basins; but it may be doubted how far this term is generally correct; for admitting that many accumulations of these beds have been deposited within depressions of the surface, it would by no means seem to follow, reasoning at least from the mode in which vegetable accumulations are now formed, that all coal-measures have been thus produced. Suppose plants to be carried down rivers, such as now happens at the mouth or delta of the Mississippi, we should ill characterize the deposit, by the term basin-shaped, which would seem to imply a hollow or depression, bounded by a circumference of nearly equal elevation.

The following analyses by Dr. Thomson exhibit the various proportions of elementary substances existing in some of the coals of England and Scotland:—

	Carbon.	Hydrogen.	Oxygen.	Nitrogen.
Newcastle Caking Coal . . .	75·28	4·18	4.58	15·96
Glasgow Splint Coal	75·00	6·25	12·50	6·25
Glasgow Cherry Coal	74·45	12·40	2·93	10·22
Cannel Coal	64·72	21·56	0·00	13·72

Carburetted hydrogen, or *fire damp* of the collieries, is well known as an abundant product of most coal-mines. Mr. Buddle observes that the discharge of this gas in the ordinary workings of a colliery much depends on the pressure of the atmo-

* Buddle, Trans. Nat. Hist. Soc. Newcastle, vol. i. p. 217.

sphere; being greatest when that pressure is least, the gas being then enabled more freely to escape from the pores and fissures of the coal. The same author remarks, that a sandstone roof (the bed immediately above the coal stratum) from being commonly split into innumerable fissures, and, consequently, receiving the gas when evolved from the coal beneath, becomes a great natural gasometer, ready to pour out the carburetted hydrogen when circumstances are favourable*. To this it may be added, that those sandstones which are sufficiently porous to permit the percolation of water downwards, would likewise allow of the escape of gas upwards, particularly if it were in a compressed state: hence a sandstone roof would probably, in many cases, be saturated with carburetted hydrogen, of greater or less density, according to circumstances. Mr. Hutton considers that this gas exists in a highly condensed, and even liquid, state in the pores of the coal. This opinion is rendered probable by the fact that small explosions from coal, technically called *eructations*, are not uncommon when the coal is struck with the pick; these explosions being apparently due to the sudden expansion of condensed gas. Mr. Buddle relates, that in a part of Jarrow colliery, near Newcastle, the *eructations* were as loud as the report of a musket, large fragments of coal being thrown off at the same time†.

Coal varies considerably in the quantity of bitumen it contains, and is more or less valuable for economical purposes according to the admixture of this substance. The quantity of coal raised in the British Isles is very considerable, and it may be said that to this substance and the iron-ore found in the same deposit, England owes a great part of her commercial prosperity; for to the abundance and cheapness of both these substances in various districts, we are indebted for a large proportion of our manufactures, the same series of beds not only furnishing fuel for working the steam-engines, but also iron for their construction.

In the present condition of the coal-measures, opportunities of observing design seem to be afforded us, even when these rocks are so disposed that at first sight such design does not appear very obvious.

* Buddle, on the Explosion in Jarrow Colliery: Trans. Nat. Hist. Soc. Newcastle, vol. i.

† Buddle, *Ibid.* The explosion which took place in Jarrow colliery, August, 1830, was caused by a volume of carburetted hydrogen being highly compressed near a small fault. When the workmen approached this volume of compressed gas, they necessarily weakened the resistance of the coal in the direction of the workings. Finally, the resistance being unequal to the pressure of the gas, the coal was driven out into the galleries, the carburetted hydrogen mixed with the common air, and exploded at the contact with flame, killing many of the miners.

The accumulation of vegetable matter at a remote epoch in the history of the world, for the consumption of creatures which should afterwards exist on its surface, must strike the least inquiring; but when the upturned, twisted, and shattered strata, so common in the districts composed of the coal-measures, are before us, design is not so apparent, more particularly when the miner complains of the dislocations (faults) which interrupt his progress*. We might therefore regard this apparent confusion as a bar to the ingenuity and industry of man in extracting the combustible so valuable to him. When, however, we look more closely into this subject, we find that the shattered and contorted condition of the rocks, though it may embarrass mining operations for a time, is in reality highly advantageous. The fractures, termed faults, frequently so cross each other, that the surface, if it could be examined without its covering of vegetation and detritus, would present much the same appearance on the great scale, as the frozen surface of a lake broken to pieces and reunited by subsequent frost. Masses of fractured strata are thus often bounded by faults which prevent the passage of subterraneous waters from one mass into the other; and the miners, in collieries situated in one particular mass, have only to contend with the waters in it; whereas if the strata were always horizontal, unbroken, and continuous, the abundance of water that would flow into the workings would render them so difficult and expensive, that the extraction of the coal must be abandoned.

It should be observed, that though the two sides of a fault often come into close contact, there is very frequently an interposed clayey substance impervious to water; and it rarely happens that water on the one side passes to water on the other, so as to form a continuous and abundant percolation in one direction. On the contrary, the water is commonly thrown out along the line of the fissure, particularly on mountain sides, in the shape of springs,—often good guides to the geologist, not only in tracing faults among the coal-measures, but in other rocks. The appearance of springs along lines of fault is what we should expect; for not only do they act as main drains to the strata which they traverse, but as Artesian wells, producing the same effects as these artificial perforations. For supposing faults to be abundant, in countries where Artesian wells are now so valuable, such countries would possess an abundant supply of water, upon the same principle that these wells now act. Knowing, therefore, that faults are abundant on the sur-

* For sections of faults in coal-measures, see Sections and Views illustrative of Geological Phenomena, pl. 5. 6. 7.; Geol. Trans. 2nd series, vol. i. pl. 32.; Transactions of the Nat. Hist. Soc. of Newcastle, vol. i.; La Richesse Minerale, by M. Heron de Villefosse, &c.

face of our planet, we may infer that no inconsiderable portion
of water is conveyed by them to that surface.

It will be obvious that in a district where the coal-measures
are greatly contorted, the relative position of strata alone would
prevent the abundant percolation of water from one situation
to another.

The following section (Fig. 97), at Jarrow colliery, near
Newcastle, will afford an idea of the manner in which the coal-
measures are sometimes fractured *.

Fig. 97.

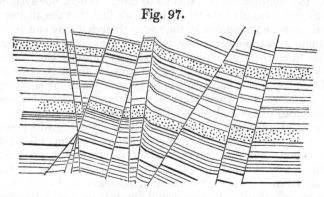

The mode in which they sometimes are contorted without
fracture, is shown in Fig. 98., a section of the coal-measures
at Little Haven, St. Bride's Bay, Pembrokeshire.

Fig. 98.

Little Haven.

The organic remains discovered in the coal-measures are
principally terrestrial plants ; with these are a few fresh-water
shells, and certain marine exuviæ, which for the most part
would rather appear to occur in beds alternating with the coal-
beds and their accompanying shales and sandstones, than
mingled with the terrestrial remains.—The following is a summary of the fossils stated to have been found in the coal-measures.

Plantæ.—Stigmaria, 9 species. Pinites, 3. Peuce, 1. Spe-
nophyllum, 10. Annularia, 7. Asterophyllites, 12. Bechera, 1. Flabellaria? 1. Nœggerathia, 2. Cannophyllites, 1.

---

* A small portion of Mr. Buddle's beautiful sections of the Newcastle
coal-field, Trans. Nat. Hist. Soc. Newcastle, vol. i. pl. 21. 22. 23.

Sternbergia, 4. Poacites, 2. Trigonocarpum, 4. Musocarpum, 3. Equisetum, 2. Calamites, 13. Sphenopteris, 32. Cyclopteris, 9. Neuropteris, 17. Odontopteris, 6. Pecopteris, 62. Lonchopteris, 1. Schizopteris, 1. Caulopteris, 1. Lycopodites, 8. Selaginites, 2. Lepidodendron, 43. Ulodendron, 1. Lepidophyllum, 2. Lepidostrobus, 2. Cardiocarpon, 5. Sigillaria, 37. Volkmannia, 4. Cyperites, 1. Polyporites, 1.

Conchifera.—Pecten, 2. Mytilus, 1. Lutricola, 3. Unio, 3. Nucula, 2. Mya? 3.

Mollusca.—Turritella, 2. Bellerophon, 2. Orthoceratites, 5. Ammonites, 5.

Pisces.—Palæothrissum, 2. Acanthæssus, 1.

Thus making: *Plantæ*, 35 genera, 310 species. *Conchifera*, 6 genera, 14 species. *Mollusca*, 4 genera, 14 species. *Pisces*, 2 genera, 3 species.—Total, 47 genera, 341 species.

Carboniferous Limestone.

This rock, in the South of England, Wales, the North of France, and Belgium, seems to possess a somewhat general character, being a compact limestone, frequently traversed by veins of calcareous spar, and at times being in a great measure composed of organic remains, while at others not a trace of these remains can be observed. The colours are mostly gray, varying in intensity of shade; other tints are however observed in it, and in some situations it affords good marble. It is occasionally of an oolitic structure, as near Bristol; and sometimes contains parts of encrinital columns in such abundance, that the rock is in great measure made up of them, whence the name *Encrinal Limestone.* It has also been known by the name of *Metalliferous Limestone*, in consequence of the quantity of lead-ore obtained from it, more particularly in the central and northern parts of England. Though these characters may suffice for a considerable portion of the carboniferous limestone, it will be seen by the sequel, that even within the distance of two or three hundred miles, the series of limestone beds known by this name becomes mixed with shale, sandstone, and even with coal.

The following is a summary of the organic remains enumerated as having been discovered in the carboniferous limestone, the plants only being omitted; as when such exuviæ occur, they are generally found to correspond with the plants of the coal-measures.

Zoophyta.—Gorgonia, 4 species. Cellepora, 1. Retepora, 1. Caryophyllia, 3. Cyathophyllum, 4. Astrea, 1. Tubipora, 1.

Syringopora, 3. Calamopora, 2. Aulopora, 1. Favosites, 3.
Lithostrotion, 3.*
 Radiaria.—Pentremites, 3. Poteriocrinites, 2. Platycri-
nites, 7. Actinocrinites, 5. Melocrinites, 1. Rhodocrinites, 1.
Cyathocrinites, 2.
 Annulata.—Serpula, 2.
 Conchifera.—Spirifer, or Delthyris, 29. Terebratula, 21.
Atripa, 1. Producta, or Leptæna, 29. Crania, 1. Inoce-
ramus, 1. Pecten, 2. Megalodon, 1. Nucula, 1. Arca, 1.
Isocardia, 1. Cardium, 4. Lucina, 2. Sanguinolaria, 1.
Solen, 2.
 Mollusca.—Pileopsis, 1. Melania, 1. Ampullaria, 2. Ne-
rita, 3. Delphinula, 1. Euomphalus, 12. Trochus, 7. Tur-
bo, 6. Rotella, 3. Helix? 1. Turritella, 10. Buccinum, 1.
Phasianella, 3. Bellerophon, 12. Conularia, 2. Orthocera-
tites, 13. Nautilus, 9. Ammonites, 2.
 Crustacea.—Asaphus, 1. Other Trilobites not determined.
 Pisces.—Ichthyodorulites and fish palates.
 Thus making: Zoophyta, 12 genera, 27 species. *Radiaria,*
7 genera, 21 species. *Annulata,* 1 genus, 2 species. *Conchi-
fera,* 15 genera, 97 species. *Mollusca,* 18 genera, 89 species.
Crustacea, 1 genus, 1 species. *Pisces,* 1 genus, 1 species.
—Total, 55 genera, 238 species.

Old Red Sandstone.

This rock is of very variable thickness, sometimes consisting of a few conglomerate beds, while at others it swells out
to the depth of several thousand feet. As might be expected,
this variation in thickness is accompanied by differences in mineralogical structure, conglomerates being abundant in some
situations, while in others they are exceedingly rare. The
sandstone possesses different degrees of induration, and is not
unfrequently schistose and micaceous, affording flag-stones
and coarse materials for roofing. The prevalent colour is red,
generally dull, which, as commonly occurs in the red marls
and sandstones of all ages, is occasionally intermixed with
different tints of greenish blue (Pembrokeshire, &c.). The
conglomerates of course vary in their contents, but pieces of
quartz are very common,—so much so in the southern parts of
England and Wales, that the greater portion of such beds
is wholly composed of them. The sandstones also are prin-

---

* This catalogue is exceedingly meagre ; for, unfortunately, British naturalists have published next to nothing on the zoophytic contents of the carboniferous limestone, though this rock is extensively developed in the British
Isles, and contains a great abundance of Polypifers.

cipally siliceous, so that if the mass be considered as wholly
of mechanical origin, it must have resulted from a considerable destruction of pre-existing siliceous rocks.

Few organic remains have been discovered in this rock, and
those that have been observed would appear, for the most part,
to be the same as in the grauwacke beneath, or the carboniferous limestone above. According to Dr. Fleming, *Orthoceratites cordiformis, O. giganteus, Nautilus bilobatus,* and *N. pentagonus,* are found in a limestone associated with the old red
sandstone of Dumfriesshire *; and M. Dumont notices *Producta
concinna* in the old red sandstone of the Liége district.

The student being now acquainted with the general zoological and botanical characters, and with the more marked mineralogical composition of the three rocks comprised within
this group, we will proceed to a more general notice of the
same rocks taken in the mass.

It has been above remarked, that the rothliegendes is considered, in certain parts of Europe, to pass into the coal-measures, so that the two rocks constitute the upper and lower
portion of the same mass. Some geologists have gone further,
and considered the coal-measures as subordinate to the rothliegendes, the carboniferous portion bearing the same relation to the general mass, which certain lignites, such for
instance as those noticed by M. Elie de Beaumont in Dauphiné and Provence, bear to the transported matter in which
they are found included.

It would appear from the observations of Von Veltheim on
the district of Wettin and Loebejun, and on the eastern Hartz,
that the coal-measures are there subordinate to the rothliegendes, and that the whole mass may be divided into three
portions:—1. The lowest, composed of red sandstone, with
slaty clays, schistose sandstone, and conglomerate, 500 feet
thick (considered equivalent to the English old red sandstone). 2. The middle, carboniferous rocks and limestone,
250 feet thick (supposed equivalent to the carboniferous limestone, coal-measures, and millstone grit of the English series).
3. The upper, formed of red sandstone, with slate-clays, conglomerates, and porphyry breccia, 2600 feet thick (referred
to the red sandstone group).

M. von Dechen observes that it is particularly difficult to
afford a better explanation of these facts, as the coal deposit of Wettin and Loebejun, wedged in between two kinds of
porphyry, is disturbed and irregular. The same author remarks, however, the important fact, that there can be no
doubt of the coal deposit being covered by the *upper beds* of

---

* Fleming, History of British Animals.

the real rothliegendes, and resting on red sandstone. The identity of the latter sandstone with the lower strata of the rothliegendes on the borders of the Hartz seems not so evident, though it would appear probable from the careful researches of M. von Veltheim *. It is certainly exceedingly desirable that the identity of these latter rocks should be more clearly proved; for as the sandstones of the coal-measures are often red, it might be considered that the lowest rocks at Wettin and Loebejun were merely such red sandstones, and that the coal-measures were covered by the upper members of the rothliegendes, because the lower beds were wanting.

A good example of coal subordinate to the red sandstone is observed in the coal series of Waldenburg and Neurode, in Lower Silesia, especially in that part of the deposit which extends into Bohemia, between Schatzlar and Nachod. In the neighbourhood of Altwasser, red sandstone and conglomerate rest in a narrow band upon grauwacke, and, in an adit, a bed of coal, with slate-clay, is seen apparently in a continuation of the red conglomerates. North from Schatzlar, above Liebau, to the environs of Waldenburg, the coal-measures rest immediately upon grauwacke, without any intermediate beds. Beneath the Bohemian coal, at Rohnow on the Metau, there is a thick mass of fine-grained red sandstone. Its beds form a saddle from Nachod to Schatzlar, and on the south-west side of it, towards the basin of Bohemia, no coal-mine is known, there being merely a continuous uninterrupted mass of red sandstone. A bed of bituminous shale, with the remains of fish, indistinct impressions of plants, and disseminated copper-ore, occurs *beneath* the coal near Eipel, not far from Saugwitz; above this occurs limestone, about 20 feet thick, and beneath the whole, white sandstone. North-west of Rohnow, red sandstone appears between the coal strata, and soon attains such thickness, that with a dip of about 20°, it occupies a breadth of $2\frac{1}{2}$ miles on the surface. It in this manner divides the coal-measures, for a length of $17\frac{1}{2}$ miles, into two clearly defined parallel lines. This sandstone, though it perfectly resembles that beneath, must necessarily be considered as an integral portion of the coal-measures.

The two lines of coal above noticed do not unite (as far as is yet known) with the coal-field of Schatzlar, but cease in the middle of the red sandstone district at Goldenelse and Teichwasser; and the upper line is also covered by red sandstone, and that uniformly for the distance above mentioned. The same red sandstone covers the coal deposit on the Silesian side. At Langwaltersdorf, the red and gray coal sandstones alternate with each

* Von Dechen, German Transl. of Manual.

other, but in general the limits of both are well defined. Conglomerates are rare in the upper sandstone, which is usually fine-grained; pebbles of quartz and flinty slate are found only in its lower portion. Red slate clays are not frequent. Single, but long continued, beds of limestone and dolomite are found in it (Grüsow, and Conradswalde); they are usually reddish gray, seldom dark gray (Ottendorf, and Scheidwinkel). It contains beautiful impressions of *Neuropteris conferta.* Impressions of fish are found in it at Ruppersdorf, which, like those of Saugwitz, are not yet determined. As the red sandstone is covered by quadersandstein (green sand), many of its relations are uncertain, but very probably it is parallel with the rothliegendes of Mansfeld. If at Waldenburg a separation of the upper red sandstone from the coal deposit is possible, because an unconformable stratification sometimes takes place, the latter cannot be separated from the lower red sandstone, as the beds of this sandstone are quite conformable with those of the coal strata *.

By consulting Mr. Weaver's observations respecting the views of the German geologists previous to those of MM. von Veltheim and Hoffmann, it will be seen he was led to the conclusion, that the old red sandstone of the English geologists might be equivalent to the lowest part of the German rothliegendes†.

Perhaps by following up the views of Mr. Weaver, and considering that the coal is not necessarily constant to a particular part of the series, and that the mutual relations of different portions of the whole mass vary materially, we may approach towards a solution of this apparent difficulty. In the first place, we obtain no help from organic remains; for it will have been observed that the general zoological character of the marine exuviæ is the same in the zechstein (above the rothliegendes), in the carboniferous limestone, and, as will be seen in the sequel, in the grauwacke series. The general character of the vegetable remains was probably also similar, as we know it was in the descending order. Assuming, therefore, (and it does not appear unphilosophical to do so,) that organic remains will not aid us in the investigation, we can only appeal to mineralogical structure and relative geological position. Our first inquiry therefore should be, Are these constantly the same, allowances being only made for smaller

---

* Von Dechen, German Transl. of Manual.

† It should be stated, while on this subject, that, according to M. Herault, the coal-measures of Littry, Normandy, which rest unconformably on grauwacke, pass into the red sandstone series above them.—*Tableau des Terrains de Calvados*: Caen, 1832.

variations? We can only reply to this question by a statement
of facts.

In the southern part of our island the three divisions of old
red sandstone, carboniferous limestone, and coal-measures, are
well marked, and there is clearly no passage of the latter into
the red sandstone (commonly termed new red sandstone) above
it; on the contrary, the coal-measures and the inferior rocks
have been upset prior to the deposition of the magnesian conglomerates and limestones with their associated red sandstones
and conglomerates; and it seems more than probable that the
lower portions of the latter series of rocks resulted from the
disturbance produced by the fracture, contortion, and elevation
of the coal-measures and older rocks. With respect also to
the carboniferous group itself, the masses of the old red sandstone, carboniferous limestone, and coal-measures, are well
separated from each other, though there may be small alternations at their contact, as the student can observe at Clifton
gorge near Bristol, and other places in that district *.

As we advance northwards into the central part of England,
we find that the lower part of the coal-measures and the upper
part of the carboniferous limestone, which in the south only
alternated at their contact in a comparatively moderate degree,
have now assumed a new character as they approach each
other, presenting a mass of shales, sandstones (most frequently
coarse), and limestones, with occasional seams of coal, the
whole being of very considerable thickness, and known as
Millstone Grit.

Prof. Sedgwick has shown, that still further north in England
the great lines of distinction between the carboniferous limestone and the coal-measures are broken up, and that the one
rock is lost in the other. As the student can have no better
or more condensed view of the subject than in Prof. Sedgwick's
own words, I shall offer no apology for inserting them here.

"On the re-appearance of the carboniferous limestone at
the base of the Yorkshire chain, we still find the same general
analogies of structure: enormous masses of limestone form
the lowest part, and the rich coal-fields the highest part of the
whole series; and we also find the millstone-grit occupying an
intermediate position. The millstone-grit, however, becomes
a very complex deposit, with several subordinate beds of coal;
and is separated from the great inferior calcareous group

* For the necessary details respecting the coal-measures and the carboniferous series generally, consult the labours of Mr. Conybeare in his Outlines
of the Geology of England and Wales; and for that of the southern part of
our island in particular, the Observations on the South-western Coal District of England, by Dr. Buckland and Mr. Conybeare, Geol. Trans., 2nd
series, vol. i.

(known in the north of England by the name of *scar limestone*), not merely by the great shale and shale-limestone, as in Derbyshire, but by a still more complex deposit, in some places not less than 1000 feet thick, in which five groups of limestone strata, extraordinary for their perfect continuity and unvarying thickness, alternate with great masses of sandstone and shale, containing innumerable impressions of coal plants, and three or four thin seams of good coal extensively worked for domestic use.

" In the range of the carboniferous chain from Stainmoor, through the ridge of Cross Fell to the confines of Northumberland, we have a repetition of the same general phænomena. On its eastern flanks, and superior to all its component groups, is the rich coal-field of Durham.   Under the coal-field we have, in regular descending order, the millstone-grit, the alternations of limestones and coal-measures nearly identical with those of the Yorkshire chain, and at the base of all is the great *scar limestone.*   The *scar limestone* begins however to be subdivided by thick masses of sandstone and carbonaceous shale, of which we had hardly a trace in Yorkshire, and gradually passes into a complex deposit, not distinguishable from the next superior division of the series.   Along with this gradual change is a greater development of the inferior coal-beds alternating with the limestone, some of which, on the north-eastern skirts of Cumberland, are three or four feet in thickness, and are now worked for domestic use, with all the accompaniments of rail-roads and steam-engines.

" The alternating beds of sandstone and shale expand more and more as we advance towards the north, at the expense of all the calcareous groups, which gradually thin off and cease to produce any impress on the features of the country.   And thus it is that the lowest portion of the whole carboniferous system, from Bewcastle Forest along the skirts of Cheviot Hills to the valley of the Tweed, has hardly a single feature in common with the inferior part of the Yorkshire chain ; but, on the contrary, has all the most ordinary external characters of a coal-formation.   Corresponding to this change is also a gradual thickening of carbonaceous matter in some of the lower groups.   Many coal-works have been opened upon this line, and near the right bank of the Tweed (almost on a parallel with the great *scar limestone*) is a coal-field with five or six good seams, some of which are worked, not merely for the use of the neighbouring districts, but also for the supply of the capital *."

---

* Sedgwick, Address to the Geological Society, 1831.   Phil. Mag. and Annals, vol. ix. pp. 286, 287.

We thus observe that a very material change has been effected in the carboniferous rocks, the limestone beds having become mixed up with, and even disappearing among, the arenaceous and shaly coal-measures. Two rocks of the series are therefore as it were amalgamated, and no line of distinction can be drawn between them. Not only have the separate characters of coal-measures and carboniferous limestone disappeared, but the remaining rock, the old red sandstone, no longer presents the common arenaceous aspect which it possesses in the South of England and in Wales. It is here a conglomerate, which, instead of offering the appearance of a passage into the grauwacke strata beneath, actually rests on the upturned edges of those strata, and is frequently absent, so that, as has been shown by Prof. Sedgwick and Mr. J. Phillips, the carboniferous limestones repose directly on the previously disturbed and upset grauwacke rocks[*].

If we now proceed to Scotland, to that mass of conglomerate and arenaceous deposits intermixed with limestone and coal described by Dr. Fleming, Prof. Jameson, Dr. MacCulloch, Mr. Bald, Dr. Boué, Prof. Sedgwick, Mr. Murchison, and other geologists, there would appear to be some difficulty in establishing distinctions such as can readily be made in the southern parts of our island; and this difficulty is increased by the presence of rocks, referrible, at least in part, to the (new) red sandstone group. In the northern English districts noticed by Prof. Sedgwick, the red sandstone rocks have clearly been deposited on the carboniferous limestone and coal-measures after the two latter rocks had suffered great disturbance and violent dislocations; but it may be questionable, at least in parts of Scotland, how far fine lines of distinction can be drawn between the upper part of the coal-measures and the lower portion of the red sandstone group. Organic remains will be of little assistance, for reasons before stated, neither is the mineralogical character of much avail, for it will have been seen that this also changes; and there is nothing, that we are aware of, which should prevent the zechstein, if produced under general similar circumstances, from assuming the character of the carboniferous limestone, such more particularly as the latter appears when divided and included in the coal-measures. The colour of the rocks is, if possible, of still less importance, for the coal-measures are not unfrequently red; so that should the whole get mixed up together, more

* See Phillips's Sections in the Geol. Trans., 2nd series, vol. iii.; and Sedgwick, Proceedings of the Geol. Soc. 1831. When the sections and descriptions of the latter author shall have been made public, geologists will be in possession of a highly valuable and illustrative series of documents on this subject.

particularly without discordant stratification, it would appear highly theoretical to distinguish the various portions by particular names, each portion being considered the decided equivalent of divisions that may be established elsewhere. It is by no means intended to infer that during any considerable deposit, such as the one under consideration, there should not be equivalents in age: such there must always have been; but the contemporaneous effects produced by different causes may have varied most materially; so that distinctions which mark particular events in one situation are not always useful when applied generally; for, perhaps without being aware that we are doing so, we theoretically consider circumstances to have been generally the same at a given time,—whereas we should, in the first place, consider them as more local, resulting from the operation of more limited causes. I am fully aware that this view may be carried too far, and that the minor divisions of rocks should be established as much as possible; but we should also avoid extremes, and not pass that point where the distinctions may admit of very great doubt, for by doing so we seem in a great measure to preclude ourselves from tracing the causes which have produced the great changes in the mineralogical and zoological characters of rocks on the surface of the earth generally.

Dr. Boué considers the conglomerates, sandstones, limestones and coal of the great arenaceous deposit of Scotland, as subordinate portions of one great whole, which he believes equivalent to the red sandstone (*grès rouge*). To what extent this opinion may be correct would not yet appear to be well decided; and it no doubt may startle English geologists to compare, in any manner, the old red sandstone with a system of rocks containing the rothliegendes: but supposing a series of conglomerate, sandstone, and other rocks of a certain common character, to have been produced, so that during the deposit the inferior should not in any manner have been disturbed, but the strata to have been laid regularly on one another, and that we obtain little or no aid from organic remains,—it would seem difficult to regard the mass in any other light than as resulting from the operation of nearly similar and uninterrupted causes. I am far from stating that this actually is the case in Scotland, being merely desirous to show that an union of the rothliegendes and zechstein with the carboniferous group may not be impossible elsewhere. It may, however, be interesting to the student to learn, while on this subject, that the *Calamites Mougeotii*, first discovered in the red sandstone group of the Vosges, is, according to Prof. Lindley and Mr. Hutton[*], also found in the Edinburgh coal-measures. Undoubtedly we

[*] Lindley and Hutton, Fossil Flora of Great Britain.

cannot hence infer the exact contemporaneous deposit of the two rocks, any more than the identity of the zechstein and carboniferous limestone, because certain fossils are stated to be common to both; but the connexion, as far as respects organic remains, is important.

In certain districts, such as Pembrokeshire, the old red sandstone passes into the grauwacke series beneath: in such situations, therefore, we may regard this rock as resulting from the continuance of causes similar to those which have produced the grauwacke; for the zoological character of the two deposits is the same, as is also their mineralogical structure; the difference between them is in the colour,—a circumstance of no importance, for red rocks are often present in the body of the grauwacke group itself.

In the north of England, the old red sandstone, as has been above noticed, rests on upturned grauwacke: causes therefore have acted violently in one situation at a given period, which have scarcely, if at all, produced marked effects in another at no very considerable distance; leading us to infer that at still greater distances the differences observable in deposits of this period may be more remarkable.

For the following sketch of the range of the old red sandstone through Great Britain, I am indebted to my friend Professor Sedgwick. It will no doubt be perused with much interest, presenting as it does this distinguished author's views on a part of a system of rocks, which, as is well known, have for many years more particularly engaged his attention.—He observes that the old red sandstone not only insensibly passes into grauwacke beneath it in the south-western parts of England, but that it may also be said to graduate into the carboniferous limestone above it in the same districts by the intervention of alternating beds of sandstone. He also observes, as a general fact, that the coal-measures, the carboniferous limestone, and the old red sandstone, are obviously affected by a common system of flexures, produced by disturbing forces, posterior to them all. The truth of this observation is evident from the various sections of Professor Buckland, Mr. Conybeare, and Mr. Weaver*.

Professor Sedgwick points out, as the next remarkable fact connected with the history of these deposits, the entire, or nearly entire, absence of the old red sandstone along the whole base of the carboniferous limestones, which, commencing at Llanymynech, ranges along the eastern skirts of Denbighshire and Flintshire, and, doubling to the N.W., runs to the Great Orms Head; thence taking its course to the Menai Straits,

* Geol. Trans., 2nd series.

and forming a small part of the interior of Anglesea. He considers that, in this long range (partly examined by Mr. Murchison and partly by himself), there is not perhaps a single trace of old red sandstone, unless we designate by that name some beds of reddish shale and sandstone, which here and there form the base of the carboniferous group, and are, he thinks, only varieties of the lowest limestone shale. This conclusion is confirmed by the observations of Professor Henslow, published about ten years since *. Professor Sedgwick believes that some part of the Isle of Anglesea, coloured by the latter as old red sandstone, is in fact a red sandstone of the grauwacke group. Prof. Henslow himself states the impossibility of separating the old red sandstone in all cases from the grauwacke; and it should be recollected that at the time his paper was written, extensive deposits of red sandstone in the grauwacke group were not generally known. Along the whole line above mentioned, where the old red sandstone is wanting, the carboniferous group rests unconformably on the grauwacke group;—a fact which seems to show that the old rocks of North Wales underwent a great movement, anterior to the period of the old red sandstone; and that, by this movement, the bottom of the neighbouring seas was raised out of those causes which produced the old red sandstones.

In no part of Denbighshire do we see the base of the carboniferous group; but on the confines of Yorkshire, Lancashire, and Westmoreland, we find, in several places, the great escarpments of the carboniferous limestone resting unconformably on the edges of the grauwacke, or separated from it by masses (sometimes of great thickness) of coarse red conglomerate. The same statement applies to the carboniferous zone wrapped round the Cumbrian mountains, and also to the chain of Cross Fell. The old red conglomerates sometimes, though rarely, alternate with green and red marls, and with red, greenish red, or white sandstone; and we occasionally find in them, not only pebbles derived from the older calcareous rocks, but calcareous concretions like the Herefordshire *Cornstone*.

On the confines of Scotland, the red conglomerates appear (though Professor Sedgwick considers rarely) at the base of the carboniferous series. He has seen them occupying this position on the flanks of the Cheviot Hills, in two or three places pointed out by Mr. Culley, of Coupland Castle. He remarks the importance of this fact, as it proves that the very old carboniferous system of the Tweed (the sandstone beds of which are generally of a red colour) is newer than the old red conglomerates last mentioned; and therefore *probably newer* than the old red conglomerates of Cumberland, Westmoreland,

---

* Transactions of the Cambridge Philosophical Society.

and Yorkshire. He does not therefore believe that the carboniferous red sandstone of the Tweed is the representative of
the old red sandstone of Herefordshire; but that it is superior
to the old red sandstone, and is of about the age of the great
scar limestone of Yorkshire and Cross Fell.

The carboniferous red sandstone, appearing here and there
in the great Caledonian trough (between the chains of the
Grampians and the grauwacke chain which stretches from St.
Abbs Head to the Mull of Galloway), is probably in no instance older than the carboniferous red sandstone of the Tweed.
Professor Sedgwick doubts whether in any part of this great
trough there be a true representative of the (new) red and variegated sandstone of central England.

In the Isle of Arran we have an old red sandstone and
conglomerate, a carboniferous series of considerable thickness
(but obscurely developed), and an upper red sandstone and
conglomerate. Guided by analogy, it was concluded by Prof.
Sedgwick and Mr. Murchison, in 1827, that the upper red
sandstone and conglomerate were equivalents of the (new) red
sandstone of England. Knowing by subsequent experience
the great development of red sandstone in the carboniferous
system of Scotland, Prof. Sedgwick now doubts the truth of
this conclusion, and thinks that the upper red sandstone and
conglomerate of the Isle of Arran may perhaps be only an
unusual development of a portion of the carboniferous series.
To settle this point it would be necessary to connect the Arran sections with those on the coast of Ayrshire, and, by a
northern traverse, to connect the Ayrshire red sandstone with
the red conglomerate system flanking the Grampians. This
task has not yet been attempted by any one well acquainted
with the English types, and at the same time with the Scotch
carboniferous deposits.

In regard to the vast masses of red sandstone and conglomerate on the shores of the Highlands, in the old oceanic valley of the Caledonian canal, and on the south flank of the
Grampians, they were referred, by Prof. Sedgwick and Mr.
Murchison, for the most part to the old red sandstone of England; and this classification has been subsequently confirmed
by the observations of Dr. Fleming and Mr. Lyell, who have
traced a part of the system under the great coal-field of Fifeshire, and thus left no doubt respecting its relative situation.

In Caithness this system is divided into (*a*) old red sandstone and conglomerate; (*b*) calcareo-bituminous schist, with
numerous impressions of fish* (*Dypteri, &c.*); and (*c*) upper red

---

* For figures of these fish, and a detailed account of the beds in which
they occur, see the paper by Prof. Sedgwick and Mr. Murchison, Geol.
Trans. 2nd series, vol. iii.

sandstone. The first and second of these divisions cannot be separated from each other, and must therefore be included in the old red sandstone,—a conclusion amply confirmed by recent observations. The upper division (c) forms the highest part of an imperfect section, and it is perhaps impossible to determine its exact place, but was hypothetically referred by Prof. Sedgwick and Mr. Murchison to the lowest division of the (new) red sandstone series. With his present views, Professor Sedgwick would wish this hypothetical adjustment of the upper groups (c) of Caithness to be changed, and to see them classed with the old red sandstone; as he considers them identical in age with the upper part of the series which descends from the southern Highlands, and passes under the carboniferous system of Fifeshire *.

The carboniferous group occupies the surface of a large portion of Ireland, the limestones being exceedingly abundant. Mr. Weaver describes sandstones and conglomerates as frequently, though not constantly, interposed between the older deposits and the carboniferous limestone, and refers them to the old red sandstone. The Gaultees mountains are mentioned as wholly composed of them. They occur along the flanks of the clay-slate districts, and isolated caps of the sandstone often rest on these older deposits. The red sandstone emerges from beneath the interior of the great limestone plain at Moat, Ballymahon, and Slievegoldry Hill. The same author notices that the strata of this sandstone deposit are most inclined as they approach, and are in contact with, the older rocks; but that as they accumulate and recede from the latter, they become more and more horizontal.

The carboniferous limestone may be considered as the prevalent rock in Ireland; for, as Mr. Weaver observes, all its counties, with the exception of Derry, Antrim, and Wicklow, are more or less composed of it. This limestone is described as coming in contact with, and sweeping round, various mountain chains, "filling up every interval and hollow between them." It supports the coal-measures, properly so called; and thus the analogy between the carboniferous series of central and southern England, and that of the corresponding portions of Ireland, is complete; the arenaceous and conglomerate deposits of the old red sandstone in the latter country being surmounted by a sheet of limestone, varying in thickness, sometimes attaining a depth of 700 or 800 feet, but generally averaging 200 or 300 feet; and being in its turn covered by coal-measures †.

* Sedgwick, MSS.
† Weaver, On the Geological Relations of the East of Ireland, Geol. Trans.

It would appear from the observations of Archdeacon Verschoyle on the north-west portions of Mayo and Sligo, that the carboniferous limestone of that part of Ireland has been largely developed; for Benbulden, 1700 feet, Knocknodie, 1025 feet, and Knocknashie, 980 feet, are stated to be wholly composed of it*.

The carboniferous rocks of the north of France and Belgium have a direction from east-north-east to west-south-west from the vicinity of Aix-la-Chapelle to and beyond Valenciennes, and rise from beneath cretaceous or newer rocks. The carboniferous limestone and coal-measures of the Boulognais can be considered only as a continuous portion of the same deposit.

According to M. de Villeneuve the coal-measures and limestone alternate at their contact with each other between Liége and Chaude Fontaine. The limestones are metalliferous, bluish and compact, and contain subordinate conglomerates of blue limestone. The alternating sandstones are sometimes reddish, and at others greenish brown; they are sometimes compact, at others fissile with mica, and the lines of cleavage are in some beds not the same with those of stratification. The upper part of the limestone and sandstone contains aluminous shale, worked for profitable purposes (Huy and other places)†.

The same author informs us that the coal-measures, which are composed of the usual mixture of sandstones, shales, and coalbeds, present at the Montagne de St. Gilles no less than sixty-one beds of the latter, varying from six feet to a few inches in thickness; and M. Dumont states that the coal-measures of Liége contain eighty-three beds of coal. The strata of the district are greatly disturbed, as is well seen at Mons, and are traversed by faults, as may be observed at St. Gilles. Coal is worked far down in the lower beds, and even amid the limestones at Mons, which circumstance, however, M. de Villeneuve attributes to the contortions of the strata.

The line of demarcation between the coal-measures and the carboniferous limestone in the north of France, Belgium, and the country extending beyond Aix-la-Chapelle to Eschweiler, is generally well defined; the occurrence of beds of coal between the beds of limestone is rare, and partly apparent, being produced by contortions of the beds. That portion of the strata which is known in England by the name of Millstone

---

vol. v.—Also consult Griffith's Account of the Connaught and Leinster Coal Districts; and Sections and Views illustrative of Geological Phænomena, pl. 29.

* Verschoyle, Proceedings of Geol. Soc., Nov. 1832.

† De Villeneuve, Ann. des Sci. Nat. t. xvi. 1829.

Grit, and in Westphalia as *Rauher sandstein*, is by no means
thick, and a bed of aluminous shale occurs in it between Huy
and Liége, as above noticed. This aluminous shale also occurs in Westphalia, at Lintdorf, and between Werden and
Velbert, as far as Schwelm. Between Namur and Huy, beds
of coal rest immediately on carboniferous limestone. On the
south of Werden the millstone-grit is not much developed,
but becomes more so to the north of Elberfeld. Very thick
beds of conglomerate here occur in it. The lower division,
immediately above the carboniferous limestone, is composed
of a series of strata, consisting of slaty clays, shales, limestones, and sandstones, and is, in essential points, absolutely
identical with the limestone shale of England. These beds
become still more developed further east, and attain a thickness, in the neighbourhood of Arnsberg, Merscede, and
Warstein, which has not yet been observed at any other
point. No old red sandstone has hitherto been observed in
Westphalia. The carboniferous limestone here rests immediately on grauwacke, the continuation of that of the slate districts of the Netherlands, the Rhine, and Westphalia.

In Belgium, especially on the Meuse, the carboniferous
limestone occurs extensively. From the undulating character
of the stratification, it is difficult to determine the various beds
of which the mass is composed. The intervening rocks have
been usually termed grauwacke and clay-slate, but probably
the red conglomerates may be considered as equivalent to the
old red sandstone. These are found on the Meuse at Lustin
and Profondeville; on the Hoyoux at Masleye, south of Huy;
and on the Vesdre at Pepinster; on the Vichtbach; &c.*

In all cases where the coal-measures come into contact with
grauwacke, without intervening beds, we should be careful to
recollect that the chances of an overlap are as great with the
coal-measures as with any other rock. We have in Pembrokeshire an instructive example of an overlap of this kind.
The coal-measures of Pembrokeshire are well known as forming the western continuation of the South Wales coal-field. On
the eastern part of Pembrokeshire they are still retained within
bands of carboniferous limestone, which separate them from
the old red sandstone to the north and south. On the western side of the same county circumstances are different; for
instead of resting on carboniferous limestone, they repose on
grauwacke, having passed over carboniferous limestone and
old red sandstone in an oblique direction. The direct continuation of the coal-measures of the eastern part of the county

* Von Dechen, German Transl. of Manual.

is seen, at Broad Haven and Little Haven, in St. Bride's Bay, to rest on grauwacke on the north. Trap rocks have disturbed its quiet relations on the south, but it is seen to rest against at least a small portion of carboniferous limestone in that direction. To the north of Broad Haven, a patch of coal-measures is observed to rest wholly on grauwacke, evidently the continuation of the coal strata on the south, though this continuity is not evident on the surface from the indentation of St Bride's Bay. This overlap of the coal-measures has not arisen from a greater development of coal-measures than of carboniferous limestone or old red sandstone, for both these rocks occur in great abundance to the south in the same district, but to a passage over these rocks in a direction different from their general range *.

To return to the carboniferous rocks of the Netherlands:— It would appear that they are continued into Germany, to the deposits between Essen, Werden, Bochum, Hattingen, Wetter and Dortmund, which repose on the north-west corner of the great exposure of grauwacke rocks in that part of Europe †. To the north of these deposits, on the northern side of the great gulf of cretaceous and supracretaceous rocks which enters easterly into Germany, and on which stands Münster, there is, according to M. Hoffmann, an outcrop of carboniferous strata at Ibbenbühren, between Osnabrück and Rheine ‡. Coal-measures occur at Seefeld, in Saxony. At Wettin, north of Halle, is another deposit; and at Saarbrück, and the

* For a description of this country, with maps and sections, see my Memoir on Southern Pembrokeshire, Geol. Trans. 2nd series, vol. ii.; and Sections and Views illustrative of Geological Phænomena, pl. 12.

† M. von Dechen remarks that the carboniferous limestone of northwestern Germany, Belgium, and the north of France, is so connected with the grauwacke group, that it has hitherto been impossible to distinguish them. The limestone of the Meuse from Namur to Visé, and from Ratingen to Arnsberg and Warstein, is decidedly carboniferous limestone. The limestone beds of the Dillenburg country, on the Lahn, from Stromberg near Bingen, may perhaps, without much objection, be considered as grauwacke limestone. But the position of the limestone from Pfaffrath and Bensberg on the right bank of the Rhine, and which may be traced eastwards to Gummersbach and Mittelacher, is wholly doubtful. The Eiffel limestone from Schönecken near Prüm to the Erft below Münstereifel, on the left bank of the Rhine, seems proved by the observations of M. Schulze merely to fill cavities in the grauwacke, and on the large scale never to be covered by it. This Eifel limestone may therefore, as far as present observations extend, be referred either to the carboniferous or grauwacke groups. —German Transl. of Manual.

‡ For the localities of the coal in the north-west of Germany, consult M. Hoffmann's map of that country ; and for descriptions, "Uebersicht der orographischen und geognostischen Verhältnisse vom Nordwestlichen Deutschland," by the same author.

neighbouring country, the coal-measures are abundant, and rest, when trappean rocks are not interposed, upon part of the grauwacke mass previously mentioned *.

The coal-measures of Saarbrück are rendered particularly interesting by the development of the upper part of the series, such upper part being apparently a passage from the coal-measures into a rock equivalent to the rothliegendes, reminding us of the connexion of this rock with the red sandstone group previously noticed. The coal-beds of the upper portion of the coal-measures, and which occur at considerable intervals from each other, are intermingled with beds of red sandstone, in many places not to be distinguished from the rothliegendes of Mansfeld. This red sandstone alternates with thin beds of limestone and dolomite, and with shale. It contains layers of nodules with impressions of fish (Lebach, and Börschweiler), and of the same ferns which are discovered in the lower coal-beds. This sandstone, which is of great thickness, would, as M. von Dechen observes, be certainly considered as rothliegendes, if it were not so decidedly connected with the coal-measures beneath it. It must not, however, be confounded with the Bunter sandstein, which rests unconformably on the coal †.

M. Pusch describes the coal-measures in Poland as extending from Hultschin to Krzeszowice, the more ancient beds passing into the grauwacke on which they rest: but the same author remarks, that in the rocky valleys of Czerna Szklary, and near Debnik, not far from Krzeszowice, a black marble, employed in the arts, supports the coal-measures. M. Pusch considers this marble as equivalent to the carboniferous limestone of the English geologists, and observes that the calcareous conglomerates which accompany the coal, sandstones and shales in the gorges of Miekina and Filipowice are referrible to the same marble beds. The same author states that the coal-measures contain the plants so commonly observed elsewhere in similar deposits, and that he has identified thirty-six species with those noticed in the works of MM. Sternberg and Ad. Brongniart ‡.

M. Kovalevski describes a very rich carboniferous deposit

---

* The student will find instructive plans and sections of the coal-mines at Werden, Essen, Eschweiler, Valenciennes, Mons, Fuchsgrube (Silesia), and Saarbrück, in the Atlas to la Richesse Minerale, by M. Heron de Ville-fosse, pl. 24, 25, 26, 27, and 28. He should also consult the geological map and sections of the countries bordering the Rhine, by MM. Oeynhausen, la Roche, and von Dechen, for the coal-measures of Saarbrück and the adjacent country. Parts of these sections are inserted in Sections and Views illustrative of Geological Phænomena, pl. 18. fig. 1. and 2.

† Von Dechen, German Transl. of Manual.

‡ Pusch, Journal de Géologie, t. ii.

as existing in the mountains, which extend for a distance of 150 wersts, on the right bank of the Donetz in Southern Russia. The exact age of this deposit, which is of considerable importance, is perhaps somewhat doubtful; but it would appear referrible either to that of the group under consideration, or to the upper part of the grauwacke. The rocks in which the coal principally occurs are described as arenaceous, and commonly red, with a mixture of argillaceous and schistose beds. The whole is described as passing into the grauwacke on which it rests. The coal, occurring in beds from a few inches to seven feet in thickness, is bituminous among the sandstones and shales, but becomes anthracitic where the rocks pass into grauwacke, the latter rock containing thin beds of limestone, and being traversed by quartz veins. Various fossil plants are discovered in this carboniferous deposit, which is principally worked in the districts of Bakmout and Shavianoserbskoi, of the Government of Ekaterineslavsk, and partly in the country of the Mious, in the military district of the Don Cossacks *.

The coal deposits of central France repose on granite, gneiss, mica-slate, &c., without the intervention of any limestones, sandstones, or slates, which can be distinctly referred to the carboniferous limestone, old red sandstone, or grauwacke : such are the coal-fields of St. Etienne, Rive de Gier, Brassac, Fuis, &c. At St. Georges-Châtellaison the coal-measures also rest on gneiss and mica-slate †.

The carboniferous deposits of the United States are, according to Prof. Eaton, of different ages ; one being contained in the argillaceous slates (argillite) of Worcester (Mass.) and Newport; another being considered equivalent to the coal-measures of Europe; and a third being of a more recent epoch, though older than certain lignites. The deposit referred to the same epoch as the carboniferous series of Europe, occurs at Carbondale, Lehigh, Lackawaxen, Wilkesbarre, and other places ‡.

Mr. Cist describes the coal of Wilkesbarre as alternating with various sandstones and shales, the latter containing a great abundance of fossil plants §, many of which it will have been seen by the lists of organic remains are identical with some discovered in the coal-measures of Europe, and all are of the

* Kovalevski, Gornoi Journal, 1829; and in Boué's Mémoires Géol. et Paléontologiques, vol. i. Other carboniferous deposits exist in the same country, but appear to be more modern.

† Mr. Grammer remarks that the coal deposit of Virginia rests on granite. —American Journal of Science, vol. i.

‡ Eaton, Amer. Journ. of Science, vol. xix.

§ Cist, Amer. Journ. of Science, vol. iv., where a map of the deposit will be seen.

same general character with those obtained from the carboniferous and grauwacke series. The sandstone beds vary from
five to one hundred feet in depth, and the coal is sometimes
from thirty to forty feet thick, its general thickness being from
twelve to fifteen feet. Prof. Silliman states that the beds at
Mauch Chunk (Pennsylvania) consist of conglomerates, sandstones, and argillaceous slate. The pebbles in the conglomerate are described as pieces of quartz rounded by attrition,
and the cementing matter of the conglomerates and sandstones
as siliceous *. According to Prof. Eaton, the limestone which
supports the strata containing the Pennsylvanian coal extends
along the foot of the Catskill Mountains, and is continued
from the southern part of Pennsylvania to Sackett's Harbour
on Lake Ontario †.

Mr. Hitchcock informs us that coal is associated with trappean rocks, fetid, siliceous, and bituminous limestones, red
and gray sandstones, and conglomerates, in Connecticut. It
is described as bituminous; whereas the Wilkesbarre coal is
frequently termed Anthracite by the American geologists ‡.
It occurs at Durham, Chatham, Berlin, Enfield, and other
places in Connecticut, and is described as passing into the so-
called old red sandstone of the country, which is composed of
a series of sandstones and conglomerates, generally of a dark
red colour. An excellent section of this coal deposit, described in great detail by Mr. Hitchcock, is exposed where
the Connecticut river cuts through it between Gill and Mont-
ague.

Fossil fish are obtained from associated bituminous shale at
Westfield (Connecticut) and Sunderland (Massachusetts); and
one species is considered referrible to the genus *Palæothrissum* of Blainville §, a genus which the student has seen noticed both under the head of zechstein, and of the coal-measures.

The relative antiquity of the coal deposit of India is not
very clearly ascertained. According to the observations of
Dr. Voysey, Mr. Calder‖, and Mr. Royle, it rests upon gneiss
and other rocks of that character. The latter author states
that the coal-field of Damuda, discovered by Mr. Jones in
1815, is composed of various beds of bituminous coal, (of

* Silliman, Amer. Journ. of Science, vol. xix.
† Eaton, ib. vol. xix.
‡ This distinction would not, in itself, appear to be of any great im-
portance; for the continuous coal deposit of South Wales is anthracitic in
Pembrokeshire, and bituminous in its eastern prolongation through Mon-
mouthshire.
§ Hitchcock, American Journal of Science, vol. vi.
‖ Calder, Trans. Phys. Class. Asiatic Soc. Bengal, vol. i. 1829.

good quality,) shales, sandstones, and conglomerates; the latter, mixed with red sandstone, constituting the upper part of
the series. The shales contain the abundant remains of *Ver-
tebraria Indica*, Royle (commonly termed by Indian geologists the Ranijung reed); of *Sphenophyllum? speciosum*, Royle;
of *Glossopteris angustifolia*, Ad. Brong.; of *Glossopteris Browniana*, Ad. Brong.; of *Pustularia Calderiana*, Royle; *Pecopteris Lindleyana*, Royle; and of various other plants *. Mr.
Jones observes that the shales and coal-beds of Damuda crop
out in many places, and that the strata are generally undulating. The coal is principally worked at Ranijung colliery,
where eight coal-beds, varying from four inches to nine feet
in thickness, are associated with shales and sandstones †.
Mr. Everest observed the effects of pseudo-volcanic action in
this coal district, which appear evidently to have been caused
by the combustion of the coal, the various shales and sandstones being more or less acted upon by the heat evolved.
The same author remarks that some of the sandstones of the
Damuda coal-field contain calcareous matter, and that the
sandstones are raised in large slabs for economical purposes
on the banks of the Adji ‡.

The coal of India is by no means confined to the Damuda
district. Mr. Jones considers that the coal of Sylhet and of
Cachar constitutes an eastern prolongation of the same deposit; and Mr. Royle infers that it is continued a considerable
distance to the westward, having been observed resting on
gneiss and other rocks of that class, as in the district first
noticed, at Goomeah, Palamow, Jubbulpoor, and Hosanha-
bad §. Capt. Franklin states that five beds of coal were found
in a part of the Palamow coal district, which, judging from
the distances of the localities mentioned by this author, seems
somewhat extensive ‖.

Coal is therefore by no means rare in India: on the contrary it extends, probably at intervals, along an east and west
line of several hundred miles. There is indeed no direct evidence, such as that of a perfect correspondence in organic remains, to show that the various deposits observed are contemporaneous, but there is sufficient evidence to make such an
inference highly probable. Although the fossil plants discovered in the shales are not specifically the same with any yet
noticed in the carboniferous rocks of Europe or America, they

* Royle, Illustrations of the Botany, &c. of the Himalayan Mountains,
London, 1833,—where further detail and figures of these plants are given.
† Jones, Trans. Phys. Class. Asiatic Soc. Bengal: Calcutta, 1829.
‡ Everest, Gleanings in Science, vol. iii.: Calcutta, 1831.
§ Royle, Illustrations of the Botany, &c. of the Himalayan Mountains.
‖ Franklin, Gleanings in Science, (Calcutta,) vol. ii. p. 217.

are still of the same general character as far as respects the temperature in which they probably flourished. The presence of *Glossopteris Browniana* in the coal district of Damuda is remarkable as it points to a connexion, with regard to date, between that deposit and the carboniferous series of Eastern Australia, where it was first observed by Dr. Robert Brown.

If we, for the moment, abstract the limestone beds, there would appear little doubt that the carboniferous series is of mechanical formation, a deposit from water varying in its transporting powers. Thus, at one time the velocities were sufficient to force forward gravels, while at others they only accomplished the transport of silt or mud. If proportional sections be made of coal deposits, it will be observed that the coal-beds occur at very unequal intervals, showing that the causes which produced them have acted irregularly. From the careful examination of the Forest of Dean by Mr. Mushet, we have a detailed list of the various beds of the coal-measures, carboniferous limestone and old red sandstone, the whole constituting a collective thickness of about 8700 feet; the coal-measures being 3060 feet in depth, and the limestone 705. The mass reposes on the grauwacke (transition) limestone of Long Hope and Huntley *. The sandstones of the old red sandstone in Gloucestershire, Somersetshire, and the neighbouring parts of England, afford us no great evidence of a quick deposit, more particularly as the conglomerates are not common; the latter are, however, sufficient to show that the velocities of the transporting waters were not constant, but liable to variation. A great change in the depositing and transporting powers subsequently took place; and instead of the siliceous and arenaceous sediment, carbonate of lime, often enveloping a variety of marine animal remains, was produced; and this not for a short time, but apparently during a long period; for the carboniferous limestone of this district bears marks of slow formation, many beds being composed of a mass of fossils, the remains of myriads of animals, which have apparently lived and died where we now find them entombed. It must however be admitted that the origin of many beds, which do not present a trace of animal exuviæ, remains obscure, and we have no direct evidence that they may not have been produced more suddenly by deposits from water, either holding carbonate of lime in chemical solution or in mechanical suspension. After a thickness of seven or eight hundred feet of calcareous rock had been formed, another great change in the matter deposited was effected; not however so suddenly but that the arenaceous sediment which afterwards became so

abundant, and the calcareous matter, were alternately produced for a comparatively limited period.   An immense mass
of sandstone, shales, and coal was then accumulated in beds
one above another, which, though irregular with regard to
the relative periods of deposit, are frequently persistent over
considerable areas.

By general consent the coal is considered as resulting from
the distribution of a body of vegetable remains over areas of
greater or less extent, upon a previously deposited surface of
sand, argillaceous silt, or mud, but principally the latter, now
compressed into shale.   After the distribution of the vegetables, other sands, silt, or mud, were accumulated upon them;
and this kind of operation was continued irregularly for a considerable time, during which there was an abundant growth
of similar vegetables at no very distant place, to be suddenly,
at least in part, destroyed and distributed over considerable
areas on the more common detritus.

Great length of time would be requisite for this accumulation, because the phænomena observed would lead us to consider the transporting power, though variable, to have been
generally moderate : moreover a very considerable growth of
vegetables requiring time, would be necessary at distinct intervals; for coal-beds only now six or ten feet thick, must,
before pressure was exerted upon them, have occupied a much
greater depth.   It is a remarkable circumstance connected
with the coal-measures of the South of England, that marine
remains have not been detected in them, which, though it
does not prove the deposit of coal to have been effected in
fresh water, does appear to show that there was something
which prevented the presence of marine animals,—a circumstance the more remarkable, as we have seen that such animals swarmed during the formation of the carboniferous limestone.

These remarks are not only applicable to the small district above noticed, but to a large extent of country,—one
stretching from Belgium, through the North of France, and
the southern parts of England and Wales, into Ireland; and
for the most part concealed beneath masses of newer rocks.
As we advance northward, however, the marked distinctions,
as before noticed, disappear; so that the causes, whatever
they were, which produced such a separation of arenaceous
and calcareous rocks on the south, became modified, and the
limestones were more intimately blended, in alternating beds,
with the sandstones and shales, affording a greater mixture
of marine with terrestrial remains.   It has long been known
that the Yorkshire coal-measures presented a bed containing
the remains of *Ammonites* and *Pectines*, and that the fossils of

the carboniferous limestone and coal-measures were detected in the millstone grit; or, in other words, that there was an alternation of terrestrial with marine remains,—showing that the causes which effected the deposit of calcareous matter and envelopment of marine remains sometimes predominated, while at others a transport of mud and sand entombed an abundance of vegetables. The occurrence of marine remains amid the coal-measures, it will be observed by reference to the foregoing lists, is not confined to Great Britain, but is also remarked in different parts of Germany; so that the same modification of circumstances which has produced a mixture, or rather alternation, of marine and terrestrial remains in Great Britain, has extended into the continent of Europe.

Mr. Buddle observes, respecting the Newcastle coal-field, that the sandstones increase in number and thickness as they rise or crop upwards, whereas the argillaceous shales become thicker in the opposite direction*. This is precisely the appearance we should expect from a deposit from water moving with moderate velocity, and which should hold detritus of different degrees of fineness in mechanical suspension. The sands would be first deposited, while the silt and mud would be carried greater distances. It would hence follow, that the sands would be more abundantly deposited in proportion as they approached the situations whence the detritus-bearing water proceeded, and consequently the resulting beds would become thinner as they receded from the same situations. The very reverse would happen with the mud deposits. So that supposing the transporting current of water to have been nearly uniform and not rapid, we should have the effects produced stated to be now observed in the Newcastle coal-field.

There is another class of appearances connected with these rocks which demands our attention. From a considerable mixture of porphyry in certain situations with the coal-measures, it has sometimes been considered that this rock was an essential and component part of the group under consideration. From all analogy it may be concluded that porphyries are of igneous origin; and for the same reason it is inferred that the coal-measures and their accompanying beds were produced by aqueous deposition. We therefore should be led *à priori* to consider, that two substances of such different origin did not necessarily constitute parts of a common whole, but that their admixture was accidental. And we may consider this at once proved, by the abundant occurrence of coal-measures without porphyry, such as is so commonly the case in England.

In the sections which M. Hoffmann has presented us of the

---

* Buddle, Trans. Nat. Hist. Soc. Newcastle, vol. i. p. 238.

coal-measures of Wettin, and other places in North-western
Germany, it is easy to conceive, although porphyry occurs
both above and beneath the coal strata, that the latter are not
necessarily of contemporaneous formation; on the contrary,
the fractured and contorted state of the beds shows that great
violence has been exercised upon them, precisely such as would
be expected if igneous rocks had burst in amidst them, when
among other accidents we should expect to find large masses
of coal-measures caught up and included in the porphyry, as
we find masses of chalk caught up and enveloped by basalt in
the North of Ireland. As we shall return to the subject of the
igneous rocks found among the carboniferous group in another place, the above notice has been introduced merely to
show that the supposed connexion of porphyry and coal strata
has not been overlooked.

From the similarity of general circumstances attendant on
the coal strata, we have reason to conclude, although the series
may contain more limestone at one place than at another, that
in Poland, Western Germany, Northern France, Belgium,
and the British Isles, there were some common causes in operation at the same epoch, producing the envelopment of a great
abundance of terrestrial vegetables, of a nature that could not,
from the want of the necessary heat, now flourish in the same
latitudes.

Proceeding to the central part of France, we find several
smaller coal deposits, which more particularly from their organic character are referred to the carboniferous epoch of
which we are treating. How far they may have been once
more extensive and continuous, and how far they may have
suffered from movements in the land, dislocations, and denudation, we are not certain; but we are certain that they were
directly deposited on granite, mica-slate, gneiss, and other rocks
of that character. The causes therefore which produced the
calcareous beds, sometimes very abundantly, in the countries
above noticed, have not extended to them. The observed
phænomena are however sufficient to show that a vegetation
similar to that of the more northern carboniferous rocks is there
entombed, though we are not quite assured to what precise
period their formation can be referred; for, as will be seen in
the sequel, vegetables of the same general character are detected in the grauwacke series, and it is also possible that they
might be discovered in the rothliegendes under the zechstein.
The precise period of any particular deposit of similar vegetables may thus be sought through a considerable lapse of time,
and it becomes hazardous to fix, without very good evidence,
on any relative portion of that time. The conglomerates usually referred to the old red sandstone in Northern England,

sometimes intervening between the upturned grauwacke rocks and the carboniferous limestone beds resting upon them, and noticed by Prof. Sedgwick and other geologists, may have been followed by a coal deposit where circumstances were favourable; and the result would have been a formation similar to the deposits of central France, except that the subjacent rocks would be, perhaps, more ancient in the latter case. It may however have also happened, that during the grauwacke deposit, to be noticed in the next section, circumstances favoured a production of rocks similar to those of St. Etienne and other places; as also that during a subsequent period, one equivalent to the lower part of the red sandstone group, a similar deposit might be effected; for as rocks may be violently disturbed in one place and not in another, so may they also be quietly formed in one situation, while a few hundred miles distant, disruptions of strata and the trituration of organic remains may have taken place, so that no trace of organic life may be left.

While on the subject of the coal series of France deposited in gneiss, mica-slate, &c., we may remark the very unequal manner in which the St. Etienne coal-field has been deposited. The rocks beneath (gneiss, mica-slate, and talc-slate) have a very irregular surface, in which the series of conglomerates, sandstones, shales, and coal are moulded; and the result is that the whole is divisible into several minor basins or cavities, as far at least as the coal is concerned. From such a state of things, we could scarcely expect that comparative even thickness of the coal-beds observable in the coal-measures of England. The coal-beds of St. Etienne suddenly swell out, and as suddenly become mere seams; sometimes being from fifty to sixty-five feet thick, at others presenting mere traces of coal. The extraordinary thickness of this coal may surprise some readers; it nevertheless rests on excellent authority [*]. The mean thickness of the beds worked is rarely beyond sixteen or twenty feet, and more rarely below three feet and a quarter.

Let us now consider the mode in which the remains of terrestrial vegetables, so abundantly preserved in the coal strata, occur. They are for the most part laid flat, and the leaves and stems parallel to the line of stratification; but there are other cases where they repose at various angles in the beds; and finally they are found vertical, with their roots downwards. The student will recollect that this is precisely the manner in which the vegetables of the submarine forests are found; and if several submarine forests, such as those which are discovered around the shores of Great Britain, occurred above each other,

---

[*] Chevalier, Annales des Mines, 3ème Series, 1832, t. ii. p. 411.

with the intervention of sandy and clay beds, we should have a series of deposits not very unlike the coal strata, so far as regards the position of the vegetable remains. If we are to consider parts of the coal-measures as in any way resulting from a series of similar deposits, we are certainly called upon to admit a very remarkable series of changes in the relative surface levels of land and water; but there are also very great difficulties attending the supposition that the vegetables have been swept by strong currents of water into the positions where we now find them; for not only have similar effects been produced over considerable areas, but the vegetables have suffered very little injury, their delicate leaves being most beautifully preserved. Now, though we know that vegetables are abundantly borne down by river floods into the sea, they by no means remain uninjured; and if they be of a soft nature, such as the bulk of the coal plants are considered to have been, the damage done them by transport is considerable, as I have had occasion to remark on the coast of Jamaica, where arborescent ferns and other tropical productions are sometimes, though very rarely, carried by floods from the neighbouring mountains into the sea. In the few cases which passed under my observation, the fern-trees were so damaged in the river-courses as to be with difficulty recognisable*.

We have now so many cases in France, Germany, and Great Britain, of the occurrence of some coal plants in a vertical position, with their roots downwards, that such cases can scarcely be considered as accidental, but, in some measure, as characteristic of the deposit in particular situations.

Mr. Witham has brought forward some good examples of vertical stems in the carboniferous rocks of Durham and Newcastle. Two stumps or stems of *Sigillariæ* are described as standing erect, with their roots imbedded in bituminous shale, in the Derwent Mines, near Blanchford, Durham: the space round them was cleared out to obtain the lead ore; and one plant is stated to have been about five feet high, and two feet in diameter. A more curious case was observed by the same author in the Newcastle district, where, in sandstone beneath the High Main coal, numbers of fossil vegetables, chiefly *Sigillariæ*, are discovered erect, their roots imbedded

---

* The height at which arborescent ferns are found, would seem much to depend on local causes. Thus, on the southern side of Jamaica they do not flourish much under an elevation of 2000 feet above the sea; while on the northern side of the same island I have seen them at not more than 400 or 500 feet above the same level. The cause would seem to be the greater moisture of the northern side. It would therefore appear that a considerably moist climate would be necessary for the abundant production of this class of plants in the low situations, such as it has been imagined the lands were which produced the mass of the coal plants.

in a small seam of coal under the sandstone, while they are all truncated on the line of the High Main coal-bed, to the formation of which their higher ends have in all probability partly contributed *.

A still more curious example of vertical stems of plants in coal-measures is noticed by Mr. Wood, as having been observed above the High Main coal, at Killingworth colliery in the same district. There were many of them, and they rose ten feet through various strata of shale and sandstone. That figured beneath (a stem of *Sigillaria pachyderma*), though not the largest, was upon the whole better exposed to view than the others. It was about two feet in diameter in the lower part, and the roots could be traced running into the shale, for about four feet from the stem. The roots of the various stems were interlaced with each other, and the interior of the plants was filled with a sandstone, resembling, not that of the lower beds through which the plant rose, but that of the upper beds. This fact proves that the interior of the plant was sufficiently hollow, when the upper beds were deposited, to permit the infiltration of the sands downwards. About thirty of these stems were visible within an area of fifty square yards. We can scarcely refuse to admit with Mr. Wood that these stems of *Sigillariæ* are exactly in the position in which they grew, the shale being the soil or mud in which they vegetated †.

The following sketch (Fig. 99.) will illustrate better than words the manner in which one of the stems was preserved ‡.

Fig. 99.

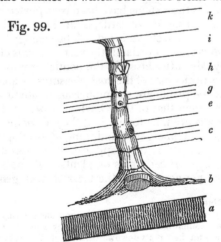

* Witham, Observations on Fossil Vegetables, 1831, p. 7, where there is an illustrative section.

† Wood, Trans. Nat. Hist. Soc. Newcastle, vol. i. p. 205, and pl 19; and Lindley and Hutton's Fossil Flora of Great Britain.

‡ Taken from Lindley and Hutton's Fossil Flora of Great Britain, pl. 54, the beds being enumerated from Mr. Wood's figure.

a, High Main coal; *b,* argillo-bituminous shale; *c,* blue
shale; *e,* compact sandstone; *g,* alternating shales and sandstones; *h,* white sandstone; *i,* micaceous sandstone; *k,* shale.
The beds dip gently to the westward.

Such cases as these, and that long since noticed by M. Ad.
Brongniart at St. Etienne*, where numerous stems are also
included upright in coal sandstone, without however being truncated by a coal- or shale-bed, are sufficient to show the very great
analogy which exists between them, certain submarine forests,
the *dirt-bed* at Portland, and the vertical stems in the Yorkshire oolite, inasmuch as they all apparently point to a quiet
submergence†.

We may have some difficulty in considering the deposition
of sand to have been effected so quietly amid the stumps of
trees as not to have washed away the substances in which they
were imbedded; but we have only to recollect, that among the
submarine forests round our shores, if once any of them were
at such a depth beneath the surface of the sea as to be sufficiently beyond the influence of the waves, they would become
quietly covered by sand; for the velocity of water sufficient to
transport this sand, would scarcely disturb the trees.

It seems impossible to come to any other conclusion, respecting the vertical stems, not only in the coal-measures, but also in
the other rocks above noticed, except that they occur in the relative situations in which they grew, were submerged quietly,
and as quietly entombed in sands, shales, or calcareous matter.
The supposition that these stems have been transported in mass
by water, with their roots downwards, is quite untenable when
we consider the various facts adduced. The analogy between
the trees in the *dirt-bed* of Portland, the vertical stems of Equiseta observed by Mr. Murchison in the oolitic series of Yorkshire, and the vertical stems of the coal-measures, is quite perfect; in all cases the plants are too numerous to admit of any
other explanation than that of their having grown in the beds
in which their roots are now found, and they all point, as before stated, to quiet submergence and tranquil envelopment.

That this explanation is only applicable to cases of large
areas, occupied by vertical stems of plants, will be at once
granted. It does not account for the mass of coal generally,

---

* Annales des Mines, 1821.

† It cannot be denied that, under particular circumstances, stems of trees
preserving a vertical position may be forced onwards by river inundations.
Thus, *snags,* or trees with their roots downwards, and only forced by the current from a vertical position, are common, so as to be very dangerous, in the
Mississippi; and trees were forced down the valley, during the debacle of
the Vallée de Bagnes, and left standing with their roots downwards, at Martigny. These facts admit of easy explanation; for if trees be suddenly detached from the soil, and their roots loaded with stones and other heavy
matter, they would naturally float with the branches upwards.

which has every appearance of having been drifted at unequal intervals. Drifts of vegetable matter now take place into lakes and estuaries; but admitting that much coal may have accumulated in such situations, there are serious difficulties attending this as a general explanation of coal accumulations, more especially where, among the marine remains contained in limestones, alternating with coal strata, we find *corals*. Now corals are generally very abundant in these alternating limestone beds, and it is well known that the creatures which construct the corals of the present day avoid fresh and brackish waters. It therefore follows, either that the coal in these alternations could not have been deposited in fresh or estuary waters, or that the habitats of corals were not then of the same kind as we now find them.

The occurrence of limestone strata, continuous over considerable areas, and alternating with the shales and sandstones, would seem to require comparative tranquillity for their production, more particularly as the marine shells entombed in them have evidently not been subjected to violence, but appear to have been imbedded at no great distances from the places where they lived and died.

The vegetable remains are often of considerable size. M. Brongniart observes that in the coal strata of Dortmund, Essen, and Bochum, stems are found in the planes of the strata, more than fifty or sixty feet long, and that they may be traced in some of the galleries for more than forty feet without observing their natural extremities [*]. Vegetables of large size have also been detected in Great Britain. Mr. Witham mentions one in Craigleith quarry as being forty-seven feet in length, from the highest part discovered to the root. The bark is described as converted into coal [†]. *Lepidodendra* have been discovered in the northern parts of England, from twenty to forty-five feet long, and four feet and a half in diameter [‡].

Respecting the general character of the vegetation of this period, such as we find it entombed in the carboniferous rocks of the northern hemisphere, M. Ad. Brongniart observes, that it is remarkable; 1. for the considerable proportion of the vascular cryptogamic plants, such as the *Equisetaceæ, Filices, Marsileaceæ,* and *Lycopodiaceæ*; 2. for the great development of the vegetables of this class, so that they have attained a magnitude far beyond those of the same class now existing; thus proving that circumstances were particularly favourable to their production during the period under consideration. It

---

[*] Brongniart, Tableau des Terrains qui composent l'Ecorce du Globe.

[†] Witham, Edinburgh Journal of Natural and Geographical Science. April, 1831.

[‡] Lindley and Hutton, Fossil Flora, vol. i. p. 17.

has been ascertained within the last few years, that dicotyledonous plants, which were once considered to be exceedingly
rare, if not altogether absent, in the coal-measures, exist in
great abundance in the same rocks. Coniferæ are sufficiently
common. It would appear that authors are by no means agreed
as to what families certain genera of fossil plants should be referred. Thus M. Adolphe Brongniart refers the genus *Stigmaria*, very common in the coal-measures, to the family of *Lycopodiaceæ*; while Prof. Lindley and Mr. Hutton consider that,
if an existing analogy must be found, it is with greater probability of accuracy referrible to *Euphorbiaceæ* or *Cacteæ*, most
probably to the former;—a difference of considerable importance, as upon it depends whether the genus in question belongs to the class *Cellulares*, or to the class *Vasculares*. No
doubt this difference of opinion arises from the obscurity of the
subject, fossil botany being beset with very great difficulties;
difficulties far beyond those which attend the study of fossil
zoology, though the latter are by no means either small or rare.
The possibility of this variety of opinion among distinguished
botanists, should teach geologists caution, and prevent them
from indulging in those hasty generalizations, which, though
often brilliant, too frequently impede the progress of the science they cultivate, not only by rendering authors satisfied
with rapid conclusions, but also by throwing doubt, when such
conclusions are found to be erroneous, upon others which are
firmly based, and which may be considered as exact as those
of any other science.

As, in the opinion of botanists, islands in warm countries
are favourable to the growth of Ferns and other plants of the
same natural class, not only from the presence of the necessary
heat, but from the moisture so congenial to them, it has been
considered by MM. Sternberg, Boué, and Ad. Brongniart,
that the vegetation of this period, such as we find it in the carboniferous deposits of Europe and North America, was the
growth of islands scattered in archipelagos.

When we come narrowly to look into the structure of the
coal-measures, the vast accumulations of shale and sandstones,
sometimes amounting to the depth of 460 feet (Forest of Dean),
do not precisely accord with mere oscillations of islands above
and beneath the level of the sea, which might in some cases
appear probable; for these accumulations of detritus require
considerable drift, and must have resulted from the destruction
of pre-existing rocks, mostly siliceous, and therefore, if solid,
requiring much time for their degradation, with the assistance
of other forces than the mere battering of the surf on clusters
of low islands, perhaps defended, like those in the Pacific, by
coral reefs.

The presence of larger masses of land, with mountains, rivers, and other physical features necessary for the production of a larger amount of detritus, would seem requisite, independent of volcanic eruptions, and other exertions of internal force, for the accumulations we observe. The oscillation of low islands is mentioned merely as the possible explanation of some of the observed phænomena, and the student must be careful to consider it only in that light. While on this subject, however, it may be as well to notice a possible explanation of some of the minor alternations of limestones with marine remains, with shales and coal containing terrestrial remains such as are found in the millstone grit; because such hints, without attributing any particular value to them, very frequently lead to further inquiry. Suppose a tract of low land covered with a dense vegetation, such as is found in the tropics, to be, by a movement in the earth,—an earthquake, for instance,—submerged a few feet beneath the sea; marine animals would establish themselves on the submerged surface, which would become in the condition of the submarine forests previously noticed; and the consequence would probably be, that not only millions of testaceous creatures would leave their exuviæ, but that the corals would also swarm, and might eventually produce coral islands, upon which vegetation might again establish itself, to be again submerged. That coral islands are sometimes raised above the sea, is what we should expect; and evidence of it has been adduced by Captain Beechey, who describes Henderson's Island (in the Pacific) as apparently upheaved by one effort of nature to the height of eighty feet. It is composed of dead coral, bounded by perpendicular cliffs, which are nearly encompassed by a reef of living coral, so that the cliffs are beyond the reach of the spray*. Now depression to this amount might as easily have taken place; in which case the vegetation of the island would have been submerged eighty feet, and the amount of destruction it would suffer would depend on the greater or less suddenness of the movement. Such movements cannot be considered great when regarded, as they always should be, with reference to the mass of the world; for we have proof that far greater have occurred, and the differences which have been produced in the relative levels of land and water are, when viewed on the great scale, of very trifling importance.

According to M. Ad. Brongniart, if we look at the mass of the coal plants, we must consider the vegetation of the carbo-

---

* Beechey, Voyage to the Pacific Ocean and Behring's Straits, p. 194. Descriptions of other coral islands, with sections of their general structure, will be found, pp. 160 and 186 of the same work.

niferous group to have been produced in climates at least as
warm as those of the tropics; and as we now find plants of the
same class increase in size as we advance towards warm latitudes, and as the coal-measure plants exceed the general size
of their existing congeners, he concludes, with much apparent
probability, that the climates in which the coal plants existed
were even warmer than those of our equinoctial regions.

This view leads us to another consideration. There certainly was a similar vegetation about the same period (for whether the American coal-measures may be, like those in parts of
Europe, somewhat older, does not alter the question,) over
parts of Europe and North America : we may therefore infer
a similar climate over a large portion of the northern hemisphere, such as we have not at present, for it was at least tropical, and very probably ultra-tropical. The question naturally arises, Is there any evidence to show that the same temperature existed at the same period in the southern hemisphere?
for if there is, there must have been some common cause to
produce such an equality of climate, at present unknown to us.
Unfortunately, the actual state of our knowledge will not permit an answer to this question; but by it we learn the importance of ascertaining the botanical character of the various
rocks in the southern hemisphere, more particularly those of
the earliest formation, such as may be considered the equivalents of the carboniferous and grauwacke groups of the North.

With respect to the testaceous remains, the limestones contain a great abundance, not only of species, but of individuals
of the genera *Spirifer* and *Producta*. Of the form of these
shells, and of the *Cardium hibernicum* and *C. alæforme*, (the
latter by no means a rare fossil in the limestones of the next
group,) the following figures will afford examples.

Fig. 100.   101.   102.

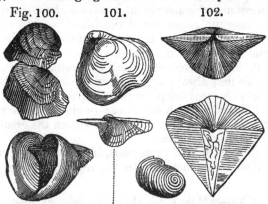

Fig. 105.   Fig. 106.   Fig. 104.   Fig. 103.

Fig. 100. *Producta Martini;* Fig. 101. *Spirifer glaber;* Fig. 102. *Spirifer attenuatus;* Fig. 103. *Spirifer cuspidatus;* Fig. 104. one of the two spiral appendages contained in *Spirifer trigonalis* *; Fig. 105. *Cardium hibernicum;* and Fig. 106. *Cardium alæforme.*

Of the vertebrated creatures which may have existed at this period our knowledge is very limited; but it may be observed that the Tritores (or palates of fish) still retain phosphate of lime; for Dr. Turner ascertained that a palate from the carboniferous limestone of Bristol, contained 24·4 per cent. of phosphate of lime, the remainder being carbonate of lime and bituminous matter, the latter abundant. A palate from the chalk, examined for the purpose of comparison, was found to contain 18·8 per cent. of phosphate of lime, the remainder being carbonate of lime, with traces of bituminous matter.

* Their position in the shell will be seen by reference to Sowerby's Mineral Conchology, pl. 265, fig. 1.

SECTION IX.

GRAUWACKE GROUP.

Syn.—Grauwacke (*Traumate*, Daubuisson). Grauwacke Slate (*Grauwacke schistoide*, Fr.; *Schiste Traumatique*, Daubuisson; *Grauwackenschiefer*, Germ.). Grauwacke Limestone (*Transition Limestone*, Engl. Authors; *Calcaire de Transition, Calcaire Intermédiaire*, Fr. Authors; *Uebergangskalkstein*, Germ. Authors).

It has been observed that the old red sandstone of some countries graduates into grauwacke; whence it may be inferred that the causes, whatever they may have been, which produced the latter deposit, were not violently interrupted in such situations, but that they were gradually modified,—if indeed it be necessary to consider the old red sandstone in any other light, taken generally, than the upper portion of the grauwacke series. That it is so, is the opinion of most continental geologists; and where the one graduates into the other, such an opinion seems well founded. Variations in the classification of the old red sandstone would appear solely to arise from its mode of occurrence in the particular countries where geologists have been accustomed to observe it. When accidents have happened to the grauwacke, throwing the strata on their edges, and a red sandstone or conglomerate deposit intervenes between the carboniferous limestone and the upturned beds, classifications made in the countries where such phænomena prevail, would naturally be framed so as to separate the (old) red sandstone from the grauwacke: but when such accidents have not happened, and the carboniferous limestone, the intervening red sandstone, and the grauwacke are so circumstanced that the two former rest conformably on the latter, and they all graduate into one another, it is altogether as natural that the old red sandstone should be pronounced the upper part of the grauwacke series. Nor should we be surprised that the carboniferous limestone should also be included in the group; for the general organic character of the whole is similar, and does not differ more, if so much, as the upper part of the oolitic group from the lower portion of the same deposit, or as the chalk from the green sand.

Viewed on the large scale, the grauwacke series consists of a large stratified mass of arenaceous and slaty rocks intermingled with patches of limestone, which are often continuous for considerable distances. The arenaceous and slate beds, considered generally, bear evident marks of mechanical origin, but that of the included limestones may be more questionable. The arenaceous rocks occur both in thick and schistose beds; the latter state being frequently owing to the presence of mica disposed in the lines of the laminæ. Their mineralogical character varies materially; and while they sometimes, though rarely, pass into a conglomerate, they very frequently graduate into slates, which become of so fine a texture as to lose the arenaceous character altogether. Roofing-slate is far from rare among the grauwacke rocks; and if we consider it of mechanical origin, like the mass of the strata among which it is included, we must suppose it to have originated from the deposition of a highly comminuted detritus.

If the size of transported substances be considered as the necessary evidence of rapid currents of water, the grauwacke rocks, taken as a mass, have been slowly deposited; for though evidences of cross currents are sufficiently abundant in the various directions of the laminæ, and in the mode in which arenaceous and slaty beds are associated with each other, the substances are generally fine-grained, rarely passing into conglomerates. A rapid current would however not appear to require large pebbles as the necessary evidences of its existence at any given period, although when large pebbles are present we infer that small currents of water could not have transported them; for the size of the substances transported by a current moving with considerable velocity will greatly depend on the surface over which it passes and the nature of the substances carried onwards. Perhaps from the absence of organic remains in a large proportion of the arenaceous part of this deposit, and their abundance in some of the included limestones, we might infer that there had been something in the transport and deposit of the sands unfavourable to their preservation, such as trituration in waters moving with rapidity. There is, however, a general appearance in the mass of the grauwacke which would lead us rather to consider a great portion of it of slow deposition.

It is by no means an uncommon circumstance for the laminæ of the slates of this group to be so arranged as to form various angles with other lines, which may be considered as those of the beds, or of stratification. Of this structure the annexed section of grauwacke slates at Bovey Sand Bay on the east side of Plymouth Sound, affords us an instructive example.

a a, curved beds of slate, the laminæ of which meet the apparent lines of stratification at various angles, being even perpendicular to them. The beds are cut off by the fault (*f*) from the slates *c*, the laminæ of which are more confusedly disposed, having however a general horizontal arrangement. The whole

is covered by a detritus (*b b*) composed of fragments of the
same kind of slate as that on which it reposes, and of the various grauwacke rocks of the hill behind.

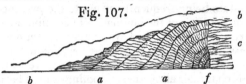

Fig. 107.

It is often interesting to remark, in countries where vertical
or highly inclined strata render such observations clear, the
various mineralogical changes which frequently take place in
minor divisions of the grauwacke group.   Thus, for instance,
in some parts of Devonshire, where quartz rocks are associated
with the grauwacke, there are often opportunities of observing
various changes of this kind within the distances of ten or
twenty miles.   At first the series of beds may consist of fine-
grained argillaceous slates.   After a range of two or three miles
the continuation of these beds will become more arenaceous.
The arenaceous structure of the rocks will then gradually increase, until they assume the character of compact sandstones.
Finally the arenaceous appearance is lost by a kind of union
of the grains with each other, and the rocks consist of beds of
quartz, in which traces of a mechanical origin can seldom or
rarely be observed.   If now we continue to follow up the same
system of beds, we shall find that, after a course of some miles
of quartz rock, the latter character gradually disappears, the
strata at first becoming arenaceous, and afterwards passing into
schistose rocks, in which the grain is more or less fine.   It not
unfrequently happens that the quartz rock suffers its first change
by the acquisition of mica, which renders it to a certain extent
schistose, though at first it only causes it to assume the character of avanturine.   In such situations mineralogical mica-
slate, that is, a slaty rock solely composed of mica and quartz,
is by no means uncommon, and is merely the quartz rock rendered highly schistose by an abundance of mica.   Of course
the change is not always sufficiently great to produce decided
quartz rocks.   It frequently arrives only at that state on which
the arenaceous rock becomes exceedingly hard, but in which its
mechanical origin and sandstone character are quite apparent.
Changes of minor importance are innumerable ; yet it not unfrequently happens that these minor divisions retain their
mineralogical characters unimpaired, or with only slight modifications, for considerable distances : thus showing that certain minor causes have varied considerably during the production of the grauwacke, and that, though the mass of this group,
viewed on the large scale, presents considerable uniformity of

structure, there is abundant evidence of much variation and change of mineralogical character on the small scale.

The mode in which the calcareous beds occur in the group under consideration is also exceedingly variable. They are at times little else than the finer grained strata, which may contain calcareous matter, such calcareous matter varying much in its proportion to the other substances of which the beds are composed. Hence we have every change from complete limestone to beds in which the calcareous matter is sparingly disseminated*. This calcareous character, though liable to be lost at unequal intervals, can often be traced for considerable distances in the general direction of the strata. Whence it may be inferred, that, during the deposit of the grauwacke, calcareous matter was often disseminated over large areas, sometimes being merely mixed with the more common substances of the deposit, while at others it was sufficiently abundant to constitute bands of limestone. Though this mode of occurrence is far from uncommon, the more sudden appearance and disappearance of carbonate of lime is by no means rare. Limestone beds will often become abundant, and constitute an important part of the series, without any variation in the mineralogical character of the grauwacke in their line of bearing, which would lead us to expect so sudden a development of calcareous matter. Examples of this fact can readily be observed in several parts of Southern Devon.

The origin of the limestones is of far more difficult explanation than the sandstones and slates in which they are included. We cannot well seek it in the destruction of pre-existing calcareous rocks; for as far as our knowledge extends, such rocks are of comparative rarity among the older strata. In fact, the quantity of calcareous matter present in the grauwacke group greatly exceeds that discovered in the older rocks; and the same remark applies to many of the newer deposits when considered with reference to the grauwacke series. If we take the mass of deposits up to the chalk inclusive, we shall find that, instead of a decrease of carbonate of lime, such as we should expect if that contained in each deposit originated solely from the destruction of pre-existing limestones, the calcareous matter is more abundant in the upper than in the lower parts of the mass; and we may hence conclude that this explanation is insufficient.

If, as has often been done with other limestones, we attribute the origin of the grauwacke limestones in a great measure

---

* It may not perhaps be generally known that many of these beds, which are little else than argillaceous limestones, with the addition of a small proportion of silica, are highly useful as water-setting limestones, being scarcely, if at all, inferior to the lias limestones, so much valued on this account.

to the exuviæ of testaceous animals and polypifers, we must grant the animals carbonate of lime with which to construct their shells and solid habitations. This they may have obtained either in their food or from the medium in which they existed. The marine vegetables are not likely to have supplied them with a greater abundance of carbonate of lime at that time than at present. Those that were carnivorous might acquire much carbonate of lime by devouring other animals more or less possessed of this substance : but the difficulty is by no means lessened by this explanation; for the creatures devoured must have procured the lime somewhere. It would appear that we should look to the medium in which testaceous animals and polypifers existed, for the greater proportion, if not all, of the carbonate of lime with which they constructed their shells and habitations. Now if we consider the mass of limestone rocks to have originated from the exuviæ of marine animals, we are called upon to consider that carbonate of lime was once far more abundant in the sea than we now find it, and that it has been gradually deprived of it. This supposition would lead us to expect, that as the sea was gradually deprived of its carbonate of lime, limestone deposits would become less and less abundant; and consequently, that calcareous rocks would be most common, when circumstances were most favourable, that is to say, during the formation of the older rocks. This, however, is precisely the reverse of what has happened. Hence we may infer that the origin of the mass of limestone deposits must be sought otherwise than in the attrition or solution of older and stratified rocks, or from the exuviæ of marine animals deriving their solid parts from a sea which has gradually been deprived of nearly all its carbonate of lime. Both these causes may have eventually produced important modifications on the surface of the earth ; but the great proportion of lime necessary for the formation of the calcareous masses covering a considerable part of it, would appear to have been otherwise obtained.

It has been usual to consider the lime of calcareous deposits as derived from limestone rocks, through which waters charged with carbonic acid percolated, the carbonic acid dissolving a certain portion of the lime, which is thus held in solution by the water until it reaches the surface, where it is thrown down in the shape of limestone. This explanation may suffice for the small deposits we observe in calcareous countries, but is insufficient for the productions of limestones generally; for it assumes that the solution of a small quantity of lime obtained from older rocks is, as previously noticed, capable of producing an immense deposit of the same substance. We know that carbonic acid is now discharged into the atmosphere from the

earth by means of volcanos, fissures, and springs, and we have
no reason to doubt that this has been the case during a long
succession of ages; indeed we have every reason to believe
that such discharge of carbonic acid formed a part of the great
economy of nature, for without this aid we should have much
difficulty in explaining the abundance of carbon and carbonic
acid now locked up in coal deposits and limestones, all of which
have clearly been produced successively on the earth's surface.
The reason why extensive tracts of carbonate of lime have
been produced at one time more than at another is not quite
so apparent; but it may be observed, as a mere conjecture,
that as this substance is not very unfrequent in volcanic regions, great disruptions of strata may have produced circumstances favourable for its deposition, and that without disturbances, carbonate of lime may have been thrown upwards in
water, through fissures, more abundantly at one time than at
another, from causes unknown to us. It is worthy of attention
that when the limestones occur, then also do the organic remains generally become more abundant, appearing as if the
calcareous rocks and the organic remains were connected with
each other. That the animals, by secreting carbonate of lime
from the medium in which they lived, sometimes contributed
considerably to the mass, we are certain, as their remains now
constitute a large portion of it; but that they were the means
through which all the carbonate of lime was derived from the
waters, may very justly be doubted, more particularly as in
certain districts not a trace of animal exuviæ can be detected
in such limestones. If carbonate of lime were present in some
situations and not in others, animals, such as the *Crinoidea*,
Testacea, and *Polyparia*, would naturally flourish more in the
former than in the latter, as they could there more readily obtain the lime necessary for them, and we should consequently
expect to find their remains more common there than elsewhere.
In limestones devoid of organic remains we appear to have
evidence of carbonate of lime being abundant in such situations
unconnected with animal life, and we may consider it derived
from the interior and dispersed through the waters over a given
space, where it has been gradually deposited. When, however,
the remains of shells and corals are present, and nearly constitute the mass of the rock, other causes may have produced
the effects required, precisely as coral reefs and accumulations
of shells now occur in one place and not in another, either in
consequence of shelter, proximity to the surface of the sea, or
other favourable circumstances.

Be the general origin of the grauwacke limestones what it
may, the causes which produced them were destined to cease
during the deposit of the grauwacke itself, and a series of sand-

stones and slates, similar for the most part to those beneath, were accumulated upon them. In some districts, such as the North of Devon, there has been a return of causes favourable to the deposit of limestone, and two bands parallel to each other have been produced.

In other districts more limestones have been formed, while in some they are nearly absent; a state of things we should expect from variations produced by local circumstances on similar general causes in operation over a considerable area.

The grauwacke sometimes assumes a red colour in the midst of beds of the usual gray and brown tints (South of Devon, Pembrokeshire, Normandy, the grauwacke district of the Moselle, &c.), and is then undistinguishable from the old red sandstone of English geologists*. This red colour seems, for the most part, due to little else than the highly oxidized state of the iron disseminated through the rock. In the red grauwacke the iron is in the state of a peroxide, while in the gray or brown grauwacke it is combined with a less proportion of oxygen. This change of colour may sometimes be seen in the vicinity of trap-rocks, which have been protruded through the grauwacke. In many cases we may suppose the iron contained in the grauwacke to have acquired an additional quantity of oxygen, converting it into a peroxide, after the formation of the rock. We may indeed consider the red tint as sometimes caused by heat, from observing that common brown grauwacke, when kept for a long time at a heat insufficient to fuse it, will become red, the iron having acquired additional oxygen from the atmosphere. It will, however, be obvious that the rock when heated beneath water would not be precisely under the same conditions; and we must be careful not to attribute the occasional red colour of the strata under consideration solely to this or a similar cause, for it will be clear that the disseminated iron may easily have been in a state of peroxide from other causes before such red grauwacke was deposited. Indeed that this has been the case is often sufficiently evident; and if the old red sandstone be included in this group, there can be little doubt that the iron, during the deposit of the upper part of the series, was most commonly in that state.

Beds and even accumulations of strata are sometimes mingled with the common grauwacke and grauwacke slate, which

---

* This circumstance renders the determination of those limestones of Southern Devonshire which are much broken by faults, greatly disturbed and contorted, or much concealed by superincumbent (new) red sandstones, exceedingly difficult. The difficulty is particularly felt in the vicinity of Tor Quay, where, however, the limestones of the southern side of Tor Bay would certainly appear to be included in the grauwacke series, as is shown by coast sections, and their prolongation to the Dart.

at least show a variation in the mode of deposit. Thus, flinty slate, sometimes associated in this series (Devonshire, &c.), is exceedingly compact, and, as its name implies, is principally composed of silica, the rock having much the appearance of a deposit from water in which silica was chemically dissolved *.

Associated with the grauwacke, more particularly in its older portions, we often find beds which, in mineralogical composition, precisely resemble certain greenstones, corneans, &c., known from their mode of occurrence to be of igneous origin. These associated beds do not cut the other strata; on the contrary they are clearly included in them, and have every appearance of being interstratified. After a course of, perhaps, a few miles, they are seen to terminate on either side, first becoming slaty; at least this frequently takes place. When this slaty condition arises, the rock is then frequently undistinguishable from hornblende slates. The most satisfactory mode of explaining these facts, seems that proposed by Prof. Sedgwick for the association of similar rocks with the slates of Cumberland. He considers that they were igneous rocks ejected during the time that such slates were depositing. This hypothesis is applicable to certain of these rocks associated with the grauwacke of Southern Devon. But care must be taken to distinguish them from other greenstones and porphyries of the same country, which occur in dykes and masses, and which have clearly been ejected at a much more recent epoch. It is also very necessary to distinguish them from the altered rocks of the same part of England, as these last are more particularly deceptive.

Another difficulty, and one by no means easy to surmount, attends the examination of such included beds, or apparent beds, of greenstone and porphyry. If a mass of schistose or stratified rock, such as grauwacke, be exposed to the action of a disruptive force, such mass would rend in the parts of least resistance. Now the lines of stratification would necessarily be those of least resistance, and hence the mass would be most likely to part in those lines, permitting the injection of igneous rocks, should these endeavour to escape through the fissures of the grauwacke. We should thus have tabular masses of greenstone, porphyries, and other rocks of that character presenting every appearance of included beds. Instructive examples of such deceptive tabular masses are well seen in Pembrokeshire and Devonshire, as they can frequently be traced to larger masses of similar rocks, which

* The reader will recollect that under the head of Deposits from Springs, siliceous beds were noticed as having been produced by deposition from thermal waters in Iceland and the Azores.

have evidently been protruded through the grauwacke after its consolidation.

Towards the lower part of the grauwacke group, its fossiliferous character disappears, and the presence of crystalline rocks, apparently of contemporaneous origin, becomes more common. This change varies much in different countries, but in general the slaty rocks gradually prevail, presenting a very great thickness of argillaceous schists. We seem to have arrived, in the descending order, at a state of the world when there was a combination of those causes which have produced fossiliferous and non-fossiliferous rocks. That there should be a transition or passage, even effected by the alternate operation of particular causes, from that condition of the world's surface when chemical action prevailed to that when mechanical action became more abundant, is what we should expect, since it is in accordance with our knowledge of rock deposits generally; for we observe, however sudden certain changes may have been produced in particular situations, that viewed on the large scale, a general change of circumstances attending rock formations has been more or less gradual.

Chlorite slates are not unfrequently associated with the lowest portions of the grauwacke series. I have often observed this to commence by small alternations of chloritic argillaceous slate with common argillaceous slate, the former gradually becoming more charged with chlorite. There are often also very ambiguous rocks, to which it is exceedingly difficult to assign names, as they are constantly varying in their mineralogical composition. Their general character is, however, far more chemical than mechanical.

The study of this part of our subject must always be attended with great difficulty; for, independently of the mixture of chlorite, talcose, and other slates with the lowest fossiliferous deposits, we have to contend with the presence of igneous rocks injected among these deposits in their line of stratification, producing, as before observed, the most deceptive appearances. No small difficulties are also caused by the alteration of strata, arising from the protrusion of granite and other igneous rocks among them, such products often causing, when the masses are large, very remarkable changes, the various grauwacke beds assuming the appearance of a great variety of older rocks. Mica-slates, gneiss, and hornblende rocks are often thus re-produced, but the changes so effected are limited in extent, and by careful examination can be easily traced. This reproduction of gneiss, mica-slate, and other rocks of the same kind,—to account for which nothing can be more simple, as I shall have occasion to show elsewhere,— has given rise, I am informed, to the hypothesis, that all the crystalline and non-fossiliferous rocks are but fossiliferous rocks out of which the organic remains have been driven by heat. This may certainly be a convenient hypothesis as far as regards a particular theory, but can scarcely be seriously entertained by those who have examined any of those vast tracts of the true inferior or non-fossiliferous rocks, so common in various parts of the world, more especially with respect to the mode in which the various mineral masses occur.

MM. Brongniart and Omalius d'Halloy long since pointed out the apparent alternation of the granitic and schistose rocks of the Cotentin and Brittany, as also that the deposits thus associated with the granitic compounds were fossiliferous *. The grauwacke of these districts certainly appears associated, more particularly in its lower parts, with rocks, the mechanical origin of which is far from evident; but while studying them we must be on our guard against granite veins, and other intrusions of the same rock, which are also observable in that country. Independently, however, of decidedly intruded rocks, there are associated crystalline rocks which render it exceedingly hazardous to affirm where the series, in which confusedly crystalline compounds prevail, may commence, or where the mechanical and fossiliferous deposits may terminate. The highly indurated sandstones also, which are clearly, like those of South Devon, associated with the fossiliferous rocks of Normandy, so pass into quartz rock, that, as M. Brongniart has observed, they often present the appearance of having been produced by confused crystallization.

The following is a summary of the various organic remains stated to have been detected in the grauwacke series; it is necessarily one which does not pretend to more than an approximation to the truth, for the catalogue on which it is founded will no doubt receive both important additions and corrections; it may, however, be found useful as affording a general view of the subject.

Plantæ.—Fucoides, 2 species. Calamites, 2. Sphenopteris, 1. Cyclopteris, 1. Pecopteris, 1. Sigillaria, 2. Lepidodendron, 1. Stigmaria, 1. Asterophyllites, 1.

Zoophyta.—Manon, 2. Scyphia, 5. Tragos, 2. Gorgonia, 2. Stromatopora, 2. Madrepora, 1. Millepora? 1. Cellepora, 2. Retepora, 3. Flustra, 1. Ceriopora, 6. Glauconome, 1. Agaricia, 1. Lithodendron, 3. Caryophyllia, 1. Fungites, 2. Anthophyllum, 1. Turbinolia, 3. Cyathophyllum, 20. Strombodes, 1. Astrea, 2. Columnaria, 1. Sarcinula, 3. Coscinopora, 1. Catenipora, 3.

Syringopora, 4. Calamopora, 9. Aulopora, 5. Favosites,
6. Mastrema, 1. Amplexus, 1. Pleurodyctium, 1. Cyclolites, 1.

Radiaria.—Apiocrinites? 2. Pentremites, 1. Pentacrinites, 1. Actinocrinites, 7. Cyathocrinites, 4. Platycrinites, 4. Rhodocrinites, 5. Melocrinites, 2. Cupressocrinites, 3. Eugeniacrinites, 1. Eucalyptocrinites, 1. Sphæronites, 4.

Annulata.—Serpula, 5.

Conchifera.—Thecidea? 1. Pentamerus, 2. Gypidia,
(Pentamerus?) 3. Spirifer, or Delthyris, 41. Terebratula,
31. Strygocephalus, 3. Calceola, 1. Atrypa, 14. Producta, or Leptæna, 22. Orbicula, 1. Crania, 1. Gryphæa,
1. Pecten, 5. Plagiostoma, 1. Inoceramus, 1. Avicula,
1. Pterinea, 9. Posodonia, 1. Arca, 1. Nucula, 5. Trigonia, 2. Megalodon, 1. Modiola, 3. Mytilus, 1. Crassatella, 1. Cardium, 9. Cardita, 4. Isocardia, 2. Venericardium, 1. Lucina, 3. Cyprina, 1. Corbula, 1. Cythere, 8. Sanguinolaria, 8. Pholadomya, 1.

Mollusca.—Patella, 5. Pileopsis, 3. Melania, 1. Natica, 1. Nerita, 2. Delphinula, 5. Cirrus, 1. Pleurotomaria, 1. Euomphalus, 16. Trochus, 6. Rotella, 1. Turbo,
2. Turritella, 7. Pleurotoma, 1. Murex? 1. Buccinum,
5. Phasianella, 3. Bellerophon, 9. Conularia, 3. Orthoceratites, 30. Cyrtoceratites, 6. Spirula, 7. Lituites, 2.
Nautilus, 9. Ammonites, 15. Aptychus, 2.

Crustacea.—Calymene, 17. Asaphus, 21. Ogygia, 4.
Paradoxides, 9. Nileus, 2. Illænus, 3. Ampyx, 1.
Agnostus, 1. Isotelus, 2.

Pisces.—At least 1 genus, and 2 or 3 species.

Thus making: *Plantæ*, 9 genera, 12 species. *Zoophyta*,
33 genera, 98 species. *Radiaria*, 12 genera, 35 species.
Annulata, 1 genus, 5 species. *Conchifera*, 35 genera, 191
species. *Mollusca*, 26 genera, 144 species. *Crustacea*, 9
genera, 60 species. *Pisces*, 1 genus, 2 species.—Total, 126
genera, 547 species.

From the above it would appear that the grauwacke series
contains a mixture of genera inhabiting the seas and oceans of
the present day, and of others which are not now known. It
may be doubtful how far all the genera have been correctly
determined; for possibly some of them may have been rather
hastily referred to those now existing, while others may have
been considered extinct without sufficient evidence. But, admitting these sources of error, some genera are certainly, as
far as our actual knowledge extends, extinct, while others do
not differ from those now existing. This catalogue also shows

that, at the early epoch when the grauwacke was produced, there was not that poverty of organic structure which was once supposed.

From the various forms of the fossils imbedded in the grauwacke, we may infer that the animals, of which they constituted the solid parts, occupied situations as different as those of the present day; some preferring deep waters, while others were fitted for shallow seas, and not a few swam freely in the open ocean; certain creatures frequenting one kind of bottom, while others sought another of a different description. The most abundant shells belong to the genera *Orthoceratites*, *Producta*, *Spirifer*, and *Terebratula*. The former often attain a large size, even reaching a yard or more in length; so that if they really once constituted a part of swimming mollusca, analogous to the Nautilus of the present day, some of such creatures must have far exceeded the size of the animals of that kind now known to us. The three latter most abundant genera constitute a natural group, which the Swedish naturalists have arranged under the heads of *Leptæna* (*Producta*), *Orthis*, *Cyrtia*, *Delthyris* (*Spirifer*), *Gypidia*, *Atrypa*, *Rhynchora*, and *Terebratula*; the characters being considered such as to justify the formation of the genera. Supposing this arrangement to be well founded, it would appear from the lists of those who have proposed it, that *Terebratulæ* are rare in the older rocks, such as those under consideration, while they are abundant in the newer strata.

Productæ are, as has been seen, common in this and the carboniferous groups, and existed during the deposit of the zechstein. Spirifers, which also abounded during the deposit of the grauwacke and carboniferous series, have been observed as high up as the lias, where three species of the genus Spirifer have been detected, one (*Spirifer Walcotii*) being a very common and characteristic shell. The Terebratulæ, which, even admitting the Swedish divisions, are found in the preceding series, if not in the higher part of this, extend upwards to the present day, many species being now known. Taking, therefore, this natural group as it existed at this early period, in which we should probably include the carboniferous limestone, and tracing it upwards through the various rocks, we find that the Productæ first disappeared, and then the Spirifers, while the Terebratulæ have been preserved through all the changes which have taken place on the surface of our planet.

The family of the Trilobites was one, the individuals of which must have swarmed in particular places during the deposit of the grauwacke. In some parts of Wales the

Asaphus Debuchii (Fig. 108.) is so abundant that the laminæ of the slates are
charged with them, so that millions have
probably lived and died not far distant
from those places where we now discover
their remains. This species has not been
confined to Wales, though it is there very
abundant, but has also been discovered
in Norway and Germany. The Trilobite long known in museums as the Dudley Trilobite, because found so commonly
at that place, is the *Calymene Blumenbachii* of M. Al. Brongniart (Fig. 109.).

Fig. 108.

This species existed over a considerable area, having not only
been discovered in England, Germany, and Sweden, but also
in North America.   Although many parts
of these creatures are found distributed in
such a manner that we may conclude they
were separated by decomposition after the
death of the animal, the perfect preservation
of others, and their frequent contracted at-
titudes, such as we should expect creatures
of this structure to assume when disturbed,
would lead us to conjecture that they had
been often suddenly destroyed, and as suddenly enveloped in that matter which subsequently became hard rock ; thus preventing
the separation of the harder parts by decom-

Fig. 109.

position.   The forms of the Trilobite family vary more considerably than might be supposed from the *Asaphus* and *Calymene* represented above, as will be seen by the annexed figure
of *Agnostus pisiformis*, Fig. 110. being the natural
size of the animal, and Fig. 111. a magnified representation of it.   The Trilobite family seem
now to have entirely disappeared from among
existing animals; and we may perhaps venture to
infer, from our present information respecting organic remains, that it became extinct before the
Productæ; and we are nearly certain it ceased to
exist long before the Spirifers, for neither in the
muschelkalk nor in the lias has the smallest trace
of them ever been detected.

Fig. 110.

Fig. 111.

Unlike the Trilobites, the Crinoidea common in this early
period are continued up to the present day, though many genera observed in the grauwacke series and in the carboniferous
group seem to have disappeared previous to the deposit of the
oolitic series, when other genera were called into existence.

The genus *Pentacrinites* being, according to M. Goldfuss, found in the rocks under consideration, and being well known in the present seas, this genus has also survived the various changes that have taken place on the earth's surface.

The discovery of the defensive fin bones, named Ichthyodorulites, in the grauwacke series is worthy of attention, as it shows that the class of animals to which they belong was among the earliest inhabitants of the globe, and that it continued to exist over what now constitutes Europe, up to the cretaceous rocks inclusive, though differing in species,. as far at least as we can judge from the various forms of the bones. The Ichthyodorulites are usually accompanied by palates; these latter have not yet been detected in the grauwacke.

Among the corals will be found several genera now existing; and it deserves notice, that throughout the series of fossiliferous rocks, wherever there is an accumulation of polypifers, such as would justify the supposition of coral banks or reefs, the genera Astrea and Caryophyllia are present,—genera which, according to the more recent observations of naturalists, in addition to Meandrina and one or two others, are the principal architects of coral reefs at the present day.

Our knowledge of the kind of vegetation existing at, and entombed during, the epoch of the grauwacke group, is insufficient to warrant any general conclusions respecting it, further than it was probably much the same as that, the remains of which are abundantly preserved in the carboniferous series. Anthracite has been long known in the grauwacke of North Devon, and may have been derived from the remains of vegetables. Where there are vegetables entombed in rocks we may expect to find accumulations of them, and there seems no good reason why grauwacke should not contain its coal-beds as well as other great deposits. It will have been observed that all the fossiliferous groups of rocks have their accumulations of vegetable matter in some part or other of the areas respectively occupied by them. In Europe these accumulations have been more abundant in our carboniferous group than at any other time, more especially in the mass of sandstones and shales thence named the Coal-measures; but it by no means follows that this should have been the case as regards the whole surface of the world. On the contrary, all analogy with other rock deposits would lead us to infer that the coal-measures would have their equivalent marine deposits, in which, if terrestrial plants occurred at all, they would be found merely scattered here and there, as they are in other rocks abounding in marine remains, and hence termed Marine deposits.

M. Élie de Beaumont observes that the grauwacke rocks

of the Bocage (Calvados), and of the south-eastern angle of the Vosges, contain vegetable impressions differing but little from those discovered in the coal-measures, as also anthracite, sometimes worked for profitable purposes *. According to M. Voltz, certain anthracitic rocks of Baden are of this age; and M. Virlet refers the coal of St. Georges Châtelaison to the grauwacke series †. Mr. Weaver considers that all the coal in the province of Munster, excepting that in the county of Clare, is of this age. He states that thin beds of anthracite, inclined at various angles from 70° to verticality, are included in the grauwacke at Knockasartnet, near Killarney, and on the north of Tralee. Mr. Weaver further remarks that this old coal is more developed in the county of Cork, particularly at Kanturk, and that large quantities of it are annually raised at Dronagh collieries. He also enumerates beds in the county of Limerick, on the left bank of the Shannon, north of Abbey-feale and at Longhill. The remains of plants, described as chiefly those of *Equiseta*, and *Calamites*, with some indications of *Fucoides*, are stated to be common ‡.

Assuming the foregoing observations to be correct, we obtain evidence that the accumulation of vegetables sufficient to produce beds of anthracitic coal commenced at the epoch of the grauwacke in Europe. Prof. Eaton states that anthracite is found in an equivalent deposit in America (Worcester, and Newport) §. If the relative age of these latter rocks be also correctly determined, it proves the existence of dry land, at different distant points, with vegetation upon it, contemporaneously, or nearly so, with the first appearance of animal life.

Although when we regard the mass of the grauwacke rocks we are struck with the minute proportion that organic remains bear to the whole, we must still perceive that the atmosphere was capable of supporting vegetation, and the seas of sustaining zoophytes, crinoidea, annulata, conchifera, mollusca, crustacea, and fish. What other creatures existed we are unable, from the absence of their remains, to judge: it may however be by no means unphilosophical to conclude that vegetation did not exist alone on dry land, but that, consistently with the general harmony of nature, it afforded food to terrestrial creatures suited to the circumstances under which they were placed.

* Elie de Beaumont, Researches on some of the Revolutions which have taken place on the Surface of the Globe; Phil. Mag. and Annals, vol. x. p. 247.
 † Bulletin de la Soc. Géol. de France, t. iii.
 ‡ Weaver, Proceedings of the Geological Society, June 4, 1830.
 § Eaton, American Journal of Science, vol. xix.

To describe the precise limits of the fossiliferous deposits, and draw fine lines of distinction between them and the non-fossiliferous rocks, is obviously impossible. We can only infer that the remains of organic life were generally entombed in deposits of mechanical origin, as well at this early period as subsequent to it. Respecting the abundance and different structures of the animals first called into existence, we shall never perhaps have any definite ideas; for the preservation of any portion of their more solid parts must always have depended on a great variety of circumstances, not likely to have been most favourable during a state of things, in which a change was effected from the formation of such rocks as gneiss, mica-slate, and others of the same description, to the deposit of those evidently of mechanical origin.

Whatever the kind of animal life may have been which first appeared on the surface of our planet, we mav be certain that it was consistent with the wisdom and design which has always prevailed throughout nature, and that each creature was peculiarly adapted to that situation destined to be occupied by it. Bearing therefore in mind this general adaptation of animals to the circumstances under which they are placed, we may be led so far to speculate at this early condition of life, as to inquire, what kind of creatures, judging from the general character of those known to us, might flourish at a period when there might have been a comparative difficulty in procuring carbonate of lime for their solid parts. It will be obvious that fleshy and gelatinous creatures, such as *Medusæ* and other animals of the like kind, might have abounded, as far as regards a comparative scarcity of this substance. Hence it would be possible to have the seas swarming with these and similar animals, while testaceous creatures and others with solid parts were rare.

These remarks are merely intended to show, that the scarcity of organic remains observed in the lowest part of the grauwacke by no means proves a scarcity of animal life at the same period, though from it we may infer that testaceous and other animals with solid parts were not abundant. Mere fleshy creatures may have existed in myriads without a trace of them having been transmitted to us. In proof of this, if any were requisite, we may inquire what portion of those myriads of fleshy animals, which now swarm in some seas, could be transmitted, as organic remains, to future ages *.

* Dr. Turner has suggested to me, that under this supposition of an abundance of Medusæ or of analogous creatures among the early inhabitants of our globe, we may perhaps account for the bituminous nature of some of the earlier limestones, more particularly of the carboniferous series, in which not a trace of solid organic remains can be observed; for the de-

It may be remarked, while on this subject, that though an
extensive distribution of carbonate of lime is essential to a
great variety of animals, it is surprising how little may supply
the wants of some, even those with vertebræ, such as sharks
and cartilaginous fish generally. To consider that there may
have been some connexion between the animals with solid
parts and a facility of procuring carbonate of lime on the surface of the globe, appears perfectly consistent with the design
manifested in the creation, because it assumes such design at
all periods, and constant harmony between the forms of creatures and their mode of existence. If we imagine a mass of
animals to be suddenly called into life, each properly provided with its solid parts, the carbonate of lime contained in
their bodies would no doubt be sufficient for a constant quantity of the same animal life during a succession of ages; for,
by devouring each other, this necessary substance would be
transmitted from one creature to another. We are however
certain that this has not been the case; for the solid parts of
animals which have been successively imbedded in various
rocks, constitute a very large proportion of certain of those
rocks, and if withdrawn from the fossiliferous deposits generally, would very considerably diminish their thickness. Therefore if the exuviæ of animals had not been entombed, and if
the supply of carbonate of lime had not been greater than
that which could have been derived from the mere destruction of one animal by another, for the purpose of food, the
surface of our planet would not have been what it now is; and
consequently, the fitness of things for the end proposed being
constant in creation, the general condition of animal and vegetable life would not have been such as we now find it.

From the advance of Geology, many districts which were
formerly considered as composed of grauwacke, are now referred to less ancient deposits, and consequently the surface
occupied by grauwacke is much less extensive than was formerly supposed. Thus large portions of the Alps and Italy
have been deprived of their supposed antiquity, which had
been founded on the mineralogical structure of the deposits.

The grauwacke group occurs in Norway, Sweden, and
Russia. It forms a portion of southern Scotland, whence it
ranges, with breaks, as far as regards the surface, formed by
newer deposits or the sea, down western England and Wales,
into Normandy and Brittany. It appears abundantly in Ireland. A large mass of it is exposed in the district constituting

---

composition of a mass of such creatures would produce much bituminous
matter, which may have entered largely into the composition of limestones
then forming.

the Ardennes, the Eifel, and the Taunus. Another mass forms a large portion of the Hartz mountains, while smaller patches emerge in other parts of Germany, on the north of Magdeburg, and other places. In all these situations there is, notwithstanding small variations, a general and prevailing mineralogical character, which points to a common mode of formation over a considerable area. From all the accounts which have been presented to us by Dr. Bigsby and the American geologists, we have every reason to consider that a deposit closely agreeing in relative antiquity, and in its general mineralogical and zoological characters, exists extensively in North America: so that there is evidence to show that some general causes were in operation during the same epoch over a large portion of the northern hemisphere, and that the result was the production of a thick and extensive deposit enveloping animals of similar organic structure over a considerable surface *.

* It was considered useless to present a long detail of the exact areas occupied by the grauwacke rocks, as the reader will comprehend more by a single glance at good geological maps of any given country,—such as Greenough's Map of England, Hoffmann's North-western Germany, Oeynhausen, La Roche, and Von Dechen's Countries near the Rhine, and De Beaumont's and Dufrénoy's France,—than by long and tedious descriptions.

Fig. 112.

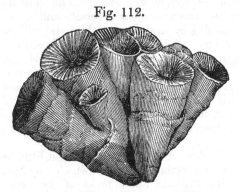

Cyathophyllum turbinatum, Goldf.

SECTION X.

INFERIOR STRATIFIED OR NON-FOSSILIFE-ROUS ROCKS.

SYN.—Clay slate (*Schiste Argilleux*, Fr.; *Phyllade*, Daubuisson; *Thonschiefer*, Germ.). Aluminous slate (*Ampelite Alumineux*, Brong. *Schiste Alumineux*, Fr.; *Alaunschiefer*, Germ.). Whetstone slate (*Schiste coticulé*, Brong.; *Wetzschiefer*, Germ.). Flinty slate (*Schiste siliceux*, Fr.; *Jaspe Schistoide*, Brong.; *Kieselschiefer*, Germ.). Chlorite slate (*Schiste Chloriteux*, Fr.; *Chloritschiefer*, Germ.). Talcose slate (*Schiste Talqueux*, Fr.; *Talkschiefer*, Germ.). Steachist. Hornblende slate (*Ampnibolite Schistoide*, Fr.; *Hornblendschiefer*, Germ.). Hornblende rock (*Amphibolite*, Daubuisson). Quartz rock (*Quartzite*, Brong.; *Quarzfels*, Germ.). Serpentine (*Ophiolite*, Brong.; *Serpentin*, Germ.). Diallage rock (*Euphotide*, Haüy; *Schillerfels*, Germ.). Whitestone, (*Eurite*, Daubuisson; *Weisstein*, Germ.). Mica slate (*Schiste Micacé*, *Micaschiste*, Fr.; *Glimmerschiefer*, Germ.). Gneiss (*Gneiss*, Fr.; *Gneuss*, Germ.). Protogine.

WE have now arrived at that early condition of our planet, when, as far as our knowledge extends, neither animal nor vegetable life existed on its surface. The student, instead of wandering in imagination amid forests and over lands and seas, surrounded by strange vegetables and still stranger animals, should now direct his attention to those laws which govern inorganic matter. This may not at first sight be so attractive as the contemplation of the varied forms of organic life and the probable conditions under which it may have existed; but it will nevertheless be found equally, if not more delightful, as the inquirer obtains more certain results, from the investigation being conducted through the medium of the exact sciences.

It must, on the outset, be confessed that little has yet been accomplished respecting the causes which may have produced gneiss, mica-slate, and other rocks of the same character. Names of the various compound and confusedly crystalline rocks we have in abundance, and if the investigation required no other aid we might sit down satisfied; but unfortunately the abundance of these names has confused the subject, and the student has more frequently contented himself with arranging and disarranging particular mineral compounds in a cabinet, than in investigating their general relations to each other, and the occurrence of the whole in the mass.

It will readily be admitted, that the difficulty of the subject is very considerable, requiring no small insight into the

exact sciences; but the subject being difficult would seem a
good reason why the more advanced cultivators of those sciences should attack it, offering as it does such an ample field
for the exertion of their abilities.

The inferior stratified rocks are of various compositions,
sometimes so passing into each other, that it is almost impossible to affix definite names to the different mixtures. The
strata rarely present a simple mineral substance constituting
a large tract of country, without the admixture of other substances, unless we consider clay slate as such. Before however we proceed further, the student should become acquainted
with the following rocks, which more particularly appear to
deserve distinguishing names.

Argillaceous or Clay Slate.

This rock, as its name implies, is schistose, and contains a
considerable portion of argillaceous matter. It varies materially as to induration, fissility, and composition; and is commonly undistinguishable, except in its geological relations, from
the argillaceous slates of the grauwacke series. Its origin
therefore becomes very ambiguous, and is not the less so from
often containing cubical and other regularly formed crystals of
iron pyrites, affording evidence that the rock was once in the
condition to permit the free arrangement of sulphuret of iron
into crystals,—a fact observable in the argillaceous deposits of
all ages, some decidedly of mechanical origin: therefore we
have no direct evidence to show that the argillaceous slates of
this epoch may not also have been mechanically produced; for
the fineness of grain will by no means assist us, the texture of
the roofing slates obtained from the grauwacke series being
altogether as fine as that of the argillaceous slates associated
with the mica slate, or gneiss. Like, also, the argillaceous
slates of the same series, the lines of cleavage are frequently
not the same with those which appear to be lines of stratification, but meet them at various angles. Argillaceous schist
passes into chlorite slate, talcose slate, and other rocks, by
gradually acquiring particular minerals, which finally replace
the matter of the argillaceous slate.

Chlorite Slate.

This is by no means an unfrequent associate of the preceding,
into which it passes on the one hand, while it graduates into
mica slate, &c. on the other. It is of course essentially composed of chlorite, which occurs alone or mixed with quartz,
felspar, hornblende or mica, in various proportions.

2 F

Talcose Slate.

This is also a rock into which argillaceous slate graduates, at first acquiring a few plates of talc, and afterwards becoming replaced by that mineral, generally associated with quartz, or quartz and felspar. There is not unfrequently a transition from this rock into mica slate.

Quartz Rock.

Quartz rock, as has been observed by Dr. Macculloch, when viewed on the large scale, so that the grauwacke series be included, sometimes appears of chemical, at others of mechanical origin. We should, however, carefully separate the quartz rock which occurs in the grauwacke series, from that associated with the rocks under consideration. That it should possess an arenaceous character in the former case, would be in accordance with the structure of grauwacke generally, though I would be far from stating that some of the grauwacke quartz rocks may not have been chemically produced. The quartz rocks usually interstratified with gneiss, mica slate, &c. are commonly either granular, or resemble common quartz. As it is a subject of much interest to determine if they really present marks of mechanical origin when associated with the true non-fossiliferous rocks, and not those which may appear such from alteration, quartz rocks should be very carefully examined, for there is much reason to believe that some of the quartz rocks stated to occur among the inferior rocks, are really not so associated. The quartz rocks, intermingled with mica slate and gneiss, are observed to pass into both those rocks, by acquiring mica in the one case and mica and felspar in the other. This rock often occupies extensive areas. It is well known in Scotland and its isles; and according to MM. Humboldt and Eschwege, it is of an extent and thickness in the Cordilleras of the Andes and in Brazil, far exceeding what we are acquainted with in Europe. Some of these Brazilian rocks are auriferous, and M. Eschwege attributes the auriferous and platiniferous deposits of that country to their decomposition or destruction.

Hornblende Rock and Slate.

Under this head are included, following the suggestions of Dr. Macculloch, all those compounds, clearly contemporaneous with the rocks among which they occur, of which hornblende constitutes an essential and prevailing ingredient. Much of this rock has been known by the names of primitive greenstone, and greenstone slate, being composed of hornblende

and felspar. The hornblende sometimes so predominates as
to exclude other minerals. As the names imply, these rocks
occur both compact and fissile; in the latter case the felspar
is frequently green. Some curious changes in the structure of
continuous beds may occasionally be observed. I have seen
thick beds of a compound consisting of nearly equal parts of
hornblende and felspar, not differing in mineralogical character
from the common unstratified greenstones, become gradually
schistose by acquiring mica, so that the compound resembles
certain varieties of gneiss. After a time the hornblende would
become scarce, and the rock would become a mixture of mica
and felspar, with probably some quartz. Changes of this kind
are innumerable, and serve to distinguish the hornblende rocks
from the greenstones, with which, without careful examination,
they may be confounded. In the southern part of Devon,
hornblende rocks insensibly become converted, in the line of
their direction, into chlorite slate. This is observable in the direction of the strata between the promontory named the Bolt
Tail and the neighbourhood of Salcombe. From the information of Mr. Royle, it appears that large tracts of country
are occupied by hornblende slate in India, particularly in the
central range of mountains. It occurs also in the Himalah
mountains, associated with gneiss and mica slate. In both
situations it often contains disseminated grains of magnetic or
titaniferous iron ore, which in the central range of mountains
is found abundantly in the river courses, being washed out,
by the rains, from the decomposed hornblende rock. In the
Himalah mountains the natives pound up this variety of hornblende rock, and obtain the iron ore by washing*. Professor
Sedgwick informs me that the menaccanite (titaniferous iron
ore), found abundantly in the bed of a stream near Tregonwell
mill, Menaccan, Cornwall, is derived from the decomposition of
a hornblende rock, composed of hornblende and felspar†. In
these various cases the titaniferous iron ore appears to form a
constituent part of the rock.

Limestone.

This rock occurs variously associated among the inferior
stratified rocks. The saccharine variety is, however, by no
means confined to them; for, as has already been noticed, it
is discovered among the fossiliferous deposits, as for instance,
amid the belemnitic rocks of the Western Alps. The limestone is of various colours, but principally white and crystalline, affording the well known statuary marbles of Greece and
Italy. It is sometimes large-grained, as, for example, that

* Royle, MS. † Sedgwick, MS.

included in mica slate on the lake of Como, which afforded the
mass of materials for the construction of the celebrated Duomo
at Milan. From a mixture of talc or mica, it sometimes becomes schistose. Some of the crystalline dolomites are associated with these marbles and others of the rocks under consideration. The limestones not only vary in their crystalline
character, but pass into compact substances, and become mixed
with various minerals, such as hornblende, augite, quartz, &c.
A remarkable compound, consisting of nearly compact limestone with small crystals of felspar, and thus forming a kind of
porphyry with a calcareous base, occurs at the Col de Bon-
homme, near Mont Blanc, constituting the *calciphyre felspa-
thique* of M. Brongniart.

Eurite.

A rock principally, and in many cases entirely, composed of
the substance named compact felspar. It does not appear to
constitute any extensive tracts in nature, but to be generally
subordinate to gneiss or mica slate.

Mica Slate.

This rock is essentially composed of mica and quartz, and
forms extensive tracts of country, as well as thin beds included
among other rocks. Mica slate sometimes contains garnets so
abundantly, that they may almost be regarded a regular component part of the rock. It graduates on the one hand into
gneiss, and on the other into talcose slate, chlorite slate, and
other compounds.

Gneiss.

This rock is either schistose or divided into beds which vary
in thickness. It is composed of quartz, felspar, mica, and
hornblende, with the occasional mixture of other minerals.
Sometimes one of these minerals is absent, sometimes another:
from this loss of either the quartz, felspar, mica, or hornblende,
and from the occasional absence of even two of them, as well
as the admixture of other substances, there results a very variable general compound. When it occurs confusedly crystallized in regular beds, the mica not being distributed in plates
parallel to the strata, as is the case in the fissile and schistose
gneiss, it is really, as far as mineralogical characters are concerned, nothing but that much disputed substance, stratified
granite. And this is rendered even more apparent, when,
as happens in the Alps, Scotland, and other situations, large
crystals of felspar are disseminated through it, precisely as in
the granite of Dartmoor, &c. When blocks have been detached

from this gneiss, as has happened with many of the erratic blocks of the Alps, they cannot be distinguished from those of true granite. Gneiss, with its variations, constitutes very considerable tracts of country.

Protogine may conveniently be arranged with gneiss, the only difference between its decidedly stratified varieties and the gneiss being the substitution of talc and steatite for the mica. Protogine is the well known granitic rock of Mont Blanc, which certainly has the appearance of graduating into a more massive compound; but in this it does not differ from gneiss, which also seems to pass into granite in a similar manner.

Although the above are the most remarkable of the inferior stratified rocks, they are far from being the whole of them. The varieties and transitions of one to the other appear endless, and, occurring in no determinate order, set classifications utterly at defiance. It was at one time considered that gneiss was the inferior rock, and was succeeded by mica slate; but this is found to be by no means the case, the two being intimately blended with each other as well as with other compounds. It must however be confessed that the mass of the gneiss frequently appears to occupy an inferior position.

All this apparent confusion, and this passage of one rock into another, though it embarrasses arrangements, may be precisely the circumstances which may lead to some knowledge of the causes that have produced the lowest stratified rocks. These irregular passages, and the possibility of discovering any given rock at the top as well as at the bottom of the series, show that the causes, whatever they may have been, which produced this variety in the substances, were secondary, and that there was some general cause upon which the formation of the whole depended.

If we also consider what minerals have entered most largely into the composition of the whole mass, we find that quartz, felspar, mica, and hornblende, are those with which it most abounds, and which impress their characters upon its various portions; chlorite, talc, and carbonate of lime, are certainly not wanting; but if we, as it were, withdraw ourselves from the earth and look down upon such parts of its surface as are geologically known, we find that these latter mineral substances constitute a very small portion of the whole. The inferior stratified rocks which form the largest part of the exposed surface of our planet are gneiss and mica slate, and when viewed on the great scale, the others are more or less subordinate to them.

Supposing this view an approximation to the truth, we arrive at another and important conclusion; namely, that the

minerals which compose the mass of these stratified rocks are
precisely those which constitute the mass of the unstratified
rocks, rocks which, from the phænomena attending them, are
referred to an igneous origin. We may here inquire what are
the circumstances which have determined the arrangement of
these minerals into stratified masses in the one instance, and
into unstratified masses in the other. This question is by no
means of easy solution in the present state of our knowledge:
but while we wait for information, it may be observed that the
conditions, under which the two classes of rocks were produced, must, to a certain extent, have been very distinct. Yet
we find, still viewing the subject in the mass, that the same
elementary substances have produced the same minerals in
both, the only difference between them being their general
difference of arrangement relatively to each other, so that they
should constitute a stratified compound in the one case, and
not in the other. Looking into the structure of gneiss, mica
slate, chlorite slate, talc slate, &c. we find, if we except the
thick-bedded gneiss or stratified granite, that it is the arrangement of the mica, chlorite, or talc in certain general planes
which has produced the fissile and schistose structure. This,
however, has not been the only cause of stratification, (if it
may be so termed, the lines of fissility not being necessarily
those of stratification,) for we find, in the thick-bedded gneiss,
the hornblende rock, the quartz rock, the eurite, and the saccharine limestone, that other causes must have produced thick
beds of confusedly crystallized substances.

There is, nevertheless, so much apparent mineralogical resemblance between these two classes of rocks, that we can
scarcely refrain from conjecturing the remote origin of the
one and of the other to be in some manner connected, modifying circumstances having impressed certain characters on
each. It must be confessed this is a mere hypothesis, and the
student must be careful only to consider it in that light; but
it may be asked, what essential difference there is between
thick-bedded gneiss, particularly that with imbedded crystals
of felspar, and granite, between some hornblende rocks and
greenstone,—except that the one occurs quietly interstratified
in beds, while the other is unstratified, even sometimes cutting
through stratified and similar compounds? We may here also
notice serpentine and diallage rock, of which there is often
good evidence (as will be seen in the next section) for considering igneous and injected rocks, cutting strata in the manner of granite and greenstone. I have never myself observed
these rocks stratified, but Dr. Macculloch appears to be certain that they are so in the Scottish Isles. *A priori*, we should
imagine that there was as much probability in finding strati-

fied rocks, whose mineralogical composition should render
them serpentine, and its common associate diallage rock, as
that we should find stratified rocks mineralogically the same
with granite and greenstone : therefore we should be disposed
to admit them into the catalogue of inferior stratified rocks,
even if we had not the direct opinions of Dr. Macculloch
and some other geologists on the subject. As the question is
one of some interest, it should be stated that the localities
where the stratification may be observed, and which are
pointed out by this author, are ;—for diallage rock, Unst, Balta,
and Fetlar; and for serpentine, also the Shetland Islands.
The stratification is described as often obscure; but the diallage rock is stated to be associated with gneiss, mica slate,
chlorite slate, and argillaceous slate, alternating with them;
and when occurring distinct, presenting the same dip and direction as the neighbouring rocks. According to Dr. Macculloch, there can be no doubt that serpentine is stratified in
Unst; as also appears to be the case in Fetlar, though the
strata are not there so regular.

Let us now cast a glance at the substances which enter into
the composition of some of the more marked inferior stratified
rocks, and see in what respect such rocks differ chemically
from each other. To do this we must search for the best existing analyses of those minerals which enter into their composition, and then calculate the relative proportions of the
constituent substances in one hundred parts of each rock.
These calculations will necessarily be little else than approximations to the truth, more particularly as we shall take the
mean of several analyses of the same mineral, and consequently
the mean of the losses in each; moreover we shall be compelled to suppose definite compounds of those things which
vary much in nature ; but it is hoped that the calculations will
be sufficiently accurate to answer the purpose for which they
are intended.

If we assume that quartz, as it occurs in these rocks, is pure
silica, we shall commit no great error as far as regards the
present inquiry. With mica, however, we shall have far more
difficulty, inasmuch as two substances, of much the same external characters, pass by that name, the one containing lithia,
the other fluoric acid. How far the one may extensively prevail over the other is not well known, but probably the fluoric
acid mica is most common in the inferior stratified rocks. Assuming this, for the sake of our inquiry, we may proceed.
The mean of fifteen analyses of mica from various parts of the
world, by Klaproth, Vauquelin, Rose, and Beudant, gives :—
Silica 46·14, alumina 26·16, potash 10·12, magnesia 4·99,

lime 0·35, peroxide of iron 8·17, oxide of manganese 0·61, fluoric acid 1·09, and water 2.

Seven analyses of felspar by Klaproth, Vauquelin, Bucholz, Rose, Berthier, and Beudant, give, for the mean composition of that mineral,—Silica 64·04, alumina 18·94, potash 13·66, lime 0·76, and oxide of iron 0·74.

A *gneiss*, therefore, composed of equal parts of quartz, felspar, and mica, would contain,

Silica	70·06	Oxide of iron	2·97
Alumina	15·03	Oxide of manganese	0·20
Magnesia	1·66	Fluoric acid	0·36
Lime	0·37	Water	0·66
Potash	7·92		

We should not forget, that instead of common felspar, albite sometimes enters into the composition of gneiss, and other of the inferior stratified rocks. The mean of four analyses of Albite from Finland, Fimbo, Arendal, and Chesterfield (United States), by Tengstrom, Eggertz, Rose, and Stromeyer, gives for the composition of that mineral,—Silica 69·45, alumina 19·44, soda 9·95, lime 0·22, magnesia 0·13, and the oxides of iron and manganese 0·27. Hence a gneiss composed of equal parts of quartz, albite, and mica, would contain,

Silica	71·86	Lime	0·25
Alumina	15·20	Fluoric acid	0·36
Potash	3·37	Oxides of iron and man-	
Soda	3·31	ganese	3·01
Magnesia	1·70	Water	0·35

The composition of the gneiss with imbedded crystals of felspar, by no means an uncommon rock, would of course differ from the varieties of gneiss above noticed, in proportion to the abundance of such crystals.

A *mica slate*, composed of equal parts of quartz and mica, would contain,

Silica	73·07	Oxide of iron	4·08
Alumina	13·08	Oxide of manganese	0·30
Magnesia	2·49	Fluoric acid	0·54
Lime	0·17	Water	1·00
Potash	5·06		

A *mica slate*, composed of equal parts of quartz, mica, and garnet, by no means an uncommon mixture, would contain (taking the mean of several analyses of garnet by Vauquelin, Hisinger, and Wachtmeister, to be,—Silica 39·69, alumina 20·19, protoxide of iron 35·99, protoxide of manganese 3·09, and lime 1·01),

Silica	61·94	Oxide of iron	14·72
Alumina	15·45	Oxide of manganese	1·23
Potash	3·37	Fluoric acid	0·36
Magnesia	1·66	Water	0·66
Lime	0·45		

The chief difference in this compound from the gneiss first noticed, would consist in a less proportion of silica (8·12) and potash (4·55), and in a larger proportion of the oxides of iron (11·75).

Hornblende appears to differ much in the quantity of iron it contains. We shall probably commit no great error if we take Bonsdorff's analysis of a hornblende from Pargas as affording a fair view of the substances contained in this mineral, more particularly as it approaches the mean of several analyses of hornblende from different places. The hornblende in question contained,—silica 45·69, alumina 12·18, lime 13·83, magnesia 18·79, protoxide of iron 7·32, protoxide of manganese 0·22, and fluoric acid 1·50.

That variety of *hornblende rock* which is almost entirely composed of confused crystals of hornblende, will necessarily consist of little else than the constituent parts of the mineral. A variety of hornblende rock composed of equal parts of hornblende and felspar, would contain,

Silica	54·86	Magnesia	9·39
Alumina	15·56	Oxide of iron	4·03
Lime	7·29	Oxide of manganese	0·11
Potash	6·83	Fluoric acid	0·75

The mean of three analyses of chlorite by Vauquelin, Berthier, and Gruner, gives for the composition of that mineral, —Silica 27·43, alumina 17·90, oxide and protoxide of iron 30·63, magnesia 14·56, potash 1·56, lime 0·50, and water 6·92. Assuming this to be a fair estimate of the substances forming chlorite, a *chlorite slate* composed of equal parts of chlorite and quartz would contain,

Silica	63·71	Oxide of iron	15·31
Alumina	8·95	Lime	0·25
Magnesia	7·28	Water	3·46
Potash	0·78		

A *talcose slate*, composed of equal parts of quartz and talc, would contain, assuming the mean of two analyses of talc by Berthier from St. Bernard and St. Foix to afford a fair estimate of the constituent parts of this mineral (viz. silica 56·9, alumina 0·8, lime 4·0, magnesia 26·4, protoxide of iron 8·1, water 3·0),

Silica	78·45	Lime	2·00
Magnesia	13·20	Alumina	0·40
Oxide of iron	4·05	Water	1·50

Protogine is a rock which occurs extensively in the Alps, and differs only from gneiss in containing, as before stated, talc or steatite instead of mica. Talc and steatite do not differ materially in their chemical contents; both probably occur in protogine, but as steatite generally prevails, we will suppose a compound of equal parts of quartz, felspar, and steatite. This protogine would contain (taking the mean of three analyses of steatite by Vauquelin, Bucholz, and Brandes, at—silica 61·68, magnesia 27·80, oxide and protoxide of iron 2·50, lime 0·25, alumina 0·83, and water 6·00),

Silica	75·24	Lime	0·33
Alumina	6·59	Oxide of iron	1·08
Potash	4·55	Water	2·00
Magnesia	9·26		

It is by no means easy to determine the chemical composition of *Eurite*. If we consider it the same with the compact felspar rock of Dr. Macculloch, the analysis of compact felspar ought to afford us the requisite information. There would appear little doubt that substances not precisely the same pass under the name of compact felspar. The author last cited states that compact felspar contains both potash and soda at the same time. According to Bucholz, compact felspar from Passau is composed of

Silica	60·00	Lime	0·75
Alumina	22·00	Loss	3·25
Potash	14·00		

According to Klaproth, compact felspar from Siebenlen contains,

Silica	51·00	Lime	11·25
Alumina	30·00	Oxide of iron	1·75
Soda	4·00	Loss	2·00

The petrosilex of some authors is sometimes also classed under the head of compact felspar. According to Berthier the petrosilex of Nantes is formed of

Silica	75·20	Lime	1·20
Alumina	15·00	Magnesia	2·40
Potash	3·40	Water	1·50

Upon the whole we can scarcely avoid agreeing with M. Beudant, who observes, that, although there are certainly varieties of compact felspar, many substances, which cannot be considered as felspars, are so called, often because it is not known what else to do with them[*]. Be this as it may, eurite is sometimes a mixture of the, so called, compact felspar with mica, and at others with quartz.

[*] Beudant, Traité de Mineralogie, 2me Edition, 1832, t. ii. p. 106.

Quartz rock, as its name implies, is, when pure, composed of little else than silica. When it is formed of equal parts of mica and quartz, it contains the same substances as has been noticed under the head of mica slate, which in fact it then is. Quartz rock, composed of equal parts of quartz and felspar, would contain,

Silica	82·02	Potash	6·83
Alumina	9·47	Oxide of iron	0·37
Lime	0·38		

The *limestones* of this age are commonly saccharine, though varieties are not wanting which are merely compact. The carbonate of lime is not always pure, but often becomes mixed with carbonate of magnesia, constituting *dolomite*. As previously observed, these rocks frequently contain disseminated minerals.

There can be little doubt that the *argillaceous slates* vary materially in their chemical composition, though they present nearly the same external appearance. Silica and alumina appear, however, to enter largely into their composition.

The reader will have observed that silica constitutes the principal ingredient of all these rocks ; for the limestones and dolomites form such an insignificant part of the whole that they may be readily omitted ; indeed, the mass of the inferior stratified rocks must be composed of more than one half of this substance. Alumina is the next abundant substance ; then follow potash, magnesia, and soda. Lime, though by no means scarce, occurs for the most part in very small quantities. Fluoric acid also occurs extensively in small proportions. The oxides of iron and manganese are also common, the former being the most abundant. It would thus appear that the enormous mass of matter constituting the inferior stratified rocks is essentially composed of a few simple substances, variously combined ; and that the greater proportion of them exists in the state of silicates.

It must not be inferred, from the small space here dedicated to the inferior stratified rocks, that they are of little importance; for they are found to occupy a large portion of the earth's surface, wherever, from denudations and disruptions of strata, or from the original absence of superincumbent rocks, they are exposed to our observation. As whenever they are observed, whether in Asia, North America, or Europe, they appear with constant general characters, we may assume that common causes have produced them over the surface of the globe, and that these common causes are principally chemical, inasmuch as the prevalent mineralogical character of the mass is confusedly crystalline.

We may therefore infer, from finding these rocks with con-

stant general characters, whenever circumstances permit us to
observe them emerging from beneath the mass of strata in
which organic remains are entombed, that general chemical
laws have been in operation contemporaneously over the surface of our planet, and previously to the existence of animal
and vegetable life upon it, producing rocks of great collective
thickness. Hence the student may always consider, that,
whatever may be the nature of the deposits on which he stands,
such strata exist beneath them, unless in cases where masses
of igneous rocks have, by protrusion, forced them asunder,
and left no stratified substances intermediate between the surface and the interior of the globe.

It would be tedious to enumerate the various situations
where these rocks may be found; it will suffice to state that
there is scarcely any very large extent of country, where from
some accident or other they are not exposed on the surface.
They abound in Norway, Sweden, and Northern Russia; they
are common in the North of Scotland, whence they stretch
over into Ireland. In the Alps and some other mountains
they occupy the central lines of elevation, as if brought to light
by the movements which have thrown up the different chains.
They abound in the Brazils, and occur extensively in the
United States. Our navigators have shown that they are sufficiently common in the various remote parts of North America
visited by them. They occupy a considerable area in central
India, and are found extensively in the great range of the
Himalah. Ceylon is in a great measure composed of them;
and they do not appear to be scarce in various other parts
of Asia. In Africa also we know that they are not wanting,
though but so small a part of that continent has been yet explored with scientific views.

SECTION XI.

UNSTRATIFIED ROCKS.

THE rocks constituting this natural group are widely distributed over the surface of the world, are found mixed with almost all the stratified rocks, and bear every mark of having been ejected from beneath. They commonly occur either as protruded masses, as overlapping masses, resulting from the spread of matter after ejection, or as veinstones filling fissures, apparently consequent on some violence to which the strata have been subjected.

The aspect of the unstratified rocks is exceedingly various as far as respects their texture, and the absence or presence of the few minerals which essentially enter into their composition. These variations would however in general appear the result of the circumstances to which they have been exposed; and not unfrequently the same mass, if of tolerable extent, will present a great variety of compounds, to which separate names might be (and indeed have been) assigned, if, instead of directing attention to the mass, the small changes in mineralogical structure are alone observed.

In the earlier days of geology, granite was considered the fundamental rock on which all others were accumulated; but this opinion, like many others, has now given way before facts; for, as will be seen in the sequel, we have examples of granite resting upon stratified and fossiliferous rocks of no very great comparative antiquity. It must however be confessed, that granite appears sometimes to alternate in considerable thickness with the inferior stratified rocks, and that the separation of it from gneiss, particularly thick-bedded gneiss, is very ambiguous. Before, however, we proceed further with the consideration of the unstratified rocks, it will be necessary to premise a sketch of their mineralogical characters, omitting those of the rocks usually termed volcanic, which have been already noticed.

Granite

Is a confusedly crystalline compound of quartz, felspar, mica, and hornblende. It is not essential that all these four minerals should be present; on the contrary, rocks have been termed granite when only felspar and mica, felspar and quartz, felspar and hornblende, and quartz and hornblende, have been the

constituent minerals.   Such an employment of the term granite must be used with much caution, as for instance in the case
of the compound of felspar and hornblende, which in fact is
mineralogical greenstone, and should not be named granite
unless it constitutes a very subordinate portion of a mass to
which the term may be more properly applied, and results
from the accidental absence of one or two of the above-named
minerals for a limited space.   The most prevalent compound
is one with quartz, felspar, and mica; when hornblende replaces the mica, it is sometimes termed sienite. Other minerals,
such as chlorite, talc, steatite, &c. are sometimes arranged with
those above enumerated in various ways and proportions; but
such compounds can only be considered as accidental varieties.
When the quartz and felspar occur alone, and the crystallization is such that the former appears disseminated in the latter,
it is termed *graphic granite*, from the supposed resemblance it
bears to antique characters.   Granite is occasionally porphyritic, as is the case in Cornwall and Devonshire, large crystals
of felspar being disseminated through the mass, showing that
however confused the general crystallization may have been,
circumstances were such as to permit the production of distinct
crystals of felspar.

*Diallage Rock (Euphotide, Haüy; Schillerfels, Germ.).   Ser-
pentine (Ophiolite, Al. Brong.; Serpentin, Germ.).*

These are so intimately connected, that to separate them
seems impossible, passing, as they sometimes do, in all directions into each other.   Diallage rock when pure is composed
of diallage and felspar.   Serpentine when pure is generally
considered as a simple mineral substance, and forms large
masses in that state, but seldom prevails to any extent without
acquiring diallage.   These rocks are sometimes blended with
compounds of the greenstone class, and apparently pass so
insensibly into them that they can only be considered as parts
of a common mass, though the serpentine and diallage rock
generally prevail in such cases.

*Greenstone (Grünstein, Germ.; Diabase, Al. Brong.), and the
other Rocks usually termed Trappean.*

These also so pass one into the other, that frequently in a
mass of inconsiderable extent a great variety may readily be
obtained.   They vary in texture from an apparently simple
rock to a confusedly crystalline compound, in which crystals
of felspar are disseminated.   It has long since been observed
by Dr. Macculloch that "the predominant substance in the
members of this family is a simple rock, of which indurated

clay or wacké may be placed at one extreme, and compact felspar at the other; the intermediate member being claystone
and clinkstone. In some cases it forms the whole mass; in
others it is mixed with other minerals, in various proportions
and in various manners; thus producing great diversities of
aspect, without any material variations in the fundamental
character*." As may be readily imagined, no exact definition
can be given of that which is constantly changing in nature.
Claystone, as its name implies, resembles clay under different
degrees of induration; and not unfrequently, when in mass,
acquires a columnar structure. Clinkstone appears an intermediate step to compact felspar, which, according to Dr. Macculloch, contains both potash and soda, while common felspar
contains potash only. I have elsewhere† applied the term
cornean to designate some of the more simple forms of that
kind of rock known as hornstone, which would appear in some
cases to be nothing else than compact felspar ; in others, however, it partakes of the characters of other minerals. Thus,
in Pembrokeshire, where there is a remarkable variety of trappean rocks, the corneans may be divided into felspathic, quartzose, and hornblendic, as those minerals appear to prevail in
the mass; the quartzose variety, which is the most rare, even
appearing like some kinds of quartz rock, with the exception
that it is unstratified. These more simple forms of trap-rock
very frequently become porphyritic by the admixture of either
quartz or felspar crystals, and sometimes of both in the same
mass, as in the red quartziferous porphyries, rocks which not
unfrequently pass into granite. Porphyries are generally
known by the name of the base or paste which includes the
disseminated crystals ; thus, we have claystone porphyry
(*Thonstein porphyr*, Germ. ; *Argillophyre*, Brongniart); felspathic porphyry (*True porphyry* of Brongniart; *Porphyre Eu-
ritique*, Fr. ; *Hornstein porphyr*, *Felspath porphyr*, Germ.);
and clinkstone porphyry (*Klingstein porphyr*).
It very frequently appears as if the elements of quartz, felspar, and hornblende composed the mass, and various circumstances determined their union in such a manner as to produce
a large proportion of the various compounds known as trap-
rocks, sometimes the hornblende being in mass, at others the
felspar, while the quartz rarely predominates. In other situations confusedly crystalline compounds have been the result;
quartz, felspar, and hornblende united, form sienite ; or felspar
and hornblende without the quartz, constitute greenstone. The

* Macculloch, Geological Classification of Rocks, 1821, p. 480.
† Geology of Southern Pembrokeshire; Geol. Trans. 2nd Series, vol. ii.

granular structure of these compounds varies materially, and
finally becomes somewhat imaginary; at least this texture is
rather inferred than seen. The compounds occasionally contain disseminated crystals of felspar, and thus become what
are commonly known as greenstone porphyries (*Diabase por-
phyroide*, Fr.; *Grunstein porphyr*, Germ.). A paste of green
hornblende cornean containing crystals of felspar constitutes
the *ophite* of Brongniart, the antique green porphyry.

Some of the rocks of this family are not unfrequently vesicular, in the manner of modern lavas, the vesicles however
being generally filled up by some mineral substances which
have since been infiltrated into them. Such substances are
not unfrequently agates, and those employed in the arts are
principally thus derived. From these cavities being frequently
of an almond shape, or rather from the appearance of their
solid contents resembling almonds in form, the term *Amygdaloid* has been applied to rocks of this class. It will be readily
understood that the base or paste of the amygdaloids is not
constantly the same, but varies materially. A trappean rock
is sometimes both amygdaloidal and porphyritic at the same
time (Devonshire, Scotland, &c.). The amygdaloidal cavities
afford the mineralogist a great abundance of siliceous, calcareous, zeolitic, and other minerals.

Other minerals than those above enumerated occur in the
trappean rocks, but cannot be considered as forming an essential part of them, with the exception of augite and hypersthene,
which with the mixture of either common, compact, or glassy
felspar, constitute the *augite* and *hypersthene rocks* of Dr. Macculloch. It would be endless to attempt a notice of the various
aspects under which these rocks present themselves; it should
however be remarked that the term basalt is applied to substances which are not precisely the same, being sometimes
given to a fine compound of augite and compact felspar, at
others to a minute mixture of hornblende and compact felspar,
sometimes to dark indurated claystones, and finally to a compound of felspar, augite, and titaniferous iron. The last mixture seems that now most commonly termed basalt.

Let us now consider how far some of the more marked of
the unstratified rocks differ chemically from each other. A
granite composed of equal parts of quartz, felspar, and mica,
would contain precisely the same substances, and in the same
proportion, as the gneiss previously noticed (p. 440.). Mica,
however, rarely constitutes a third part of a large mass of
granite; it is usually in smaller proportions. A granite composed of two fifths of quartz, two fifths of felspar, and one fifth
of mica, would appear much more common. Such a rock

would contain (taking the composition of felspar and mica to be the same as noticed under the head of the inferior stratified rocks),

Silica	74·84	Lime	0·37
Alumina	12·80	Oxide of iron	1·93
Potash	7·48	Oxide of manganese	0·12
Magnesia	0·99	Fluoric acid	0·21

When granite contains disseminated crystals of felspar,— by no means a rare circumstance,—felspar may be considered as constituting at least one half of the compound. Assuming that in such a rock the felspar $= \frac{3}{6}$, the quartz $= \frac{2}{6}$, and the mica $= \frac{1}{6}$, the chemical composition would be,

Silica	73·04	Lime	0·44
Alumina	13·83	Oxide of iron	1·73
Potash	8·51	Oxide of manganese	0·10
Magnesia	0·83	Fluoric acid	0·18

In this estimate of the contents of such a granite, we must not forget that the disseminated crystals are sometimes those of albite: if we assume that such disseminated crystals are $= \frac{1}{6}$, the quartz $= \frac{2}{6}$, the common felspar $= \frac{2}{6}$, and the mica $= \frac{1}{6}$, we should have for the chemical composition of such a rock (taking the constituent parts of albite to be the same as those noticed under the head of the inferior stratified rocks),

Silica	73·94	Lime	0·35
Alumina	13·92	Oxide of iron	1·64
Potash	6·24	Oxide of manganese	0·11
Soda	1·66	Fluoric acid	0·18
Magnesia	0·85		

A granite composed of quartz, felspar, and hornblende, in that case usually termed *sienite*, would contain (supposing the minerals to be in equal parts, and the chemical constituents of hornblende to be as noticed under the head of the inferior stratified rocks),

Silica	69·91	Magnesia	6·26
Alumina	10·37	Oxide of iron	2·69
Potash	4·55	Oxide of manganese	0·07
Lime	4·86	Fluoric acid	0·50

A granite composed of quartz, felspar, mica, and hornblende is by no means a very common variety; it does, however, occasionally occur, even in considerable masses. Such a rock, supposing the four minerals to be in equal parts, would contain,

Silica	63·96	Magnesia	5·94
Alumina	14·32	Oxide of iron	4·06
Potash	5·94	Oxide of manganese	0·21
Lime	3·73	Fluoric acid	0·65

A granite formed of quartz and felspar in equal parts would

afford, upon analysis, the same result as the substance noticed under the head of quartz rock, and supposed to be similarly composed.

A rock, formed of schorl and quartz, and hence termed *schorl rock*, is often found in granite districts; and as it sometimes constitutes important masses, should be noticed.

Schorl is now generally admitted to be only one of the black varieties of tourmaline. These vary in their constituent parts, but agree in containing silica and alumina in larger proportions than any other integral substance, and in the presence of boracic acid. The mean of six analyses by Gmelin of black tourmalines from Rabenstein, Saint Gothard, Greenland, Bovey, Eibenstock, and Käringbrika, gives as the composition of this mineral,—Silica 36·03, alumina 35·82, potash 0·71, soda 1·96, lime 0·28, magnesia 4·44, oxide of iron 13·71, oxide of manganese 1·62, and boracic acid 3·49.

The relative proportion of the minerals in this rock, like all those now under consideration, varies materially; but supposing, for the sake of our inquiry, that the quartz and schorl occur in equal parts, the rock would contain,

Silica	68·01	Magnesia	2·22
Alumina	17·91	Oxide of iron	6·85
Potash	0·35	Oxide of manganese	0·81
Soda	0·98	Boracic acid	1·79
Lime	0·14		

A *greenstone*, composed of equal parts of felspar and hornblende, would present the same results as the hornblende rock previously noticed, and similarly constituted. For the sake of comparison it may be useful to repeat the calculation, which is as follows:

Silica	54·86	Magnesia	9·39
Alumina	15·56	Oxide of iron	4·03
Potash	6·83	Oxide of manganese	0·11
Lime	7·29	Fluoric acid	0·75

A *porphyritic greenstone*, assuming the disseminated crystals of felspar to form one third of the mass, the confusedly crystallized felspar one third, and the hornblende one third, would contain,

Silica	57·92	Magnesia	6·26
Alumina	16·69	Oxide of iron	2·93
Potash	9·10	Oxide of manganese	0·07
Lime	5·11	Fluoric acid	0·50

The exact composition of hypersthene being little known, that of *hypersthene rock*, assuming any proportions of the constituent minerals which we may consider a fair average view of its chemical contents, can necessarily be but imperfectly calculated. The hypersthene from Labrador contains, according to Klaproth,—Silica 54·25, magnesia 14·00, oxide of iron 24·50, lime 1·50, alumina 2·25, water 1·00. If we suppose the rock to be composed of equal parts of hypersthene, thus constituted, and common felspar, it would contain,

Silica	59·14	Magnesia	7·00
Alumina	10·59	Oxide of iron	12·62
Potash	6·83	Water	0·50
Lime	1·13		

If we assume that the same rock is composed of equal parts of hypersthene and albite, it would contain,

Silica	61·85	Magnesia	7·06
Alumina	10·84	Oxide of iron	12·32
Soda	4·97	Oxide of manganese	0·06
Lime	0·86	Water	0·50

When quartz enters into the composition of hypersthene rock, the proportion of silica is necessarily increased, while that of the other constituent substances is diminished, according to the relative amount of quartz.

From the want of good analyses, it becomes exceedingly difficult to arrive at any approximative estimate of the chemical contents of a great variety of porphyries. In the first place, the base or paste in which the crystals are imbedded varies most materially; and in the second, the disseminated crystals are those of different minerals, though they principally consist of common felspar, albite, hornblende, and augite. Some of the bases of porphyries have the character of that rock which is commonly termed compact felspar; and we might proceed to estimate the chemical contents of porphyries with this base, if it were clear that several substances, differing from each other in composition, were not known by the name of compact felspar. Porphyries, for the most part, seem to consist principally of variable mixtures of silica, alumina, potash, soda, lime, iron, and manganese. The imbedded crystals are most frequently those of either common felspar or albite; and, accordingly, as the one or the other prevails in the mass, should we expect to find potash or soda most common.

In like manner the chemical contents of those various trappean compounds, which I have elsewhere termed corneans, are exceedingly difficult to estimate. Silica certainly sometimes prevails in them to a great extent; so that, probably, if the circumstances under which they have been produced had been such as to allow of confused crystallization, the rock would be one in which quartz would form a principal constituent mineral. That they are silicates of various substances of the same kind as those contained in granites, greenstones, and the like, can be scarcely doubted. Very frequently we find circumstances have been such, in various parts of their mass, that crystallization has taken place to a greater or less extent; the products being porphyries of different kinds, greenstones, or sienites.

Various dark coloured rocks of a close fine-grained texture having been termed basalts, there is again difficulty under this head. Certain of these basalts are but exceedingly fine-grained greenstones, and would therefore have the same chemical composition. If, as it is supposed by some, true basalt is a compound of augite, felspar, and titaniferous iron, we can scarcely attempt a calculation, inasmuch as the analyses of minerals which have been termed augite differ much from each other; and we scarcely know what proportions to assume as a fair estimate of the contained minerals. According to Phillips, a *basalt* from Saxony was composed of

Silica	44·50	Soda	2·60
Alumina	16·75	Oxide of iron	20·00
Lime	9·50	Oxide of manganese	0·12
Magnesia	2·25	Water	2·00

A basalt from Baulieu afforded, on analysis, to M. Beudant,

Silica	59·5	Lime	1·3
Alumina	11·5	Soda	5·9
Peroxide of iron	0·5	Potash	1·6
Protoxide of iron	19·7		

Trachyte is again another igneous rock, the chemical composition of which is most difficult to estimate. Silica certainly sometimes prevails more than at others. Both the potash and soda felspars apparently enter largely into it.

Our difficulties are by no means diminished when we arrive at the *lavas.* They appear of a very variable composition. Still, however, the mass of the substances of which they are formed consists chiefly of silicates. Potash or soda commonly constitutes a portion of their general composition.

The analyses of *pitchstone* from Newry, by Knox, and of the same rock of Meissen, by Duménil, do not, as will be seen beneath, differ very materially from each other; and are remarkable not only for exhibiting this general similarity in their chemical contents, but also for showing that bitumen may enter somewhat largely into the composition of an igneous rock.

	Newry *Pitchstone.*	*Meissen* *Pitchstone.*
Silica	72·80	73·00
Alumina	11·50	10·84
Soda	2·86	1·48
Lime	1·12	1·14
Oxide of iron	3·03	1·90
Bituminous matter	8·50	9·40

It would be useless to estimate the contents of *obsidian*. It
appears little else than the vitreous condition of various melted
rocks, as is indeed shown by the analyses which have been
made of it; for these differ very considerably.

The analyses of *serpentine*, though they vary as to the numbers and proportions of the substances which enter into the
composition of this rock, always afford silica, magnesia, and
water as the principal constituents, as will readily be seen
beneath (*a*. Serpentine from Germantown, by Nutall; *b*. from
Skyttgrufa, by Hisinger; *c*. by John*, from a place not mentioned):

	a.	b.	c.
Silica	42·00	43·07	42·50
Magnesia	33·00	40·37	38·63
Alumina	0·00	0·25	1·00
Lime	3·50	0·50	0·25
Oxide of iron	7·00	1·17	1·50
Oxide of manganese	0·00	0·00	1·62
Oxide of chrome	0·00	0·00	1·25
Water	13·00	12·45	15·20

We may here remark, that, though chrome does not always
constitute one of the ingredients of serpentine, chromate of iron
is exceedingly common in it and some diallage rocks. The
chrome of commerce, so extensively used in the preparation
of certain paints, seems almost entirely to be thus derived.

It is by no means easy to calculate the approximative composition of *diallage rock*, from the variable nature both of the
diallage and felspar of which it is composed. Minerals much
resembling each other in appearance, but differing in the relative proportions of their constituent substances, have been
named diallage. The felspar is sometimes compact, at others
common, and is not unfrequently an albite. According to
Berthier, the diallage from La Spezia contains,—Silica 47·2,
magnesia 24·4, lime 13·1, protoxide of iron 7·4, alumina 3·7,
and water 3·2. Albite is frequently mixed with the diallage
in the same country, forming diallage rock. Assuming that
the rock is composed of two fifths of diallage and three fifths
of albite, which is not an uncommon proportion in the diallage
rock of La Spezia, we should have for the chemical contents
of this rock,

Silica	60·55	Lime	5·37
Alumina	13·14	Oxides of iron	3·03
Soda	5·97	Oxides of manganese	0·07
Magnesia	9·83	Water	1·28

If, instead of albite, we take common felspar as mixed
with diallage in diallage rock, in the proportion of two

* Mohs's Mineralogy, by Haidinger.

thirds of the former mineral to one third of the latter, we
obtain,

Silica	58·42	Magnesia	8·13
Alumina	13·86	Oxide of iron	2·96
Potash	9·10	Water	1·06
Lime	4·87		

Though the foregoing calculations are merely approximative and incomplete, they nevertheless afford data for many
important inferences. To enter at length into this subject
would not accord with the plan of the present work. It may
however be remarked, that, like the inferior stratified rocks,
the unstratified rocks are composed of a few simple substances,
variously combined. The silicates of alumina, potash, soda,
lime, and magnesia, with the oxides of iron and manganese,
constitute the chief ingredients of the rocks under consideration. In the granites without hornblende the silicates of alumina, potash, and soda prevail, but principally the two former.
When hornblende enters into the composition of granite, the
silicates of lime and magnesia are not without their importance,
particularly when hornblende constitutes at least one third of
the mass. The importance of the silicates of lime and magnesia is increased in the greenstones, and in a large portion
of the basalts. The silicate of magnesia is in excess in the
serpentines.

The relative fusibility of the rock is necessarily determined
by the relative proportions of the above noticed ingredients
contained in it. Thus, when silica greatly prevails, it is refractory; and the like seems to occur when silicate of magnesia is in excess (as takes place in the serpentines) without
the presence of a sufficient quantity of any substance, such as
silicate of lime, which may act as a flux. When silicate of
lime is in tolerable quantity, the rock is then readily fusible.
This takes place when either hornblende or augite are sufficiently abundant; indeed the augitic rocks, which contain a
large amount of silicate of lime*, may be considered as the
most fusible.

Although a large proportion of the unstratified rocks may
be considered as having been ejected directly from beneath
all stratified rocks, it by no means follows that some of them
do not consist of the stratified rocks fused and thrown up.
In fact, the composition of some lavas would not be opposed
to this inference. That the inferior stratified rocks could
readily be converted, by fusion, into many of the unstratified
rocks, is sufficiently apparent, their general composition being
the same. The conversion of some of the fossiliferous rocks

* The proportion of lime in the minerals known under the common name
of augite, varies in the different analyses from 13 to 24 per cent.

into lavas and substances of that character is not less simple. If we abstract a large proportion of carbon and lime, the fossiliferous rocks appear, in a great measure, derived from the degradation or chemical destruction of the inferior stratified or unstratified rocks. The unequal dispersion of the materials so derived will of course produce unequal accumulations of them ; so that a collection of grains of quartz, forming a siliceous sandstone, would be highly refractory. But suppose, as very often happens, that the grains of quartz are cemented by calcareous matter (no consequence whence derived), the compound would be exceedingly fusible. Pumice is clearly often nothing else than a schistose rock sufficiently heated to produce the necessary vesicular character, the heat not being intense enough to cause fusion. The pumice of the volcanic district of the Rhine is probably the grauwacke schists of that country which have been thus circumstanced.

Such are the rocks commonly considered unstratified. It will have been seen that they so pass into one another that distinctions are not easily established between them. A common passage of diallage rock into greenstone, the reason for which does not at first sight seem apparent, will be seen, by comparing the foregoing calculations respecting these two rocks, to be produced by a very small difference in the proportions of the substances composing them. Mineralogical granite passes through various stages, and graduates into the compounds named greenstone, and others of the trappean class *.

---

* Dr. Hibbert notices the passage of granite into one of those compounds named basalt, (in this case formed of an intimate mixture of hornblende with a small proportion of felspar,) as taking place in the Shetland Islands. As this author's account is illustrative of such changes in general, it may advantageously find a place here. The basalt extends from the Island of Mickle Voe northwards to Roeness Voe, a distance of twelve miles. On the west of this is a considerable mass of granite. The transition is thus described : " Not far from the junction we may find, dispersed through the basalt, very minute particles of quartz. This is the first indication of an approaching change in the nature of the rock. In again tracing it still nearer the granite, we find the particles of quartz dispersed through the basalt becoming still more distinct, more numerous, and larger, an increase of magnitude even extending to every other description of particles. The rock may now be observed to consist of separate ingredients of quartz, hornblende, felspar, and greenstone; the latter substance (greenstone) being a homogeneous commixture of hornblende and felspar. Again, as we approach still nearer the granite, the disseminated portions of greenstone disappear, their place being supplied by an additional quantity of felspar and quartz. The rock now consists of three ingredients, felspar, quartz, and hornblende. The last change which takes place results from the still increasing accumulation of quartz and felspar, and from the proportionate diminution of hornblende. The hornblende eventually disappears, and we have a well characterized granite, consisting of two ingredients of felspar and quartz." Hibbert, Brewster's Edin. Journal of Science, vol. i. p. 107. The same author also notices a passage of felspar porphyry into granite near Hillswick Ness.

Instead also of being solely mixed with rocks of the oldest date, it is found, among the Montagnes de l'Oisans (Western Alps), cutting through and superincumbent upon deposits, referrible, according to M. Elie de Beaumont, to the oolitic series[*].

Observations of the same kind have been made by MM. Hugi and Studer on part of the Swiss Alps. The annexed figure is a section, by M. Hugi, of the Bötzberg.

a, limestones and slates, referred to the lias; *b*, mica slate; *c*, gneiss; *d*, granitic rock. Assuming the section to be correct, the super-position of the crystalline rocks, in this case, is evident. But it may have been produced in two ways: the rock (*a*) considered as lias

Fig. 113.

may have been quietly deposited upon the mica-slate (*b*), which in its turn reposed on the gneiss (*c*), also resting on the granitic rock (*d*), and the whole may have been thrown over,—a circumstance not so rare as may be supposed in great mountain chains; or the granitic rock, supposing it to belong to the class now under consideration, may have been ejected from beneath, thus overflowing the fossiliferous beds, and altering those nearest to it.

If we have recourse to the observations of M. Studer on this part of the Alps, we shall find that the former opinion is most probably the true one, at least in the greater number of cases. The annexed figure represents his section of the Jungfrau.

a, limestones and slates, referred to rocks of the oo-litic group; *b b*, gneiss. This gneiss is described as composed of talcose mica, white or brown felspar, with little quartz; and is stated to os-cillate between gneiss and granite, appearing on the large scale to belong rather to the former than the latter. In fact it seems to belong to that great system of gneiss

Fig. 114.

Jungfrau.

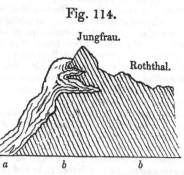

Roththal.

a　　　　*b*　　　　*b*

with steatite, talc, or talcose mica, constituting a large portion of the central range of the Alps, and not unfrequently known as *Protogine*. As a mass, the rocks equivalent to those of the oolitic group, rest unconformably on the mica slate and

* Elie de Beaumont, sur les Montagnes de l'Oisans; Mém. de la Soc. d'Hist. Nat. de Paris, tom. v.; as also Sections and Views illustrative of Geological Phænomena, pl. 15.

gneiss of the Alps, though of course there must be numerous sections where the planes of the beds of the two rocks are parallel to each other. Such being the case, a deposit of limestones and slates on an unequal surface of gneiss would necessarily mould itself on the latter, filling up the cavities, as often happens with other rocks similarly circumstanced. It hence follows that if the gneiss be upraised by a force acting along the central range of the Alps, the limestones and slates would be tilted up at the same time, so that the whole, when viewed as it now appears, would have the aspect of gneiss and fossiliferous limestones reciprocally protruded into each other. If the reader will place the section of the Jungfrau before him, in such a manner that the limestone and slates become horizontal, he will perceive how easily they may have been moulded on the gneiss. If another very interesting section by M. Studer, representing five horizontal and reciprocal protrusions of gneiss and limestones into each other at the Gstellihorn, be also so turned that the gneiss is beneath and the limestone above, the former rise in peaks into the latter. We must here remark the uncertainty of many sections of this kind, however clearly the lines of separation may be exhibited on the face of a huge precipice, as this is; for it is well known to all accustomed to examine disturbed districts, where unconformable rocks have been tilted up together, that natural lines of section often cause a large mass of inferior rock to appear included in the superior beds, when in fact such appearance is entirely deceptive. This arises from a portion of the older rock projecting into the newer rock, having been accidentally cut through in the line of section. If fortunate sections were made through the upper uneven surface of the chalk in many parts of England and France, where this rock projects in pinnacles and ridges into the plastic clay, or beds of that character, we should have such deceptive appearances; and if both rocks were tilted up at right angles to their former position, the result would be a section not differing in appearance, though greatly in magnitude, from the Alpine sections above noticed.

M. Studer considers these appearances to have arisen from the breaks and contortions of the limestones and gneiss when the whole was upraised. This, which is also a probable explanation of the phænomena, might be ascertained by a careful examination of the relative stratification of the two rocks*.

While, however, we take this view of the subject, we must not neglect facts which may be thought to support the hypothesis, that the crystalline rocks have been protruded in a

* Studer, Bulletin de la Société Géologique de France, t. ii. p. 53. pl. 1. figs. 2 and 3.

heated state among the calcareous strata. The calcareous beds intermingled with the gneiss of the Tossenhorn are observed to be dolomitic, and in certain cases crystalline *. The beds which rest on the gneiss at the Jungfrau occur in the following ascending order:—1. compact dolomite, 30 feet; 2. quartz rock, associated with variegated argillaceous schist, 15 feet; 3. oolitic iron ore; 4. limestone, generally black, gray, and schistose. The limestone and iron ore contain numerous organic remains, considered as referrible to those of the oolitic group. It might be supposed that the lowest of the calcareous beds were to a certain extent altered, but this can scarcely be considered as a sufficient explanation. Moreover, M. Studer remarks that in the Roththal, where the gneiss covers the limestone series, there is, though the contact of the two rocks is well exposed, no appearance whatever of their having produced any effect on each other.

Whether or not granite is to be considered as having been intruded among the limestones of the Bernese Alps, it has already been seen that it covers the chalk of Weinbohla, whence it may be inferred that granite was produced at the supracretaceous epoch. Assuming therefore that the evidence is good, we should expect to find granitic rocks traversing or superincumbent upon beds of all ages, from the inferior stratified to the cretaceous inclusive. The superposition of granitic rocks to fossiliferous limestone has long since been remarked by Von Buch in Norway, and by Dr. Macculloch in the Isle of Sky. Similar rocks have also been noticed as incumbent on those of the age of either the oolitic or cretaceous series at Predazzo. Respecting the latter, Sir J. Herschel observes, that where the dolomite plunges beneath the granitic rock at Canzocoli, at an angle of 50° to 60°, both rocks appear altered, and that there are laminæ of serpentine between the two†.

The contact of the granite with the oolitic rocks of Brora is attributed by Prof. Sedgwick and Mr. Murchison to the elevation of the granite in mass, which is supposed to have turned up the edges of the oolitic deposit‡. The same authors have also remarked a very curious occurrence of granite and limestone on the north coast of Caithness, near Sandside, where the granite appears thrust up among the limestones, and a breccia has been produced containing fragments of limestone and granite. The cement of this breccia is described as generally granitic, though it is calcareous in some places, and approaches a sandstone in others. One great block of limestone

---

* M. von Dechen states that the limestone of these protruding wedges contains laminæ of talc.—German Transl. of Manual, p. 552.

† Herschel, Edinburgh Journal of Science, vol. iii.

‡ Sedgwick and Murchison, Geol. Trans. vol. ii. pl. 34.

is noticed as apparently entangled in the granite. The limestone beds on the eastern side are stated not to be much disturbed, while those on the western side are in the utmost confusion, and, which are important circumstances, crystalline and cellular*.

Thus far we have only seen granite rising through and covering other rocks in considerable masses; but we have also evidence in granite veins, that the matter of the rock was in such a state of fusion, as to penetrate into thin clefts opened in stratified and older rocks by some violence, such as probably resulted from the upburst of the igneous matter accompanied by elastic vapours. If we imagine fractures to be suddenly produced in contact with a mass of rock in fusion, such as we may presume granite to have been, the natural result would be the injection of the substance in fusion into all the crevices, in consequence of the great pressure exerted on one side; the intruding substance breaking off and including in it all loose fragments, and those projecting portions which opposed the fury of the injection. This is precisely the condition of granite veins, which, though much doubted during the reign of the Wernerian theory, are now known to be abundant in nature.

Glen Tilt, which is reported to have produced such delight in Hutton when viewed by him for the first time, presents excellent examples of the intrusion of granite veins into other and stratified rocks. The great features consist of a mass of granite on the northern side of the glen, and of schist and limestone on the southern; from the former, veins issue in all directions, disturbing and intermingling with the latter in such a complicated manner, as to render a description useless without the aid of maps and sections, for which, and for a detail of the various singular phænomena observable in Glen Tilt, the student must be referred to the memoirs of Lord Webb Seymour, Prof. Playfair†, and Dr. Macculloch‡.

Granite veins traversing the stratified rocks are now known in various parts of the world. Some fine examples are to be observed in the district of the Land's End; among them one at Cape Cornwall shows that there has been a shift or fault in the slate rocks, for a quartz vein has been cut through, and elevated more on one side than the other, thus proving that force has been employed§. At Mousehole the veins can be seen to

---

* Sedgwick and Murchison, Geol. Trans. vol. iii. p. 132.
† Trans. of Royal Soc. of Edinburgh, vol. vii.
‡ Geol. Transactions, 1st Series, vol. iii.
§ Oeynhausen and Von Dechen, Phil. Mag. and Annals of Philosophy, 1829; also Sections and Views illustrative of Geological Phænomena, pl. 17. fig. 4.

proceed from the main body of the granite *. In the Alps they also proceed from masses of granite, which appear to have much influenced the present position of strata in parts of those mountains, as has been shown by M. Necker de Saussure†. They traverse gneiss in the Vallée de Vallorsine, as also at the head of the lake of Como. It is curious that in the Hartz, once considered as beautifully illustrating the Wernerian theory of granite, masses of granite cut through the direction of the clay slate and grauwacke nearly at right angles, sending off veins. At the Rosstrappe and at the Rehberg Graben, on the south side of the Brocken, granite veins clearly run into the clay slate and grauwacke. Few circumstances can more clearly prove how a leading theory may pervert the judgement, and thus cause the misrepresentation of facts. According to M. Dufrénoy, the granite veins in the vicinity of St. Paul de Fenouillet (Pyrenees) occur in limestones referrible to the age of the cretaceous group. These limestones become more and more crystalline as they approach the granitic masses‡.

Granite veins are not confined to Europe, but are found cutting and including portions of slate rocks at the Cape of Good Hope, as has been shown by Captain Basil Hall and Dr. Clarke Abel§. In America also they have been observed by Mr. Hitchcock traversing mica slate, hornblende slate, limestone (described as of a peculiar character), gneiss, and granite in Connecticut; the veins frequently branching out in various directions ‖. Granite veins therefore cannot be considered as rare; on the contrary, they would appear sufficiently common when circumstances permit good sections of the junctions of the granitic mass, and of the rocks among which they appear intruded. We should expect these veins to be of various dates, and accordingly we find that masses of granite are themselves traversed by veins, also of granite.

The exact composition of the granite in these veins must naturally vary, depending much on local circumstances; for if we suppose a substance in fusion to be injected into fissures of rocks, such injected matter will be subjected to different conditions. Where the fused substance cooled more suddenly, as was likely to be the case in the distant and smaller fissures,

---

* Sections and Views illustrative of Geological Phænomena, fig. 5.
† Necker de Saussure, sur le Vallée de Vallorsine; Mém. de la Soc. de Physique et d'Hist. Nat. de Genève.
‡ Dufrénoy, Bull. de la Soc. Géol. de France, t. ii. p. 71.
§ Basil Hall, Transactions of the Royal Soc. of Edinburgh, vol. vii.; and Clarke Abel's Voyage to China.
‖ Hitchcock, On the Geology of Connecticut, American Journal of Science, vol. vi.

the result would be less crystalline; while in the wider clefts, and near the great heated mass, the crystallization would be more perfect, and bear the greatest resemblance to the parent mass. Consequently, in a system of granite veins we should expect a great diversity in the aspect of the granitic matter, which generally appears to be the case.

The trappean rocks, though there is much difficulty in separating them from the granitic, may for convenience be considered separately from them. They also form considerable masses, and constitute dykes and veins. When considered in the mass, they may be regarded as containing much less mica than the granitic rocks, while hornblende has become much more abundant; they also, when viewed on the large scale, appear more abundantly among the comparatively modern deposits than the granites, though it cannot be denied that they run into the latter in a remarkable manner. If this opinion of the greater prevalence of the granitic rocks over the trappean at the earliest periods be correct, it would seem to point to a certain condition of things at such periods, which subsequently became so modified that the igneous eruptions became altered. Of what that condition of things may have been, we do not as yet appear to have any very definite ideas, and we obtain little help on the subject from the phænomena of modern volcanos, granite never having been known to flow from them. We however learn from this circumstance that igneous eruptions into the atmosphere are not favourable to the production of granites; and we may consequently infer, that the conditions under which granite was produced were not similar to those which we now observe on the surface of the earth, at least so far as relates to those phænomena which occur in the atmosphere. We do not exactly see why a difference of pressure beneath water, or any cause of that kind, should render mica less abundant, or increase the quantity of hornblende; and therefore we may infer that there was something in the then condition of our planet's surface, which permitted the production of that great abundance of granite so commonly associated with the earliest stratified rocks with which we are acquainted, and which frequently differ from it only in being stratified, or having the component minerals arranged in laminæ.

Admitting this prevalence of granitic compounds at the earliest periods, their production at more recent epochs shows that the conditions necessary for their formation continued up to such epochs, though they may have been infinitely more rare, having in a great measure given place to those under which the more common trappean rocks were produced.

Trappean rocks, under their various modifications, are so common in nature, that to attempt a notice of localities would be entirely useless. They occur mingled with the stratified rocks in every possible way,—injected among the beds for considerable distances, so that sections are exhibited in which the igneous rock appears quietly interstratified with the aqueous deposits; constituting caps of hills, thus appearing like a stratified and quiet deposit on other beds, the continuous parts that once connected these caps into a sheet of matter, ejected from beneath, having been removed by denudation; or as dykes or veins filling fissures, previously produced, in some instances showing that the igneous matter entered the fissure with such force as to tear away portions of the sides, while at others it seems to have risen more slowly, gradually filling the rent.

There would appear no more convenient place for observing all these modes of occurrence, or indeed the various mineral aspects of the rocks themselves, than the coasts and islands of Scotland, which have been described by Dr. Macculloch [*] and other geologists, and where the student possesses the great advantage of innumerable coast sections, those invaluable aids in all geological investigations.

Apparent interstratifications of igneous rocks with beds which have had a different origin may be observed in many places, but may be well studied in High Teesdale, where the igneous matter has been injected among strata of limestone, sandstone, and shale, forming part of the carboniferous limestone series, in such a manner that a great apparent bed, commonly known as the Great Whin Sill, was considered as constituting a regularly stratified portion of a common whole, before the investigations of Prof. Sedgwick showed that it had evidently been injected among the aqueous deposits, and was connected with a mass of igneous matter which had disturbed and altered a continuation of the same rocks [†]. In Derbyshire, trappean rocks, generally known by the provincial term *toad-stones*, from the aspect of the prevailing amygdaloid, are apparently interstratified with the carboniferous limestone. These we may, from all analogy, consider as injected among the limestones, the strata of which would easily be separated by the application of the proper force, in the manner already noticed under the head of Volcanic Rocks (p. 140). The student will find ample details of this association of trap-rocks and limestones in Mr. Conybeare's account of the rocks of Derbyshire [‡].

---

[*] Macculloch's Western Islands.

[†] Sedgwick, Trans. of the Cambridge Phil. Soc. vol. ii. p. 139; and Sections and Views illustrative of Geological Phænomena, pl. 13. In some situations the limestone and slate have been turned up by the trap, and the former has become granular, and the latter indurated.

[‡] Outlines of the Geology of England and Wales, Book III. chap. v.

According to Mr. Aikin, a good example of the apparent interstratification of greenstone with the coal-measures is observable at Birch Hill colliery, Staffordshire. The bed seems to be connected with a mass of trap on one side, whence it has been injected among the coal strata, altering the coal where it covers it, by depriving it of its bitumen*.

The connexion of trap rocks with coal-measures has often been insisted on, and certainly in some countries they are much associated; but when the facts are narrowly examined, it generally appears that the igneous rocks have been introduced among the sandstones, shales, and coal, subsequently to the deposit, and even consolidation, of the latter. It does not however necessarily follow that igneous eruptions and coal deposits may not have been contemporaneous; on the contrary, we may inquire if the violent movements of land, which probably accompanied such igneous eruptions, did not assist in the destruction of the vegetation, by depressing it beneath the waters; and even if accompanying and violent agitations of the atmosphere, similar to that which happened during the great eruption of Sumbawa, might not also contribute something towards the transport of the different parts of plants in particular cases†. Difficulties very frequently arise from the want of good natural sections; for it is clear that a mass of injected trap may be so spread over, or injected among, the coal strata, in a district wanting such sections, and only explored by means of the miner's galleries, that ambiguous appearances abound, more particularly when the whole mass has been traversed by faults, as is often the case.

Among the various trap dykes noticed by Mr. Winch as traversing the coal-measures in the vicinity of Newcastle, there is one described by Mr. Hill as occurring at Walker colliery, which has converted the coal contiguous to it into coke. This dyke is stated not to alter the level of the coal-measures, though it cuts through them; but in the plan which accompanies the memoir there is a fault marked on the south side of the dyke, parallel to it, and on the eastern part, producing a dislocation to the amount of nine feet, so that the fracture does not appear to have been quite simple‡.

Trap dykes are to be found in all parts of the world, the

* Aikin, Geol. Trans. vol. iii. In the section which illustrates this memoir, a fault is seen to have traversed the beds after the injection of the trap, for it is dislocated with the rest;—a fact also observable in Prof. Sedgwick's sections of High Teesdale, where dislocations are represented to have affected all the rocks equally.

† During tropical hurricanes among islands, such as those in the West Indies, plants, more particularly their lighter parts, are carried abundantly out to sea.

‡ Geol. Trans. 1st Series, vol. iv.

composition of the rock varying materially, even in the dyke itself, as we might expect from differences in the cooling and pressure, so that the central parts are not unfrequently more crystalline than the sides *.

There is good evidence of the great mechanical force which has been exerted on the stratified rocks, as has been already pointed out in the North of Ireland, where large disrupted masses have been caught up in the igneous matter; similar phænomena have also been noticed by Dr. Macculloch and Mr. Murchison in the Western Islands of Scotland, the disrupted and included rocks being in the latter cases of an older date than those fractured in Northern Ireland†.

The annexed figure will illustrate a considerable fracture and alteration in the limestones at the d Black Head, Babbacombe Bay, Devon, effected by the eruption of greenstone, which, though it overlies the limestones in this section, occurs beneath them not far distant.

Fig. 115.

$a \quad b \quad f \quad c \quad e \quad f \quad b$

a, argillo-calcareous slate, traversed by veins of calcareous spar, and occasionally indurated. $b\,b$, limestones which have become semi-crystalline; they have also, judging from lines of colour, been once more fissile than at present. c, a slate, with a thin bed of reddish limestone (e). This slate is apparently much altered. $d\,d$, greenstone and its varieties, constituting the mass of the hill, and traversed by calcareous veins near the limestones. $f\,f$, lines of fracture which divide the limestones and slates into three masses. The slates and limestones have evidently suffered, not only from the mechanical action of the erupted greenstone, but also chemically

* One of the longest dykes with which we are acquainted is that described by Prof. Sedgwick in his memoir on those of Yorkshire and Durham. It is very probably continued from "High Teesdale to the confines of the eastern coast, a distance of more than sixty miles." During this traverse it cuts the coal-measures, red sandstone, and lias.—Cambridge Phil. Trans., vol. ii. p. 31; and Sections and Views illustrative of Geological Phænomena, pl. 14.

Archdeacon Verschoyle notices many east and west trap dykes in the counties of Sligo and Mayo, one of which can be traced for sixty or seventy miles, and probably extends to greater length. These dykes cut through all the stratified rocks in that part of Ireland, from the gneiss to the carboniferous limestone inclusive.—Proceedings of the Geol. Soc., Nov. 21, 1832.

† Macculloch, On the Western Islands of Scotland. Several of the sections contained in this work are copied into Sections and Views illustrative of Geological Phænomena, for the purpose of exhibiting the various modes in which trappean are associated with stratified rocks in the Hebrides. Mr. Murchison has represented fragments of the oolitic series of the same islands as caught up in the trap of the southern side of Mull.—Geol. Trans. 2nd Series, vol. ii. pl. 35.

from the proximity of the mass in a state of igneous fusion. Notwithstanding the general pressure preventing the disengagement of the carbonic acid contained in the limestones, some elementary substances have been given off, and filled crevices and cracks in the trappean mass above with carbonate of lime. The alterations of limestone in contact with trappean rocks is sufficiently common, producing a greater or less amount of crystallization, in accordance with the well-known experiments of Sir James Hall*, who has proved that carbonate of lime when subjected to great heat beneath sufficient pressure, does not part with its carbonic acid, but that it is fused and is rendered crystalline,—a fact previously doubted.

The alteration of limestones and trap at their contact would not always appear to be confined to a crystalline arrangement in the parts of the former; for Dr. Macculloch has observed trap to become changed into serpentine at the point of contact with limestone, at Clunie in Perthshire. A trap vein traverses limestone, and the rock is noticed as a kind of greenstone, the greater part of which is moderately coarse, passing into lamellar towards the sides. This (lamellar) structure "gradually becomes more distinct towards the edges of the vein, where it frequently splits off by the process of decomposition into laminæ, resembling, on a cursory view, black slate. These laminæ are often intersected by other cross fissures, dividing the whole into cuboidal masses, which sometimes decompose still further into spheroidal forms, during the approximation to the calcareous boundary; and whether the laminæ are present or not, the texture becomes gradually finer and softer, the rock still retaining its black colour, or sometimes assuming a greenish cast. At length the observer finds the vein converted under his hand into serpentine, without being at first aware that any change has been brought about†." Thus the transition from greenstone to serpentine can gradually be traced, but only where the vein traverses the limestone; for where the continuation of the same vein cuts through schist and conglomerate, no such change is observable. The trap is much entangled, in the small scale, with the limestone, and the minuter ramifications from the vein are entirely composed of serpentine. The limestone does not pass into the serpentine; on the contrary, the line of separation is described as well defined. Veins of green asbestos and steatite occur in the serpentine. Dr. Macculloch also states that the trap veins traversing the calcareous sandstone at Strathaird abound in steatite, found in the outer parts of the vein, and approaching the calcareous

---

* Sir James Hall, Trans. of the Royal Soc. of Edinburgh, vol. vi.
† Macculloch, Brewster's Edin. Journal of Science, vol. i. p. 1.

rock. The same author observes, that the trap vein which traverses the white marble of Strath passes into serpentine at its outer edges, as at Clunie. " At the line of contact, a zone of transparent serpentine of a fine oil-green colour, is found intermixed with the limestone *."

The above is sufficient to show that trap under certain conditions may pass into serpentine. We have now to consider dykes and masses of serpentine and diallage rock which occur under circumstances analogous to those of the trap rocks. Mr. Lyell has described a serpentine dyke which cuts through a sandstone (equivalent either to grauwacke or the old red sandstone), near West Balloch Farm, in Forfarshire. The phænomena can be well observed where the dyke traverses the Carity. The serpentine dyke is ninety feet thick, nearly vertical, and ranges east and west. It is stated to be flanked in part by a hard compact rock, about three yards thick, standing vertically and forming a parting wall between the sandstone and the serpentine. " This rock consists of equal parts of green serpentine and an indurated brick-coloured rock, harder than serpentine, and sometimes passing into jasper." The serpentine is also described as bounded, on the left bank of the Carity, by " a vertical mass of sandstone conglomerate, evidently much altered, about five yards thick. Some parts of this rock approach jasper in hardness and appearance." But the most interesting fact connected with this altered conglomerate is, that the quartz pebbles contained in it have been fractured and reunited,—a circumstance also noticed by Mr. Lyell in a conglomerate flanking a greenstone dyke on the Isla, also in Forfarshire. This fracture of the quartz pebbles is precisely what we should expect from a sudden application of heat, and would speak strongly in favour of the once igneous fusion of the serpentine in the dyke, if any evidence were wanting. That common association of serpentine with greenstone is here also observable, the dyke being bordered on the right side of the Carity by a fine-grained rock of that kind. The dyke can be traced at intervals for at least fourteen miles, stretching in a straight line from Cortachie to Banff†.

The serpentine and diallage rocks of Liguria are particularly instructive, as they occur under a variety of forms, and appear connected with the disturbed condition of the strata in that country. These rocks pass into each other in all directions, and into those of a trappean character (Levanto). Between Braco and Matanara the student may observe them with perfect ease, on the high road from Genoa to Florence, cut-

<hr />

* Macculloch, Brewster's Edin. Journ. of Science, vol. i.
† Lyell, Brewster's Edin. Journ. of Science, vol. iii.

ting through the limestone and slate as a dyke, insinuated between the strata, so as to seem interstratified, and constituting an enormous mass, apparently thrust upwards. The whole district is full of interesting facts of this nature.

If it has been correctly determined that the limestones of La Spezia are of the age of the oolitic groups of England, France, and Germany, the serpentine and diallage rocks of southern Liguria have been erupted since that period; for the La Spezia limestones and their associated deposits have been upheaved, contorted, and cut through by them. Possibly also the date of their intrusion may even be later, and of the supracretaceous period, for the lignite deposits of Caniparola, near Sarzana, are thrown into a vertical position, and I did not detect any serpentine or diallage rock pebbles among their associated conglomerates; this latter date, however, must be considered uncertain, for the serpentine rocks are not observed to be actually intruded among the supracretaceous deposits.

At Capo Mesco, between Levanto and Monte Rosso, gray schist and a compact calcareo-siliceous sandstone (one of the Italian *macignos*) are broken into faults by a mass of serpentine and diallage rock, which branches from a larger mass at Levanto. The valley of Rochetta, near Borghetto, has attracted much attention since it was noticed by M. Brongniart*. It shows the intrusion of the serpentine and diallage rocks (which here also pass into each other in various ways) among stratified rocks, similar to those which are observed at Capo Mesco. At the entrance into the valley, the sandstone is seen dipping at a considerable angle and resting upon gray limestone and schist, which are supported by serpentine. The serpentine then passes over contorted gray limestone and schist, and occupies a considerable portion of the valley, mixed with diallage rock, until the latter predominating to the exclusion of the serpentine, the mass rests on beds of red and green jasper, having the same dip as the sandstones at the entrance of the valley. These jasper beds repose on contorted gray limestone and schist, on the left bank of the river and opposite Rochetta. The jasper beds have sometimes been considered as a subordinate part of the serpentine: that it may be an altered rock is very possible; but I do not imagine it can be regarded as a portion of the unstratified mass of the diallage rock and serpentine, more particularly as similar jaspers occur among the limestones in the gulf of La Spezia, not far from Lerici, interstratified with them and distant from either serpentine or diallage rock.

The mass of serpentine and diallage rock which constitutes

* Annales des Mines, 1821.

2 H 2

the Monte Ferrato, north of Prato, in Tuscany, rests also upon stratified jasper on the west, and this again upon a schistose rock based on limestone: this also appears an accidental circumstance, for jasper is interstratified with brown shale at Paciana on the opposite side of the mountain, where it is not in contact with the serpentine. The diallage rock and serpentine here also pass into each other in all directions, and one variety of the former is worked for millstones. The whole seems a mass ejected from beneath, which has overflowed the stratified rocks, appearing to cut through them on the northward, beyond the north-west knoll, where there is a good section of the serpentinous mass resting on the jaspers, slates, and limestones.

At the Lizard, Cornwall, there is a well known mass of serpentine, which seems intimately connected with greenstones: unfortunately, however, from its position, we do not obtain any clear idea of its relative date*.

The volcanic rocks, at least such as have been considered the products of what are commonly termed modern and extinct volcanos, have already been noticed; therefore a statement of their general characters need not be repeated.

If we regard these various igneous products as a mass of matter which has successively, and during the lapse of all that time comprehended between the earliest formation of the stratified rocks and the present day, been ejected from the interior of the earth, we shall be struck with certain differences of these rocks on the great scale, which has led to their practical arrangement under the heads of granitic, trappean, serpentinous, and volcanic products, as above noticed. The two former and the last occur most abundantly, whilst the third is comparatively more scarce, though sufficiently common in nature.

It has been generally considered that the mineralogical character of igneous rocks has changed during the deposit of the stratified rocks, through which they have more or less forced their way; that is, we do not find granite and serpentine flowing from modern volcanos, nor trachite nor leucitic lavas intimately associated with the oldest strata in such a manner, that their relative differences of age could not be very considerable. Admitting that true mineralogical granite may be reckoned among the products of the supracretaceous period, the mass of granite is associated with the oldest rocks, even omitting all consideration of the gneiss, composed of the same minerals, and probably of exactly the same amount of elementary substances. The same with those igneous compounds into which

* For descriptions of this district, consult the memoirs of Prof. Sedgwick, Cambridge Phil. Trans. vol. i.; of Mr. Magendie, Trans. Geol. Soc. of Cornwall, vol. i.; and of Mr. Rogers, same work, vol. ii.

augite largely enters, which abound in the more recent products, while certainly they are scarce, if they be not altogether
absent, among the older rocks of an igneous origin; and we
have no stratified rocks of similar mineralogical composition
constituting extensive districts, as is the case with gneiss. We
are compelled therefore to admit, that the conditions, under
which the two kinds of igneous rocks have been formed, have
not been the same. What those conditions may have been is
a separate question, and one, as above noticed, requiring investigation; but it will be at once obvious, that the ejection
of a mass, in a state of fusion, into the atmosphere, would
be likely to have its constituent parts arranged in a different
manner from those in a similar mass forced out beneath great
pressure, such as we may consider to exist beneath deep seas
or a great mass of superincumbent rock. Independently,
however, of this consideration, there appears to have been
something in the condition of the world at the earliest times,
causing certain compounds to be formed in great abundance,
which does not now continue in such force as to permit the
production of similar compounds.

We cannot conclude this sketch of the unstratified rocks
without adverting to the concretionary and columnar structure
which they frequently assume. The most familiar examples
of the columnar structure are those of the basalt in the Giant's
Causeway and at Staffa, in the latter place constituting the sides
of the justly celebrated Fingal's Cave *. The concretionary or
globular structure is often visible in the decomposition of trappean and volcanic rocks, and is remarkable in a solid rock
named the orbicular granite of Corsica (*diorite orbiculaire*,
Al. Brong.), in which balls or spheroids of concentric and
alternate coats of hornblende and compact felspar are disseminated in the mass of the rock.

We are indebted to Mr. Gregory Watt for our first great
advance towards a knowledge of the circumstances which have
produced this structure. This author fused seven hundred
weight of an amorphous basalt named Rowley Rag, described
as fine-grained and of a confused crystalline texture; the fire
was maintained for more than six hours, and the fused mass
was suffered to cool very gradually, so that eight days elapsed
before it was removed from the furnace. The fused mass was
then three feet and a half long, two feet and a half wide, about
four inches thick at one end, and above eighteen inches at the
other. This irregularity of form, resulting from the shape of
the furnace, was highly advantageous, showing the arrange-

* See Macculloch's Western Islands of Scotland; and Sections and Views
illustrative of Geological Phænomena, pl. 11. and 19.

ment of the bodies passing from a vitreous to a stony state. A portion taken out while the basalt was in fusion became perfect glass. The most important result observed was the formation of spheroids, sometimes extending to a diameter of two inches. They were radiated with distinct fibres, the latter also forming concentric coats, when circumstances were only favourable to such an arrangement; but this structure gradually disappeared when the temperature was sufficiently continued, the centres of most of the spheroids becoming compact before they attained the diameter of half an inch. This structure gradually pervaded the whole body of the spheroid. " A continuation of the temperature favourable to arrangement speedily induces another change. The texture of the mass becomes more granular, its colour rather more gray, and the brilliant points larger and more numerous; nor is it long before these brilliant molecules arrange themselves in regular forms, and finally the whole mass becomes pervaded by thin crystalline laminæ which intersect it in every direction, and form projecting crystals in the cavities."

Mr. Gregory Watt applied the facts here noticed in explanation of the globular structure of many decomposing basaltic rocks, in which, after a certain stage of disintegration, the included balls resist decomposition with great obstinacy. He moreover extended his remarks to the columnar structure, and observed, that when in his experiments " two spheroids came into contact, no penetration ensued, but the two bodies became mutually compressed and separated by a plane, well defined and invested with a rusty colour;" and when several met, they formed prisms. His inferences from this arrangement were as follows :—

" In a stratum composed of an indefinite number in superficial extent, but only one in height, of impenetrable spheroids, with nearly equidistant centres, if their peripheries should come in contact on the same plane, it seems obvious that their mutual action would form them into hexagons; and if these were resisted below and there was no opposing cause above them, it seems equally clear that they would extend their dimensions upwards, and thus form hexagonal prisms, whose length might be indefinitely greater than their diameters. The further the extremities of the radii were removed from the centre, the nearer would be their approach to parallelism; and the structure would be finally propagated by nearly parallel fibres, still keeping within the limits of the hexagonal prism with which their incipient formation commenced; and the prisms might thus shoot to an indefinite length into the undisturbed central mass of the fluid, till their structure was deranged by the superior influence of a counter-acting cause[*].

Basaltic columns are often curved, and sometimes there is a somewhat confused arrangement of them, so that the disturbing causes have been considerable. They are also frequently articulated, which Mr. Watt considered might be produced by the same cause which determined the concentric fractures of the fibres of the spheroids. Supposing the general theory of the formation of columns correct, the irregularities of the prisms would obviously depend upon the unequal distances of the centres of the spheroids, and the consequent unequal pressure. Mr. Watt accounts for the horizontal arrangement of basaltic columns in some perpendicular dykes, such for example as those at the Giant's Causeway, from the refrigerating or absorbing cause operating on each side of the vein, so that columns should strike out from it, but would not coincide so as to form continuous prisms across the vein, as there would be confusion where the two sets of columns met, if indeed circumstances were sufficiently favourable to produce a meeting.

The columnar arrangement is not confined to basalt, but is more or less observed in all the trappean rocks, the magnitude of the columns being often very considerable. Granite also assumes a prismatic form, as has already been remarked by Mr. Carne respecting that of the western part of Cornwall[†], where it is well seen near the Land's End; but instead of assuming an hexagonal arrangement, such as we might presume to be the figure, if the theory respecting basalt was altogether applicable to it, it is quadrangular, and so divided into joints that the resulting solids are parallelopipeds and even cubes. If the student should pass round the Land's End, Cornwall, in a boat, he will be particularly struck with the general arrangement of granite into columns, and the picturesque effect is considerably heightened by the varied disintegration of the blocks from the united action of the sea and atmosphere.

While on the subject of the columnar structure of rocks, it may be remarked, that some of the stratified rocks, more particularly the arenaceous, are rendered columnar by the action of heat upon them. Some sandstones, when kept in our furnaces at a heat insufficient to fuse them, take this structure.

---

[*] Gregory Watt, Observations on Basalt, and on the Transition from the vitreous to the stony Texture which occurs in the gradual Refrigeration of melted Basalt; Phil. Trans. 1804.

[†] Carne on the Granite on the western part of Cornwall; Geol. Trans. of Cornwall, vol. iii. p. 208.

From the want of a good material for road-making between Halifax and Huddersfield, a coal-measure sandstone is thrown into kilns and burnt, when it frequently becomes columnar, the columns varying in the number of their sides. They are generally curved, and about half an inch in diameter.

Dr. Macculloch was, I believe, the first to connect this altered structure of a sandstone in a furnace with the columnar character of masses of sandstone in nature. He observed in a hearth-stone, taken down from a blast furnace, at the Old Park Iron Works, near Schiffnall, and which had been in constant work for sixteen or eighteen years, that its structure was prismatic. The prisms sometimes extended through the whole thickness of the stone, about ten inches, while in other instances they did not penetrate so deep. The prismatic structure was considered to have been produced by the long-continued action of considerable heat upon the slab of sandstone. Reasoning from this fact, the same author explains the occurrence of the columnar sandstone beneath basaltic rock at the hill of Scuirmore, in Rum, as also the columnar rocks at Dunbar, by the action of heat *.

Mr. Yates informs me that at about a mile N.E. of Bad-Ems, in the duchy of Nassau, where clay ironstone is in contact with a mass of trap, the former exhibits the red colour and columnar structure of the Staffordshire clay ironstone after it has been *roasted*, and previously to smelting†. Capt. Belcher notices a columnar sandstone on the Rio Nuñez, west coast of Africa, which appears, from the circumstances connected with it, to have assumed its columnar structure from the effect of heat acting on it after consolidation‡.

* Macculloch, Quarterly Journal of Science, 1829.
† Yates, MSS.
‡ Belcher, Journal of the Geographical Society, vol. ii. p. 282.

Section XII.

On the Mineralogical Differences in Contemporaneous Rocks, either Original, or resulting from Alteration after Deposition.

Among the variety of stratified rocks which have been noticed above, it will have been observed that there was much difference as well in the mineralogical composition as in the zoological character of the deposits. Some rocks are evidently formed from the destruction of others; some are chemical; while others appear as if they had suffered alteration subsequently to deposition. The rocks which have been produced by deposit from water, in which mud, sand, gravel, and great blocks were for a time mechanically suspended, have already been sufficiently discussed; and that they should not precisely resemble each other over considerable areas is only what would have been expected, as we cannot imagine any detritus so uniform as to be the same over considerable spaces, for it assumes a perfect uniformity in the transporting power, a constant, equal, and uniform supply of detritus, and a surface over which the transported substances were carried, so constituted as to offer an equal resistance throughout.

When a stratified rock is crystalline, it has evidently been chemically and not mechanically produced; but it remains to inquire whether such structure be original, or consequent on circumstances which have permitted an alteration of the rock. This investigation is one of considerable difficulty, as we cannot always obtain the data necessary for decision, since it is obvious that the same substance may often be obtained in different ways: thus crystalline carbonate of lime may either be produced directly by deposition from an aqueous solution of that substance, or common limestone may be fused by heat under pressure, and the results be similar. The same with many other substances. It therefore becomes a very difficult though always interesting question to discover, whether such stratified and crystalline substances have been produced in the one way or the other.

There are certain generalities on which we may base our investigations. If crystalline and stratified substances occur as sheets of matter, included among beds evidently of mechanical origin, and igneous rocks are not intimately connected with the whole, and there has been no violent disturbance of

strata, permitting the possible influence of gaseous compounds on the beds, we may fairly infer that the crystalline
rock was formed by aqueous chemical deposition, and that its
occurrence among decidedly mechanical compounds, merely
shows a difference in the condition of the medium from which
they have both resulted; there being solution in the one case,
and mere mechanical suspension in the other.

When we find uncrystalline strata assuming a crystalline
structure in the immediate vicinity of igneous rocks, so that
the crystalline and uncrystalline portions constitute different
parts of a common whole, the question assumes another character, and we have to inquire if this difference arises from an
alteration of a part of the whole, subsequently to deposition,
or whether it is the result of certain causes which have operated only on parts of the same whole during deposition.

It has been seen that dolomite, a crystalline compound of
carbonate of lime and carbonate of magnesia, occurs in the
oolitic series of Poland and Germany; therefore we should
not be surprised that it occurred in the same series in the
Alps, and apparently among the same rocks in Dalmatia and
Greece. From this presence of a particular crystalline compound in a given rock, and over a considerable area, we should
be led to consider, that circumstances existed during the formation of the rock which produced this compound over the
area, and consequently, that it was original, and did not result from the subsequent application of heat, or any other
chemical agent.

Supposing compounds of this nature to have resulted from
an aqueous solution of the carbonates of lime and magnesia,
we should not be surprised at the absence of organic remains;
for it would scarcely be a mixture in which animals would
flourish. Organic remains are however not absent from dolomite, though they are rare in it, for I have seen them in the
dolomite of Nice, and they have been noticed elsewhere. The
occurrence of organic remains in dolomite does not, it must
be confessed, well accord with the supposition that it has been
a limestone on which chemical agents have subsequently so
acted that it became crystalline and charged with magnesia;
for we cannot well understand in the new arrangement of particles how the form of the organic remains could be preserved,
more particularly when they are of the same substance with
the rock, or solely carbonate of lime. The fossiliferous dolomites would therefore appear to be excluded from the altered rocks, and reduced to those of original and chemical
deposition. There are however masses of dolomite not so
easily reconcileable with the supposition of aqueous deposition, which occur in patches among limestones, and not far

removed from igneous rocks, in such a manner that Von Buch considers them to have resulted from the action of chemical agents on the limestones, subsequently to the deposition and consolidation of the latter, and at the time when igneous rocks were intruded among the stratified mass. To convert a series of beds into a crystalline mass to a certain distance from a rock in a state of fusion, provided there was sufficient pressure to prevent the escape of the carbonic acid, would, we know, not be difficult: but it is difficult to obtain the magnesia necessary to produce the dolomite, unless it was insinuated into the altered mass at the time when the various particles were arranging themselves conformably to the laws of crystallization; in fact, when all the elementary substances were so circumstanced that they could freely unite according to their proper affinities. Von Buch considers this was effected by the escape of magnesia from the augite porphyries or *melaphyres**, at the period when such porphyries were protruded through limestones, as in the Tyrol and other places. He is of opinion that the gas evolved at the time these igneous rocks were upheaved, entered among the fissures of the limestone, and converted a considerable proportion of it into dolomite. This celebrated author adduces the mountain of San Salvador, on the lake of Lugano, as confirming the truth of his theory. A red conglomerate, of a similar character to that which occurs on the lake of Como, separates the mica slate on which Lugano is situated from the limestones and dolomite. "These beds dip rapidly at 70° to the south, and form a promontory on which the chapel of San Martino is built. This rock appears in place for about ten minutes walk, the dip of the beds diminishing to 60°. It is then covered by a compact smoke-gray limestone, in beds about a foot thick. These dip as the beds on which they rest, and have the same inclination on the side of the mountain; but in their prolongation towards the lake the dip continually diminishes, until, at its level, it is scarcely 20°. The beds as they rise describe a curve that somewhat resembles a parabola. The further we advance on the road, the more we find these beds traversed by small veins, the sides of which are covered by rhombs of dolomite. Similar crystals are also observable in small cavities of the rocks. As we advance, the rock appears divided into fissures, and the stratification ceases to be

---

* If we consider with M. Rose, that augite and hornblende constitute one mineral, we shall find, judging at least from published analyses, that the magnesia contained in it varies materially. The extremes are the hornblende from Pargas, which, according to Bonsdorff, contains 18·79 per cent. of magnesia, and the augite of Taberg, containing, according to H. Rose, 4·99 per cent. of the same substance.

distinct. Lastly, where the face of the mountain becomes
nearly perpendicular, it is found to be entirely formed of dolomite. There is no marked separation between the limestone
and the latter rock. By the increase of the veins and geodes
the calcareous rock entirely disappears, and pure dolomite
occurs in its place. * * * * As we advance along the high
road, the purer we find the dolomite, and at the same time
the more white and granular. * * * * From hence to beyond
Melide the mountains are composed of dark augite porphyry
mixed with epidote, the same as it appears at Campione, Bis-
sone, and Rovio *."

This is undoubtedly a remarkable case, as the mass of augitic rock is on the side of the dolomite, and as crystals of
dolomite are found in the cracks of the limestones, because
the latter circumstance shows that such crystals were not contemporaneous with the deposition of the limestone, but were
formed subsequently, after cracks had been produced in it,
while the former circumstance is precisely in accordance with
the theory. According to Von Buch's sections, however, a
small quantity of red porphyry and mica slate intervenes between the mass of the augite porphyry and the dolomite: it
certainly does not follow that they should constantly intervene,
because they may do so in one situation and not in another,
therefore this is no great objection; indeed, according to the
map, they do not always separate the dolomite from the igneous
rock. Other masses of dolomite occur round a large patch of
granite extending westward from the south-western branch of
the Lago di Lugano, on which are situated Casco al Monte
and Porto. One of these masses at Monte Schieri is connected
with tufaceous augite rock; while others are only in contact
with the granite, as far at least as regards the surface: but
this proves little, for the augite porphyry may be beneath
them, as it is seen to cut through the granite at Brincio.

From its vicinity to these places the dolomites of the lakes
of Como and Lecco acquire considerable interest, although
augite rocks have not yet been detected among them. They
certainly occur intermingled with the limestones, which are
compact and gray, while at other times they also appear the
prolongation of limestone beds which have gradually lost
their compact texture, and at the same time have acquired
magnesia and become crystalline. In countries like these,
where so much confusion prevails, and where we may expect
to find extensive faults, a perfect continuance of a given series

* Von Buch, Ann. des Sci. Nat. 1827, where there are sections and a
map of the district of the lakes Orta, Maggiore, and Lugano; as also in
Sections and Views illustrative of Geological Phænomena, pl. 8. fig. 2. and
pl. 30.

of beds is most difficult to trace; but with ample allowance for these difficulties, there seems every probability that the continuations of some of the limestone beds become dolomite*.

The north part of the lake of Como is composed of gneiss and mica slate, which correspond on either shore, and dip southwards; the lake of Lecco and the southern part of that of Como are formed of limestones and dolomite. Between the two masses of rock are conglomerates and sandstones which have the same dip and direction as the gneiss and mica slate. On the south of these latter rocks the sides of the lake cease to correspond. Thus, on the eastern shore, after passing a small portion of dolomite, we find limestones, among which are the black marbles of Varenna, and these continue as far as opposite to Bellaggio Point, while on the other and western side dolomite prevails during the whole corresponding distance, with the exception of a few limestone beds on the south of Menaggio. Here then is no correspondence; on the contrary we have limestones on the one side and dolomite on the other, the latter containing a mass of gypsum at Nobiallo. If we proceed down the lake of Como, from Bellaggio to Como, we find nothing but limestones, producing the fine scenery for which this lake is so celebrated, after having passed the promontory of Dosso d'Albido and the opposite shores of the Croci Galle: but if we go down the lake of Lecco to the town of the same name, also from Bellaggio, there is scarcely anything to be seen but dolomite, if we except a mass of gypsum included in it at Limonta, and a long strip of limestone between Lierna and Mandello. Here also we have no correspondence, though the general direction of the beds in both lakes would lead us to suspect, that those in the one were continued from those in the other. And this view is strengthened if we ascend the Monte San Primo, a mountain already noticed as strewed over with thousands of erratic blocks, for the highest crest is composed of limestone ranging W.N.W. and E.S.E, with a dip to the S.S.W. If we follow the direction of these beds to the lake of Lecco on the east, we find dolomite, so that the change in this place appears somewhat sudden.

Notwithstanding this apparent conversion of limestone into dolomite in the direction of the beds, which might lead us to suppose that some cause had produced a change in the rock after its consolidation, it must be confessed that there is also an interstratification of the dolomite with the limestone, and moreover, the dolomite rests upon limestone in the lake of

* See Geological Map and Sections of the shores of the lake of Como, in Sections and Views illustrative of Geological Phænomena, pl. 31. and 32.

Lecco; facts which are at variance with the supposition that all the dolomites of this part of Italy have been altered rocks. Some of them at least appear to have been original deposits; and this supposition, as far as it affects the rocks which are apparently limestones in one place and dolomites in another, involves a curious question; for if we admit a contemporaneous deposition of the two to have taken place, it follows, that carbonate of lime was thrown down in one place, while a mixture of the carbonates of lime and magnesia was deposited close to it in another, and that the two depositions were so far influenced by circumstances, that the one was compact while the other was either wholly or semicrystalline. These observations do not apply to the alternations, as there we have to suppose a change of circumstances over the same place, and this by no means sudden; for the calcareous beds seem gradually to acquire magnesia, as may be seen on the western shores of the lake of Lecco.

Some good examples of a mixture of dolomite and limestone may be observed near Nice, where again the continuation of limestone beds becomes dolomitic, the dolomite here, as elsewhere, generally losing its stratification when most pure, while the less pure semicrystalline compounds are distinctly stratified. This is however not a general rule, for I have seen some nearly pure beds stratified; and supposing such beds to be original deposits, their division into strata is no more remarkable, than that the saccharine marble of Carrara should be stratified.

In the vicinity of Nice, gypsum also accompanies the dolomite, and the connexion of the two is so intimate that the gypsum of Sospello contains rhombs of dolomite, a circumstance also observed in the gypsum accompanying the dolomites in the Tyrol. This frequent association of gypsum with dolomite has not yet been satisfactorily explained.

Gypsum is not a rare accompaniment of rocks, even those of a decidedly mechanical origin, yet it must be considered either as a chemical deposit, or an altered rock; its occurrence therefore in such situations shows that other causes were in force than a mere drift of detritus; and when gypsum is to a certain extent characteristic of a deposit over a considerable area, it proves that the operation of such causes has not been local, but that during such period, and over such area, the circumstances permitting those deposits prevailed extensively. Gypsum has been considered characteristic of the upper part of the red sandstone series, generally known as red or variegated marls. Without however asserting that it is a necessary part of the rock, or that it is constantly present, it is remarkable how very frequently it is discovered in this deposit, in

England, France, and Germany, proving that circumstances were then favourable to its production, if it proves nothing more.

When we recollect that the intrusion of igneous rocks has been sufficient to convert chalk into granular limestone in the north of Ireland, we need not be surprised that other rocks have been altered by the intrusion of similar substances. The slates for instance in many parts of the country surrounding the granite of Dartmoor, Devon, have suffered from its intrusion, some being simply micaceous, others more indurated and with the characters of mica slate and gneiss, while others again appear converted into a hard zoned rock strongly impregnated with felspar.

The alteration of the rocks in this case is of very easy explanation. The grauwacke, which is for the most part the altered rock, is, when taken in the mass, only the consolidated detritus of more ancient and crystalline rocks, composed of a few simple substances. These substances have necessarily been variously accumulated in different beds, so that their relative proportions would also vary. If long-continued heat, insufficient to produce fusion, be now applied to the ends of these beds, thus differently constituted, the results will necessarily be dissimilar. The various substances would have a tendency to resume their original state, at least that state in which they existed in the crystalline rock, whence they have been derived; and consequently we should have compounds resembling various crystalline stratified rocks. The granite of Dartmoor has evidently been intruded among the grauwacke of the same district, sending veins into it. It to a certain extent cuts through the line of direction of the grauwacke, twisting and contorting the strata, which come more immediately in contact with it, to various distances. Such strata must have been exposed to long-continued heat, insufficient to fuse them, and hence their altered character.

M. von Dechen remarks, that the grauwacke of the Hartz is in like manner altered by granite, which converts the former into flinty slate, quartz rock, greenstone, &c. Transitions of this kind, even into coarse-grained grauwacke, may be observed. The slates of the Vosges and Swartzwald are also altered by granite.

It might at first sight appear necessary that complete fusion should take place before the rock could be converted into substances like hornblende rocks, &c., and that therefore the stratified character of the rock would be destroyed. The common experiment of throwing a fragment of green bottle glass into a fire, and keeping it for a long time at a heat insufficient to fuse it, shows that the internal particles of a body may, under certain circumstances, arrange themselves in a crystalline form without any alteration of the external appearance of the same body. In like manner a bed of grauwacke composed of comminuted hornblende and felspar, probably resulting from the destruction of some hornblende rocks, would, when exposed to a long-continued heat, insufficient to produce fusion, become converted into a greenstone or hornblende rock, arranged in beds.

Cases of induration and alteration of rocks in contact with igneous products are so common that it would be useless to enumerate them; but the student must carefully distinguish between igneous rocks which have evidently been intruded among the others, and those which are the older rocks, upon which the others have been deposited; for it may happen that the older rocks have been disintegrated previous to deposition of the superincumbent substance: in which case, if the latter be arenaceous, there may be an apparent passage of the arenaceous into the igneous rock, producing the false appearance of an altered substance.

There can be little doubt that many rocks resulting from the deposit of detritus have, more particularly when such detritus has been in a highly comminuted state, suffered alteration by the chemical action of various substances on each other. Changes have also been produced by the percolation of water, charged with various substances, through rocks. Matter soluble in water is thus conveyed from a superior to an inferior rock. The cavities or vesicles of igneous rocks filled with carbonate of lime and other minerals, constituting an amygdaloid, afford a well-known illustration of this percolation.

The change in the mineralogical character of certain calcareous rocks, at different points of the area which they may happen to cover, has been previously adverted to, and it has been shown that in the oolitic group there was a probability of a portion having been produced at the bottom of a deep sea, while other parts were formed in shallower waters. The physical circumstances under which the different parts of the deposit would be placed, could scarcely do otherwise than influence the product; but what that precise influence may have been we are as yet ignorant: it may, however, be anticipated that the differences of pressure, and the liability to be disturbed by currents in one situation, while the latter might be scarcely felt, if not entirely absent, in another, would alone cause a great variation in the mineralogical texture.

It might also happen, that in a deep part of an ocean successive depositions were effected during periods when frequent changes were produced in other and remote situations, so that though contemporaneous there should be no mineralogical agreement between them; and if in the course of events, the continuous and quiet deposits were upheaved, and a continent be the result, the difficulty of identifying clear divisions in the one place with the mass in the other would be insurmountable. It is more than probable that this supposition has been realized on the surface of our planet, and that eventually geologists will show less determination in identifying deposits, more particularly those of moderate comparative antiquity, over very considerable distances. It is much more desirable, for instance, that India should be described with reference to itself,— so that when its geology shall have been sufficiently advanced, Europe may be fairly compared with it,—than that there should be a determination to find nothing but European equivalents in that quarter of the world.

On the Elevation of Mountains.

Although the direction of the various chains of mountains has long engaged the attention of geologists and geographers, and although the direction of upturned strata has also long been noticed by the former, and found generally to coincide with that of the mountain chains which they constitute, it has only been recently and since the labours of M. Elie de Beaumont, that the subject has acquired a new interest, and will henceforth form an important branch of geological investigation, whether the theory of this distinguished geologist shall eventually be found tenable to the extent supposed, or require very material modifications.

Von Buch appears some time since to have discovered that the mountain-systems of Germany were not contemporaneous, but were of distinct dates; and geologists had long been in the habit of noticing the unconformable position of strata, an older and inferior rock having been upheaved, while a newer rock rested upon the edges of the upturned strata. Here, however, the subject seemed to rest, until M. Elie de Beaumont, from a series of very exact observations in certain parts of France and the Alps, remarked, that the dislocations of the strata were not only referrible to distinct epochs, but that there was a parallelism between dislocations and upheaved mountains of the same date; and he further considered that these events produced breaks in the rocks then in the course of deposition, so that those subsequently formed rested unconformably on the disturbed strata of the older rocks.

To determine the general unconformable position of two rocks sometimes requires very great care, though at first sight it may appear extremely easy to observe whether one rock rests on the upturned edges of another, or not; and so it undoubt-

edly is in many cases; but when they meet at small angles*,
or the one rests on the contortions of the other, the inquiry
becomes more difficult, and it requires numerous observations
to be certain of the general fact.   The difficulty in the case of
contortions will be seen in the annexed cut.

If a section be obtained on
the left or right, it will very   Fig. 116.
easily be seen that the beds
a a rest upon the contorted
strata *b b*; but if one were
only observed at *c*, the unconformable position of the two
rocks might be doubtful. The student may perhaps consider,
that from a little research in the same place he would soon
find evidence of the disturbed condition of the strata beneath;
and if the contortions were always on the small scale, and natural sections common, such would be the case: but when the
latter are scarce, or the undulations and contortions on the
large scale, the great bends being measured by miles instead
of fathoms, the subject is not so easy.   It may be stated as an
example, that the mass of the calcareous Alps is considered
to rest unconformably on the mass of those composed of protogine, gneiss, &c.; but the situations where the contrary
opinion may be formed are very numerous, the sections there
exposing perfect conformability.   It also requires great care
in tracing strata up to a mountain range, for the purpose of
ascertaining its relative antiquity, to distinguish between those
beds which have been decidedly upturned subsequently to deposition, and those which may have originally taken a small
angle during their formation on the flanks of a chain previously elevated to a certain extent.

Perhaps the annexed diagram may assist the student in comprehending the manner in which the relative age of mountain
ranges is determined from the position of strata.

If the rocks *a a* are found resting quietly on the upturned
strata *b b*, it is inferred that *b b* have been disturbed previous

Fig. 117.

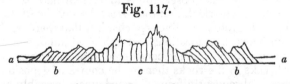

* A general unconformability does not always prove a movement in the
inferior rocks prior to the deposition of the superior; for supposing a given
series to be so produced that the newer rocks may be formed within successively diminishing areas, and another deposit to cover the whole, it is evident that the higher mass will so far rest unconformably on the inferior
rocks that it will cover them all in succession.   Now this is what has hap-

to the deposition of *a a*; and consequently, if *a a* be a known rock occupying a certain place in the series, we obtain a relative date for the elevation of *b b*, so far as relates to *a a*; but if it should chance that there are no commonly known deposits absent between them, we obtain the exact relative date of the elevation of so much of the mountains as do not exhibit any other unconformability of strata, and of the whole range, if no such unconformability can be detected. When however this does occur, as in the above diagram, we learn, that more than one elevation of strata has taken place in the same chain, for *b b* resting in discordant stratification on *c*, proves that *c* was tilted up prior to the deposition of *b b*; so that in this case there would be evidence of two disturbances in the same chain, and the relative dates of both would be obtained on the same principles. It will be obvious if two lines of elevated strata cross each other, there would be much confusion where they traverse; and should great violence have been employed, so that the strata be even thrown over, it will require much caution to determine the relative age of the fractures at those points.

From the obliging communications of M. Elie de Beaumont, it appears that he now considers he can distinguish twelve systems of mountains in Europe, each regarded as characterized by the relative direction and elevation of its strata, and each elevation as corresponding with a solution observed in the continuity of the sedimentary deposits, also of Europe*. These various systems are as follows, commencing with that of the greatest relative antiquity:

I. *System of Westmoreland and the Hundsruck.*—The direction of the slate rocks in Westmoreland is, according to Prof. Sedgwick, N.E. by E. and S.W. by W.; and that of the slates and grauwacke of the Eifel, the Hundsruck, and of the Nassau Mountains, about N.E. and S.W. The slates, grauwacke, and grauwacke limestones of the northern and central part of the Vosges have the same direction. The carboniferous rocks rest on the upturned rocks of the North of England; coal-measures repose upon the edges of those of the Vosges; and the carbo-

---

pened with the chalk and oolite groups in England, the former overlapping the various members of the latter as they successively fine off. See Sections and Views illustrative of Geological Phænomena, for an overlap of the chalk on the coasts of Dorset and Devon. The angles at which the cretaceous and other rocks meet is there so small, that their unconformability could scarcely be determined at any particular point, though in the mass it is evident.

* M. Elie de Beaumont's communication to me, of which the accompanying sketch is a brief notice, will be found in the Phil. Mag. and Annals of Philosophy, for October 1831.

On the Elevation of Mountains.

niferous rocks of Belgium and Saarbruck were probably deposited at the foot of the Eifel, Hundsruck, &c.

II. *System of the Ballons (Vosges), and of the Hills of the Bocage (Calvados)*.—It is observed under this head, that the first system only shows that the slates and grauwacke of Westmoreland and the Hundsruck have been elevated before the deposition of the carboniferous series; but it would also appear that there has been an elevation of strata before the more recent transition strata were deposited, so that the latter have not been elevated in a N.E. and S.W. direction, but would on the contrary seem to have been formed on upturned beds of the former. Such are the calcareous, marly, and arenaceous deposits, with *Orthoceratites, Trilobites*, &c., in Podolia, of the environs of St. Petersburg, and of Sweden, where they are but slightly removed from their original horizontal position. Such are also the transition beds of Dudley and Gloucestershire; and possibly also the transition beds of the South of Ireland may be included in the same list, which M. Elie de Beaumont considers may also contain certain slate and grauwacke beds with anthracite, forming the south-east angle of the Vosges. When these beds are not horizontal, they are dislocated in directions the most marked of which is comprised between an E. and W. line and one E. 15° S. and W. 15° N.

III. *System of the North of England*.—This is the north and south range of the carboniferous series, noticed by Prof. Sedgwick. It is considered to have been produced immediately previous to the deposit of the red conglomerate (rothliegendes). M. Elie de Beaumont remarks; "The elevation of the chain in the North of England is not probably an isolated phænomenon; but if we glance at the geological map of England by Mr. Greenough, or that accompanying the memoir of Prof. Buckland and Mr. Conybeare on the environs of Bristol, we are naturally led to remark, that the problematical rocks which penetrate and dislocate the coal deposits of Shrewsbury and Colebrooke Dale, and those which form the Malvern Hills, appear to be connected with a series of fractures which run nearly N. and S., and are prolonged across the more recent transition and the carboniferous rocks to the environs of Bristol." It is also considered probable that the form of the west coast of the department of La Manche, which runs nearly N. and S., may be due to a fracture of this age.

IV. *System of the Netherlands and South Wales*.—This is the great E. and W. range of the carboniferous rocks from the environs of Aix-la-Chapelle to St. Bride's Bay, Pembrokeshire, which, whenever visible from beneath other deposits, exhibits this general direction for about 400 miles. It is considered that the beds of the (new) red sandstone series which repose on this dislocation are not so ancient as those noticed in the previous group*.

V. *System of the Rhine.*—The Vosges and the Swarzwald terminate opposite one another in two long cliffs parallel to each other and to the course of the Rhine. These are apparently due to great faults, having a direction S. 15° W. and N. 15° E. These fractures preceded the deposit of the rocks in the basin of Alsace, among which are the red or variegated sandstone (grès bigarré), the muschelkalk, and the variegated marls.

VI. *System of the S.W. coast of Brittany and of La Vendée, of Morvan, of the Böhmerwaldgebirge, and of the Thuringerwald.* —The general direction of this system is from N.W. to S.E.; and while the red or variegated marls (marnes irisées), the muschelkalk, and all older strata have been thrown out of their original positions, the oolitic series, comprehending the lias and its lower sandstone, have remained undisturbed in these various situations.

VII. *System of the Pilas, the Côte d'Or, and the Erzgebirge.* —In this system, which also contains a portion of the Cevennes, the strata are disturbed up to the oolitic rocks inclusive, while the cretaceous series (green sand and chalk) remain apparently in the position in which they were deposited. The direction of this system is considered to be N.E. and S.W.

VIII. *System of Mont Viso.*—" The French Alps and the south-west extremity of the Jura, from the environs of Antibes to those of Pont d'Ain, and Lons le Saulnier, present a series of crests and dislocations with a direction towards the N.N.W., in which the older rocks of the Wealden formation, the green sand, and chalk, are found upheaved, as well as those of the oolitic series. The pyramid of primæval rocks composing Mont Viso is traversed by enormous faults, which, from their direction, evidently belong to this system of fractures. The eastern crests of the Devolny, on the north of Gap, are formed of the oldest beds of the green sand and chalk thrown up in the direction in question, and raised more than 4700 English feet above the sea. At the feet of these enormous escarpments are, horizontally deposited and at more than 2000 feet

---

* It should be here noticed that some recent observations among the Mendip Hills have shown me that they have been dislocated in N. and S. lines, subsequently to their E. and W. elevation. These faults are the cross courses so well known to the miners of the district, and exhibit disturbances parallel to the lines of dislocation in System III. Faults with a general N. and S. direction are also observable in the Blackdown Hills, to the south of the Western Mendips. These latter were clearly produced after the deposit of the chalk, and perhaps of certain supracretaceous rocks: whether they may or may not be contemporaneous with the N. and S. Mendip faults is not yet determined.

lower down, those upper beds of the cretaceous system which are distinguished from the rest by the presence of *Nummulites, Cerithia, Ampullariæ,* and other shells, the genera of which were long considered as exclusively found in the tertiary (supracretaceous) rocks. Thus it was between the two portions of that which is commonly termed the series of the Wealden formation, green sand, and chalk, that the beds of the Mont Viso system have been thrown up."

IX. *Pyreneo-Apennine System.*—" This includes the whole chain of the Pyrenees, the northern and some other ridges of the Apennines, the calcareous chains on the north-east of the Adriatic, those of the Morea, nearly the whole Carpathian chain, and a great series of inequalities continued from that chain through the north-east escarpment of the Hartz Mountains to northern Germany." The general direction of this system is about W.N.W. and E.S.E.

X. *System of the Islands of Corsica and Sardinia.*—This elevation is considered to have taken place during the supracretaceous period, and it is remarked that the north and south direction of the system in Corsica and Sardinia is observed "in many small valleys and ridges of mountains in the Apennines, and in Istria, and in the disposition of many volcanic masses and metalliferous sites of Hungary."

" It is worthy of remark that the directions of the system of the Pilas and the Côte d'Or, of that of the Pyrenees, and that of the islands of Corsica and Sardinia, are respectively nearly parallel to those of the system of Westmoreland and the Hundsruck, of the system of the Ballons and of the Bocage, and of the system of the North of England. The corresponding directions differ but a small number of degrees, and the corresponding systems of the two series have succeeded each other in the same order; leading to the supposition that there has been *a kind of periodical recurrence* of the same, or nearly the same, directions of elevation."

XI. *System of the Western Alps.*—The mean direction of this system is about N.N.E. and S.S.W., and the elevation is considered to have taken place after the deposit of those recent supracretaceous beds named Shelly Molasse (*molasse coquillière*), beds contemporaneous with the *fahluns* of Touraine.

The direction of strata is of a complicated character where this system and that to be next mentioned cross each other, as they do around the Mont Blanc, Mont Rose, and Finsteraarhorn, at about an angle of 45° or 50°.

XII.—*System of the principal Chain of the Alps (from the Valais into Austria), comprising also the Chains of the Ventoux, the Lebaron and the Sainte Baume (Provence).*—The direction of this system is about E. $\frac{1}{4}$ N.E. and W. $\frac{1}{4}$ S.W., and the strata are considered to have been elevated previous to the dispersion of the erratic blocks and those gravels which have been termed *diluvium*, but which in the vicinity of the Alps are found to have been deposited upon other gravels, often of considerable thickness.

M. Elie de Beaumont concludes with the following observations: "The independence of successive sedimentary formations is the most important result obtained from the study of the superficial beds of our globe; and one of the principal objects of my researches has been to show, that this great fact is the consequence and even a proof of the independence of mountain-systems having different directions.

"The fact of a general uniformity in the direction of all beds upheaved at the same epoch, and consequently in the crests formed by these beds, is perhaps as important in the study of mountains, as the independence of successive formations is in the study of superimposed beds. The sudden change of direction in passing from one group to another has permitted the division of European chains into a certain number of distinct systems, which penetrate and sometimes cross each other without becoming confounded. I have recognised from various examples, of which the number now amounts to twelve, that there is a coincidence between the sudden changes established by the lines of demarcation observable in certain consecutive stages of the sedimentary rocks, and the elevation of the beds of the same number of mountain-systems.

"Pursuing the subject as far as my means of observation and induction will permit, it has appeared to me, that the different systems, at least those which are at the same time the most striking and recent, are composed of a certain number of small chains, ranged parallel to the semi-circumference of the earth's surface, and occupying a zone of much greater length than breadth; and of which the length embraces a considerable fraction of one of the great circles of the terrestrial sphere. It may be observed respecting the hypothesis of each of these mountain-systems being the product of a single epoch of dislocation, that it is easier geometrically to conceive the manner in which the solid crust of the globe may be elevated into ridges along a considerable portion of one of its great circles, than that a similar effect may have been produced in a more restricted space.

"However well established it may be by facts, the assemblage of which constitutes positive geology, that the surface of the globe has presented a long series of tranquil periods, each separated from that which followed it by a sudden and violent convulsion, in which a portion of the earth's crust was dislocated,—that, in a word, this surface was ridged at intervals in different directions; the mind would not rest satisfied if it did not perceive, among those causes now in action, an element, fitted from time to time to produce disturbances different from the ordinary march of the phænomena which we now witness.

" The idea of *volcanic action* naturally presents itself when we search, in the existing state of things, for a term of comparison with these great phænomena. They nevertheless do not appear susceptible of being referred to volcanic action, unless we define it, with M. Humboldt, as being *the influence exercised by the interior of a planet on its exterior covering during its different stages of refrigeration.*

" Volcanos are frequently arranged in lines following fractures parallel to mountain chains, and which originate in the elevation of such chains; but it does not appear to me that we can thence regard the elevation of the chains themselves as due to the action of *volcanic foci*, taking the words in their ordinary and restricted sense. We can easily conceive how a *volcanic focus* may produce accidents circularly and in the form of rays from a central point, but we cannot conceive how even many united *foci* could produce those ridges which follow a common direction through several degrees.

" Volcanic action, such as it is commonly understood, could not therefore be itself the first cause of these great phænomena; but volcanic action appears to be related (and this is a subject which has long occupied M. Cordier, though he has considered it under another point of view,) with the high temperature now existing in the interior of the globe.

" Now the secular refrigeration, that is to say, the slow diffusion of the primitive heat to which the planets owe their spheroidal forms and the generally regular disposition of these beds from the centre to the circumference, in the order of specific gravity,—the secular refrigeration, on the march of which M. Fourier has thrown so much light, does offer an element to which these extraordinary effects may be referred. This element is the relation which a refrigeration so advanced as that of the planetary bodies establishes between the capacity of their solid crusts and the volume of their internal masses. In a given time, the temperature of the interior of the planets is lowered by a much greater quantity than that on their surfaces, of which the refrigeration is now nearly insensible. We are, undoubtedly, ignorant of the physical properties of the matter composing the interior of these bodies; but analogy leads us to consider, that the inequality of cooling above noticed would place their crusts under the necessity of continually diminishing their capacities, notwithstanding the nearly rigorous constancy of their temperature, in order that they should not cease to embrace their internal masses exactly, the temperature of which diminishes sensibly. They must therefore depart in a slight and progressive manner from the spheroidal figure proper to them, and corresponding to a maximum of capacity; and the gradually increasing tendency to revert to that figure, whether it acts alone, or whether it combines with other internal causes of change which the planets may contain, may, with great probability, completely account for the ridges and protuberances which have been formed at intervals on the external crust of the earth, and probably also of all the other planets*."

From this sketch of M. Elie de Beaumont's theory, in which his views respecting the connexion of distant mountains with those of Europe have been omitted, it will be evident that it will require much time and very exact observation in various parts of the world before we can fairly ascertain what are exceptions and what the general rules. It will have been observed that M. Elie de Beaumont has already remarked on the near parallelism of three particular systems respectively with three other particular systems of European mountains, leading to the presumption that parallelism is alone insufficient to determine the relative age of an elevated range of strata; a conclusion that may be still further strengthened by observing certain lines of disturbed strata in the British Isles, which, when we regard the general surface of the world, are not far distant from each other.

The disturbed strata in the Isle of Wight range east and west, as do those also in the Weymouth district, in the Mendip Hills, in a large part of Devonshire, and in South Wales. The date of the elevation of the Isle of Wight beds was certainly posterior to the deposition of the London clay, and there would appear little reason to doubt that the disturbed and fractured condition of the Weymouth district was effected at the same time. But when we continue our researches into Devonshire, we find that the east and west arrangement of a large proportion of the grauwacke in that country was produced anterior to the deposit of the (new) red sandstone series, since the latter rests upon the upturned edges of the former †. If we now proceed northwards to the carboniferous rocks of the Mendips and South Wales, we find they also have suffered an elevation in an east and west direction, prior to the formation of the (new) red sandstone; so that in the southern parts of England the strata have been twice elevated in a given direction at dif-

---

* Elie de Beaumont, MSS.

† Both the one and the other have been subsequently fractured, and many of the faults have somewhat of east and west direction.

ferent dates; and if we continue our researches, we find, from the observations of Mr. Weaver, that the grauwacke in the South of Ireland was elevated previous to the deposition of the old red sandstone, also in an east and west direction; thus affording three elevations of strata (not far distant from each other) in the same direction, but at different dates *.

In offering these observations it is by no means intended to combat the general principle of the possibly contemporaneous elevation of strata at various and distant places, resulting from the gradual refrigeration of the globe; beds of various kinds having been subsequently and quietly deposited over large areas, on such disturbed strata; but simply to remark, that, though parallelism may frequently exist, it may not be a necessary condition of contemporaneously elevated strata; for perhaps by laying too much stress upon this point, we not only are in danger, in the present state of our knowledge, of permitting theory to take the lead of facts, but of shutting ourselves out from a consideration of other possible lines which contemporaneously elevated strata may follow. And should it eventually be discovered that contemporaneously disturbed beds are by no means parallel though still in straight lines, it does not appear that the main principle of M. Elie de Beaumont's theory would be affected by it. What lines may eventually be found to prevail, will, as previously remarked, require much time and great patience to discover; but let the event be as it may, geologists will not the less have reason to feel thankful to M. Elie de Beaumont for having rescued the subject from the state in which he found it; it being impossible but that the investigations, to which this theory will necessarily give rise, must end in the most important additions to geological knowledge †.

It has already been noticed by Prof. Sedgwick, that the change in the zoological character of deposits has not always coincided with disruptions of strata; and the student will have collected from the foregoing pages, as indeed is also remarked by Prof. Sedgwick, that there was, in Europe, no important change in the general zoological character of deposits up to the zechstein inclusive; the first great alteration, as far as we

* It has been considered that the north and south line of the carboniferous rocks in Northern England was elevated at a different epoch from the east and west line of South Wales and Somersetshire, but it must be confessed that this point is far from being proved.

† It is but right to inform the student that various objections have been made to this theory by different geologists, more particularly as relates to some of the lines of elevation and their relative dates. Dr. Boué, who was among the first to notice that mountain-masses were elevated at different dates, has inserted a series of objections in the Journal de Géologie, tom. iii. p. 338.

can at present see our way, being observed in the remains entombed in the variegated sandstone (grès bigarré), and muschelkalk. It has already been observed, but may be conveniently repeated here, that the effect produced on animal and vegetable life by an upburst of a line of rocks sufficient to produce a range of mountains, might destroy all terrestrial animals and even a large proportion, if not the whole, of the vegetation within the influence of the disturbing cause, not only by producing a deluge over the land, which might wash off the animals and carry away a great proportion of the vegetation, but by elevating such vegetation into colder regions of the atmosphere where it could no longer exist. In this case we suppose land so situated as to produce plants and to support terrestrial creatures; but it will be evident that if we admit a contemporaneous disruption of strata at different points, it would take place under various conditions. In one place it may be effected in the atmosphere, in another in shallow seas, and in a third beneath a great depth and pressure of the ocean; consequently the resulting phænomena would be as various as the conditions under which the disruption and elevation of strata were produced: but it will be obvious that the destruction of marine life would be very difficult, and we can scarcely, from known facts, consider that there has been a disruption of the rocks composing the earth's surface so general as to annihilate all marine creatures at a given time, even with every allowance for the operations of powerful and destructive currents; though we can conceive that near the centres of every great disturbance there might be a very great, and as far as related to certain areas complete destruction of marine creatures.

On the Occurrence of Metals in Rocks.

To enter fully into this subject would require a volume; the following notice is therefore solely intended to call the attention of the student to a few circumstances which may be generally interesting.

Metals occur in rocks either disseminated; in bunches; in a net-work of strings or small veins; in beds; or in veins filling fissures, which traverse beds or masses of rock. When metals are disseminated through a rock, as tin often is in granite, and iron pyrites in many trap rocks and clay slates, there can be little doubt that they constituted original portions of the rock, and that they were chemically separated from the mass during consolidation. When metals occur in bunches, as the copper at Ecton, Staffordshire, or the lead in the Sierra Nevada, in Spain, there is a difficulty in considering them otherwise than contemporaneous with the rocks in which they are included.

The occurrence also of metals in strings or small veins crossing each other in all directions, so that in a section they appear like net-work, reminds us strongly of the small strings or veins of carbonate of lime in many limestones, as has already been observed by Mr. Weaver respecting those of copper in Ross Island, Lake of Killarney; so that if not precisely contemporaneous with the original formation of the including rock, they were, like the calcareous veins in the limestone, secreted from the rock into small cracks possibly produced during consolidation. The occurrence of metals in beds has been much disputed or commented on, but it must be admitted that iron ore frequently occurs in beds, and we must regard the copper slate of Thuringia and other adjacent countries as to a certain extent a metallic bed, though it does not strictly come under the head of a bed of solid ore. The appearance of metals in beds is often deceptive, being nothing more than a continuation of a vein laterally between strata; thus in the rich copper mine of Allihies, in the South of Ireland, " the ore occurs in a large quartz vein, which generally intersects the slaty rocks of the country from north to south, but in some cases runs parallel to the stratification *." Mr. Taylor informs me that the lead at Nent Head in Alston Moor, Cumberland, shoots out laterally among the strata, and that the same fact is observable in different mines in Yorkshire and Flintshire.

The most common occurrence of metals is however in veins, or, as they are termed in Cornwall, *lodes*. These are in part filled up, but in various proportions, with metallic substances, and have the general appearance of fissures. They dip at various angles, not unfrequently approaching a vertical position. It was at one time much disputed whether these fissures had been filled from above or beneath; but from facts that have been noticed within a few years, more particularly by Mr. Taylor and Mr. Carne, there is much difficulty in considering that either hypothesis is generally correct. It now appears that the mineral character of a metalliferous vein greatly depends upon the rock which it traverses, that is, when a vein traverses two rocks, as for instance granite and slate, the contents of the vein are not generally the same in the two rocks, but will be different in the one and the other.

Mr. Carne has observed respecting the metalliferous veins of Cornwall, that it is a rare circumstance when a vein which has been productive in one rock continues rich long after it has entered into another. The same author has also remarked that a similar change will be observed even in the same rock, should such rock become harder or softer, more slaty or more

* Weaver, Proceedings of the Geol. Soc. June 4, 1830.

compact. He admits that such changes are sometimes small, but states that the general fact is sufficiently apparent, and often very striking *.

Such facts are not confined to Cornwall, but have been observed elsewhere; thus the lead veins traversing the carboniferous limestone of Derbyshire, which is in some places much associated with trap rocks, are found to be so altered in their passage through the trap, which, from the mode of association, presents the appearance of interstratification, that it was once considered the trap cut off the lead veins; this is however now well known not to be the case.

This fact of the alteration of metallic veins in their passage from one kind of rock to another, or in the same rock, should that become changed, would lead us to consider, with Mr. Fox, that their formation has been in a great measure due to the silent though powerful influence of electricity. This inquiry may yet be considered in its infancy; but the experiments of Mr. Fox on the electro-magnetic properties of the metalliferous veins of Cornwall will be read with great interest †.

That many of these veins are fissures produced by dislocations similar to those which are commonly found in various countries, and are supposed to abound more in the coal-measures only because opportunities of detecting them are there more frequent, seems highly probable; indeed if veins are of different ages, and by cutting one another shift each other, as has been shown to be frequently the case in Cornwall, we can scarcely doubt it. The following is, according to Mr. Carne, the relative ages of the veins in Cornwall: 1. oldest tin lodes; 2. the more recent tin lodes; 3. the oldest east and west copper lodes; 4. the contra copper lodes; 5. cross courses; 6. the more recent copper lodes; 7. the cross flukans (clay veins); and 8. the slides (faults with clay in the fissures)‡.

Now if this relative antiquity of veins be generally correct as far as respects Cornwall, it becomes a curious question, why, if similar causes have produced them, similar results should

---

* Carne, Trans. Geol. Soc. of Cornwall, vol. iii. p. 81.

† Fox, Philosophical Transactions, 1830, p. 399. This author considers that the relative power of conducting galvanic electricity is in the following order in some of the metalliferous minerals. *Conductors:* Copper nickel, purple copper, yellow sulphuret of copper, vitreous copper, sulphuret of iron, arsenical pyrites, sulphuret of lead, arsenical cobalt, crystallized black oxide of manganese, Tennantite, Fahlerz. *Very imperfect conductors:* Sulphuret of molybdenum, sulphuret of tin, or rather bell-metal ore. *Non-conductors:* Sulphuret of silver, sulphuret of mercury, sulphuret of antimony, sulphuret of bismuth, cupriferous bismuth, realgar, sulphuret of manganese, sulphuret of zinc, and mineral combinations of metals with oxygen, and with acids.

‡ On the relative Age of the Veins in Cornwall; Carne, Geol. Trans. of Cornwall, vol. ii.

not be the consequence. If we admit the possibility of secreting the contents of veins from the rocks by electrical means, we cannot so readily understand why different metals should fill the veins in the same rocks, though the direction of the veins might have considerable influence on the conditions and mineralogical combinations of the *same metal.* While again if we consider them ejected from beneath, we are at a loss to understand why the metallic veins should be so much altered in their passage through different rocks. We are certainly not prepared to say what effect may have been produced on the vein, and on the including rocks, from the continued passage of electricity through the vein during an immense lapse of time, or from the arrangement of rocks on the large scale, producing, when properly connected, the effects of a grand galvanic battery; but as the information at present stands, the history of metalliferous veins is anything but clear. It is quite certain from the dissemination of metals in rocks, that they may constitute an original portion of them; the small strings also which cross each other, and are unconnected with great veins, have all the appearance of chemical separations from the including rock; therefore a given rock may contain the necessary elements for secreting substances into a fissure, in the same manner that carbonate of lime frequently fills fissures in limestones, and quartzose veins are common in rocks where silica is abundant.

If the theory of internal heat be well founded, it will be evident that the two ends of a metallic vein will be differently heated, and therefore we should have a thermo-electrical apparatus on the large scale, producing effects which, though slow, might be very considerable. How far such really exist in nature remains questionable; but it may be observed that the experiments of Mr. Fox show the possibility of their occurrence; and should further researches in this highly interesting subject merely so divide it, that some of its present apparent complexity may disappear, a great advance will be made in this now obscure branch of geological inquiry.

SECTION XIII.

ORGANIC REMAINS.

Organic Remains of the Supracretaceous Group.

[THE reader will find a great number of organic exuviæ noticed, in connexion with the various rocks which contain them, under the head of the Supracretaceous Group. The following are lists which were considered too long to be inserted with the descriptions of the beds in which they are stated to have been discovered.]

A. *Organic Remains in the Supracretaceous Blue Marls of the South of France, according to M. Marcel de Serres*.*

LENTICULITES complanata, *Defr.*, Italy, Bordeaux, and C.

VAGINELLA depressa, *Bast.*, Bordeaux; BULLA ampulla, *Lam.*, Italy; B. striata? *Lam.*, Italy, Bord.; B. hydatis? *Lam.*, Italy; B. truncatula, *Broc.*, Italy, Bord.; B. lignaria, *Lam.*, Bord., Italy, Paris, England; TESTACELLA haliotidea, *Draparnaud*, an analogue.

PLANORBIS minutus, *Favjas de St. Fond.*

AURICULA Pisum, *M. de S.*, Italy; Au. (species resembling Voluta myotis, *Broc.*), Italy; Au. myosotis, *Draparnaud*, Italy.

TORNATELLA fasciata, *Lam.*, analogous with the existing species, T. allegata, *Desh.*, Paris; T. inflata, *Férussac*, Bord., Paris.

PALUDINA Brardii. AMPULLARIA Faujasii.

MELANOPSIS lævigata, *Lam.*; M. deperdita, *M. de S.*

MELANIA ventricosa, *Fauj. de St. Fond;* M. pyramidata, *Fauj. de St. Fond.*

NERITA Plutonis, *Bast.*, Bordeaux.

NATICA epiglotina, *Al. Brong.*, Italy and C.; N. patula, Italy, England; N. cruentata? *Lam.*, Italy; N. vitellus? *Lam.*, Italy; N. Guilleminii, *Payrandeau*, analogous to the living species, N. Olla, *M. de S.*, Italy; N. helicina, *Broc.*, Italy.

DELPHINULA solaris (Trochus solaris, *Broc.*), Italy.

TURBO rugosus, *Broc.*, Italy; T. tuberculatus, *M. de S.;* numerous opercula of the Turbo.

TROCHUS cingulatus, *Broc.*, Italy; T. striatus, *Broc.*, Italy; T. magus, *Lam.;* T. conulus, *Lam.;* T. Matonii, *Payrandeau* (analogous with the species now living in the Mediterranean); T. Fermonii, *Payrandeau;* T. zizyphinus, *Lam.*, an analogue; Trochus, resembling T. moniliferus, *Lam.;* T. patulus, *Broc.*, Bord., Italy; T. agglutinans, *Broc.*, Italy; T. granulatus, *M. de S.*

PHASIANELLA pulla, *Payrandeau* (analogous with the existing species); Ph. lævis, *M. de S.*

SOLARIUM sulcatum, *Lam.*, Paris; Solarium, very near S. lævigatum, *Lam.*

SCALARIA Textorii, *M. de S.* (Turbo pseudo-scalaris, *Broc.*), Italy and C.; Sc. cancellata (Turbo cancellatus, *Broc.*), Italy; Sc. lamellosa (Turbo lamellosus, *Broc.*), Italy.

* The names which follow those of the authors who have named the species, point out the other localities, or supracretaceous basins as they are termed, in which the same fossil is considered to be found. When the letter C. is appended, it shows that it is also discovered in the calcaire moellon of the South of France.

TURRITELLA rotifera, *Lam.*, in the marine sands, the calcareous marls, and the calcaire moellon; T. terebralis, *Lam.*, Bord., Italy and C.; T. terebra, *Lam.*, (analogue of the existing species), Italy and C.; T. turris, *Bast.*, Italy and C.; T. tricarinata (Turbo tricarinatus, *Broc.*), Italy; T. varricosa (Turbo varricosus, *Broc.*), Italy; T. cathedralis, *Al. Brong.*, Italy; T. cochleata (Turbo cochleatus, *Broc.*), Italy; T. Archimedis, *Al. Brong.*, Italy; T. serrata (Trochus serratus, *Broc.*), Italy; T. marginalis (Turbo marginalis, *Broc.*), Italy; T. muricata (Turbo muricatus, *Broc.*), Italy; T. imbricata? *Lam.*, Paris and C.; T. duplicata (Turbo duplicatus, *Broc.*), Italy; T. perforata? *Lam.*, Paris; T. acutangula (Turbo acutangulus, *Broc.*), Italy; T. triplicata (Turbo triplicatus, *Broc.*), Italy and C.; T. vermicularis, *Al. Brong.*, Italy and C.; T. fuscata, *Lam.*, (analogous to the species existing in the Mediterranean and Ocean); T. Proto, *Bast.*, Bordeaux, Italy, and C.; T. replicata (Turbo replicatus, *Broc.*), Italy; T. quadriplicata, *Bast.*, Bordeaux; T. lata, *M. de S.*; T. corona, *M. de S.*

CERITHIUM marginatum, *Bruguière*, Italy and C.; C. prismaticum, *Al. Brong.*, Italy; C. cinctum, *Bast.* (not C. cinctum, *Bruguière*), Bordeaux; C. cinctum, *Bruguière* (C. lemniscatum, *Al. Brong.*), Italy; C. pictum, *Bast.*, Bord., Italy; C. sulcatum, *Bruguière* (C. plicatum, *Bast.*), Bord., Italy; C. doliolum (Murex doliolum, *Broc.*), Italy; C. plicatum, *Bruguière* (not *Bast.*), Bordeaux; C. papaveraceum, *Bast.*, Bordeaux; C. subgranosum, *Lam.*, Paris; C. tuberculosum? *Lam.*, Paris; C. umbilicatum, *Lam.*, Paris; C. Castellini, *Al. Brong.*, Italy; C. Lima, *Bruguière* (Murex scaber, *Broc.*), Italy; C. mutabile, *Lam.*, Paris; C. bicarinatum, *Lam.*, Paris; C. turbinatum (Murex turbinatus, *Broc.*), Italy; C. vulgatum antiquum, Italy; C. multisulcatum, *Al. Brong.*, Italy; C. calcaratum, *Al. Brong.*, Italy; C. multigranulatum, *M. de S.*; C. alucaster (Murex alucaster, *Broc.*), Italy; C. baccatum, *Al. Brong.*, Italy; C. ampullosum? *Al. Brong.*, Italy.

PLEUROTOMA turricula (Murex turricula, *Broc.*), Italy; P. dimidiata (Murex dimidiata, *Broc.*), Italy; P. muricata, *M. de S.*, Italy; P. auricula (Murex auricula, *Broc.*), Italy; P. textile (Murex textile, *Broc.*), Italy; P. oblonga (Murex oblongus, *Broc.*), Italy; P. contigua (Murex contiguus, *Broc.*), Italy; P. mitræformis (Murex mitræformis, *Broc.*), Italy; P. multinoda, *Lam.*, Paris; P. spiralis, *M. de S.*; P. subulata (Murex subulatus, *Broc.*), Italy; P. Farinensis, *M. de S.*; P. harpula (Murex harpula, *Broc.*), Italy; P. clathrata, *M. de S.*; P. Pannus, *Bast.*, Bordeaux.

FUSUS lignarius, *Payrandeau* (analogue to the existing species, common in the Mediterranean), Italy; F. subcarinatus, *Al. Brong.*, Italy; F. subulatus (Murex subulatus, *Broc.*), Italy; Fusus, a species between F. Syracusanus of *Lamarck*, and another and undescribed species of the Mediterranean; F. polygonus, *Al. Brong.*, Italy; F. rugosus, *Lam.*, Paris; F. longirostris (Murex longirostris, *Broc.*), Italy; F. uniplicatus, *Lam.*, Paris.

CANCELLARIA clathrata, *Lam.*, Paris.

PYRULA transversalis, *M. de S.*; P. ficoides, *Lam.*, (analogue of a living species), Italy; P. clathrata, *Lam.*, Italy; P. clathroïdes, *M. de S.*

RANELLA marginata, *Al. Brong.* (Buccinum marginatum, *Broc.*), Bordeaux, Italy; R. ranina, *Lam.* (an analogue).

MUREX brandaris, *Lam.*, Italy; M. anguliferus, *Lam.* (apparently an analogue of the living species), Italy; Murex Motacilla, *Lam.* (an analogue of the living species), Italy; M. craticulatus, *Broc.*, Italy; Murex approaching M. trunculus, Italy; M. intermedius, *Broc.*, Italy; M. calcitrapoïdes, *Lam.*, Paris; M. Blainvillii, *Payrandeau* (so like the living species in the Mediterranean, that it cannot be distinguished from it); Murex cornutus, *Lam.* (apparently the analogue of the existing species), Italy; M. Haustellum, *Lam.* (resembles the living species); M. brevispina, *Lam.* (an analogue of an existing species); M. tenuispina, *Lam.* (an analogue), Bordeaux, Italy; M. crassispina, *Lam.* (analogous to a living species), Italy; M. rarispina, *Lam.* (a complete analogue), Italy; Murex, approaching M. heptagonus of

Brocchi, Italy; M. tripterus, *Lam.* (var.); M. cristatus, *Broc.*, Italy; M. decussatus, *Broc.*, Italy; M. transversalis, *M. de S.*; M. rostratus, *Broc.*, Italy; M. oblongus, *Broc.*, Italy.

TURBINELLA infundibulum? *Lam.*, analogous to the existing species.

TRITON corrugatum, *Lam.* (an analogue), Italy; T. pileare, *Lam.* (analogous to a species now living in the Mediterranean), Italy; T. doliare, *Broc.*, Bordeaux, Italy; T. personatum, *M. de S.*; T. intermedium (Murex intermedium, *Broc.*), Italy; T. Chlorostoma, *Lam.*, (an analogue.)

ROSTELLARIA Pes Pelicani (Strombus Pes Pelicani), Italy, Bordeaux.

STROMBUS pugilis, *Lam.* (a species completely analogous with that now existing in the Mediterranean); S. tuberculiferus, *M. de S.*

CASSIDARIA echinophora (Buccinum echinophorum, *Broc.*) (an analogue), Italy.

CASSIS Rondeleti, *Bast.*, Bordeaux; C. marginatus, *M. de S.*; C. diluvii, *M. de S.*; C. striatus, *M. de S.*; C. inflatus, *M. de S.*

DOLIUM, casts of.

NASSA gibba (Buccinum gibbum, *Broc.*), Italy; N. Caronis, *Al. Brong.*, Italy; N. semi-striata, *Al. Brong.*, Italy.

BUCCINUM asperulum, *Broc.*, Italy; B. semi-striatum, *Broc.*, Italy; B. transversale, *M. de S.*; B. corrugatum, *Broc.*, Italy; B. semi-costatum, *Broc.*, Italy; B. Calmeilii, *Payrandeau* (altogether resembling the species so common in the Mediterranean); B. prismaticum, *Broc.*, Italy; B. Lacepedii, *Payrandeau*, C.; Buccinum, apparently approaching B. gemmulatum of *Lamarck*, C.; B. polygonum, *Broc.*, Italy; B. flexuosum, *Broc.*, Italy; B. clathratum, *Lam.*, Italy; B. gibbum, *Broc.*, Italy; B. Miga, *Lam.* (closely approaches the living species); B. angulatum, *Broc.*, Italy; B. reticulatum, *Lam.* (analogous to the existing species), Bordeaux, Italy; B. olivaceum, *Lam.* (apparently an analogue of the living species); B. Turbinellus, *Broc.*, Italy; B. politum, *Bast.*, Bordeaux, Italy; B. mutabile (completely analogous to the living species), Italy; B. crenulatum, *Lam.* (apparently an analogue of the existing species), Italy; B. Carcassonii, *M. de S.*; B. costulatum, *Broc.*, Italy; B. parvulum, *M. de S.*; B. gibbosulum, *Broc.*, Italy; B. pusillum, *M. de S.*, Italy.

TEREBRA duplicata, *Lam.* (an analogue of an existing species), Bordeaux, Italy; T. Vulcani, *Al. Brong.*, Italy; T. pertusa, *Bast.*, Bordeaux, Italy; T. dimidiata (an analogue of the existing species); T. plicaria, *Bast.*, Bordeaux.

MITRA scrobiculata (Voluta scrobiculata, *Broc.*), Italy; M. Brocchii, *M. de S.*, Italy; M. Gervilii, *Payrandeau*, C.; M. pyramidella (Voluta pyramidella, *Broc.*), Italy.

PURPURA Lassaignei, *Bast.*, Bordeaux; P. bicostalis, *Lam.* (analogue of the living species); P. undata, *Lam.* (also an analogue of an existing species.)

VOLUTA varricosa, *Broc.*, Italy; V. piscatoria, *Broc.*, Italy; V. citharella, *Al. Brong.*, Italy; V. buccinea, *Broc.*, Italy; V. pyramidella, *Broc.*, Italy; V. tornatilis, *Broc.*, Italy.

RISSOA Cimex, *Bast.*, Bordeaux, Italy; R. cancellata, *Lam.*; R. pusilla, (Turbo pusillus, *Broc.*), Italy; R. cochlearella, *Lam.*, Bordeaux, Italy, Paris.

MARGINELLA cypræola (Voluta cypræola, *Broc.*), Italy; M. buccinea (Voluta buccinea, *Broc.*), Italy.

CYPRÆA Amygdalum, *Broc.*, Italy; C. Mus, *Lam.* (analogous to the living species), C.; C. Coccinella, *Bast.*, Bordeaux, C. elongata, *Broc.*, Italy, C.

ANOPLAX inflata, *Al Brong.*, Italy.

OVULA carnea, *Lam.* (an analogue to the existing species), C.

CONUS betulinoides, *Lam.*, Paris; C. virginalis, *Broc.*, Italy; C. Pyrula, *Broc.*, Italy; C. avellana, *Lam.*, Italy; C. turricula, *Broc.*, Italy; C. Aldrovandi, *Broc.*, Italy ; C. Pelagicus, *Broc.*, Italy ; C. Mercati, *Bast.*, Bord., Italy ; C. canaliculatus, *Broc.*, Italy and C. ; C. deperditus, *Broc.*, Bordeaux, Italy, Paris ; C. mediterraneus, *Lam.* (analogous to the existing species), C.

Sigaretus costatus (Nerita costata, *Broc.*), Italy, C. ; S. striatus, *M. de S.*

Pileopsis Paretti, *M. de S.*

Calyptræa lævigata, *Desh.*, Paris ; C. muricata (Patella muricata, *Broc.*), Italy, C.

Crepidula unguiformis, *Bast.*, Bordeaux, Italy.

Patella vulgata? *Lam.* ; P. Bonardii, *Payrandeau* (analogous to the existing species), C. ; P. Umbella, *Lam.* (also an analogue), C. ; P. glabra, *Desh.*, Paris.

Fissurella græca, *Desh.* (Patella græca, *Broc.*), Italy, Paris.

Emarginula, a species closely approaching E. fissura of *Lamarck*, and E. reticulata of *Sowerby.*

Avicula, species not determined.

Perna mytiloides, *Lam.*, Bordeaux, Italy.

Lima bullata, *Payrandeau* (analogue) ; L. Breislaki, *Bast.*, Bordeaux, Italy ; L. mutica, *Lam.*, Italy ; L. nivea, (Ostrea nivea, *Renieri, Broc.*), Italy.

Pecten laticostatus, *Lam.*, Italy and C. ; P. benedictus, *Lam.*, Bordeaux, Italy and C. ; P. Plica (Ostrea Plica, *Broc.*), Italy ; P. scabrellus, *Bast.*, Bord., Italy and C. ; P. dubius (O. dubia, *Broc.*), Italy and C. ; P. multiradiatus, Bord., Italy ; P. plebeius (O. plebeia, *Broc.*), Italy ; P. arcuatus (O. arcuata, *Broc.*), Italy ; P. turgidus, *Lam.*, apparently approaches the species found in the American seas ; P. lepidolaris, *Lam.*, Italy and C. ; P. striatulus, *Lam.*, Italy and C. ; P. striatus (O. striata, *Broc.*), Italy ; P. inæquicostalis? *Lam.*, Italy ; P. Pusio, *Lam.*, Italy and C. ; P. scutularis? *Lam.*, Italy and C. ; P. unicolor, *Lam.*, Italy and C. ; P. flabelliformis (O. flabelliformis, *Broc.*), Italy ; P. palmatus, *Lam.*, Bordeaux ; P. solarium, *Lam.*, Italy ; P. terebratulæformis, *M. de S.*, Italy and C. ; P. Tournalii, *M. de S.* ; P. Phaseolus? *Lam.*, Italy ; P. seniensis, *Lam.*, Italy and C. ; P. jacobæoides, *M. de S.*, Italy ; P. pusioides, *M. de S.*, Italy.

Spondylus gæderopus, *Broc.*, Bord., Italy ; S. rastellum, *Lam.*, Italy and C.

Hinnites Brussonii, *M. de S.* ; H. Leufroyi, *M. de S.*

Plicatula, species not determined.

Ostrea canalis, *Lam.*, Paris and C. ; O. crassissima, *Lam.*, C. ; O. undata, *Lam.*, Bord., Italy and C. ; O. virginica, *Al. Brong.*, Italy ; O. edulina, *Lam.*, Italy and C. ; O. colubrina, *Lam.*, Paris ; O. scabrella, *M. de S.*, Italy ; O. anomialis, *Lam.*, Italy, Paris ; O. flabellula, *Lam.*, Bord., Italy, Paris ; O. frondosa, *M. de S.* ; O. crenulatoides, *M. de S.* ; O. cristata, *Lam.*, apparently analogous to the existing species, Italy ; O. corrugata, *Broc.*, Italy.

Anomia Ephippium, *Broc.*, analogous to the existing species, Italy ; A. costata, *Broc.*, Bord., Italy ; A. sulcata, *Broc.*, analogous to the species now living in the Mediterranean, Italy ; A. radiata, *Broc.*, Italy ; A. cepa, *Lam.*, analogue of the existing species, Italy ; A. sinistrorsa, *M. de S.* ; A. electrica, *Lam.*, analogue, Italy and C. ; A. Lens, *Lam.*, closely approaches the living species, Italy and C. ; A. Pellis Serpentis, *Broc.*, Italy and C.

Pinna subquadrivalvis, Italy ; P. augustana? *Lam.* ; P. tetragona, *Broc.*, Italy ; P. pectinata, *Lam.*, approaches the living species.

Arca barbata, *Lam.*, analogue ; A. Gaymardi, *Payrandeau*, apparently analogous to the living species ; A. antiquata, *Lam.*, analogue of the existing species, Italy ; A. Diluvii, *Bast.*, Bord., Italy and C. ; A. aurita, *Broc.*, Italy ; A. biangula, *Lam.*, Italy ; A. lactea, *Lam.*, analogue of the living species, Italy ; A. Quoyi, *Payrandeau*, analogous to the living species ; A. cardiiformis, *Bast.*, Bordeaux, Italy ; A. Breislaki, *Bast.*, Bord., Italy ; A. pectinata, *Broc.*, Italy ; A. clathrata, *Bast.*, Bordeaux.

PECTUNCULUS violacescens, *Lam.*, analogous to the living species; P. num-
marius (Arca nummaria, *Broc.*), C. ; P. pygmæus, *Lam.*, Paris; P. sub-
concentricus, *Lam.*, Paris; P. pulvinatus, Bord., Italy, Paris, England,
and C.

NUCULA minuta (Arca minuta, *Broc.*), Italy; N. pella, *Lam.*, analogue
of the living species, Italy ; N. nicobarica, *Lam.*, Italy ; N. rostrata, *Lam.*
(an analogue), Italy; N. margaritacea, *Lam.*, analogous to the existing
species, Bord., Italy.

MODIOLA discrepans, *Lam.*, C.; M. Semen, analogous to the existing
species; M. subcarinata? *Lam.*; Mytilus edulis, *Bast,* (*Broc.*), Bord., Italy.

UNIO, species undetermined.

ANODONTA, perhaps many species.

CYPRICARDIA, many species, not determined.

CARDITA ajar, *Lam.*; C. trapezia, *Lam.*, an analogue ; C. sinuata, *Payrandeau,* an analogue.

CRASSATELLA latissima, *Lam.*

ISOCARDIA Cor, *Lam.*, exactly resembling the living species, Bord., Italy.

CHAMA intermedia, *Broc.*, approaches Cardita of *Lamarck,* Italy ; C.
pectinata, *Broc.*, Italy ; C. gryphoides, *Lam.*, analogous to the existing
species, Bordeaux, Italy.

CARDIUM hians, *Broc.*, Italy; C. punctatum, *Broc.*, Italy; C. ciliare,
Broc., Italy ; C. oblongum? *Broc.*, Italy ; C. serratum? *Lam.*, Italy ; C.
rusticum, *Broc.*, Italy ; Cardium, approaching C. tuberculatum, *Lam.*, Italy ;
C. rhomboides? *Lam.*, Italy and C. ; C. scrobinatum, *Lam.*, C. ; C. distans?
Lam., Italy ; C. lævigatum, analogous to the existing species, Italy ; C.
edule, *Broc.* (*Bast.*), an analogue, widely spread, from Antibes to the Pyrenees, Italy and C. ; C. glaucum, *Bruguière,* an analogue; C. fragile, *Broc.*,
Italy ; C. striatulum, *Broc.*, Italy; C. planatum, *Broc.*, Italy ; C. echinatum, *Broc.*, Bord, Italy.

TELLINA stricta, *Broc.*, Italy ; T. carinulata, *Desh.*, Paris and C. ; T.
zonaria, *Lam.*, Bordeaux; T. tenui-stria, *Desh.*, Italy, Paris ; T. pellucida,
Broc., Italy ; T. rudis, *Lam.*, Paris ; T. subrotunda, *Desh.*, Paris ; T. elegans,
Bast., Bordeaux ; T. depressa, *Lam.*, analogous to the existing species,
Italy ; T. elliptica, *Broc.*, Italy ; T. strigosa, *Lam.*, analogous to the exist-
ing species, Italy ; T. compressa, Italy and C. ; T. pulchella, C. ; T. planata,
Lam., C. ; T. striatella, *Broc.*, Italy and C. ; T. rostralina, *Desh.* ; T. nitida,
Lam., analogous to the existing species.

LUCINA lactea, *Lam.*, analogous to the existing species, C. ; L. Scopulorum, *Bast.*, Italy and C. ; L. Saxorum, *Desh.*, Paris ; L. concentrica, *Lam.*,
Paris and C.

CORBIS lamellosa, *Lam.*, Paris ; C. ventricosa, *M. de S.*

CYRENA, many species, not determined.

CYCLAS, perhaps many species.

CYPRINA islandicoides, *Lam.*, in the marine sands, the calcaire moellon,
the blue marls, and in the supracretaceous basins of Bordeaux, Italy, Paris,
and England.

CYTHEREA exoleta, *Lam.*, analogous to the existing species; C. erycinoides, *Bast.*, Bordeaux ; C. Lincta, *Lam.*, an analogue, Bordeaux, Italy ; C.
Chione (*Lam.*) (*Broc.*), Italy ; C. elegans, *Lam.* (*Desh.*), Paris ; C. erycinoides, *Lam.*, Bordeaux, Italy ; C. mactroides, *Lam.* ; C. Cypria? (Venus
Cypria, *Broc.*), Italy ; C. Deshayesiana, *Bast.*, Bordeaux ; C. nitidula, *Lam.*,
Paris ; C. Aphrodite, *M. de S.*, Italy ; C. undata, *Bast.*, Bordeaux ; C. semisulcata, *Lam.*, Paris ; C. incrassata (Venus incrassata, *Broc.*), Italy ; C.
globulosa?? *Desh.*, Paris.

VENERICARDIA Jouannetii, *Bast.*, Bord.; V. Lauræ, *Al. Brong.*, Italy ;
V. planicosta, *Lam.*, Paris ; V. pinnula, *Bast.*, Bordeaux.

VENUS impressa, *M. de S.* ; V. angula, *M. de S.* ; V. senilis, *Broc.*, Italy ;
V. Pullastra, *Lam.*, an analogue, Italy ; V. Dysera, *Broc.*, Bord., Italy and
C. ; V. gallina, *Lam.*, an analogue ; V. rugosa, *Broc.*, Italy ; V. cassinoides,

Bast., Bordeaux; V. Pectunculus, *Broc.*, Italy; V. radiata, *Broc.*, Italy; V. circinnata, *Broc.*, Italy; V. Lupinus, *Broc.*, Italy. DONAX nitida, *Lam.*, Paris; D. Basterotina, *Desh.*, Paris; D. Fabagella, *Payrandeau*, an analogue. MYA conglobata, *Broc.*, Italy. CORBULA revoluta, *Bast.*, Bord., Italy. PETRICOLA, striata, *Lam.* LUTRARIA elliptica? *Lam.*, Italy; L. piperatra, *Lam.*; L. solenoides? *Lam.*, Italy. MACTRA triangula, *Bast.*, also in the fahluns of Touraine; M. crassatella, *Lam.*, England; M. lactea, C. SOLEN Vagina, *Lam.*, Italy; S. siliqua? *Lam.*, Italy and C.; S. strigillatus, *Lam.*, Bord., Italy; S. candidus, *Broc.*, Bord., Italy; S. coarctatus, *Broc.*, Italy; all these species of Solen have their existing analogues. PSAMMOBIA Labordei, *Bast.*, Bordeaux; Ps. pulchella, *Lam.*, Italy; Ps. vespertina, *Lam.*, an analogue. PANOPÆA Faujasii, *Menard de la Groye*, Bord., Italy and C. SANGUINOLARIA, species not determined. GASTROCHÆNA cuneiformis, *Lam.*, analogous to the existing species, C. TEREBRATULA ampulla, *M. de S.* (Anomia ampulla, *Broc.*), Italy. PHOLAS Branderi, *Bast.*, Bordeaux. CIRRIPEDA. BALANUS Tintinnabulum, *Lam.*, also in the marine sands and C.; B. miser? *Lam.*, also in marine sands and C.; B. semiplicatus? *Lam.*, also in marine sands and C.; B. perforatus, *Lam.*, also in marine sands and C.; B. patellaris, *Lam.*, also in marine sands, C., and in Italy; all the above Balani are analogues; B. pustularis, *Lam.*, marine sands, Italy; B. crispatus, *Lam.*, marine sands, C., Italy. ANNULATA. SERPULA quadrangularis? *Lam.*; S. arenaria, *Lam.*, Italy; S. contortuplicata, *Lam.*; S. spirorbis? *Broc.*, analogous to the existing species, Italy; S. spirulæa, *Lam.*; S. ammonoides, *Broc.*, Italy; S. annulata? *Lam.*; S. protensa, *Lam.*, analogous to the existing species in the Mediterranean, Italy. DENTALIUM elephantinum, *Lam.*, apparently analogous to the existing species; D. sexangulum, *Broc.*; D. triquetrum, *Broc.*; D. entalis, *Lam.*, apparently analogous to the existing species, Italy; D. coarctatum, *Lam.*, Italy; D. Tarentinum, *Lam.*, Italy; D. striatum, *Lam.*, Italy, Paris. RADIARIA. Species of the Echinite family are not stated to occur in the blue marls; but the following are found, according to M. Marcel de Serres, in the calcaire moellon, or calcareous marls. ECHINUS miliaris, *Lam.*, calc. moel.; E. granularis? *Lam.*, perhaps analogous to the existing species, calc. moel. SCUTELLA striatula, *M. de S.*, calc. moel.; S. gibercula, *M. de S.*, calc. moel. GALERITES pustulata, *M. de S.*, calcareous marls. CLYPEASTER altus, *Lam.*, calc. moel., and in Italy; C. marginatus, *Lam.*, calc. moel., and also Bord., Italy; C. politus? *Lam.*, calc. moel., also Italy; Clypeaster, closely approaching C. oviformis of Lamarck, calc. moel.; C. excentricus, *Lam.*, calc. moel.; C. hemisphericus, *Lam.*, calc. moel., also Italy; C. stelliferus, *Lam.*, calc. moel., also Italy; C. gibbosus, *M. de S.*, calc. moel.; C. scutellatus, *M. de S.*, calc. moel. SPATANGUS canaliferus, *Lam.*, calc. moel.: the specimens of this fossil found highly preserved in the calc. moel. of Barcelona appear quite analogous to the existing species; Sp. lævis? *Deluc*, calc. moel.; Sp. arcuarius, *Lam.*, analogous to the existing species, calc. moel.; Sp. retusus? *Lam.*, calc. moel. CRUSTACEA. PODOPHTHALMUS Defrancii, *Desmarest.* (This is the only species noticed by M. Marcel de Serres in the blue marls ; but he states that the Atelecylus rugosus, *Desmarest*, is found in the calcaire moellon, near Montpellier. Remains of the genus *Portunus* are also mentioned.)

B. *Fossil Shells contained in the Supracretaceous Rocks of Bordeaux and Dax, and enumerated by M. De Basterot* *.

NAUTILUS Aturi, *Bast.*, Dax., Houdan ; not considered identical with the N. Pompilius existing in the Eastern seas.

LENTICULITES complanata, *Defr.*, Dax, Léognan, Antwerp, Pontoise, Montpellier, and Italy; common at Saucats.

NUMMULITES lævigata, *Lam.*, common in many supracretaceous deposits ; N. complanata, *Lam.*, Dax, Soissons ?

LYCOPHORIS lenticularis, *Defr.*, common near Bordeaux, Claudiopolis in Transylvania.

VAGINELLA depressa, *Bosc.* Léognan, Saucats.

BULLA lignaria, *Linn.* (var.), analogous to the existing species, Dax, Léognan, Piedmont, England, Paris ; B. cylindrica, *Brug.*, Grignon, Piedmont, Vienna, Dax, Bordeaux ; B. Utriculus, *Broc.*, analogous to the existing species, Piacenza, Bordeaux, Dax ; B. Labrella, *Fér.*, Dax ; B. clathrata, *Defr.*, Dax ; B. truncatula, *Brug.*, analogous to the existing species, environs of Paris, Sienna, Riluogo, Dax.

BULLINA Lajonkaireana, *Bast.*, abundant at Saucats, Léognan, and Mérignac.

HELIX nemoralis, analogous to the existing species, fresh-water limestone at Saucats ; H. variabilis, *Drap.*, analogous to the existing species, fresh-water limestone, Saucats.

BULIMUS? terebellatus, *Lam.*, analogous to the existing species, Grignon, Placentine, Dax.

PLANORBIS corneus, *Drap.*, analogous to the existing species, Saucats.

LYMNEA palustris, *Drap.*, analogous to the existing species, in many supracretaceous deposits, Saucats.

AURICULA ringens, *Lam.*, analogous to the existing species, Paris, Nice, Italy, Touraine, Bordeaux, Dax ; A. hordeola, *Lam.*, Grignon, Léognan.

TORNATELLA sulcata (Auricula sulcata, *Lam.*), Grignon, Dax, Bordeaux ; T. inflata, *Fér.*, Champagne, Dax ; T. semistriata, *Defr.*, Léognan ; T. punctulata, *Fér.*, Léognan, Saucats, Dax ; T. papyracea, *Bast.*, Dax ; T. Dargelasi, *Bast.*, Léognan, Saucats.

PYRAMIDELLA Mitrula, *Fér.*, Léognan, Mérignac ; P. terebellata (Auricula terebellata, *Lam.*), Grignon, Volterra, Bordeaux, Dax.

TURBO Parkinsoni, *Bast.*, Dax ; T. Fittoni, *Bast.*, Dax ; T. Lachesis, *Bast.*, common at Bordeaux and Dax.

DELPHINULA marginata, *Lam.*, Grignon, Dax ; D. Scobina (Turbo Scobina, *Al. Brong.*), Castelgomberto, Dax, and near Valognes ; D. sulcata, *Lam.*, Grignon, var. at Léognan ; D. trigonostoma, *Bast.*, Dax.

TURRITELLA terebralis, *Lam.*, common at Dax, Léognan, and Saucats ; T. Archimedis, *Al. Brong.* (var. Burdigalensis, *Bast.*), Ronca, var. α Bassano, var. β Anjou, var. γ Bordeaux ; T. asperula, *Al. Brong.*, Ronca, Dax ; T. Turris, *Bast.*, analogous to the existing species, Dax ; T. quadriplicata, *Bast.*, above the fresh-water limestone at Saucats ; T. cathedralis, *Al. Brong.*, Turin, Léognan, Saucats ; T. Proto, *Bast.*, Saucats ; T. Desmarestina, *Bast.*, Dax.

SCALARIA communis (var.), analogous to the existing species, Placentine, Volterra, Bramerton, Dax ; S. acuta, *Sow.*, Barton, Dax ; S. multilamella, *Bast.*, Parnes, Léognan.

CYCLOSTOMA Lemani, *Bast.*, fresh-water limestone, Saucats, Dax, and Tongres, near Maëstricht.

PALUDINA pusilla (Bulimus pusillus, *Al. Brong.*), analogous to the existing species, Paris, Bordeaux.

MONODONTA elegans, *Faujas de St. F.*, Léognan, rare at Bordeaux ; M.

---

* Description Géologique du Bassin Tertiaire du Sud-Ouest de la France: Mém. de la Soc. d'Hist. Nat. de Paris, t. ii.

Modulus, *Lam.*, analogous to the existing species, Dax; M. Araonis, *Bast.*, analogous to the existing species? Mérignac, Touraine, Dax.

Trochus Benetti, *Sow.*, Stubbington, Turin, Léognan, Saucats; T. patulus, *Broc.*, Placentine, Bologna, Vienna, Bordeaux, Dax; T. Boscianus, *Al.* Brong., Castelgomberto, Dax; T. Labarum, *Bast.*, Dax; T. turgidulus? *Broc.*, Italy, Mérignac; T. Bucklandi, *Bast.*, above the fresh-water limestone, Saucats; T. Audebardi, *Bast.*, Léognan.

Rotella Defrancii, *Bast.*, Léognan.

Solarium, carocollatum, *Lam.*, Léognan, Dax.

Ampullaria compressa, *Bast.*, Dax; A. crassatina, *Lam.*, Pontchartrain, var. at Dax.

Melania costellata, *Lam.*, Grignon, Ronca, Sangonini, Dax; M. subulata, Volterra, Léognan, Dax; M. hordeacea, *Lam.*, Houdan, Pierrelaye, Beauchamp, Isle of Wight, var. at Saucats; M. clathrata, *Bast.*, Dax; M. nitida, *Lam.*, Grignon, Placentine, Parnes, Dax; M. distorta (Turbo politus, *Montagu*), analogous to the existing species, Thorigne, Bordeaux.

Melanopsis Dufourii, *Fér.*, Dax.

Rissoa Cochlearella (Melania Cochlearella, *Lam.*), Grignon, Mérignac, var. at Dax; R. Cimex (Turbo Cimex, *Broc.*), analogous to the existing species, Bologna, Isle of Ischia, Mérignac, var. at Dax; R. varicosa, *Bast.*, Mérignac; R.? Grateloupi, *Bast.*, Mérignac.

Phasianella turbinoides, *Lam.*, Grignon, Mérignac, Dax; P. Prevostina, *Bast.*, Léognan, Saucats.

Natica Canrena, *Broc.*, analogous to the existing species, Italy, England, Léognan, Saucats, Dax; N. glaucina, *Broc.*, analogous to the existing species, Italy, Léognan, Dax.

Nerita Plutonis, *Bast.*, Mérignac.

Neritina fluviatilis, *Lam.*, analogous to the existing species, Tuscany, Mérignac, Dax (often preserves its colours).

Conus deperditus, *Lam.* (analogous to the existing species at Owhyhee), Grignon, Ronca, Turin, Bordeaux, Dax; C. alsiosus, *Al.* Brong., Ronca, Dax, Bordeaux; C. Mercati, Vienna, San Miniato, Saucats.

Cypræa Coccinella, *Lam.*, Grignon, Suffolk, Angers, Nantes, Dax; Cy. annulus, *Broc.*, analogous to the existing species, Piedmont, Ronca, Bordeaux; Cy. annularia, *Al.* Brong., Turin, Bordeaux; Cy. leporina, *Lam.*, Dax; Cy. lyncoides, *Al.* Brong., Turin, Bordeaux; Cy. Duclosiana, *Bast.*, Dax.

Oliva plicaria, *Lam.*, Mérignac, Léognan, Dax, Saucats; O. Clavula, *Lam.*, Mérignac, Dax; O. Dufresnii, *Bast.*, Mérignac, Dax, Saucats.

Ancillaria canalifera, *Lam.* (A. turrellata, *Sow.*), Grignon, Barton, Dax, Bordeaux; A. inflata (Anolax inflata, *Al. Brong.*), Turin, Vienna, Léognan, Mérignac, Dax, Saucats.

Voluta Lamberti, *Sow.*, analogous to the existing species, Suffolk, Anjou, Léognan; V. rarispina, *Lam.*, Dax, Bordeaux; V. affinis, *Broc.*, Ronca, Turin, Léognan.

Marginella cypræola, Placentine, Touraine, Dax.

Mitra Dufresnii, *Bast.*, rare at Léognan; M. scrobiculata, Placentine, Piedmont, Sienna, var. at Bordeaux; M. incognita, *Bast.*, Mérignac, Dax.

Cancellaria acutangula (C. acutangularis, *Lam.*), Léognan, Saucats; C. trochlearis, *Lam.*, Léognan, Saucats; C. doliolaris, *Bast.*, rare, Léognan, C. Geslinii, *Bast.*, Léognan, Saucats; C. buccinula, *Lam.*, Vienna, Crépy in Valois, Bordeaux; C. contorta, *Bast.*, Italy, Saucats; C. cancellata, *Lam.*, anal. exist. species; Piedmont, Placentine, Sienna, Bordeaux.

Buccinum Veneris, *Faujas de St. F.*, Léognan, Saucats; B. baccatum, *Bast.*, Saucats, Léognan, Mérignac, var. α Dax. var. β Saucats, var. γ Vienna; B. politum, *Bast.*, Piedmont, Saucats.

Eburna spirata (Buccinum spirata, *Brug.*), anal. exist. species, Rennes, Dax, Saucats.

NASSA reticulata (Buccinum reticulatum, *Broc.*), anal. exist. species, San Miniato, Castel-Arquato, Sienna, var. *α* Dax, var. *β* Saucats and Léognan; N. asperula, *Broc.*, Placentine, Sienna, var. *α* Dax, var. *β* Léognan and Saucats; N. angulata, Volterra, Saucats; N. columbelloides, *Bast.*, Vienna, Angers, Touraine, Dax, Léognan, Saucats (approaches a living species); N. Desnoyersi, *Bast.*, Dax, Saucats; N. cancellaroides, *Bast.*, Dax; N. Andrei, *Bast.*, Bordeaux.

PURPURA costata (Nerita costata, *Broc.*), Placentine, Dax, Bordeaux; P. Lassaignei, *Bast.*, Léognan.

CASSIS Saburon (Cassidea Saburon, *Brug.*), analogous to the existing species, Calabria, Placentine, Vienna, Léognan, Saucats, Dax; C. Rondeleti, *Bast.*, Léognan, Dax.

CASSIDARIA Cythara, Italy, Bordeaux.

TEREBRA plicaria, analogous to the existing species, Saucats; T. plicatula, *Lam.*, Grignon, Saucats, Léognan, Dax; T. cinerea (T. aciculina, *Lam.*), analogous to the existing species, Piedmont, Léognan, Saucats; T. striata, analogous to the existing species, Saucats; T. duplicata, *Lam.*, analogous to the existing species? Sienna, Piedmont, Rome; T. pertusa (var.), analogous to the existing species, Saucats; T. murina, Dax.

CERITHIUM margaritaceum, *Al. Brong.*, Sienna, Mayence; C. corrugatum, *Al. Brong.*, Ronca, Saucats; C. inconstans, *Bast.*, Saucats; C. ampullosum, *Al. Brong.*, Castelgomberto, Vienna, Mérignac, Dax; C. plicatum, *Lam.*, Montpellier, Pontchartrain, Mayence, Castelgomberto, Saucats; C. cinctum, *Lamb.*, Montpellier, Pontchartrain, Beynes, Houdan, Saucats; C. Charpentieri, *Bast.*, Dax; C. papaveraceum, *Bast.*, Touraine, Mérignan; C. lemniscatum, *Al. Brong.*, Ronca, Dax; C. Salmo, *Bast.*, Léognan, Mérignac; C. pictum, *Bast.*, Vienna, Mérignac, Saucats; C. lamellosum, *Lam.*, Courtagnon, Grignon, var. Dax; C. angulosum, Grignon, Saucats; C. Diaboli, *Al. Brong.*, the Diablerets, Switzerland, Dax; C. resectum, *Defr.*, Hauteville, Dax, Mérignac; C. calculosum, *Bast.*, Dax, Léognan; C. pupæforme, *Bast.*, rare, Mérignac; C. granulosum (Murex granulosus, *Broc.*), analogous to the existing species, Volterra, Mérignac; C. scaber, analogous to the existing species, Italy, Mérignac, Léognan.

MUREX Pomum, *Linn.*, analogous to the existing species, Placentine, Saucats, Mérignac; M. sublavatus, *Bast.*, rare, Mérignac, Léognan, Saucats; M. Lingua-Bovis, *Bast.*, Saucats, Léognan; M. suberinaceus, *Bast.*, Bordeaux.

TYPHIS tubifer (Murex tubifer, *Lam.*), analogous to the existing species, Grignon, Barton, Highgate, Léognan.

TRITON doliare (Murex doliaris, *Broc.*), Placentine, Pisa, Sienna, Léognan.

RANELLA marginata, *Al. Brong.*, Piedmont, Pisa, Placentine, Volterra, Turin, Léognan, Mérignac; R. leucostoma, *Lam.*, analogous to the existing species, Placentine, Bordeaux.

FUSUS lavatus, *Sow.*, Barton, Paris, Léognan, Saucats, Dax; F. buccinoides, *Bast.* (Buccinum subulatum, *Broc.*), Placentine, Saucats, Mérignac (a Mediterranean shell approaches this species); F. rugosus, *Lam.*, Grignon, Valognes, Dax (different from the Fusus rugosus of Sowerby); F. clavatus, Placentine, var. Bordeaux.

PLEUROTOMA tuberculosa, *Bast.*, Vienna, Saucats, Léognan; P. Pannus, *Bast.*, Léognan, Saucats, Dax; P. denticulata, *Bast.*, Touraine, Saucats, Léognan, Mérignac, Dax; P. ramosa, *Bast.*, Thorigné, Angers, Léognan, Saucats; P. Borsoni, *Bast.*, Saucats, Léognan, Mérignac; P. plicata, *Lam.*, Grignon, Dax; P. undata, *Lam.*, Grignon, Epernay, Dax; P. multinoda, *Lam.*, Bordeaux; P. Turrella, *Lam.*, Grignon, var. Dax; P. crenulata, *Lam.*, Grignon, var. Léognan; P. cataphracta, Placentine, Sienna, Bologna, Bordeaux; P. purpurea, *Bast.*, analogous to the existing species, Léognan; P. terebra, *Bast.*, Léognan, Saucats, Dax; P. costellata, *Lam.*, Grignon, Léognan, Dax; P. cheilotoma, *Bast.*, Bordeaux.

FASCIOLARIA Burdigalensis, *Defr.*, Léognan, Saucats, Mérignac; F. uniplicata (Fusus uniplicatus, *Lam.*), Grignon, Epernay, Dax.

PYRULA condita, *Al. Brong.*, Turin, Léognan, Saucats; P. Clava, *Bast.*, Dax, Bordeaux; P. Lainei, *Bast.*, Saucats, Léognan, Mérignac, Dax; P. Melongena, analogous to the existing species, Courtagnon, Saucats, Dax, Mérignac; P. rusticula, *Bast.*, Dax, Bordeaux.

TURBINELLA Lynchi, *Bast.*, rare, Léognan.

STROMBUS decussatus, Dax; S. Bonelli, *Al. Brong.*, Turin, Dax.

ROSTELLARIA Pes-Pelicani, analogous to the existing species, common in many supracretaceous deposits, Léognan, Dax; R. curvirostris, *Lam.*, analogous to the existing species, Dax.

SIGARETUS canaliculatus, *Sow.* Hordwell, Paris, Bordeaux, Dax.

CAPULUS (Pileopsis) sulcosus (Nerita sulcosa, *Broc.*), Placentine, Mérignac.

CREPIDULA unguiformis (C. Italica, *Defr.*), analogous to the existing species, Placentine, Sienna, Vienna, Saucats; C. cochlearia, *Bast.*, analogous to the existing species, Mérignac.

FISSURELLA costaria, *Desh,*, Grignon? Dax.

CALYPTRÆA deformis, *Lam.*, Dax, Bordeaux; C. depressa, *Lam.*, abundant at Bordeaux; C. muricata (Patella muricata, *Broc.*), analogous to the existing species, Piedmont, Placentine, Castel-Arquato, Léognan, Saucats; C. ornata, *Bast.*, Dax.

HIPPONYX granulatus, *Bast.*, Dax.

OSTREA flabellula, *Lam.*, Grignon, Hordwell, Barton, Brussels, Saucats, Léognan; O. undata, *Lam.*, Dax, Bordeaux; O. Cymbula, *Lam.*, Grignon, Barton, Saucats.

PECTEN scabrellus, *Lam.* (Ostrea dubia, *Broc.*), Val Andone, Piedmont, Saucats; P. Burdigalensis, *Lam.*, Saucats, approaches P. Pleuronectes, P. obliteratus, and P. Laurenti; P. multiradiatus, *Lam.*, Italy, Saucats.

SPONDYLUS, fragments.

PERNA Ephippium, *Lam.*, analogous to the existing species, Bordeaux.

AVICULA phalænacea, *Lam.*, Léognan.

PINNA, fragments.

ARCA biangula, *Lam.*, Grignon, Léognan; A. scapulina, *Lam.*, Grignon, Mérignac; A. clathrata, *Defr.*, analogous to the existing species, Angers, Thorigné, Nice, Mérignac; A. Diluvii, *Lam.* (A. Pectinata, *Broc.*), Houdan, Touraine, Placentine, Sienna, Turin, Vienna, Bordeaux; A. Breislaki, *Bast.*, Dax.

PECTUNCULUS Cor, *Lam.*, Saucats, Mérignac, Léognan; P. pulvinatus, *Lam.*, Paris, Touraine, var. α Dax and Bordeaux, var. β Léognan; M. de Basterot considers this shell the same with that found at Walton.

NUCULA emarginata, *Lam.*, Léognan, Saucats; N. margaritacea, *Lam.* (Arca Nucleus, *Broc.*), analogous to the existing species, Grignon, Placentine, Barton, Highgate, Léognan, Dax.

MYTILUS antiquorum, *Sow.*, Suffolk, var. Saucats, Mérignac; M. Brardii, *Al. Brong.*, Mayence, Dax, Mérignac; M. edulis, *Linn.*, analogous to the existing species, Piedmont, Placentine, Sienna, Volterra, Saucats.

MODIOLA cordata, *Lam.*, Paris, Domfront, Saucats.

CARDITA hippopea, *Bast.*, Saucats.

VENERICARDIA Pinnula, *Bast.*, beds above the fresh-water limestone, Saucats, Dax; V. Jouanneti, *Bast.*, Italy, Vienna, Bordeaux; V. intermedia (Cardita intermedia, *Lam.*), Placentine, Sienna, Dax.

ERYCINA elliptica, *Lam.*, Ecouen, Senlis, Saucats.

CHAMA gryphoides, *Broc.*, analogous to the existing species, Piedmont, Placentine, Sienna, Dax, Léognan, Saucats, Mérignac.

CARDIUM edule, *Linn.*, analogous to the existing species, Placentine, Piedmont, Sienna, Bramerton, Ipswich, Dax; C. Burgalinum, *Lam.*, Dax, Bordeaux; C. serrigerum, *Lam.*, Grignon, Bordeaux; C. echinatum,

Brug., analogous to the existing species, Placentine, Touraine, Bordeaux, var. Vienna; C. Pallassianum, *Bast.*, Dax; C. multicostatum, *Broc.*, Placentine, var. Léognan; C. discrepans, *Bast.*, bed above the fresh-water limestone, Saucats, Dax.

Donax anatinum, *Lam.*, analogous to the existing species; var. Dax, Bordeaux; D. elongata, *Lam.*, analogous to the existing species, Mérignac; D. triangularis, *Bast.*, approaches an existing species, Saucats; D. irregularis, *Bast.*, Dax; D.? difficilis, *Bast.*, Dax.

Cyrena Brongniartii, *Bast.*, Ronca, Mérignac, Saucats; C. Sowerbii, *Bast.*, Paris, Saucats.

Tellina zonaria, *Lam.*, Dax, Saucats, Léognan, Mérignac (preserves its colours); T. elegans, *Desh.*, Hauteville, Grignon, above the fresh-water beds, Saucats; T. bipartita, *Bast.*, Saucats; T. biangularis, *Desh.*, Paris, var. Dax.

Lucina Columbella, *Lam.*, Touraine, Léognan, Saucats, Dax, Mérignac; L. divaricata, *Lam.*, analogous to the existing species, Grignon, Léognan, Mérignac, common at Hordwell and Saucats; L. scopulorum, *Al. Brong.*, Ronca, Turin, Mérignac, Saucats; L. dentata, *Bast.*, Dax, Saucats; L. digitalis, *Lam.*, analogous to the existing species, rare, Saucats; L. hiatelloides, *Bast.*, rare, Léognan; L. gibbosula, *Lam.*, analogous to the existing species; Grignon, Dax; L. renulata, *Lam.*, analogous to the existing species, Grignon, Bordeaux; L. neglecta, *Bast.*, Dax, Bordeaux.

Venus Dysera, *Linn.*, analogous to the existing species, Piedmont, Placentine, Dax, Saucats; V. casinoides, *Lam.*, Vienna, Léognan, Saucats; V. vetula, *Bast.*, Saucats, Léognan; V. radiata, *Broc.*, analogous to the existing species, Italy, Saucats, Léognan, Dax.

Cytherea erycinoides, *Lam.*, analogous to the existing species, Paris, Turin, Rome, Saucats, Léognan, Dax; C. Deshayesiana, *Bast.*, Saucats, Léognan ; C. tincta, *Lam.*, perfect resemblance to existing species, Saucats, C. leonina, *Bast.*, Léognan, Saucats; C. undata, *Bast.*, Saucats, abundant at Mérignac ; C. nitidula, *Lam.* (Venus transversa, *Sow.*), Grignon, Barton, Saucats.

Cyprina Islandicoides, *Lam.* (Venus æqualis, *Sow.*), Suffolk, Placentine, Antwerp, Dax, Bordeaux.

Venerupis Faujasii, *Bast.*, Bordeaux.

Petricola peregrina, *Bast.*, in large madrepores, Mérignac.

Saxicava anatina, *Bast.*, in the holes which it has bored in the fresh-water limestone, when the latter was covered by the waters of an ancient sea, Saucats.

Clotho? unguiformis, *Bast.*, in holes which it has pierced in the marine and fresh-water limestones, Saucats.

Corbula revoluta, *Sow.*, Barton, Italy, Dax, Léognan, Mérignac, Saucats; C. striata, *Lam.*, Grignon, var. Angers, var. Bordeaux.

Mactra striatella, *Lam.*, analogous to the existing species? Saucats; M. deltoides, *Lam.*, Grignon, Saucats; M. triangula, *Broc.*, analogous to the existing species, Placentine, Saucats.

Lutraria Sanna, *Bast.*, Saucats. Mya ornata, *Bast.*, Dax.

Panopæa Faujasii, *Mesnard de la Groye*, Parma, Sienna, Pisa, San Miniato (Reggio), Placentine, Piedmont, Léognan.

Psammobia Labordei, *Bast.*, approaches Ps. vespertina, *Leach*, a living species, Saucats.

Solen strigillatus, *Linn.*, analogous to the existing species, Placentine, Piedmont, Vienna, Grignon, Léognan, Dax ; S. Vagina, *Linn.*, analogous to the existing species, Placentine, Grignon, Saucats; S. Legumen, *Linn.*, Saucats.

Pholas Branderi, *Bast.*, in the rolled stones and corals, Touraine and Mérignac.

Clavagella coronata, *Desh.*, Meaux, Pauliac, nine leagues from Bordeaux.

506

Gosau Fossils.

C. *Gosau Fossils.*

The following is a list, according to Prof. Sedgwick and Mr. Murchison, of the organic remains detected by them either in the Gosau deposit, or its equivalents in the Alps. (G., Gosau; Z., Zlam; W., Flanks of the Wand; M., Marzoll; R., Hinter Reutter; T., Bavarian Traunstein.) *Zoophyta.*—Tragos. Nullipora. Madrepora. Cellepora, G. Lithodendron granulosum, *Goldf.*, G. (also Castell' Arquato, *Supracretaceous*). Fungia radiata, *Goldf.*, G. Fung. discoidea, G. Fung. polymorpha, *Goldf.*, G., Z. (also Bassano and Dauphiné, *Supracretaceous*). Fung. undulata, *Goldf.*, G. Diploctenium cordatum, *Goldf.*, G. (also Maestricht.) Turbinolia complanata, *Goldf.*, G. Turb. duodecimcostata, G. (also Castell' Arquato, *Supracretaceous*). Turb. lineata, *Goldf.*, G. Turb. cuneata, *Goldf.*, G. (also Castell' Arquato, *Supracretaceous*). Turb. aspera, *Sow.*, G. Cyathophyllum rude, *Sow.*, G. Cy. compositum, *Sow.*, G. Meandrina agaricites, *Goldf.*, G. Astrea striata, *Goldf.*, G. Ast. formosa, *Goldf.*, G. Ast. reticulata, *Goldf.*, G. Ast. agaricites, *Goldf.*, G. Ast. grandis, *Sow.*, G. Ast. media, *Sow.*, G. Ast. formosissima, *Sow.*, G. Ast. ambigua, *Sow.*, G. Ast. tenera, *Sow.*, G. Ast. ramosa, *Sow.*, G.

Annulata.—Serpula.

Conchifera.—Teredo, G. Solen, G. Panopæa plicata? *Sow.*, G. (also Sandgate, *Lower Green Sand*). Anatina, G. Crassatella impressa, *Sow.*, G. Corbula angustata, *Sow.*, G. Sanguinolaria Hollowaysii?? *Sow.*, G. (also Bracklesham Bay, *London Clay*). Lucina, G. Astarte macrodonta, *Sow.*, G. Cyclas cuneiformis? *Sow.*, G., W. (also Woolwich, *Plastic Clay*). Cytherea lævigata, *Lam.*, G. (also Grignon, *Calcaire Grossier*). Venus, G. Venericardia, G. Cardium productum, *Sow.*, G., M. Isocardia, G. Cucullæa carinata, *Sow.*, G. (Blackdown, *Green Sand*). Arca, G. Pectunculus Plumsteadiensis, *Sow.*, G. (also Plumstead, near Woolwich, *Plastic Clay*). Pect. brevirostris, *Sow.*, G. (also Bognor, *London Clay*). Pect. pulvinatus, *Lam.*, G. (also Grignon, Bracklesham, *Calcaire Grossier*, and *London Clay*). Pect. calvus, *Sow.*, G., M., W. Nucula amygdaloides, *Sow.*, G. (also Southend, *London Clay*). Nuc. concinna, *Sow.*, G., R. Trigonia alæformis, *Sow.*, G. (also Parham Park and Black Down, *Green Sand*). Modiola, G. Inoceramus Cripsii, *Mant.*, G., W. (also Hamsey, *Chalk Marl*). Avicula, G. Pecten quinquecostatus, *Sow.*, G. (*common cretaceous fossil*). Plicatula aspera, *Sow.*, G., W. Gryphæa elongata, *Sow.*, G. Gryphæa expansa, *Sow.*, G. Exogyra, G. Ostrea, G. Terebratula dimidiata? *Sow.*, G. (also Haldon Hill, *Green Sand*). Axinus? G., W. Trigonellites, G., W.

Mollusca.—Dentalium grande? *Desh.*, G., M. (also *Calcaire Grossier*). Calyptræa? G. Auricula decurtata, *Sow.*, G. Aur. simulata, *Sow.*, G., M. (also Barton Cliff, *London Clay*). Melania, G. Melanopsis, G. Natica Ambulacrum, *Sow.*, G. (also Barton Cliff, *London Clay*). Nat. lyrata, *Sow.*, G. Nat. angulata, *Sow.*, G. Nat. bulbiformis, *Sow.*, G., Z. Nerita, G. Solarium quadratum, *Sow.*, G. Trochus spiniger, *Sow.* Turbo arenosus, *Sow.*, G. Turritella angusta, *Desh.*, G. Tur. biformis, *Desh.* G., T. Tur. rigida, *Desh.*, G. Tur. læviuscula, *Desh.*, G. Tornatella gigantea, *Desh.*, G., Z., Meyersdorf, &c. Torn. Lamarckii, *Desh.*, Gams-Gebirge. Nerinea flexuosa, *Desh.*, G. Cerithium reticosum, *Desh.*, G. Cer. conoideum, *Desh.*, G., Z., T. Cer. pustulosum, *Desh.*, G. Pleurotoma prisca, *Sow.*, G., M. (also Barton Cliff, *London Clay*). Pleur. fusiforme, *Sow.*, G. Pleur. spinosum, *Sow.*, G. Fasciolaria elongata, *Sow.*, G. Fusus intortus, *Lam.*, G. (also Grignon and Ronca, *Supracretaceous*). Fus. heptagonus, *Sow.*, G. Fus. carinella, *Sow.*, G. Fus. muricatus, *Sow.*, G. Fus. abbreviatus, *Sow.*, G. Fus. cingulatus, *Sow.*, G. Rostellaria plicata, *Sow.*, G. Rost. costata, *Sow.*, G. Rost. granulata, *Sow.*, G., M. Rost. lævigata, *Sow.*, G. Nassa carinata, *Sow.*, G. Nas. affinis, *Sow.*, G. Mitra pyramidella? *Broc.*, G. (*Supracretaceous*). Mit. cancellata, *Sow.*, G. Voluta coronata? *Broc.*, G.

(also *Supracretaceous*). Vol. citharella? *Al. Brong.*, G. (also Turin, *Supra-cretaceous*). Vol. acuta, *Sow.*, G. Terebra coronata, *Sow.*, G. Volvaria lævis, *Sow.*, G. Baculites or Hamites, G.

Organic Remains of the Cretaceous Group.

PLANTÆ.

Confervæ.

1. Confervites fasciculata, *Ad. Brong.* Arnager, Bornholm, *Ad. Brong.* Chalk, Sussex, *Mant.*
2. ――― ægagropiloides, *Ad. Brong.* Arnager, Bornholm, *Ad. Brong.* ―――, species not determined. Chalk, Sussex, *Mant.*

Algæ.

1. Fucoides Orbignianus, *Ad. Brong.* Isle d'Aix, Rochelle, *Ad. Brong.*
2. ――― strictus, *Ad. Brong.* Isle d'Aix, Rochelle, *Ad. Brong.*
3. ――― tuberculosus, *Ad. Brong.* Isle d'Aix, Rochelle, *Ad. Brong.*
4. ――― difformis, *Ad. Brong.* Bidache, Bayonne, *Ad. Brong.*
5. ――― intricatus, *Ad. Brong.* Bidache, *Ad. Brong.*
6. ――― Lyngbianus, *Ad. Brong.* Arnager, Bornholm, *Ad. Brong.*
7. ――― Brongniarti, *Mant.* Chalk, Sussex, *Mant.*
8. ――― Targioni, *Ad. Brong.* Chalk, Sussex, *Mant.*
9. ――― canaliculatus, *Ad. Brong.*, Env. of Bayonne; Bidache; Green Sand, Rochefort, *Dufr.*
――――, species not determined. Chalk, Gault, Sussex, *Mant.*

Naïades.

1. Zosterites cauliniæfolia, *Ad. Brong.* Isle d'Aix, *Ad. Brong.*
2. ――― lineata, *Ad. Brong.* Isle d'Aix, *Ad. Brong.*
3. ――― Bellovisana, *Ad. Brong.* Isle d'Aix, *Ad. Brong.*
4. ――― elongata, *Ad. Brong.* Isle d'Aix, *Ad. Brong.*

Cycadeæ.

1. Cycadites Nilssonii, *Ad. Brong.*, Chalk, Scania.
1. Thuites aliena, *Sternb.* Smetschna (Rakonitzer Kreis), *G. T.*
Dicotyledonous wood, perforated by some boring shell; Chalk, Sussex, *Mant.*; Green Sand, Lyme Regis, *De la B.*
Cones of Coniferæ, Green Sand, Lyme Regis, *De la B.* Green Sand? Köpinge, Scania, *Nils.*
Ferns? Green Sand, Lyme Regis, *De la B.*
―― leaves, between Platanus and Lyriodendron, *Sternb.* Green Sand, Tetschen; Blankenburg; Wernigerode; Quedlinburg, *G. T.*

ZOOPHYTA.

1. Achilleum glomeratum, *Goldf.* Maestricht, *Goldf.*
2. ――― fungiforme, *Goldf.* Maestricht, *Goldf.*
3. ――― Morchella, *Goldf.* Cretaceous Rocks, Essen, Westphalia, *Sack.*
1. Manon capitatum, *Goldf.* Maestricht, *Goldf.*
2. ――― tubuliferum, *Goldf.* Maestricht, *Goldf.*
3. ――― pulvinarium, *Goldf.* Maestricht; Essen, Westphalia, *Goldf.*
4. ――― Peziza, *Goldf.* Maestricht; Cretaceous Rocks, Essen, Westphalia, *Goldf.*
5. ――― stellatum, *Goldf.* Cretaceous Rocks, Essen, *Goldf.*
6. ――― pyriforme, *Goldf.* Chalk, Coesfeld, *Goldf.*
7. ――― verticillites, . Maestricht, *G. T.*
1. Scyphia mammillaris, *Goldf.* Essen, Westphalia, *Goldf.*
2. ――― furcata, *Goldf.* Cretaceous Rocks, Essen, *Goldf.*

3. Scyphia infundibuliformis, *Goldf.* Essen, *Goldf.*
4. ——— foraminosa, *Goldf.* Cretaceous Rocks, Essen, *Goldf.*
5. ——— Sackii, *Goldf.* Essen, Westphalia, *Sack.*
6. ——— tetragona, *Goldf.* Essen, *Goldf.*
7. ——— fungiformis, *Goldf.* Green Sand, Coesfeld, Westphalia, *Goldf.*
8. ——— Mantellii, *Goldf.* Coesfeld, *Goldf.*
9. ——— Dechenii, *Goldf.* Coesfeld, *Goldf.*
10. ——— Oeynhausii, *Goldf.* Green Sand, Darup, Westphalia, *Goldf.*
11. ——— Murchisonii, *Goldf.* Darup, *Goldf.*
12. ——— verticillites, *Goldf.* Maestricht; Chalk, Nehou, *Goldf.*
1. Spongia ramosa, *Mant.* Chalk, Sussex, *Mant.;* Chalk? Yorkshire, *Phil.;* Noirmoutier, *Al. Brong.*
2. ——— lobata, *Flem.* Chalk, Sussex, *Mant.*
3. ——— plana, *Phil.* Chalk, Yorkshire, *Phil.*
4. ——— capitata, *Phil.* Chalk, Yorkshire, *Phil.*
5. ——— osculifera, *Phil.* Chalk, Yorkshire, *Phil.*
6. ——— convoluta, *Phil.* Chalk, Yorkshire, *Phil.*
7. ——— marginata, *Phil.* Chalk, Yorkshire, *Phil.*
8. ——— radiciformis, *Phil.* Chalk, Yorkshire, *Phil.*
9. ——— terebrata, *Phil.* Chalk, Yorkshire, *Phil.*
10. ——— lævis, *Phil.* Chalk, Yorkshire, *Phil.*
11. ——— porosa, *Phil.* Chalk, Yorkshire, *Phil.*
12. ——— cribrosa, *Phil.* Chalk, Yorkshire, *Phil.*
1. Spongus Townsendi, *Mant.* Chalk, Sussex, *Mant.;* Chalk, Rouen, *Pas.*
2. ——— labyrinthicus, *Mant.* Chalk, Sussex, *Mant.;* Chalk, Rouen, *Pas.*
1. Tragos Hippocastanum, *Goldf.* Maestricht, *Goldf.*
2. ——— deforme, *Goldf.* Cretaceous Rocks, Essen, *Goldf.*
3. ——— rugosum, *Goldf.* Cretaceous Rocks, Essen, Westphalia, *Sack.*
4. ——— pisiforme, *Goldf.* Cretaceous Rocks, Essen, Westphalia, *Goldf.*
5. ——— stellatum, *Goldf.* Cretaceous Rocks, Essen, *Goldf.*
1. Alcyonium globulosum, *Defr.* Chalk, Beauvais; Meudon; Amiens; Tours; Gien; Baculite Limest., Normandy, *Desn.*
——? pyriforme, *Mant.* Chalk, Sussex, *Mant.*
———, species not determined. Chalk, Sussex, *Mant.;* Upper Green Sand, Warminster, *Lons.*
1. Choanites subrotundus, *Mant.* Chalk, Sussex, *Mant.*
2. ——— Königi, *Mant.* Chalk, Sussex; Warminster, *Mant.*
3. ——— flexuosus, *Mant.* Chalk, Sussex, *Mant.*
1. Ventriculites radiatus, *Mant.* Chalk, Sussex, *Mant.;* Chalk, Moen, *Al. Brong.*
2. ——— alcyonoides, *Mant.* Chalk, Sussex; Warminster, *Mant.*
3. ——— Benettiæ, *Mant.* Chalk, Sussex, *Mant.;* Chalk, Yorkshire, *Phil.*
1. Siphonia Websteri, *Mant.* Chalk, Sussex, *Mant.*
2. ——— cervicornis, *Goldf.* Chalk, Haldern, Westphalia, *Goldf.*
3. ——— Ficus, *Goldf.* Green Sand, Quedlinburg, *Goldf.*
4. ——— punctata, *Goldf.* Green Sand, Goslar, *G. T.*
1. Hallirhoa costata, *Lam^x.* Green Sand, Normandy, *De la B.;* Upper Green Sand, Warminster, *Lons.*
1. Serea pyriformis, *Lam.* Green Sand, Normandy, *Al. Brong.*
1. Gorgonia bacillaris, *Goldf.* Maestricht, *Goldf.*
1. Nullipora racemosa, *Goldf.* Maestricht, *Goldf.*
1. Millepora Fittoni, *Mant.* Chalk, Sussex, *Mant.*
2. ——— Gilberti, *Mant.* Chalk, Sussex, *Mant.*
3. ——— antiqua? *Defr.* Baculite Limest., Normandy, *Desn.*
4. ——— madreporacea, *Goldf.* Maestricht, *Goldf.*
5. ——— compressa, *Goldf.* Maestricht, *Goldf.*
——— species not determined. Chalk, Meudon, *Al. Brong.*
1. Eschara cyclostoma, *Goldf.* Maestricht, *Goldf.* Kjuge, Sweden, *His.*

2. Eschara pyriformis, *Goldf.* Maestricht, *Goldf.*
3. ——— stigmatophora, *Goldf.* Maestricht, *Goldf.*
4. ——— sexangularis, *Goldf.* Maestricht, *Goldf.*
5. ——— cancellata, *Goldf.* Maestricht, *Goldf.*
6. ——— arachnoidea, *Goldf.* Maestricht, *Goldf.*
7. ——— dichotoma, *Goldf.* Maestricht, *Goldf.*
8. ——— striata, *Goldf.* Maestricht, *Goldf.*
9. ——— filograna, *Goldf.* Maestricht, *Goldf.*
10. ——— disticha, *Goldf.* Meudon, *Goldf.*
1. Cellepora ornata, *Goldf.* Maestricht, *Goldf.*
2. ——— Hippocrepis, *Goldf.* Maestricht, *Goldf.*
3. ——— Velamen, *Goldf.* Maestricht, *Goldf.*
4. ——— dentata, *Goldf.* Maestricht, *Goldf.*
5. ——— crustulenta, *Goldf.* Maestricht, *Goldf.*
6. ——— bipunctata, *Goldf.* Maestricht, *Goldf.*
7. ——— escharoides, *Goldf.* Cretaceous Rocks, Essen, Westphalia, *Goldf.*
1. Coscinopora infundibuliformis, *Goldf.* Coesfeld, *G. T.*
2. ——— macropora, *Goldf.* Stormede, Münster, *G. T.*
1. Retepora clathrata, *Goldf.* Maestricht, *Goldf.*
2. ——— lichenoides, *Goldf.* Maestricht, *Goldf.*
3. ——— truncata, *Goldf.* Maestricht, *Goldf.*
4. ——— disticha, *Goldf.* Maestricht, *Goldf.*
5. ——— cancellata, *Goldf.* Maestricht, *Goldf.*
1. Flustra utricularis, *Lam.* Chalk, Sussex, *Mant.*
2. ———? reticulata, *Desm.* Baculite Limestone, Normandy, *Desn.*
3. ——— flabelliformis, *Lam.* Baculite Limestone, Normandy, *Desn.*
——, species not determined. Chalk, Sussex, *Mant.*
1. Cœloptychium lobatum, *Goldf.* Chalk, Coesfeld, *Goldf.*
2. ——— acaule, *Goldf.* Maestricht; Münster, *Goldf.*
3. ——— agaracoides, *Goldf.* Coesfeld, *G. T.*
1. Ceriopora micropora, *Goldf.* Maestricht, *Goldf.;* Essen, *G. T.*
2. ——— cryptopora, *Goldf.* Maestricht, *Goldf.*
3. ——— anomalopora, *Goldf.* Maestricht, *Goldf.*
4. ——— dichotoma, *Goldf.* Maestricht, *Goldf.*
5. ——— milleporacea, *Goldf.* Maestricht, *Goldf.;* Mörby; Kjuge, &c. Sweden, *His.*
6. ——— madreporacea, *Goldf.* Maestricht, *Goldf.*
7. ——— tubiporacea, *Goldf.* Maestricht, *Goldf.;* Kjuge, &c., Sweden, *His.*
8. ——— verticillata, *Goldf.* Maestricht, *Goldf.*
9. ——— spiralis, *Goldf.* Maestricht, *Goldf.*
10. ——— pustulosa, *Goldf.* Maestricht, *Goldf.*
11. ——— compressa, *Goldf.* Maestricht, *Goldf.*
12. ——— stellata, *Goldf.* Maestricht; Cretaceous Rocks, Essen, *Goldf.*
13. ——— Diadema, *Goldf.* Maestricht, *Goldf.*
14. ——— polymorpha, *Goldf.* Cretaceous Rocks, Essen, Westphalia, *Goldf.*
15. ——— gracilis, *Goldf.* Cretaceous Rocks, Essen, *Goldf.*
16. ——— spongites, *Goldf.* Cretaceous Rocks, Essen, *Goldf.*
17. ——— clavata, *Goldf.* Essen, Westphalia, *Goldf.*
18. ——— trigona, *Goldf.* Cretaceous Rocks, Essen, *Goldf.*
19. ——— Mitra, *Goldf.* Cretaceous Rocks, Essen, *Goldf.*
20. ——— venosa, *Goldf.* Cretaceous Rocks, Essen, *Goldf.*
21. ——— cribrosa, *Goldf.* Cretaceous rocks, Essen, *Goldf.*
1. Lunulites cretacea, *Defr.* Maestricht; Tours; Baculite Limestone, Normandy, *Desn.*
1. Orbulites lenticulata, *Lam.* Chalk, Sussex, *Mant.;* Green Sand, Perte du Rhone, *Al. Brong.;* Chalk, Bray, Normandy, *Pas.*

1. Lithodendron gibbosum, *Munst.*   Green Sand, Bochum, Westphalia, *Munst.*
2. —— gracile, *Goldf.*   Green Sand, Quedlinburg, *Goldf.*
1. Caryophyllia centralis, *Mant.*   Chalk, Sussex, *Mant.;* Chalk, Yorkshire, *Phil.;* Baculite Limestone, Normandy, *Desn.;* Chalk, Duclair; Dieppe, *Pas.*
2. —— Conulus, *Phil.*   Speeton Clay, Yorkshire, *Phil.*
1. Antophyllum proliferum, *Goldf.*   Faxoe, *G. T.*
1. Turbinolia mitrata, *Goldf.*   Aix-la-Chapelle, *Goldf.*
2. —— Kœnigi, *Mant.*   Gault, Sussex, *Mant.*
1. Fungia radiata, *Goldf.*   Cretaceous Sand, Aix-la-Chapelle, *Goldf.*
2. —— cancellata, *Goldf.*   Maestricht, *Goldf.*
3. —— coronula, *Goldf.*   Cretaceous Rocks, Essen, Westphalia, *Goldf.*
1. Chenendopora fungiformis, *Lam.*   Upper Green Sand, Warminster, *Lons.;* Havre; Rouen, *Pas.*
1. Hippalimus fungoides, *Lam.*   Upper Green Sand, Warminster, *Lons.*
1. Diploctenium cordatum, *Goldf.*   Maestricht, *Goldf.*
2. —— Pluma, *Goldf.*   Maestricht, *Goldf.*
1. Meandrina reticulata, *Goldf.*   Maestricht, *Goldf.*
1. Astrea flexuosa, *Goldf.*   Maestricht, *Goldf.*
2. —— geometrica, *Goldf.*   Maestricht, *Goldf.*
3. —— clathrata, *Goldf.*   Maestricht, *Goldf.*
4. —— escharoides, *Goldf.*   Maestricht, *Goldf.*
5. —— textilis, *Goldf.*   Maestricht, *Goldf.*
6. —— velamentosa, *Goldf.*   Maestricht, *Goldf.*
7. —— gyrosa, *Goldf.*   Maestricht, *Goldf.*
8. —— elegans, *Goldf.*   Maestricht, *Goldf.*
9. —— angulosa, *Goldf.*   Maestricht, *Goldf.*
10. —— geminata, *Goldf.*   Maestricht, *Goldf.*
11. —— arachnoides, *Schröter.*   Maestricht, *Goldf.*
12. —— Rotula, *Goldf.*   Maestricht, *Goldf.*
13. —— macrophthalma, *Goldf.*   Maestricht, *Goldf.*
14. —— muricata, *Goldf.*   Chalk, Meudon, *Goldf.*
15. —— stylophora, *Goldf.*   Meudon, *Goldf.*
1. Pagrus Proteus, *Defr.*   Meudon; Tours; Baculite Limestone, Normandy, *Desn.*
Polypifers, genera not determined.   Green Sand, Grand Chartreuse, *Beaum.;* Green Sand, Maritime Alps, *De la B.;* Lower Green Sand, Isle of Wight, *Sedg.;* Gourdon, S. of France, *Dufr.*

RADIARIA.

1. Apiocrinites ellipticus, *Miller.*   Chalk, Sussex, *Mant.;* Chalk, Yorkshire, *Phil.;* Chalk, Touraine; Baculite Limestone, Normandy, *Desn.;* Westphalia; Maestricht, *Goldf.;* Chalk, Dieppe, *Pas.*
1. Pentacrinites, species not determined.   Chalk, Sussex, *Mant.;* Speeton Clay, Yorkshire, *Phil.*
1. Marsupites ornatus, *Miller.*   Chalk, Sussex, *Mant.;* Chalk, Yorkshire, *Phil.*
1. Glenotremites paradoxus, *Goldf.*   Marly Chalk, Speldorf, between Duisberg and Mühlheim, *Goldf.*
1. Asterias quinqueloba, *Goldf.*   Chalk, North Fleet; Chalk, Maestricht; Rinkerode near Münster, *Goldf.*
——, species not determined.   Chalk, Paris; Rouen; *Al. Brong.;* Baculite Limstone, Normandy, *Desn.;* Chalk, England.
1. Cidaris cretosa, *Mant.*   Chalk, Sussex, *Mant.*
2. —— variolaris, *Al. Brong.*   Chalk, Sussex, *Mant.;* Green Sand,

Hâvre; Green Sand, Perte du Rhone, *Al. Brong.*; Cretaceous
Rocks, Koesfeld and Essen, Westphalia; Cretaceous Rocks,
Saxony, *Goldf.*
3. Cidaris claviger, *König.* Chalk, Sussex, *Mant.*; Rouen, *Pas.*
4. ——— vulgaris. Chalk, Poland, *Al. Brong.*
5. ——— regalis, *Goldf.* Maestricht, *Goldf.*
6. ——— vesiculosa, *Goldf.* Cretaceous Rocks, Essen, Westphalia, *Goldf.*
7. ——— scutiger, *Munst.* Cretaceous Rocks, Kelheim, Bavaria, *Goldf.*
8. ——— crenularis, *Lam.* Chalk, France, *Goldf.*
9. ——— granulosa, *Goldf.* Chalk, Aix-la-Chapelle; Maestricht; Cretaceous Rocks, Essen, Westphalia, *Goldf.*
———, species not determined. Chalk, Speeton Clay, Yorkshire, *Phil.*
1. Echinus regalis, *Hœninghaus.* Cretaceous Rocks, Essen, Westphalia, *Goldf.*
2. ——— alutaceus, *Goldf.* Cretaceous Rocks, Essen, *Goldf.*
3. ——— granulosus, *Munst.* Cretaceous Sandstone, Kelheim, Bavaria, *Munst.*
4. ——— areolatus, *Wahl.* Balsberg, Scania, *Nils.* Green Sand, Wilts; Lyme Regis, *König.*
5. ——— Benettiæ, *König.* Green Sand, Chute, Wilts, *König.*
———, species not determined. Green Sand, M. de Fis, *Al. Brong.*; Baculite Limestone, Normandy, *Desn.*; Upper Green Sand, Warminster, *Lons.*
1. Galerites albo-galerus, *Lam.* Chalk, Sussex, *Mant.*; Chalk, Yorkshire, *Phil.*; Chalk, Dieppe, *Al. Brong.*; Chalk, Quedlinberg and Aix-la-Chapelle, *Goldf.* Chalk, Lublin, Poland, *Pusch.* Chalk, Lyme Regis, *De la B.*
2. ——— vulgaris, *Lam.* Chalk, Sussex, *Mant.*; Chalk, Dreux, &c., *Al. Brong.*; Quedlinberg; Aix-la-Chapelle, *Goldf.* Chalk, Lyme Regis, *De la B.*
3. ——— subrotundus, *Mant.* Chalk, Sussex, *Mant.*; Chalk, Yorkshire, *Phil.*
4. ——— Hawkinsii, *Mant.* Chalk, Sussex, *Mant.*
5. ——— abbreviatus, *Lam.* Cretaceous Rocks, Quedlinberg; Aix-la-Chapelle, *Goldf.*
6. ——— canaliculatus, *Goldf.* Cretaceous Rocks, Büren and Brencken, Westphalia, *Goldf.*
7. ——— Subuculus, *Linnæus.* Cretaceous Rocks, Koesfeld and Essen, Westphalia, *Goldf.*; Hâvre, *Pas.*
8. ——— sulcato-radiatus, *Goldf.* Maestricht, *Goldf.*
9. ———? depressus, *Lam.* Green Sand, M. de Fis, *Al. Brong.*
———, species not determined. Chalk, Upper Green Sand, Warminster, *Lons.*
Clypeus, species not determined. Upper Green Sand, Warminster, *Lons.*
1. Clypeaster Leskii, *Goldf.* White Chalk, Maestricht, *Goldf.*
2. ——— fornicatus, *Goldf.* Cretaceous Rocks, Münster, Westphalia, *Goldf.*
3. ——— oviformis, *Lam.* Green Sand, Mans, *Desn.*
1. Echinoneus subglobosus, *Goldf.* Maestricht, *Goldf.*
2. ——— Placenta, *Goldf.* Maestricht, *Goldf.*
3. ——— Lampas, *De la B.* Green Sand, Lyme Regis, *De la B.*
4. ——— peltiformis, *Wahl.* Balsberg, Scania, *Wahl.*
1. Nucleolites Ovulum, *Lam.* Mæstricht, *Goldf.*; Rouen, *Pas.*
2. ——— scrobicularis, *Goldf.* Maestricht, *Goldf.*
3. ——— Rotula, *Al. Brong.* Chalk, Rouen; Green Sand, M. de Fis, *Al. Brong.*
4. ——— castanea, *Al. Brong.* Green Sand, M. de Fis, *Al. Brong.*; Rouen, *Pas.*

5. Nucleolites patellaris, *Goldf.* Maestricht, *Goldf.*
6. ——— pyriformis, *Goldf.* White Chalk, Maestricht and Aix-la-Chapelle, *Goldf.*
7. ——— lacunosus, *Goldf.* Cretaceous Rocks, Essen, Westphalia, *Goldf.*
8. ——— cordatus, *Goldf.* Cretaceous Rocks, Essen, *Goldf.*
9. ——— carinatus, *Goldf.* Chalk, Aix-la-Chapelle and Hildesheim; Cretaceous Rocks, Essen, Westphalia, *Goldf.*
10. ——— Lapis Cancri, *Goldf.* Aix-la-Chapelle; Maestricht, *Goldf.*; Upper Green Sand, Warminster, *Lons.*
11. ——— depressa, *Al. Brong.*, Rouen, *Pas.*
12. ——— heteroclita, *Defr.*, Chalk, Beauvais, *Pas.*
——— , species not determined. Baculite Limestone, Normandy; Lower Chalk, Tours; Rouen, *Desn.*
1. Ananchytes ovata, *Lam.* Chalk, Sussex, *Mant.*; Chalk, Yorkshire, *Phil.*; Chalk, Moen; Meudon, *Al. Brong.*; Baculite Limestone, Normandy, *Desn.*; Limhamn, Sweden, *Nils.*; Cretaceous Rocks, Coesfeld, Westphalia, *Goldf.*; Chalk, Lublin, Poland, *Pusch*; Chalk, Rouen, *Pas.*
2. ——— hemisphærica, *Al. Brong.*; Chalk, Yorkshire, *Phil.*; Chalk, Etretat; Duclair, Normandy, *Pas.*
3. ——— intumescens,   . Chalk, Yorkshire, *Phil.*
4. ——— pustulosa, *Lam.* Chalk, Joigny; Paris; Rouen; and Moen, *Al. Brong.*; Chalk, Norwich, *Woodward.*
5. ——— conoidea, *Goldf.* Cretaceous Rocks, Aubel, Belgium, *Goldf.*
6. ——— striata, *Lam.* Maestricht; Aix-la-Chapelle; Quedlinburg, *Goldf.*
7. ——— sulcata, *Goldf.* Chalk, Aix-la-Chapelle; Maestricht, *Goldf.*
8. ——— Corculum, *Goldf.* Cretaceous Rocks, Coesfeld, Westphalia, *Goldf.*
——— , species not determined. Chalk, Warminster, *Lons.*
1. Spatangus Cor-anguinum, *Lam.* Chalk, Sussex, *Mant.*; Chalk, Yorkshire, *Phil.*; Chalk, Meudon; Joigny; Dieppe; Green Sand, M. de Fis, *Al. Brong.*; Baculite Limestone, Normandy, *Desn.*; Torp, Scania, *Nils.*; Chalk, Dorset and Devon, *De la B.*; Marly Chalk, Paderborn; Bielefeld; Münster; Coesfeld; Aix-la-Chapelle, *Goldf.*; Planerkalk, Saxony, *Munst.*; Chalk, Lublin, Poland, *Pusch*; Mont-Ferrand; Pic de Bugarach, Pyrenees, *Dufr.*
2. ——— rostratus, *Mant.* Chalk, Sussex, *Mant.*; Chalk, Joigny, *Al. Brong.*
3. ——— planus, *Mant.* Chalk, Sussex, *Mant.*; Chalk, Yorkshire, *Phil.*
4. ——— retusus, *Park.* Upper Green Sand, Wiltshire, *Lons.*
5. ——— cordiformis, *Mant.* Chalk, Sussex, *Mant.*
6. ——— suborbicularis, *Defr.* Green Sand, Dives, Normandy, *Al. Brong.*; Marly Chalk, Maestricht, *Goldf.*; Rouen, *Pas.*
7. ——— punctatus, *Lam.* Upper Green Sand, Warminster, *Lons.*
8. ——— granulosus, *Goldf.* Maestricht, *Goldf.*
9. ——— subglobosus, *Leske.* White Chalk, Quedlinburg, Cretaceous Rocks, Büren, Paderborn, *Goldf.*; Rouen, *Pas.*
10. ——— nodulosus, *Goldf.* Cretaceous Rocks, Essen, Westphalia, *Goldf.*
11. ——— radiatus, *Lam.* Maestricht, *Goldf.*
12. ——— truncatus, *Goldf.* White Chalk, Maestricht, *Goldf.*
13. ——— ornatus, *Cuv.* Chalk, Aix-la-Chapelle, *Goldf.*; Env. of Bayonne, *Dufr.*; Chalk, Dieppe; Rouen, *Pas.*
14. ——— Bucklandii, *Goldf.* Cretaceous Rocks, Essen, *Goldf.*
15. ——— Bufo, *Al. Brong.* Chalk, Meudon, Hâvre, *Al. Brong.*; Chalk, Sussex, *Mant.**; Baculite Limestone, Normandy, *Desn.*; Chalk, Aix-la-Chapelle; Maestricht, *Goldf.*

* *Sp. Prunella* of Mantell, according to Brongniart.

16. Spatangus arcuarius, *Lam.*   White Chalk, Maestricht, *Goldf.*
17. ——— Prunella, *Lam.*   Marly Chalk, Maestricht, *Goldf.*
18. ——— Amygdala, *Goldf.*   Chalk, Aix-la-Chapelle, *Goldf.*
19. ——— gibbus, *Lam.*   Cretaceous Rocks, Paderborn, Westphalia, *Goldf.*
20. ——— Cor-testudinarium, *Goldf.*   White Chalk, Maestricht and Qued-
     linburg; Cretaceous Rocks, Coesfeld, Westphalia, *Goldf.*
21. ——— Bucardium, *Goldf.*   Chalk, Aix-la-Chapelle, *Goldf.*
22. ——— lacunosus, *Linnæus.*   Chalk, Quedlinburg and Aix-la-Chapelle,
     Goldf.
23. ——— Murchisonianus, *Kœnig.*   Upper Green Sand, Sussex, *Murch.*,
     Mant.
24. ——— hemisphæricus, *Phil.*   Chalk, Yorkshire, *Phil.*
25. ——— argillaceus, *Phil.*   Speeton Clay, Yorkshire, *Phil.*
26. ——— lævis, *Defr.*   Green Sand, Perte du Rhône, *Al. Brong.;* Havre;
     Rouen, *Pas.*
27. ——— acutus, *Desh.*, S. of France; Rouen, *Desh.*
28. ——— Ambulacrum, *Desh.*, Pyrenees, *Desh.*
29. ——— crassissimus, *Defr.*, Hâvre, *Pas.*
——, species not determined.   Gault and Lower Green Sand, Sussex,
     Mant.; Green Sand, Grande Chartreuse, *Beaum.;* Chalk,
     Warminster, *Lons.*

ANNULATA.

1. Serpula ampullacea, *Sow.*   Chalk, Sussex, *Mant.;* Chalk, Norfolk,
     Barnes.
2. ——— Plexus, *Sow.*   Chalk, Sussex, *Mant.*
3. ——— Carinella, *Sow.*   Green Sand, Blackdown, *Sow.*
4. ——— antiquata, *Sow.*   Green Sand, Wilts, *Sow.*
5. ——— rustica, *Sow.*   Upper Green Sand, Folkstone, *Goodhall.*
6. ——— articulata, *Sow.*   Upper Green Sand, Folkstone, *Sow.*
7. ——— obtusa, *Sow.*   Chalk, Norfolk, *Rose.*
8. ——— fluctuata, *Sow.*   Chalk, Norfolk, *Barnes.*
9. ——— ? macropus, *Sow.*   Chalk, Norfolk, *Leathes.*
10. ——— Trachinus, *Goldf.*   Green Sand, Essen, Westphalia, *Goldf.*
11. ——— lophioda, *Goldf.*   Green Sand, Essen, *Goldf.*
12. ——— lævis, *Goldf.*   Green Sand, Essen, *Goldf.*
13. ——— triangularis, *Münst.*   Gault? Rinkerode, Münster, *Münst.*
14. ——— draconocephala, *Goldf.*   Chalk Marl, Maestricht, *Goldf.*
15. ——— depressa, *Goldf.*   Green Sand, Essen, *Goldf.*
16. ——— Rotula, *Goldf.*   Green Sand, Regensburg, *Goldf.*
17. ——— quadricarinata, *Münst.*   Green Sand, Regensburg, *Goldf.*
18. ——— cincta, *Goldf.*   Green Sand, Essen; Green Sand, Coesfeld; Aix-
     la-Chapelle, *Goldf.*
19. ——— arcuata, *Münst.*   Green Sand, Regensburg, *Goldf.*
20. ——— subtorquata, *Münst.*   Cretaceous Blue Marl, Rinkerode near
     Münster, *Goldf.*
21. ——— sexangularis, *Münst.*, Rinkerode, *Goldf.*
22. ——— Nöggerathii, *Münst.*, Rinkerode, *Goldf.*
23. ——— erecta, *Goldf.*   Cretaceous Marl, Maestricht, *Goldf.*
24. ——— Amphisbœna, *Goldf.*   Green Sand, Bochum, Westphalia; Cre-
     taceous Marl, Maestricht, *Goldf.*
25. ——— spirographis, *Goldf.*   Green Sand, Essen, *Goldf.*
26. ——— parvula, *Münst.*   Green Sand, Essen, *Goldf.*
27. ——— subrugosa, *Münst.*   Blue Cretaceous Marl, Baumberg, near
     Münster, *Goldf.*
28. ——— crenato-striata, *Münst.*, Baumberg, *Goldf.*
29. ——— vibicata, *Münst.*   Blue Cretaceous Marl, Rinkerode, *Goldf.*
30. ——— gordialis, *Schlot.*, Münster; Paderborn; Essen; Osnabrück;

Maestricht; Regensburg; Strehla and Perna, near Dresden, *Goldf.*

Serpula, species not determined. Red Chalk, Speeton Clay, Yorkshire, *Phil.*; Chalk, Paris, *Al. Brong.*; Charlottenlund; Köpinge, Scania, *Nils.*

CIRRIPEDA.

1. Pollicipes sulcatus, *Sow.* Chalk, Sussex, *Mant.*
2. ———— maximus, *Sow.* Chalk, Norfolk, *Barnes.*

CONCHIFERA.

1. Magas pumilus, *Sow.* Chalk, Norwich, *Taylor;* Chalk, Meudon, *Al. Brong.*; Chalk, Dieppe, *Pas.*
1. Thecidea radians, *Defr.* Chalk, Maestricht, *Fauj. de St. Fond;* Baculite Limestone, Normandy, *Desn.*; Chalk, Dieppe, *Pas.*
2. ———— recurvirostra, *Defr.* Maestricht; Baculite Limestone, Normandy, *Desn.*
3. ———— hieroglyphica, *Defr.* Chalk, Essen, *Hœn.*
1. Terebratula subrotunda, *Sow.* Chalk, Sussex, *Mant.*; Green Sand, Bochum, *Hœn.*
2. ———— carnea, *Sow.* Chalk, Sussex, *Mant.*; Chalk, Meudon, *Al. Brong.*; Green Sand, Bochum, *Hœn.*; Chalk, Rouen; Dieppe, *Pas.*
3. ———— ovata, *Sow.* Chalk, Lower Green Sand, Sussex, *Mant.*; Köpinge, Scania, *Nils.*; Green Sand, Bochum, *Hœn.*; Chalk, Rouen, *Pas.*
4. ———— undata*, *Sow.* Chalk, Sussex, *Mant.*; Chalk, Rouen, *Pas.*
5. ———— elongata, *Sow.* Chalk, Sussex, *Mant.*
6. ———— plicatilis, *Sow.* Chalk, Sussex, *Mant.*; Chalk, Meudon, Moen; M. de Fis, *Al. Brong.*; Green Sand, Grande Chartreuse, *Beaum.*; Chalk, Gravesend, *Sow.*; Jonsac; Cognac, *Dufr.*
7. ———— subplicata, *Mant.* Chalk, Sussex, *Mant.*; Chalk, Yorkshire, *Phil.*; Chalk, Maestricht; Tours; Beauvais; Bac. Limestone, Normandy, *Desn.*
8. ———— curvirostris, *Nils.* Köpinge, Scania, *Nils.*
9. ———— Mantelliana †, *Sow.* Chalk, Sussex, *Mant.*
10. ———— Martini ‡, *Mant.* Chalk, Sussex, *Mant.*
11. ———— rostrata, *Sow.* Chalk, Sussex, *Mant.*
12. ———— squamosa, *Mant.* Chalk, Sussex, *Mant.*
13. ———— biplicata, *Sow.* Upper Green Sand, Sussex, *Mant.*; Upper Green Sand, Cambridge, *Sedg.*; Rouen; Hâvre, *Pas.*; Green Sand, Calvados, *Her.*
14. ———— lata, *Sow.* Lower Green Sand, Sussex, *Mant.*; Green Sand, Devizes, *Sow.*; Upper Green Sand, Warminster, *Lons.*; Gourdon, *Dufr.*
15. ———— subundata, *Sow.* Chalk, Speeton Clay, Yorkshire, *Phil.*; Chalk, Rouen, *Al. Brong.*
16. ———— pentagonalis, *Phil.* Chalk, Yorkshire, *Phil.*
17. ———— lineolata, *Phil.* Speeton Clay, Yorkshire, *Phil.*
18. ———— Defrancii §, *Al. Brong.* Chalk, Meudon, *Al. Brong.*; Chalk, Sussex, *Mant.*; Speeton Clay, Yorkshire, *Phil.*; Balsberg, Mörby, Sweden, *Nils.*; Maestricht, *Hœn.*; Chalk, Rouen, *Pas.*
19. ———— intermedia, *Sow.* Upper Green Sand, Warminster, *Sow.*
20. ———— alata, *Lam.* Chalk, Meudon, *Al. Brong.*; Köpinge; Mörby, Sweden, *Nils.*; Cognac, *Dufr.*; Chalk, Rouen, *Pas.*

---

* *T. subundata, T. intermedia,* and *T. semiglobosa,* according to Mantell.
† *T. sulcata* Mant. ‡ *T. Pisum* of Sowerby. § *T. striatula* of Mantell.

21. Terebratula octoplicata, *Sow.* Chalk, Sussex, *Mant.* ; Chalk, Dieppe, *Al. Brong.* ; Balsberg ; Ignaberga, Sweden? *Nils.* ; Green Sand, Quedlinburg, *Hœn.* ; Jonsac ; Cognac, *Dufr.* ; Chalk, Rouen, *Pas.*
22. ——— Gallina, *Al. Brong.* Green Sand, Perte du Rhône, *Al. Brong.* ; Baculite Limestone, Normandy, *Desn.* ; Rouen ; Hâvre, *Pas.*
? 23. ——— ornithocephala, *Sow.* Green Sand, Perte du Rhône ; M. de Fis, *Al. Brong.*
24. ——— pectita, *Sow.* Baculite Limestone, Normandy, *Desn.* ; Ignaberga, Scania? *Nils.* ; Hâvre, *Al. Brong.* ; Upper Green Sand, Wilts, *Meade* ; Maestricht, *Hœn.*
25. ——— recurva, *Defr.* Maestricht ; Baculite Limestone, Normandy, *Desn.*
26. ——— lævigata, *Nils.* Köpinge, Scania, *Nils.*
27. ——— triangularis, *Wahl.* Köpinge, Scania, *Nils.*
28. ——— longirostris, *Wahl.* Balsberg ; Kjuge, Sweden, *Nils.*
29. ——— Lyra, *Sow.* Upper Green Sand, Warminster, *Lons.* ; Hâvre, *Pas.*
30. ——— rhomboidalis, *Nils.* Kjuge ; Mörby, Sweden, *Nils.*
31. ——— semiglobosa, *Sow.* Charlottenlund, Sweden, *Nils.* ; Chalk, Moen, *Al. Brong.* ; Green Sand, Bochum, *Hœn.* ; Chalk, Yorkshire, *Phil.*
32. ——— obtusa, *Sow.* Upper Green Sand, Cambridge, *Sedg.* ; Green Sand, Quedlinburg, *Hœn.*
33. ——— obesa, *Sow.* Chalk, Warminster, *Lons.* ; Chalk, Bray, Normandy, *Pas.*
34. ——— dimidiata, *Sow.* Green Sand, Haldon, *Sow.*
35. ——— aperturata, *Schlot.* Chalk, Essen, *Hœn.*
36. ——— chrysalis, *Schlot.* Maestricht, *Hœn.*
37. ——— curvata, *Schlot.* Green Sand, Quedlinburg, *Hœn.*
38. ——— dissimilis, *Schlot.* Green Sand, Bochum; Chalk, Speldorf, *Hœn.*
39. ——— lacunosa, *Schlot.* Green Sand, Quedlinburg, *Hœn.*
40. ——— microscopica, *Fauj. de St. F.*, Maestricht.
41. ——— nucleus, *Defr.* Green Sand, Bochum ; Quedlinburg, *Hœn.*
42. ——— ovoidea*, *Sow.* Green Sand, Bochum, *Hœn.*
43. ——— peltata, . Maestricht, *Hœn.*
44. ——— semistriata, *Lam.* Green Sand, Bochum, *Hœn.*
45. ——— striatula, *Sow.* Green Sand, Bochum, *Hœn.*
46. ——— varians, . Chalk, Essen, *Hœn.*
47. ——— vermicularis, *Schlot.* Maestricht, *Hœn.*
48. ——— minor, *Nils.* Kjuge, *Nils.*
49. ——— pulchella †, *Nils.* Scania, *Nils.*
50. ——— costata, *Nils.* Kjuge, *Nils.*
51. ——— Lens, *Nils.* Charlottenlund, Sweden, *Nils.*
52. ——— depressa, *Lam.* Gourdon, S. of France, *Dufr.*
53. ——— Gibbsiana, *Sow.* Green Sand, Folkstone, *Sow.*
54. ——— rigida, *Sow.* Norfolk, *Sow.*
1. Crania Parisiensis, *Defr.* Chalk, Meudon, *Al. Brong.* ; Chalk, Brighton, *Sow.* ; Chalk, Dieppe, *Pas.*
2. ——— antiqua, *Defr.* Baculite Limestone, Normandy, *Desn.* ; Chalk, Schlenacken, *Hœn.*
3. ——— striata, *Defr.* Baculite Limestone, Normandy, *Desn.* ; Balsberg, &c. Sweden, *Nils.*
4. ——— stellata, *Defr.* Baculite Limestone, Normandy, *Desn.*

---

* According to Von Dechen very like *T. lata* of Sowerby.
† *T. pumila*, Lam.

516   *Organic Remains of the Cretaceous Group.*

5. Crania spinulosa, *Nils.* Kjuge; Mörby, Sweden, *Nils.*; Maestricht, *Hœn.*
6. —— tuberculata, *Nils.* Scania, *Nils.*
7. —— Nummulus, *Lam.* Balsberg; Kjuge; Ifö, in Scania, *Nils.*
8. —— nodulosa, *Hœn.* Maestricht, *Hœn.*
Orbicula, species not determined. Lower Green Sand, Sussex, *Martin;* Speeton Clay, Yorkshire, *Phil.*
1. Hippurites radiosa, *Des M.* Cendrieux, Périgord, *Des M.*
2. ——— Cornu Pastoris, *Des M.* Pyles, Périgueux, *Jouannet.*
3. ——— striata, *Defr.* Alet, Aude; Manbach, Berne, *Des M.*
4. ——— sulcata, *Defr.* Alet, Aude, *Des M.*
5. ——— dilatata, *Defr.* Alet, Aude, *Des M.*
6. ——— bioculata, *Lam.* Alet, Aude, *Des M.*
7. ——— Fistulæ, *Defr.* Alet, Aude, *Des M.*
8. ——— resecta, *Defr.* Marseille; Dauphiné; Regensburg, Untersberg, *G. T.*
———, species not determined. Cretaceous Rocks, South of France, *Beaum.*; Pyrenees; Jonsac (very large), *Dufr.*; Western Alps, *Lill von Lillienbach;* Chalk, Sussex, *Mant.*
1. Sphærulites dilatata, *Des M.* Chalk, Royan and Talmont, mouth of the Gironde, *Des M.*
2. ——— Bournonii, *Des M.* Royan and Talmont; Vallée de la Couze, Dordogne, *Des M.*
3. ——— ingens, *Des M.* Royan and Talmont, *Des M.*
4. ——— Hœninghausii, *Des M.* Royan and Talmont; Chalk, Languais, Dordogne, *Des M.*
5. ——— foliacea, *Lam.* Isle d'Aix, *Fleurian de Bellevue.*
6. ——— Jodamia, *Des M.* Mirambeau, Charente-Inférieure, *Defr.*
7. ——— Jouannetti, *Des M.* Vallée de la Couze, Périgord, *Des M.*
8. ——— crateriformis, *Des M.* Royan; Languais, Dordogne, *Des M.*
9. ——— cylindracea, *Des M.* Vallée de la Couze, *Des M.*
10. ——— ventricosa, *Des M.* Vallée de la Couze, *Des M.*
11. ——— turbinata, *Des M.* Vallée de la Couze, *Des M.*
12. ——— cristata, *Des M.* Department of the Var, *Des M.*
13. ——— bioculata, *Des M.* Var, *Des M.*
14. ——— imbricata, *Des M.* Var, *Des M.*
15. ——— calceoloides, *Des M.* Vallée de la Couze, *Des M.*
1. Ostrea vesicularis, *Lam.* Chalk, Sussex, *Mant.*; Chalk, Périgueux, Meudon, *Al. Brong.*; Chalk, Maestricht, *Fauj. de St. F.*; var. Baculite Limestone, Normandy, *Desn.*; Köpinge; Kjuge, Sweden, *Nils.*; Chalk, Dieppe, *Pas.*
2. —— semiplana, *Mant.* Chalk, Sussex, *Mant.*
3. —— canaliculata, *Sow.* Chalk, Sussex, *Mant.*; Aix-la-Chapelle, *G. T.*
4. —— carinata, *Lam.* Upper Green Sand, Sussex, *Mant.*; Green Sand, Normandy, *De la B.*; Green Sand, Grasse, (Dep. of the Var,) *Martin de Martigues;* Green Sand, Bochum; Chalk, Essen, *Hœn.*
5. —— serrata, *Defr.* Chalk, Sweden; Dreux, *Al. Brong.*; Green Sand, Grasse, Var; Maestricht, *Hœn.*; Jonsac; Cognac; Angoulême; Coustouge, *Dufr.*; Chalk, Rouen, *Pas.*
6. —— lateralis, *Nils.* Köpinge; Ifö, Scania, *Nils.*; Chalk, Essen, *Hœn.*
7. —— clavata, *Nils.* Mörby, Sweden, *Nils.*
8. —— Hippopodium, *Nils.* Ifö; Carlshamn, Sweden, *Nils.*
9. —— curvirostris, *Nils.* Ifö; Kjuge, Scania, *Nils.*
10. —— acutirostris, *Nils.* Ifö; Scania, *Nils.*
11. —— flabelliformis, *Nils.* Kjuge, Mörby, Sweden, *Nils.*; Chalk, Essen, *Hœn.*
12. —— pusilla, *Nils.* Köpinge, Scania, *Nils.*

13. Ostrea diluviana?* *Lam.* Balsberg; Kjuge; Mörby; Carlshamn, Sweden,
    Nils.; Orcher, Rouen, *Pas.*
14. —— lunata, *Nils.* Ähus, Yngsjö, Scania, *Nils.*
15. —— truncata, *Goldf.* Green Sand, Griesenbeck, *Hœn.*
16. —— incurva, *Nils.* Kjuge; Oppmanna, *Nils.*
17. ——? plicata, *Nils.* Kjuge, Sweden, *Nils.*
18. —— biauricularis, . Jonsac; Cognac; Angoulême, *Dufr.*
19. —— pectinata, *Lam.* Green Sand, Orcher, Rouen, *Pas.*; Bochum,
    G. T.
20. —— Rotomagensis, *Defr.* Rouen, *Pas.*
21. —— pectinoïdes, *Defr.* Rouen, *Pas.*
22. —— auriculata, *Defr.* Rouen, *Pas.*
1. Exogyra digitata, *Sow.* Green Sand, Lyme Regis, *De la B.*
2. —— conica, *Sow.* Green Sand, Sussex; Upper Green Sand, Wilts;
    Green Sand, Blackdown, *Sow.*; Köpinge, *Nils.*; Green Sand,
    Haldon Hill, *Baker.*
3. —— undata, *Sow.* Green Sand, Blackdown, *Goodhall.*
4. —— haliotoidea, *Sow.* Upper Green Sand, Warminster, *Lons.*; Chalk,
    Essen, *Hœn.*; Kjuge; Balsberg; Mörby, *Nils.*; Lillebonne,
    Pas.
5. —— lævigata, *Sow.* Green Sand, N. of Ireland, *Sow.*
1. Gryphæa vesiculosa, *Sow.* Upper Green Sand, Sussex, *Mant.*; Green
    Sand, Warminster, *Bennet*; Green Sand, Bouches du Rhône,
    Hœn.; Bourg St. Andréol, Env. of Pont St. Esprit; Gourdon,
    Dufr.
2. —— sinuata, *Sow.* Speeton Clay, Yorks., *Phil.*; Green Sand, Grande
    Chartreuse, *Beaum.*, Lower Green Sand, Isle of Wight, *Sedg.*;
    Pic de Bugarach; Bourg St. Andréol, *Dufr.*; Bray, Norm., *Pas.*
3. —— auricularis, *Al. Brong.* Chalk, Périgueux, *Al. Brong.*; Green
    Sand, Grande Chartreuse, *Beaum.*; Chalk, Kazimirz, Poland,
    Pusch; Green Sand, Apt, Vaucluse, *Hœn.*; Jonsac; Cognac,
    Dufr.
4. —— Aquila, *Al. Brong.* Green Sand, Perte du Rhône, *Al. Brong.*;
    Pic de Bugarach, Pyrenees; Bourg St. Andréol; Jonsac;
    Cognac, *Dufr.*; Rouen, *Pas.*
5. —— Columba, *Lam.* Green Sand, Normandy; Green Sand, Mari-
    time Alps, *De la B.*; Chalk, Kazimirz, Poland, *Pusch*; Re-
    genburg; Pirna; Königstein, *Holl*; Chalk, Saumur; Mans,
    Hœn.; Env. of Pont St. Esprit; Angoulême, *Dufr.*; Plessis-
    Grimoult, Calvados, *Desl.*
6. —— truncata, *Goldf.* Maestricht, *Hœn.*
7. —— secunda, . Env. of Pont St. Esprit; Jonsac; Cognac;
    Gourdon; Pic de Bugarach, Pyrenees, *Dufr.*
8. —— canaliculata, *Sow.* Upper Green Sand, Wilts, *Sow.*
    ——, a small species in the baculite limestone and chalk of other parts
    of France, *Desn.*
1. Sphæra corrugata, *Sow.* Lower Green Sand, Isle of Wight, *Sedg.*
1. Podopsis obliqua, *Mant.* Chalk, Sussex, *Mant.*
2. —— striata, *Sow.* Chalk, Yorks., *Phil.*; Chalk, Hâvre, *Al. Brong.*;
    Chalk, Essen; Bochum, *Hœn.*; Chalk, Dieppe, *Pas.*; Chalk,
    Sussex, *Mant.*
3. —— truncata, *Lam.* Chalk, Normandy, Touraine, *Al. Brong.*;
    Balsberg and other places in Sweden, *Nils.*; Lyme Regis,
    De la B.
4. —— lamellata, *Nils.* Kjuge, Mörby, Sweden, *Nils.*

---

* M. Brongniart considers that this shell, cited by M. Nilsson as *O. diluviana,*
may be the *O. serrata* of Defrance.

5. Podopsis spinosa,   .   Coustouge, *Dufr.*
———, species not determined. Gourdon, *Dufr.*
1. Spondylus? strigilis, *Al. Brong.; * Green Sand, Perte du Rhône, *Al.*
     Brong.
1. Plicatula inflata, *Sow.* Chalk, Sussex, *Mant.;* Chalk, Cambridge, *Sedg.*
2. ——— pectinoides, *Sow.* Chalk, Sussex, *Mant.;* Gault, Cambridge, *Sedg.*
3. ——— radiata, *Goldf.* Coesfeld, *G. T.*
4. ——— spinosa, *Mant.* Orcher, Norm., *G. T.*
1. Pecten quinquecostatus, *Sow.* Chalk, Sussex, *Mant.;* Chalk, Meudon,
     Al. Brong.; Green Sand, Perte du Rhone, *Al. Brong.;* Ba-
     culite Limestone, Normandy, *Desn.;* Köpinge, and other
     places in Sweden, *Nils.;* Green Sand, Blackdown, *Sow.;*
     Green Sand, Lyme Regis, *De la B.;* Upper Green Sand,
     Warminster, *Lons.;* Green Sand, Coesfeld, Osterfeld; Chalk,
     Saumur, *Hœn.;* Env. of Pont St. Esprit; Cognac; Mont-
     Ferrand; Pic de Bugarach, Pyrenees; Env. of Bayonne, *Dufr.;*
     Green Sand, Calvados, *Her.;* Maestricht, *G. T.*
2. ——— Beaveri, *Sow.* Chalk, Sussex, *Mant.;* Büren; Quedlinburg, *G. T.;*
     Lillebonne, *Pas.*
3. ——— triplicatus, *Mant.* Chalk, Sussex, *Mant.*
4. ——— orbicularis, *Sow.* Chalk, Gault, Lower Green Sand, Sussex, *Mant.;*
     Köpinge, Sweden? *Nils.;* Green Sand, Aix-la-Chapelle, *Hœn.;*
     Lillebonne, *Pas.*
5. ——— quadricostatus, *Sow.* Lower Green Sand, Sussex, *Mant.;* Chalk,
     Maestricht; Baculite Limestone, Normandy, *Desn.;* Green
     Sand, Grande Chartreuse, *Beaum.;* Green Sand, Haldon,
     Baker; Upper Green Sand, Warminster, *Lons.*
6. ——— obliquus, *Sow.* Lower Green Sand, Sussex, *Mant.;* Green Sand,
     Calvados, *Her.;* Lillebonne, *Pas.*
7. ——— cretosus, *Defr.* Chalk, Meudon, *Al. Brong.;* Chalk, Lublin, Po-
     land, *Pusch;* Chalk, Angers.
8. ——— arachnoides, *Defr.* Chalk, Meudon and Normandy, *Al. Brong.;*
     Chalk, Lublin, Poland, *Pusch.*
9. ——— extextus *, *Al. Brong.* Chalk, Hâvre; Baculite Limestone, Nor-
     mandy, *Desn.;* Chalk, Angers, *Hœn.*
10. ——— serratus, *Nils.* Balsberg; Köpinge, Sweden, *Nils.*
11. ——— septemplicatus, *Nils.* Balsberg, Kjuge, Sweden, *Nils.;* Maestricht,
     G. T.
12. ——— multicostatus, *Nils.* Balsberg, Sweden, *Nils.*
13. ——— undulatus, *Nils.* Köpinge; Käserberga, Scania, *Nils.*
14. ——— subaratus, *Nils.* Balsberg; Kjuge, Sweden, *Nils.*
15. ——— pulchellus, *Nils.* Köpinge; Balsberg, Sweden, *Nils.*
16. ——— lineatus, *Nils.* Köpinge; Mörby, Sweden, *Nils.*
17. ——— virgatus, *Nils.* Balsberg; Mörby, *Nils.*
18. ——— membranaceus, *Nils.* Köpinge, and other places, Sweden, *Nils.*
19. ——— lævis, *Nils.* Köpinge; Yngsjoe, Sweden, *Nils.;* Aix-la-Chapelle,
     Hœn.
20. ——— inversus, *Nils.* Köpinge, Sweden, *Nils.*
21. ——— asper, *Lam.* Upper Green Sand, Warminster, *Lons.;* Chalk,
     Lublin, Poland, *Pusch;* Green Sand, Bochum; Chalk, Hat-
     teren, *Hœn.;* Green Sand, Calvados, *Her.;* Lillebonne, *Pas.;*
     Maestricht, *G. T.*
22. ——— asperrimus, *Hœn.* Green Sand, Hardt, *Hœn.*
23. ——— gryphæatus,   . Green Sand, Aix-la-Chapelle, *Hœn.*
24. ——— nitidus, *Sow.* Chalk, Sussex, *Mant.;* Green Sand, Aix-la-Chapelle,
     Hœn.; Chalk, Rouen; Dieppe, *Pas.*

* M. Hœninghaus considers this shell the same with *P. serratus,* Nilsson.

25. Pecten versicostatus, . Green Sand, Aix-la-Chapelle; Green Sand, Minden, *Hœn.*

?26. —— corneus, *Sow.* Köpinge? *Nils.*

27. —— dentatus, *Nils.* Balsberg, *Nils.*

28. —— dubius, *Defr.* Chalk, Rouen; Dieppe, *Pas.*

——, species not determined. Chalk, Sussex, *Mant.* ; Speeton Clay, Yorks., *Phil.* ; Green Sand, Maritime Alps, *De la B.*

1. Lima pectinoides, *Hœn.* Maestricht, *Hœn.*

2. —— striata, *Goldf.* Maestricht, *G. T.*

3. —— muricata, *Goldf.* Maestricht, *G. T.*

1. Plagiostoma spinosum *, *Sow.* Chalk, Sussex, *Mant.* ; Chalk, Meudon, Dieppe, Rouen, Périgueux, Poland, *Al. Brong.* ; Köpinge, Sweden, *Nils.* ; Chalk, Dorset and Devon, *De la B.* ; Chalk, Weinbohla, Saxony, *Weiss*; Quedlinburg, *Holl* ; Osterfeld, *Hœn.* ; Env. of Pont St. Esprit; Coustouge, *Dufr.*

2. —— Hoperi, *Mant.* ; Chalk, Sussex, *Mant.* ; Orcher; Rouen, *Pas.*

3. —— Brightoniense, *Mant.* ; Chalk, Sussex, *Mant.*

4. —— elongatum, *Sow.* ; Chalk, Sussex, *Mant.*

5. —— asperum, *Mant.* ; Chalk, Sussex, *Mant.* ; Coustouge, *Dufr.*

6. —— ovatum, *Nils.* Balsberg and Kjuge, Sweden, *Nils.*

7. —— semisulcatum, *Nils.* Balsberg and other places, Sweden, *Nils.* ; Chalk, Künder, Saumur, *Hœn.*

8. —— Mantelli, *Al. Brong.* Chalk, Dover ; Moen, Denmark, *Al. Brong.* ; Chalk, Dieppe, *Pas.*

9. —— granulatum, *Nils.* Köpinge, Kjuge, Sweden, *Nils.*

10. —— elegans, *Nils.* Balsberg, Mörby, Sweden, *Nils.*

11. —— pusillum, *Nils.* Balsberg, Köpinge, Sweden, *Nils.*

12. —— turgidum, *Lam.* Chalk, Saintes; Green Sand, Osterfeld, *Hœn.*

13. —— denticulatum, *Nils.* Ignaberga, Kjuge, *Nils.*

14. —— squamatum, *Goldf.* Maestricht, *G. T.*

15. —— Juliobonæ, *Pas.* Lillebonne, *Pas.*

——, species not determined :—Upper Green Sand, Sussex, *Mant.*

1. Avicula cœrulescens, *Nils.* Köpinge, Kaseberga, Sweden, *Nils.*

2. —— triptera, *Bronn.* Maestricht, *G. T.*

——, species not determined. Chalk, Sussex, *Mant.* ; Maestricht? *Hœn.* ; Gourdon, *Dufr.*

1. Inoceramus Cuvieri, *Sow.* Chalk, Sussex, *Mant.* ; Chalk, Yorks., *Phil.* ; Chalk, Meudon, *Al. Brong.* ; Balsberg; Ignaberga, Kjuge, Sweden, *Nils.* ; Jonsac; Cognac ; Gourdon, *Dufr.* ; Chalk, Rouen; Dieppe, *Pas.*

2. —— Brongniarti, *Mant.* Chalk, Sussex, *Mant.* Chalk, Yorks., *Phil.* ; Käseberga, Köpinge, Sweden, *Nils.* ; Chalk, Czarkow, Poland, *Pusch* ; Quedlinburg, *Hœn.*

3. —— Lamarckii †, *Mant.* Chalk, Sussex, *Mant.* ; Rouen, *Pas.*

4. —— mytiloides, *Mant.* Chalk, Sussex, *Mant.* ; Chalk, Warminster, *Lons.* ; Quedlinburg ; Pirna, Königstein, *Holl* ; Env. of Pont St. Esprit, *Dufr.* ; Chalk, Dieppe ; Duclair, *Pas.*

5. —— cordiformis, *Sow.* Chalk, Sussex, *Mant.* ; Chalk, Gravesend, *Sow.*

6. —— latus, *Mant.* Chalk, Sussex, *Mant.* ; Rouen ; Meulers, *Pas.*

---

* *Pachites spinosa* of Defrance. According to M. Deshayes, the species of *Plagiostoma* which have been named *Pachites* by M. Defrance, are referrible to the genus *Spondylus*, while the remaining species of the same supposed genus belong to the genus *Lima*.

† According to M. Deshayes, *Inoceramus (Catillus) Lamarckii* and *I. Brongniarti* are the same shells. All the *Inocerami* of the chalk are *Catilli* according to Deshayes.

7. Inoceramus Websteri, *Mant.* Chalk, Sussex, *Mant.*
8. ——— striatus, *Mant.* Chalk, Sussex, *Mant.;* Quedlinburg; Dresden, *G. T.*
9. ——— undulatus, *Mant.* Chalk, Sussex, *Mant.*
10. ——— involutus, *Sow.* Chalk, Sussex, *Mant.;* Chalk, Norfolk, *Rose.*
11. ——— tenuis, *Mant.* Chalk, Sussex, *Mant.*
12. ——— Cripsii, *Mant.* Chalk, Sussex, *Mant.;* Chalk, Dieppe, *Pas.*
13. ——— concentricus, *Park.* Gault, Sussex, *Mant.;* Green Sand, Perte du Rhône, M. de Fis, *Al. Brong.;* Chalk, Warminster, *Lons.;* Green Sand, Quedlinburg, Bochum, and Essen, *Hœn.;* Chalk, Rouen; Meulers, *Pas.*
14. ——— sulcatus, *Park.* Gault, Sussex, *Mant.;* Green Sand, Perte du Rhône; M. de Fis, *Al. Brong.;* Köpinge, Scania, *Nils.;* Green Sand? Nice, *De la B.;* Rouen; Meulers, *Pas.*
15. ——— gryphæoides, *Sow.* Gault, Sussex, *Mant.;* Green Sand, Lyme Regis, *De la B.*
16. ——— pictus, *Sow.* Chalk, Surrey, *Murch.*
17. ——— rugosus, . Quedlinburg, *Hœn.*
18. ——— fornicatus, *Goldf.* Westphalia, *G. T.*
19. ——— cardissoides, *Goldf.* Quedlinburg, *G. T.*
———, species not determined. Lower Green Sand, Sussex, *Martin;* Baculite Limestone, Normandy, *Desn.*
1. Pachymya Gigas, *Sow.* Lower Chalk, Lyme Regis, *De la B.*
1. Meleagrina approximata, *Bronn.* Maestricht, *G. T.*
1. Gervillia aviculoides, *Sow.* Lower Green Sand, Sussex, *Mant.;* Green Sand, Lyme Regis, *De la B.;* Quedlinburg, *Holl;* Lower Green Sand? Isle of Wight, *Sedg.*
2. ——— solenoides, *Defr.* Lower Green Sand, Sussex, *Mant.;* Baculite Limestone, Normandy, *Desn.;* Green Sand, Lyme Regis, *De la B.;* Upper Green Sand, Warminster, *Lons.;* Maestricht, *Hœn.;* Upper Green Sand, Aix-la-Chapelle, *Dum.*
3. ——— acuta, *Sow.* Lower Green Sand, Sussex, *Mant.*
1. Pinna gracilis, *Phil.* Speeton Clay, Yorks., *Phil.*
2. ——— tetragona, *Sow.* Upper Green Sand, Devizes, *Gent;* Maestricht; Bochum; Aix-la-Chapelle; Green Sand, Pirna, *G. T.*
3. ——— restituta, . Chalk, Valkenburg, *Hœn.*
? 4. ——— subquadrivalvis, *Lam.* Cotentin; Saumur, *Hœn.*
1. Mytilus lanceolatus, *Sow.* Lower Green Sand, Sussex, *Mant.;* Green Sand, Blackdown, *Sow.*
2. ——— lævis, *Defr.* Chalk, Bougival, *Al. Brong.;* Rouen, *Pas.*
3. ——— edentulus, *Sow.* Green Sand, Blackdown. *Sow.*
4. ——— problematicus, . Green Sand, Bochum, *Hœn.*
5. ——— simplex, *Pas.* Chalk, Rouen, *Pas.*
1. Modiola æqualis, *Sow.* Lower Green Sand, Sussex, *Mant.*
2. ——— bipartita, *Sow.* Lower Green Sand, Sussex, *Mant.;* Env. of Pont St. Esprit. *Dufr.*
1. Chama Cornu Arietis, *Nils.* Kjuge; Mörby, Sweden, *Nils.*
2. ——— laciniata, *Nils.* Kjuge; Balsberg; Mörby, Sweden, *Nils.*
———, species not determined. Chalk, Sussex, *Mant.*
1. Trigonia Dædalea, *Park.* Lower Green Sand, Sussex, *Mant.;* Green Sand, Haldon? *Baker;* Lower Green Sand, Isle of Wight, *Sedg.;* Env. of Pont St. Esprit, *Dufr.*
2. ——— aliformis, *Sow.* Lower Green Sand, Sussex, *Mant.;* Blackdown, *De la B.;* Upper Green Sand? Eddington, *Lons.;* Lower Green Sand, Isle of Wight, *Sedg.;* Gourdon, *Dufr.;* Aix-la-Chapelle; Quedlinburg, *G. T.;* Orcher, Normandy, *Pas.*
3. ——— spinosa, *Sow.* Lower Green Sand, Sussex, *Martin;* Green Sand, Blackdown, *Steinhauer;* Lillebonne, *Pas.*

4. Trigonia rugosa, *Lam.* Green Sand, Perte du Rhône, *Al. Brong.* Rouen, *Pas.*
5. —— scabra, *Lam.* Green Sand, Perte du Rhône, *Al. Brong.*; Baculite Limestone? Normandy, *Desn.*; Plessis-Grimoult, Calvados, *Her.*; Lillebonne; Rouen, *Pas.*
6. —— pumila, *Nils.* Köpinge, Scania, *Nils.*
7. —— eccentrica, *Sow.* Green Sand, Blackdown, *Steinhauer.*
8. —— nodosa, *Sow.* Lower Green Sand, Hythe, Kent, *Sow.*
9. —— spectabilis, *Sow.* Green Sand, Blackdown, *Goodhall.*
10. —— arcuata, *Lam.* Aix-la-Chapelle, *Hœn.*
11. —— alata, . Env. of Pont St. Esprit; Pic de Bugarach, Pyrenees, *Dufr.*
—— , species not determined. Lower Green Sand, Wiltshire, *Lons.*
1. Nucula pectinata, *Mant.* Gault, Sussex, *Mant.*; Blue Marl, Bray, Normandy, *Pas.*
2. —— ovata, *Mant.* Gault, Sussex, *Mant.*; Speeton Clay, Yorkshire, *Phil.*
3. —— impressa, *Sow.* Lower Green Sand, Sussex, *Mant.*; Green Sand, Blackdown, *Sow.*
4. —— subrecurva, *Phil.* Speeton Clay, Yorkshire, *Phil.*
5. —— ovata, *Nils.* Köpinge; Käseberga, Scania, *Nils.*
6. —— truncata, *Nils.* Käseberga, Scania, *Nils.*
7. —— panda, *Nils.* Käseberga, Scania, *Nils.*
8. —— producta, *Nils.* Käseberga, Scania, *Nils.*
9. —— antiquata, *Sow.* Green Sand, Blackdown, *Sow.*
10. —— angulata, *Sow.* Green Sand, Blackdown, *Sow.*
11. —— undulata, *Sow.* Gault, Folkestone, *Sow.*
12. —— siliqua, *Goldf.* Maestricht, *G. T.*
1. Pectunculus lens, *Nils.* Balsberg; Köpinge, Sweden, *Nils.*
2. —— sublævis, *Sow.* Green Sand, Blackdown, *Sow.*; Paderborn; Quedlinburg, *G. T.*
3. —— umbonatus, *Sow.* Green Sand, Blackdown, *Sow.*; Rouen, *Pas.*
1. Arca carinata, *Sow.* Upper Green Sand, Sussex, *Mant.*
2. —— exaltata, *Nils.* Carlshamn, Sweden, *Nils.*; Green Sand? Aix-la-Chapelle, *Hœn.*
3. —— rhombea, *Nils.* Balsberg, Sweden, *Nils.*; Aix-la-Chapelle, *G. T.*
4. —— clathrata, *Lam.* Chalk, Angers; Saumur, *Hœn.*
5. —— ovalis, *Nils.* Köpinge, Scania, *Nils.*; Aix-la-Chapelle, *G. T.*
6. —— subacuta, *Hœn.* Maestricht, *Hœn.*
—— , species not determined. Chalk, Gault, Sussex, *Mant.*
1. Cucullæa decussata, *Sow.* Lower Green Sand, Sussex, *Mant.*; Chalk, Rouen, *Al. Brong.*
2. —— glabra, *Sow.* Green Sand, Blackdown, *Sow.*; Upper Green Sand, Warminster, *Lons.*; Rouen, *Pas.*
3. —— carinata, *Sow.* Green Sand, Blackdown, *Sow.*; Aix-la-Chapelle, *G. T.*; Rouen, *Pas.*
4. —— fibrosa, *Sow.* Green Sand, Blackdown, *Hill.*
5. —— costellata, *Sow.* Green Sand, Blackdown, *Sow.*
6. —— crassatina, *Lam.* Chalk, Beauvais, *Hœn.*
—— , species not determined. Chalk, Sussex, *Mant.*; Speeton Clay, Yorkshire, *Phil.*; Gourdon, *Dufr.*
1. Cardita Esmarkii, *Nils.* Köpinge, Scania, *Nils.*
2. —— Modiolus, *Nils.* Käseberga, Scania, *Nils.*
3. —— tuberculata, *Sow.* Upper Green Sand, Devizes, *Gent.*
4. —— crassa, *Lam.* Chalk, Doué, *Hœn.*
—— , species not determined. Upper Green Sand, Sussex, *Mant.*
1. Cardium decussatum, *Sow.* Chalk, Sussex, *Mant.*

2. Cardium Hillanum, *Sow.* Green Sand, Blackdown, *Hill;* Env. of Pont
   St. Esprit; Gourdon, *Dufr.*
3. ——— proboscideum, *Sow.* Green Sand, Blackdown, *Hill.*
   Venericardia, species not determined. Chalk, Sussex, *Mant.*
1. Astarte striata, *Sow.* Green Sand, Blackdown, *Sow.;* Upper Green
   Sand, Devizes, *Lons.*
   ———, species not determined. Chalk, Sussex, *Mant.;* Lower Green
   Sand, Wilts, *Lons.*
1. Thetis minor, *Sow.* Lower Green Sand, Sussex, *Mant.;* Green Sand,
   Lyme Regis, *De la B.*
2. ——— major, *Sow.* Upper Green Sand, Devizes, *Gent;* Green Sand,
   Blackdown, *Hill.*
1. Venus Ringmeriensis, *Mant.* Chalk, Sussex, *Mant.*
*2. ——— parva, *Sow.* Lower Green Sand, Sussex, *Mant.;* Green Sand,
   Lyme Regis, *De la B.;* Green Sand, Isle of Wight, *Sow.*
*3. ——— angulata, *Sow.* Lower Green Sand, Sussex, *Mant.;* Green Sand,
   Blackdown, *Hill.*
*4. ——— Faba, *Sow.* Lower Green Sand, Sussex, *Mant.;* Green Sand,
   Blackdown; Green Sand, Isle of Wight, *Sow.*
5. ——— ovalis, *Sow.* Lower Green Sand, Sussex, *Mant.*
*6. ——— lineolata, *Sow.* Green Sand, Blackdown, *Hill;* Green Sand, Bochum, *Hœn.*
7. ——— plana, *Sow.* Green Sand, Blackdown, *Hill.*
*8. ——— caperata, *Sow.* Green Sand, Lyme Regis, *De la B.;* Green Sand,
   Blackdown, *Hill.*
9. ———? exerta, *Nils.* Köpinge, *Nils.*
1. Lucina sculpta, *Phil.* Speeton Clay, Yorkshire, *Phil.*
1. Tellina æqualis, *Mant.* Lower Green Sand, Sussex, *Mant.*
2. ——— inæqualis, *Sow.* Lower Green Sand, Sussex, *Mant.;* Green
   Sand, Blackdown, *Sow.*
3. ——— striatula, *Sow.* Green Sand, Blackdown, *Sow.*
   ———, species not determined. Speeton Clay, Yorkshire, *Phil.*
1. Corbula striatula, *Sow.* Lower Green Sand, Sussex, *Mant.*
2. ——— Punctum, *Phil.* Speeton Clay, Yorkshire, *Phil.*
3. ——— gigantea, *Sow.* Green Sand, Blackdown, *Hill.*
4. ——— lævigata, *Sow.* Green Sand, Blackdown, *Hill.*
5. ——— ovalis, *Nils.* Köpinge, *Nils.*
6. ——— caudata, *Nils.* Köpinge, *Nils.*
   ———, species not determined, Rouen, *Pas.*
1. Crassitella latissima, *Hœn.* Maestricht, *Hœn.*
2. ——— tumida, . Coustouge, *Dufr.*
1. Cythere subdeltoidea, *Munst.* Chalk, Rinkerode; Strehla; Haldem;
   Maestricht, *G. T.*
2. ——— compressa, *Munst.;* Haldem, *G. T.*
1. Lutraria Gurgitis, *Al. Brong.;* Green Sand, Perte du Rhône, *Al. Brong.;*
   Köpinge, Mörby, Sweden, *Nils.*
2. ———? carinifera, *Sow.* Chalk, Lyme Regis, *De la B.;* Rouen, *Pas.*
   ———, species not determined. Speeton Clay? Yorkshire, *Phil.*
1. Panopæa plicata, *Sow.* Green Sand, Osterfeld, *Hœn.;* (var.?) Lower
   Green Sand, Sussex, *Mant.;* Coustouge, *Dufr.*
1. Mya mandibula, *Sow.* Lower Green Sand, Sussex, *Martin;* Gault, Isle
   of Wight, *Fitton;* Gourdon, *Dufr.;* Rouen, *Pas.*
2. ——— phaseolina, *Phil.* Speeton Clay, Yorkshire, *Phil.*
   ———? Chalk, near Calne, *Lons.*

---

* *Cytherea* of Lamarck, according to Lonsdale.

Teredo, species not determined. Maestricht, *Hœn.*
1. Pholas? constricta, *Phil.* Speeton Clay, Yorkshire, *Phil.*
1. Fistulana pyriformis, *Mant.* Gault, Sussex, *Mant.*

MOLLUSCA.

1. Dentalium striatum, *Sow.* Gault, Sussex, *Mant.*
2. ——— ellipticum, *Sow.* Gault, Sussex, *Mant.*; Rouen, *Pas.*
3. ——— decussatum, *Sow.* Gault, Sussex, *Mant.*
4. ——— nitens, *Hœn.* Maestricht, *Hœn.*
———, species not determined. Lower Green Sand, Sussex, *Mant.*
1. Patella ovalis, *Nils.* Balsberg, Scania, *Nils.*
———, species not determined. Lower Green Sand, Sussex, *Mant.*; Lower Green Sand, Wiltshire, *Lons.*
1. Emarginula Sanctæ Catherinæ, *Pas.*; Rouen, *Pas.*
2. ——— pelagica, *Pas.*; Rouen, *Pas.*
Pileopsis, species not determined. Lower Green Sand, Sussex, *Mant.*
1. Helix Gentii, *Sow.* Upper Green Sand, Devizes, *Gent.*
1. Auricula incrassata, *Sow.* Chalk, Sussex, *Mant.*; Green Sand, Blackdown, *Hill.*
2. ——— obsoleta, *Phil.* Speeton Clay, Yorkshire, *Phil.*
3. ——— avellana, *Mant.*; Rouen, *Pas.*
Melania, species not determined. Speeton Clay? Yorkshire, *Phil.*
1. Paludina extensa, *Sow.* Green Sand, Blackdown, *Hill.*
1. Ampullaria canaliculata. Gault, Sussex, *Mant.*; Rouen, *Pas.*
2. ——— spirata, *Hœn.* Maestricht, *Hœn.*
———, species not determined. Green Sand, M. de Fis, *Al. Brong.*
1. Nerita rugosa, *Hœn.* Maestricht, *Hœn.*
1. Natica canrena, *Park.* Lower Green Sand, Sussex, *Mant.*
2. ——— spirata, . Green Sand, Aix-la-Chapelle, *Hœn.*
———, species not determined. Gault, Sussex, *Mant.*; Lower Green Sand, Wiltshire, *Lons.*; Env. of Pont St. Esprit, *Dufr.*
1. Vermetus polygonalis, *Sow.* Lower Green Sand, Hythe, Kent, *Lord Greenock.*
2. ——— umbonatus, *Mant.* Chalk, Sussex, *Mant.*
3. ——— Sowerbii, *Mant.* Chalk, Sussex, *Mant.*; Speeton Clay, Yorkshire, *Phil.*
4. ——— concavus, *Sow.* Lower Green Sand, Sussex, *Mant.*; Upper Green Sand, Wilts, *Lons.*
———, species not determined. Lower Green Sand, Isle of Wight, *Sedg.*
Delphinula, species not determined. Speeton Clay, Yorkshire, *Phil.*
1. Solarium tabulatum? *Phil.* Speeton Clay, Yorkshire, *Phil.*
1. Cirrus depressus, *Mant.* Chalk, Sussex, *Mant.*
2. ——— perspectivus, *Mant.* Chalk, Sussex, *Mant.*
3. ——— granulatus, *Mant.* Chalk, Sussex, *Mant.*
4. ——— plicatus, *Sow.* Gault, Sussex, *Mant.*
1. Pleurotomaria Rhodani, *Al. Brong.* Rouen, *Pas.*
2. ——— depressa, *Sow.* Rouen, *Pas.*
3. ——— perspectiva, *Sow.* Chalk, Rouen, *Pas.*
———, species not determined. Maestricht, *Hœn.*; Gourdon; Bourg St. Andréol, *Dufr.*
1. Trochus Basteroti, *Al. Brong.* Chalk, Sussex, *Mant.*; Köpinge, Scania, *Nils.*; Rouen, *Pas.*
2. ——— linearis, *Mant.* Chalk, Sussex, *Mant.*
3. ——— Rhodani, *Al. Brong.* Upper Green Sand, Sussex, *Mant.*; Green Sand, Perte du Rhône, *Al. Brong.*; Lower Chalk, Lyme Regis, *De la B.*; Green Sand, Essen; Green Sand, Osterfeld, *Hœn.*
4. ——— bicarinatus, *Sow.* Upper Green Sand? Sussex, *Mant.*

5. Trochus Gurgitis, *Al. Brong.* Green Sand, Perte du Rhône, *Al. Brong.*;
   Green Sand, Bochum, *Hœn.*; Rouen, *Pas.*
6. ———? Cirroides, *Al. Brong.* Green Sand, Perte du Rhône, *Al. Brong.*;
   Rouen, *Pas.*
7. ——— lævis, *Nils.* Köpinge, Scania, *Nils.*
8. ——— onustus, *Nils.* Köpinge, Scania, *Nils.*
   ———, species not determined. Green Sand, M. de Fis, *Al. Brong.*
1. Turbo pulcherrimus, *Bean.* Speeton Clay, Yorkshire, *Phil.*
2. ——— sulcatus, *Nils.* Chalk, Köpinge, Scania, *Nils.*
3. ——— moniliferus, *Sow.* Green Sand, Blackdown, *Sow.*
4. ——— carinatus, *Sow.* Green Sand, Coesfeld, *Hœn.*; Rouen, *Pas.*
1. Turritella terebra, *Broc.* Green Sand, Weddersleben, *Hœn.*
   ———, species not determined. Speeton Clay? Yorkshire, *Phil.*
1. Cerithium excavatum, *Al. Brong.* Green Sand, Perte du Rhône, *Al.*
   Brong.; Green Sand, Aix-la-Chapelle, *Hœn.*
   ———, species not determined. Green Sand, M. de Fis, *Al. Brong.*
1. Pyrula planulata, *Nils.* Chalk, Köpinge, Scania, *Nils.*
2. ——— minima, *Hœn.* Green Sand, Aix-la-Chapelle, *Hœn.*
1. Fusus quadratus, *Sow.* Green Sand, Blackdown, *Sow.*
1. Murex Calear, *Sow.* Green Sand, Blackdown, *Sow.*
1. Pterocera maxima, *Hœn.* Martigues, *Hœn.*
1. Rostellaria Parkinsoni, *Mant.* Chalk, Lower Green Sand, Sussex,
   Mant.; Green Sand, Bochum; Coesfeld, *Hœn.*; Lillebonne,
   Pas.
2. ——— carinata, *Mant.* Gault, Sussex, *Mant.*; Rouen, *Pas.*
3. ——— calcarata, *Sow.* Lower Green Sand, Sussex, *Mant.*; Green
   Sand, Blackdown, *Sow.*
4. ——— anserina, *Nils.* Chalk, Köpinge, Scania, *Nils.*
5. ——— inflata, *Pas.*; Rouen, *Pas.*
   ———, species not determined. Lower Green Sand, Isle of Wight,
   Sedg.
1. Strombus papilionatus,   Chalk, Maestricht, Aix-la-Chapelle,
   Hœn.
1. Cassis avellana, *Al. Brong.* Chalk, Sussex, *Mant.*; Chalk, Rouen; M.
   de Fis, *Al. Brong.*
1. Dolium nodosum, *Sow.* Chalk, Sussex, *Mant.*
   Eburna, species not determined. Green Sand, Perte du Rhône, *Al.*
   Brong.; Chalk? Sussex, *Mant.*
1. Nummulites, species not determined. Green Sand, Alps of Savoy,
   Dauphiny, and Provence, *Beaum.*; Maritime Alps, *De la B.*;
   Chalk, Weinbohla, Saxony, *Klipstein*; Cretaceous rocks, South
   of France; Pyrenees, *Dufr.*
1. Lenticulites Comptoni, *Sow.* Green Sand, Earlstoke, Wilts, *Sow.*; Green
   Sand, Scania, *Nils.*
2. ——— cristella, *Nils.* Chalk, Charlottenlund, Sweden, *Nils.*
1. Lituolites nautiloidea, *Lam.* Chalk, Paris, *Al. Brong.*; Rouen, *Pas.*
2. ——— difformis, *Lam.* Chalk, Paris, *Al. Brong.*
   Miliolites,   . S. of France; Pyrenees, *Dufr.*
1. Planularia elliptica, *Nils.* Charlottenlund, Sweden, *Nils.*
2. ——— angusta, *Nils.* Köpinge, Scania, *Nils.*
1. Nodosaria sulcata, *Nils.* Chalk and Green Sand, Scania, *Nils.*
2. ——— lævigata, *Nils.* Scania, *Nils.*
1. Belemnites mucronatus, *Schlot.* Chalk, Sussex, *Mant.*; Chalk, Yorkshire, *Phil.*; Green Sand, Sweden, *Nils.*; Chalk, Meudon,
   &c., *Al. Brong.*; Baculite limestone, Normandy, *Desn.*;
   Chalk, Lublin, Poland, *Pusch*; Maestricht, Aix-la-Chapelle,
   Schlot.; Haldem; Rinkerode, *Münst.*; Dieppe, *Pas.*
2. ——— granulatus, *Defr.* Chalk, Sussex, *Mant.*

3. Belemnites lanceolatus *, *Schlot.* Chalk, Sussex, *Mant.* ; Quedlinburg, *Holl.*
4. —— minimus, *Lister.* Gault, Sussex, *Mant.;* Red Chalk, Yorkshire, *Phil.*
5. —— attenuatus, *Sow.* Gault, Sussex, *Mant.*
6. —— mamillatus †, *Nils.* Chalk, Scania, *Nils.*
——, species not determined. Speeton Clay, Yorkshire, *Phil.;* Green Sand, Perte du Rhône, *Al. Brong.*
1. Actinocamax verus, *Miller.* Chalk, Kent, *Miller.*
1. Nautilus elegans, *Sow.* Chalk, Sussex, *Mant.;* Chalk, Rouen, *Al. Brong.*
2. —— expansus, *Sow.* Chalk, Sussex, *Mant.*
3. —— inæqualis, *Sow.* Gault, Sussex, *Mant.;* Rouen, *Pas.*
4. —— obscurus, *Nils.* Chalk, Scania, *Nils.*
5. —— simplex, *Sow.* Upper Green Sand, Warminster, *Berrett.* Lyme Regis, *De la B.* Rouen, *Al. Brong.*; Green Sand? Aix-la Chapelle, *Hœn.*
6. —— Listeri, *Mant.*, Gault, Sussex, *Mant.;* Quedlinburg, *G. T.;* Green Sand, Hâvre, *Pas.*
7. —— undulatus, *Sow.* Upper Green Sand, Nutfield, *Sow.* Green Sand, Griesenbruch, near Bochum, *Hœn.*
——, species not determined. Lower Green Sand, Sussex, *Martin ;* Speeton Clay, Yorkshire, *Phil.;* Green Sand, M. de Fis, *Al. Brong.;* Baculite limestone, Normandy, *Desn.*
1. Scaphites striatus, *Mant.* Chalk, Sussex, *Mant.;* Chalk, Rouen; Mont de Fis, *Al. Brong.*
2. —— costatus, *Mant.* Chalk, Sussex, *Mant.;* Chalk, Rouen, *Al. Brong.*
—— species not determined. Baculite limestone, Normandy, *Desn.;* Köpinge, *Nils.*
1. Ammonites varians, *Sow.* Chalk, Sussex, *Mant.;* Chalk, Rouen; M. de Fis, *Al. Brong.;* Baculite limestone, Normandy, *Desn.;* Chalk and Upper Green Sand, Wiltshire, *Lons.;* Green Sand, Bochum, *Hœn.*
2. —— Woollgari, *Mant.* Chalk, Sussex, *Mant.;* Rouen, *Pas.*
3. —— navicularis, *Mant.* Chalk, Sussex, *Mant.;* Rouen, *Pas.*
4. —— catinus, *Mant.* Chalk, Sussex, *Mant.*
5. —— Lewesiensis, *Mant.* Chalk, Sussex, *Mant.;* Chalk, Essen, *Hœn.;* Rouen, *Pas.*
6. —— peramplus, *Mant.* Chalk, Sussex *Mant.*
7. —— rusticus, *Sow.* Chalk, Lyme Regis, *Buckl.;* Chalk, Sussex, *Mant.;* Green Sand, Bochum, *Hœn.*
8. —— undatus, *Sow.* Chalk, Sussex, *Mant.*
9. —— Mantelli, *Sow.* Chalk, Sussex, *Mant.;* Hanover, *Holl;* Green Sand, Bochum ; Chalk, Saumur, *Hœn.*
10. —— Rhotomagensis‡, *Al. Brong.* Chalk, Sussex, *Mant.;* Baculite limestone, Normandy, *Desn.;* Rouen, *Al. Brong.;* Chalk, Wilts, *Sow.*
11. —— cinctus, *Mant.* Chalk, Sussex, *Mant.*
12. —— falcatus, *Mant.* Chalk, Sussex, *Mant.;* Chalk, Rouen, *Al. Brong.*
13. —— curvatus, *Mant.* Chalk, Sussex, *Mant.*
14. —— complanatus, *Mant.* Chalk, Sussex, *Mant.*
15. —— rostratus, *Sow.* Chalk, Sussex, *Mant.;* Chalk, Oxfordshire, *Buckl.*

---

* B. semicanaliculatus, *Blainv.*   † B. Scaniæ, *Blainv.*
‡ According to Sowerby, *Am. Rhotomagensis* and *Am. Sussexiensis* are the same shell.

16. Ammonites tetrammatus, *Sow.* Chalk, Sussex, *Mant.*
17. ——— planulatus, *Sow.* Upper Green Sand, Sussex, *Mant.*
18. ——— Catillus, *Sow.* Upper Green Sand, Sussex, *Mant.*
19. ——— splendens, *Sow.* Gault, Sussex, *Mant.*; Green Sand, Sénéfontaine, Norm., *Pas.*
20. ——— auritus, *Sow.* Upper Green Sand, Devizes, *Gent.*; Gault, Sussex, *Mant.*; Hâvre, *Pas.*
21. ——— planus, *Mant.* Gault, Sussex, *Mant.*; Speeton Clay? Yorkshire, *Phil.*
22. ——— lautus, *Park.* Gault, Sussex, *Mant.*
23. ——— tuberculatus, *Sow.* Gault, Sussex, *Mant.*
24. ——— Goodhalli, *Sow.* Lower Green Sand, Sussex, *Mant.*; Green Sand, Blackdown, *Goodhall;* Green Sand, Lyme Regis, *De la B.*
25. ——— venustus, *Phil.* Speeton Clay, Yorkshire, *Phil.*
26. ——— concinnus, *Phil.* Speeton Clay, Yorkshire, *Phil.*
27. ——— Rotula, *Sow.* Speeton Clay, Yorkshire, *Phil.*
28. ——— trisulcosus, *Phil.* Speeton Clay, Yorkshire, *Phil.*
29. ——— marginatus, *Phil.* Speeton Clay, Yorkshire, *Phil.*
30. ——— parvus, *Sow.* Speeton Clay? Yorkshire, *Phil.*
31. ——— hystrix, *Phil.* Speeton Clay, Yorkshire, *Phil.*
32. ——— fissicostatus, *Phil.* Speeton Clay, Yorkshire, *Phil.*
33. ——— curvinodus, *Phil.* Speeton Clay, Yorkshire, *Phil.*
34. ——— inflatus, *Sow.* Green Sand, I. of Wight, *Buckl.*; Green Sand, Perte du Rhône; Rouen; Hâvre; M. de Fis, *Al. Brong.*; Upper Green Sand, Wilts, *Lons.*
35. ——— Deluci, *Al. Brong.* Green Sand, Perte du Rhône; M. de Fis, *Al. Brong.*; Puits de Meulers, Norm., *Pas.*
36. ——— subcristatus, *De Luc* Green Sand, Perte du Rhône, *Al. Brong.*
37. ——— Beudanti*, *Al. Brong.* Green Sand, Perte du Rhône; M. de Fis, *Al. Brong.*
38. ——— clavatus, *De Luc.* Green Sand, M. de Fis, *Al. Brong.*; Rouen, *Pas.*
39. ——— Selliguinus, *Al. Brong.* Green Sand, M. de Fis, *Al. Brong.*; Chalk, Lublin, Poland, *Pusch;* Chalk, Essen, *Hœn.*; Gault, Sussex, *Mant.*; Rouen, *Pas.*
40. ——— Gentoni, *Defr.* Baculite Limestone, Normandy, *Desn.*; Gault, Sussex, *Mant.*; Chalk, Rouen, *Al. Brong.*
41. ——— constrictus, *Sow.* Baculite Limestone, Normandy, *Desn.*; Chalk, Lublin, Poland, *Pusch.*
42. ——— Stobæi, *Nils.* Chalk, Scania, *Nils.*
43. ——— varicosus, *Sow.* Green Sand, Blackdown, *Sow.*
44. ——— Hippocastanum, *Sow.* Chalk with quartz grains, Lyme Regis, *De la B.*; Green Marl, Puits de Meulers, Norm., *Pas.*
45. ——— Benettianus, *Sow.* Gault, Warminster, *Lons.*
46. ——— denarius, *Sow.* Green Sand, Blackdown, *Goodhall.*
47. ——— Nutfieldiensis, *Sow.* Chalk, near Calne, *Lons.*
48. ——— virgatus, *Goldf.*, G. S. Moskau, *G. T.*
49. ——— Coupei, *Al. Brong;* Rouen, *Pas.*
50. ——— canteriatus, *Al. Brong.;* Rouen, *Pas.*
1. Turrilites costatus, *Sow.* Chalk, Sussex, *Mant.*; Chalk, Rouen; Hâvre, *Al. Brong.*; Chalk, near Calne, *Lons.*
2. ——— undulatus, *Sow.* Chalk, Sussex, *Mant.*; Rouen, *Pas.*
3. ——— tuberculatus, *Sow.* Chalk, Sussex, *Mant.*

* According to Sowerby, *Am. Beudanti, Am. Selliguinus,* and *Am. lævigatus* are the same shell.

Organic Remains of the Cretaceous Group. 527

4. Turrilites Bergeri, *Al. Brong.*; Green Sand, Perte du Rhône; M. de Fis, *Al. Brong.*
5. ———? Babeli, *Al. Brong.* Green Sand, M. de Fis, *Al. Brong.*; Rouen, *Pas.*
6. ——— acutus, *Pas.*; Rouen, *Pas.*
——, species not determined. Green Sand, Maritime Alps, *Risso.*
1. Baculites Faujasii, *Lam.* Chalk, Sussex, *Mant.*; Chalk, Norfolk, *Rose*; Maestricht, *Desm.*; Chalk, Sweden, *Nils.*; Bochum, Aix-la-Chapelle, *Hœn.*
2. ——— obliquatus, *Sow.* Chalk, Sussex, *Mant.*; Scania, *Nils.*; Rouen, *Pas.*
3. ——— vertebralis, *Defr.* Chalk, Maestricht, *Fauj. de St. Fond*; Baculite Limestone, Normandy, *Desm.*
4. ——— anceps, *Lam.* Chalk, Scania, *Nils.*
5. ——— triangularis, *Desm.* Maestricht, *Desm.*
1. Hamites armatus, *Sow.* Chalk, Sussex, *Mant.*; Chalk, Oxfordshire, *Buckl.*; Rouen, *Pas.*
2. ——— plicatilis, *Mant.* Chalk, Sussex, *Mant.* Speeton Clay? Yorkshire, *Phil.*; Rouen, *Pas.*
3. ——— alternatus, *Mant.* Chalk, Sussex, *Mant.*; Speeton Clay, Yorkshire, *Phil.*
4. ——— ellipticus, *Mant.* Chalk, Sussex, *Mant.*; Baculite Limestone? Normandy, *Desn.*
5. ——— attenuatus, *Sow.* Chalk, Gault, Sussex, *Mant.*; Speeton Clay, Yorkshire, *Phil.*; Rouen, *Pas.*
6. ——— maximus, *Sow.* Gault, Sussex, *Mant.*; Speeton Clay, Yorkshire, *Phil.*
7. ——— intermedius, *Sow.* Gault, Sussex, *Mant.*; Speeton Clay, Yorkshire, *Phil.*; Green Sand, Aix-la-Chapelle, *Hœn.*; Rouen, *Pas.*
8. ——— tenuis, *Sow.* Gault, Sussex, *Mant.*; Rouen, *Pas.*
9. ——— rotundus, *Sow.* Gault, Sussex, *Mant.*; Speeton Clay, Yorkshire, *Phil.*; Green Sand, Perte du Rhône, *Al. Brong.*; Green Sand, Aix-la-Chapelle, *Hœn.*; Rouen, *Pas.*
10. ——— compressus, *Sow.* Gault, Sussex, *Mant.*; Green Sand, Nice, *Risso.*
11. ——— raricostatus, *Phil.* Speeton Clay, Yorkshire, *Phil.*
12. ——— Beanii, *Y. & B.* Speeton Clay, Yorkshire, *Phil.*
13. ——— Phillipsii, *Bean.* Speeton Clay, Yorkshire, *Phil.*
14. ——— funatus, *Al. Brong.* Green Sand, Perte du Rhône; M. de Fis, *Al. Brong.*; Rouen, *Pas.*
15. ——— canteriatus, *Al. Brong.* Green Sand, Perte de Rhône, *Al. Brong.*
16. ——— virgulatus, *Al. Brong.* Green Sand, M. de Fis, *Al. Brong.*
17. ——— cylindricus, *Defr.* Baculite Limestone, Normandy, *Desn.*
18. ——— spinulosus, *Sow.* Green Sand, Blackdown, *Miller.*
19. ——— grandis, *Sow.* Lower Green Sand, Kent, *Buckl.*
20. ——— Gigas, *Sow.* Lower Green Sand, Hythe, Kent, *G. E. Smith.*
21. ——— spiniger, *Sow.* Gault, Folkestone, *Gibbs.*

CRUSTACEA.

1. Astacus Leachii, *Mant.* Chalk, Sussex, *Mant.*
2. ——— Sussexiensis, *Mant.* Chalk, Sussex, *Mant.*
3. ——— ornatus, *Phil.* Speeton Clay, Yorkshire, *Phil.*
4. ——— longimanus, *Sow.* Green Sand, Lyme Regis, *De la B.*
——, species not determined. Gault, Sussex, *Mant.*
1. Pagurus Faujasii, *Desm.* Chalk? Sussex, *Mant.*; Maestricht.
1. Scyllarus Mantelli, *Desm.* Chalk, Sussex, *Mant.*

1. Orythia Labechii, *Desl.* ; Green Sand, Calvados, *Desl.*
   Eryon, species not determined. Chalk, Sussex, *Mant.*
   Arcania, species not determined. Gault, Sussex, *Mant.*
   Etyæa, species not determined. Gault, Sussex, *Mant.*
   Coryster, species not determined. Gault, Sussex, *Mant.*

PISCES.

1. Squalus Mustelus? Chalk, Sussex, *Mant.*
2. —— Galeus? Chalk, Sussex, *Mant.*
3. —— pristodontes, *Bronn.* Aix-la-Chapelle, *G. T.*
1. Muræna, Lewesiensis, *Mant.* Chalk, Sussex, *Mant.*
1. Zeus Lewesiensis, *Mant.* Chalk, Sussex, *Mant.*
1. Salmo? Lewesiensis, *Mant.* Chalk, Sussex, *Mant.*
1. Esox Lewesiensis, *Mant.* Chalk, Sussex, *Mant.*
1. Amia? Lewesiensis, *Mant.* Chalk, Sussex, *Mant.*
   Fish, genera not determined. Speeton Clay, Yorkshire, *Phil.* ; Chalk, Paris, *Al. Brong.* ; Chalk, Lyme Regis, *De la B.* ; Upper Green Sand, Wilts, *Lons.* Gault, Isle of Wight, *Fitton ;* Chalk, Troyes, *Clement-Mullet.*
   —— teeth and palates; common in England and France, *var. authors ;* Bochum; Aix-la-Chapelle, *Hœn. ;* Scania, *Nils.*

REPTILIA.

1. Mososaurus Hoffmanni, Maestricht, *Fauj. de St. Fond ;* Chalk, Sussex, *Mant.*
1. Crocodile of Meudon, *Cuv. ;* Chalk, Meudon, *Al. Brong.*
   Reptiles, genera not determined. Speeton Clay, Yorkshire, *Phil.*

Organic Remains of the Wealden Rocks of England.

PLANTÆ.

1. Sphenopteris Mantelli, *Ad. Brong.* Hastings Sands, Sussex, *Mant.*
1. Lonchopteris Mantelli, *Ad. Brong.* Hastings Sands, Sussex, *Mant.*
   Lycopodites? species not determined. Hastings Sands, Sussex, *Mant.*
1. Clatharia Lyellii, *Mant.* Hastings Sands, Sussex, *Mant.*
1. Carpolithus Mantelli, *Ad. Brong.* Hastings Sands, Sussex, *Mant.*
   Lignite, and undescribed vegetables. Hastings Sands, Sussex, *Mant.*

CONCHIFERA AND MOLLUSCA.

Corbula, species not determined. Pounceford, *Fitton.*
Tellina, species not determined. Pounceford, *Fitton.*
Mytilus, species not determined. Pounceford, *Fitton.*
Ostrea, species not determined. Weald Clay, Isle of Wight, *Sedg. ;* Purbeck Beds, near Weymouth, *Buckl.* and *De la B.*
1. Cyclas membranacea, *Sow.* Weald Clay, Hastings Sands, Ashburnham Beds, Sussex, *Mant. ;* Weald Clay? Swanage Bay, *Fitton.*
2. —— media, *Sow.* Weald Clay, Hastings Sands, Ashburnham Beds, Sussex, *Mant. ;* Weald Clay, Isle of Wight, Swanage Bay ; Hastings Sands, Isle of Wight; Sussex, *Fitton.*
   —— species not determined. Weald Clay, Isle of Wight; Swanage Bay, *Fitton.*
1. Unio porrectus, *Sow.* Hastings Sands, Sussex, *Mant.*
2. —— compressus, *Sow.* Hastings Sands, Sussex, *Mant.*
3. —— antiquus, *Sow.* Hastings Sands, Ashburnham Beds, Sussex, *Mant.*
4. —— aduncus, *Sow.* Hastings Sands, Sussex, *Mant.*
5. —— cordiformis, *Sow.* Hastings Sands, Sussex, *Mant.*
   Bulla, small species. Tilgate beds, *Fitton.*

Melanopsis, species not determined. Pounceford, *Fitton.*
1. Paludina vivipara, *Lam.* Weald Clay, Hastings Sands, Ashburnham
   Beds, Sussex, *Mant.;* Purbeck Beds, Purbeck, *Conyb.*
2. —— elongata, *Sow.* Weald Clay, Hastings Sands, Ashburnham
   Beds, Sussex, *Mant.;* Weald Clay, Isle of Wight; Swanage
   Bay, *Fitton.*
3. —— carinifera, *Sow.* Weald Clay, Sussex, *Mant.*
4. —— Sussexiensis, . Wealden rocks, *Fitton.*
Potamides, species not determined. Weald Clay, Sussex, *Mant.*
Neritina, species not determined. Tilgate Beds, *Fitton.*

PISCES.

Lepisosteus, species not determined. Hastings Sands, Sussex, *Mant.*
Silurus, species not determined. Hastings Sands, Sussex, *Mant.*
Remains of Fish, genera not determined. Weald Clay, Ashburnham
   Beds, Sussex, *Mant.;* Purbeck Beds, Purbeck, *De la B.;*
   Hastings Sands, Isle of Wight, *Fitton.*

CRUSTACEA.

1. Cypris faba, *Desm.* Weald Clay, Isle of Wight; Swanage Bay, &c.
   Fitton; Weald Clay, Hastings Sands, Sussex, *Mant.*

REPTILIA.

1. Crocodilus priscus, Hastings Sands, Sussex, *Mant.*
   ——, species not determined. Ashburnham Beds, Sussex, *Mant.;*
   Purbeck Beds, Purbeck, *Conyb.;* Weald Clay, Swanage Bay,
   Fitton.
1. Iguanodon Mantelli, *v. Meyer.* Hastings Sands, Sussex, *Mant.*
Megalosaurus ——. Hastings Sands, Ashburnham Beds, Sussex,
   Mant.
Hylæosaurus ——. Hastings Sands, Sussex, *Mant.*
Reptiles of the genera Trionyx, Emys, Chelonia, Plesiosaurus, and Ptero-
   dactylus? Hastings Sands, Sussex, *Mant.*
Tortoise, Purbeck Beds, Purbeck, *Conyb.**

Organic Remains of the Oolitic Group.

PLANTÆ.

Algæ.

1. Fucoides furcatus, *Ad. Brong.* Stonesfield slate, *Ad. Brong.*
2. —— Stockii, *Ad. Brong.* Solenhofen, *Ad. Brong.*
3. —— encelioides, *Ad. Brong.* Solenhofen, *Ad. Brong.*

Equisetaceæ.

1. Equisetum columnare, *Ad. Brong.* Lower carbonaceous series, Yorkshire, *Phil.;* Brora, *Murch.*

Filices.

1. Pachypteris lanceolata, *Ad. Brong.* Coal, shale, &c. between inferior
   and great oolite, Yorkshire, *Phil.*
2. —— ovata, *Ad. Brong.* Coal, shale, &c. between inferior and great
   oolite, Yorkshire, *Phil.*
1. Pecopteris Reglei, *Ad. Brong.* Forest marble, Mamers, *Desn.*

* In this list, the sands, sandstones, and clays, grouped by Mr. Mantell under
me head of Tilgate Beds, are given as Hastings Sands, although this arrangement
thay perhaps clash with one or two local divisions.

2. Pecopteris Desnoyersii, *Ad. Brong.* Forest marble, Mamers, *Desn.*
3. —— polypodioides, *Ad. Brong.* Coal, shale, &c. between cornbrash and great oolite, Yorkshire, *Phil.*
4. —— denticulata, *Ad. Brong.* Coal, shale, &c. between cornbrash and great oolite, Yorkshire, *Phil.*
5. —— Phillipsii, *Ad. Brong.* Coal, &c. of the oolitic series, Yorkshire, *Ad. Brong.*
6. —— Whitbiensis, *Ad. Brong.* Coal, shale, &c. between cornbrash and great oolite, Yorkshire, *Phil.*
1. Sphænopteris hymenophylloides, *Ad. Brong.* Stonesfield slate, *Buckl.*; Coal, shale, &c. between great and inferior oolite, Yorkshire, *Phil.*
2. ——? macrophylla, *Ad. Brong.* Stonesfield slate, *Buckl.*
3. —— Williamsonis, *Ad. Brong.* Coal, &c. of the oolitic series, Yorkshire, *Ad. Brong.*
4. —— crenulata, *Ad. Brong.* Coal, &c. of the oolitic series, Yorkshire, *Ad. Brong.*
5. —— denticulata, *Ad. Brong.* Coal, &c. of the oolitic series, Yorkshire, *Ad. Brong.*
1. Tæniopteris latifolia, *Ad. Brong.* Coal, shale, &c. between cornbrash and great oolite, Yorkshire, *Phil.*
2. —— vittata, *Ad. Brong.* Coal, shale, &c. between cornbrash and great oolite, Yorkshire, *Phil.*
1. Cyclopteris Beanii, *L. & H.* Coal, shale, &c. of the oolitic series, Yorkshire, *Williamson.*
2. —— digitata, *L. & H.* Sandstone, near Scarborough, *Geol. Soc. Mus.*
1. Glossopteris Phillipsii, *Ad. Brong.* Shale, near Scarborough, *Bean.*
1. Neuropteris recentior, *L. & H.* Gristhorpe, Scarborough, *Bean.*
2. —— ligata, *L. & H.* Gristhorpe, Scarborough, *Bean.*

Lycopodiaceæ.

1. Lycopodites falcatus, *L. & H.* Clougton, Yorkshire, *Bean.*

Cycadeæ.

1. Pterophyllum Williamsonis. Coal, shale, &c. between cornbrash and great oolite, Yorkshire, *Phil.*
2. —— comptum, *L. & H.* Gristhorpe, Scarborough, *Bean.*
3. —— minus, *L. & H.* Near Scarborough, *Mus. Geol. Soc.*
4. —— Nilsoni, *L. & H.* Near Scarborough, *Bean.*
1. Zamia pectinata, *Ad. Brong.* Stonesfield slate, *Buckl.*
2. —— patens, *Ad. Brong.* Stonesfield slate, *Ad. Brong.*
3. —— longifolia, *Ad. Brong.* Coal, shale, &c. between cornbrash and great oolite, Yorkshire, *Phil.*
4. —— pennæformis, *Ad. Brong.* Coal, shale, &c. between great and inferior oolite, Yorkshire, *Phil.*
5. —— elegans, *Ad. Brong.* Coal, shale, &c. between great and inferior oolite, Yorkshire, *Phil.*
6. —— Goldiæi, *Ad. Brong.* Coal, &c. of the oolitic series, Yorkshire, *Ad. Brong.*
7. —— acuta, *Ad. Brong.* Coal, &c. of the oolitic series, Yorkshire, *Ad. Brong.*
8. —— lævis, *Ad. Brong.* Coal, &c. of the oolitic series, Yorkshire, *Ad. Brong.*
9. —— Youngii, *Ad. Brong.* Coal, shale, &c. between great and inferior oolite, Yorkshire, *Phil.*
10. —— Fcneonis, *Ad. Brong.* Coal, &c. of the oolitic series, Yorkshire, *Ad. Brong.*

11. Zamia Mantelli, *Ad. Brong.* Coal, shale, &c. between great and inferior oolite, Yorkshire, *Phil.*
1. Zamites Bechii, *Ad. Brong.* Forest marble, Mamers, *Desn.*; Lias, Lyme Regis, *De la B.*
2. —— Bucklandii, *Ad. Brong.* Forest marble, Mamers, *Desn.*; Lias, Lyme Regis, *De la B.*
3. —— Lagotis, *Ad. Brong.* Forest marble, Mamers, *Desn.*
4. —— hastata, *Ad. Brong.* Forest marble, Mamers, *Desn.*

Coniferæ.

1. Thuytes divaricata, *Sternb.* Stonesfield slate, *Buckl.*; Solenhofen, *G. T.*
2. —— expansa, *Sternb.* Stonesfield slate, *Buckl.*
3. —— acutifolia, *Ad. Brong.* Stonesfield slate, *Buckl.*
4. —— cupressiformis, *Sternb.* Stonesfield slate, *Buckl.*
1. Taxites podocarpoides, *Ad. Brong.* Stonesfield state, *Buckl.*

Lilia.

1. Bucklandia squamosa, *Ad. Brong.* Stonesfield, *Buckl.*

Class uncertain.

1. Mamillaria Desnoyersii, *Ad. Brong.* Mamers, *Desn.*
Many undescribed vegetables. Lias, Lyme Regis, *De la B.*

ZOOPHYTA.

1. Achilleum dubium, *Goldf.* Solenhofen, *Goldf.*
2. —— cheirotonum, *Goldf.* Oolitic rocks, Baireuth, *Munst.*
3. —— muricatum, *Goldf.* Streitberg, *Munst.*
4. —— tuberosum, *Munst.* Hattheim, *Munst.*
5. —— cancellatum, *Munst.* Hattheim, *Munst.*
6. —— costatum, *Munst.* Streitberg, *Munst.*
1. Manon Peziza, *Goldf.* Streitberg; Hattheim; Giengen; Regensberg, *Goldf.*
2. —— marginatum, *Munst.* Streitberg: Muggendorf, *Munst.*
3. —— impressum, *Munst.* Muggendorf, *Munst.*
1. Scyphia cylindrica, *Goldf.* Muggendorf, *Munst.*
2. —— elegans, *Goldf.* Thurnau; Baireuth, *Goldf.*
3. —— calopora, *Goldf.* Thurnau; Baireuth, *Goldf.*
4. —— pertusa, *Goldf.* Streitberg; Baireuth, *Goldf.*
5. —— texturata, *Goldf.* Giengen, Wurtemberg, *Goldf.*
6. —— texata, *Goldf.* Legerberg, Switzerland; Streitberg, *Goldf.* Calc. Grit, Bernese Jura, *Thur.*
7. —— polyommata, *Goldf.* Baireuth & Switzerland, *Goldf.*
8. —— clathrata, *Goldf.* Streitberg; Baireuth, *Goldf.*
9. —— milleporata, *Goldf.* Baireuth, *Goldf.*
10. —— parallela, *Goldf.* Streitberg, *Munst.*
11. —— psilopora, *Goldf.* Muggendorf, *Goldf.*
12. —— obliqua, *Goldf.* Muggendorf, *Munst.*; Oxford Clay, Bernese Jura, *Thur.*
13. —— rugosa, *Goldf.* Streitberg, *Munst.*
14. —— articulata, *Goldf.* Muggendorf, *Goldf.*
15. —— pyriformis, *Goldf.* Streitberg, *Munst.*
16. —— radiciformis, *Goldf.* Streitberg, *Goldf.*
17. —— punctata, *Goldf.* Streitberg, *Munst.*
18. —— reticulata, *Goldf.* Streitberg, *Goldf.*
19. —— dictyota, *Goldf.* Streitberg, *Munst.*
20. —— procumbens, *Goldf.* Baireuth, *Goldf.*
21. —— paradoxa, *Munst.* Streitberg & Amberg, *Munst.*
22. —— empleura, *Munst.* Streitberg, *Munst.*

23. Scyphia striata, *Munst.* Streitberg & Muggendorf, *Munst.*
24. —— Buchii, *Munst.* Streitberg, *Munst.*
25. —— Munsteri, *Goldf.* Regensberg ; Streitberg, *Goldf.*
26. —— propinqua, *Munst.* Streitberg ; Muggendorf, *Munst.*
27. —— cancellata, *Munst.* Streitberg ; Muggendorf, *Munst.*
28. —— decorata, *Munst.* Muggendorf, *Munst.*
29. —— Humboldtii, *Munst.* Muggendorf, *Munst.*
30. —— Sternbergii, *Munst.* Streitberg, *Munst.*
31. —— Schlotheimii, *Munst.* Thurnau ; Streitberg, *Munst.*
32. —— Schweiggeri, *Goldf.* Baireuth, *Goldf.*
33. —— secunda, *Munst.* Heiligenstadt; Streitberg, *Munst.;* Calc. Grit, Bernese Jura, *Thur.*
34. —— verrucosa, *Goldf.* Streitberg & Wurgau, *Goldf.*
35. —— Bronnii, *Munst.* Wurtemberg & Baireuth, *Munst.;* Calc. Grit, Bernese Jura, *Thur.*
36. —— milleporacea, *Munst.* Thurnau ; Aufsees ; Streitberg, *Munst.*
37. —— pertusa, *Goldf.* Streitberg & Amberg, *Goldf.*
38. —— intermedia, *Munst.* Hattheim ; Streitberg, *Munst.*
39. —— Neesii, *Goldf.* Streitberg, *Goldf.*
40. —— turbinata, *Goldf.* Streitberg, *Goldf.*
41. —— tenuistriata, *Goldf.* Streitberg, *Goldf.*
1. Tragos pezizoides, *Goldf.* Muggendorf, *Goldf.*
2. —— Palella, *Goldf.* Wurtemberg & Switzerland ; Rabenstein ; Heiligenstadt, *Goldf.*
3. —— sphærioides, *Goldf.* Sigmaringen, Wurtemberg, *Goldf.*
4. —— tuberosum *, *Goldf.* Inferior Oolite, Rabenstein; Streitberg, *Munst.*
5. —— acetabulum, *Goldf.* Streitberg ; Randen, *Goldf.*
6. —— radiatum, *Munst.* Streitberg, *Munst.*
7. —— rugosum, *Munst.* Streitberg, *Munst.*
8. —— reticulatum, *Munst.* Streitberg, *Munst.*
9. —— verrucosum, *Munst.* Streitberg, *Munst.*
1. Spongia floriceps, *Phil.* Coral Oolite, Yorkshire, *Phil.*
2. —— clavaroides, *Lam.* Great Oolite, Wiltshire, *Lons.*
—— , species not determined. Lower Calcareous Grit, Yorkshire, *Phil.;* Inferior Oolite, Middle and South of England, *Conyb.;* Forest Marble, Wiltshire, *Lons.*
Alcyonium, species not determined. Forest Marble, Normandy, *De Cau.;* Great Oolite? Wilts, *Lons.*
1. Cnemidium lamellosum, *Goldf.* Randen, Switzerland, *Goldf.*
2. —— stellatum, *Goldf.* Randen, Switzerland, *Goldf.*
3. —— striato-punctatum, *Goldf.* Randen, *Goldf.*
4. —— rimulosum, *Goldf.* Randen, *Goldf.*
5. —— mammillare, *Goldf.* Streitberg, *Goldf.*
6. —— Rotula, *Goldf.* Thurnau, *Goldf.*
7. —— granulosum, *Munst.* Streitberg, *Munst.*
8. —— astrophorum, *Munst.* Hattheim ; Regensberg, *Munst.*
9. —— capitatum, *Munst.* Amberg, *Munst.*
1. Limnorea mammillaris †, *Lam^z.* Forest Marble, Normandy, *De Cau.*
1. Siphonia pyriformis, *Goldf.* Streitberg, *Goldf.*
1. Myrmecium hemisphæricum, *Goldf.* Thurnau, *Goldf.*
1. Gorgonia dubia, *Goldf.* Glücksbrunn ; Thuringia, *Goldf.*
1. Millepora dumetosa, *Lam^x.* Forest Marble, Normandy, *De Cau.*
2. —— corymbosa, *Lam^x.* Forest Marble, Normandy, *De Cau.*
3. —— conifera, *Lam^z.* Forest Marble, Normandy, *De Cau.*

---

* *Limnorea lamellosa* of Lamouroux according to M. Goldfuss.
† Is this *Limnorea mammillosa*, Lam.? If it be, it is the *Cnemidium tuberosum* of Goldfuss.

4. Millepora pyriformis, *Lam*. Forest Marble, Normandy, *De Cau.*
5. —— macrocaule, *Lam*. Forest Marble, Normandy, *De Cau.*
6. —— straminea, *Phil.* Great Oolite and Cornbrash, Yorkshire, *Phil.*
—— , species not determined. Cornbrash and Forest Marble, North of France, *Bobl.*; Forest Marble, Mamers, Normandy, *Desn.*; Forest Marble and Great Oolite, Wiltshire, *Lons.*
1. Madrepora limbata, *Goldf.* Heidenheim, *G. T.*
1. Cellepora orbiculata, *Goldf.* Streitberg, *Munst.*; Oxford Clay, Haute Saone, *Thir.*
2. —— echinata, *Goldf.* Inferior Oolite, Haute Saone, *Thir.*
—— , species not determined. Inferior Oolite, Midland and Southern England, *Conyb.*
Retepora?——. Great Oolite, Yorkshire, *Phil.*
Flustra, species not determined. Great Oolite, Wiltshire, *Lons.*
1. Ceriopora radiciformis, *Goldf.* Thurnau, Baireuth, *Goldf.*
2. —— striata, *Goldf.* Streitberg; Thurnau, *Munst.*
3. —— angulosa, *Goldf.* Thurnau, *Munst.*
4. —— alata, *Goldf.* Thurnau, *Munst.*
5. —— crispa, *Goldf.* Thurnau, *Munst.*
6. —— favosa, *Goldf.* Streitberg; Thurnau, *Munst.*
7. —— radiata, *Goldf.* Thurnau, *Munst.*
8. —— compressa, *Munst.* Thurnau, *Munst.*
9. —— orbiculata, *Goldf.* Inferior Oolite, Haute Saone, *Thir.*; Calc. Grit, Inferior Oolite, Bernese Jura, *Thur.*
1. Agaricia rotata, *Goldf.* Randenberg, Switzerland, *Goldf.*
2. —— crassa, *Goldf.* Randen, Switzerland, *Goldf.*
3. —— granulata, *Munst.* Bâle; Hattheim, *Munst.*
1. Lithodendron elegans, *Munst.* Wurtemberg, *Munst.*
2. —— compressum, *Munst.* Heidenheim, Wurtemberg, *Munst.*
3. —— Raaracum, *Thur.* Coral Rag, Bernese Jura, *Thur.*
1. Caryophyllia cylindrica, *Phil.* Coralline Oolite, Yorks., *Phil.*
2. —— truncata, *Lam*. Forest Marble, Normandy, *De Cau.*
3. —— Brebissonii, *Lam*. Forest Marble, Normandy, *De Cau.*
4. —— convexa, *Phil.* Inferior Oolite, Yorkshire, *Phil.*
5. —— like C. cespitosa, *Ellis.* Coral Oolite, Yorks., *Phil.*; Great Oolite, Mid. and S. of England, *Conyb.*
6. —— like C. flexuosa, *Ellis.* Coral Oolite, Yorkshire, *Phil.*; Great Oolite, Midland and Southern England, *Conyb.*
7. —— approaching C. Carduus, *Park.* Coral Rag, Great Oolite, Middle and South of England, *Conyb.*
—— , species not determined. Inferior Oolite, North of France, *Bobl.*; Rochelle Beds, *Dufr.*; Forest Marble, Mamers, Normandy, *Desn.*; Forest Marble, Bradford Clay, and Great Oolite, Wiltshire, *Lons.*
1. Anthophyllum turbinatum, *Munst.* Hattheim; Heidenheim, *Munst.*
2. —— obconicum, *Munst.* Hattheim; Heidenheim, *Munst.*; Calc. Grit, Bernese Jura, *Thur.*
3. —— decipiens, *Goldf.* Alsace, *Goldf.*
1. Fungia orbiculites, *Lam* . Forest Marble, Normandy, *De Cau.*; Cornbrash, Wiltshire, *Lons.*
2. —— lævis, *Goldf.* Calc. Grit, Bernese Jura, *Thur.*
—— , species not determined. Inferior Oolite, Midland and Southern England, *Conyb.*
1. Turbinolia dispar, *Phil.* Coral Oolite, Yorkshire, *Phil.*
2. —— dydyma, *Goldf.* Coral Rag, Bernese Jura, *Thur.*
—— , species not determined. Inferior Oolite and Lias, North of France, *Bobl.*
1. Turbinolopsis ochracea, *Lam*. Forest Marble, Normandy, *De Cau.*

534   *Organic Remains of the Oolitic Group.*

1. Cyathophyllum Tintinnabulum, *Goldf.* Banz; Staffelstein; Bamberg, *Goldf.*
2. —— Mactra, *Goldf.* Banz; Bamberg, *Goldf.*
3. —— quadrigeminum, *Goldf.* Coral Rag, Bernese Jura, *Thur.*
4. —— ceratites, *Goldf.* Calc. Grit, Bernese Jura, *Thur.*
5. —— plicatum, *Goldf.* Calc. Grit, Bernese Jura, *Thur.*
6. —— vermiculare, *Goldf.* Calc. Grit, Bernese Jura, *Thur.*
1. Meandrina Sœmmeringii, *Munst.* Hattheim; Heidenheim, *Munst.*
2. —— astroides, *Goldf.* Coral Rag, Haute Saone, *Thir.*; Giengen, *Goldf.*
3. —— tenella, *Goldf.* Giengen, *Goldf.*; Coral Rag, Bernese Jura, *Thur.*
4. —— magna, *Thur.* Coral Rag, Bernese Jura, *Thur.*
5. —— foliacea, *Thur.* Coral Rag, Bernese Jura, *Thur.*
   ——, species not determined. Inferior Oolite and Coral Oolite, Yorks., *Phil.*; Inferior Oolite? Midl. and Southern England, *Conyb.*; Kimmeridge Clay, Haute Saone, *Thir.*; Great Oolite, Wilts, *Lons.*
1. Astrea Microconos, *Goldf.* Biberbach, near Muggendorf, *Goldf.*
2. —— limbata, *Goldf.* Giengen, *Goldf.*
3. —— concinna, *Goldf.* Giengen, *Goldf.*
4. —— pentagonalis, *Munst.* Hattheim; Heidenheim, *Munst.*
5. —— gracilis, *Munst.* Boll, Wurtemberg, *Munst.*
6. —— explanata, *Munst.* Wurtemberg, *Munst.*
7. —— tubulosa, *Goldf.* Wurtemberg, *Goldf.*; Coral Rag, Haute Saone, *Thir.*; Coral Rag, Bernese Jura, *Thur.*
8. —— oculata, *Goldf.* Giengen, *Goldf.*; Coral Rag, Haute Saone, *Thir.*
9. —— alveolata, *Goldf.* Heidenheim, Wurtemberg, *Goldf.*
10. —— helianthoides, *Goldf.* Heidenheim; Giengen, *Goldf.*; Inferior Oolite, Coral Rag, Haute Saone, *Thir.*; Coral Rag, Inferior Oolite, Bernese Jura, *Thur.*
11. —— confluens, *Goldf.* Heidenheim; Giengen, *Goldf.*; Coral Rag, Haute Saone, *Thir.*; Coral Rag, Bernese Jura, *Thur.*
12. —— caryophylloides, *Goldf.* Giengen, *Goldf.*; Coral Rag, Haute Saone, *Thir.*; Coral Rag, Bernese Jura, *Thur.*
13. —— cristata, *Goldf.* Giengen; Heidenheim, *Goldf.*; Coral Rag, Bernese Jura, *Thur.*
14. —— sexradiata, *Goldf.* Giengen, *Goldf.*
15. —— favosioides, *Smith.* Coral Oolite, Yorkshire, *Phil.*; Coral Rag and Great Oolite, Midland and Southern England, *Conyb.*
16. —— inæqualis, *Phil.* Coral Oolite, Yorkshire, *Phil.*
17. —— micastron, *Phil.* Coral Oolite, Yorkshire, *Phil.*
18. —— arachnoides, *Flem.* Coral Oolite, Yorkshire, *Phil.*
19. —— tubulifera, *Phil.* Coral Oolite, Yorkshire, *Phil.*
20. —— macropthalma, *Goldf.* Kim. Clay, Porrentruy, Bernese Jura, *Thur.*
21. —— textilis, *Goldf.* Coral Rag, Bernese Jura, *Thur.*
22. —— geminata, *Goldf.* Coral Rag, Bernese Jura, *Thur.*
23. —— velamentosa, *Goldf.* Coral Rag, Bernese Jura, *Thur.*
24. —— geometrica, *Goldf.* Coral Rag, Bernese Jura, *Thur.*
   ——, species not determined. Coral Rag, Normandy, numerous, *De Cau.*; Great Oolite, Midland and Southern England, *Conyb.*; Lias, Hebrides, *Murch.*; Great Oolite, Wiltshire, *Lons.*
1. Thamnasteria Lamourouxii, *Le Sauvage.* Coral Rag, Norm., *De Cau.*
1. Aulopora compressa, *Goldf.* Rabenstein; Grafenberg, *Munst.*
2. —— dichotoma, *Goldf.*, Streitberg, *Goldf.*
3. —— intermedia, *Münst.* Streitberg, *Münst.*

1. Entalophora cellarioides, *Lam*. Forest Marble, Normandy, *De Cau.*
Favosites, species not determined. Forest Marble, Mamers, Normandy, *Desn.*
1. Spiropora tetragona, *Lam*. Forest Marble, Normandy, *De Cau.*
2. ———— cæspitosa, *Lam*. Forest Marble, Normandy, *De Cau.; * Great Oolite, Wiltshire, *Lons.*
3. ———— elegans, *Lam*. Forest Marble, Normandy, *De Cau.*
4. ———— intricata, *Lam*. Forest Marble, Normandy, *De Cau.*
1. Eunomia radiata, *Lam*. Forest Marble, Normandy, *De Cau.; * Great Oolite, Wiltshire, *Lons.*
1. Chrysaora damæcornis, *Lam*. Forest Marble, Normandy, *De Cau.; * Great Oolite, Wiltshire, *Lons.*
2. ———— spinosa, *Lam*. Forest Marble, Normandy, *De Cau.*
1. Theonoa clathrata, *Lam*. Forest Marble, Normandy, *De Cau.; * Great Oolite, Wiltshire, *Lons.*
1. Idmonea triquetra, *Lam*. Forest Marble, Normandy, *De Cau.; * Great Oolite, Wiltshire, *Lons.*
1. Alecto dichotoma, *Lam*. Great Oolite, Wiltshire, *Lons.; * Forest Marble, Normandy, *De Cau.*
————, species not determined. Inferior Oolite, Midland and Southern England, *Conyb.*
1. Berenicea diluviana, *Lam*. Great Oolite, Wiltshire, *Lons.; * Forest Marble, Normandy, *De Cau.*
————, species not determined. Great Oolite, Haute Saone, *Thir.; * Forest Marble, Wiltshire, *Lons.*
1. Terebellaria ramosissima, *Lam*. Forest Marble and Great Oolite, Somerset, *Lons.; * Forest Marble, Normandy, *De Cau.*
2. ———— Antilope, *Lam*. Forest Marble, Normandy, *De Cau.*
1. Cellaria Smithii, *Phil.* Cornbrash, Yorkshire, *Phil.*
1. Sarcinula astroites, *Goldf.* Coral Rag, Bernese Jura, *Thur.*
1. Intricaria Bajocensis, *Defr.* Inferior Oolite, Bernese Jura, *Thur.*
Explanaria, species not determined. Great Oolite, Wilts, *Lons.*
Polypifers, genera not determined. Lias (rare), Lyme Regis, *De la B.; * Lias (rare), Yorkshire, *Phil.; * Lias (rare), Normandy, *De Cau.; * Coral Rag (numerous), North of France, *Bobl.; * Coral Rag (abundant), Burgundy, *Beaum.; * Coral Rag (abundant), South of France, *Dufr.; * Inferior Oolite, Calvados, *Her.*

Radiaria.

1. Cidaris florigemma, *Phil.* Coral Oolite, Yorkshire, *Phil.*
2. ———— intermedia, *Park.* Coral Oolite, Yorkshire, *Phil.*
3. ———— monilipora, *Y. & B.* Coral Oolite, Yorkshire, *Phil.*
4. ———— vagans, *Phil.* Calcareous Grit, Cornbrash, and Great Oolite, Yorkshire, *Phil.*
5. ———— crenularis, *Lam.* Coral Rag, Midland and Southern England, *Conyb.; * Calc. Grit, Bernese Jura, *Thur.*
6. ———— ornata, . Bradford Clay, North of France, *Bobl.*
7. ———— globata, *Schlot.* Coral Rag, North of France, *Bobl.*
8. ———— maxima, *Munst.* Baireuth; Hohenstein, Saxony, *Munst.*
9. ———— Blumenbachii, *Munst.* Thurnau, Muggendorf, Pretzfeld and Theta, *Goldf.; * Calc. Grit, Bernese Jura, *Thur.*
10. ———— nobilis, *Munst.* Baireuth, *Munst.*
11. ———— elegans, *Munst.* Baireuth, *Munst.; * Kelloway Rock, Haute Saone, *Thir.*
12. ———— marginata, *Goldf.* Regensburg, Heidenheim, *Goldf.*
13. ———— coronata, *Goldf.* Coral Rag, Midland and Southern England, *Conyb.; * Streitberg, Thurnau, Staffelstein, Heidenheim, Randen, *Goldf.; * Calc. Grit, Bernese Jura. *Thur.*

14. Cidaris propinqua, *Munst.* Streitberg, *Munst.* ; Kim. Clay, Calc. Grit,
    Bernese Jura, *Thur.*
15. —— glandifera, *Goldf.* Altdorf, Bavaria; Wurtemberg; Randen,
    Goldf. ; Calc. Grit, Bernese Jura, *Thur.*
16. —— Schmidelii, *Munst.* Dischingen, Switzerland, *Munst.*
17. —— subangularis, *Goldf.* Thurnau; Muggendorf, *Goldf.* ; Kim.
    Clay, Bernese Jura, *Thur.*
18. —— variolaris, *Al. Brong.* Streitberg, Regensberg, Heidenheim,
    Goldf.
—— , species not determined. Inf. Oolite, Yorkshire, *Phil.* ; Lias,
    Lyme Regis, *De la B.* ; Cornbrash, Bradford Clay, Great
    Oolite, Inferior Oolite and Lias, Midland and Southern En-
    gland, *Conyb.* ; Coral Rag, Forest Marble, Normandy, *De
    Cau.* ; Forest Marble, Great Oolite, Wiltshire, *Lons.*
—— , spines of. Great Oolite and Lias, Yorkshire, *Phil.* ; Lias, Mid.
    and South of England, *Conyb.* ; Oolite beds, Lower System,
    South of France, *Bobl.* ; Coral Rag, Normandy, *Desn.* ; Coral
    Rag, Haute Saone, *Thir.*
1. Echinus germinans, *Phil.* Coral Oolite, Calcareous Grit, and Great
    Oolite, Yorkshire, *Phil.*
2. —— lineatus, *Goldf.* Regensburg, Bâle, *Goldf.* ; Calc. Grit, Bernese
    Jura, *Thur.*
3. —— excavatus, *Leske.* Regensburg, *Goldf.* ; Calc. Grit, Bernese
    Jura, *Thur.*
4. —— nodulosus, *Munst.* Baireuth, *Munst.*
5. —— hieroglyphicus, *Goldf.* Regensburg; Thurnau, *Goldf.* ; Calc.
    Grit, Bernese Jura, *Thur.*
6. —— sulcatus, *Goldf.* Thurnau; Streitberg; Muggendorf; Heiden-
    heim, *Goldf.*
—— , species not determined. Coral Rag, North of France, *Bobl.*
1. Galerites depressus, *Lam.* Wurtemberg; Bavaria, *Goldf.* ; Coral Oolite,
    Calcareous Grit, Cornbrash, Yorkshire, *Phil.* ; Oxford Clay,
    Normandy, *Desn.* ; Oxford Clay, Haute Saone, *Thir.* ; Ho-
    henstein, Saxony, *Munst.* ; Oxford Clay, Compound Great
    Oolite, Bernese Jura, *Thur.*
2. —— speciosus, *Munst.* Heidenheim, Wurtemberg, *Munst.*
3. —— Patella, .   Oxford Clay, Normandy, *Desn.*
1. Clypeaster pentagonalis, *Phil.* Calcareous Grit, Yorks., *Phil.*
—— , species not determined :—Coral Rag, Normandy, *De Cau.*
    Kimmeridge Clay, Haute Saone, *Thir.*
1. Nucleolites scutatus, *Lam.* Oxford Clay, Normandy, *Desn.* ; Oxford
    Clay, Haute Saone, *Thir.* ; Oxford Clay, Compound Great
    Oolite, Bernese Jura, *Thur.*
2. —— columbarius, .   Cornbrash, Forest Marble, North of France,
    Bobl.
3. —— granulosus, *Munst.* Amberg; Streitberg; Würgau, *Munst.*
4. —— semiglobus, *Munst.* Pappenheim; Monheim; Bavaria, *Munst.*
5. —— excentricus, *Munst.* Kehlheim, Bavaria, *Munst.*
6. —— canaliculatus, *Munst.* Blaubeuren, Wurtemberg, *Munst.*
—— , species not determined :—Oxford Clay, North of France, *Bobl.*
1. Ananchytes bicordatus, *Lam.* Oxford Clay, Normandy, *Desn.* ; Calc.
    Grit, Bernese Jura, *Thur.*
1. Spatangus ovalis, *Park.* Coral Oolite, Calcareous Grit, Kelloway Rock,
    Yorkshire, *Phil.*
2. —— intermedius, *Munst.* Blaubeuren, Wurtemberg, *Munst.*
3. —— carinatus, *Goldf.* Baireuth, Wurtemberg, *Goldf.*
4. —— capistratus, *Goldf.* Baireuth, *Goldf.* ; Oxford Clay, Haute Saone,
    Thir. ; Oxford Clay, Bernese Jura, *Thur.*

Spatangus, species not determined :—Cornbrash, Forest Marble, North of France, *Bobl.*
1. Clypeus sinuatus, *Park.* Coral Oolite, Yorkshire, *Phil.* ; Coral Rag, Cornbrash, Great Oolite, Inferior Oolite, Mid. and Southern England, *Conyb.* ; Forest Marble, Normandy, *De Cau.*
2. —— emarginatus, *Phil.* Coralline Oolite, Yorkshire, *Phil.*
3. —— clunicularis, *Smith.* Coral Oolite, Cornbrash, Yorkshire, *Phil.*; Coral Rag, Cornbrash, Great Oolite, Inferior Oolite, Midland and Southern England, *Conyb.* ; Forest Marble, Normandy, *De Cau.* ; Coral Rag, Weymouth, *Sedg.*
4. —— dimidiatus, *Phil.* Coral Oolite, Yorkshire, *Phil.*
5. —— semisulcatus, *Phil.* Coralline Oolite, Yorkshire, *Phil.*
6. —— orbicularis, *Phil.* Cornbrash, Yorkshire, *Phil.*
——, species not determined :—Cornbrash, Great Oolite, Wiltshire, *Lons.*
Echinites, genera not determined. Inferior Oolite, Normandy, *De Cau.*
——, spines of. Coral Rag, Burgundy, *Beaum.* ; Coral Rag, North of France, *Bobl.* ; Forest Marble, Mamers, *Desn.* ; Mauriac beds, South of France, *Dufr.*
1. Eugeniacrinites caryophyllatus, *Goldf.* Baireuth; Wurtemberg; Switzerland, *Goldf.* ; Calc. Grit, Bernese Jura, *Thur.*
2. —— mutans, *Goldf.* Streitberg; Muggendorf, *Goldf.* ; Calc. Grit, Bernese Jura, *Thur.*
3. —— pyriformis, *Munst.* Randen, *Goldf.*
4. —— moniliformis, *Munst.* Thurnau; Streitberg; Switzerland, *Goldf.*
5. —— Hoferi, *Munst.* Switzerland; Streitberg, *Goldf.*
6. —— compressus, *Munst.* Baireuth; Wurtemberg, *Munst.*
1. Apiocrinites rotundus, *Miller.* Forest Marble, Normandy, *De Cau.* ; Bradford Clay, Great Oolite, Mid. and S. England, *Conyb.* ; Forest Marble, *Buckl.* ; Great Oolite, Alsace, *Al. Brong.* ; Forest Marble, Normandy, *De Cau.* ; Forest Marble, Wiltshire; Great Oolite, Somerset, *Lons.* ; Germany; Alsace, *Goldf.* ; Calc. Grit, Bernese Jura, *Thur.*
2. —— Prattii, *Gray.* Great Oolite, Somerset, *Lons.*
3. —— elongatus, *Miller.* Bâle; Soleure; Elsas, near Béfort, Alsace; Forest Marble, Normandy, *Goldf.*
4. —— rosaceus, *Schlot.* Canton Soleure; Elsas; Muggendorf, *Goldf.* ; Calc. Grit, Bernese Jura, *Thur.*
5. —— mespiliformis, *Schlot.* Heidenheim; Giengen, *Goldf.* ; Kim. Clay, Haute Saone, *Thur.*
6. —— Milleri, *Schlot.* Wurtemberg, *Goldf.* ; Calc. Grit, Oxford Clay, Bernese Jura, *Thur.*
7. —— flexuosus, *Goldf.* Wurtemberg, *Goldf.*
8. —— subconicus, *Goldf.* Bath, *Goldf.*
1. Pentacrinites vulgaris, *Schlot.* Cornbrash, Coral Oolite, and Lias, Yorks., *Phil.* ; Inf. Oolite, and Lias, Midl. and S. England, *Conyb.* ; Lias, Alsace, Gundershofen, Figeac, *Al. Brong.*
2. —— subangularis, *Miller.* Inferior Oolite and Lias, Midland and Southern England, *Conyb.* ; Lias, Banz; Boll, *Goldf.*
3. —— Briareus, *Miller.* Lias, Midland and Southern England,*Conyb.*; Lias, Yorkshire, *Phil.* ; Lias, Banz; Boll, *Goldf.*
4. —— basaltiformis, *Miller.* Lias, Midland and Southern England, *Conyb.* ; Lias, Alsace, *Voltz* ; Baireuth; Banz; Boll, *Goldf.*
5. —— tuberculatus, *Miller.* Lias, Midland and Southern England, *Conyb.* ; Lias, Alsace, *Voltz.*
6. —— subteres, *Goldf.* (Var.) Oxford Clay, Haute Saone, *Thir.*
7. —— Jurensis, *Munst.* Coral Rag, Haute Saone, *Thir.*
8. —— scalaris, *Goldf.* Baireuth ; Banz ; Boll, *Goldf.*

9. Pentacrinites cingulatus, *Munst.* Streitberg; Thurnau, *Goldf.*
10. ———— pentagonalis, *Goldf.* Streitberg; Thurnau; Boll, *Goldf.; Ox-*
ford Clay, Bernese Jura, *Thur.*
11. ———— moniliferus, *Munst.* Lias, Baireuth, *Goldf.*
12. ———— subsulcatus, *Munst.* Lias, Baireuth, *Goldf.*
13. ———— subteres, *Munst.* Streitberg, *Goldf.*
14. ————? paradoxus, *Goldf.* Baireuth; Wurtemberg, *Goldf.*
————, species not determined. Forest Marble, Normandy, *De Cau.;*
Bradford Clay, North of France, *Bobl.;* Cornbrash, Forest
Marble, Great Oolite, Midland and Southern England, *Conyb.;*
Inferior Oolite, Wotton-under-Edge, Forest Marble, Great
Oolite, Somerset, *Lons.*
1. Solanocrinites costatus, *Goldf.* Giengen; Heidenheim, Wurtemberg,
Goldf.
2. ———— scrobiculatus, *Munst.* Streitberg; Thurnau, *Goldf.*
3. ———— Jaegeri, *Goldf.* Baireuth, *Goldf.*
1. Rhodocrinites echinatus, *Schlot.* Amberg; Wurtemberg; Switzerland;
Berrach, *Goldf.*
1. Comatula pinnata, *Goldf.* Solenhofen, *Goldf.*
2. ———— tenella, *Goldf.* Solenhofen, *Goldf.*
3. ———— pectinata, *Goldf.* Solenhofen, *Goldf.*
4. ———— filiformis, *Goldf.* Solenhofen, *Goldf.*
1. Ophiura Milleri, *Phil.* Lias, Yorkshire, *Phil.;* Inferior Oolite sands,
Bridport, *De la B.*
2. ———— speciosa, *Munst.* Solenhofen, *Goldf.*
3. ———— carinata, *Munst.* Solenhofen, *Goldf.*
1. Asterias lumbricalis, *Schlot.* Walzendorf, Coburg; Lichtenfels, Bamberg, *Goldf.*
2. ———— lanceolata, *Goldf.* Walzendorf; Lichtenfels, *Goldf.*
3. ———— arenicola, *Goldf.* Porta Westphalica, *Goldf.*
4. ———— Jurensis, *Munst.* Hattheim, Wurtemberg; Baireuth, *Goldf.*
5. ———— tabulata, *Goldf.* Streitberg, *Goldf.*
6. ———— scutata, *Goldf.* Streitberg; Heiligenstadt, *Goldf.*
7. ———— stellifera, *Goldf.* Streitberg, *Goldf.*
8. ———— prisca, *Goldf.* Wasseralfingen, *Schübber.*
————, species not determined. Coral Rag, Bernese Jura, *Thur.*

ANNULATA.

1. Lumbricaria Intestinum, *Munst.* Solenhofen, *Goldf.*
2. ———— Colon, *Munst.* Solenhofen, *Goldf.*
3. ———— recta, *Munst.* Solenhofen, *Goldf.*
4. ———— gordialis, *Munst.* Solenhofen, *Goldf.*
5. ———— coniugata, *Munst.* Solenhofen, *Goldf.*
6. ———— Filaria, *Munst.* Solenhofen, *Goldf.*
1. Serpula squamosa, *Bean.* Coral Oolite, Yorkshire, *Phil.*
2. ———— lacerata, *Phil.* Calcareous Grit, and Great Oolite, Yorkshire,
Phil.; Calc. Grit, Bernese Jura, *Thur.*
3. ———— intestinalis, *Phil.* Oxford Clay, and Cornbrash, Yorkshire,
Phil.
4. ———— deplexa, *Bean.* Inferior Oolite, Yorkshire, *Phil.*
5. ———— capitata, *Phil.* Lias, Yorkshire, *Phil.*
6. ———— quadrangularis, *Lam.* Oxford Clay, Normandy, *Desn.;* Calc.
Grit, Bernese Jura, *Thur.*
7. ———— sulcata, *Sow.* Calcareous Grit, Oxford, *Sow.*
8. ———— tricarinata, *Sow.* Calcareous Grit, Oxford; Coral Rag, Steeple
Ashton, Wilts, *Sow.;* Oxford Clay, Haute Saone, *Thir.*
9. ———— triangulata, *Sow.* Bradford Clay or Great Oolite, Bradford, *Sow.*
10. ———— runcinata, *Sow.* Coral Rag, Oxford, *Sow.*

11. Serpula tricristata, *Goldf.* Lias, Banz, *Goldf.*
12. ——— quinque-cristata, *Munst.* Lias, Banz, *Goldf.*
13. ——— quinque-sulcata, *Munst.* Lias, Theta, Baireuth, *Goldf.*
14. ——— circinnalis, *Munst.* Lias, Banz, *Goldf.*
15. ——— complanata, *Goldf.* Lias, Theta, *Munst.*
16. ——— grandis, *Goldf.* Ferrug. Oolite, Baireuth; Wurtemberg; Coral Oolite, Haute Saone; Upper Jura Limestone, Heidenheim, *Goldf.*
17. ——— Limax, *Goldf.* Ferrug. Oolite, Baireuth, *Goldf.*
18. ——— conformis, *Goldf.* Alsace, *Goldf.; Kim.* Clay, Bernese Jura, *Thur.*
19. ——— convoluta, *Goldf.* Ferrug. Oolite, Wasseralfingen; Baireuth, *Goldf.*
20. ——— lituiformis, *Munst.* Ferrug. Oolite, Gräfenberg, Baireuth, *Goldf.*
21. ——— Delphinula, *Goldf.* Thurnau; Streitberg, *Goldf.*
22. ——— capitata, *Goldf.* Streitberg, *Goldf.;* Calc. Grit, Bernese Jura, *Thur.*
23. ——— limata, *Munst.* Streitberg, *Goldf.*
24. ——— plicatilis, *Munst.* Gräfenberg; Streitberg, *Goldf.*
25. ——— gibbosa, *Goldf.* Muggendorf, *Goldf.*
26. ——— nodulosa, *Goldf.* Streitberg, *Goldf.*
27. ——— Spirolinites, *Munst.* Streitberg, *Goldf.*
28. ——— tricarinata, *Goldf.* Ferrug. Oolite, Rabenstein; Baireuth; Alsace, *Goldf.*
29. ——— pentagona, *Goldf.* Streitberg, *Goldf.*
30. ——— quinquangularis, *Goldf.* Kim. Clay, Largue, Sundgau; Normandy, *Goldf.;* Calc. Grit, Bernese Jura, *Thur.*
31. ——— quadrilatera, *Goldf.* Ferrug. Oolite, Rabenstein; Buxweiler, *Goldf.*
32. ——— vertebralis, *Sow.* Buxweiler, *Goldf.*
33. ——— prolifera, *Goldf.* Streitberg, *Goldf.*
34. ——— planorbiformis, *Munst.* Thurnau; Streitberg, *Goldf.*
35. ——— trochleata, *Munst.* Streitberg, *Goldf.*
36. ——— macrocephala, *Goldf.* Thurnau, *Goldf.*
37. ——— heliciformis, *Goldf.* Neuburg; Doubs, *Goldf.*
38. ——— quadristriata, *Goldf.* Berrach, Burgundy; Amberg, *Goldf.*
39. ——— convoluta, *Munst.* Streitberg, *Goldf.*
40. ——— canaliculata, *Munst.* Streitberg, *Goldf.*
41. ——— Deshayesii, *Munst.* Streitberg, *Goldf.*
42. ——— volubilis, *Munst.* Ferrug. Oolite, Rabenstein, *Goldf.*
43. ——— spiralis, *Munst.* Muggendorf; Hattheim; Heidenheim, *Goldf.*
44. ——— cingulata, *Munst.* Streitberg, *Goldf.*
45. ——— Flagellum, *Munst.* Streitberg, *Goldf.*
46. ——— substriata, *Munst.* Ferrug. Oolite, Rabenstein, *Goldf.*
47. ——— flaccida, *Munst.* Ferrug. Oolite, Rabenstein; Bâle; Elsas, *Goldf.*
48. ——— gordialis, *Schlot.* Streitberg; Heidenheim; Buxweiler, *Goldf.;* Coral Rag, Bernese Jura, *Thur.*
49. ——— intercepta, *Goldf.* Streitberg; Culmbach, *Goldf.*
50. ——— Ilium, *Goldf.* Thurnau; Streitberg, *Goldf.;* Calc. Grit, Bernese Jura, *Thur.*
51. ——— Filaria, *Goldf.* Ferrug. Oolite, Gräfenberg; Streitberg, *Goldf.*
52. ——— socialis, *Goldf.* Bavaria; Swabia; Burgundy, *Goldf.;* Calc. Grit, Bernese Jura, *Thur.*
53. ——— problematica, *Munst.* Solenhofen, *Goldf.*
——, species undetermined. Coral Rag, Oxford Clay, Cornbrash, Forest Marble, Bradford Clay, Great Oolite, Mid. and South of England, *Conyb.;* Oxford Clay, Inferior Oolite, Haute Saone, *Thir.;* Cornbrash, Forest Marble, Bradford Clay, Great Oolite, Fuller's Earth, Wiltshire, *Lons.*

CONCHIFERA.

1. Spirifer Walcotii, *Sow.* Lias, Yorkshire, *Phil.*; Lias, Bath, Lyme
    Regis, *De la B.*; Lias, Normandy, *De Cau.*; Lias, South o
    France, *Dufr.*; Lias, Western Islands, Scotland, *Murch.*
1. Delthyris * verrucosa, *Von Buch.* Lias, Bahlingen, Wurtemberg, *Von
    Buch.*
2. ———— rostrata, *Schlot.* Lias, Wurtemberg, *Von Buch.*; Oxford Clay,
    Bernese Jura, *Thur.*
1. Terebratula intermedia, *Sow.* Coral Oolite, and Great Oolite, Yorks.,
    Phil.; Cornbrash, Mid. and S. England; Inferior Oolite,
    Dundry, *Conyb.*; Kim. Clay, Bernese Jura, *Thur.*
2. ———— globata, *Sow.* Coral Oolite? Great Oolite, Yorkshire, *Phil.*;
    Forest Marble, Normandy, *De Cau.*; Oolite, Env. of Bath;
    Sow.; Fuller's Earth, Env. of Bath, Great Oolite, Haute
    Saone, *Thir.*
3. ———— ornithocephala, *Sow.* Coralline Oolite, and Kelloway Rock,
    Yorkshire, *Phil.*; Kelloway Rock, Cornbrash, Lias? Mid.
    and South of England; Inferior Oolite, Dundry, *Conyb.*;
    Oxford Clay and Lias, Normandy, *De Cau.*; Inferior Oolite,
    Uzer, South of France, *Dufr.*; Kimmeridge Clay, Great
    Oolite, Haute Saone, *Thir.*; Inferior Oolite, Wiltshire, *Lons.*;
    Soleure, Buxweiler, *Hœn.*; Oxford Clay, Bernese Jura, *Thur.*
? 4. ———— ovata, *Sow.* Coralline Oolite? Yorkshire, *Phil.*; Inferior Oolite,
    Mid. and South of England, *Conyb.*; Coral Rag, Haute Soane,
    Thir.
5. ———— obsoleta, *Sow.* Coralline Oolite? Inferior Oolite, Yorkshire,
    Phil.; Cornbrash, Bradford Clay, Great Oolite, and Inferior
    Oolite, Mid. and South of England, *Conyb.*; Great Oolite,
    Normandy, *De Cau;* Lias and Inferior Oolite, South of
    France, *Dufr.*; Forest Marble, Wiltshire, *Lons.*; Oxford
    Clay, Bernese Jura, *Thur.*
6. ———— socialis, *Phil.* Calcareous Grit, and Kelloway Rock, York-
    shire, *Phil.*
7. ———— ovoides, *Sow.* Cornbrash? Yorkshire, *Phil.*; Inferior Oolite,
    Normandy, *De Cau.*; Rubbly Limestone, &c. Braambury
    Hill, Brora, *Murch.*; Inf. Oolite, Calvados, *Her.*; Leisacker;
    Neuburg; Neresheim, *G. T.*
8. ———— digona, *Sow.* Cornbrash, Yorks., *Phil.*; Cornbrash and Brad-
    ford Clay, Mid. and S. England; Inferior Oolite, Dundry,
    Conyb.; Forest Marble, Normandy, *De Cau.*; Bradford Clay
    and Coral Rag? North of France, *Bobl.*; Forest Marble,
    Bradford Clay, Great Oolite, Wilts, *Lons.*
9. ———— spinosa, *Townsend and Smith.* Great Oolite, Yorkshire, *Phil.*;
    Inf. Oolite, Bath, *Lons.*; Oxford Clay, Bernese Jura, *Thur.*
*9. ———— spinosa, *Schlot.* Inferior Oolite, Southern Germany, *Munst.*
10. ———— trilineata, *Y. & B.* Inferior Oolite and Lias, Yorkshire, *Phil.*
11. ———— bidens, *Phil.* Inferior Oolite and Lias, Yorks., *Phil.*; Lias,
    Boll, *G. T.*
12. ———— punctata, *Sow.* Lias, Yorkshire, *Phil.*; Inferior Oolite, Mid.
    and South of England, *Conyb.*; Lias, Western Islands, Scot-
    land, *Murch.*; Inf. Oolite, Southern Germany, *Munst.*
13. ———— resupinata, *Sow.* Lias, Yorkshire, *Phil.*; Inferior Oolite, Mid.
    and South of England, *Conyb.*; Inferior Oolite, Bärendorf;
    Thurnau, *Munst.*

---

* The genus *Delthyris*, Dalman, is the same with the genus *Spirifer*, Sowerby;
both names have been retained above for the purpose of more easy reference.

14. Terebratula acuta, *Sow.* Lias, Yorkshire, *Phil.;* Inferior Oolite and Lias, Mid. and South of England, *Conyb.;* Lias, Normandy, *De Cau.;* Fuller's Earth, Frome, *Lons.;* Lias, Wurtemberg, *Von Buch.*

15. ———— triplicata, *Phil.* Lias, Yorkshire, *Phil.;* Lias, Wurtemberg, *Von Buch.*

16. ———— tetraëdra, *Sow.* Lias, Yorkshire, *Phil.;* Inferior Oolite, Mid. and South of England, *Conyb.;* Lias, South of France, *Dufr.;* Forest Marble? Mauriac, South of France, *Dufr.;* Lias and Micaceous Sandstone, Western Islands, Scotland, *Murch.;* Echterdingen, Buxweiler, *Hœn.;* Lias, Waldhausen, Tubingen, *G. T.*

17. ———— subrotunda, *Sow.* Cornbrash and Inferior Oolite, Mid. and South of England, *Conyb.;* Cornbrash and Forest Marble, North of France, *Bobl.;* Forest Marble? Mauriac, South of France, *Dufr.;* Inferior Oolite, Env. of Bath, *Lons.*

18. ———— obovata, *Sow.* Cornbrash, Mid. and South of England, *Conyb.;* Inferior Oolite, Env. of Bath, *Lons.*

19. ———— reticulata, *Sow.* Bradford Clay, Mid. and South of England, *Conyb.;* Forest Marble, Normandy, *De Cau.*

20. ———— media, *Sow.* Inferior Oolite, Dundry, *Conyb.;* Inferior Oolite, Great Oolite, and Bradford Clay, North of France, *Bobl.;* Dunrobin Oolite, Scotland, *Murch.;* Fuller's Earth, Inferior Oolite, Env. of Bath, *Lons.*

21. ———— crumena, *Sow.* Inf. Oolite and Lias? Mid. and S. England, *Conyb.;* Echterdingen, *Hœn.*

22. ———— concinna, *Sow.* Fuller's Earth, Mid. and South of England, *Conyb.;* Inferior Oolite, Normandy, *De C.;* Forest Marble? Mauriac, South of France, *Dufr.;* Fuller's Earth, Frome; Inferior Oolite, Env. of Bath, *Lons.*

? 23.———— biplicata, *Sow.* Oxford Clay, Forest Marble, Great Oolite, Normandy, *Her.;* Soleure, *Hœn.;* Kim. Clay, Bernese Jura, *Thur.;* Papenheim, *G. T.*

24. ———— tetrandra, . Forest Marble, Normandy, *De C.*

25. ———— coarctata, *Park.* Forest Marble, Normandy, *De C.;* Bradford Clay, North of France, *Bobl.;* Bradford Clay, Bath, *Loscombe.*

26. ———— plicatella, *Sow.* Inf. Oolite, Bridport, *De la B.;* Forest Marble, Normandy, ·*De C.;* Bavaria; Hohenstein, Saxony, *Munst.*

27. ———— serrata, *Sow.* Lias, Lyme Regis, *De la B.;* Lias, Waldhaenserhof, *G. T.*

28. ———— bullata, *Sow.* Inferior Oolite, Normandy, *De C.;* Inferior Oolite, Bridport, Dorset, *Sow.;* Cornbrash, Wiltshire; Fuller's Earth, Env. of Bath, *Lons.*

29. ———— sphæroïdalis, *Sow.* Inferior Oolite, Normandy, *De C.;* Inferior Oolite, Dundry, *Braikenridge.*

30. ———— emarginata, *Sow.* Inferior Oolite, Normandy, *De C.;* Inferior Oolite, Env. of Bath, *Lons.*

31. ———— quadrifida, . Inf. Oolite, Calvados, *Desl.*

32. ———— numismalis, *Lam.* Lias, Bahlingen; Gonningen, *Von Buch.;* Inf. Oolite, Calvados, *Desl.*

33. ———— perovalis, *Sow.** Inf. Oolite, Dundry, *Braikenridge;* Forest Marble? Mauriac, and Kim. Clay, Cahors, South of France; Rochelle Limestone, *Dufr.;* Oxford Clay, Kell. Rock, Haute Saône, *Thir.;* Inf. Oolite, Calvados, *Desl.*

34. ———— maxillata, *Sow.* Inf. Oolite, Env. of Bath, *Sow.;* Forest Marble, Wilts, *Lons.*

* *T. bisuffarcinata,* Schlot., according to Count Munster.

35. Terebratula flabellula, *Sow.* Great Oolite, Ancliff, near Bradford, Wilts, *Cookson.*
36. —— furcata, *Sow.* Great Oolite, Ancliff, *Cookson.*
37. —— orbicularis, *Sow.* Lias, Bath, *Sow.*
38. —— hemisphærica, *Sow.* Great Oolite, Ancliff, *Cookson.*
39. —— inconstans, *Sow.* Shelly Limestone and Calc. Grit, Portgower, &c. N. of Scotl., and Shell Limestone, Beal, Isle of Sky, *Murch.*; Coral Rag, Weymouth, *Sedg.*; Bavaria; Wurtemberg; Porta Westphalica; Hohenstein, Saxony, *Munst.*
40. —— avicularis, *Munst.* Inf. Oolite, Southern Germany, *Munst.*
41. —— loricata, *Schlot.* Baireuth, *Hœn.*; Thurnau; Streitberg, *G. T.*
42. —— plicata, *Lam.* Thurnau, *G. T.*
43. —— spinosa, *Schlot.* Compound Great Oolite, Bernese Jura, *Thur.*; Blomberg; Wurtemberg; Donaueschingen, *G. T.*
44. —— vulgaris, *Schlot.* Porta Westphalica, *Hœn.*
45. —— Defrancii, *Al. Brong.* Amberg, *Hœn.*
46. —— Hœninghausii, *Blain.* Baireuth, *Hœn.*
47. —— rimosa, *Von Buch.* Lias, Bahlingen, Wurtemberg, *Von Buch.*
48. —— bicanaliculata, *Sow.* Hohenstein, Saxony; Ferruginous Oolite, Bavaria and Wurtemberg, *Munst.*
49. —— cornuta, *Sow.* Inf. Oolite, Ilminster, *Sow.*; Bavaria; Hohenstein, *Munst.*
50. —— trilobata, *Munst.* Bavaria; Porta Westphalica; Hohenstein, *Munst.*
51. —— lacunosa, *Schlot.* Calc. Grit, Oxford Clay, Bernese Jura, *Thur.*
52. —— varians, *Schlot.* Oxford Clay, Compound Great Oolite, Bernese Jura, *Thur.*
53. —— depressa, *Sow.* Oxford Clay, Bernese Jura, *Thur.*
54. —— variabilis, *Schlot.* Oxford Clay, Bernese Jura, *Thur.*
55. —— personata, *Her.* Inf. Oolite, Calvados, *Desl.*
56. —— sella, *Sow.* Lochen; Bahlingen, *G. T.*
57. —— impressa, . Hohenzollern; Stufenberg; Thurnau, *G. T.*
58. —— nucleolata, *Schlot.* Streitberg, *G. T.*
59. —— tegularis, *Schlot.* Kelheim; Heidenheim; Amberg, *G. T.*
60. ——? alata, *Lam.* Bahlingen; Locherberg; Hohenzollern, *G. T.*
1. Orbicula? radiata, *Phil.* Coral. Oolite, Yorks., *Phil.*
2. —— granulata, *Sow.* Great Oolite, Ancliff, Wilts, *Cookson.*
——, species not determined. Inferior Oolite, Yorks., *Phil.*
1. Lingula Beanii, *Phil.* Inferior Oolite, Yorks., *Phil.*
1. Ostrea gregarea, *Sow.** Coral Rag, Yorks., Wilts, &c.; Calc. Grit and Great Oolite? Yorks., *Phil.*; Coral Rag, Mid. and S. of Eng.; Inf. Oolite, Dundry, *Conyb.*; Coral Rag and Oxford Clay, Norm., *De C.*; Oxford Clay and Coral Rag, N. of Fr., *Bobl.*; Kim. Clay, Hâvre, *Phil.*; Coral Rag, Weymouth, *Sedg.*; Great Oolite, Calvados, *Desl.*
2. —— solitaria, *Sow.* Coral Rag and Inf. Oolite, Yorks., Oxon, &c. *Phil.*; Kim. Clay, Haute Saone, *Thir.*; Coral Rag, Weymouth, *Sedg.*; Kim. Clay, Bernese Jura, *Thur.*
3. —— duriuscula, *Bean.* Coralline Oolite, Yorks., *Phil.*
4. —— inæqualis, *Phil.* Oxford Clay, Yorks., *Phil.*
5. —— undosa, *Bean.* Kell. Rock, Yorks., *Phil.*
6. —— archetypa, *Phil.* Kell. Rock, Yorks., *Phil.*
7. —— Marshii †, *Sow.* Kell. Rock, Cornb., and Great Oolite, Yorks., *Phil.*; Cornb. and Fuller's E., Mid. and S. of Eng., *Conyb.*; Oxford Clay, Forest Marb., and Inf. Oolite, Norm., *De C.*; Cornb., Wilts, *Lons.*; Coral Rag, Weymouth, *Sedg.*; Oxford

* Query, *Ostrea Crista-Galli*, Smith.   † Ostrea flabelloides, *Lam.*

Clay, N. of France, *Bobl.;* Wiesgoldingen; Fürstenberg; Stufenberg; Baireuth, *G. T.*

8. Ostrea sulcifera, *Phil.* Great Oolite, Yorks., *Phil.;* Inf. Oolite, Haute Saone, *Thir.*

9. ——— deltoidea, *Sow. and Smith.* Kim. Clay, Yorks., *Phil.;* Oxford Clay, N. of Fr. *Bobl.;* Kim. Clay, S. and Mid. of England, *Conyb.;* Shell Limestone and Calc. Grit? Portgower, &c. Scotl., *Murch.;* Kim. Clay, Hâvre, *Phil.;* Sandst., Limest., and Shale, Inverbrora, Scotl., *Murch.;* upper part of Coral Rag, Weymouth, *Sedg.*

10. ——— expansa, *Sow.* Portland Stone, *Conyb.*

11. ——— palmetta, *Sow.* Oxford Clay, Mid. and S. of Eng., *Conyb.;* Oxf. Clay and For. Marb., Norm., *De C.;* Baireuth; Wurtemberg, *G. T.*

12. ——— acuminata, *Sow.* Bradford Clay and Inf. Oolite, Mid. and S. of Eng., *Conyb.;* Great Oolite and Brad. Clay, N. of Fr., *Bobl.;* Great Oolite, Haute Saone, *Thir.;* Fuller's E., Inf. Oolite, Environs of Bath, *Lons.;* Comp. Great Oolite, Bernese Jura, *Thur.;* Inf. Oolite? Calvados, *Desl.*

13. ——— rugosa, *Sow.* Inf. Oolite, Mid. and S. of Eng., *Conyb.*

14. ——— minima, *Desl.* Coral Rag and Oxford Clay, Norm., *De C.*

15. ——— plicatilis. Oxford Clay, Norm., *De C.*

16. ——— costata*, *Sow.* Brad. Clay, N. of Fr., *Bobl.;* Great Oolite, Ancliff, near Bath, *Cookson.;* Comp. Great Oolite, Bernese Jura, *Thur.*

17. ——— pectinata. Oxford Clay, N. of Fr., *Bobl.*

18. ——— pennaria. Oxford Clay, N. of Fr., *Bobl.*

19. ——— læviuscula, *Sow.* Lias, Eng., *Sow.*

20. ——— obscura, *Sow.* Great Oolite, Ancliff, Wilts, *Cookson.*

21. ——— Meadii, *Sow.* Inf. Oolite, Env. of Bath, *Lons.*

22. ——— colubrina, *Lam.* Nattheim, *G. T.*

23. ——— carinata, *Lam.* Calc. Grit, Bernese Jura, *Thur.*

24. ——— irregularis, *Munst.* Lias, Amberg, *G. T.*

25. ——— Ungula, *Munst.* Lias, Amberg; Banz, *G. T.*

26. ——— læviuscula, *Munst.* Lias, Amberg, *G. T.*

27. ——— Synama, *Munst.* Lias, Baireuth, *G. T.*

28. ——— semiplicata, *Munst.* Lias, Baireuth, *G. T.*

1. Exogyra Bruntrutana, *Thur.* Portland Beds, Kim. Clay, Porrentruy, Bernese Jura, *Thur.*

2. ——— spiralis, *Goldf.* Elligser Brink, *G. T.*

3. ——— reniformis, *Goldf.* Westphalia, *G. T.*

———, species not determined. Kim. Clay, Haute Saone, *Thir.* Forest Marble? Wilts, *Lons.*

1. Gryphæa chamæformis, *Phil.* Calc. Grit, Yorks.; and Oolite, Sutherland, *Phil.*

2. ——— bullata, *Sow.* Coral. Oolite? Calc. Grit? *Phil.;* Oxford Clay, Lincolnshire, *Sow.;* Oolite of Braambury Hill, Brora, *Murch.*

3. ——— inhærens, *Phil.* Calc. Grit, Yorks., *Phil.*

4. ——— dilatata, *Sow.* Kell. Rock, Yorks., *Phil.;* Oxford Clay, Mid. and S. of Eng., *Conyb.;* Oxford Clay and Lias, Norm., *De C.;* Oxford Clay, N. of Fr., *Bobl.;* Oxford Clay, Burgundy, *Beaum.;* Great Arenaceous Formation, Western Islands, Scotl. *Murch.;* Oxford Clay, Haute Saone, *Thir.;* Lower part of Coral Rag, Weymouth, *Sedg.;* Oxford Clay, Beggingen, Schafhausen, *Von Buch.*

5. ——— incurva, *Sow.* Lias, Yorks., *Phil;* Lias, Mid. and S. Eng.

* Ostrea Knorrii, *Voltz.*

544   *Organic Remains of the Oolitic Group.*

Conyb. ; Lias, Norm., *De C.* ; Lias and Inf. Oolite, N. of Fr.,
Bobl. ; Lias, S. of Fr., *Dufr.* ; Lias, Metz, Salins, Amberg,
Al. Brong. ; Lias, Western Islands, Scotl. ; Lias, Ross and
Cromarty, Scotl., *Murch.* ; Göppingen, Bahlingen, *Hœn.*
6. Gryphæa nana, *Sow.* Kim. Clay, Oxford, *Sow.* ; Shale and Grit, Dunrobin Reefs, Scotl., *Murch.* ; Lias and Oxford Clay? N. of
Fr., *Bobl.*
7. ——— Maccullochii, *Sow.* ; Lias, Western Islands, Scotl., *Murch.* ;
Lias, Yorks., *Phil.* ; Oxford Clay, Norm., *De C.* ; Lias, S. of
Fr., *Dufr* ; Lias, Env. of Bath, *Lons.*
8. ——— depressa, *Phil.* Lias, Yorks., *Phil.*
9. ——— obliquata, *Sow.* Lias, Mid. and S. Eng., *Conyb.* ; Lias, S. of Fr.,
Dufr. ; Lias, Western Islands, Scotl., *Murch.* ; Lias, Env.
of Bath, *Lons.*
10. ——— Cymbium, *Lam.* Inf. Oolite, N. of Fr., *Bobl.* ; Lias, S. of France;
Inf. Oolite, Villefranche, S. of France, *Dufr.* ; Inf. Oolite,
Haute Saone, *Thir.* ; Lias, Bahlingen, *Von Buch* ; Inf. Oolite,
Calvados, *Desl.* ; Lias, Boll; Wasseralfingen; Ellwangen, *G.T.*
11. ——— lituola, *Lam.* Brad. Clay, Cornb., and For. Marb., N. of Fr.,
Bobl.
12. ——— gigantea, *Sow.* Lias, S. of Fr., *Dufr.* ; Lias, Ross and Cromarty,
Scotl.; Great Arenaceous Formation, Western Islands, Scotl.,
Murch. ; Porta Westphalica ; Hohenstein, Saxony, *Munst.*
13. ——— minuta, *Sow.* Great Oolite, Ancliff, Wilts, *Cookson.*
14. ——— virgula*, *Defrance.* Kim. Clay, Hâvre, *Al. Brong.* ; Kim. Clay,
Burgundy, *Beaum.* ; Kim. Clay, S. of Fr. *Dufr.* ; Kim. Clay,
Weymouth, *Buckl. & De la B.* ; Kim. Clay, Haute Saone,
Thir. ; Portland Beds, Kim. Clay, Bernese Jura, *Thur.*
15. ——— lævis, *Schlot.* Lias, Malsch, near Heidelberg, *Bronn.*
1. Plicatula spinosa, *Sow.* Lias, Yorks., *Phil.* ; Lias, Mid. and S. Eng.,
Conyb. ; Lias, Norm., *De C.* ; Inf. Oolite, N. of Fr. *Bobl.* ;
Great Arenaceous Formation, Western Islands, Scotl., *Murch.* ;
Lias, Gundershoffen, *Voltz.* ; Lias, Wittberg, *G. T.*
2. ——— tubifera,   Calc. Grit, Bernese Jura, *Thur.*
3. ——— pectinoides, *Desl.* Inf. Oolite, Calvados, *Desl.*
4. ——— squamosa, *Goldf.* Elligser Brink, *G. T.*
1. Pecten abjectus, *Phil.* Coral Rag, Yorks. and Oxon ; Calc. Grit, Great
Oolite, and Inf. Oolite, Yorks., *Phil.*
2. ——— inæquicostatus, *Phil.* Coralline Oolite, Yorks. ; Calc. Grit,
Oxon, *Phil.* ; Coral Rag, Bernese Jura, *Thur.*
3. ——— cancellatus, *Bean.* Coralline Oolite, Yorks. ; Oolite, Sutherland? *Phil.*
4. ——— demissus, *Phil.* Coralline Oolite, Kell. Rock, Cornbrash, and
Great Oolite, Yorks., *Phil.*
5. ——— Lens, *Sow.* Coralline Oolite, Kell. Rock, Great Oolite, Inf.
Oolite, and Lias, Yorks., *Phil.* ; Coral Rag, Mid. and S. Eng.;
Inf. Oolite, Dundry, *Conyb.* ; Coral Rag and Oxford Clay,
Norm., *De C.* ; Cornb. and For. Marb., N. of Fr., *Bobl.* ; Inf.
Oolite, Alsace, and Stranen near Luxembourg, *Al. Brong.* ;
Sandst., Limest., and Shale, Inverbrora, Scotl., *Murch.* ; Inf.
Oolite, Haute Saone, *Thir.* ; Elligser Brink, *G. T.* ; Comp.
Great Oolite, Bernese Jura, *Thur.*
6. ——— vagans, *Sow.* Coral Rag, Yorks. and Oxford; Calc. Grit, Yorks.,
Phil. ; For. Marb., Norm., *De C.* ; Sandst. and Rubbly Limest.,
Braambury Hill, Brora, *Murch.* ; For. Marb., Wilts, *Lons.* ;
Oxf. Clay, Comp. Great Oolite, Bernese Jura, *Thur.*

* Exogyra virgula, *Voltz.*

7. Pecten fibrosus, *Sow.* Kell. Rock and Cornbrash, Yorks., *Phil.;* Coral
Rag, Kell. Rock, Cornb., For. Marb., Brad. Clay, and Inf,
Oolite, Mid. and S. Eng., *Conyb.;* Coral Rag, Norm.? *De C.;*
Cornb. and For. Marb., N. of Fr., *Bobl.;* For. Marb.? Mau-
riac, S. of Fr., *Dufr.;* Rubbly Limestone, &c., Braambury
Hill, Brora, *Murch.;* For. Marb., Wilts, *Lons.;* Soleure,
Hœn.; Culmbach, *G. T.*
8. —— virguliferus, *Phil.;* Inferior Oolite, Yorks., *Phil.*
9. —— sublævis, *Y. & B.* Lias, Yorks., *Phil.*
10. —— æquivalvis, *Sow.* Lias, Yorks., *Phil.;* Inf. Oolite, Mid. and S.
Eng., *Conyb.;* Lias, S. of Fr., *Dufr.;* Lias, Western Islands,
Scotl., *Murch.;* Inf. Oolite, Env. of Bath, *Lons.;* Inf. Oolite,
Calvados, *Desl.*
11. —— lamellosus, *Sow.* Portland Stone, *Conyb.*
12. —— arcuatus, *Sow.* Coral Rag, Mid. and S. Eng., *Conyb.;* Portland
Beds, Kim. Clay, Haute Saone, *Thir.;* Kim. Clay, Bernese
Jura, *Thur.*
13. —— similis, *Sow.* Coral Rag, Mid. and S. Eng., *Conyb.;* Coral
Rag, Norm.? *De C.;* Great Oolite, Haute Saone, *Thir.*
14. —— laminatus, *Sow.* Cornb., Mid. and S. Eng., *Conyb.*
15. —— barbatus, *Sow.* Inf. Oolite, Dundry, *Conyb.;* Lias, Norm.,
De C.; Inf. Oolite, Lias, Env. of Bath, *Lons.*
16. —— vimineus, *Sow.* Oxford Clay, For. Marb., and Inf. Oolite,
Norm., *De C.;* Forest Marble, Malton, *Sow.;* Rubbly Limestone, &c., Braambury Hill, Brora, *Murch.;* Coral Rag, Haute
Saone, *Thir.;* Coral Rag, Yorksh. and Oxon, *Phil.;* Calc.
Grit, Oxf. Clay, Bernese Jura, *Thur.*
17. —— obscurus, *Sow.* Stonesfield, *Sow.;* For. Marb.? Mauriac, S. of
Fr., *Dufr.*
18. —— annulatus, *Sow.* Cornb., Felmersham, *Marsh.*
19. —— marginatus, Wasseralfingen, *Hœn.*
20. —— squamosus, *Von Buch.* Lias, Weissenburg, *G. T.*
21. —— Phillipsii, *Voltz.* Comp. Great Oolite, Bernese Jura, *Thur.*
22. —— paradoxus, *Munst.* Inf. Oolite, Bernese Jura, *Thur.*
23. —— striatus, *Sow.* Inf. Oolite, Bernese Jura, *Thur.*
24. —— corneus, . Great Oolite; Inf. Oolite, Calvados, *Desl.*
25. —— personatus, *Goldf.* Mistelgau; Wasseralfingen, *G. T.*
26. —— dentatus, *Sow.* Lias, Bollerbad, *G. T.*
27. —— contrarius, *Von Buch.* Lias, Wittberg; Metzingen, *G. T.*
28. —— canaliculatus, *Goldf.* Lias, Culmbach, *G. T.*
1. Monotis salinaria, *Bronn.* Regensberg, *G. T.*
2. —— similis, *Munst.* Pappenheim, *G. T.*
3. —— decussata, *Munst.* Hildesheim; Bükkeberg; Suntel; Wett-
bergen, *G. T.*
4. —— concinnus, *Goldf.* Minden; Wurtemberg, *G. T.*
1. Plagiostoma, læviusculum, *Sow.* Coralline Oolite, Yorks.; Coral Rag
and Calcareous Grit, Oxon, *Phil.;* Coral Rag, Marthon, S. of
Fr., *Dufr.*
2. —— rigidum, *Sow.* Coralline Oolite, Yorks.; Coral Rag, Oxon,
Phil.; Inf. Oolite, Dundry, *Conyb.;* Coral Rag, N. of Fr.,
Bobl.; Coral Rag, Haute Saone, *Thir.*
3. —— rusticum, *Sow.* Coralline Oolite, Yorks.; Calc. Grit, Oxon,
Phil.
4. —— duplicatum, *Sow.* Coralline Oolite, Oxford Clay, and Kell.
Rock, Yorks., *Phil.;* Inf. Oolite, Norm., *De C.;* Dunrobin
Oolite, Scotl., *Murch.;* Lias, Env. of Bath, *Lons.*
5. —— rigidulum, *Phil.* Cornbrash, Yorks., *Phil.*
6. —— interstinctum, *Phil.* Cornb. and Great Oolite, Yorks., *Phil.*

2 N

7. Plagiostoma cardiiforme, *Sow.*   Petty France, Gloucestershire, *Stein-hauer;* Great Oolite, Yorks., *Phil.;* Cornb. and For. Marb., N. of France, *Bobl.*
8. —— giganteum, *Sow.**   Inf. Oolite and Lias, Yorks., *Phil.;* Inf. Oolite, Dundry? Lias, Mid. and S. Eng., *Conyb.;* Lias, Norm., *De. C.;* Lias, N. of Fr., *Bobl.;* Lias, Western Islands, Scotl., *Murch.;* Inf. Oolite, Haute Saone, *Thir.;* Bahlingen, *Hœn.;* Lias, Malsch, near Heidelberg, *Bronn.*
9. —— obscurum, *Sow.* Kell. Rock, Mid. and S. Eng., *Conyb.*
10. —— pectinoïde, *Sow.* Lias, Yorks., *Phil.;* Shale and Grit, Reefs at Dunrobin, Scotl., *Murch.;* Lias, Vachingen; Waldhaenserhof, *G. T.*
11. —— punctatum, *Sow.* Inf. Oolite, Dundry. Lias, Mid. and S. England, *Conyb.;* For. Marb. and Inf. Oolite, Norm., *De C.;* Lias, N. of France, *Bobl.;* Lias, S. of France, *Dufr.;* Lias, Western Islands, Scotl., *Murch.;* Inf. Oolite, Bärendorf; Thurnau, *Munst.*
12. —— sulcatum. Lias, S. of France, *Dufr.*
13. —— ovale, *Sow.* For. Marb.? Mauriac, S. of France, *Dufr.*
14. —— Hermanni, *Voltz.* Lias, Alsace, *Voltz;* Lias, Env. of Bath, *Lons.;* Lias, Lyme Regis, *De la B.*
15. —— obliquatum, *Sow.* Sandstone and Limestone, Braambury Hill, Brora, Sandst., Limest., and Shale, Inverbrora, Scotl., *Murch.*
16. —— acuticostatum, *Sow.* Sandst., Limest., and Shale, Inverbrora, Scotl., *Murch.*
17. —— concentricum, *Sow.* Lias, Ross and Cromarty, Scotl., *Murch.*
18. —— transversum, *Von Buch.* Nipf, Bopfingen; Stufenberg, *G. T.*
——, species not determined. Bradford Clay and Great Oolite, Mid. and S. Eng., *Conyb.;* Lias, Gundershofen, *Voltz;* Kim. Clay, Bernese Jura, *G. T.*
1. Posidonia Bronni, *Goldf.* Lias, Ubstadt, near Bruchsal, *Hœn.*
1. Lima rudis, *Sow.* Coralline Oolite, Calc. Grit, Kell. Rock, and Great Oolite, Yorks., *Phil.;* Coral Rag, Mid. and S. Eng., *Conyb.;* Coral Rag, N. of Fr., *Bobl.;* Rubbly Limestone, &c., Braambury Hill, Brora, *Murch.*
2. —— proboscidea, *Sow.* Inf. Oolite? Yorks., *Phil.;* Inf. Oolite, Dundry, *Conyb.;* Oxford Clay, For. Marb., and Inf. Oolite, Norm., *De C.;* Inf. Oolite, Haute Saone, *Thir.;* Soleure, Bâle, *Hœn.;* Coral Rag, Weymouth, *Sedg.;* Inf. Oolite, Bärendorf; Thurnau, *Munst.;* Calc. Grit, Bernese Jura, *Thur.*
3. —— gibbosa, *Sow.* Cornb. and Inf. Oolite, Mid. and S. Eng., *Conyb.;* Great Oolite, Calvados, *Her.*
4. —— antiqua, *Sow.* Lias, Mid. and S. Eng., *Conyb.;* Lias, S. of France, *Dufr.;* Inf. Oolite, Haute Saone, *Thir.;* Elligser Brink, *G. T.*
5. —— heteromorpha, *Desl.* Inf. Oolite, Calvados, *Her.*
——, species not determined. Great Oolite, Wilts, *Lons.*
1. Avicula expansa, *Phil.* Coralline Oolite, Oxford Clay? Kell. Rock and Great Oolite, Yorks., *Phil.*
2. —— ovalis, *Phil.* Coralline Oolite and Calc. Grit, Yorks., *Phil.*
3. —— elegantissima, *Bean.* Coralline Oolite, Yorks., *Phil.*
4. —— tonsipluma, *Y. & B.* Coralline Oolite, Yorks., *Phil.*
5. —— Braamburiensis, *Sow.* Sandstone, Braambury Hill, Brora, *Murch.;* Kell. Rock, Great Oolite, and Inf. Oolite, Yorks., *Phil.;* Inf. Oolite, Bernese Jura, *Thur.*

---

* *Lima gigantea,* Deshayes. The same author considers that the genus *Plagiostoma* of Sowerby and Lamarck should be suppressed, the species composing it being referrible either to *Spondylus* or to *Lima.*

6. Avicula inæquivalvis, *Sow.* Inf. Oolite and Lias, Yorks., *Phil.*; Great
   Oolite and Inf. Oolite, Norm., *De C.*; Lias, S. of Fr., *Dufr.*;
   Great Arenaceous Formation, Western Islands; and Shell
   Limest. and Grit, Portgower, Scotland, *Murch.*; Lias, Lyme
   Regis, *De la B.*; Bahlinghen, *Hœn.*; Lias, Gandershofen,
   Voltz; Full. E., Inf. Oolite, and Lias, Env. of Bath, *Lons.*;
   Lias, Calvados, *Her.*; Wisgoldingen; Nipf, Bopfingen; Banz;
   Lias, Mögglingen; Baireuth, *G. T.*
7. ———— echinata, *Sow.* Lias? Yorks., *Phil.*; Cornb., Mid. and S. Eng.,
   Conyb.; For. Marb., Norm., *De C.*; Brad. Clay, Cornb., and
   For. Marb., N. of Fr., *Bobl.*; Great Oolite, Haute Saone,
   Thir.; Full. E., Env. of Bath, *Lons.*; Inf. Oolite, Bernese
   Jura, *Thur.*
8. ———— cygnipes, *Y. & B.* Lias, Yorks., *Phil.*; Lias, Western Islands,
   Scotl., *Murch.*
9. ———— costata, *Sow.* Cornb. and Brad. Clay, Mid. and S. Eng., Inf.
   Oolite, Dundry, *Conyb.*; For. Marb., Norm, *De C.*
10. ———— lanceolata, *Sow.* Lias, Lyme Regis, *De la B.*
11. ———— ovata, *Sow.* Stonesfield Slate, *Sow.*
12. ———— Munsteri, *Bronn.* Lias, Malsch, Heidelberg, *Bronn.*
1. Inoceramus dubius, *Sow.* Lias, Yorks., *Phil.*; Lias, Osnabrück; Gr.
   Gschaid, *G. T.*
1. Gervillia aviculoides, *Sow.* Coralline Oolite, Yorks., Calcareous Grit,
   Oxfordshire, *Phil.*; Oxford Clay, Mid. and S. Eng., Inf.
   Oolite, Dundry Hill, *Conyb.*; Oxford Clay, Norm., *De la B.*;
   Sandst., Limest., and Shale, Inverbrora, Scotl., *Murch.*; Lias,
   Gundershofen, *Voltz;* Coral Rag, Weymouth, *Sedg.*; Inf.
   Oolite, Bärendorf; Thurnau, *Munst.*; Calc. Grit, Bernese
   Jura, *Thur.*; Boll; Neuhausen, Germs; Graefenberg, Nürnberg, *G. T.*
2. ————? acuta, *Sow.* Collyweston, *Sow.*; Great Oolite, Yorks., *Phil.*
3. ———— lata, *Phil.* Inf. Oolite, Yorks., *Phil.*
4. ———— pernoides, *Desl.* Oxford Clay, For. Marb., Great Oolite, and
   Inf. Oolite, Norm., *De C.*
5. ———— siliqua, *Desl.* Oxf. Clay and For. Marb., Norm., *De C.*
6. ———— monotis, *Desl.* For. Marb., Norm., *De C.*
7. ———— costellata, *Desl.* For. Marb., Norm., *De C.*
————, species not stated. Coral Rag, Norm., *De C.*; Kim. Clay and
   Inf. Oolite, Haute Saone, *Thir.*; Kim. Clay, Bernese Jura,
   Thur.
1. Perna quadrata, *Sow.* Coralline Oolite, Kell. Rock, and Great Oolite,
   Yorks., *Phil.*; Cornb., Bulwick, *Sow.*
2. ———— mytiloides, *Lam.* Lias, Gundershofen, *Voltz;* Oxford Clay, Dives,
   Normandy, *Desh.*; Neuhausen, Germs; Wisgoldfingen; Bopfingen; Inf. Oolite, Metzingen; Kahleberg, Echte, *G. T.*
3. ———— plana, *Thur.* Kim. Clay, Bernese Jura, *Thur.*
————, species not determined. Oxford Clay, Yorks., *Phil.*
1. Crenatula ventricosa, *Sow.* Husband Bosworth, Leicestershire, *Conyb.*;
   Gloucestershire, *Sow.*; Lias, Yorks., *Phil.*
————, species not determined. Portland Stone, *Conyb.*
1. Trigonellites antiquatus, *Phil.* Coral. Oolite, Yorks., *Phil.*
2. ———— politus, *Phil.* Oxford Clay, Yorks., *Phil.*
1. Pinna lanceolata, *Sow.* Coralline Oolite and Calcareous Grit, Yorks.,
   Phil.; Inf. Oolite, Dundry, *Conyb.*; Lias, Norm., *De C.*;
   Oxford Clay, N. of Fr., *Bobl.*; Coral Rag, Weymouth, *Sedg.*;
   Soleure, *G. T.*
2. ———— mitis, *Phil.* Oxford Clay and Kell. Rock? Yorks., *Phil.*
3. ———— cuneata, *Bean.* Cornbrash and Great Oolite, Yorks., *Phil.*

4. Pinna Folium, *Y. & B.* Lias, Yorks., *Phil.*
5. —— pinnigena, . Coral Rag, For. Marb., and Inf. Oolite, Norm., *De C.*
6. —— granulata, *Sow.* Kim. Clay, Weymouth, *Sedg.*; Kim. Clay, Cahors, S. of Fr., *Dufr.*; Lias, Skye, *Murch.*
7. —— diluviana, . Lias, Kaltenthal; Plieningen, Stuttgardt, *G. T.*
   ——, species not determined. Inf. Oolite, Env. of Bath, *Lons.*
1. Mytilus cuneatus, *Phil.* Inf. Oolite, Yorks., *Phil.*
2. —— amplus, . Great Oolite, Norm., *De C.*
3. —— pectinatus, *Sow.* Kim. Clay, Weymouth, *Sedgwick;* Rochelle Limestone, *Dufr.*
4. —— sublævis, *Sow.* Cornb., Eng., *Sow.*
5. —— solenoïdes, . Kim. Clay, Cahors, S. of Fr., *Dufr.*
6. —— Jurensis, *Merian.* Kim. Clay, Bernese Jura, *Thur.*
   ——, species not determined. Coral Rag and Inf. Oolite, Mid. and S. Eng., *Conyb.*; Coral Rag, Norm., *De C.*; Portland beds, Haute Saone, *Thir.*
1. Modiola imbricata, *Sow.* Coralline Oolite? and Great Oolite, Yorks., *Phil.*; Cornb., Mid. and S. Eng., *Conyb.*; Cornb., Wilts, *Lons.*
2. —— ungulata, *Y. & B.* Coralline Oolite, Great Oolite, and Inf. Oolite, Yorks., *Phil.*
3. —— bipartita, *Sow.* Calc. Grit, Yorks., *Phil.*; Sandstone and Limestone, Braambury Hill, Brora, *Murch.*
4. —— cuneata, *Sow.* Oxford Clay, Kell. Rock? and Cornb., Yorks., *Phil.*; Inf. Oolite, Mid. and S. Eng., *Conyb.*; Lias, Norm., *De C.*; Lias, Western Islands, Scotl.; Sandst., Limest., and Shale, Inverbrora, Scotl., *Murch.*; Hohenstein; Ferriferous Oolite, Bavaria and Wurtemberg, *Munst.*; Comp. Great Oolite, Bernese Jura, *Thur.*
5. —— pulchra, *Phil.* Kell. Rock, Yorks., *Phil.*; Oolite, Sutherland.
6. —— plicata, *Sow.* Inf. Oolite, Yorks., *Phil.*; Cornb., Mid. and S. Eng., Inf. Oolite, Dundry, *Conyb.*; Portland Beds, Haute Saone, *Thir.*; Full. E., Somerset, *Lons.*; Portland Beds, Kim. Clay, Bernese Jura, *Thur.*
7. —— aspera, *Sow.* Inf. Oolite, Yorks., *Phil.*; Cornb. Mid. and S. Eng., *Conyb.*
8. —— Scalprum, *Sow.* Lias, Lyme Regis; Lias, Yorks., *Phil.*; Lias, S. of Fr., *Dufr.*
9. —— Hillana, *Sow.* Lias, Yorks., *Phil.*; Lias, Mid. and S. Eng., *Conyb.*; Full. E.? Env. of Bath, *Lons.*
10. —— lævis, *Sow.* Lias, Mid. and S. Eng., *Conyb.*
11. —— depressa, *Sow.* Lias, Mid. and S. Eng., *Conyb.*
12. —— minima, *Sow.* Lias, Mid. and S. Eng., *Conyb.*
13. —— tulipea, *Lam.* Oxford Clay, N. of Fr., *Bobl.*
14. —— pallida, *Sow.* Shale and Grit, Dunrobin Reefs, &c., Scotl., *Murch.*
15. —— gibbosa, *Sow.* Inf. Oolite, Env. of Bath, *Lons.*
16. —— livida, *Goldf.* Metz, *G. T.*
17. —— ventricosa, *Goldf.* Soleure, *Hœn.*
18. —— Thirriæ, *Voltz.* Kim. Clay, Bernese Jura, *Thur.*
19. —— striolaris, *Mérian.* Kim. Clay, Bernese Jura, *Thur.*
20. —— elegans, . Great Oolite, Calvados, *Her.*
21. —— æquiplicata, *Von Strombeck.* Kahleberg, Echte, *G. T.*
    ——, species not determined. Lias, Gundershofen, *Voltz;* Lias, Bath, *Lons.*
1. Lithodomus Sowerbii, *Thur.* Coral Rag, Bernese Jura, *Thur.*
   ——, species not determined. Inf. Oolite, N. of Fr., *Bobl.*; Inf. Oolite, Env. of Bath, *Lons.*

1. Chama mima, or Gryphæa mima, *Phil.* Coral. Oolite and Calc. Grit., Yorks., *Phil.*
2. ———? crassa, *Smith.* Bradford Clay, Mid. and S. Eng., *Conyb.*
3. ——— Berno-jurensis, *Thur.* Calc. Grit, Bernese Jura, *Thur.* ———, species not determined. For. Marb., Cornb., and Brad. Clay, Wilts, *Lons.*
1. Unio peregrinus, *Phil.* Cornb., Yorks., *Phil.*
2. ——— abductus, *Phil.* Inferior Oolite and Lias, Yorks., *Phil.*
3. ——— concinnus, *Sow.* Lias, Yorks., *Phil.;* Inf. Oolite, Mid. and S. Eng., *Conyb.;* Inf. Oolite, Lias, Env. of Bath, *Lons.;* Lias, Mögglingen, Gmündt, *G. T.*
4. ——— crassiusculus, *Sow.* Lias, Yorks., *Phil.*
5. ——— Listeri, *Sow.* Lias, Yorks., *Phil.;* Inf. Oolite, Mid. and S. Eng., *Conyb.*
6. ——— crassissimus, *Sow.* Lias, Mid. and S. Eng., *Conyb.;* Lias, Norm., *De C.;* For. Marb.? Mauriac, and Inf. Oolite, Uzer, S. of Fr., *Dufr.*
1. Trigonia costata, *Sow.* Coralline Oolite, Great Oolite, and Inf. Oolite, Yorks., *Phil.;* Cornb., For. Marb., and Brad. Clay, Mid. and S. Engl., Inf. Oolite, Dundry, *Conyb.;* Oxford Clay, For. Marb., and Inf. Oolite, Norm., *De C.;* Oxford Clay, N. of Fr., *Bobl.;* Kim. Clay and Inf. Oolite, Haute Saone, *Thir.;* Lias, Gundershofen, *Voltz;* Inf. Oolite, Env. of Bath, *Lons.;* Coral Rag, Weymouth, *Sedg.;* Porta Westphalica; Hohenstein, *Munst.;* Banz, *G. T.*
2. ——— clavellata, *Sow.* Coralline Oolite, Kell. Rock, and Cornb., Yorks., *Phil.;* Portland Stone and Cornb., Mid. and S. Engl., Inf. Oolite, Dundry, *Conyb.;* Oxford Clay, Norm., *De la B.;* Oxford Clay, N. of Fr., *Bobl.;* Kim. Clay? Angoulême, *Dufr.;* Sandst., Shale, &c., Inverbrora, Scotl., *Murch.;* Coral Rag and Inf. Oolite, Haute Saone, *Thir.;* Coral Rag, Weymouth, *Sedg.;* Kim. Clay, Calc. Grit, Bernese Jura, *Thur.;* Wisgoldingen; Stufenberg; Ehningen Sandstone, *G. T.*
3. ——— conjungens, *Phil.* Great Oolite, Yorks., *Phil.*
4. ——— striata, *Sow.* Inferior Oolite, Yorks., *Phil.;* Inf. Oolite, Dundry, *Conyb.;* Inf. Oolite, Norm., *De C.;* Lias, S. of Fr., *Dufr.*
5. ——— angulata, *Sow.* Inf. Oolite, Yorks., *Phil.;* Inf. Oolite, near Frome, *Sow.*
6. ——— literata, *Y. & B.* Lias, Yorks., *Phil.*
7. ——— gibbosa, *Sow.* Portland Stone, *Conyb.;* Forest Marb., Norm., *De C.*
8. ——— duplicata, *Sow.* Inf. Oolite, Mid. and S. Eng., *Conyb.;* For. Marb., Norm., *De C.*
9. ——— elongata, *Sow.* Oxford Clay, Norm., *De C.;* Oxford Clay, Eng., *Sow.;* Great Oolite, Alsace, *Voltz;* Cornb., Wilts, *Lons.*
10. ——— imbricata, *Sow.* Great Oolite, Ancliff, Wilts, *Cookson.*
11. ——— cuspidata, *Sow.* Great Oolite, Ancliff, *Cookson;* var. Coral Rag, Haute Saone, *Thir.*
12. ——— Pullus, *Sow.* Great Oolite, Ancliff, *Cookson.*
13. ——— navis, *Lam.* Lias, Gundershofen, *Voltz.*
14. ——— incurva, *Benett.* Portland Beds, Tisbury, Wilts, *Benett.*
15. ——— nodulosa, *Lam.* Kim. Clay, Hâvre; Bray, *Pas.* ———, species not determined. Coral Rag, Midl. and S. Eng., *Conyb.;* Coral Rag, Norm., *De C.*
1. Nucula elliptica, *Phil.* Oxford Clay, Yorks., *Phil.*
2. ——— nuda, *Y. & B.* Oxford Clay, Yorks., *Phil.*
3. ——— variabilis, *Sow.* Great Oolite and Inf. Oolite, Yorks., *Phil.;* Great Oolite, Ancliff, near Bath, *Cookson.*

4. Nucula Lachryma, *Sow.*   Great Oolite and Inf. Oolite, Yorks., *Phil.;*
     Great Oolite, Ancliff, *Sow.*
5. —— axiniformis, *Phil.*   Inferior Oolite, Yorks., *Phil.*
6. —— Ovum, *Sow.*   Lias, Yorks., *Phil.*
7. —— pectinata, *Sow.*   Oxford Clay, Norm., *De C.;* Brad. Clay?
     Wilts, *Lons.*
8. —— clariformis,   .   Lias, S. of Fr., *Dufr.*
9. —— mucronata, *Sow.*   Great Oolite, Ancliff, Wilts, *Cookson;* Banz,
     G. T.
10. —— Stahlii, *Bronn.*   Lias, Ubstadt, near Heidelberg, *Bronn.*
11. —— acuminata, *Mérian.*   Oxf. Clay, Bernese Jura, *Thur.*
12. —— medio-jurensis, *Thur.*   Oxf. Clay, Bernese Jura, *Thur.*
13. —— Hammeri, *Defr.*   Gundershofen, *G. T.*
14. —— lobata, *Von Buch.*   Metzingen; Nipf., Bopfingen, *G. T.*
15. —— subovalis, *Goldf.*   Wasseralfingen, *G. T.*
16. —— rostrata, *Goldf.*   Wasseralfingen, *G. T.*
17. —— elongata, *Goldf.*   Wasseralfingen, *G. T.*
18. —— arcacea, *Goldf.*   Banz; Baireuth, *G. T.*
—— , species not determined.   Coralline Oolite, Yorks., *Phil.;* Inf.
     Oolite, Dundry; Lias, Mid. and S. Eng., *Conyb.;* Kim. Clay,
     Bernese Jura, *Thur.;* Lias, Bahlingen, *G. T.*
1. Pectunculus minimus, *Sow.*   Great Oolite, Ancliff, Wilts, *Cookson.*
2. —— oblongus, *Sow.*   Great Oolite, Ancliff, Wilts, *Cookson.*
1. Arca quadrisulcata, *Sow.*   Coral Rag, Malton, *Sow.*   Coralline Oolite,
     Yorks., *Phil.*
2. —— æmula, *Phil.*   Coralline Oolite, Yorks., *Phil.*
3. —— pulchra, *Sow.*   Great Oolite, Ancliff, Wilts, *Cookson;* Rochelle
     Limestone, *Dufr.*
4. —— trigonella,   .   Wasseralfingen, *Hœn.*
5. —— elongata,   .   Wasseralfingen, *Hœn.*
6. —— rostrata,   .   Wasseralfingen, *Hœn.*
7. —— medio-jurensis, *Thur.*   Oxf. Clay, Bernese Jura, *Thur.*
—— , species not determined.   Lias, Mid. and S. Eng., *Conyb.;* Brad.
     Clay, Wilts; Full. E., Inf. Oolite, Env. of Bath, *Lons.;* Kim.
     Clay, Bernese Jura, *Thur.;* Lias, near Heidelberg, *Bronn.*
1. Cucullæa oblonga, *Sow.*   Coralline Oolite, Yorks., *Phil.;* Inf. Oolite,
     Dundry, *Conyb.;* Inf. Oolite, Bärendorf; Thurnau, *Munst.*
2. —— contracta, *Phil.*   Coralline Oolite, Yorks., *Phil.*
3. —— triangularis, *Phil.*   Coralline Oolite, Yorks., *Phil.*
4. —— pectinata, *Phil.*   Coralline Oolite, Yorks., *Phil.*
5. —— elongata, *Sow.*   Coralline Oolite? and Great Oolite, Yorks.,
     Phil.; Rochelle limestone, *Dufr.*   Cross Hands, Gloucester-
     shire, *Steinhauer.;* Winzingen, Wisgoldingen, *G. T.*
6. —— concinna, *Phil.*   Oxford Clay and Kell. Rock? Yorks., *Phil.*
7. —— imperialis, *Bean.*   Great Oolite, Yorks., *Phil.*
8. —— cylindrica, *Phil.*   Great Oolite, Yorks., *Phil.*
9. —— cancellata, *Phil.*   Great Oolite, Yorks., *Phil.*
10. —— reticulata, *Bean.*   Inf. Oolite, Yorks., *Phil.*
11. —— minuta, *Sow.*   Great Oolite, Ancliff, Wilts, *Cookson.*
12. —— rudis, *Sow.*   Great Oolite, Ancliff, Wilts, *Cookson.*
13. —— parvula, *Munst.*   Oxf. Clay, Bernese Jura, *Thur.*
?14. —— decussata, *Sow.*   Mistelbach, Baireuth, *G. T.*
—— , species not determined.   Oxford Clay, Haute Saone, *Thir.* Lias,
     Yorks., *Phil.;* Lias, Mid. and S. Eng., *Conyb.*
1. Hippopodium ponderosum, *Sow.*   Coralline Oolite and Lias, Yorks.,
     Phil.; Lias, Mid. and S. Eng., *Conyb.;* Inf. Oolite, Calva-
     dos, *Desl.*
1. Isocardia rhomboidalis, *Phil.*   Coralline Oolite, Yorks., *Phil.*

2. Isocardia tumida, *Phil.*   Calc. Grit., Yorks., *Phil.*
3. ———— minima, *Sow.* Cornb. and Great Oolite? Yorks., *Phil.; Cornb.*,
    Wilts, *Lons.*
4. ———— concentrica, *Sow.* Great Oolite and Inf. Oolite, Yorks., *Phil.;*
    Cornb., Northamptonshire, *Sow.;* Full. E., Somerset, *Lons.;*
    Inf. Oolite, Calvados, *Desl.*
5. ———— angulata, *Phil.* Great Oolite? Yorks., *Phil.*
6. ———— rostrata, *Sow.* Gloucestershire, *Sow.* Inf. Oolite, Yorks., *Phil.;*
    Comp. Great Oolite, Bernese Jura, *Thur.*
7. ———— striata, *D'Orb.* Kim. Clay, Portland Beds, Haute Saone, *Thir.;*
    Portland Beds, Kim. Clay, Bernese Jura, *Thur.*
8. ———— excentrica, *Voltz.* Portland Beds, Kim. Clay, Bernese Jura,
    Thur.
9. ———— inflata, *Voltz.* Portland Beds, Kim. Clay, Bernese Jura, *Thur.*
10. ———— carinata, *Voltz.* Kim. Clay, Bernese Jura, *Thur.*
11. ———— costulata, *Voltz.* Kim. Clay, Bernese Jura, *Thur.*
———, species not determined.   For. Marb., Norm., *De C.*
1. Cardita similis, *Sow.* Coralline Oolite, Great Oolite, and Inf. Oolite,
    Yorks., *Phil.;* Inf. Oolite, Dundry, *Conyb.*
2. ———— lunulata, *Sow.* Inf. Oolite, Dundry, *Conyb.;* Inf. Oolite, Norm.,
    De C.
3. ———— striata, *Sow.* Lias, Norm.? *De C.* Inf. Oolite, Bath, *Sow.*
———, species not determined.   Portland Stone, *Conyb.*
1. Cardium lobatum, *Phil.* Coralline Oolite, Yorks., *Phil.*
2. ———— dissimile, *Sow.* Kell. Rock, Yorks., *Phil.;* Portland Stone,
    Portland, *Sow.;* Rocks of the Oolite series, Braambury Hill,
    Brora, *Murch.*
3. ———— citrinoideum, *Phil.* Cornb., Yorks., *Phil.*
4. ———— cognatum, *Phil.* Great Oolite, Yorks., *Phil.*
5. ———— acutangulum, *Phil.* Great Oolite and Inf. Oolite, Yorks., *Phil.*
6. ———— semiglabrum, *Phil.* Great Oolite, Yorks., *Phil.*
7. ———— incertum, *Phil.* Inf. Oolite, Yorks., *Phil.*
8. ———— striatulum, *Sow.* Sandst., Limest. and Shale, Inverbrora, Scot-
    land, *Murch.;* Inferior Oolite, Yorks., *Phil.*
9. ———— gibberulum, *Phil.* Inf. Oolite, Yorks., *Phil.*
10. ———— truncatum, *Sow.* Lias, Yorks., *Phil.;* Sandst., Limest., &c.,
    Inverbrora, *Murch.*
11. ———— multicostatum, *Bean.* Lias, Yorks., *Phil.*
1. Myoconcha crassa, *Sow.* Inf. Oolite, Dundry, *Brackenridge;* Inf. Oolite,
    Normandy, *De C.*
1. Astarte * cuneata, *Sow.* Portland Stone, S. Eng.; Inf. Oolite? Dundry,
    Conyb.
2. ———— excavata, *Sow.* Inf. Oolite, Dundry, *Conyb.;* Inf. Oolite, Norm.,
    De C.; Bopfingen; Lauchheim, Banz, *G. T.*
? 3. ———— planata, *Sow.* Inf. Oolite, Norm., *De C.;* Bradf. Clay, N. of
    Fr., *Bobl.*
4. ———— trigonalis, *Sow.* Inf. Oolite, Dundry, *Johnstone.*
5. ———— orbicularis, *Sow.* Great Oolite, Ancliff, Wilts, *Cookson.*
6. ———— pumila, *Sow.* Great Oolite, Ancliff, Wilts, *Cookson;* Rochelle
    Limestone, *Dufr.*
7. ———— Voltzii, *Hœn.* Fullon, near Vesoul, *Hœn.;* Banz, *G. T.*
8. ———— medio-jurensis, *Thur.* Oxf. Clay, Bernese Jura, *Thur.*
9. ———— modiolaris, *Goldf.* Normandy; Wurtemberg, *G. T.*
10. ———— ovata, *Smith.* Coralline Oolite, Wilts; Oxon; Yorks., *Phil.*
11. ———— elegans, *Sow.* Coralline Oolite and Inf. Oolite, Yorks., *Phil.;*
    Rochelle Limest., *Dufr.;* Shell Limest. and Calc. Grit, Port-

---

* *Crassina* of Phillips is the *Astarte* of Sowerby.

gower, &c., *Murch.;* Limest., Shale, Sandst., Inverbrora, *Murch.*

12. Astarte aliena, *Phil.* Coralline Oolite, Yorks., *Phil.*
13. —— extensa, *Phil.* Coralline Oolite, Yorks., *Phil.*
14. —— carinata, *Phil.* Calc. Grit, Oxford Clay, and Kell. Rock, Yorks., *Phil.*
15. —— lurida, *Sow.* Inf. Oolite, Dundry, *Conyb.;* Oxford Clay, Yorks., *Phil.*
16. —— minima, *Phil.* Great Oolite, Inf. Oolite, and Lias, Yorks., *Phil.;* Kim. Clay, and Coral Rag, Haute Saone, *Thir.*
—— , species not determined. Lias, Mid. and S. Eng., *Conyb.;* Coral Rag and Kim. Clay, Haute Saone, *Thir.;* Cornb., Wilts, *Lons.*
1. Venus varicosa, *Sow.* Felmersham, *Sow.*
—— , species not determined. Coral. Oolite, Calc. Grit and Lias, Yorks., *Phil.;* Portland Stone, *Smith;* Coral Rag, Norm., *De C.;* Sandst., Shale, &c., Inverbrora, Scotl., *Murch.*
1. Cytherea dolabra, *Phil.* Great Oolite, Yorks., *Phil.*
2. —— trigonellaris, *Voltz.* Lias, Gundershofen, *Voltz.*
3. —— lucinea, *Voltz.* Lias, Gundershofen, *Voltz.*
4. —— cornea, *Voltz.* Lias, Gundershofen, *Voltz.*
—— , species not determined. Coralline Oolite, Yorks., *Phil.;* Lias, N. of Fr., *Bobl.;* Kim. Clay, Bernese Jura, *Thur.*
1. Pullastra recondita, *Phil.* Great Oolite, Yorks., *Phil.*
2. —— oblita, *Phil.* Inf. Oolite, Yorks., *Phil.*
—— , species not determined. Lias, Yorks., *Phil.*
1. Donax Alduini, *Al. Brong.* Inf. Oolite? N. of Fr., *Bobl.;* Kim. Clay, Hâvre and the Jura, *Al. Brong.;* Kim. Clay, Bernese Jura, *Thur.;* Nipf, Bopfingen; Rautenberg, Scheppenstadt; Kahleberg, Echte, *G. T.*
2. —— Saussurii, *Al. Brong.* Kim. Clay, Bernese Jura, *Thur.;* Kahleberg, Echte, *G. T.*
1. Corbis lævis, *Sow.* Coralline Oolite? Kell. Rock? Yorks., *Phil.;* Marsham Field, Oxford, *Smith.*
2. —— ovalis, *Phil.* Kell. Rock, Yorks., *Phil.*
3. —— uniformis, *Phil.* Lias, Yorks., *Phil.*
1. Tellina ampliata, *Phil.* Coralline Oolite, Yorks., *Phil.*
2. —— incerta, *Thur.* Kim. Clay, Bernese Jura, *Thur.*
1. Psammobia lævigata, *Phil.* Coralline Oolite, Great Oolite, and Inf. Oolite, Yorks., *Phil.*
1. Lucina crassa, *Sow.* Sandstone and Rubbly Limestone, Braambury Hill, Brora; Great Arenaceous Formation, Western Islands, Scotl., *Murch.;* Calc. Grit, Yorks., *Phil.;* Lincolnshire, *Sow.*
2. —— lyrata, *Phil.* Kell. Rock, Yorks., *Phil.*
3. —— despecta, *Phil.* Great Oolite, Yorks., *Phil.*
4. —— Elsgaudiæ, *Thur.* Kim. Clay, Bernese Jura, *Thur.*
—— , species not stated. Coral Rag and For. Marb., Norm., *De Cau.;* Inf. Oolite, Yorks., *Phil.;* Shale, &c. Inverbrora, Scotl., *Murch.*
1. Sanguinolaria undulata, *Sow.* Sandst., Limest., and Shale, Inverbrora, Scotl., *Murch.;* Calc. Grit, Oxford Clay, and Cornbrash, Yorks., *Phil.*
2. —— elegans, *Phil.* Lias, Yorks., *Phil.*
—— , species not determined. Lias, Ross and Cromarty, Scotl., *Murch.;* Lias, Yorks., *Phil.*
1. Corbula curtansata, *Phil.* Coralline Oolite and Kell. Rock, Yorks., *Phil.*
2. —— depressa, *Phil.* Great Oolite, Yor ks., *Phil.*

3. Corbula? cardioides, *Phil.* Lias, Yorks., *Phil.;* Lias, Ofterdingen, *G. T.*
4. —— obscura, *Sow.* Brora, *Murch.*
——, species not determined. For. Marb., Wilts, *Lons.;* Kim. Clay, Bernese Jura, *Thur.*
1. Mactra gibbosa, For. Marb., Norm, *De C.*
1. Amphidesma decurtatum, *Phil.* Cornb. and Great Oolite, Yorks, *Phil.;* Kim. Clay? and Great Oolite? Haute Saone, *Thir.*
2. —— recurvum, *Phil.* Coralline Oolite? and Kell. Rock, Yorks., *Phil.;* Kim. Clay, Hâvre, *Phil.*
3. —— securiforme, *Phil.* Cornb., Inf. Oolite, Yorks., *Phil.;* Kim. Clay, Hâvre, *Phil.*
4. —— donaciforme, *Phil.* Lias, Yorks., *Phil.*
5. —— rotundatum, *Phil.* Lias, Yorks., *Phil.;* Lias, Bahlingen; Waldhaenserhof, *G. T.*
1. Lutraria Jurassi, *Brong.* For. Marb., Ligny, Meuse, *Brong.*
1. Gastrochæna tortuosa, *Sow.* Inf. Oolite, Yorks., *Phil.*
1. Mya literata, *Sow.* Coralline Oolite, Calc. Grit, Oxford Clay, Kelloway Rock, Cornb., Inf. Oolite, and Lias, Yorks., *Phil.;* Shale, Sandstone, and Limestone, Inverbrora, Scotl., *Murch.*
2. —— depressa, *Sow.* Oxford Clay? Yorks., *Phil.;* Kim. Clay? Angoulême, *Dufr.;* Kim. Clay, Hâvre, *Phil.;* Shale, Limestone, and Sandstone, Inverbrora, Scotl., *Murch.*
3. —— calceiformis, *Phil.* Kell. Rock, Great Oolite, aud Inf. Oolite, Yorks., *Phil.*
4. —— dilata, *Phil.* Inf. Oolite, Yorks, *Phil.*
5. —— æquata, *Phil.* Inf. Oolite, Yorks., *Phil.*
6. —— V scripta, *Sow.* Inf. Oolite, Dundry, *Conyb.;* Great Oolite, Alsace, *Brong.;* Micaceous Sandstone, Western Islands, Scotl., *Murch.*
7. —— Mandibula, *Sow.* Kim. Clay? Env. of Angoulême, *Dufr.*
8. —— angulifera, *Sow.* Great Oolite, Haute Saone, *Thir.;* Lias, Alsace, *Voltz;* Fuller's Earth, Environs of Bath, *Lons.;* Calc. Grit, Bernese Jura, *Thur.*
1. Pholadomya Murchisoni, *Sow.* Sandstone, Limestone, and Shale, Inverbrora, Scotl., *Murch.;* Coralline Oolite? and Cornbrash, Yorks., *Phil.;* Inf. Oolite, Normandy, *De Cau.;* Kim. Clay, Bernese Jura, *Thur.;* Nipf, Bopfingen; Metzingen; Porta Westphalica, *G. T.*
2. —— simplex, *Phil.* Calc. Grit, Yorks., *Phil.*
3. —— deltoidea, *Sow.* Calc. Grit, Yorks., *Phil.;* Kell. Rock and Cornbrash, Midl. and S. Engl., *Conyb.*
4. —— obsoleta, *Phil.* Oxford Clay and Kell. Rock, Yorks., *Phil.*
5. —— ovalis, *Sow.* Cornbrash, Yorks., *Phil.;* Portland Stone, *Conyb.;* Oxford Clay, Normandy, *DeC.;* Kim. Clay? Angoulême; Rochelle Limestone, *Dufr.*
6. —— acuticostata, *Sow.* Great Oolite, Yorks, *Phil.;* Kim. Clay, Cahors, S. of Fr., *Dufr.;* Kim. Clay? Angoulême, *Dufr.;* Kim. Clay, Haute Saone, *Thir.;* Brora, *Farey;* Portland Beds, Kim. Clay, Bernese Jura, *Thur.*
7. —— nana, *Phil.* Great Oolite, Yorks., *Phil.*
8. —— producta, *Sow.* Great Oolite? Yorks., *Phil.;* Cornb. and Inf. Oolite, Midl. and S. Engl., *Conyb.;* Cornb., Wilts, *Lons.*
9. —— obliquata, *Phil.* Great Oolite, Inf. Oolite, and Lias, Yorks., *Phil.*
10. —— fidicula, *Sow.* Inf. Oolite, Yorks., *Phil.;* Cornb., Mid. and S. of Eng.; Inf. Oolite, Dundry, *Conyb.;* Lias, Norm., *De C.;* Cornb., Wilts; Full. E., Env. of Bath, *Lons.;* Soleure, *Hœn.;* Inf. Oolite, Haute Saone, *Thir.*

11. Pholadomya obtusa, *Sow.* Inf. Oolite, Dundry, *Conyb.*
12. ———— ambigua, *Sow.* Inf. Oolite, Dundry, *Conyb.;* Oxford Clay, Norm., *De C.;* Lias, S. of Fr., *Dufr.;* Lias, Alsace, *Voltz;* Lias, Bath, *Lons.;* Lias, Soleure; Lias Bahlingen, *Von Buch.*
13. ———— æqualis, *Sow.* Weymouth, *Sow.;* Inf. Oolite, Norm., *De C.;* Lübbecke, Minden, *G. T.*
14. ———— gibbosa, . Lias, Norm., *De C.;* Soleure, *Hœn.*
15. ———— Protei, *Brong.* Rochelle Limest., *Dufr.;* Kim. Clay, Hâvre and the Jura, *Brong.;* Portland Beds and Kim. Clay, Haute Saone, *Thir.;* Kim. Clay, Bernese Jura, *Thur.;* Kahleberg, Echte, *G. T.*
16. ———— clathrata, *Munst.* Bavaria; Hohenstein, Saxony, *Munst.*
17. ———— angustata, *Sow.* Kim. Clay, Calc. Grit, Bernese Jura, *Thur.*
18. ———— cardiiformis, *Goldf.* Mammers, *G. T.*
19. ———— concentrica, *Goldf.* Soleure, *G. T.*
20. ———— decussata, *Goldf.* Wurtemberg, *G. T.*
————, species not determined. Oxford Clay, Haute Saone, *Thir.;* Oxford Clay, Bernese Jura, *Thur.*
1. Panopæa gibbosa, *Sow.* Great Oolite? Yorks., *Phil.;* Inf. Oolite, Dundry, *Conyb.*
1. Pholas recondita, *Phil.* Coralline Oolite, Yorks., *Phil.*
2. ————? compressa, *Sow.* Kim. Clay, Oxford, *G. E. Smith.*

MOLLUSCA.

1. Dentalium giganteum, *Phil.* Lias, Yorks., *Phil.*
2. ———— cylindricum, *Sow.* Lias, Mid. and S. Eng., *Conyb.*
————, species not determined. Calc. Grit, Yorks., *Phil.*
1. Patella latissima, *Sow.* Oxford Clay, Yorks., *Phil.;* Oxford Clay, Mid. and S. of Eng., *Conyb.*
2. ———— rugosa, *Sow.* For. Marb., Mid. and S. of Eng., *Conyb.;* For. Marb., Norm., *De C.*
3. ———— lævis, *Sow.* Lias, Mid. and S. of Eng., *Conyb.*
4. ———— lata, *Sow.* Stonesfield Slate, *Sow.*
5. ———— ancyloïdes, *Sow.* Great Oolite, Ancliff, Wilts, *Cookson.*
6. ———— nana, *Sow.* Great Oolite, Ancliff, Wilts, *Cookson.*
7. ———— discoides, *Schlot.* Lias, Gundershofen, *Voltz.*
8. ———— papyracea, *Goldf.* Lias, Banz, *G. T.*
1. Emarginula scalaris, *Sow.* Great Oolite, Ancliff, Wilts, *Cookson.*
1. Pileolus plicatus, *Sow.* Great Oolite, Wilts, *Lons.*
1. Bulla elongata, *Phil.* Coral. Oolite, Yorks., *Phil.*
1. Helicina polita, *Sow.** Inf. Oolite, Cropredy, *Conyb.*
2. ———— expansa, *Sow.* Lias, Mid. and S. of Eng., *Conyb.*
3. ———— solarioides, *Sow.* Lias, Mid. and S. of Eng., *Conyb.*
1. Auricula Sedgvici, *Phil.* Inf. Oolite, Yorks., *Phil.*
1. Melania Heddingtonensis, *Sow.* Coral. Oolite, Cornb., Great Oolite and Inf. Oolite, Yorks., *Phil.;* Coral Rag, Mid. and S. of Eng.; Inf. Oolite, Dundry, *Conyb.;* Coral Rag and Inf. Oolite, Norm., *De C.;* Rubbly Limest., &c., Braambury Hill, Brora, *Murch.;* Kim. Clay, Hâvre, *Phil.;* Inf. Oolite? Haute Saone, *Thir.;* Coral Rag, Weymouth, *Sedg.;* Kelheim; Kahleberg, Echte, *G. T.*
2. ———— striata, *Sow.* Coral. Oolite and Great Oolite? Yorks., *Phil.;* Coral Rag and Lias, Mid. and S. of Eng., *Conyb.;* Coral Rag, N. of Fr., *Bobl.;* Kim. Clay, Hâvre, *Phil.;* Coral Rag, Weymouth, *Sedg.*
3. ———— vittata, *Phil.* Cornb., Yorks., *Phil.*

* *Turbo callosus*, Deshayes.

4. Melania lineata, *Sow.* Inf. Oolite, Yorks., *Phil.*; Inf. Oolite, Dundry,
   Conyb.; Inf. Oolite, Norm., *De C.*
5. —— medio-jurensis, *Thur.* Oxf. Clay, Bernese Jura, *Thur.*
   ——, species not determined. Great Oolite, Mid. and S. of Eng.,
   Conyb.
   Paludina, species not determined. Portland Beds, Haute Saone,
   Thir.
1. Ampullaria Gigas, *Von Strombeck.* Kahleberg, Echte, *G. T.*
   ——, species not determined. Coral Rag, Cornb., and Inf. Oolite,
   Mid. and S. Eng., *Conyb.*; Coral Rag, Norm., *De C.*;
   Brad. Clay, N. of Fr., *Bobl.*; Kim. Clay, Bernese Jura,
   Thur.
1. Nerita costata, *Sow.* Inf. Oolite, Yorks., *Phil.*; Great Oolite, Ancliff,
   Wilts, *Cookson.*
2. —— sinuosa, *Sow.* Portland Stone, *Conyb.*
3. —— lævigata, *Sow.* Inf. Oolite, Dundry, *Conyb.*; Shell Limestone
   and Calc. Grit, Portgower, &c., Scotland, *Murch.*
4. —— minuta, *Sow.* Great Oolite, Ancliff, Wilts, *Cookson.*
1. Natica arguta, *Smith.* Coral. Oolite, Yorks., *Phil.*
2. —— nodulata, *Y. & B.* Coral. Oolite, Yorks., *Phil.*
3. —— cincta, *Phil.* Coral. Oolite, Yorks., *Phil.*
4. —— adducta, *Phil.* Great Oolite and Inf. Oolite, Yorks., *Phil.*
5. —— tumidula, *Bean.* Inf. Oolite, Yorks., *Phil.*
   ——, species not determined. Lias, Yorks., *Phil.*; Kim. Clay, Bernese Jura, *Thur.*
1. Vermetus compressus, *Y. & B.* Coral. Oolite, Inf. Oolite, Yorks.,
   Phil.
2. —— Nodus, *Phil.* Cornb., Great Oolite, Yorks., *Phil.*
   ——, species not determined. Cornbrash, Wilts, *Lons.*
   Delphinula, species not determined. Coral. Oolite and Great Oolite,
   Yorks., *Phil.*
1. Solarium Calix, *Bean.* Inf. Oolite, Yorks., *Phil.*
2. —— conoideum, *Sow.* Portland Stone, *Conyb.*
1. Cirrus cingulatus, *Phil.* Calc. Grit, Yorks., *Phil.*
2. —— depressus, *Sow.* Kell. Rock, Yorks., *Phil.*
3. —— nodosus, *Sow.* Inf. Oolite, Dundry, *Conyb.*
4. —— Leachii, *Sow.* Inf. Oolite, Dundry, *Conyb.*
5. —— carinatus, *Sow.* Inf. Oolite, Wilts, *Lons.*
   ——, species undetermined. Lias, N. of Fr., *Bobl.*; Oxford Clay, Haute
   Saone, *Thir.*
1. Pleurotomaria conoidea, *Desh.* Normandy, *Desh.*
2. —— ornata, *Defr.* Inf Oolite, Bayeux, *Desh.*; Inf. Oolite, Dundry,
   Conyb.; Inf. Oolite, Norm., *De C.*; Lias, N. of France, *Bobl.*
3. —— decorata, *Von Buch.* Neuhausen, *G. T.*
4. —— compressa, *Sow.* Lias, Southern England, *Conyb.*
1. Trochus arenosus, *Sow.* Coral. Oolite, Calc. Grit, Cornb., and Inf.
   Oolite, Yorks., *Phil.*; Inf. Oolite, Dundry, *Conyb.*; Inf. Oolite,
   Norm., *De C.*
2. ——? tornatilis, *Phil.* Coral. Oolite, Yorks., *Phil.*
3. —— Tiara, *Sow.* Calc. Grit, Yorks., *Phil.*; Coral Rag, Mid. and S.
   Eng., Inf. Oolite, Dundry, *Conyb.*; Inf. Oolite, Norm., *De C.*
4. —— guttatus, *Phil.* Kell. Rock, Yorks., *Phil.*
5. —— monilitectus, *Phil.* Great Oolite, Yorks., *Phil.*
6. —— bisertus, *Phil.* Inf. Oolite, Yorks., *Phil.*
7. —— pyramidatus, *Bean.* Inf. Oolite, Yorks., *Phil.*
8. —— Anglicus, *Sow.* Lias, Yorks., *Phil.*; Lias, Mid. and S. Eng.,
   Conyb.; Inf. Oolite, Haute Saone, *Thir.*; Stufenberg; Hei-
   ningen, *G. T.*

9. Trochus angulatus, *Sow.* Inf. Oolite, Mid. and S. Eng., *Conyb.;* Inf.
Oolite, Norm.
10. ——— dimidiatus, *Sow.* Inf. Oolite, Mid. and S. Eng., *Conyb.*
11. ——— duplicatus, *Sow.* Inf. Oolite, Mid. and S. Eng., *Conyb.;* Inf.
Oolite, Haute Saone, *Thir.;* Lias, Gundershofen, *Voltz.;*
Banz, *G. T.*
12. ——— elongatus, *Sow.* Inf. Oolite, Dundry, *Conyb.;* For. Marb. and
Inf. Oolite, Norm., *De C.*
13. ——— punctatus, *Sow.* Inf. Oolite, Dundry, *Conyb.;* Inf. Oolite,
Norm., *De C.*
14. ——— abbreviatus, *Sow.* Inf. Oolite, Dundry, *Conyb.;* Inf. Oolite,
Norm., *De C.*
15. ——— fasciatus, *Sow.* Inf. Oolite, Dundry, *Conyb.;* Inf. Oolite, Norm.,
De C.
16. ——— prominens, *Sow.* Inf. Oolite, Dundry, *Conyb.;* Inf. Oolite,
Norm., *De C.*
17. ——— imbricatus, *Sow.* Lias, Mid. and S. Eng., *Conyb.;* Inf. Oolite,
Norm., *De C.;* Lias, S. of Fr., *Dufr.;* Soleure, *Hœn.*
18. ——— reticulatus, *Sow.* Inf. Oolite, Norm., *De C.;* Coral Rag, Weymouth, *Sedg.*
19. ——— rugatus, *Benett.* Portland Beds, Tisbury, Wilts, *Benett.*
20. ——— speciosus, *Munst.* Hohenstein, Saxony; Inf. Oolite, Bavaria,
Munst.
21. ——— niloticiformis, *Stahl.* Stufenberg, *G. T.*
———, species not determined. Portland Stone and Bradford Clay,
Mid. and S. Eng., *Conyb.;* Coral Rag, Norm., *De C.;* Oxford
Clay, Coral Rag, and Great Oolite, Haute Saone, *Thir.;* Kim.
Clay, Coral Rag, Oxford Clay, Bernese Jura, *Thur.*
1. Rissoa lævis, *Sow.* Great Oolite, Ancliff, Wilts, *Cookson.*
2. ——— acuta, *Sow.* Great Oolite, Ancliff, *Cookson.*
3. ——— obliquata, *Sow.* Great Oolite, Ancliff, *Cookson.*
4. ——— duplicata, *Sow.* Great Oolite, Ancliff, *Cookson.*
1. Turbo muricatus, *Sow.* Coral. Oolite, Great Oolite, and Inf. Oolite,
Yorks., *Phil.;* Coral Rag, Mid. and S. Eng., *Conyb.;* Coral
Rag, Weymouth, *Sedg.*
2. ——— funiculatus, *Phil.* Coral. Oolite, Yorks., *Phil.*
3. ——— sulcostomus, *Phil.* Kell. Rock, Yorks., *Phil.*
4. ——— unicarinatus, *Bean.* Inf. Oolite, Yorks., *Phil.*
5. ——— lævigatus, *Phil.* Inf. Oolite, Yorks., *Phil.*
6. ——— undulatus, *Phil.* Lias, Yorks., *Phil.*
7. ——— ornatus, *Sow.* Inf. Oolite, Mid. and S. Eng., *Conyb.;* Inf. Oolite,
Norm., *De C.;* Lias, Gundershofen, *Voltz.*
8. ——— obtusus, *Sow.* Great Oolite, Ancliff, *Cookson.*
———, species not determined, Cornb. and Great Oolite, Norm., *De C.*
1. Phasianella cincta, *Phil.* Great Oolite, Yorks., *Phil.*
2. ——— angulosa, *Sow.* Porta Westphalica, *G. T.*
1. Turritella muricata, *Sow.* Coral. Oolite, Calc. Grit, Kell. Rock, and
Inf. Oolite, Yorks., *Phil.;* Rochelle Limestone, *Dufr.;* Shell
Limestone and Grit, Portgower, &c., Scotland, *Murch.*
2. ——— cingenda, *Sow.* Coral. Oolite? Great Oolite, and Inf. Oolite,
Yorks., *Phil.*
3. ——— quadrivittata, *Phil.* Inf. Oolite, Yorks., *Phil.*
4. ——— concava, *Sow.* Portland Stone, Tisbury, *Benett.*
5. ——— echinata, *Von Buch.* Banz, Langheim, *Von Buch.*
———, species not determined. Portland Stone, Coral Rag? Cornb., For.
Marb., and Brad. Clay, Mid. and S. Eng., *Conyb.;* Brad. Clay,
N. of Fr., *Bobl.;* Portland Beds and Coral Rag, Haute Saone,
Thir.; Lias, Bath, *Lons.;* Oxford Clay, Bernese Jura, *Thur.*

1. Nerinæa tuberculata, *Blain.* Bailly, near Auxerre, *Hœn.*
2. —— Mosæ, *Desh.* St. Mehiel (Meuse), *Desh.*
3. —— Bruckneri, *Thur.* Kim. Clay, Bernese Jura, *Thur.*
4. —— Bruntrutana, *Thur.* Coral Rag, Bernese Jura, *Thur.*
5. —— elegans, *Thur.* Coral Rag, Bernese Jura, *Thur.*
6. —— pulchella, *Thur.* Coral Rag, Bernese Jura, *Thur.*
——, species not determined. Coral Rag and For. Marb., Norm., *De C.;* Brad. Clay, N. of Fr., *Bobl.;* Coral Rag, Inf. Oolite, Haute Saone, *Thir.;* Rochelle, Nancy, *Desh.;* Neufchatel; Kelheim; Kahleberg, Echte, *Von Buch.*
1. Cerithium intermedium (var.). Böhlhorst, near Minden, *Hœn.*
2. —— muricatum, Mühlhausen, Bas Rhin, *Hœn.*
3. —— quinquangulare, *Thur.* Bernese Jura, *Thur.*
——, species not determined. Lias, Gundershofen, *Voltz.*
1. Murex Haccanensis, *Phil.* Coral. Oolite, Yorks., *Phil.*
2. —— rostellariformis, *Von Buch.* Coral Rag, Randen, Schafhausen, *Von Buch.*
1. Rostellaria bispinosa, *Phil.* Calc. Grit? and Kell. Rock, Yorks., *Phil.*
2. —— trifida, *Bean.* Oxford Clay, Yorks., *Phil.*
3. —— composita, *Sow.* Sandst., Limest., and Shale, Inverbrora, Scotl., *Murch.;* Great? and Inf. Oolite, Yorks., *Phil.;* Oxford Clay, Weymouth, *Sow.;* Kim. Clay, Hâvre, *Phil.*
——, species not determined. Lias, Yorks., *Phil.;* Oxford Clay, Kell. Rock, Cornb., Forest Marb., and Inf. Oolite, Mid. and S. Eng., *Conyb.;* Oxford Clay, Norm., *De C.;* Coral Rag, Bernese Jura, *Thur.*
1. Pteroceras Oceani, *Al. Brong.* Kim. Clay, Hâvre and the Jura, *Al. Brong.;* Portland Beds, Kim. Clay? Haute Saone, *Thir.;* Portland Beds, Kim. Clay, Bernese Jura, *Thur.;* Kahleberg, Echte, *G. T.*
2. —— Ponti, *Al. Brong.* Kim. Clay, Hâvre and the Jura, *Al. Brong.;* Kim. Clay, Haute Saone, *Thir.*
3. —— Pelagi, *Al. Brong.* Kim. Clay, Hâvre and the Jura, *Al. Brong.*
1. Actæon retusus, *Phil.* Calc. Grit, Yorks., *Phil.*
2. —— glaber, *Bean.* Great Oolite and Inf. Oolite, Yorks., *Phil.*
3. —— humeralis, *Phil.* Inf. Oolite, Yorks., *Phil.*
4. —— cuspidatus, *Sow.* Great Oolite, Ancliff, Wilts, *Cookson.*
5. —— acutus, *Sow.* Great Oolite, Ancliff, Wilts, *Cookson.*
——, species not determined. Lias, Yorks., *Phil.*
1. Buccinum unilineatum, *Sow.* Great Oolite, Ancliff, Wilts, *Cookson.*
——, species not determined. Shale, Sandst., and Limest., Inverbrora, Scotl., *Murch.*
1. Terebra melanoïdes, *Phil.* Coral. Oolite, Yorks., *Phil.*
2. ——? granulata, *Phil.* Coral. Oolite and Cornb., Yorks., *Phil.*
3. —— vetusta, *Phil.* Great Oolite and Inf. Oolite, Yorks., *Phil.*
4. —— sulcata, . Coral Rag, N. of Fr., *Bobl.;* Oxford Clay, Bernese Jura, *Thur.*
1. Belemnites sulcatus, *Mill.* Coral. Oolite? Calc. Grit, Oxford Clay, and Kell. Rock, Yorks., *Phil.;* Shale, Sandst., and Limest., Inverbrora, Scotl., *Murch.;* Lias, S. of France, *Dufr.*
2. —— fusiformis, *Mill.* Coral. Oolite? Yorks., *Phil.*
3. —— gracilis, *Phil.* Oxford Clay, Yorks., *Phil.*
4. —— abbreviatus, *Mill.* Great Oolite, Yorks., *Phil.;* Lias, Ross and Cromarty, Scotland ; and Micaceous Sandstone, Western Islands, Scotland, *Murch.*
5. —— elongatus, *Miller.* Lias, Yorks., *Phil.;* Lias, Ross and Cromarty, Scotl., *Murch.;* Fer. Oolite, Wasseralfingen, *Zieten.*
6. —— trisulcatus, *Blain.* Inferior Oolite, N. of Fr., *Bobl.*

7. Belemnites compressus, *Blain.* Fuller's E., N. of Fr., *Bobl.* ; Inf. Oolite
     Yorks., *Sow.* ; Lias, Gundershofen, *Voltz* ; Culmbach ; Witt-
     berg ; Metzingen, *G. T.*
8. —— dilatatus, *Blain.* Fuller's E., N. of Fr., *Bobl.* ; Theta, Baireuth,
     G. T.
9. —— apicicurvatus, *Blain.* Lias, S. of Fr., *Dufr.* ; Lias, Alais, *Al.
     Brong.*
10. —— pistilliformis, *Blain.* Lias, S. of Fr., *Dufr.* ; Lias, Gunder-
     shofen, *Voltz.*
11. —— brevis, *Blain.* Lias, Alais, *Brong.* ; Goppingen, *G. T.*
12. —— longissimus, *Miller.* Lias, Bath, *Lons* ; Lias, Boll, *Zieten.*
13. —— canaliculatus, *Schlot.* Oxford Clay and Inf. Oolite, Haute Saone,
     Thir. ; Inf. Oolite, Southern Germany, *Münst.* ; Stufenberg,
     Zieten.
14. —— ellipticus, *Miller.* Inf. Oolite, Haute Saone, *Thir.*
15. —— longus, *Voltz.* Great Oolite, Haute Saone, *Thir.*
16. —— ferruginosus, *Voltz.* (var.) Oxford Clay, Haute Saone, *Thir.* ;
     Oxford Clay, Bernese Jura, *Thur.* ; Swabia ; Bavaria, *G. T.*
17. —— aduncatus, *Miller.* Lias, Bath, *Lons.*
18. —— subclavatus, *Voltz.* Lias, Gundershofen ; Lias, Boll, *Voltz.*
19. —— tenuis, *Stahl.* Lias, Gundershofen, *Voltz.* ; Lias, Altdorf, *G. T.*
20. —— subdepressus, *Voltz.* Lias, Gundershofen, *Voltz.*
21. —— subaduncatus, *Voltz.* Lias, Gundershofen, *Voltz* ; Lias, Boll,
     Zieten.
22. —— digitalis, *Biguet.* Lias, Gundershofen, *Voltz.*
23. —— breviformis, *Voltz.* Lias, Gundershofen, *Voltz* ; Lias, Boll,
     Zieten.
24. —— ventroplanus, *Voltz.* Lias, Béfort, Haut Rhin, *Voltz.*
25. —— paxillosus, *Schlot.* Lias, Béfort ; Lias, Boll, *Voltz.* ; Lias,
     Ubstadt, near Heidelberg, *Bronn.*
26. —— longisulcatus, *Voltz.* Lias, Wurtemberg, *Voltz.*
27. —— trifidus, *Voltz.* Lias, Gundershofen, *Voltz.*
28. —— comprimatus, *Voltz.* Lias, Bahlingen, *Von Buch.*
29. —— Aalensis, *Voltz.* Inf. Oolite, Nipf, Bopfingen ; Baireuth, *G. T.*
30. —— grandis, *Schübler.* Inf. Oolite, Stufenberg, *Zieten.*
31. —— quinquesulcatus, *Blain.* Schlatt, Wurtemberg, *Zieten* ; Inf.
     Oolite, Baireuth, *G. T.*
32. —— tumidus, *Zieten.* Inf. Oolite, Stufenberg, *Zieten.*
33. —— teres, *Stahl.* Lias, Gosbach, Wurtemberg, *Zieten.*
34. —— lævigatus, *Zieten.* Lias, Boll, *Zieten.*
35. —— crassus, *Voltz.* Lias, near Göppingen, Wurtemberg, *Zieten* ;
     Besancon, *Voltz.*
36. —— semihastatus, *Blain.* Lias, Gamelshausen, Wurtemberg, *Zieten* ;
     Inf. Oolite, Baireuth, *G. T.*
37. —— incurvatus, *Hehl.* Lias, Boll, *Zieten* ; Lias, Banz, *G. T.*
38. —— pyramidatus, *Schübler.* Lias, Gross-Eislingen, Wurtemberg,
     Zieten.
39. —— rostratus, *Zieten.* Lias, Boll, *Zieten.*
40. —— papillatus, *Plieninger.* Lias, Boll, *Zieten.*
41. —— acuminatus, *Schübler.* Inf. Oolite, Stufenberg.
42. —— subhastatus, *Zieten.* Inf. Oolite, Stufenberg, *Zieten.*
43. —— oxyconus, *Heyl.* Lias, Boll, *Zieten.* ; Lias, Banz ; Altdorf, *G.T.*
44. —— carinatus, *Heyl.* Lias, Boll, *Zieten.*
45. —— pygmæus, *Zieten.* Lias, Boll, *Zieten.*
46. —— unisulcatus, *Hartmann.* Geisslingen, &c., Wurtemberg, *Zieten.*
47. —— bisulcatus, *Hartmann.* Lias, Boll, *Zieten.*
48. —— quadrisulcatus, *Hartmann.* Lias, Gross-Eislingen, near Göp-
     pingen, *Zieten.*

49. Belemnites pyramidalis, *Munst.* Lias, Wurtemberg, *Zieten.*; Lias, Banz, *G. T.*
50. ——— bipartitus, *Hartmann.* Gruibingen, Wurtemberg, *Zieten.*
51. ——— unicanaliculatus, *Hartmann.* Donzdorf, Wurtemberg, *Zieten.*
52. ——— bicanaliculatus, *Hartmann.* Ganslosen, Wurtemberg, *Zieten.*
53. ——— tricanaliculatus, *Hartmann.* Lias, Stufenberg, *Zieten.*
54. ——— quadricanaliculatus, *Hartmann.* Stufenberg, *Zieten.*
55. ——— quinquecanaliculatus, *Hartmann.* Lias, near Göppingen, *Zieten.*
56. ——— semisulcatus, *Munst.* Upper part of Oolite Group, Southern Germany, (Staffelberg; Lichtenfels; Solenhofen, &c.), *Munst.*; Oxford Clay, Bernese Jura, *Thur.*
57. ——— pusillus, *Munst.* Streitberg, *Munst.*
58. ——— acuarius, *Schlot.* Lias, Banz, *Munst.*; Lias, Altdorf, *G. T.*
59. ——— latesulcatus, *Voltz.* Oxford Clay, Bernese Jura, *Thur.*
60. ——— deformis, *Munst.* Southern Germany, *G. T.*
61. ——— gladius, *Blain.* Metzingen, Baireuth, *G. T.*
62. ——— Blainvillii, *Voltz*.* Swabia, *G. T.*
63. ——— hastatus, *Blain.* Inf. Oolite, Baireuth; Metz; Banz, *G. T.*
64. ——— tripartitus, *Schlot.* Lias, Altdorf, *G. T.*
65. ——— clavatus, *Blain.* Lias, Boll; Amberg; Banz; Lyme Regis, *G. T.*
——— , species not determined. Kim. Clay and Inf. Oolite, Yorks., *Phil.* Kim. Clay, Coral Rag, Oxford Clay, Kell. Rock, Stonesfield Slate, Bradford Clay, and Inf. Oolite, Mid. and S. England, *Conyb.*; Oxford Clay, For. Marb., Great Oolite, Inf. Oolite, and Lias, Norm., *De C.*; Lias, N. of Fr., *Bobl.*
1. Orthoceratites? elongatus, *De la B.* Lias, Lyme Regis, *De la B.*
1. Nautilus hexagonus, *Sow.* Kell. Rock? Yorks., *Phil.*; Calc. Grit, Oxford, *Sow.*
2. ——— lineatus, *Sow.* Inf. Oolite and Lias, Yorks., *Phil.*; Inf. Oolite, Dundry, *Conyb.*; Inf. Oolite? Haute Saone, *Thir.*; Lias, Bath, *Lons.*
3. ——— astacoides, *Y. & B.* Lias, Yorks., *Phil.*
4. ——— annularis, *Phil.* Lias, Yorks., *Phil.*
5. ——— obesus, *Sow.* Inf. Oolite, Mid. and S. Eng., *Conyb.*; Inf. Oolite, Norm., *De C.*
6. ——— sinuatus†, *Sow.* Inf. Oolite, Mid. and S. Eng., *Conyb.*; Oxford Clay, Norm.? *De la B.*
7. ——— intermedius, *Sow.* Lias, Mid. and S. Eng., *Conyb.*; Wurtemberg, *G. T.*
8. ——— striatus, *Sow.* Lias, Mid. and S. Eng., *Conyb.*; Lias, Alsace, *Brong.*
9. ——— truncatus, *Sow.* Lias, Mid. and S. Eng., *Conyb.*; For. Marb. and Lias, Norm., *De Cau.*
10. ——— angulosus, *D'Orbigny.* Portland Stone, Isle d'Aix, *Brong.*
——— , species not stated. Great Oolite, Yorks., *Phil.*; Kim. Clay, Coral Rag, Oxford Clay, Kell. Rock, and Stonesfield Slate, Mid. and S. Eng., *Conyb.*; Coral Rag, Norm., *De Cau.*; Fuller's Earth, N. of Fr., *Bobl.*
Hamites, species not determined. Lias, Zell, near Boll, *Zieten*; Inf. Oolite, Bayeux, *Desh. & Majendie‡.*
1. Scaphites bifurcatus, *Hartmann.* Lias, Göppingen, Wurtemberg, *Zieten.*
2. ——— refractus §, Gamelshausen, *G. T.*

---

* *B. acutus* and *B. apiciconus,* Blain.   † *Nautilus aganaticus,* Schlot.
‡ It should also be noticed, that M. Deshayes (Desc. des Coquilles Caracteristiques des Terrains) describes and figures a Hamite, by the name of *Hamites annulatus,* as found in the ferruginous oolite, but unfortunately does not mention the locality.
§ *Ammonites refractus,* Rein.

Scaphites, species not determined. Lias, S. of England, *Conyb.*
Turrilites Babeli, *Al. Brong.* Coral Rag? N. of Fr., *Bobl.*

1. Ammonites perarmatus, *Sow.* Coral Rag, Malton, *Sow.; * Coral. Oolite,
Calc. Grit, and Kel. Rock, Yorks., *Phil.;* Oolitic Rocks, Bra-
ambury Hill, Brora, *Murch.;* Coral Rag, Wilts, *Lons.;* Coral
Rag, Randen, *Von Buch.;* Oxford Clay, Bernese Jura, *Thur.;*
Mordberg, Nürnberg, *G. T.*
2. —— plicomphalus, *Sow.* Bolingbroke, Lincolnshire, *Sow.;* Kim.
Clay? Yorks., *Phil.;* Oxford Clay, Norm., *De C.*
3. —— triplicatus, *Sow.* Coral. Oolite, Yorks, *Phil.;* Inf. Oolite, Norm.,
De C.; Coral Rag, Randen, *Von Buch.*
4. —— plicatilis, *Sow.* Coral. Oolite and Kell. Rock, Yorks., *Phil.;*
Coral Rag, Mid. and S. Eng., *Conyb.;* Oxford Clay and Kell.
Rock, Haute Saone, *Thir.;* Coral Rag, Randen, *Von Buch.*
5. —— Williamsoni, *Phil.* Coral. Oolite, Yorks., *Phil.*
6. —— Sutherlandiæ*, *Sow.* Sandstone, Braambury Hill, Brora, *Murch.;*
Coral Oolite and Calc. Grit, Yorks., *Phil.;* Randen; Thurnau; Staffelberg, *G. T.*
7. —— sublævis, *Sow.* Coral. Oolite and Kell. Rock, Yorks., *Phil.;*
Full. E., Env. of Bath, *Lons.;* Oxford Clay, Beggingen, Schaf-
hausen, *Von Buch.;* Kell. Rock, Mid. and S. Eng., *Conyb.;*
Oxford Clay, Norm., *De la B.*
8. —— lenticularis, *Phil.* Coral. Oolite? Kell. Rock, and Lias, Yorks.,
Phil.
9. —— vertebralis and cordatus, *Sow.* Coral. Oolite, Calc. Grit, and
Oxford Clay, Yorks., *Phil.;* Coral Rag, Mid. and S. Eng.,
Conyb.; Oolite of Braambury Hill, Brora, *Murch.;* Kim.
Clay and Oxford Clay, Haute Saone, *Thir.;* Coral Rag, Wilts,
Lons.
10. —— instabilis, *Phil.* Calc. Grit., Yorks., *Phil.*
11. —— solaris, *Phil.* Calc. Grit, Yorks., *Phil.*
12. —— oculatus, *Phil.* Oxford Clay, Yorks., *Phil.*
13. —— Vernoni, *Bean.* Oxford Clay, Yorks., *Phil.*
14. —— Athleta, *Phil.* Oxford Clay and Kell. Rock, Yorks., *Phil.*
15. —— Kœnigi, *Sow.* Kell. Rock, Yorks., *Phil.;* Kell. Rock, Kelloways, Wilts; Lias, Charmouth, *Sow.;* Micaceous Sandst.,
Western Islands, Scotl., *Murch.;* Gammelshausen, *Zieten.;*
Hohenzollern, *G. T.*
16. —— bifrons, *Phil.* † Kell. Rock, Yorks., *Phil.*
17. —— Gowerianus, *Sow.* Shale, Sandst. and Limest., Inverbrora,
Scotl., *Murch.;* Kell. Rock, Yorks., *Phil.*
18. —— Calloviensis‡, *Sow.* Kell. Rock, Yorks., *Phil.;* Kell. Rock,
Kelloways, *Sow.*
19. —— Duncani, *Sow.* St. Neots, Huntingdonshire, *Duncan;* Kell.
Rock, Yorks., *Phil.;* Oxford Clay, Mid. and S. Eng., *Conyb.;*
Oxford Clay, Norm., *De Cau.;* Oxford Clay, Haute Saone,
Thir.
20. —— gemmatus, *Phil.* Kell. Rock, Yorks., *Phil.*
21. —— Herveyi, *Sow.* Spalden, Lincolnshire; Bradford; Knowles
Hill, Somerset, *Sow.;* Kell. Rock? and Cornb., Yorks., *Phil.;*
Inf. Oolite, Mid. and S. Eng., *Conyb.;* Inf. Oolite, Wasseralfingen, Wurtemberg, *Zieten.*
22. —— flexicostatus, *Phil.* Kell. Rock, Yorks., *Phil.*
23. —— funiferus, *Phil.* Kell. Rock, Yorks., *Phil.*

---

* *Am. inflatus*, Rein.

† This Ammonite must be distinguished from *A. bifrons*, Bruguière, which is,
according to M. Deshayes, the *A. Walcotii* of Sowerby.   ‡ *Am. Jason*, Rein.

24. Ammonites terebratus, *Phil.* Cornb., Yorks., *Phil.*
25. —— Blagdeni, *Sow.* Great Oolite, Yorks.,*Phil.*; Inf. Oolite, Dundry, *Conyb.*; Inf. Oolite, Norm., *De Cau.*; Spaichingen; Metzingen, *G. T.*
26. —— striatulus, *Sow.* Inf. Oolite and Lias, Yorks.,*Phil.*; Lias, Wasseralfingen, *Zieten.*
27. —— heterophyllus, *Sow.* Lias, Yorks., *Phil.*; Lias, Midland and Southern England, *Conyb.*
28. —— subcarinatus, *Y. & B.* Lias, Yorks., *Phil.*
29. —— Henleii, *Sow.* Lias, Yorks., *Phil.*; Lias, Mid. and S. Eng., *Conyb.*
30. —— heterogeneus, *Y. & B.* Lias, Yorks., *Phil.*
31. —— crassus, *Y. & B.* Lias, Yorks., *Phil.*
31.*—— crassus, *Montf.* Kim. Clay, Hécourt, Norm., *Pas.*
32. —— communis, *Sow.* Lias, Yorks., *Phil.*; Lias, Mid. and S. Eng., *Conyb.*; Lias, Western Islands, Scot.,*Murch.*; Soleure, *Hœn.*; Lias, Wurtemberg, *Zieten.*
33. —— angulatus, *Sow.* Lias, Yorks., *Phil.*; Lias, Mid. and S. Eng., *Conyb.*
34. —— annulatus, *Sow.* Lias, Yorks.,*Phil.*; Inferior Oolite and Lias, Midl. and S. Eng., *Conyb.*; Oxford Clay, For. Marb., and Inf. Oolite, Norm., *De C.*; Inf. Oolite, Uzer, S. of Fr.; Rochelle Limestone, *Dufr.*; Inf. Oolite and Lias, Montdor, Lyon, *Al. Brong.*; Coral Rag, Inf. Oolite, Wilts, *Lons.*; Coburg, *Holl*; Inf. Oolite, Gamelshausen, Wurtemberg, *Zieten.*; Moritzberg, Nürnberg, *G. T.*
35. —— fibulatus, *Sow.* Lias, Yorks., *Phil.*
36. —— subarmatus, *Sow.* Lias, Yorks., *Phil.*
37. —— maculatus, *Y. & B.* Lias, Yorks., *Phil.*
38. —— gagateus, *Y. & B.* Lias, Yorks., *Phil.*
39. —— planicostatus*, *Sow.* Maston Magna, Yeovil, Somerset, *Sow.*; Lias, Yorks., *Phil.*; Lias, Mid. and S. Eng., *Conyb.*; Lias, Bath, *Lons.*; Kahlefeld, Hartz; Amberg, Altdorf, *Holl*; Lias, Bahlingen, *Von Buch.*
40. —— balteatus, *Phil.* Lias, Yorks., *Phil.*
41. —— arcigerens, *Phil.* Lias, Yorks., *Phil.*
42. —— brevispina, *Sow.* Lias, Western Islands, Scotl., *Murch.*; Lias, Yorks., *Phil.*
43. —— Jamesoni, *Sow.* Lias, Western Islands, Scotl., *Murch.*; Lias, Yorks., *Phil.*
44. —— erugatus, *Bean.* Lias, Yorks., *Phil.*
45. —— fimbriatus, *Sow.*† Lias, Lyme Regis, *Buckl.*; Lias, Yorks., *Phil.*; Lias, Mid. and S. Eng., *Conyb.*; Lias, Norm., *De C.*; Lias, Wurtemberg, *Zieten*; Lias, Mende, Lozere, Banz; Randen, *Von Buch*; Inf. Oolite, Calvados, *Desl.*; Lias, Gr. Gschaid; Culmbach; Rautenberg, Scheppenstadt, *G. T.*
46. —— nitidus, *Y. & B.* Lias, Yorks., *Phil.*
47. —— anguliferus, *Phil.* Lias, Yorks., *Phil.*
48. —— crenularis, *Phil.* Lias, Yorks., *Phil.*
49. —— Clevelandicus, *Y. & B.* Lias, Yorks., *Phil.*
50. —— Turneri, *Sow.* Lias, Watchet; Wymondham Abbey, *Sow.*; Lias, Yorks., *Phil.*; Lias, South of France, *Dufr.*; Lias, Wurtemberg, *Zieten.*
51. —— geometricus, *Phil.* Lias, Yorks., *Phil.*

---

* *Am. Capricornus,* Schlot.
† According to Von Buch, this Ammonite is the same with *A. lineatus* and *A. hircinus* of Schlotheim.

562   *Organic Remains of the Oolitic Group.*

52. Ammonites vittatus, *Y. & B.* Lias, Yorks., *Phil.*
53. —— sigmifer*, *Phil.* Lias, Yorks., *Phil.*; Inf. Oolite, Haute Saone, *Thir.*; Lias, Wurtemberg, *Voltz.*
54. —— Hawkserensis, *Y. & B.* Lias, Yorks., *Phil.*
55. —— Conybeari, *Sow.* Lias, Yorks., *Phil.*; Lias, Mid. and S. Eng., *Conyb.*; Lias, Gundershofen and Buxweiller, *Al. Brong.*; Lias, Western Islands, Scotl., *Murch.*; Lias, Southern Germany, *G. T.*
56. —— Bucklandi, *Sow.*† Lias, Yorks., *Phil.*; Lias, Mid. and S. Eng., *Conyb.*; Lias, Norm., *De Cau.*; Lias, Malsch, near Heidelberg, *Bronn*; Inf. Oolite, Calvados, *Desl.*
57. —— obtusus, *Sow.* Lias, Yorks., *Phil.*; Lias, Mid. and S. Eng., *Conyb.*
58. —— Walcotii, *Sow.* Lias, Yorks., *Phil.*; Inf. Oolite and Lias, Mid. and S. Eng., *Conyb.*; Lias, S. of Fr., *Dufr.*; Lias, Béfort, Haut Rhin; Lias, Boll, *Voltz.*; Achelberg, *Hœn.*; Inf. Oolite, Calvados, *G. T.*
59. —— ovatus, *Y. & B.* Lias, Yorks., *Phil.*
60. —— Mulgravius, *Y. & B.* Lias, Yorks., *Phil.*; Lias, Boll, *G. T.*
61. —— exaratus, *Y. & B.* Lias, Yorks., *Phil.*
62. —— Lythensis, *Y. & B.* Lias, Yorks., *Phil.*
63. —— concavus, *Sow.* Lias? Yorks., *Phil.*; Inf. Oolite, Mid. and S. Eng., *Conyb.*; Coburg, *Holl*; Inf. Oolite, Calvados, *G. T.*
64. —— elegans‡, *Sow.* Lias? Yorks., *Phil.*; Inf. Oolite, Dundry, *Conyb.*; Lias, Norm., *De C.*; Inf. Oolite, Uzer, S. of Fr., *Dufr.*; Lias, Wurtemberg, *Zieten.*; Inf. Oolite, Calvados, *G. T.*
65. —— discus, *Sow.* Inf. Oolite, Dundry, Cornb., Mid. and S. Eng., *Conyb.*; Inf. Oolite, Norm., *De C.*; Cornb., Wilts., *Lons.*; Inf. Oolite, Wasseralfingen, *Zieten.*; Comp. Great Oolite, Bernese Jura, *Thur.*; Spaichingen, *G. T.*
66. —— Banksii, *Sow.* Inf. Oolite, Dundry, *Conyb.*
67. —— Braikenridgii, *Sow.* Inferior Oolite, Dundry, *Conyb.*; Inf. Oolite, Norm., *De C.*; Gammelshausen, *Zieten.*
68. —— Brocchii, *Sow.* Inf. Oolite, Dundry, *Conyb.*; Inf. Oolite, Haute Saone, *Thir.*
69. —— Sowerbii, *Miller.* Inf. Oolite, Dundry, *Conyb.*
70. —— falcifer, *Sow.* Inf. Oolite, Dundry, *Conyb.*; Lias, Norm., *De C.*; Lias, S. of Fr., *Dufr.*; Lias, Wurtemberg, *Zieten.*; Inf. Oolite, Bärendorf; Thurnau, *Munst.*; Inf. Oolite, Bernese Jura, *Thur.*
71. —— Brownii, *Sow.* Inf. Oolite, Dundry, *Conyb.*
72. —— læviusculus, *Sow.* Inf. Oolite, Dundry, *Braikenridge*; Inf. Oolite, Norm., *De C.*
73. —— acutus, *Sow.* Oxford Clay, Inf. Oolite, Norm., *De C.*; Lias, Western Islands, Scotl., *Murch.*; Inf. Oolite, Haute Saone, *Thir.*; Wasseralfingen, *Zieten.*
74. —— contractus, *Sow.* Inf. Oolite, Dundry, *Sow.*; Inf. Oolite, Norm., *De C.*
75. —— giganteus, *Sow.* Portland Stone, Coral Rag, and Lias, Mid. and S. Eng., *Conyb.*; Portland Stone, Isle d'Aix, *Brong.* (var.); Inf. Oolite, Haute Saone, *Thir.*
76. —— Lamberti, *Sow.* Portl. Stone, *Conyb.*; Rochelle Limest., *Dufr.*; Coburg; Heinberg; Bamberg, *Holl.*; Oxf. Clay, Bernese Jura, *Thur.*

   * *Am. costulatus,* Rein.
   † This Ammonite is, according to M. Deshayes, the *A. bisulcatus* of Bruguière, and the *A. Arietis* of Schlotheim.
   ‡ *Am. radians,* Rein.

77. Ammonites excavatus, *Sow.*   Coral Rag, Mid. and S. Eng., *Conyb.;* Oxford Clay, Norm., *De la B.;* Lias, Norm., *De C.;* Altdorf, *G. T.*
78. —— armatus, *Sow.* Oxford Clay and Lias, Mid. and S. Eng., *Conyb.;* Oxford Clay, Norm., *De la B.;* Oxford Clay, Haute Saone, *Thir.;* Lias, Bath, *Lons.;* Oxf. Clay, Bernese Jura, *Thur.*
79. —— modiolaris, *Smith.* Fuller's Earth? Mid. and S. Eng., *Conyb.*
80. —— jugosus, *Sow.* Inf. Oolite, Mid. and S. Eng., *Conyb.*
81. —— Stokesii, *Sow.** Inf. Oolite, Mid. and S. Eng., *Conyb.;* Lias, S. of Fr. *Dufr.;* Inf. Oolite, Haute Saone, *Thir.;* Lias, Wurtemberg, *Zieten.;* Lias, Ubstadt, near Heidelberg, *Bronn;* Oxf. Clay, Inf. Oolite, Bernese Jura, *Thur.;* Inf. Oolite, Calvados, *Desl.*
82. —— Strangwaysii, *Sow.* Inf. Oolite, Mid. and S. Eng., *Conyb.;* Lias, Norm., *De C.*
83. —— Brookii, *Sow.* Lias, Lyme Regis, *Buckl.;* Lias, Göppingen, *G. T.*
84. —— Bechii, *Sow.* Inf. Oolite and Lias, Mid. and S. Eng., *Conyb.;* Lias, Lyme Regis, *De la B.;* Coburg, *Holl.;* Lias, Rottweil; Bahlingen, *G. T.*
85. —— stellaris, *Sow.* Lias, Mid. and S. Eng., *Conyb.;* Lias, Norm., *De C.;* Lyme Regis, *De la B.*
86. —— Greenovii, *Sow.* Lias, Mid. and S. Engl., *Conyb.;* Lias, Lyme Regis, *De la B.;* Halsbach; Dünkelsbühl, *G. T.*
87. —— Loscombi, *Sow.* Lias, Mid. and S. Engl., *Conyb.;* Lias, Lyme Regis, *De la B.*
88. —— Birchii, *Sow.* Lias, Mid. and S. Engl., *Conyb.;* Lias, Lyme Regis, *De la B.;* Lias, Göppingen, *G. T.*
89. —— omphaloides, *Sow.* Portland Stone, *Sow.;* Oxford Clay, Norm., *De la B.;* Gt. Arenaceous Formation, Western Islands, Scotl., *Murch.;* Oxf. Clay, Bernese Jura, *G. T.*
†90 —— quadratus, *Sow.* Inf. Oolite, Norm., *De C.*
91. —— Gervillii, *Sow.* Inf. Oolite, Norm., *De C.*
92. —— Brongniartii, *Sow.* Inf. Oolite, Norm., *De C.*
†93.—— biplex, *Sow.* Inf. Oolite, Norm., *De C.;* Lias, Ross and Cromarty, Scotl., *Murch.;* Oxford Clay, Haute Saone, *Thir.;* Solenhofen, *Hœn.;* Calc. Grit, Oxf. Clay, Bernese Jura, *Thur.;* Randen; Rathhausen; Streitberg; Altdorf, *G. T.*
94. —— rotundus, *Sow.* Inf. Oolite, Norm., *De C.;* Kim. Clay, Purbeck, *Sow.*
†95.—— decipiens. Hohenstein, Saxony; Solenhofen, *Munst.*
96. —— Deslongchampi . Inf. Oolite, N. of Fr., *Bobl.*
97. —— vulgaris . Bradford Clay, N. of Fr., *Bobl.*
98. —— coronatus . Oxford Clay? N. of Fr., *Bobl.*
99. —— Humphresianus‡, *Sow.* Lias, S. of Fr., *Dufr.;* Inf. Oolite, Sherborne, *Sow.;* Lias, Boll, *Zieten.*
100. —— Parkinsoni, *Sow.* Lias, Yeovil, *Sow.;* Inf. Oolite, Bayeux, *Majendie;* Inf. Oolite, Hohenstein; Bärendorf; Thurnau, *Munst.;* Wasseralfingen; Wisgoldingen; Bopfingen, *G. T.*
101. —— Gulielmii, *Sow.* Oxford Clay, S. Engl., *Sow.*
102. —— Davæi, *Sow.* Lias, Lyme Regis, *De la B.;* Lias, Wasseralfingen, Wurtemberg, *Zieten.*
103. —— planorbis, *Sow.* Lias, Watchet, Somerset, *Sow.*

* *A. Amaltheus.*
† Found, according to Sowerby, in the Suffolk gravel.
‡ *Am. Bollensis*, Zieten.

104. Ammonites Johnstonii, *Sow.* Lias, Watchet, Somerset, *Sow.*; Lias, Bath, *Lons.*
105. —— corrugatus, *Sow.* Inf. Oolite, Dundry, *Braikenridge.*
106. —— rotiformis, *Sow.* Lias, Yeovil, *Sow.*; Lias, Bath, *Lons.*
107. —— multicostatus, *Sow.* Lias, Bath, *Sow.*
108. —— lævigatus, *Sow.* Lias, Lyme Regis, *De la B.*
109. —— latæcostata, *Sow.* Lias, Lyme Regis, *Murch.*
110. —— Murchisonæ, *Sow.* Micaceous Sandst., Holm Cliff, Western Islands, Scotl., *Murch.*; Inf. Oolite, Allington near Bridport, *Murch.*; Wasseralfingen; Gundershofen; Wisgoldingen; Goslar, *G. T.*
111. —— serpentinus*, *Rein.* Inf. Oolite, Haute Saone, *Thir.*; Lias, Gundershofen, *Voltz.*; Lias, Ubstadt, near Heidelberg, *Bronn*; Lias, Altdorf; Boll, *G. T.*
112. —— cristatus, *Defr.* Weymouth, *Bryer*; Oxford Clay, Haute Saone, *Thir.*; Oxf. Clay, Bernese Jura, *Thur.*
113. —— interruptus, *Schlot.* Oxford Clay, Haute Saone, *Thir.*; Thirnau, *Holl.*; Oxf. Clay, Bernese Jura, *Thur.*
114. —— opalinus, *Reinecke.* Lias, Gundershofen, *Voltz.*
115. —— latina, *Sow.* Coral Rag, Wilts, *Lons.*
116. —— ammonius, *Schlot.* Lias, Gundershofen, *Voltz*; Altdorf, *Holl.*
117. —— comptus†, *Reinecke.* Lias, Gundershofen, *Voltz*; Donzdorf, *Zieten.*
118. —— planulatus, *De Haan.* Baireuth, *Holl.*
119. —— Knorrianus, *De Haan.* Boll, Wurtemberg, *Holl.*
120. —— Reineckii, *Holl.* Coburg, *Holl.*
121. —— pustulatus, *Rein.* Coburg; Thurnau, *Holl.*
122. —— granulatus, *Brug.* Coburg, *Holl.*
123. —— bifurcatus, *Brug.* Coburg; Baireuth, *Holl.*; Coral Rag, Germany, *Von Buch.*
124. —— trifurcatus, *De Haan.* Coburg, *Holl.*
125. —— macrocephalus‡, *Schlot.* Aarau; Coburg, *Holl.*; Inf. Oolite, Southern Germany, *Munst.*; Vaches Noires, Calvados, *G. T.*
126. —— Planula, *Heyl.* Donzdorf, *Holl.*
127. —— Fonticola §, *Mencke.* Ferruginous Beds, Thurnau; Langheim; *Von Buch*; Gamelshausen, *Zieten*; Oxford Clay, Haute Saone, *Thir.*; Oxf. Clay, Bernese Jura, *Thur.*
128. —— scutatus, *Von Buch.* Lias, Banz, near Bamberg, *Von Buch.*
129. —— canaliculatus, *Munst.* Wöschnau, Aarau; Furstenberg; Lochenberg, Bahlingen, *Von Buch.*
130. —— flexuosus, *Munst.* Coral Rag, Streitberg, near Erlangen; Donzdorf, Swabia; Rathhausen, near Bahlingen; summit of Mont Randen, near Schafhausen, *Von Buch.*
131. —— crenatus, *Rein.* Coral Rag, Germany, *Von Buch.*
132. —— sublævis, *Munst.* Donzdorf, *Zieten.*
133. —— hecticus, *Rein.* Inf. Oolite, Gamelshausen, *Zieten*; Oxf. Clay, Bernese Jura, *Thur.*
134. —— Pollux, *Rein.* Inf. Oolite, Gamelshausen, *Zieten*; Vaches Noires, Calvados; Goslar; Thurnau, *G. T.*
135. —— æquistriatus, *Munst.* Lias, Wurtemberg, *Zieten.*
136. —— inæqualis, *Merian.* Bâle, *Merian.*
137. —— tenuistriatus, *Munst.* Solenhofen, *Hœn.*
138. —— dubius, *Schlot.* Gamelshausen, *Zieten*; Oxf. Clay, Bernese Jura, *Thur.*

* *Am. Strangwaysii,* Sow., according to G. T.
† *Am. gracilis,* Munst.   ‡ *Am. tumidus,* Rein.
§ According to Von Buch this Ammonite is figured as *A. Lunula* by M. Zieten.

139. Ammonites Kridion, *Rein.* Lias, Stutgard, *Zieten.*
140. ——— Jason, *Rein.* Lias, Gamelshausen, *Zieten.*
141. ——— alternans, *Von Buch.* Coral Rag, Muggendorf, Gailenreuth, &c. *Von Buch.*
142. ——— Gigas, *Zieten.* Riedlingen on the Danube, *Zieten.*
143. ——— denticulatus, *Zieten.* Lias, Boll, *Zieten;* Oxf. Clay, Bernese Jura, *Thur.*
144. ——— raricostatus, *Zieten.* Lias, Boll, *Zieten.*
145. ——— decoratus, *Zieten.* Inf. Oolite, Guttenberg, Wurtemberg, *Zieten.*
146. ——— bipartitus, *Zieten.* Inf. Oolite, Guttenberg, *Zieten.*
147. ——— torulosus, *Schübler.* Lias, Stuifenberg, *Zieten.*
148. ——— oblique-costatus, *Zieten.* Lias, Kaltenthal, near Stuttgart, *Zieten.*
149. ——— insignis, *Schübler.* Lias, Reichenbach, *Zieten.*
150. ——— oblique-interruptus, *Schübler.* Lias, Wasseralfingen, *Zieten.*
151. ——— polygonius, *Zieten.* Lias, Zell, near Boll, *Zieten.*
152. ——— discoides, *Zieten.* Lias, Reichenbach, *Zieten.*
153. ——— bispinosus, *Zieten.* Wasseralfingen, *Zieten.*
154. ——— biarmatus, *Sow.* Hohenstein, Saxony; Bavaria, Wurtemberg, Switzerland; *Munst.*
155. ——— lævis, *Schlot.* Inf. Oolite, Southern Germany, *Munst.;* Lias, near Heidelberg, *Bronn.*
156. ——— colubrinus, *Rein.* Oxf. Clay, Bernese Jura, *Thur.*
157. ——— lævigatus, *Schlot.* Oxf. Clay, Bernese Jura, *Thur.*
158. ——— anceps, *Rein.* Oxf. Clay, Bernese Jura, *Thur.*
159. ——— inflatus, *Rein.* Oxf. Clay, Bernese Jura, *Thur.*
160. ——— Deluci, *Al. Brong.* Neuhausen, *G. T.*
161. ——— Comensis, *Von Buch.* Neuhausen, *G. T.*
162. ——— alternans, *Von Buch.* Coral Rag, Bamberg, *G. T.*
163. ——— cristatus, *Sow.* Guttenberg, Streitberg, *G. T.*
164. ——— polygyratus, *Rein.* Donzdorf; Randen, *G. T.*
165. ——— tripartitus, *Sow.* Randen, *G. T.*
166. ——— multiradiatus, *Reng.* Willibaldsburg, Eichstadt, *G. T.*
167. ——— longidorsatus, *Von Buch.* Lias, Moutiers, Caen, *G. T.*
168. ——— asper, *Mérian.* Haute Rive, Neufchatel, *G. T.*
169. ——— planorbiformis, *Munst.* Lias, Bavaria, *G. T.*
170. ——— colubratus, *Montf.* Lias, Vaichingen; Dünkelsbühl, *G. T.*
171. ——— angulatus, *Schlot.* Lias, Neckar Thailfingen; Wellersen, Scheppenstadt, *G. T.*
172. ——— natrix, *Schlot.* Lias, Bahlingen; Gr. Brunsrode; Altdorf, *G. T.*
173. ——— funicularis, *Von Buch.* Lias, near Strasburg, *G. T.*
  1. Aptychus * lævis, *Meyer.* Solenhofen; Stafelstein, Bavaria, *Meyer;* Stufenberg; Banzberg, Amberg, *G. T.*
  2. ——— imbricatus, *Meyer.* Solenhofen, *Meyer;* Lias, Banzberg, *G. T.*
  3. ——— bullatus, *Meyer.* Lias, Banz, *Meyer;* Lias, Boll, *G. T.*
  4. ——— Elasma, *Meyer.* Lias, Banz, *Meyer;* Lias, Boll, *G. T.*
  1. Onychoteuthis angusta †, *Munst.* Solenhofen, *Rüppell.*
  1. Sepia antiqua ‡, *Munst.* Solenhofen, *Rüppell.*
  ———, remains of, with ink-bags preserved, Lias, Lyme Regis, *Buckl.*
Rhyncolites, or Sepia beaks, Lias, Lyme Regis, *De la B.;* Lias, near Bristol, *Miller.*

---

  * *Trigonellites,* Parkinson; *Tellinites,* Schlotheim.
  † *Loligo prisca,* Rüppell.
  ‡ *Sepia hastæformis,* Rüppell.

CRUSTACEA.

1. Pagurus mysticus, *Germar.* Solenhofen, *Holl.*
1. Eryon Cuvieri, *Desm.* Solenhofen; Erchstadt, Pappenheim, *Holl.*
2. —— muticus, *Germ.* Solenhofen, *G. T.*
3. —— propinquus *, *Germ.* Solenhofen, *Holl.*
4. —— spinimanus, *Germ.* Solenhofen, *G. T.*
1. Scyllarus dubius, *Holl.·* Solenhofen, *Holl.*
1. Palæmon spinipes, *Desm.* Pappenheim, Solenhofen, *Holl.*
2. —— longimanatus, *Holl.* Solenhofen, *Holl.*
3. —— Walchii, *Holl.* Pappenheim, *Holl.*
1. Astacus modestiformis, *Holl.* Solenhofen, *Holl.*
2. —— minutus, *Holl.* Solenhofen, *Holl.*
3. —— rostratus, *Phil.* Kelloway Rock and Coral. Oolite, Yorks., *Phil.*
4. —— leptodactylus, *Germ.* Solenhofen, *G. T.*
5. —— spinimanus, *Germ.* Solenhofen, *G. T.*
6. —— fusiformis, *Holl.* Solenhofen, *Holl.*
——, species not determined. Oxford Clay and Lias, Yorks., *Phil.*
Crustacea, not yet determined. Lias, Midl. and S. Engl., *Conyb.*; Lyme Regis, *De la B.*; Forest Marble, Normandy, *De Cau.*; Stonesfield Slate, *Conyb.*; Bradford Clay, North of France, *Bobl.*

INSECTA.

Insects of the families Libellula; Æschna; Agrion; Myrmeleon? Sirax? Solpaga? Solenhofen, *Munst., Murch., G. T.*
Elytra of coleopterous insects. Stonesfield Slate, *Leach, Buckl.*

PISCES.

1. Dapedium politum, *De la B.* Lias, Lyme Regis, *De la B.*; Lias, and Oxford Clay of Normandy, *De Cau.*
1. Clupea sprattiformis, *Blain.* Solenhofen, *Holl.*
2. —— dubia, *Blain.* Solenhofen, *G. T.*
3. —— Knorrii, *Blain.* Solenhofen, *G. T.*
4. —— Salmonea, *Blain.* Solenhofen, *G. T.*
5. —— Davilei, *Blain.* Solenhofen, *G. T.*
1. Esox avirostris, *Germ.* Solenhofen, *G. T.*
2. —— acutirostris, *Blain.* Solenhofen, *G. T.*
1. Uræus gracilis, *Ag.* Lias, Wurtemberg, *Ag.*
1. Sauropsis latus, *Ag.* Lias, Wurtemberg, *Ag.*
1. Ptycholepis Bollensis, *Ag.* Lias, Boll, *Ag.*
1. Semionotus leptocephalus, *Ag.* Lias, Zell; Boll, *Ag.*
1. Lepidotes Gigas, *Ag.* Lias, Ohmden, Boll, *Ag.*
2. —— frondosus, *Ag.* Lias, Zell; Boll, *Ag.*
3. —— ornatus, *Ag.* Lias, Wurtemberg, *Ag.*
1. Leptolepis Bronnii, *Ag.* Lias, Neidingen, *Ag.*
2. —— Jægeri, *Ag.* Lias, Zell; Boll, *Ag.*
3. —— longus, *Ag.* Lias, Zell; Boll, *Ag.*
1. Tetragonolepis heteroderma, *Ag.* Lias, Zell; Boll, *Ag.*
2. —— semicinctus, *Bronn.* Lias, *G. T.*
3. —— pholidotus, *Ag.* Lias, Zell; Boll, *Ag.*
4. —— Traillii, *Ag.* Lias, England, *G. T.*
Fish, species not yet determined. Several in the Lias, Lyme Regis, *De la B.*; Barrow, Leicestershire, *Conyb.*; Portland Beds, Tisbury, Wilts, *Benett.*
Ichthyodorulites, *Buckl. & De la B.* Different kinds. Lias, Lyme Regis, and elsewhere in Southern and Midland England,

* *Eryon Schlotheimii,* Holl.

Conyb. & De la B.; Kimmeridge Clay, near Oxford, *Buckl.;*
    Stonesfield Slate, *Buckl.* In the Great Oolite, Normandy,
    De Cau.
Fish palates and teeth. Lias, Lyme Regis, and Somersetshire, &c.
    Conyb.; Stonesfield Slate, *Buckl.;* Great Oolite, Normandy,
    De Cau.; Cornbrash and Forest Marble, North of France,
    Bobl.; Coral. Oolite, Oxford Clay, Yorks., *Phil.;* Portland
    Beds, Tisbury, Wilts, *Benett.*

Reptilia.

1. Pterodactylus macronyx, *Buckl.* Lias, Lyme Regis, *Buckl.;* Lias,
    Banz, Bavaria, *Meyer.*
2. —— longirostris, *Cuv.* Aichstadt, *Collini.*
3. —— brevirostris, *Cuv.* Aichstadt, *Cuv.*
4. —— grandis, *Cuv.* Solenhofen, *Holl.*
5. —— crassirostris, *Goldf.* Solenhofen, *Goldf.*
6. —— medius, *Munst.* Monheim, *Schnitzlein.*
7. —— Munsteri, *Goldf.* Monheim, *Goldf.*
Pterodactylus, species not known. Stonesfield Slate, *Buckl.*
1. Macrospondylus Bollensis*, *Von Meyer.* Lias, Boll, *Jäg.*
1. Crocodilus cylindrirostris, *Cuv.* Kim. Clay, Hâvre, *Al. Brong.;* Alt-
    dorf, *G. T.*
2. —— brevirostris, *Cuv.* Kim. Clay, Hâvre, *Al. Brong.;* Altdorf,
    G. T.
3. Crocodile of Mans, *Cuv.* Great Oolite, *Al. Brong.*
—— remains, species not determined. Lias, Yorks., *Phil.;* Lias?
    Lyme Regis, *De la B.;* Cornbrash, Engl., *Conyb.;* Stones-
    field Slate, *Buckl.;* Coral. Oolite, Yorks., *Phil.;* Inf. Oolite,
    Calvados, *Her.*
1. Teleosaurus Cadomensis, *Geoffroy St. Hilaire.* Great Oolite, Caen,
    De Cau.
1. Megalosaurus Bucklandi. Stonesfield Slate, *Buckl.*
——, species not known. Great Oolite, Normandy, *De Cau.*
1. Geosaurus Bollensis, *Jäg.* Lias, Boll, *Jäg.*
2. —— Sömmeringii, *Cuv.* Monheim, *Sömmering.*
1. Lacerta Neptunia, *Goldf.* Monheim, *Goldf.*
2. —— gigantea, *Munst.* Monheim, *G. T.*
1. Rhacheosaurus gracilis, *Meyer.* Daiting, Solenhofen, *Meyer.*
1. Ælodon priscus†, *Von Meyer.* Monheim, *Sömmering.*
1. Pleurosaurus Goldfussii, *Meyer.* Daiting, *Meyer.*
1. Plesiosaurus dolichodeirus, *Conyb.* Lias, Lyme Regis, &c.
2. —— recentior, *Conyb.* Kim. Clay, Engl., *Conyb.;* Kim. Clay, Hon-
    fleur, *Al. Brong.*
3. —— carinatus, *Cuv.* Great Oolite, Boulogne, *Al. Brong.*
4. —— pentagonus, *Cuv.* Great Oolite, Ballon and Chaufour, *Al. Brong.*
5. ——? trigonus, *Cuv.* Great Oolite, Calvados, *Al. Brong.*
6. —— macrocephalus, *Conyb.* Lias, Lyme Regis, *De la B.*
——, species not determined. Oxford Clay, Stenay, *Bobl.;* Oxford
    Clay, Calvados, *De la B.;* Lias, N. of Ireland, *Bryce;* Lias,
    Whitby, *Dunn.*
1. Ichthyosaurus communis, *De la B.;* Lias, Lyme Regis, &c. Engl.,
    Conyb., &c.; Lias, Boll, Wurtemberg, *Jäg.;* Banz; Fried-
    richsgemünd, *G. T.*
2. —— platyodon, *De la B.* Lias, Lyme Regis, &c. Engl., *Conyb.,* &c.;
    Lias, Boll, *Jäg.*

---

* *Crocodilus Bollensis,* Jäg.   † *Crocodilus priscus,* Sömmering.

3. Ichthyosaurus tenuirostris, *De la B.*　Lias, Lyme Regis, &c.　*Conyb.,*
　　&c.; Lias, Boll, *Jäg.*
4. ———— intermedius, *Conyb.*　Lias, Lyme Regis, &c.　*Conyb.,* &c.;
　　Lias, Boll, *Jäg.*
———, species not determined.　Lias and Inferior Oolite, Normandy,
　　De Cau.; Lias, Yorks., *Phil.;* Oxford Clay, England, *Conyb.;*
　　Oxford Clay, Normandy, *De la B.;* Great Oolite, Reugny,
　　Brong.; Coral. Oolite, Yorks., *Phil.;* Calc. Grit, Midl. Eng.,
　　Conyb.; Kim. Clay, Oxford, *Buckl.;* Kim. Clay, Weymouth,
　　De la B.; Kim. Clay, Honfleur, *Brong.*
Saurian bones occur in the Kelloway Rock and Bath Oolite, Yorks.,
　　Phil.; in the Portland Stone, *Buckl. & De la B.;* Lias, Cal-
　　vados, *Her.*
Tortoise.　Stonesfield Slate, *Buckl.;* Lias? Engl. *Conyb.;* Solenhofen,
　　Munst.

MAMMALIA.

1. Didelphis Bucklandi, *Broderip.*　Stonesfield Slate, *Buckl.*

Organic Remains of the Red Sandstone Group.

Variegated Marls.

PLANTÆ.

1. Equisetum Meriani, *Ad. Brong.*　Neue Welt, Basle, *Ad. Brong.*
2. ———— arenaceum*, *Bronn.*　Near Heidelberg, *Bronn;* Wurtemberg;
　　France, *G. T.*
3. ———— columnare, *Ad. Brong.*　Lorraine; Alsace; Wurtemberg, *Rozet.*
1. Pecopteris Meriani, *Ad. Brong.*　Neue Welt, *Ad. Brong.*
1. Tæniopteris vittata, *Ad. Brong.*　Neue Welt, *Ad. Brong.;* Wurtem-
　　berg, *G. T.*
1. Filicites Stuttgardiensis, *Ad. Brong.*　Wurtemberg, *Rozet, G. T.*
2. ———— lanceolata, *Ad. Brong.*　Wurtemberg, *Rozet.*
1. Marantoidea arenacea, *Jæg.*　Stuttgart, *G. T.*
1. Pterophyllum longifolium, *Ad. Brong.*　Neue Welt, *Ad. Brong.*
2. ———— Meriani, *Ad. Brong.*　Neue Welt, *Ad. Brong.*
3. ———— Jægeri, *Ad. Brong.*　Wurtemberg; France, *Rozet.*

RADIARIA.

Ophiura, species undetermined.　Vosges, *Rozet.*

CONCHIFERA.

1. Plagiostoma lineatum, *Bronn.*　Wurtemberg, *G. T.*
1. Cardium pectinatum, *Von Alb.*　Wurtemberg, *G. T.*
1. Trigonia vulgaris, *Schlot.*　Louisburg, *G. T.*
2. ———— curvirostris, *Schlot.*　Louisburg; Schwenningen, *G. T.*
3. ———— sulcata, *Goldf.*　Villengen, *G. T.*
1. Mya musculoides, *Schlot.*　Sulz on the Neckar, *G. T.*
2. ———— elongata, *Schlot.*　Sulz on the Neckar, *G. T.*
1. Avicula socialis†, *Desh.*　Sulz, *G. T.*
2. ———— subcostata, *Goldf.*　Sulz, *G. T.*
3. ———— lineata, *Goldf.*　Sulz, *G. T.*
1. Posodonia Keuperina, *Voltz.*　Swabia; Hall, *G. T.*
2. ———— minuta, *Von Alb.*　Rottweil, *G. T.*
1. Modiola minuta, *Goldf.*　Rottweil, *G. T.*

* *Calamites arenaceus,* Jæg.　　　　　† *Mytilus socialis,* Schlot.

1. Venericardia Goldfussii, *Von Alb.* Rottweil, *G. T.*
1. Lingula tenuissima, *Bronn.* Rottweil, *G. T.*
1. Saxicava Blainvillii, *Hœn.* Ballbron, *Hœn.*

MOLLUSCA.

1. Buccinum turbilinum *, . Sulz on the Neckar, *G. T.*

PISCES.

Genera not determined. Seidmannsdorf; Neuses; Seidingstadt, Coburg, *G. T.*

REPTILIA.

1. Phytosaurus cylindricodon, *Jœg.* Boll, *Jœg.*
2. ——— cubicodon, *Jœg.* Boll, *Jœg.*
1. Mastodonsaurus Jægeri, *Holl.* Gaildorf, *G. T.*
1. Ichthyosaurus Lunevillensis, . Wurtemberg, *G. T.*
Plesiosaurus, species not determined. Dürrheim, *Hœn.*

Muschelkalk.

PLANTÆ.

1. Neuropteris Gailliardoti, *Ad. Brong.* Lunéville, *Ad. Brong.*
1. Mantellia cylindrica, *Ad. Brong.* Lunéville, *Ad. Brong.*

ZOOPHYTE.

1. Astrea pediculata, *Desh.* Locality not stated.

RADIARIA.

1. Cidaris grandævis, *Goldf.* Wurtemberg, *G. T.*
1. Ophiura prisca, *Munst.* Baireuth, *Goldf.*
2. ——— loricata, *Goldf.* Schwenningen, Wurtemberg, *G. T.*
1. Asterias obtusa, *Goldf.* Marbach, Villengen, *G. T.*
1. Encrinites moniliformis, *Mill.* Göttingen; Wurtemberg; Poland; France, &c., *Var. Auth.*
1. Pentacrinites dubius, *Goldf.* Rüdersdorf, *G. T.*

ANNULATA.

1. Serpula valvata, *Goldf.* Baireuth, *Goldf.*
2. ——— colubrina, *Goldf.* Baireuth, *Goldf.*

CONCHIFERA.

1. Terebratula communis †, *Bosc.* Göttingen, *Hœn.;* Wurtemberg; Lunéville; Toulon, *Al. Brong.;* Richen, Basle, *G. T.*
2. ——— perovalis, *Schlot.* Jena, *Hœn.*
3. ——— sufflata, *Schlot.* Jena, *Hœn.*
4. ——— orbiculata, *Schlot.* Dornberg, Jena, *Hœn.*
1. Delthyris semicircularis, *Goldf.* Villengen, *G. T.*
1. Lingula tenuissima, *Bronn.* Rottweil, *G. T.*
1. Ostrea placunoides, *Munst.* Baireuth, *G. T.*
2. ——— subanomia, *Munst.* Baireuth, *G. T.*
3. ——— reniformis, *Munst.* Baireuth, *G. T.*
4. ——— difformis, *Schlot.* Baireuth; Wurtemberg, *G. T.*
5. ——— multicostata, *Munst.* Würzberg, *G. T.*
6. ——— complicata, *Goldf.* Baireuth; Villengen, *G. T.*
7. ——— decemcostata, *Munst.* Baireuth, *G. T.*
8. ——— spondyloides, *Schlot.* Quedlinberg, *Hœn.;* Göttingen; Lunéville; Toulon, *Al. Brong.;* Baireuth; Brombach, *G. T.*

---

* *Helix turbilinum*, Schlot. † *T. vulgaris,* and *T. subrotunda,* Schlot.

9. Ostrea comta, *Goldf.* Rottweil, *G. T.*
1. Gryphæa prisca, *Goldf.* Villengen, *G. T.*
1. Pecten reticulatus, *Schlot.* Göttingen, *Hœn.;* Gotha, *G. T.*
2. —— Albertii, *Goldf.* Villengen, Rüdersdorf, *G. T.*
3. —— lævigatus, *Goldf.* Wurtemberg; Hagen; Bromberg; Baireuth; Gotha, *G. T.*
4. —— discites, *Schlot.* Wurtemberg; Richen, Basle; Rüdersdorf; Poland, *G. T.*
1. Plagiostoma lineatum *, *Bronn.* Michelstadt, *Hœn.;* Göttingen, *Al. Brong.;* Mosbach, Heidelberg, *Bronn;* Wurtemberg; Baireuth; Weimar, *G. T.*
2. —— striatum, *Schlot.* Germany; France; Poland, *G. T.*
3. —— rigidum, *Schlot.* Rauhthal, Jena, *Hœn.;* Göttingen, *Al.Brong.*
4. —— lævigatum, *Schlot.* Mosbach, *Hœn.*
5. —— punctatum, *Schlot.* Göttingen; Gotha; Toulon, *Al. Brong.;* Weimar; Baireuth, *G. T.*
1. Avicula socialis, *Desh.* Gotha; Sachsenburg, *Schlot.;* Weimar, *Hœn.;* Göttingen; Mont Meisner; Wurtemberg; Lunéville, *Al. Brong.;* Ibbenbühren; Rüdersdorf; Nischwitz; Wehrau; Kalinowitz, *G. T.;* Heidelberg, *Bronn.*
2. —— costata, *Bronn.* Würtemberg; Baireuth, *G. T.*
3. —— crispata, *Goldf.* Friedrichshall, *G. T.*
4. —— Bronnii, *Von Alberti.* Villingen, *G. T.*
1. Mytilus vetustus †, *Goldf.* Göttingen; Lunéville, *Al. Brong.;* Wurtemberg; Hagen; Baireuth, *G. T.*
1. Trigonia vulgaris, *Schlot.* Weimar, *Hœn.;* Göttingen, *Al. Brong.;* Wurtemberg; Riedern, Waldshut; Baireuth, *G. T.*
2. —— Pes-anseris, *Schlot.* Lunéville; Mosbach, *Hœn.;* Göttingen, *Al. Brong.*
3. —— curvirostris, *Schlot.* Wurtemberg, *G. T.*
4. —— cardissoides, *Goldf.* Wurtemberg, *G. T.*
5. —— lævigata, *Goldf.* Marbach, *G. T.*
6. —— Goldfussii, *Von Alberti.* Marbach, *G. T.*
1. Arca inæquivalvis, *Goldf.* Freudenstadt, Schwarzwald, *G. T.*
1. Cardium striatum, *Schlot.* Wurtemberg; Göttingen, *Al. Brong.*
2. —— pectinatum, *Von Alb.* Wurtemberg, *G. T.*
1. Mya musculoides, *Schlot.* Weimar; Wurtemberg; Upper Silesia; Poland, *G. T.*
2. —— elongata, *Schlot.* Wurtemberg, *Al. Brong.;* Seewangen, Waldshut; Upper Silesia; Poland, *G. T.*
3. —— ventricosa, *Schlot.* Lunéville, *Al. Brong.;* Wurtemberg, *G. T.*
4. —— mactroides, *Schlot.* Marbach; Upper Silesia; Poland, *G. T.*
5. —— rugosa, *Von Alberti.* Rottweil, *G. T.*
1. Venus nuda, *Goldf.* Marbach, *G. T.*
1. Mactra? trigona, *Goldf.* Marbach, *G. T.*
1. Cuccullæa minuta, *Goldf.* Villengen, *G. T.*

MOLLUSCA.

1. Calyptræa ‡ discoides, *Schlot.* Villengen, *G. T.*
1. Capulus mitratus, *Goldf.* Villengen, *G. T.*
1. Dentalium torquatum, *Schlot.* Göttingen, *Al. Brong.*
2. —— læve, *Schlot.* Göttingen, *Al. Brong.;* Alpirsbach; Baireuth, *G. T.*
1. Trochus Albertinus, *Goldf.* Rottweil, *G. T.*
1. Turritella obsoleta §,   . Göttingen, *Hœn.;* Weimar, *G. T.*

---

* *Chama lineata,* Schlot.
‡ Patellites.

† *Mytilus eduliformis,* Schlot.
§ *Buccinum obsoletum,* Schlot.

2. Turritella deperdita, *Goldf.* Weimar, *G. T.*
3. —— detrita, *Goldf.* Culmbach, *G. T.*
4. —— scalata*, . Würtemberg; Rüdersdorf, *G. T.*
5. ——? terebralis, *Schlot.* Weimar, *Hœn.*
1. Buccinum gregarium, *Schlot.* Rüdersdorf, *G. T.*
2. —— turbilinum †, . Wurtemberg; Seewangen, Waldshut; Rüdersdorf, *G. T.*
1. Strombus denticulatus, *Schlot.* Rüdersdorf, *G. T.*
1. Natica Gailliardoti, *Lefroy.* Wurtemberg, *G. T.*
2. —— Pulla, *Goldf.* Rottweil, *G. T.*
1. Turbo dubius, *Munst.* Hässel, *Bronn;* Seewangen; Riedern, *G. T.*
2. —— giganteus, *Schlot.* Seewangen, *G. T.*
1. Nautilus bidorsatus, *Schlot.* Weimar, *Hœn.;* Wurtemberg, *Al. Brong.;* Göttingen; Rüdersdorf; Lunéville, *G. T.*
2. —— nodosus, *Munst.* Germany, *Munst.*
1. Ammonites nodosus, *Schlot.* Weimar, *Hœn.;* Göttingen; Wurtemberg; Toulon, *Al. Brong.;* Lorraine, *Beaum.;* Tarnowitz, *G. T.*
2. —— bipartitus, *Gailliardot.* Lunéville, *Al. Brong.*
Rhyncolites. Jena; Göttingen; Wurtemberg; Lunéville; Rehainvillers, *G. T.*

CRUSTACEA.

1. Palinurus Sueurii, *Desm.* Dürrheim, Villingen, *Hœn.;* Blittersdorf, Saarbrück, *G. T.*

PISCES.

Fish, and Fish Teeth. Baireuth; Wurtemberg; Rudersdorf, *Munst.*, *G. T.*

REPTILIA.

Plesiosaurus, species not determined. Wurtemberg, *Jæg.;* Baireuth; Rüdersdorf, *G. T.*
1. Ichthyosaurus Lunevillensis, . Lunéville; Wurtemberg, *G. T.*
Great Saurian, genus not determined. Lunéville, *Al. Brong.*
Crocodilus, species not determined. Rüdersdorf, *G. T.*
Chelolia, species not determined. Lunéville; Bindlocherberg; Leineckerberg, *G. T.*

Red or Variegated Sandstone.

PLANTÆ.

1. Equisetum columnare, *Ad. Brong.* Sulz-les-Bains, *G. T.*
1. Calamites arenaceus, *Ad. Brong.* Wasselonne; Marmoutier (Bas-Rhin), *Ad. Brong.*
2. —— remotus, *Ad. Brong.* Wasselonne, *Ad. Brong.*
1. Anomopteris Mougeotii, *Ad. Brong.* Wasselonne; Sulz-les-Bains, *Ad. Brong.*
1. Neuropteris Voltzii, *Ad. Brong.* Sulz-les-Bains, *Ad. Brong.*
2. —— elegans, *Ad. Brong.* Sulz-les-Bains, *Ad. Brong.*
1. Sphenopteris Myriophyllum, *Ad. Brong.* Sulz-les-Bains, *Ad. Brong.*
2. —— palmetta, *Ad. Brong.* Sulz-les-Bains, *Ad. Brong.*
1. Filicites scolopendrioides, *Ad. Brong.* Sulz-les-Bains, *Ad. Brong.*
1. Voltzia brevifolia, *Ad. Brong.* Sulz-les-Bains, *Ad. Brong.*
2. —— elegans, *Ad. Brong.* Sulz-les-Bains, *Ad. Brong.*
3. —— rigida, *Ad. Brong.* Sulz-les-Bains, *Ad. Brong.*

---

* *Strombus scalatus*, Schlot. † *Helix turbilinus*, Schlot.

4. Voltzia acutifolia, *Ad. Brong.* Sulz-les-Bains, *Ad. Brong.*
5. —— heterophylla, *Ad. Brong.* Sulz-les-Bains, *Ad. Brong.*
1. Convallarites erecta, *Ad. Brong.* Sulz-les-Bains, *Ad. Brong.*
2. —— nutans, *Ad. Brong.* Sulz-les-Bains, *Ad. Brong.*
1. Paleoxyris regularis, *Ad. Brong.* Sulz-les-Bains, *Ad. Brong.*
1. Echinostachys oblongus, *Ad. Brong.* Sulz-les-Bains, *Ad. Brong.*
1. Æthophyllum stipulare, *Ad. Brong.* Sulz-les-Bains, *Ad. Brong.*

CONCHIFERA.

1. Plagiostoma lineatum, *Schlot.* Sulz-les-Bains.
2. —— striatum, *Schlot.* Sulz-les-Bains.
1. Avicula socialis, *Desh.* Sulz-les-Bains; Domptail, *Voltz.*
2. —— costata, *Bronn.* Sulz-les-Bains, *G. T.*
1. Mytilus vetustus, *Goldf.* Domptail; Sulz-les-Bains.
1. Trigonia vulgaris, *Schlot.* Domptail.
1. Mya musculoides, *Schlot.* Sulz-les-Bains.
2. —— elongata, *Schlot.* Sulz-les-Bains.

MOLLUSCA.

1. Natica Gailliardoti, *Lefroy.* Domptail.
1. Turritella scalata,   . Domptail; Sulz-les-Bains.
2. —— Schoteri,   . Sulz-les-Bains.
1. Buccinum antiquum, *Goldf.* Sulz-les-Bains, *G. T.*

Zechstein.

PLANTÆ.

1. Fucoides Brardii *, *Ad. Brong.* Cop. Slate, Frankenberg, *Ad. Brong.*
2. —— selaginoides, *Ad. Brong.* Cop. Slate, Mansfeld, *Ad. Brong.*
3. —— lycopodioides, *Ad. Brong.* Cop. Slate, Mansfeld, *Ad. Brong.*
4. —— frumentarius, *Ad. Brong.* Cop. Slate, Mansfeld, *Ad. Brong.*
5. —— pectinatus, *Ad. Brong.* Cop. Slate, Mansfeld, *Ad. Brong.*
6. —— digitatus, *Ad. Brong.* Cop. Slate, Mansfeld, *Ad. Brong.*
1. Pecopteris arborescens, *Ad. Brong.* Mont Muse, Autun, *G. T.*
2. —— abbreviata, *Ad. Brong.* Mont Muse, Autun, *G. T.*
1. Lycopodites Hœninghausii, *Ad. Brong.* Eisleben, *G. T.*
1. Asterophyllites? bulbosa,   . Thuringia, *G. T.*

ZOOPHYTA.

1. Gorgonia anceps, *Goldf.* Glücksbrunn, Thüringerwald, *G. T.*
2. —— antiqua, *Goldf.* Glüksbrunn, *G. T.*
3. —— infundibuliformis, *Goldf.* Glücksbrunn, *G. T.*
1. Calamopora spongites, (var.) *Goldf.* Glücksbrunn, *G. T.*
1. Retepora flustracea, *Phil.* Shelly Mag. Limest., Durham, *Sedg.*
2. —— virgulacea, *Phil.* Shelly Mag. Limest., Durham, *Sedg.*
Polypifera, genera not determined. Durham; Northumberland, *Sedg.*

RADIARIA.

1. Encrinus ramosus, *Schlot.* Glüksbrunn, *Al. Brong.*
1. Cyathocrinites planus, *Mil.* Mag Limest., Durham; Northumberland, *Sedg.*
Crinoidea, genera not determined. Durham; Northumberland, *Sedg.*

CONCHIFERA.

1. Spirifer † trigonalis, *Sow.* Röpsen, Gera, *Hœn.*
2. —— undulatus, *Sow.* Midderidge; Humbleton, *Sedg.*

* *Cupressus Ulmanni,* Bronn.   † Delthyris.

3. Spirifer multiplicatus, *Sow.* Humbleton, *Sedg.*
4. ———— minutus, *Sow.* Humbleton, *Sedg.*
1. Terebratula cristata, *Schlot.* Röpsen, *Hœn.*
2. ———— elongata, *Schlot.* Schmerbach, *Al. Brong.*
3. ———— complanata, *Schlot.* Gera, *G. T.*
4. ———— intermedia, *Schlot.* Röpsen, *Hœn.*
5. ———— inflata, *Schlot.* Röpsen, *Hœn.;* Schmerbach, *Al. Brong.*
6. ———— lacunosa, *Schlot.* Cop. Slate, Schmerbach; Zechstein, Röpsen, *Hœn.*
7. ———— paradoxa, *Schlot.* Schmerbach, *Al. Brong.*
8. ———— pelargonata, *Schlot.* Schmerbach, *Al. Brong.*
9. ———— pygmæa, *Schlot.* Leimstein near Schmalkalden, *Al. Brong.*
———, species not determined. Durham, *Sedg.*
1. Producta* aculeata, *Al. Brong.* Bödingen, *Hœn.;* Gera; Thalitter; Goddelsheim, Logan on the Queiss, *G. T.;* Durham; Northumberland, *Sedg.*
2. ———— rugosa, *Schlot.* Röpsen, Gera, *Hœn.*
3. ———— speluncaria, . Röpsen, *Hœn.;* Glückbrunn, *Al. Brong.*
4. ———— antiquata, *Sow.* Midderidge, *Sedg.*
5. ———— calva, *Sow.* Humbleton; Midderidge, &c., *Sedg.*
6. ———— spinosa, *Sow.* Humbleton, &c., *Sedg.*
? 7. ———— longispina, *Sow.* Cop. Slate, Schmerbach, *Hœn.*
1. Orbicula speluncaria, *Schlot.* Glücksbrunn, *G. T.*
1. Axinus obscurus, *Sow.* Durham, *Sedg.*
Ostrea, species not determined. Northumberland, *Sedg.*
Pecten, species not determined. Humbleton, *Sedg.*
Plagiostoma? species not determined. Humbleton, *Sedg.*
1. Avicula gryphæoides, *Sow.* Humbleton (abundant), *Sedg.*
1. Mytilus keratophagus, *Schlot.* Glücksbrunn, *G. T.*
2. ———— striatus, *Schlot.* Glücksbrunn, *G. T.*
3. ———— squamosus, *Sow.* Ferrybridge, *Sedg.;* Hasel near Goldberg, *G. T.*
1. Modiola acuminata, *Sow.* Black Rocks, Durham, *Sedg.*
1. Arca tumida, *Sow.* Humbleton, Durham, *Sedg.*
1. Cucullæa sulcata, *Sow.* Humbleton, *Sedg.*
Astarte? ———, . Whitley, Northumberland, *Sedg.*
Venus? ———, . Humbleton, *Sedg.*

MOLLUSCA.

Turbo? ———, . Mag. Limest., Marr; Hickleton, *Sedg.*
Pleurotomaria? ———, . Humbleton, *Sedg.*
Melania? ———, . Hawthorn Hive, *Phil.*
Ammonites, species not determined. Humbleton, *Sedg.*

PISCES.

1. Palæothrissum macrocephalum, *Blain.* Cop. Slate, Mansfeld, *Al. Brong.;* Marl Slate, Midderidge; East Thickley, *Sedg.*
2. ———— magnum, *Blain.* Mansfeld, *Al. Brong.;* Midderidge; East Thickley, *Sedg.*
3. ———— inæquilobum, *Blain.* Cop. Slate, Rothenburg, *G. T.;* Bit. Slate, Mont Muse, Autun, *Al. Brong.*
4. ———— macropterum, *Bronn.* Börschweiler, Thuringia, *Hœn.*
5. ———— parvum, *Blain.* Bit. Slate, Autun, *Al. Brong.*
6. ———— blennioides, *Holl.* Mansfeld, *G. T.*
7. ———— elegans, . Marl Slate, Midderidge; East Thickley, *Sedg.*
8. ———— Freieslebense, *Blain.* Mansfeld; Hessia, *G. T.*

---

* Leptæna.

574 *Organic Remains of the Coal Measures.*

Palæothrissum, species not determined. Marl Slate, East Thickley, *Sedg.*; Mag. Limest., Pallion, *Winch.*
1. Stromateus major, *Blain.* Cop. Slate, Hessia, *G. T.*
2. ——— gibbosus, *Blain.* Cop. Slate, North Germany, *G. T.*
1. Clupea Lametherii, *Blain.* Cop. Slate, Eisleben, *G. T.*

REPTILE.

1. Monitor antiquus, *Cuv.* Cop. Slate, Mansfeld; Rothenburg on the Saale; Glücksbrunn; Memmingen, &c. *Al. Brong.*

Organic Remains of the Coal Measures.

PLANTÆ.

[The following list of Plants, discovered fossil in the coal measures, is compiled from the labours of Adolphe Brongniart, Sternberg, Lindley, Hutton, Schlotheim, and other authorities. To have abridged it would have deprived the student not only of a valuable catalogue of localities, but also of an idea of the situations where plants of a similar general character probably existed. The names of the plants, when not otherwise noticed, are those assigned to them by M. Adolphe Brongniart.]

VASCULARES.

Subclass 1.—DICOTYLEDONS.

Euphorbiaceæ.

STIGMARIA *reticulata*, England; *S. Weltheimiana*, Madgeburg; *S. intermedia*, St. Georges-Châtellaison; Montrelais; Wilkesbarre (N. America); *S. ficoides*, Bristol; Dudley; Leeds; Newcastle; St. Georges-Châtellaison; Montrelais; St. Etienne; Liége; Charleroi; Valenciennes; Muhlheim, near Dusseldorf; Bavaria; Silesia; *S. tuberculosa*, Montrelais; Wilkesbarre; *S. rigida*, Anzin, near Valenciennes; *S. minima*, Anglesea, Charleroi; *S. Mosana* (Sauv.), Lüttich; *S. gigantea* (Sauv.), Lüttich.

Coniferæ.

PINITES *Brandlingi* (L. & H.), Newcastle; *P. Withami* (L. & H.), Craigleith, Edinburgh; *P. medullaris* (L. & H.), Craigleith, Edinburgh.
PEUCE *Withami* (L. & H.), Durham.

Doubtful Coniferæ.

SPHENOPHYLLUM *Schlotheimii*, Waldenburg, Silesia; Somerset; *Sph. emarginatum*, Env. of Bath; Wilkesbarre; *Sph. truncatum*, Somerset; *Sph. dentatum*, Newcastle; Anzin; Geislautern; *Sph. quadrifidum*, Terrasson; *Sph. dissectum*, Montrelais; *Sph. pusillum* (Saur.), Lüttich; *Sph. quadriphyllum* (Sauv.), Lüttich; *Sph. multifidum* (Sauv.), Lüttich; *Sph. erosum* (L. & H.), Newcastle.

Dicotyledonous Plants of Doubtful Affinity.

ANNULARIA *minuta*, Terrasson; *A. brevifolia*, Alais; Geislautern; *A. fertilis*, Env. of Bath; St. Etienne; Wilkesbarre; *A. floribunda*, Saarbruck (Sternb.); *A. longifolia*, Env. of Bath; Geislautern; Silesia; Alais; Wilkesbarre; (*Var.*) Charleroi; Terrasson; *A. spinulosa*, Saxony (Sternb.); *A. radiata*, Saarbruck.
ASTEROPHYLLITES *equisetiformis*, Mannebach; Saxony; Rhode Island; *As. rigida*, Alais; Valenciennes; Charleroi; Bohemia; *As. hippuroides*, Alais; *As. longifolia*, Eschweiler (Sternb.); Newcastle; *As. tenuifolia*, Newcastle; Silesia; *As. delicatula*, Charleroi; Anzin; *As. Brardii*, Terrason; *As. diffusa*, Radnitz; *As. elegans* (Sauv.), Belgium; *As. tuber-*

culata (L. & H.), Newcastle; *As. grandis* (L. & H.), Newcastle; *As. ga-lioides* (L. & H.), Newcastle.
BECHERA *grandis* (Sternb.), Newcastle.

Subclass 2. MONOCOTYLEDONS.

Palmæ.

FLABELLARIA? *borassifolia,* Swina.
NŒGGERATHIA *foliosa,* Bohemia; *N. flabellata* (L. & H.), Newcastle.

Cannæ.

CANNOPHYLLITES *Virletii,* St. Georges-Châtellaison.

Monocotyledons of Doubtful Affinity.

STERNBERGIA *angulosa,* Yorkshire; *St. approximata,* Langeac; St. Etienne; *St. distans,* Edinburgh; *St. Volkmanni* (Sauv.), Lüttich.
POACITES *equalis,* Terrasson; *P. striata,* Terrasson.
CYPERITES *bicarinata (L. & H.),* Leebatwood, Shrewsbury.
TRIGONOCARPUM *Parkinsoni,* England and Scotland; *Tr. Nœggerathii,* Langeac; *Tr. ovatum,* Langeac; *Tr. cylindricum,* Langeac.
MUSOCARPUM *prismaticum,* Langeac; *M. difforme,* Langeac; *M. contractum,* Oldham, Lancashire.

CELLULARES.

Equisetaceæ.

EQUISETUM *infundibuliforme* (Bronn.), Saarbruck; *E. dubium,* Wigan, Lancashire.
CALAMITES *Suckowii,* Newcastle; Saarbruck; Liége; Wilkesbarre, Pennsylvania; Richmond, Virginia; *C. decoratus,* Yorkshire; Saarbruck; *C. undulatus,* Yorkshire; Radnitz; Bohemia; *C. ramosus* (Artis), Yorkshire; Mannebach; Wettin, Germany; *C. cruciatus* (Sternb.), Litry; Saarbruck; *C. Cistii,* Montrelais; Saarbruck; Wilkesbarre, Pennsylvania; *C. dubius* (Artis), Yorkshire; Zanesville, Ohio; *C. cannæformis* (Fig.118.), Langeac, Haute-Loire; Alais; Yorkshire; Mannebach; Wettin; Radnitz, Germany; *C. Pachyderma,* St. Etienne; Ireland; *C. nodosus* (Schlot.), Newcastle; Le Lardin, Dordogne; *C. approximatus* (Sternb.), Alais; Liége; St. Etienne; Kilkenny; *C. Steinhaueri,* Yorkshire; *C. Mougeoti,* Edinburgh.

Fig. 118.

Filices.

SPHENOPTERIS *furcata,* Newcastle; Charleroi; Silesia; Saarbruck; *Sp. elegans,* Waldenburg in Silesia; *Sp. stricta* (Sternb.), Northumberland; Glasgow; *Sp. arte-misiæfolia* (Sternb.), Newcastle; *Sp. delicatula* (Sternb.), Saarbruck; Radnitz; *Sp. dissecta,* Montrelais; St. Hippolyte, Vosges; *Sp. linearis* (Sternb.), Swina, Bohemia; Engl.; *Sp. Brardii,* Le Lardin; *Sp. trifoliolata,* Anzin near Valenciennes; Mons; Silesia; Yorkshire; *Sp. Schlotheimii* (Sternb.), Doutweiler near Saarbruck; Waldenburg and Breitenbach, Silesia; *Sp. fragilis,* Breitenbach; *Sp. Hœninghausii,* Newcastle; Werden; *Sp. Du-buissonis,* Montrelais; *Sp. distans* (Sternb.), Ilmenau, Silesia; *Sp. gracilis,* Newcastle; *Sp. latifolia,* Newcastle; Saarbruck; *Sp. Verleti,* St. Georges-Châtellaison; *Sp. Gravenhorstii,* Silesia; Anglesea;

Sp. Loshii, Newcastle; *Sp. tenuifolia*, St. Georges-Châtellaison; *Sp. rigida*, Waldenburg; *Sp. acuta*, Werden; *Sp. trichomanoides*, Anzin; *Sp. tenella*, Yorkshire; *Sp. alata*, Geislautern; *Sp. conferta*, Waldenburg; *Sp. multifida* (Sauv.), Lüttich; *Sp. crenata* (L. & H.), Newcastle; *Sp. affinis* (L. & H.), Newcastle; *Sp. crithmifolia* (L. & H.), Newcastle; *Sp. dilatata* (L. & H.), Newcastle; *Sp. caudata* (L. & H.), Newcastle.

CYCLOPTERIS* *orbicularis*, St. Etienne; Liége; *C. trichomanoides*, St. Etienne; *C. obliqua*, Yorkshire; *C. flabellata*, Berghaupten; *C. gibbosa* (Sauv.), Lüttich; *C. semicordata* (Sauv.), Lüttich; *C. cycloidea* (Sauv.), Lüttich; *C. reniformis* (Sauv.), Lüttich; *C. tendulata* (Sauv.), Lüttich.

NEUROPTERIS *acuminata*, Klein-Schmalkalden; Newcastle; *N. Villersii*, Alais, Gard; *N. rotundifolia*, Plessis, Calvados; Yorkshire; *N. Loshii*, Gloucestershire; Newcastle; Anzin; Liége; Wilkesbarre; *N. tenuifolia* (Sternb.), Saarbruck; Miereschau; Bohemia; Waldenburg, Silesia; Montrelais; *N. heterophylla*, Saarbruck; Valenciennes; Newcastle; *N. flexuosa* (Sternb.), Env. of Bath; Saarbruck; *N. gigantea* (Sternb.), Saarbruck; Schatzlar; Wettin; Newcastle; *N. oblongata*, Paulton, Somerset; *N. cordata*, Leebotwood, Shrewsbury; Alais; St. Etienne; *N. Scheuchzeri* (Hoffman), England; Osnabruck; Wilkesbarre; *N. angustifolia*, Env. of Bath; Wilkesbarre; *N. acutifolia*, Env. of Bath; Wilkesbarre; *N. crenulata*, Saarbruck; *N. macrophylla*, Dunkerton, Somerset; *N. auriculata*, St. Etienne; *N. Soretii*, Newcastle.

ODONTOPTERIS *Brardii*, Le Lardin and Terrasson, Dordogne; St. Etienne; *O. crenatula*, Terrasson; *O. minor*, Le Lardin; St. Etienne; *O. obtusa*, Terrasson; Leebotwood, Shrewsbury; *O. Schlotheimii*, Mannebach; Wettin; *O. appendiculata* (Sauv.), Lüttich.

PECOPTERIS *blechnoides*, Werden near Dusseldorf; St. Priest, Loire; *P. Candolliana*, Alais, Gard; *P. cyathea*, St. Etienne; *P. arborescens*, St. Etienne; Aubin, Aveyron; Anzin; Mannebach; *P. platyrachis*, St. Etienne; *P. polymorpha*, St. Etienne; Alais; Litry; Wilkesbarre; *P. oreopteridis* (Sternb.), Le Lardin; Mannebach; Wettin; *P. Bucklandi*, Env. of Bath; *P. aquilina* (Sternb.), Mannebach and Wettin (Schlot.); *P. Schlotheimii*, Mannebach (Schlot.); Geislautern; *P. pteroides*, Mannebach; Aubin; *P. Davreuxii*, Liége; Valenciennes; *P. Mantelli*, Newcastle; Liége; *P. conchitica*, Newcastle; Namur; Saarbruck; Silesia; *P. Serlii*, Env. of Bath; St. Etienne; Geislautern; Wilkesbarre; *P. Grandini*, Geislautern; *P. crenulata*, Geislautern; *P. marginata*, Alais; *P. gigantea*, Abascherhütte; Treves; Liége; Saarbruck; Wilkesbarre; *P. nervosa*, Wales; Liége; Rolduc; Waldenburg; *P. obliqua*, Valenciennes; *P. Brardii*, Le Lardin; *P. Defrancii*, Saarbruck; *P. ovata*, St. Etienne; *P. Plukenetii*, Alais; St. Etienne; *P. arguta* (Sternb.), St. Etienne; Saarbruck (Schlot.); Rhode Island, United States; *P. cristata*, Saarbruck; *P. aspera*, Montrelais; *P. Miltoni* (Artis), Yorkshire; Saarbruck; *P. abbreviata*, Valenciennes; *P. microphylla*, Saarbruck; *P. æqualis*, Fresnes and Vieux-Condé near Valenciennes; Silesia; *P. acuta*, Saarbruck; Ronchamp, Haute Saone; *P. unita*, Geislautern; St. Etienne; *P. debilis*, Ronchamp; *P. dentata*, Valenciennes; Doutweiler; *P. angustissima* (Sternb.), Swina, Bohemia; Saarbruck; *P. gracilis*, Geislautern; Valenciennes; *P. pinnæformis*, Fresnes and Vieux-Condé; Saarbruck; *P. triangularis*, Fresnes and Vieux-Condé; *P. pectinata*, Geislautern; *P. plumosa*, Yorkshire (Artis); Saarbruck; Valenciennes; *P. pannonica* (Sauv.), Lüttich; *P. heterophylla* (Sauv.), Lüttich; *P. amœna* (Sauv.), Lüttich; *P. chnophoroides* (Sauv.), Lüttich; *P. distans* (Sauv.), Lüttich; *P. adiantoides* (L. & H.), Newcastle; *P. heterophylla* (L. & H.), Newcastle; *P. rigida* (Sauv.), Lüttich †.

---

* *Otopteris*, Sauveur.

† To this list should be added the other species enumerated by Count Sternberg, which however may, as M. Ad. Brongniart remarks, be the same with some of those

LONCHOPTERIS *Tournailii*, Valenciennes.
SCHIZOPTERIS *anomala*, Saarbruck.
CAULOPTERIS *primæva* (L. & H.), Radstock, Bath.

Lycopodiaceæ.

LYCOPODITES *piniformis*, Saxe-Gotha; St. Etienne; *L. Gravenhorstii*, Silesia; *L. Hœninghausii*, Eisleben; *L. imbricatus*, St. Georges-Châtellaison; *L. phlegmarioides*, Newcastle; Silesia; *L. tenuifolius*, St. Georges-Châtellaison; *L.? filiciformis*, Wettin; *L.? affinis*, Wettin.
SELAGINITES *patens*, Edinburgh; *S. erectus*, Mont Jean near Angers.
LEPIDODENDRON *selaginoides*, Bohemia; Silesia; Newcastle; *Lep. elegans*, Swina, Bohemia; *Lep. Bucklandi*, Colebrooke Dale; *Lep. Ophiurus*, Newcastle; Charleroi; *Lep. rugosum*, Charleroi; Valenciennes; *Lep. Underwoodii*, Anglesea; *Lep. taxifolium*, Ilmenau; *Lep. insigne*, St. Ingbert, Bavaria; *Lep. Sternbergii*, Swina; Newcastle; *Lep. longifolium*, Swina; *Lep. ornatissimum** (Sternb.), Edinburgh; Yorkshire; Silesia; Durham; *Lep. tetragonum* (Sternb.), Newcastle; *Lep. venosum*, Waldenburg; *Lep. transversum*, Glasgow; *Lep. Volkmannianum* (Sternb.), Silesia; *Lep. Rhodianum* (Sternb.), Yorkshire; Valenciennes; Silesia; *Lep. cordatum*, Durham; *Lep. obovatum* (Sternb.), Newcastle; Radnitz, Bohemia; Silesia; Fresnes and Vieux-Condé; *Lep. dubium*, Newcastle; *Lep. læve*, county of La Marck; *Lep. pulchellum*, Alais; Liége; *Lep. cœlatum*, Yorkshire; *Lep. varians*, Saarbruck; Wilkesbarre; *Lep. carinatum*, Montrelais; St. Georges-Châtellaison; *Lep. crenatum* (Sternb.), Bohemia; Eschweiler; Essen; Zanesville; *Lep. aculeatum* (Sternb.), Essen; Bohemia; Silesia; Wilkesbarre; *Lep. distans*, St. Etienne; *Lep. laricinum* (Sternb.), Bohemia; Silesia; *Lep. rimosum* (Sternb.), Bohemia; *Lep. undulatum* (Sternb.), Bohemia; *Lep. confluens* (Sternb.), Silesia; Eschweiler; *Lep. imbricatum* (Sternb.), Eschweiler, Wettin; *Lep. majus*, Geislautern; *Lep. lanceolatum*, Montrelais; *Lep. Boblayi*, Valenciennes; *Lep. trinerve*, Montrelais; *Lep. lineare*, Alais; *Lep. ornatum*, Shropshire; *Lep. undulatum*, England; *Lep. emarginatum*, Yorkshire; *Lep. acerosum* (L. & H.), Newcastle; *Lep. dilatatum* (L. & H.), Newcastle; *Lep. gracile* (L. & H.), Newcastle.
ULODENDRON *majus* (L. & H.), Newcastle.
LEPIDOPHYLLUM *lanceolatum* (L. & H.), Newcastle; *L. intermedium* (L. & H.), Leebotwood, Shrewsbury.
LEPIDOSTROBUS *variabilis* (L. & H.), Newcastle; *L. ornatus*, Newcastle.
CARDIOCARPON *majus*, St. Etienne; Langeac; *C. Pomieri*, Langeac; *C. cordiforme*, Langeac; *C. ovatum*, Langeac; *C. acutum*, Langeac.

Plants of Uncertain Affinity.

SIGILLARIA *punctata*, Bohemia; *S. appendiculata*, Bohemia; Yorkshire; *S. peltigera*, Alais; *S. lævis*, Liége; *S. canaliculata*, Saarbruck; *S. Cortei*, Essen; *S. elongata*, Charleroi; Liége; *S. reniformis*, Newcastle; Mons; Essen; *S. Hippocrepis*, Mons; *S. Davreuxii*, Liége; *S. Candollii*, Alais; *S. oculata*, Newcastle; Bohemia; *S. orbicularis*, St. Etienne; Saarbruck; *S. tessellata*, Env. of Bath; Alais; Eschweiler; Wilkesbarre; *S. Boblayi*, Anzin; *S. Knorrii*, Saarbruck; *S. elliptica*, St. Etienne; *S. transversalis*, Eschweiler near Aix-la-Chapelle; *S. subrotunda*, Doutsweiler, near Saar-

above enumerated. *Pecopteris orbiculata*, Swina, Bohemia; *P. discreta*, Swina; *P. cordata*, Swina; *P. varians*, Swina; *P. obtusata*, Radnitz, Bohemia; *P. undulata*, Radnitz; *P. repanda*, Radnitz; *P. antiqua*, Radnitz; *P. crenata*, Minitz, Bohemia; *P. elegans*, Schatzlar, Bohemia; *P. incisa*, Waldenburg, Silesia; Schatzlar; *P. dubia,* Bohemia.
* *Ulodendron minus*, L. & H.

2 P

bruck; *S. cuspidata,* St. Etienne; *S. notata,* Liége, Saarbruck; Silesia;
S. Tournailii, Charleroi; Valenciennes; *S. trigona,* Radnitz, Bohemia;
S. mammillaris, Charleroi; *S. alveolaris,* Saarbruck; *S. hexagona,* Eschweiler;
Bochum; *S. elegans,* Bochum; *S. Brardii,* Terrasson; *S. lævigata,* Mon-
trelais; *S. Serlii,* Paulton, Somerset; *S. pachyderma,* Newcastle; *S. alter-
nans* (L. & H.), Northumberland; Eschweiler; *S. catenulata* (L. & H.),
Newcastle; *S. alternans,* (Sauv.), Lüttich; *S. contigua* (Sauv.), Lüttich;
S. antiqua (Sauv.), Lüttich; *S. minuta* (Sauv.), Lüttich.
  VOLKMANNIA *polystachya,* Waldenburg, Silesia; *V. distachya,* Swina;
V. erosa, Terrasson; *V. grandis* (Sauv.), Belgium*.
  POLYPORITES *Bowmanni* (L. & H.), Denbighshire.

CONCHIFERA†.

1. Pecten papyraceus, *Sow.*   Werden, *Hœn.;* Bradford, *Hailstone.*
2. ——— dissimilis, *Flem.*   Locality not stated.
1. Mytilus crassus, *Flem.*   Scotland, *Flem.;* Werden? *Hœn.*
1. Lutricola truncata, *Goldf.*   Niederstaufenbach, Cusel; Werden, *G. T.*
2. ——— Blainvillii, *Goldf.*   Werden, *G. T.*
3. ——— acuta‡, *Goldf.*   Werden; Lüttich; Kammerberg, Ilmenau;
        Bochum, *G. T.*
1. Unio Urii, *Flem.*   Rutherglen, Scotl., *Flem.*
2. ——— uniformis, *Sow.*   Wettin; Loebejun, *G. T.*
3. ——— subconstrictus, *Sow.*   Derbyshire, &c., *Sow.*
1. Nucula attenuata, *Flem.*   Rutherglen, *Flem.*
2. ——— gibbosa, *Flem.*   Rutherglen, *Flem.*
1. Mya? tellinaria,   . Lüttich, *Hœn.*
2. ———? ventricosa,   . Lüttich, *Hœn.*
3. ———? minuta,   . Kammerberg near Ilmenau, *Hœn.*

MOLLUSCA.

1. Turritella Urii, *Flem.*   Rutherglen, Scotl., *Flem.*
2. ——— elongata, *Flem.*   Rutherglen, *Flem.*
1. Bellerophon decussatus, *Flem.*   Linlithgowshire, *Flem.*
2. ——— striatus, *Flem.*   Linlithgowshire, *Flem.*
1. Orthoceratites Steinhaueri, *Sow.*   Coal Measure Limestone, Choquier
        near Lüttich, *Hœn.;* Yorkshire, *Sow.*
2. ——— cylindraceus, *Flem.*   Linlithgowshire, *Flem.*
3. ——— attenuatus, *Flem.*   Linlithgowshire, *Flem.*
4. ——— sulcatus, *Flem.*   Linlithgowshire, *Flem.*

---

  * Whether or not certain of the coal strata of N. America be precisely identical
with those of Europe, or may, like some in Ireland described by Mr. Weaver, be of
the grauwacke series, it appears that many of the same plants are found, according
to the foregoing list, in both Europe and America. These are: Calamites 3 species,
Neuropteris 3, Pecopteris 4, Sigillaria 1, Sphenophyllum 1, Lepidodendron 3, Stigmaria 2, Annularia 2, Asterophyllites 1.
  Vegetables discovered as yet only in America.—*Neuropteris Cistii,* Wilkesbarre;
N. Grangeri, Zanesville; *N. macrophylla,* Wilkesbarre; *Sigillaria Cistii,* Wilkesbarre; *S. rugosa,* Wilkesbarre; *S. Sillimanni,* Wilkesbarre; *S. obliqua,* Wilkesbarre;
S. dubia, Wilkesbarre; *Lycopodites Sillimanni,* Hadley, Connecticut; *Lepidodendron
mammillare,* Wilkesbarre; *Lep. Cistii,* Wilkesbarre; *Poacites lanceolata,* Zanesville.
Pecopteris punctulata, discovered at Wilkesbarre, is, according to M. Ad. Brongniart,
found at the Montagne des Rousses, in Oisans.
  † How far all the marine remains noticed in this list may be really discovered in
the body of the coal measures, and not in the alternations with the inferior rocks,
may perhaps be questionable; some of them seem certainly to occur in the body of
the coal measures.
  ‡ *Unio acutus,* Sow.

5. Orthoceratites undatus, *Flem.* Linlithgowshire, *Flem.*
1. Ammonites Listeri, *Sow.* Bit. Shale, Werden, *Hœn.*; Yorkshire, *Steinhauer;* Liége; Melin, *Munst.*
2. ———— subcrenatus, *Schlot.* Werden, *Munst.*
3. ———— Diadema, *Haan.* Choquier, *Munst.*
4. ———— sphæricus, *Goldf.* Choquier, *Munst.*
5. ———— carbonarius, *Goldf.* Lüttich; Werden; Wetter, *G. T.*

Pisces.

1. Palæothrissum macropterum, *Bronn.* Lebach, Börschweiler, Saarbruck, *G. T.*
2. ———— Freiesbebense, *Blain.* Münsterappel, Saarbruck, *G. T.*
1. Acanthæssus Bronnii, *Ag.* Lebach; Börschweiler; Lüttich, *G. T.*
Ichthyodorulites, *Buckl. & De la B.* Shale, Felling Colliery, Durham, *Taylor;* Rutherglen, *Ure;* Sunderland, *Sow.*
Fish palates, Coal. Tong, near Leeds, *George *.*

Organic Remains of the Carboniferous Limestone.

Zoophyta.

1. Gorgonia ripisteria, *Goldf.* Tournay, *G. T.*
2. ———— infundibuliformis, *Goldf.* Lindlar; Gimborn; Arnsberg, *G. T.*
3. ———— antiqua, *Goldf.* Arnsberg, *G. T.*
4. ———— anceps, *Goldf.* Arnsberg, *G. T.*
1. Cellepora Urii, *Flem.* Rutherglen, *Flem.*
1. Retepora elongata, *Flem.* Rutherglen, *Flem.*
1. Caryophyllia duplicata, *Mart.* Derbyshire, *Mart.*
2. ———— affinis, *Mart.* Derbyshire, *Mart.*
3. ———— juncea, *Flem.* Rutherglen, *Flem.*
1. Cyathophyllum excentricum, *Goldf.* Ratingen near Dusseldorf, *Goldf.*
2. ———— flexuosum †, *Goldf.* King's County; Limerick, *Weav.*
3. ———— pentagonum, *Goldf.* Namur, *G. T.*
4. ———— Ananas, *Goldf.* Namur, *G. T.*
1. Astrea undulata, . Bristol, *Park.*
1. Tubipora tubularia, *Lam.* Theux near Liége, *Al. Brong.*
1. Syringopora ramulosa, *Goldf.* Olne, Limburg, Belgium, *G. T.*
2. ———— reticulata, *Goldf.* Olne, *G. T.*
3. ———— filiformis, *Goldf.* Arnsberg, *G. T.*
1. Calamopora polymorpha, *Goldf.* Namur, *Goldf.*
2. ———— spongites, *Goldf.* Schwelm; Sundwig, *G. T.*
1. Aulopora compressa, *Goldf.* Ratingen, *G. T.*
1. Favosites Gothlandica, *Lam.* Dublin, *Al. Brong.*
2. ———— septosus, *Flem.* Scotland, *Flem.*
3. ———— depressus, *Flem.* Scotland, *Flem.*
1. Lithostrotion striatum, *Park.* Wales, *Park.*
2. ———— floriforme, *Mart.* Bristol, *Woodward.*
3. ———— marginatum, *Flem.* Scotland, *Flem.*
Polypifers, genera undetermined. Very numerous in the British Isles.

Radiaria.

1. Pentremites Derbiensis, *Sow.* Derbyshire, *Watson.*

———————
* Zoological Journal, vol. ii. pl. 2. fig. 2. † *Amplexus coralloides*, Mill.

2. Pentremites ellipticus, *Sow.* Preston, Lancashire, *Kenyon.*
3. —— ovalis, *Goldf.* Ratingen, Dusseldorf, *Goldf.*
1. Poteriocrinites crassus, *Miller.* Somerset; Yorkshire, *Miller.*
2. —— tenuis, *Miller.* Mendip Hills; Bristol, *Miller.*
1. Platycrinites lævis, *Miller.* Dublin; Bristol, *Miller;* Ratingen; Namur, *Goldf.*
2. —— rugosus, *Miller.* Mendip Hills; Caldy Island, *Miller.*
3. —— tuberculatus, *Miller.* Mendip Hills, *Miller.*
4. —— granulatus, *Miller.* Mendip Hills, *Miller.*
5. —— striatus, *Miller.* Bristol, *Miller.*
6. —— pentangularis, *Miller.* Mendip Hills; Bristol, *Miller.*
7. —— depressus, *Goldf.* Ratingen, *Goldf.*
1. Actinocrinites triacontadactylus, *Miller.* Yorkshire; Bristol; Mendip Hills, *Miller.*
2. —— polydactylus, Mendip Hills; Caldy Island, *Miller.*
3. —— lævis, *Miller.* Ratingen, *Goldf.*
4. —— granulatus, *Goldf.* Ratingen, *Goldf.*
5. —— tesseratus, *Goldf.* Schweln, *Goldf.*
1. Melocrinites hieroglyphicus, *Goldf.* Stellberg, Aix-la-Chapelle, *Goldf.*
1. Rhodocrinites verus, *Miller.* Bristol; Mendip Hills, *Miller.*
1. Cyathocrinites planus, *Miller.* Clevedon, Bristol, *Miller.*
2. —— quinquangularis, *Miller.* Bristol, *Miller.*

ANNULATA.

1. Serpula compressa, *Sow.* Lothian.

CONCHIFERA.

1. Spirifer* ambiguus, *Sow.* Ratingen, *Hœn.;* Derbyshire, *Watson.*
2. —— bisulcatus, *Sow.* Visé, *Hœn.;* Dublin, *Sow.;* Liége, *Dum.*
3. —— glaber, *Sow.* Ratingen, *Hœn.;* Derbyshire, *Martin;* Ireland, *Sow.;* Liége, *Dum.*
4. —— oblatus, *Sow.* Visé, *Hœn.;* Derbyshire; Flintshire, *Farey;* Lüttich, *G. T.*
5. —— obtusus, *Sow.* Ratingen, *Hœn.;* Yorkshire, *Ducket.*
6. —— pinguis, *Sow.* Dublin, *Sow.;* Liége, *Dum.*
7. —— plicatus, *Hœn.* Ratingen, *Hœn.*
8. —— rotundatus, *Sow.* Limerick, *Wright;* Visé, *Hœn.*
9. —— trigonalis †, *Sow.* Ratingen; Visé, *Hœn.;* Derbyshire, *Martin;* Rutherglen, *Flem.*
10. —— triangularis ‡, *Sow.* Derbyshire, *Martin.*
11. —— striatus, *Sow.* Derbyshire, *Martin;* Namur, *Al. Brong.;* Liége, *Dum.;* Ratingen, *G. T.*
12. —— attenuatus, *Sow.* Dublin, *Sow.;* Liége, *Dum.*
13. —— distans, *Sow.* Dublin, *Sow.*
14. —— resupinatus, *Sow.* Derbyshire, *Martin;* Rutherglen, *Flem.*
15. —— Martini, *Sow.* Derbyshire, *Sow.*
16. —— Urii, *Flem.* Rutherglen, *Ure.*
17. —— exaratus, *Flem.* West Lothian, *Flem.*
18. —— cuspidatus §, *Sow.* Bristol, *Beeke;* Derbyshire, *Martin.* Ratingen, *G. T.*

* *Delthyris,* Dalm.
† *Producta trigonalis,* Deshayes. According to the same author the genus *Spirifer* should be suppressed, the species composing it being referrible either to *Terebratula* or *Producta.*
‡ *Delthyris compressa,* Goldf.   § *Delthyris elevata,* Dalm.

19. Spirifer minimus, *Sow.* Derbyshire, *Watson.*
20. —— octoplicatus *, *Sow.* Derbyshire, *Martin.*
1. Delthyris dorsata, *Goldf.* Arnsberg, *G. T.*
2. —— bisulcata, *Goldf.* Lüttich; Dublin, *G. T.*
3. —— polymorpha, *Goldf.* Ratingen, *G. T.*
4. —— incisa, *Goldf.* Ratingen, *G. T.*
5. —— biplicata, *Goldf.* Ratingen, *G. T.*
6. —— radiata, *Goldf.* Ratingen, *G. T.*
7. —— thecaria, *Goldf.* Ratingen, *G. T.*
8. —— striatula, *Goldf.* Ratingen, *G. T.*
9. —— concentrica, *Goldf.* Ratingen, *G. T.*
1. Terebratula acuminata, *Sow.* Ratingen, *Hœn.;* Yorkshire; Derbyshire, *Sow.;* Rutherglen, *Flem.;* (var.) Clitheroe, Lancashire, *Stokes;* (var.) Ireland, *Sow.*
2. —— crumena, *Sow.* Derbyshire, *Martin.*
3. —— hastata, *Sow.* Dublin ; Limerick, *Wright;* Bristol, *Sow.*
4. —— resupinata †, *Sow.* Ratingen, *Hœn.;* Derbyshire, *Martin.*
5. —— affinis, *Sow.* ‡ Derbyshire, *Martin;* Liége, *Dum.*
6. ——? lineata, *Sow.* Derbyshire, *Martin;* Liége, *Dum.*
7. ——? imbricata § *Sow.* Derbyshire; Yorkshire, *Sow.;* Liége, *Dum.*
8. —— Sacculus, *Sow.* Derbyshire, *Martin;* Rutherglen, *Flem.*
9. —— lateralis, *Sow.* Dublin, *Moore.*
10. —— Wilsoni ‖, *Sow.* Mordeford, E. S. E. of Hereford, *Sow.;* Liége, *Dum.*
11. —— Mantiæ, *Sow.* Ireland, *Sow.*
12. —— cordiformis, *Sow.* Ireland, *Sow.*
13. —— platyloba, *Sow.* Clitheroe, *Stokes.*
14. —— Pugnus, *Sow.* Derbyshire, *Martin;* Ireland, *Sow.*
15. —— Fimbria, *Sow.* Gloucestershire, *Taylor.*
16. —— reniformis, *Sow.* Dublin, *Sow.*
17. —— lateralis, *Sow.* Dublin, *Weav.*
18. —— diodonta, *Dalm.* Ratingen, *G. T.*
19. —— plicatella, *Dalm.* Ratingen, *G. T.*
20. —— Amygdala, *Goldf.* Visé, *G. T.*
21. —— complicata, *Goldf.* Iserlohn, *G. T.*
1. Atripa cassidea, *Dalm.* Ratingen, *G. T.*
1. Producta ¶ antiquata, *Sow.* Visé; Ratingen, *Hœn.;* Derbyshire, *Sow.;* Cloghran, Dublin, *Humphreys.*
2. —— comoides, *Sow.* Visé; Ratingen, *Hœn.;* Llangeveni, Anglesea, *Farey.*
3. —— concinna, *Sow.* Visé; Ratingen, *Hœn.;* Derbyshire; Yorkshire, *Sow.*
4. —— fimbriata, *Sow.* Visé; Ratingen, *Hœn.;* Derbyshire, *Stokes.*
5. —— fornicata. Ratingen, *Hœn.*
6. —— hemisphærica, *Sow.* Ratingen, *Hœn.;* Caermarthenshire, *Taylor;* Liége, *Dum.*
7. —— humerosa, *Sow.* Ratingen, *Hœn.*
8. —— latissima, *Sow.* Ratingen ; Visé, *Hœn.;* Tydmawr, Anglesea, *Farey;* Liége, *Dum.*
9. —— lobata, *Sow.* Ratingen ; Visé, *Hœn.;* Northumberland ; Derbyshire, *Sow.;* Arran, *Leach;* Liége, *Dum.*
10. —— Martini, *Sow.* Ratingen ; Visé, *Hœn.;* Derbyshire, *Martin;* Yorkshire, *Danby;* Liége, *Dum.*
11. —— personata, *Sow.* Ratingen, *Hœn.;* Derbyshire, Kendal, *Sow.*

12. Producta plicatilis, *Sow.*   Ratingen; Visé, *Hœn.*; Derbyshire, *Stokes*;
    Liége, *Dum.*
13. —— punctata, *Sow.*   Visé; Ratingen, *Hœn.*; Derbyshire, *Martin*;
    Rutherglen, *Flem.*; Liége, *Dum.*
14. —— spinulosa, *Sow.*   Ratingen; Visé, *Hœn.*; Linlithgowshire, *Flem.*
15. —— sulcata, *Sow.*   Visé, *Hœn.*; Derbyshire, *Salt*; Liége, *Dum.*
16. —— transversa,   .   Visé, *Hœn.*
17. —— Flemingii, *Sow.*   Rutherglen, *Ure*; Linlithgow, *Flem.*
18. —— longispina *, *Sow.*   Linlithgow, *Flem.*
19. —— crassa, *Flem.*   Derbyshire, *Martin.*
20. —— aculeata, *Sow.*   Derbyshire, *Martin.*
21. —— scabricula, *Sow.*   Derbyshire, *Martin*; Visé, *Al. Brong.*; Liége,
    Dum.
22. —— spinosa, *Sow.*   Scotland, *Flem.*; Ratingen, *G. T.*
23. —— Scotica, *Sow.*   Scotland, *Flem.*; Liége, *Al. Brong.*
24. —— gigantea, *Sow.*   Derbyshire, *Martin*; Yorks., *Sow.*
25. —— costata, *Sow.*   Glasgow, *Murch.*; Visé, *G. T.*
26. —— depressa,   .   Liége, *Dum.*
 1. Leptæna lævis, *Goldf.*   Visé, *G. T.*
 2. —— corrugata, *Goldf.*   Ratingen; Visé, *G. T.*
 3. —— conoidea, *Goldf.*   Visé, *G. T.*
 1. Crania prisca, *Hœn.*   Ratingen, *Hœn.*
 1. Inoceramus vetustus,   .   Ratingen, *G. T.*
 1. Pecten granosus, *Sow.*   Queen's County, Ireland, *Sow.*
 2. —— plicatus, *Sow.*   Queen's County, *Sow.*
 1. Megalodon cucullatus, *Sow.*   Liége, *Dum.*
 1. Nucula Palmæ, *Sow.*   Derbyshire, *Martin.*
 1. Arca cancellata, *Sow.*   Derbyshire, *Martin.*
 1. Isocardia lineata, *Goldf.*   Visé, *G. T.*
 1. Cardium elongatum, *Sow.*   Ratingen, *Hœn.*; Derbyshire, *Martin.*
 2. —— hibernicum, *Sow.*   Ratingen, *Hœn.*; Queen's County, *Sow.*;
    Limerick, *Weav.*; Namur, *Om. d'Halloy.*
 3. —— alæforme, *Sow.*   Queen's County, *Sow.*
 4. —— carinatum, *Goldf.*   Ratingen, *G. T.*
 1. Lucina lineata, *Goldf.*   Ratingen, *G. T.*
 2. —— cincta, *Goldf.*   Ratingen, *G. T.*
 1. Sanguinolaria gibbosa, *Sow.*   Queen's County, *Sow.*
 1. Solen pelagicus, *Goldf.*   Ratingen, *G. T.*
 2. —— vetustus, *Goldf.*   Ratingen, *G. T.*

MOLLUSCA.

 1. Pileopsis vetusta, *Sow.*   Ratingen, *G. T.*
 1. Melania constricta, *Sow.*   Derbyshire, *Martin.*
 1. Ampullaria helicoides, *Sow.*   Queen's County, *Sow.*
 2. —— nobilis, *Sow.*   Queen's County, *Sow.*
 1. Nerita striata †, *Flem.*   Corry, Arran, *Flem.*
 2. —— spirata, *Sow.*   Bristol, *Beeke*; Derbyshire, *Sow.*; Liége, *Dum.*;
    Ratingen; Visé, *G. T.*
 3. —— lineata, *Goldf.*   Ratingen, *G. T.*
 1. Delphinula tuberculata, *Flem.*   West Lothian, *Flem.*
 1. Euomphalus nodosus, *Sow.*   Ratingen, *Hœn.*; Derbyshire, *Martin.*
 2. —— angulosus, *Sow.*   Benthnall Edge, *Flem.*
 3. —— catillus, *Sow.*   Ratingen, *Hœn.*; Derbyshire, *Martin*; Liége, *Dum.*
 4. —— delphinularis‡, *Hœn.*   Ratingen, *Hœn.*

---

* Considered the same with *P. Flemingii* by Dr. Fleming.
† According to Dr. Fleming this shell closely approaches the recent *Nerita polita.*
‡ *Cirrus delphinularis,* Goldfuss; *Helicites delphinularis,* Schlotheim.

5. Euomphalus pentangulatus, *Sow.* Ratingen, *Hœn.*; Dublin, *Weav.*;
   Namur, *Al. Brong.*; Liége, *Dum.*
6. ———— coronatus, . Visé; Ratingen, *Hœn.*
7. ———— rotundatus, . Ratingen; Visé, Pfaffrath, *Hœn.*
8. ———— rugosus, *Sow.* Colebrooke Dale, *Sow.*
9. ———— discus, *Sow.* Colebrooke Dale, *Sow.*; Ratingen, *G. T.*
10. ———— æqualis*, . Kendal, *Sow.*
11. ———— lævis, *Goldf.* Ratingen, *G. T.*
12. ———— Dionysii†, *Goldf.* Ratingen; Visé, *G. T.*; Yorkshire, *Sow.*;
    Liége, *Dum.*
———, species not determined. Argenteau, Belgium; Visé, *Al. Brong.*
1. Trochus catenulatus, *Goldf.* Ratingen, *G. T.*
2. ———— cingulatus, *Goldf.* Ratingen, *G. T.*
3. ———— scalaris, *Goldf.* Ratingen, *G. T.*
4. ———— carinatus, *Goldf.* Ratingen? *G. T.*
5. ———— acutus‡, *Goldf.* Derbyshire, *Martin;* Visé, *G. T.*
6. ———— alutaceus, *Goldf.* Visé, *G. T.*
7. ———— tæniatus, *Goldf.* Ratingen, *G. T.*
1. Turbo carinatus§, *Hœn.* Visé; Ratingen, *Hœn.*; Yorkshire, *Sow.*;
   Liége, *Dum.*
2. ——— Tiara, *Sow.* Preston, Lancashire, *Gilbertson.*
3. ——— striatus‖, *Goldf.* Visé, *Hœn.*; Derbyshire, *Martin.*
4. ——— lineatus, *Goldf.* Visé, *G. T.*
5. ——— canaliculatus, *Goldf.* Ratingen, *G. T.*
6. ——— cingulatus, *Goldf.* Ratingen, *G. T.*
1. Rotella helicinæformis ¶, *Goldf.* Iserlohn, *G. T.*
2. ——— lævigata, *Goldf.* Ratingen, *G. T.*
3. ——— elliptica, *Goldf.* Ratingen, *G. T.*
1. Helix? cirriformis, *Sow.* Derbyshire, *Martin.*
1. Turritella constricta**, *Flem.* Derbyshire, *Martin.*
2. ——— spinata, *Goldf.* Ratingen, *G. T.*
3. ——— angulata, *Sow.* Ratingen, *G. T.*
4. ——— costata, *Goldf.* Ratingen, *G. T.*
5. ——— cancellata, *Goldf.* Iserlohn, *G. T.*
6. ——— angustata, *Goldf.* Ratingen, *G. T.*
7. ——— conoidea, *Goldf.* Ratingen, *G. T.*
8. ——— tenuis, *Goldf.* Ratingen, *G. T.*
9. ——— acuminata, *Goldf.* Ratingen, *G. T.*
10. ——— prisca, *Goldf.* Ratingen, *G. T.*
1. Buccinum acutum, *Sow.* Queen's County, Ireland, *Sow.*; Liége, *Dum.*
1. Phasianella auricularis, *Goldf.* Ratingen, *G. T.*
2. ——— striata, *Goldf.* Ratingen, *G. T.*
3. ——— neritoidea, *Goldf.* Ratingen, *G. T.*
1. Bellerophon hiulcus, *Sow.* Ratingen, *G. T.*; Derbyshire, *Martin.*
2. ——— apertus, *Sow.* Kendal; Bristol; Yorks., *Sow.*; Liége, *Dum.*
3. ——— tenuifascia, *Sow.* Visé; Ratingen, *Hœn.*; Kendal; Derbyshire,
   and Yorkshire, *Sow.*; Liége, *Dum.*
4. ——— costatus, *Sow.* Visé; Ratingen, *G. T.*; Derbyshire, *Sow.*
5. ——— depressus, *Montfort.* Ratingen, *Hœn.*
6. ——— Cornu-Arietis, *Sow.* Kendal, *Sow.*; Linlithgowshire, *Flem.*
7. ——— Urii, *Flem.* Rutherglen, *Ure.*
8. ——— vasulites, *Montf.* Namur, *Holl.*
9. ——— imbricatus, *Goldf.* Ratingen; Visé, *G. T.*

* *Planorbis æqualis,* Sow.   † *Cirrus rotundatus,* Sow.
‡ *Cirrus acutus,* Sow.   § *Helix carinatus* of Sowerby.
‖ Probably *Helix striatus* of Sowerby.   ¶ *Helicina,* Lam.
** *Melania constricta* of Sowerby.

10. Bellerophon sulcatus, *Goldf.* Ratingen, *G. T.*
11. —— undulatus, *Goldf.* Chimay; Schwelm, *G. T.*
12. —— Hüpschii, *Defr.* Chimay, *G. T.*
 1. Conularia quadrisulcata, *Miller.* Bristol, *Miller ;* Rutherglen, *Flem.*
 2. —— teres, *Sow.* Scotland, *Sow.*
 1. Orthoceratites undulatus, *Sow.* Scalebar, Yorkshire, *Ducket, (Schlot.)* ;
     Visé, near Liége, *Al. Brong.*
 2. —— Breynii, *Sow.* Ashford, Derbyshire, *Sow.*
 3. —— annulatus, *Sow.* Colebrooke Dale, Shropshire, *Sow.;* King's
     County, *Weav.*
 4. —— paradoxicus, *Sow.* Ireland, *Ogilby.*
 5. —— fusiformis, *Sow.* Queen's County, Ireland, *Sow.;* Lancashire,
     Gilbertson.
 6. —— cinctus, *Sow.* Preston, Lancashire, *Moore.*
 7. —— Gesneri, . Derbyshire, *Martin.*
 8. —— lævis, *Flem.* Linlithgowshire, *Flem.*
 9. —— pyramidalis, *Flem.* Linlithgowshire, *Flem.*
10. —— convexus, *Flem.* Linlithgowshire, *Flem.*
11. —— annularis, *Flem.* Linlithgowshire, *Flem.*
12. —— rugosus, *Flem.* Linlithgowshire, *Flem.*
13. —— angularis, *Flem.* Linlithgowshire, *Flem.*
 1. Nautilus globatus, *Sow.* Ratingen, *G. T.*
 2. —— discus, *Sow.* Kendal, *Sow.*
 3. —— ingens, . Derbyshire, *Martin.*
 4. —— marginatus, *Flem.* Bathgate, Scotland, *Flem.*
 5. —— quadratus, *Flem.* West Lothian, *Flem.*
 6. —— biangulatus, *Sow.* Bristol, *Beeke.*
 7. —— sulcatus, *Sow.* Derbyshire, *Sow.*
 8. —— Woodwardii, *Sow.* Derbyshire, *Martin.*
 9. —— excavatus, *Flem.* Limerick, *Wright.*
 1. Ammonites striatus, *Sow.* Derbyshire, *Sow.;* Ratingen; Visé, *G. T.*
 2. —— Listeri, *Sow.* Chockier, *G. T.*

CRUSTACEA.

 1. Asaphus Dalmanni, *Goldf.* Ratingen, *G. T.*
    Trilobites, genera not determined. Bristol, *De la B.;* Llangeveni,
    Anglesea, *Farey ;* Linlithgowshire, *Flem.;* Liége, *Dum.*

PISCES.

Ichthyodorulites, *Buckl. & De la B.* Bristol, *De la B.*
Tritores, or Fish Palates. Bristol, *De la B.;* Northumberland, *Phil.*

Organic Remains of the Grauwacke Group.

PLANTÆ.

Algæ.

 1. Fucoides antiquus, *Ad. Brong.* Christiania, Sweden, *Ad. Brong.*
 2. —— circinatus, *Ad. Brong.* Kinnekulle, Sweden, *Ad. Brong.*
    ——, species not determined. South of Ireland, *Weav.;* Rudelstadt,
    Silesia; Schweighof, Swarzwald, *G. T.*

Equisetaceæ.

 1. Calamites radiatus, *Ad. Brong.* Bitschweiler, Haut Rhin, *Ad. Brong.*
 2. —— Voltzii, *Ad. Brong.* Bitschweiler, Haut Rhin, *Ad. Brong.*
    ——, species not determined. S. of Ireland, *Weav.;* Val St. Amarin,
    Haut Rhin, *Hœn.*

Filices.
1. Sphenopteris dissecta, *Ad. Brong.* Berghaupten, Baden, *Ad. Brong.*
1. Cyclopteris flabellata, *Ad. Brong.* Berghaupten, Baden, *Ad. Brong.*
1. Pecopteris aspera, *Ad. Brong.* Berghaupten, *Ad. Brong.*

Lycopodiaceæ.
Lepidodendron, several species not determined. Berghaupten and Bitschweiler, *Ad. Brong.*

Euphorbiaceæ.
1. Stigmaria ficoides, *Ad. Brong.* Bitschweiler, *Ad. Brong.*

Dicotyledonous plant of doubtful affinity.
1. Asterophyllites pygmæa, *Ad. Brong.* Berghaupten, *Ad. Brong.*

Plants of uncertain affinity.
1. Sigillaria tessellata, *Ad. Brong.* Berghaupten, *Ad. Brong.*
2. ——— Voltzii, *Ad. Brong.* Zundsweiler, *Ad. Brong.*

ZOOPHYTA.

1. Manon cribrosum, *Goldf.* Rebinghausen, Eifel, *Goldf.*
2. ——— favosum, *Goldf.* Eifel, *Goldf.*
1. Scyphia conoidea, *Goldf.* Nieder-Ehe, Eifel, *Goldf.*
2. ——— costata, *Goldf.* Eifel, *Goldf.*
3. ——— turbinata, *Goldf.* Eifel, *Goldf.*
4. ——— clathrata, *Goldf.* Eifel, *Goldf.*
5. ——— empleura, *Munst.* Gottland, *His.*
1. Tragos acetabulum, *Goldf.* Keldenich, Eifel, *Goldf.*
2. ——— capitatum, *Goldf.* Bensberg, Prussian Prov., *Goldf.*
1. Gorgonia antiqua, *Goldf.* Eifel; Ural, *Goldf.*
2. ——— infundibuliformis, *Goldf.* Lindlar; Gimborn; Eifel; Ural, *G. T.*
1. Stromatopora concentrica, *Goldf.* Eifel, *Goldf.*
2. ——— polymorpha*, *Goldf.* Eifel; Bensberg, *Goldf.*
Madrepora, species not determined. Gloucestershire; Herefordshire;
S. of Ireland, *Weav.*
1. Millepora? repens, *Wahl.* Gottland, *His.*
1. Cellepora antiqua, *Goldf.* Heisterstein, Eifel, *Goldf.*
2. ——— favosa, *Goldf.* Eifel; Dudley, *Goldf.*
———, species not determined. Gloucestershire; Herefordshire, *Weav.*
1. Retepora antiqua, *Goldf.* Heisterstein, Eifel, *Goldf.*
2. ——— prisca, *Goldf.* Eifel, *Goldf.*
3. ——— clathrata, *Goldf.* Gottland, *His.*
———, species not determined. Gloucestershire; Herefordshire; S. of
Ireland, *Weav.*
1. Flustra lanceolata, *Goldf.* Gottland, *His.*
———, species not determined. Gloucestershire; Herefordshire; S. of
Ireland, *Weav.*
1. Ceriopora verrucosa, *Goldf.* Bensberg, Pruss. Prov., *Goldf.*
2. ——— affinis, *Goldf.* Eifel; Dudley, *Goldf.*
3. ——— punctata, *Goldf.* Eifel; Dudley, *Goldf.*
4. ——— granulosa, *Goldf.* Eifel; Dudley, *Goldf*
5. ——— oculata, *Goldf.* Eifel; Dudley, *Goldf.*
6. ——— prisca, *Munst.* Regnitzlosau, *G. T.*
1. Glauconome disticha, *Goldf.* Eifel; Dudley, *G. T.*
1. Agaracia lobata, *Goldf.* Eifel, *Goldf.*

---
* *Tragos capitatum,* Goldf.; and *Ceriopora verrucosa,* Goldf., according to German Transl. of Manual.

1. Lithodendron cæspitosum, *Goldf.* Bensberg, *Goldf.*
2. ――― bicostatum, *Goldf.* Eifel, *G. T.*
3. ――― denticulatum, *Goldf.* Niagara, *G. T.*
1. Caryophyllia explanata, *His.* Gottland, *His.*
――――, species not determined. Gloucestershire; Herefordshire, *Weav.*
1. Fungites patellaris, *Lam.* Gottland, *His.*
2. ――― rimosus, *His.* Gottland, *His.*
1. Anthophyllum bicostatum, *Goldf.* Heisterstein, Eifel, *Goldf.*
1. Turbinolia turbinata, *Lam.* Gottland, *His.*
2. ――― mitrata, *Goldf.* Gottland; Ostrogothia, *His.*
3. ――― pyramidalis, *His.* Gottland, *His.*
――――, species not determined. Gloucestershire; Herefordshire; S. of Ireland, *Weav.*
1. Cyathophyllum Dianthus, *Goldf.* Eifel, *Goldf.;* Gottland, *His.*
2. ――― radicans, *Goldf.* Eifel, *Goldf.*
3. ――― marginatum, *Goldf.* Bensberg, *Goldf.*
4. ――― explanatum, *Goldf.* Bensberg, *Goldf.*
5. ――― turbinatum, *Goldf.* Eifel, *Goldf.;* Vestrogothia; Gottland, *His.*
6. ――― hypocrateriforme, *Goldf.* Eifel, *Goldf.*
7. ――― Ceratites, *Goldf.* Bensberg; Eifel, *Goldf.;* Gottland, *His.*
8. ――― flexuosum *, *Goldf.* Eifel, *Goldf.;* Gottland, *His.*
9. ――― vermiculare, *Goldf.* Eifel, *Goldf.;* Gottland, *His.*
10. ――― vesiculosum, *Goldf.* Eifel, *Goldf.*
11. ――― secundum, *Goldf.* Eifel, *Goldf.*
12. ――― lamellosum, *Goldf.* Eifel, *Goldf.*
13. ――― placentiforme, *Goldf.* Eifel, *Goldf.*
14. ――― quadrigeminum †, *Goldf.* Eifel; Bensberg, *Goldf.*
15. ――― cæspitosum, *Goldf.* Bensberg; Eifel, *Goldf.;* Gottland, *His.*
16. ――― hexagonum, *Goldf.* Bensberg; Eifel, *Goldf.*
17. ――― helianthoides, *Goldf.* Eifel; Lake Huron, *Goldf.*
18. ――― plicatum, *Goldf.* Kentucky, *G. T.*
19. ――― articulatum, *Wahl.* Gottland, *His.*
20. ――― Ananas, *Goldf.* Gottland, *His.*
1. Strombodes pentagonus, *Goldf.* Drummond Island, Lake Huron, *Goldf.*
1. Astrea porosa, *Goldf.* Eifel; Bensberg, *Goldf.;* Gottland, *His.*
2. ――― favosa, *Linn.* Gottland, *His.*
――――, species not determined. Gloucestershire; Herefordshire; S. of Ireland, *Weav.*
1. Columnaria alveolata, *Goldf.* Senekasee, New York, *Goldf.*
1. Sarcinula organum, *Goldf.* Gottland, *His.*
2. ――― microphthalma, *Goldf.* Eifel? *G. T.*
3. ――― auloticon, *Goldf.* Linich, *G. T.*
1. Coscinopora Placenta, *Goldf.* Eifel, *Goldf.*
1. Catenipora escharoides, *Lam.* Eifel; Norway; Drummond Isl., *Goldf.;* Ratoska, Gov. of Moscow, *Fischer;* Gottland, *His.*
2. ――― labyrinthica, *Goldf.* Groningen; Drummond Island, *Goldf.;* Gottland, *His.*
3. ――― tubulosa, *Lam.* Christiania, *Al. Brong.*
――――, species not determined. Gloucestershire; Herefordshire, *Weav.*
1. Syringopora verticillata, *Goldf.* Drummond Isl., *Goldf.;* Gottland, *His.*
2. ――― cæspitosa, *Goldf.* Pfaffrath, *Goldf.*
3. ――― ? fascicularis, *Wahl.* Gottland, *His.*
4. ――― ? Serpula, *Wahl.* Gottland, *His.*
Tubipora, species not determined. Gloucestershire; Herefordshire, *Weav.*
1. Calamopora alveolaris, *Goldf.* Eifel, *Goldf.*

---

* *Amplexus corralloides,* Mill., according to G. T.
† *Manon favosum,* Goldf., and *Columnare sulcata,* Goldf., according to G. T.

2. Calamopora favosa, *Goldf.* Drummond Isl., *Goldf.*
3. ———— Gothlandica, *Goldf.* Eifel, *Goldf.*; Gottland, *His.*
4. ———— basaltica, *Goldf.* Eifel; Gothland; Lake Erie, *Goldf.*
5. ———— infundibulifera, *Goldf.* Eifel; Besenberg, *Goldf.*
6. ———— polymorpha, *Goldf.* Eifel; Bensberg, *Goldf.*; Ural; Ems; Harz, *G. T.*; Gottland, *His.*
7. ———— spongites, *Goldf.* Eifel; Bensberg; Sweden; Dudley, *Goldf.*; Ural, *G. T.*; Gottland, *His.*
8. ———— fibrosa, *Goldf.* Eifel; Bensberg, *Goldf.*
9. ———— dubia, *Von Meyer.* Kaub on the Rhine, *G. T.*
1. Aulopora serpens, *Goldf.* Eifel, *Goldf.*; Christiania, *Al. Brong.*; Gottland, *His.*
2. ———— tubæformis, *Goldf.* Eifel, *Goldf.*; Gottland, *His.*
3. ———— spicata, *Goldf.* Eifel; Bensberg, *Goldf.*
4. ———— conglomerata, *Goldf.* Bensberg, *Goldf.*
5. ———— sarmentacea, *Goldf.* Eifel, *G. T.*
1. Favosites Gothlandica, *Lam.* Sloeben-Aker; Christiania; Eifel; Catskill; Batavia, New York, *Al. Brong.*
2. ———— Bromelli, *Ménard de la Groye.* Nehou, *Al. Brong.*
3. ———— truncata, *Rafinesque.* Kentucky, *Al. Brong.*
4. ———— Kentuckensis, *Raf.* Kentucky, *Al. Brong.*
5. ———— Boletus, *Ménard de la Groye.* Christiania, *Al. Brong.*
6. ———— Alcyonium, *Defr.* Gottland, *G. T.*
1. Mastrema pentagona, *Raf.* Garrard, Kentucky, *Al. Brong.*
1. Amplexus coralloides, *Miller.* South of Ireland, *Weav.*; Montechaton, near Coutances; Catskill, New York, *Al. Brong.*
———, species not determined. Plymouth, *Hennah.*
1. Pleurodyctium problematicum, *Goldf.* Abentheur, Hündsrück; Ems; Braubach, *G. T.*
1. Cyclolites nummismalis, *Lam.* Gottland, *His.*

RADIARIA.

1. Apiocrinites? scriptus, *His.* Gottland, *His.*
2. ————? punctatus, *His.* Gottland, *His.*
1. Pentremites ovalis, *Goldf.* (Grauwacke) Ratingen, *G. T.*
1. Pentacrinites priscus, *Goldf.* Eifel, *Goldf.*
1. Actinocrinites moniliformis, *Miller.* S. of Ireland, *Weav.*
2. ———— triacontadactylus, *Miller.* South of Ireland, *Weav.*; Eifel, *Goldf.*
3. ———— cingulatus, *Goldf.* Eifel, *Goldf.*
4. ———— muricatus, *Goldf.* Eifel, *Goldf.*
5. ———— nodulosus, *Goldf.* Eifel, *Goldf.*
6. ———— moniliferus, *Goldf.* Eifel, *Goldf.*
7. ———— granulatus, *Goldf.* Regnitzlosau, *G. T.*
———, species not determined. Gloucestershire; Herefordshire, *Weav.*
1. Cyathocrinites tuberculatus, *Miller.* South of Ireland, *Weav.*; Dudley, *Miller*; Rhenish Prov., *Goldf.*
2. ———— rugosus, *Miller.* Shropshire; Herefordshire; Isl. of Oeland; Dalecarlia, *Miller*; Eifel, *Goldf.*
3. ———— geometricus, *Goldf.* Eifel, *Goldf.*
4. ———— pinnatus*, *Goldf.* Eifel, *Goldf.*; Lindlar; Ems; Martenberg, Waldeck; Oberweiler, Schwarzwald, *G. T.*
———, species not determined. Gloucestershire; Herefordshire, *Weav.*
1. Platycrinites lævis, *Miller.* Cork, *Weav.*
2. ———— pentangularis, *Miller.* Dudley; Dinevar Park, Wales, *Miller.*
3. ———— rugosus, *Miller.* Regnitzlosau, Baireuth, *Goldf.*
4. ———— ventricosus, *Goldf.* Eifel, *Goldf.*

---

* *Act. moniliformis,* Mill., according to G. T.

1. Rhodocrinites verus, *Miller.* Dudley, *Miller;* Eifel, *Goldf.*
2. ———— gyratus, *Goldf.* Eifel, *Goldf.*
3. ———— quinquepartitus, *Goldf.* Eifel, *Goldf.*
4. ———— canaliculatus, *Goldf.* Eifel, *Goldf.*
5. ———— crenatus, *Goldf.* Eifel, *Goldf.*
1. Melocrinites lævis, *Goldf.* Regnitzlosau, Baireuth, *Goldf.*
2. ———— gibbosus, *Goldf.* Eifel, *Goldf.*
1. Cupressocrinites crassus, *Goldf.* Eifel, *Goldf.*
2. ———— gracilis, *Goldf.* Eifel, *Goldf.*
3. ———— tesseratus *, *Goldf.* Eifel, *G. T.*
1. Eugeniacrinites mespiliformis, *Goldf.* Eifel, *Goldf.*
1. Eucalyptocrinites rosaceus, *Goldf.* Eifel, *Goldf.*
1. Sphæronites† Pomum, *Wahl.* Isl. of Oeland; Kinnekulle, in Vestrogothia; Dalecarlia, *His.;* Tzarko-Sselo, near St. Petersburgh, *Al. Brong.*
2. ———— Aurantium, *Wahl.* Mösseburg, Vestrogothia, *His.*
3. ———— granatum, *Wahl.* Furudal, Dalecarlia; Boedahamn, Isl. of Oeland, *His.*
4. ———— Wahlenbergii, *Esmark.* Gulf of Christiania, *Al. Brong.*

Annulata.

1. Serpula epithonia, *Goldf.* Bensberg, *Goldf.*
2. ———— ammonia, *Goldf.* Eifel, *Goldf.*
3. ———— omphalodes, *Goldf.* Bensberg; Eifel, *Goldf.*
4. ———— socialis, *Goldf.* Eifel, *Goldf.*
5. ———— Lithuus, *Schlot.* Gottland, *His.*

Conchifera.

1. Thecidea? antiqua, *Hœn.* Gerolstein, *Hœn.*
1. Pentamerus Knightii, *Sow.* Shropshire, *Murch.;* Eifel, *G. T.*
2. ———— lævis, *Sow.* Shropshire, *Aikin.*
1. Gypidia‡ gryphoides, *Goldf.* Pfaffrath, *G. T.*
2. ———— lævis, *Goldf.* Pfaffrath, *G. T.*
3. ———— Conchidium, *Dalm.* Gottland, *His.;* Pokroi, *Von Buch.*
1. Spirifer§ speciosus, *Bronn.* Eifel, *Holl.*
2. ———— cuspidatus, *Sow.* Eifel, *Holl.;* South of Ireland, *Weav.;* Bensberg; Blankenheim, *Hœn.;* Plymouth, *Hennah.*
3. ———— glaber, *Sow.* South of Ireland, *Weav.;* Plymouth? *Hennah.*
4. ———— obtusus, *Sow.* South of Ireland, *Weav.*
5. ———— striatus, *Sow.* South of Ireland, *Weav.*
6. ———— pinguis, *Sow.* South of Ireland, *Weav.*
7. ———— intermedius‖, *Schlot.* Gloucestershire; Herefordshire, *Weav.;* Eifel; Alleghany Mountains, *Al. Brong.*
8. ———— alatus, *Sow.* Env. of Coblentz, *Al. Brong.*
9. ———— sarcinulatus¶, *Schlot.* Coblentz; Malmö; Mösseberg, Sweden; Catskill, New York, *Al. Brong.*
10. ———— rotundatus, *Sow.* Cork, *Wright;* Newton Bushel? Devon, *De la B.*
11. ———— lineatus, *Sow.* Dudley, *Stokes.*
12. ———— ambiguus, *Sow.* Blankenheim, *Hœn.*
13. ———— attenuatus, *Sow.* Eifel; Coblenz, *G. T.*
14. ———— minimus, *Sow.* Blankenheim, *Hœn.*
15. ———— Sowerbii, . Eifel, *Hœn.*
16. ———— decurrens, *Sow.* Newton Bushel, Devon, *De la B.*

* *Actinocrinites tesseratus,* Goldf.
† *Sphæronites,* Hisinger; *Echinosphærites,* Wahlenberg.
‡ *Pentamerus?* Sow.   § *Delthyris,* Dalm.
‖ *Terebratula,* Schlotheim.   ¶ *Ibid.*

17. Spirifer distans, *Sow.* Plymouth, *Hennah.*
18. —— octoplicatus, *Sow.* Plymouth, *Hennah.*
1. Delthyris elevata *, *Dalm.* Gottland, *His.*
2. —— cyrtæna, *Dalm.* Gottland, *His.; * Eifel; Bensberg, *G. T.*
3. —— crispa †, *Dalm.* Gottland, *His.; * Eifel, *G. T.*
4. —— sulcata, *His.* Gottland, *His.*
5. —— subsulcata, *Dalm.* Oeland, *His.*
6. —— ptycodes, *Dalm.* Gottland, *His.*
7. —— cardiospermiformis, *His.* Gottland, *His.; * Dudley, *G. T.*
8. ——? pusio, *His.* Gottland, *His.*
9. ——? Psittacina, *Wahl.* Dalecarlia, *His.*
10. ——? jugata, *Wahl.* Dalecarlia, *His.*
11. —— exporrecta‡, *Wahl.* Gottland, *His.*
12. —— trapezoidalis §, *His.* Gottland, *His.*
13. —— læviiosta, *Goldf.* Bensberg; Eifel; Coblenz, *G. T.*
14. —— microptera ‖, *Goldf.* Siebengebirge; Lindlar, Siegen, *G. T.*
15. —— compressa ¶, *Goldf.* Bensberg, *G. T.*
16. —— heteroclyta **, *Goldf.* Blankenheim, *G. T.*
17. —— macroptera ††, *Goldf.* Eifel; Coblenz; Ems; Lahnstein; Lindlar; Kaisersteimel; Catskill Mountains, *G. T.*
18. —— ceptoptera, *Goldf.* Eifel; Lindlar, *G. T.*
19. —— pachyoptera, *Goldf.* Hudson, New York, *G. T.*
20. —— canalifera ‡‡, *Goldf.* Bensberg, *G. T.*
21. —— canaliculata, *Goldf.* Bensberg, *G. T.*
22. —— curvata §§, *Goldf.* Gerolstein, Eifel, *G. T.*
23. —— striatula ‖‖, *Goldf.* Bensberg; Christiania; Trenton Falls, *G. T.*
1. Terebratula crumena, *Sow.* S. of Ireland, *Weav.; * Eifel; Lindlar; Lake Huron, *G. T.*
2. —— cordiformis, *Sow.* S. of Ireland, *Weav.*
3. —— Pugnus ¶¶, *Sow.* S. of Ireland, *Weav.; * Plymouth, *Hennah ; * Newton Bushel, *De la B.; * Bensberg; Gottland ; Ostrogothia, *His.*
4. —— affinis, *Sow.* Dudley, *Ryan ; * Eifel, *Hœn.*
5. —— lævigata, *Schlot.* S. of Ireland, *Weav.*
6. —— elongata, *Schlot.* S. of Ireland, *Weav.*
7. —— acuminata, *Sow.* Cork, *Wright.*
8. —— lateralis, *Sow.* Cork, *Wright.*
9. —— reniformis, *Sow.* Cork, *Sow.*
10. —— imbricata, *Sow.* Plymouth, *Hennah.*
11. —— Mantiæ, *Sow.* Ems, *G. T.*
12. —— porrecta, *Sow.* Newton Bushel, Devon, *De la B.*
13. —— platyloba (jun.), *Sow.* Plymouth, *Hennah.*
14. —— concentrica, *Bronn.* Eifel, *G. T.*
15. —— heterotypa, *Bronn.* Eifel, *G. T.*
16. —— triloba, *Goldf.* Eifel, *G. T.*
17. —— canaliculata, *Goldf.* Eifel, *G. T.*
18. —— quinquelatera, *Goldf.* Eifel, *G. T.*
19. —— dichotoma, *Goldf.* Eifel, *G. T.*
20. —— lacunosa ***, *Dalm.* Gottland, *His.; * Eifel, *G. T.*
21. —— pentagona, *Goldf.* Eifel, *G. T.*

* *Spirifer cuspidatus*, Sow. (G. T.) † *Spirifer octoplicatus*, Sow. (G. T.)
‡ *Cyrtia exporrecta*, Dalm. (G. T.) § *Cyrtia trapezoidalis*, His. (G. T.)
‖ *Terebratula intermedia*, Schlot. (G. T.) ¶ *Spirifer triangularis*, Sow. (G. T.)
** *Calceola heteroclyta*, Defr. †† *Terebratula speciosa*, & *Ter. paradoxa*, Schlot.
‡‡ *Terebratula aperturata*, Schlot., & *Ter. canalifera*, Sow.
§§ *Terebratula curvata*, Schlot. ‖‖ *Terebratula striatula*, & *Ter. excisa*, Schlot.
¶¶ *Ter. lacunosa*, Schlot; *Ter. plicatella*, Dalm. *** *Terebratula Wilsoni*, Sow.

22. Terebratula Wahlenbergii, *Goldf.* Eifel, *G. T.*
23. ——— diodonta, *Dalm.* Gottland, *His.* ; Eifel, *G. T.*
24. ——— subglobosa, *Goldf.* Eifel, *G. T.*
25. ——— bifida, *Goldf.* Eifel, *G. T.*
26. ——— clavata, *Goldf.* Eifel, *G. T.*
27. ——— cuneata, *Dalm.* Gottland, *His.*
28. ——— bidentata, *His.* Gottland, *His.*
29. ——— marginalis, *Dalm.* Gottland, *His.*
30. ——— didyma, *Dalm.* Gottland, *His.*
1. Strygocephalus Burtini *, *Defr.* Bensberg, *Hœn.*
2. ——— elongatus, *Goldf.* Bensberg, *Hœn.*
3. ——— striatus, *Goldf.* Eifel, *G. T.*
1. Atrypa reticularis †, *Wahl.* S. of Ireland, *Weav.*; Bensberg, *Schlot.*; Plymouth, *Hennah;* Vestrogothia; Gottland, *His.*; Eifel; Lindlar; Dudley, *G. T.*
2. ——— aspera, *Schlot.* Gottland, *His.*; Eifel; Bensberg, *G. T.*
3. ——— canaliculata, *Dalm.* Ostrogothia, *His.*; Pokroi, *Von Buch.*
4. ——— dorsata, *His.* Oeland, *His.*
5. ——— galeata, *Dalm.* Gottland, *His.*; Eifel, *G. T.*
6. ——— Nucella, *Dalm.* Ostrogothia, *His.*
7. ———? crassicostis, *Dalm.* Vestrogothia, *His.*
8. ———? lenticularis, *Wahl.* Vestrogothia, *His.*
9. ——— Prunum, *Dalm.* Gottland, *His.*
10. ——— tumida, *Dalm.* Gottland, *His.*
11. ———? tumidula, *His.* Gottland, *His.*
12. ——— cassidea, *Dalm.* Ostrogothia, *His.*
13. ———? micula, *Dalm.* Fogelsang, Scania, *His.*
14. ——— nitida, *Goldf.* Lake Simkoe, *G. T.*
1. Calceola sandalina, *Lam.* Eifel,*Bronn.*; Gerolstein, Blankenheim, *Hœn.*
1. Producta ‡ Scotica, *Sow.* S. of Ireland, *Weav.*; Eifel, *Hœn.*; Isle of Man, *Henslow.*
2. ——— Martini, *Sow.* S. of Ireland, *Weav.*
3. ——— concinna, *Sow.* S. of Ireland, *Weav.*
4. ——— lobata, *Sow.* S. of Ireland, *Weav.*
5. ——— longispina, *Sow.* Blankenheim, *Hœn.*
6. ——— punctata, *Sow.* Blackrock, Cork, *Wright.*
7. ——— fimbriata, *Sow.* S. of Ireland, *Weav.*
8. ——— depressa, *Sow.* S. of Ireland, *Weav.*; Dudley, *Sow.*; Plymouth, *Hennah.*; Gottland, *His.*; Pokroi, *Von Buch*; Eifel, *G. T.*
9. ——— hemisphærica, *Sow.* Eifel; Catskill Mountains; Albany, Lexington, *Hœn.*
10. ——— rostrata, *Sow.* Bensberg, *Hœn.*
11. ——— sarcinulata, *Goldf.* Eifel, Catskill Mountains, *Hœn.*
12. ——— sulcata, *Sow.* Catskill Mountains, *Hœn.*
13. ——— scabricula, *Sow.* Eifel, *G. T.*
1. Leptæna rugosa, *His.* Gottland; Ostrogothia; Vestrogothia, *His.*; Catskill Mountains, *G. T.*
2. ——— euglypha, *Dalm.* Gottland, *His.*; Eifel, *G. T.*
3. ——— transversalis, *Wahl.* Gottland; Dalecarlia, *His.*
4. ——— convoluta, *Goldf.* Eifel, *G. T.*
5. ——— furcata, *Goldf.* Eifel, *G. T.*
6. ——— capillata, *Goldf.* Eifel, *G. T.*
7. ——— striata, *Goldf.* Coblenz, *G. T.*
8. ——— pectinata, *Goldf.* Eifel; Coblenz; Kaisersteimel, *G. T.*

---

* *Terebratula rostrata,* Schlot.
† *Terebratula prisca,* & *Ter. explanata,* Schlot.; *Ter. affinis,* Sow.
‡ *Leptæna,* Dalm.

9. Leptæna minuta, *Goldf.* Eifel, *G. T.*
1. Orbicula concentrica, *Von Buch.* Martenberg, Waldeck, *G. T.*
1. Crania prisca, *Hœn.* Ratingen (Grauwacke), *G. T.*
   Gryphæa, species not determined. Keswick, near Kirby Lonsdale, *Phil.*
1. Pecten primigenius, *Meyer.* Wisenbach, Herborn, *Meyer.*
2. —— Munsteri, *Meyer.* Wisenbach, Herborn, *Meyer.*
3. —— grandævis, *Goldf.* Geistlecher Berg, Herborn, *G. T.*
4. —— Oceani, *Goldf.* Harz, *G. T.*
5. —— Neptuni, *Goldf.* Eifel, *G. T.*
   ——, species not determined. Keswick, *Phil.*; Plymouth, *Hennah;*
      S. of Ireland, *Weav.*; Pokroi, Lithuania, *Von Buch.*
   Plagiostoma, species not determined. Keswick, *Phil.*
1. Inoceramus vetustus, *Sow.* Elbersreuth, *G. T.*
1. Avicula obsoleta, *Goldf.* Abentheuer, Hundsruck, *G. T.*
2. —— lepida, *Goldf.* Geistlecher Berg, Herborn, *G. T.*
1. Pterinea ventricosa, *Goldf.* Kemmenau; Ems; Altenahr, *G. T.*
2. —— costata, *Goldf.* Ems; Siebengebirge, *G. T.*
3. —— lineata, *Goldf.* Kemmenau; Ems, *G. T.*
4. —— lamellosa, *Goldf.* Siegen; Harz, *G. T.*
5. —— reticulata, *Goldf.* Iserlohn, *G. T.*
6. —— radiata, *Goldf.* Eifel; Iserlohn, *G. T.*
7. —— carinata, *Goldf.* Pfaffendorf; Coblenz; Oneida County; Lewis
      Town, *G. T.*
8. —— plana, *Goldf.* Kemmenau; Ems, *G. T.*
9. —— trigona, *Goldf.* Kemmenau; Ems, *G. T.*
1. Posodonia Becheri, *Bronn.* Frankenberg, Hesse, *Meyer;* Geistlecher
      Berg, Herborn; Ründeroth, *G. T.*
1. Arca prisca, *Goldf.* Kloster Bruck, Solingen, *G. T.*
1. Nucula antiqua, *Goldf.* Harz; Ems, *G. T.*
2. —— subnoides, *Goldf.* Harz, *G. T.*
3. —— fornicata, *Goldf.* Olpe, *G. T.*
4. —— securiformis, *Goldf.* Ems, *G. T.*
5. —— pinguis, *Goldf.* Ems, *G. T.*
1. Trigonia sulcata, *Goldf.* Lindlar, *G. T.*
2. —— concentrica, *Goldf.* Lindlar, *G. T.*
   ——, species not determined. Keswick, *Phil.*
1. Megalodon cucullatus, *Sow.* Newton Bushel, Devon, *De la B.*
1. Modiola Goldfussii, *Hœn.* Eifel, *G. T.*
2. —— antiqua, *Goldf.* Ems, *G. T.*
3. —— Gothlandica, *His.* Gottland, *His.*
1. Mytilus vetustus, *Goldf.* Dillenberg; Upper Canada, *G. T.*
1. Crassatella obsoleta, *Goldf.* Wipperfürth, *G. T.*
1. Cardium costellatum, *Munst.* Elbersreuth; Prague, *Hœn.*
2. —— hybridum, *Munst.* Elbersreuth, *Hœn.*
3. —— lineare, *Munst.* Elbersreuth, *Hœn.*
4. —— priscum, *Munst.* Elbersreuth; Prague, *Hœn.*
5. —— striatum, *Munst.* Elbersreuth, *Hœn.*
6. —— alæforme, *Sow.* Scarlet, Isle of Man, *Henslow;* Plymouth,
      Hennah; Newton Bushel, Devon, *De la B.*
7. —— fasciculatum, *Goldf.* Kemmenau; Ems, *G. T.*
8. —— marginatum, *Goldf.* Kemmenau, Ems, *G. T.*
9. —— carpomorphum, *Dalm.* Ostrogothia, *G. T.*
1. Cardita costellata, *Munst.* Elbersreuth, *Hœn.*
2. —— gracilis, *Munst.* Elbersreuth, *Hœn.*
3. —— plicata, *Munst.* Elbersreuth, *Hœn.*
4. —— tripartita, *Munst.* Elbersreuth, *Hœn.*
1. Isocardia Humboldtii, *Hœn.* Wisenbach, near Dillenburg, *Hœn.*
2. —— oblonga, *Sow.* Cork, *Flem.*

1. Venericardium retrostriatum, *Von Buch.* Martenberg, Waldeck, *G. T.*
1. Lucina Proacia*, *Goldf.* Eifel; Bensberg, *G. T.*
2. —— lineata, *Goldf.* Eifel, *G. T.*
3. —— rugosa, *Goldf.* Eifel, *G. T.*
1. Cyprina minuta, *Goldf.* Eifel, *G. T.*
1. Corbula zonaria, *Goldf.* Eifel, *G. T.*
1. Cythere Okeni, *Munst.* Regnitzlosau, *G. T.*
2. —— suborbiculata, *Munst.* Regnitzlosau, *G. T.*
3. —— inflata, *Munst.* Regnitzlosau, *G. T.*
4. —— Hisingeri, *Munst.* Regnitzlosau, *G. T.*
5. —— elongata, *Munst.* Regnitzlosau, *G. T.*
6. —— bilobata, *Munst.* Regnitzlosau, *G. T.*
7. —— subcylindrica, *Munst.* Regnitzlosau, *G. T.*
8. —— intermedia, *Munst.* Regnitzlosau, *G. T.*
1. Sanguinolaria gibbosa, *Sow.* Altenahr, *G. T.*
2. —— undulata, *Sow.* Siebengebirge, *G. T.*
3. —— concentrica, *Goldf.* Eifel, *G. T.*
4. —— lamellosa, *Goldf.* Eifel, *G. T.*
5. —— dorsata, *Goldf.* Eifel; Altenahr, *G. T.*
6. —— truncata, *Goldf.* Eifel, *G. T.*
7. —— phaseolina, *Goldf.* Eifel, *G. T.*
8. —— solenoides, *Goldf.* Siebengebirge; Altenahr, *G. T.*
1. Pholadomya radiata, *Goldf.* Eifel, *G. T.*

MOLLUSCA.

1. Patella Neptuni, *Goldf.* Eifel; Olpe, *G. T.*
2. —— primigena, *Goldf.* Pfaffrath, *G. T.*
3. ——? conica, *Wahl.* Kinnekulle, Westrogothia, *His.*
4. ——? pennicostis, *Wahl.* Ulanda, Westrogothia, *His.*
5. ——? concentrica, *Wahl.* Mösseberg, &c., Westrogothia, *His.*
   ——, species not determined. Keswick, near Kirby Lonsdale, *Phil.*
1. Pileopsis vetusta, *Sow.* South of Ireland, *Weav.* Plymouth, *Hennah.*
2. —— prisca, *Goldf.* Eifel, *G. T.*
3. —— compressa, *Goldf.* Eifel, *G. T.*
1. Melania constricta, *Sow.* South of Ireland, *Weav.*
1. Natica, species not determined. Plymouth, *Hennah.* Newton Bushel?
   De la B.
1. Nerita spirata? *Sow.* Plymouth, *Hennah.*
2. —— subcostata†, *Goldf.* Pfaffrath; Ems, *G. T.*
   ——, species not determined. Herefordshire; Gloucestershire; South
   of Ireland, *Weav.*
1. Delphinula æquilatera, *Wahl.* Westrogothia, *His.*
2. —— alata, *Wahl.* Gottland, *His.*
3. —— catenulata, *Wahl.* Gottland, *His.*
4. —— Cornu Arietis, *Wahl.* Gottland, *His.*
5. —— subsulcata, *His.* Gottland, *His.*
1. Cirrus acutus‡, *Sow.* S. of Ireland, *Weav.* Plymouth, *Hennah.*
1. Pleurotomaria cirriformis, *Sow.* Plymouth, *Hennah.*
1. Euomphalus catillus, *Sow.* S. of Ireland, *Weav.* Lake Erie, *Hœn.*
   Dudley, *G. T.*
2. —— centrifugus, *Wahl.* Wikarby, Dalecarlia, *His.*
3. —— funatus, *Sow.* Dudley, *Johnstone.*
4. —— nodosus, *Sow.* Eifel; Bensberg, *G. T.*
5. —— spinosus, *Goldf.* Bensberg, *G. T.*

---

* *Venus orbiculata,* Schlot.   † *Buccinum subcostatum,* Schlot.
‡ *Trochus acutus,* Goldf.

6. Euomphalus lævis, *Goldf.* Bensberg, *G. T.*
7. —— radiatus, *Goldf.* Eifel, *G. T.*
8. —— striatus, *Goldf.* Eifel, *G. T.*
9. —— articulatus, *Goldf.* Eifel, *G. T.*
10. —— depressus, *Goldf.* Eifel, *G. T.*
11. —— delphinuloides*, *Goldf.* Bensberg; Eifel; Dillenberg, *G. T.*
12. —— trigonalis, *Goldf.* Bensberg; Eifel, *G. T.*
13. —— carinatus, *Goldf.* Bensberg; Eifel, *G. T.*
14. —— angulatus, *Wahl.* Gottland, *His.*
15. —— substriatus, *His.* Gottland, *His.*
16. —— costatus, *His.* Gottland, *His.*
—— , species not determined. Newton Bushel, Devon, *De la B.*
1. Trochus ellipticus, *Hisinger.* Furudal, Dalecarlia, *His.*
2. —— exaltatus, *Goldf.* Eifel, *G. T.*
—— , species not determined. Pokroi, *Von Buch.*
1. Turbo bicarinatus, *Wahl.* Wikarby, Dalecarlia; Borenshult, Ostrogothia, *His.*
2. —— Tiara, *Sow.* Plymouth, *Hennah.*
3. —— armatus, *Goldf.* Eifel, *G. T.*
4. —— nodosus, *Goldf.* Eifel, *G. T.*
5. —— cœlatus, *Goldf.* Eifel, *G. T.*
6. —— porcatus, *Goldf.* Bensberg, *G. T.*
1. Rotella helicinæformis †, *Goldf.* Pfaffrath; Eifel, *G. T.*
—— , species not determined. Pokroi, *Von Buch.*
1. Turritella abbreviata, *Sow.* Newton Bushel, Devon, *De la B.*
2. —— prisca, *Munst.* Elbersreuth, *Munst.*
3. —— cingulata, *His.* Gottland, *His.*
4. —— bilineata ‡, *Goldf.* Bensberg; Pfaffrath; Eifel, *G. T.*
5. —— coronata, *Goldf.* Pfaffrath, *G. T.*
6. —— striata, *Goldf.* Eifel, *G. T.*
7. —— obsoleta, *Goldf.* Eifel, *G. T.*
—— , species not determined. Beckfoot, near Kirby Lonsdale, *Phil.*
Pleurotoma, species not determined. Newton Bushel, Devon, *De la B.*
1. Murex? Harpula, *Sow.* Newton Bushel, *De la B.;* Plymouth, *Hennah.*
1. Buccinum spinosum §, *Sow.* Plymouth, *Hennah;* Newton Bushel, *De la B.;* Bensberg, *G. T.*
2. —— acutum, *Sow.* Plymouth, *Hennah;* Bensberg; Pfaffrath, *G. T.*
3. —— breve, *Sow.* Newton Bushel, Devon, *De la B.*
4. —— imbricatum, *Sow.* Newton Bushel, *De la B.;* Plymouth, *Hennah;* Bensberg, *G. T.*
5. —— arcuatum, *Schlot.* Bensberg; Lustheide, *G. T.*
1. Phasianella ventricosa, *Goldf.* Eifel, *G. T.*
2. —— buccinoides, *Goldf.* Eifel, *G. T.*
3. —— fusiformis, *Goldf.* Eifel, *G. T.*
1. Bellerophon tenuifascia, *Sow.* S. of Ireland, *Weav.;* Newton Bushel, Devon, *De la B.*
2. —— ovatus, *Sow.* S. of Ireland, *Weav.*
3. —— hiulcus ||, *Sow.;* Bensberg; Blankenheim, *G. T.*
4. —— Hüpschii, *Defr.* Chimay; Pfaffrath, *G. T.*
5. —— nodulosus, *Goldf.* Bensberg; Harz, *G. T.*
6. —— Cornu Arietis, *Sow.* Catskill Mountains, *Hœn.*
7. —— apertus, *Sow.* Plattsburg. New York, *Hœn.;* Eifel, *G. T.*
8. —— costatus, *Sow.* Plymouth, *Hennah;* Pokroi, *Von Buch.*
9. —— undulatus, *Goldf.* Blankenheim, *G. T.*
—— , species not determined. Plymouth, *Hennah.*

* *Helix delph.*, Schlot.   † *Helix helicinæformis*, Schlot.
‡ *Murex turbinatus*, Schlot.   § *Turritella spinosa*, Goldf.
|| *B. striatus*, Goldf.

2 Q

594   *Organic Remains of the Grauwacke Group.*

1. Conularia quadrisulcata, *Miller.* Gloucestershire, *Weav.;* Borenshult, Ostrogothia, *His.;* Montmorency Falls, Quebec, *Hœn.*
2. ———— pyramidata,   . May, near Caen, *Deslongchamps.*
3. ———— teres, *Sow.* Lockport, N. America, *Hœn.;* Eifel, *G. T.*
———, species not determined. May, Calvados, *Deslong.*
1. Orthoceratites striatus, *Sow.* S. of Ireland, *Weav.;* Malmoe, Christiania, *Al. Brong.;* Trenton Falls, New York, *Hœn.*
2. ———— undulatus, *Sow.* S. of Ireland, *Weav.* Czarko-Szelo, near St. Petersburg, *Al. Brong.*
3. ———— paradoxicus, *Sow.* S. of Ireland, *Weav.*
4. ———— circularis, *Sow.* Gloucestershire; Herefordshire, *Weav.;* Plymouth, *Hennah.*
5. ———— annulatus, *Sow.* Gloucestershire, *Weav.;* Gerolstein, Eifel, *Schlot.;* Gottland, *His.*
6. ———— flexuosus, *Schlot.* Oeland; Gerolstein, Eifel, *Holl.* Black River, New York, *Hœn.*
7. ———— communis, *Wahl.* Common in Sweden, *His.*
8. ———— duplex, *Wahl.* Kinnekulle, Sweden, *His.;* Black River, New York, *Hœn.*
9. ———— trochlearis, *Dalman.* Solleroe, Dalecarlia, *His.*
10. ———— turbinatus, *Dalman.* Dalecarlia; Isle of Oeland, *His.*
11. ———— centralis, *Dalman.* Solleroe, Dalecarlia, *His.*
12. ———— gracilis, *Schlot.* Hellenburg, Nassau, *Al. Brong.;* Wissenbach, *Hœn.*
13. ———— crassiventris *, *Wahl.* N.W. side of Lake Huron, *Hœn.;* Gottland, *His.;* Eifel, *G. T.*
14. ———— rectus, *Bosc.* Kuchel, near Prague, *Hœn.*
15. ———— regularis, *Schlot.* Oeland, *Hœn.;* Elbersreuth, Bavaria, *Munst.*
16. ———— giganteus, *Sow.* Gerolstein, *Hœn.;* Elbersreuth; Regnitzlosau, Bavaria, *Munst.*
17. ———— striolatus, *Meyer.* Herborn, Dillenberg, *Meyer.*
18. ———— acuarius, *Munst.* Elbersreuth, *Munst.*
19. ———— striatopunctatus, *Munst.* Elbersreuth, *Munst.*
20. ———— cingulatus, *Munst.* Elbersreuth, *Munst.*
21. ———— torquatus, *Munst.* Elbersreuth, *Munst.*
22. ———— Steinhaueri, *Sow.* Elbersreuth, *Munst.*
23. ———— carinatus, *Munst.* Elbersreuth, *Munst.*
24. ———— linearis, *Munst.* Elbersreuth, *Munst.*
25. ———— irregularis, *Munst.* Elbersreuth, *Munst.*
26. ———— excentrica, *Goldf.* Bensberg; Gladbach, *G. T.*
27. ———— nodulosa, *Goldf.* Eifel, *G. T.*
28. ———— imbricatus, *Wahl.* Gottland, *His.*
29. ———— angulatus, *Wahl.* Gottland, *His.*
30. ———— undulatus, *His.* (not *Sow.*). Gottland, *His.*
1. Cyrtoceratites semilunaris, *Goldf.* Bensberg, *G. T.*
2. ———— depressus, *Goldf.* Gerolstein, Eifel, *G. T.*
3. ———— compressus, *Goldf.* Eifel, *G. T.;* Oeland, *Holl.*
4. ———— ornatus, *Goldf.* Bensberg, *G. T.*
5. ———— annulatus, *Goldf.* Eifel, *G. T.*
6. ———— lineatus, *Goldf.* Eifel, *G. T.*
1. Spirula compressa, *Goldf.* Eibach, Dillenberg, *G. T.*
2. ———— nodosa, *Goldf.* Eifel, *G. T.*
3. ———— costata, *Goldf.* Eifel, *G. T.*
4. ———— annulata, *Goldf.* Eifel, *G. T.*
5. ———— carinata, *Goldf.* Eifel, *G. T.*
6. ———— dorsata, *Goldf.* Eifel, *G. T.*

* *Orthoceratites inflatus,* Goldf.

7. Spirula constricta, *Goldf.* Montmorency Falls, Canada, *G. T.*
——, species not determined. Gloucestershire; Herefordshire, *Weav.;*
    Plymouth, *Hennah;* Env. of St. Petersburg, *Strangways.*
1. Lituites perfectus, *Wahl.* Mösseberg, Sweden, *Al. Brong.;* Revel, *Hœn.*
2. —— imperfectus, *Wahl.* Jungby, Sweden, *Al. Brong.*
1. Nautilus globatus*, *Sow.* S. of Ireland, *Weav.;* Trenton Falls, *G. T.*
2. —— multicarinatus, *Sow.* S. of Ireland, *Weav.*
3. —— complanatus, *Sow.* Scarlet, Isle of Man, *Henslow.*
4. —— cariniferus, *Sow.* Black Rock, Cork, *Sow.*
5. —— divisus, *Munst.* Geistlicher Berg, near Herborn, *Hœn.*
6. —— Wrightii, *Flem.* Cork, *Wright.*
†7. —— funatus, *Flem.* Cork, *Sow.*
†8. —— compressus, *Flem.* Cork, *Sow.*
†9. —— ovatus, *Flem.* S. of Ireland, *Weav.;* Hof; Schleitz, *Munst.*
1. Ammonites Henslowi, *Sow.* I. of Man, *Henslow.*
2. —— subnautilinus, *Schlot.* Wissenbach, near Dillenberg, *Hœn.*
3. —— expansus, *Von Buch.* Geistlicher Berg, *G. T.*
4. —— evexus, *Von Buch.* Gerolstein, Eifel, *G. T.*
5. —— Nöggerathi, *Goldf.* Eifel; Wissenbach, *G. T.*
6. —— primordialis, *Schlot.* Winterberg, Harz, *G. T.*
7. —— Becheri, *Goldf.* Eibach, Dillenberg, *G. T.*
8. —— Hœninghausii, *Von Buch.* Bensberg, *G. T.*
9. —— Munsteri, *Von Buch.* Elbersreuth, *G. T.*
10. —— simplex, *Von Buch.* Rammelsberg; Goslar, *G. T.*
11. —— inæquistriatus, *Munst.* Elbersreuth, *G. T.*
12. —— semistriatus, *Munst.* Elbersreuth, *G. T.*
13. —— speciosus, *Munst.* Elbersreuth, *G. T.*
14. —— retrorsus, *Von Buch.* Martenberg, Waldeck, *G. T.*
15. —— Dalmanni, *His.* Gottland, *His.*
——, species not determined. Gloucestershire; Herefordshire; S.
    of Ireland, *Weav.;* Newton Bushel, *De la B.;* Eifel; Hof;
    Frankenberg, Herborn, *Munst.*
1. Aptychus antiquus, *Goldf.* Geistlicher Berg, Herborn, *G. T.*
2. —— lævigatus, *Goldf.* Eifel, *G. T.*

CRUSTACEA.

1. Calymene Blumenbachii, *Al. Brong.* Dudley; Lebanon, Ohio; New-
    port, .Utica, United States, *Al. Brong.;* Gloucestershire;
    Herefordshire, *Weav.;* Skartofta, Scania; Gottland; Ostro-
    gothia, *His.;* Blankenheim, *Hœn.*
2. —— macrophthalma, *Al.Brong.* United States; Cromford, near Dus-
    seldorf, *Al. Brong.;* Dudley, *Weav.;* Shropshire, *Stokes;*
    Dillenberg, *Hœn.*
3. —— variolaris, *Al. Brong.* Dudley, *Stokes;* Gloucestershire; Here-
    fordshire, *Weav.*
4. —— Tristani, *Al. Brong.* Breuville, Cotentin; Falaise; La Hunan-
    dière; Bain, near Rennes, *Al. Brong.*
5. —— bellatula, *Dalman.* Husbyfjoel, Ostrogothia, *His.*
6. —— ornata, *Dalman.* Husbyfjoel, Ostrogothia, *His.*
7. —— verrucosa, *Dalman.* Varving, near the mountain of Billingen,
    Westrogothia, *His.*
8. —— polytoma, *Dalman.* Ljung, Ostrogothia, *His.*
9. —— actinura, *Dalman.* Berg, Ostrogothia, *His.*
10. —— Sclerops, *Dalman.* Furudal, Dalecarlia; Ostrogothia, *His.*
11. —— Schlotheimi, *Bronn.* Gerolstein, Eifel, *G. T.*

* Dr. Fleming considers that this shell may probably be his *Nautilus Wrightii.*
† *Ellipsolites* of Sowerby.

2 Q 2

12. Calymene latiferus, *Bronn.* Gerolstein, Eifel, *G. T.*
13. ———? æqualis, *Meyer.* Herborn; Dillenberg, *Meyer.*
14. ——— punctata, *Dalm.* Gottland, *His.*
15. ——— concinna, *Dalm.* Gottland, *His.*
16. ——— lævigata, *Goldf.* Gerolstein, Eifel, *G. T.*
17. ——— arachnoides, *Goldf.* Gerolstein, *G. T.*
1. Asaphus cornigerus, *Al. Brong.* Env. of St. Petersburg, *Al. Brong.*;
    Revel, *Schlot.*; Blankenheim, *Hœn.*
2. ——— cordigerus, *Al. Brong.* Dudley, *Stokes.*
3. ——— Hausmanni, *Al. Brong.* Nehou (La Manche); Prague, *Al.*
    Brong.; Canada; Catskill Mountains, Karlstein; Kugel,
    Hœn.
4. ——— de Buchii, *Al. Brong.* Dinevawr Park, Wales; Cyer, Norway,
    Al. Brong.
5. ——— Brongniartii, *Deslongchamps.* May; Nehou, Normandy; Eifel,
    Al. Brong.
6. ——— extenuatus, *Wahl.* Husbyfjoel, Heda, Ostrogothia, *His.*
7. ——— granulatus, *Wahl.* Varving, Olleberg, Westrogothia; Furudal,
    Dalecarlia, *His.*
8. ——— angustifrons, *Dalman.* Husbyfjoel, Ostrogothia, *His.*
9. ——— Heros, *Dalman.* Kinnekulle, Westrogothia; Vikarby, Dale-
    carlia, *His.*
10. ——— expansus, *Wahl.* Common in Sweden, *His.*
11. ——— platynotus, *Dalman.* Westrogothia, *His.*
12. ——— frontalis, *Dalman.* Ljung, Ostrogothia, *His.*
13. ——— læviceps, *Dalman.* Husbyfjöl, Ostrogothia, *His.*
14. ——— palpebrosus, *Dalman.* Husbyfjöl, Ostrogothia, *His.*
15. ——— crassacanda, *Wahl.* Husbyfjöl; Christiania; Bain; Czarko-
    Szelo, *Al. Brong.*
16. ——— brevicaudatus, . May, Calvados, *Her.*
17. ——— incertus, . May, Calvados, *Her.*
18. ——— mucronatus, *Al. Brong.* Ostrogothia; Vestrogothia, *His.*
19. ——— caudatus, *Al. Brong.* Gottland, *His.*; Dudley, *G. T.*
20. ——— Bucephalus, *Goldf.* Eifel, *G. T.*
21. ——— armatus, *Goldf.* Eifel, *G. T.*
1. Ogygia Guettardii, *Al. Brong.* Angers, *Al. Brong.*
2. ——— Desmaresti, *Al. Brong.* Angers, *Al. Brong.*
3. ——— Wahlenbergii, *Al. Brong.* Angers, *Al. Brong.*
4. ——— Sillimani, *Al. Brong.* Banks of the Mohawk, near Schenec-
    tady, *Al. Brong.*
1. Paradoxides * Tessini, *Al. Brong.* Olstorp, Westrogothia, *Al. Brong.*;
    Ginez, Bohemia, *Hœn.*
2. ——— spinulosus †, *Al. Brong.* Andrarum, Scania, *Al. Brong.*;
    Westrogothia, *His.*
3. ——— gibbosus ‡, *Al. Brong.* Kinnekulle, *Al. Brong.*
4. ——— scaraboides §, *Al. Brong.* Falköping, *Al. Brong.*; Ostrogothia;
    Westrogothia, *His.*
1. Olenus Hoffii, *Goldf.* Braatz, near Ginez, Bohemia, *Al, Brong.*
2. ——— Bucephalus, *Wahl.* Olstorp, Vestrogothia, *His.*
3. ——— macrocephalus, *Goldf.* Eifel, *G. T.*
4. ——— flabellifer, *Goldf.* Eifel, *G. T.*
5. ——— Sulzeri ‖, *Goldf.* Ginetz, Bohemia, *G. T.*
1. Nileus Armadillo, *Dalman.* Husbyfjöl and Skarpäsen, Ostrogothia;
    Tomarp, Scania; Furudal, Dalecarlia, *His.*

---

* *Olenus*, Dalm.   † *Olenus spinulosus*, Wahlenberg.
‡ *Olenus gibbosus*, Wahlenberg.   § *Olenus scaraboides*, Wahlenberg.
‖ *Trilobites Sulzeri*, Schlot.

2. Nileus Glomerinus, *Dalman.*   Husbyfjöl, Ostrogothia, *His.*
1. Illænus Centaurus, *Dalman.*   Isle of Oeland, *His.*
2. —— centrotus, *Dalman.*   Husbyfjöl, Ostrogothia, *His.*
3. —— laticauda, *Wahl.*   Osmundsberg, Delecarlia, *His.*
1. Ampyx nasutus, *Dalman.*   Skarpäsen and Husbyfjöl, Ostrogothia;
     Varving, Westrogothia, *His.*
1. Agnostus pisiformis, *Al. Brong.* Kinnekulle, Mösseberg; Westrogothia,
     Al. Brong.
1. Isotelus Gigas, *Dekay*.*   Trenton Falls.
2. —— planus, *Dekay.*   Trenton Falls.
Trilobites, species not determined. Env. of St. Petersburg, *Strangways;*
     Isle of Man, *Henslow;* Brixham, Devon, *De la B.;* Newton
     Bushel, *Radley* and *De la B.;* Elbersreuth, *Munst.*

Pisces.

Ichthyodorulites, *Buckl.* and *De la B.*   Dudley, *Clayfield;* Hereford-
     shire, *Phil.;* Shropshire, *Murch.*
Fish bones and a tooth.   Whitefield quarry and Skeay's Grove, Tort-
     worth, Gloucestershire, *Weav.*
Casts referrible to the vertebræ of fish, S. of Ireland, *Weav.*†

        * *Asaphus platycephalus,* Stokes.

† It should be observed that there are several hitherto undescribed fossils of the grauwacke limestone of Plymouth, in the collection of the Rev. R. Hennah; among these Mr. Sowerby has noticed the following:

Conchifera.

1. Spirifer reticulatus, *Sow. MS.,* also from Ireland, *Sow.*
2. —— pentagonus, *Sow. MS.*
1. Terebratula Hennahiana, *Sow.*   Oblong, rather square, convex, and smooth ; a
     wide furrow runs along the middle of the larger valve, the beak of
     which is much produced.
2. —— gigantea, *Sow.*   Oval, the front rather straight; valves equally con-
     vex, a little flattened towards the front; beak of the large valve
     moderately produced, not incurved.   Five and a half inches long,
     four inches wide.
3. —— rotundata, *Sow.*   Globose, smooth; beaks large, touching.
4. —— like *T. affinis,* but has finer striæ, a produced beak to the larger valve,
     and a greater length; it is also rather more flat.
5. —— Lachryma, *Sow. MS.*
1. Producta anomala, *Sow. MS.,* also from Ireland and Preston, *Sow.*

Mollusca.

1. Turbo cirriformis, *Sow.*   Spire short, of three very convex whorls, smooth ;
     length and breadth equal.
1. Natica? ——.   Nearly globose; spire pointed; whorls few ; the last large ;
     smooth.
1. Terebra Hennahiana, *Sow.*   Turrited, sides nearly straight; whorls flat, crossed
     by slightly curved deep striæ.   Also from Preston, *Sow.*

APPENDIX.

A. *On some of the Terms employed in Geology.*

East.   Fig. 119.   West.

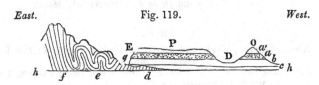

Stratum.—ALTHOUGH, perhaps, this term should only be applied to a bed of rock the upper and under surfaces of which are parallel planes, it is also employed to designate beds the upper and under surfaces of which are irregular. Hence rocks are termed stratified, even when the planes of the beds are not precisely parallel to each other.

Seam.—This term is employed to designate a thin stratum.

Dip.—Strata are said to dip when they form an angle with the horizon; the point towards which they plunge being considered the dip. The amount of the dip is estimated by the size of the angle. In Fig. 119. the strata *f* dip at a considerable angle to the west, because they form a considerable angle with the horizontal line *hh* ; the strata *d* do the same towards the east. The strata *a b c* are nearly horizontal, having only a slight dip to the west.

As it is evident that mere vertical sections, viewed only in one direction, may afford a false idea of the real dip, and as the planes of the strata may be irregular, the student must be careful to ascertain the general and real dip of such planes.

Direction.—A term applied to the course which strata take at right angles to their line of dip.

Strike.—A term recently introduced into English geological writings, from the German 'Streich,' and also applied to the bearing of strata at right angles to their dip.

Anticlinal line—is that line from which strata dip on either side : the ridge of a house-top will convey an idea of this line, the slope of the roof representing the dip of the strata. This line is often extremely useful in tracing disturbances of strata over a country.

Contorted Strata.—Strata are said to be contorted when they are twisted and bent, as at *e*, Fig. 119. These contortions are sometimes on the large scale, as for example in the Alps, where whole mountains are thus twisted.

Conformable Strata.—Strata are termed conformable when their general planes are parallel to each other. Thus *a* rests conformably on *b*.

Unconformable Strata.—Strata are said to be unconformable when they rest as *a b* and *c* do on the edges of the beds *d*, Fig. 119.

Outcrop.—Beds are stated to crop-out when they make their appearance on the surface from beneath others. Thus the strata *d* crop-out at *g*, as do also the beds at *a b* at the same place.

Outlier.—Strata are said to form outliers, when they constitute a portion of country, detached from a main mass of similar beds, of which they have evidently once formed a continuous part. Thus the beds *a'* *a* constitute the outlier O of the strata forming the plateau P, for they have evidently once been continuous, and the continuity has been interrupted by the valley D.

Escarpment.—Strata are said to terminate in an escarpment, when they end abruptly, as *a'* *a* and *b'* do at E.

Fault—is such a dislocation of strata that not only is their continuity destroyed, but the mass of beds on the one side of the fracture or on the other, and sometimes both, are heaved out of their original position. Thus the beds in Fig. 120. have been dislocated or broken into

Fig. 120.

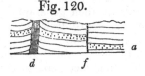

a fault at *f*, the parts of the stratum *a* being no longer on the same plane.

Dyke.—This is a wall of rock intermediate between two sides of a dislocation, interrupting the continuity of the beds on either side. Sometimes the latter exhibit marks of having been shoved up by the intrusion of the rock in the dyke, as in Fig. 120, where the dyke *d* has turned up the edges of the beds which it traverses. At other times, there has been a simple fracture and separation, permitting the presence of the matter in the dyke.

Rock.—This term is used by geologists not only for the hard substances usually thus termed, but also for sands, clays, &c. It is also employed to express a general collection of such substances; thus the expression " the rocks of a country;" or a particular series of mineral substances, such as " the carboniferous rocks," " the cretaceous rocks," &c.

Formation.—A certain series of rocks supposed to have been produced under similar general circumstances, and at about the same epoch.

B. *On Geological Maps and Sections, and the Geological Examination of a Country.*

It is of the very first importance that the geologist should, before he proceeds to the examination of a country, be provided with the best physical map that can be procured; so that his observations may be recorded on that which will not deceive him. It must be admitted that such maps are sufficiently rare; but this defect may now be considered as being gradually removed; for in many of those recently published in this country and on the Continent, much attention has been generally paid to the real physical features of the country, more particularly to the exact delineation of the mountain masses. Formerly geographers contented themselves with running a range of high land between the water-sheds, with little regard to relative heights; so that a real depression between two ranges of mountains was not unfrequently converted into a high connecting ridge, merely because the streams of water flowed in different directions in consequence of a very trifling degree of elevation.

Respecting the maps of our own country, too much praise cannot be given to those published by the Ordnance, remarkable for their general fidelity. With these maps in his hands the geologist feels that his time is not thrown away, and by noticing various minute circumstances upon them, he is subsequently enabled to soar, as it were, above the country he has examined; and by combining his various observations, he may arrive at general conclusions, with which he might not otherwise feel satisfied, and to which he might never have been led without an exact document of this nature.

As the geologist is frequently engaged on a country of which there are no good physical maps, it is quite necessary that he should be able to construct such a map, or at least one in which the hills, rivers, villages, and remarkable objects, should be placed with sufficient accuracy. To do this, it is obvious he must possess a general knowledge of trigonometrical surveying. What is commonly termed *military drawing* will be found of the most essential service; indeed without a knowledge of it, the geologist will often find himself much embarrassed, and be unable to record that which he has observed in nature.

It may be generally assumed that, in the maps of most countries, the towers, steeples, or domes of churches and monasteries, remarkably situated chapels, or very striking objects, are determined and placed with a fair approach to accuracy. If we now suppose that a geologist, acquainted with military drawing, has observed some remarkable facts in a district of which there is only a common map, and that he is desirous to record them, he can, by assuming the church towers or other striking objects in their places, readily construct a general physical map, more particularly as regards the mountains, hills, and river courses, to him the most important parts of a map. For this purpose the student will find Kater's or Schmalkalder's compasses extremely useful, as they are made very portable, and will supply the place of the common compass if it be desirable.

We will now suppose the student in possession of a good physical map, either constructed by others, or by himself. An experienced person will find no great difficulty in advancing the geological and physical maps at the same time; but we will assume that the physical map exists by itself prior to any geological observations, and that the geological student is about to work upon it. He has to trace all the lines which separate one rock from another as they appear on the surface, and thus to represent the superficial areas respectively occupied by each rock. He has also to exhibit the direction or strike of the strata or beds, and their dip, and thus to convey a general idea of their mode of superposition. The dislocations or faults must also be shown, as well as the direction of the mineral veins, where these can be observed. To trace the lines of separation between two rocks, he must carefully observe all the natural or artificial sections he can find, and he will more readily accomplish this part of his labours: if observing that the line of bearing of the strata is in any particular direction, he should follow a zig-zag course across the lines of direction, as far as he conveniently can. Among the artificial sections narrow neglected lanes will often be found valuable; as from the friction of the waggons and carts the rock is frequently laid bare or cut into. Even field ditches must be consulted with care, more particularly when theoretical conclusions are to be deduced from the mode of occurrence of two rocks in contact with each other. Let us suppose, for instance, that a stratified and an unstratified rock, such as a limestone and a trap rock, are in contact or associated with each other; it is evidently important clearly to ascertain whether the strata of the one be cut through by the other, or if the trap merely occurs between the beds of the limestone, without any appearance of having broken or cut the beds. Some countries are much covered superficially with pieces of rock transported from various distances, and which are not known, in place, in the neighbourhood; in such cases it would be obviously unsafe to consider such fragments as affording any information respecting the nature of the rocks beneath them. Other districts alone present on the surface fragments of the subjacent rocks, and much information may then be obtained by examining the surface of ploughed fields, the sides of roads, &c., attention being paid to the slopes of the hills, on the lower flanks of which fragments washed from the heights may be found. For instance, the line separating the slate from the granite on the flanks of Dartmoor, Devon, is often high above the places almost covered by granite blocks, which have fallen from, or been transported over, the slates of the lower slopes. In districts where there are no erratic blocks, large blocks of rock employed in the construction of cottages, or of common field inclosures, inform the student, that such rocks may be successfully sought for in the immediate vicinity.

The student, when he finds himself on a line separating two rocks, should make a dot on his physical map, corresponding with his true position at the time, which can very readily be done, more particularly when roads are correctly laid down, as they are upon the Ordnance Maps. By continuing to dot the points where the rocks come into contact, he will, by connecting the dots by a line, eventually obtain the true line of separation between the respective rocks. Where strata dip at a small angle, the lines of separation are necessarily cut into by the various falls of land, and thus stretch out in directions corresponding to the differences of levels. Isolated hills horizontally catching, as it were, the prolongation of any given rock, the lines of separation often circle round such hills. In countries where rocks, inclined at a small angle, rest on each other, and are composed of substances of different degrees of hardness, the harder often rise in escarpments above the softer, and thus the physical features of the hills correspond with the lines separating the different rocks. This affords great facilities in the construction of geological maps; but at the same time, if a fault occur in such a district, the separating lines are again difficult to trace, and are frequently very complicated.

To express the direction of the dip an arrow is generally employed, the arrow pointing to the direction of the dip: thus the arrow *a* (Fig. 121.) points to the south, supposing the bottom of the page to be south. The amount of the angle at which a rock may dip, is shown, either by writing such amount on one side of the arrow, or by drawing a line at right angles to the arrow, and writing the same amount above such line. I have employed other signs, which I have found useful when reconsidering a country that I have examined. Thus, the arrow *b* (Fig. 121.) shows that, though the strata undulate in the small scale, they dip in the mass to a given point. The sign *c* shows that the strata are perpendicular, the longest line exhibiting their direction or strike. An anticlinal line is shown by *d*, the long line being the direction of the anticlinal line, and the arrow-heads pointing to the dip on either side. *e* represents greatly contorted strata, or when the confusion prevails to such an extent, that they are twisted in all directions. Horizontal strata are shown by the cross *f*, formed of two lines of equal length.

Various instruments have been constructed to take the angle of dip with accuracy, but in practice they are found to be of little value. If strata were really arranged in parallel planes to any considerable distances, and were of equal thickness throughout, these instruments might be useful; but strata are rarely of an equal thickness beyond short distances, and a dip seldom continues long at the same angle. The student having determined the direction of the dip, must estimate its amount by averaging the various angles of dip around him. This certainly requires much caution, and can only be acquired by practice; but, fortunately, extreme accuracy in this point is only required in those rare cases, where an error of 2° or 3° would be important.

The value of exact geological maps is daily becoming more apparent, and it is by no means difficult to foresee that many geological problems will eventually be solved by their accumulation. Already a great change has been effected in this department, but much more remains to be accomplished; and it is exceedingly desirable that even general lines in sketches should not be hastily run. The best geological maps which have been, or will soon be, published, are Greenough's Geological Map of England and Wales, second edition; Elie de Beaumont and Dufrénoy's France; Hoffman's North-Western Germany; and Oeynhausen, La Roche, and Von Dechen's Rhine. Smaller maps of greater or less interest are sufficiently common, and will be found in various scientific works, more particularly in the Transactions of the Geological Society of London.

A geological section represents a vertical cut, at right angles to the surface of a country, towards the interior of the earth. Natural sections are afforded by cliffs either inland, on the shores of the sea, or on the banks of rivers. If these be perpendicular, a true representation of the mode in which the rocks composing them rest upon each other, is a geological section, (as far as the particular direction is concerned,) derived from direct evidence. It must, however, be always understood that the lines drawn solely represent the contact of the respective rocks, their mode of stratification, if they be stratified, or other circumstances strictly geological, to the exclusion of perspective or picturesque

drawing. Exact representations of deep canal or road excavations, the sides of tunnels, mines, and the like, also afford geological sections derived from direct evidence. The
greater part of published geological sections are, however, so
far ideal, that they are formed by the prolongation of lines
observed on the surface of a country, and considered from
analogy to be continued under ground in given directions.
To construct these last requires far more care than is usually
taken. Too much stress cannot be laid on the importance of
rendering them as conformable to nature as circumstances will
admit; that is, the perpendicular elevations and base lines
should be as much as possible on the same scale. Without
this necessary precaution such sections are little better than
caricatures of nature, and are frequently much more mischie-
vous than useful, even leading those who make them to erroneous conclusions, from the distortion and false proportions
of the various parts; gentle sloping valleys being converted
into deep ravines, moderate mountains into enormous elevations, while the possibility of conjecturing the kind of surface
upon which any particular deposit was thrown down, and the
relative importance of the deposit itself, is entirely destroyed.
It will at once be admitted that the proportional thickness of
a deposit is sometimes so trifling when compared to its length,
that it could not be conveniently represented on paper; but
as the relative importance of such a deposit is precisely one of
the circumstances that should be exhibited in a section, it will
be obvious that, though it may be necessary to quit exact proportion, the section should be kept as nearly to it as possible.
The cases, however, in which exact, or nearly exact, proportion can be kept are sufficiently numerous; and, unless it be
desirable to convey false impressions, it is clearly in the interest of science that geological sections should be, what they
pretend to be, miniature representations of nature*.

Let us suppose a student desirous of making a geological
section of a district which he has examined. He must first
determine the direction in which the section should be made.
This will necessarily depend upon a great variety of circumstances which no general rule can meet; but the direction
should always be such as will give most correctly the various
phænomena observed. When, indeed, a series of rocks has
the same, or nearly the same, strike or direction, sections
should be preferred which cut such strike or direction at right
angles. The line of section being settled, he has next to determine the scale on which the section itself should be constructed. If the section be intended to accompany a geologi-

* For the differences between correct and caricature sections, see Sections
and Views illustrative of Geological Phænomena, pl. 2.

cal map, it is desirable that the scale should bear some convenient proportion to the map, in order that both may be more readily understood. A base line should now be drawn representing some given level. The sea level is most frequently taken as a base line, because heights are most commonly measured above it. The undulations of the country above this level have next to be sketched in. This is readily accomplished by erecting perpendiculars of different lengths upon the base line, corresponding in their proportions and distances from each other with the scale adopted both for heights and distances, it being assumed that the section is to be strictly proportional. Let *a b* (Fig. 122.) be the base line, then upon

Fig. 122.

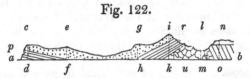

it erect the perpendiculars, *c d, e f, g h, i k, l m,* and *n o,* corresponding, at the proper distances, with the relative heights of the land at those points. With these lines as guides for heights, let the student draw the line which expresses the general outline of the land with its various rises and falls. This portion of the section being accomplished, he must proceed to mark the lines which separate the rocks, their dips, &c. We will suppose that some height up the cliff *c,* one rock rests upon another conformably, and that the dip is inland and correctly ascertained. At the proper proportional height, such as *p,* the line *p f* is to be drawn, so that the angle *p f a* should correspond with the angle of dip in the proper direction. We will now assume that upon the hill *i,* the contact of the same rocks is again seen, but that the dip is in the reverse direction of that formerly observed: in this case the student must proceed as before, so far as regards the angle, making the angle *r h b* correspond with the observed dip. When a section of stratified rocks is given, it is usual to mark the lines of stratification on them, if the scale adopted be not too small, as the conformability or unconformability of rocks in contact with each other is thus clearly shown. At the hill *n* a horizontal rock is represented as reposing unconformably on a rock beneath, it being assumed that the student has observed facts which lead to this conclusion, and that the relative height at which the contact has been observed, has been fairly estimated and represented in the section. It will be obvious that the value of such lines will only depend upon their accurately representing the proper amount of dip of the strata composing each rock respectively.

Geological sections are usually coloured, so that each rock
having a given tint their differences are easily seen. In those
which are not coloured, it is not uncommon to give each different
rock some particular shading in order to distinguish it. Lines,
however, which interfere with lines of stratification should be
avoided, and therefore dots and small hatches are preferable.
The rock *u* (Fig. 122.) is marked in a manner to show that it
is not stratified: in this case it is merely represented as up-
raising the rocks to the right and left without cutting their
lines of stratification; but as unstratified rocks often do cut
stratified rocks, care must be taken to represent such cases,
when they occur, continuing to mark hypothetically downwards the course which the student may consider to be the
true mode of intrusion of the unstratified rocks. In these
and other cases of difficulty it is advisable to dot the supposed
lines of separation, in order to express uncertainty on the subject.

To discuss the subject of geological maps and sections at
length, and to point out the various objects of geological interest which may be observed in various countries, would far
exceed our limits; but it is hoped that the foregoing hints may
be found useful to the student when engaged in the examination of a country, and assist him in the construction of geological maps and sections.

C. *Dr. Daubeny's Views respecting the Origin of Thermal Springs.*

Dr. Daubeny considers that thermal springs will, in most
cases at least, be found to occur in one of three positions; viz.
either in the vicinity of active or extinct volcanos; or, 2ndly, in
that of some uplifted chain of mountains; or, 3rdly, in some
spot which affords either in its own aspect, or in that of the
surrounding country, traces of having been affected by certain
physical convulsions.

Of the first class abundant evidence has already been adduced; the second mentioned may be illustrated in most extensive ranges of mountains, and in none better than by that
of the Pyrenees, which presents a succession of hot springs on
its northern declivity, corresponding nearly with the direction
from west to east, in which the chain itself extends. Thus,
in the Department des Landes, we have the springs of Dax,
the hottest of which has a temperature of 140° F.—Dep. des
Gers, Barbotan, temp. 104° F.—Dep. des Basses Pyrenees,
Cambo, temp. 68°; Eaux Bonnes, 92°; Eaux Chaudes, temp.
100°.—Dep. des Hautes Pyrenees, Cauterets, temp. 144°; St.
Sauveur, 95°; Barèges, 112°; Bagneres de Bigorre, 160°.—
Dep. de Haute Garonne, Encausse, temp. 80°; Bagneres de

Luchon,149°.—Dep. de l'Arriege, Ussat, near Tarascon, 100°; Ax, 158°.—Dep. des Pyrenées Orientales, Olette, 190°; Arles, 160°; Thuez, 170°.—Dep. de l'Aude, Rennes, 111°; Alette, 81°; Campagne, 80°; St. Paul de Fenouilhedes, near Caudies, 82°; &c. &c.

Dr. Daubeny has endeavoured to show, that these springs are for the most part situated, either near the line at which the elevation of the mountains appears to have commenced, or else near the central axis of the chain. The former is considered to be exemplified by the position of Bagneres de Bigorre, Dax, Oleron, Capvern, and Encausse; the latter by that of Barège, Cauterets, and St. Sauveur, which are situated in the heart of the mountains.

In other cases, as in the instances of Alette, Rennes, and Campagne, a change of dip is observed in the rocks near the place whence the spring issues; and in others, as at St. Paul, near Caudies, the thermal water occurs in a cleft separating an elsewhere continuous line of hills, where the occurrence of a fault in the strata affords an additional presumption of some violent action having taken place.

Dr. Daubeny remarks, that many hot springs which occur apart, both from volcanos, and from any leading system of elevations, are nevertheless situated in spots which exhibit proofs of violent convulsions in their vicinity. Von Hoff has supplied us with an instance of this kind in his account of the hot springs of Carlsbad in Bohemia. These waters gush out from a valley, which lies at right angles to all those in its vicinity,—which is evidently more abrupt than they are,—and which exhibits other marks of having been acted on by violence. Dr. Daubeny shows, that many hot springs in Switzerland, such as those of Pfeffers, Weissenburg, and Loueche, are placed in spots which afford similar indications. He also points out the contiguity of the thermal waters of Clifton (Bristol), Matlock, Buxton, and Bath to derangement of strata, and infers, that in the great majority of instances the origin of thermal waters may be traced to volcanic processes, which are either now proceeding, or have formerly taken place. He considers this inference as confirmed by examining the gaseous products of thermal waters, which seem for the most part identical with those given off by volcanos; remarking, that many of these hot springs contain sulphuretted hydrogen, a common volcanic product, especially in cases where the action is languid.

Carbonic acid, also emitted from volcanos, is noticed still more commonly as being evolved from hot springs; and Dr. Daubeny is inclined to attribute, in some cases, the evolution of this gas from crevices in rocks, and from springs possessing only an ordinary temperature, to volcanic processes. This appears to him probable from the more frequent occurrence of this phænomenon, in the vicinity of volcanos, either active or extinct, or in the midst of rocks which appear to have been uplifted, as in valleys of elevation. He further observes, that thermal carbonated and cold carbonated springs are commonly associated, and therefore probably derive their carbonic acid from the same cause.

Lastly, Dr. Daubeny points out, that nitrogen, a gas very generally given off from warm springs, may be inferred also to be a volcanic product from the ammoniacal salts so copiously emitted by many volcanos. Nitrogen has been detected, by Longchamp in almost all the thermal waters of the Pyrenees, by Dr. Ure at Loueche in Switzerland, and by Dr. Daubeny in many of the thermal waters of Savoy and France. In our own country it has long been known to occur in the waters of Bath and Buxton; and Dr. Daubeny has lately detected it at Bakewell and Middleton in Derbyshire, and at Taafe's Well, near Cardiff, South Wales. This evolution of nitrogen being therefore one of the most general phænomena of volcanic action in all its various degrees of intensity, and being in his opinion not attributable, all the circumstances considered, to any other source except the disengagement of atmospheric air deprived of its oxygen by some process of combustion, Dr. Daubeny is thence led to adopt the theory which attributes volcanic action to a process of oxidation, in preference to that which refers it merely to the contraction of the crust of the earth upon an internal fluid nucleus.

He proceeds to remark, that if this mode of accounting for the nitrogen be admitted, the next inquiry will be, what are the materials which by their oxidation produce the effect alluded to? Here it appears to him most natural to search for them, in the first instance, amongst those bodies which actually exist in the lavas and other substances usually ejected by volcanos. It is observable, that the bases of all of these are capable, either at ordinary or at high temperatures, of combining with oxygen, forming with it some fixed product, whilst those of several even decompose water under the same circumstances, and by their union with its oxygen give rise to a sufficient evolution of heat to cause the combustion of such other inflammable bodies as may be present. If the above existed in the interior of the globe in their unaltered condition, they might, by combining with oxygen, cause that evolution of nitrogen which has been so commonly observed.

Dr. Daubeny therefore contends, that the hypothesis originally suggested, in consequence of Sir H. Davy's discovery of the metallic bases of the earths and alkalies, is adequate to explain all the phænomena connected with volcanic action which up to this time have been discovered. It is only necessary to assume, that water, and afterwards air, find occasional admission to certain parts of the interior of the globe, containing the bases of silica, alumina, lime, potass, and soda, together with some of the metals met with on the surface, in such proportions as are indicated by the actual composition of lavas *.

D. *On the Upper Portion of the Grauwacke Group in Shropshire, Herefordshire, and Wales.*

Very important additions to our knowledge of the upper part of the grauwacke group, more particularly as it exists in Shropshire, Herefordshire, and Wales, have lately been made by Mr. Murchison. This author divides the upper part of the grauwacke of that district into six portions, or sub-groups, to which he has assigned the names of (1.) Upper Ludlow Rock; (2.) Wenlock Limestone; (3.) Lower Ludlow Rock; (4.) Shelly Sandstones; (5.) Black Trilobite Flagstone; and (6.) Red Conglomerate, Sandstone, and Slaty Schist.

The *Upper Ludlow Rock*, equivalent to the grauwacke sandstone of Tortworth, Gloucestershire, is so named because the Castle of Ludlow stands upon it. This sub-group is described as having a maximum thickness of 1000 feet, and as being principally composed of thin-bedded sandstone, often highly calcareous. The upper beds are characterized by two species of Strophomena or Leptæna, an Orbicula, a plicated Terebratula, &c., all of undescribed species. The lower beds contain the abundant remains of a small Terebratula having a gryphoid form. Trilobites of the genera Homonolotus and Calymene are also found. The upper Ludlow rock passes upwards into old red sandstone.

The *Wenlock limestone* is considered the equivalent of the Dudley limestone, and thus the true relative geological place of the latter, so long celebrated for its organic contents, is determined. Mr. Murchison remarks that in the course of this limestone between the rivers Oney and Lug, it is chiefly characterized by one species of Pentamerus.

The *Lower Ludlow Rock*, also known as the ' Die Earth,' is principally composed of incoherent, grayish, argillaceous schist, seldom micaceous. The higher beds contain Orthoceratites, Lituites, *Asaphus caudatus, &c.* A thin calcareous zone occurs at the base of this sub-group in Shropshire, containing *Pentamerus lævis,* as also another species of the same genus. Thickness, upwards of 2000 feet.

The *Shelly Sandstones* are of variegated colours, red and

* For the detail of this theory, consult Dr. Daubeny's Description of Volcanos, London, 1826, p. 389, &c.; and for the same author's Remarks on Thermal Springs, an article in the London Review, No. 2, 1829, on Mineral Springs, and a memoir in the New Edinburgh Phil. Journal, 1831.

green predominating. Calcareous bands, almost made up of Productæ, Leptæna, Spiriferi, and crinoidal remains, differing from those in the upper sub-groups, are associated with the sandy beds. Thickness, from 1500 to 1800 feet.

The *Black Trilobite Flagstone* contains the remains of the *Asaphus Buchii*, and it is remarked that this and the associated trilobites are not found in the upper sub-groups. Thickness supposed to exceed that of any one of the superior sub-groups.

The *Red Conglomerates, Sandstones,* and *Slaty Schist* constitute a deposit several thousand feet thick, in which organic remains have not been observed.

Mr. Murchison remarks that the Wenlock limestone thins out a short distance to the S.W. of Aymestrey, and that then the Upper and Lower Ludlow rocks come into contact, and generally occupy the same lofty escarpment in their course through South Wales*.

E. *Intensity of Terrestrial Magnetism in Mines.*

Mr. Henwood has been recently engaged in investigating the intensity of terrestrial magnetism at various depths and elevations in Cornwall. His experiments were made *in vacuo*, with the apparatus contrived for that purpose by Mr. Harris†, and seem to have been conducted with much care. The stations were Carn Brea Castle, near Redruth, a granitic ridge, 760 feet above the sea; the surface of Dolcoath Mine, on slate, 300 feet above the sea; and at 1200 feet in the same mine (or 900 feet beneath the level of the sea). At each of these stations he has made two sets of horary observations, continued at each place 24 hours each time. The results of these experiments indicate no appreciable difference in the intensity of terrestrial magnetism at each of these stations.

F. *Tables for calculating Heights by the Barometer.*

The following Tables are those of M. Oltmanns, which are generally admitted as among the most convenient hitherto published. Being calculated for the metrical barometer, they were useless to persons employing that graduated according to English inches and their decimal parts. To render them applicable to our barometers, a table (A) has been prefixed, in which the equivalent of every millimetre of the metrical barometer is given in English inches and the thousandth parts of inches.

To reduce the metres used in these tables into English feet, a table (E) is appended, where the number of English feet corresponding to any number of metres up to 10,000 will be immediately obtained.

Abstraction being made of table A prefixed, and table E appended, the march of operations is as follows:

* Proceedings of the Geol. Soc., April 17, 1833.
† Trans. of the British Association, vol. i. p. 559.

Let h be the height of the barometer at the lower station expressed in millimetres; h' that of the higher station; T and T' the temperature of the barometer at the different stations according to the centigrade thermometer; t and t' that of the air.

We search in table B for the number which corresponds to h; let us call it a: we likewise search in the same table for that which corresponds to h'; let this be named b: let us call c, the generally very small number which, in table C, faces $T-T'$; the approximate height will be $a-b-c$. (If $T-T'$ is negative, it should be written $a-b+c$.) In order to apply the correction necessary for the strata of air, it will suffice to multiply the thousandth part of the approximate height by the double sum $2(t+t')$ of the detached thermometers; the correction will be either positive or negative, according as $t+t'$ is itself either positive or negative.

The second and last correction, that for the latitude and the diminution of weight, is obtained by taking, in table D, the number which corresponds vertically to the latitude, and horizontally to the approximate height: this correction, which can never exceed 28 metres, is always added.

In order to understand the calculation of a height by means of these tables, and those prefixed and appended, let us suppose that in latitude $= 44°$ we had, at the level of the sea, the barometer $= 30\cdot040$ English inches, temperature of the instrument $= 22°\cdot5$ centigrade, and of the air $= 22°$. At the top of a mountain, the barometer $= 26\cdot575$ English inches, temperature of the instrument $= 17°\cdot5$, and of the air $= 17°$.

In order to obtain the equivalents of the English inches in millimetres, search in table A; where the number of millimetres corresponding to $30\cdot040$ inches observed at the sea will be 763, and that of $26\cdot575$ observed on the mountain will be 675. Having obtained these equivalents, the calculation proceeds:

	Mill.	Metres.	
Barometer at sea level	$= 763 =$	6182·0 ⎱	Table B.
Barometer on the mountain	$= 675 =$	5206·1 ⎰	
		975·9	
Diff. of attached thermometers $= 5° =$		7·4	Table C.
Apparent height		968·5	
Double the sum of the detached thermometers multiplied by the thousandth part of 968·5		75·5	
		1044·	
Correction for latitude		3·1	Table D.
Height of the mountain		1047·1	
Height in English feet		3435	Table E.

2 R 2

When the height of the barometer, graduated according to English inches and their parts, does not precisely correspond with a certain number of millimetres, and when great accuracy is required, it will be obvious, that instead of taking the next nearest number to it in the tables, as might otherwise be done, it will be necessary to calculate the difference.

Table A.

Inches.	Mil.	Inches.	Mil.	Inches.	Mil.	Inches.	Mil.	Inches.	Mil.	Inches.	Mil.
14·56	370	16·142	410	17·717	450	19·292	490	20·866	530	22·441	570
·606	371	·181	411	·756	451	·331	491	·901	531	·481	571
·646	372	·221	412	·795	452	·370	492	·945	532	·520	572
·685	373	·260	413	·835	453	·410	493	·984	533	·559	573
·725	374	·299	414	·874	454	·449	494	21·025	534	·599	574
·764	375	·339	415	·914	455	·488	495	·063	535	·638	575
·803	376	·378	416	·953	456	·528	496	·102	536	·678	576
·843	377	·417	417	·992	457	·567	497	·142	537	·717	577
·882	378	·457	418	18·032	458	·607	498	·181	538	·756	578
·921	379	·496	419	·071	459	·646	499	·220	539	·796	579
·961	380	·536	420	·110	460	·685	500	·260	540	·835	580
15·000	381	·575	421	·150	461	·725	501	·300	541	·874	581
·040	382	·614	422	·189	462	·764	502	·339	542	·914	582
·079	383	·654	423	·229	463	·803	503	·378	543	·953	583
·118	384	·693	424	·268	464	·843	504	·418	544	·992	584
·156	385	·733	425	·307	465	·882	505	·457	545	23·032	585
·197	386	·772	426	·347	466	·922	506	·496	546	·071	586
·236	387	·811	427	·386	467	·961	507	·536	547	·111	587
·276	388	·851	428	·425	468	20·000	508	·575	548	·150	588
·315	389	·890	429	·465	469	·040	509	·615	549	·189	589
·355	390	·929	430	·504	470	·079	510	·654	550	·229	590
·394	391	·969	431	·543	471	·118	511	·693	551	·268	591
·433	392	17·008	432	·583	472	·158	512	·733	552	·307	592
·473	393	·047	433	·622	473	·197	513	·772	553	·347	593
·512	394	·087	434	·662	474	·237	514	·812	554	·386	594
·551	395	·126	435	·701	475	·276	515	·851	555	·426	595
·591	396	·166	436	·740	476	·315	516	·890	556	·465	596
·630	397	·205	437	·780	477	·355	517	·929	557	·504	597
·670	398	·244	438	·819	478	·394	518	·969	558	·544	598
·709	399	·284	439	·859	479	·433	519	22·008	559	·583	599
·748	400	·323	440	·898	480	·473	520	·048	560	·622	600
·788	401	·362	441	·937	481	·512	521	·087	561	·662	601
·827	402	·402	442	·977	482	·551	522	·126	562	·701	602
·866	403	·441	443	19·016	483	·590	523	·166	563	·741	603
·906	404	·481	444	·055	484	·630	524	·205	564	·780	604
·945	405	·520	445	·095	485	·670	525	·244	565	·819	605
·985	406	·559	446	·134	486	·709	526	·284	566	·859	606
16·024	407	·599	447	·174	487	·748	527	·323	567	·898	607
·063	408	·638	448	·213	488	·788	528	·363	568	·937	608
·102	409	·677	449	·252	489	·827	529	·402	569	·977	609

TABLE A. (continued.)

Inches.	Mil.	Inches.	Mil.	Inches.	Mil.	Inches.	Mil.	Inches.	Mil.	Inches.	Mil.
24·016	610	25·197	640	26·378	670	27·560	700	28·741	730	29·922	760
·055	611	·237	641	·418	671	·600	701	·780	731	·961	761
·095	612	·276	642	·457	672	·639	702	·819	732	30·000	762
·134	613	·315	643	·496	673	·678	703	·859	733	·040	763
·174	614	·355	644	·536	674	·718	704	·898	734	·079	764
·213	615	·394	645	·575	675	·757	705	·937	735	·119	765
·252	616	·433	646	·615	676	·796	706	·977	736	·158	766
·292	617	·473	647	·654	677	·836	707	29·016	737	·197	767
·331	618	·512	648	·693	678	·875	708	·056	738	·237	768
·370	619	·551	649	·733	679	·915	709	·095	739	·276	769
·410	620	·590	650	·772	680	·954	710	·134	740	·316	770
·449	621	·630	651	·811	681	·993	711	·174	741	·355	771
·489	622	·670	652	·851	682	28·032	712	·213	742	·394	772
·528	623	·709	653	·891	683	·071	713	·252	743	·433	773
·567	624	·749	654	·931	684	·111	714	·291	744	·473	774
·607	625	·788	655	·970	685	·150	715	·331	745	·512	775
·646	626	·827	656	27·010	686	·189	716	·371	746	·552	776
·685	627	·866	657	·049	687	·229	717	·410	747	·591	777
·725	628	·906	658	·089	688	·268	718	·449	748	·630	778
·764	629	·945	659	·128	689	·308	719	·489	749	·670	779
·804	630	·985	660	·167	690	·347	720	·528	750	·709	780
·843	631	26·024	661	·206	691	·386	721	·567	751	·749	781
·882	632	·063	662	·246	692	·426	722	·607	752	·788	782
·922	633	·103	663	·285	693	·465	723	·646	753	·827	783
·961	634	·142	664	·324	694	·504	724	·686	754	·867	784
25·000	635	·182	665	·363	695	·544	725	·725	755	·906	785
·040	636	·221	666	·403	696	·583	726	·764	756	·945	786
·079	637	·260	667	·442	697	·622	727	·804	757	·985	787
·119	638	·300	668	·482	698	·662	728	·843	758	31·024	788
·158	639	·339	669	·521	699	·701	729	·882	759	·064	789

TABLE B.

Mil.	Metr.	Mil.	Metr.	Mil.	Metr.	Mil.	Metr.	Mil.	Metr.	Mil.	Metr.
370	418·5	384	714·3	398	999·5	412	1274·8	426	1540·8	440	1798·4
371	440·0	385	735·0	399	1019·5	413	1294·1	427	1559·5	441	1816·5
372	461·5	386	755·6	400	1039·4	414	1313·3	428	1578·2	442	1834·5
373	482·9	387	776·2	401	1059·3	415	1332·5	429	1596·8	443	1852·5
374	504·2	388	796·8	402	1079·1	416	1351·7	430	1615·3	444	1870·4
375	525·4	389	817·3	403	1098·9	417	1370·8	431	1633·8	445	1888·3
376	546·6	390	837·8	404	1118·6	418	1389·9	432	1652·2	446	1906·2
377	567·8	391	858·2	405	1138·3	419	1408·9	433	1670·6	447	1924·0
378	588·9	392	878·5	406	1157·9	420	1427·9	434	1689·0	448	1941·8
379	609·9	393	898·8	407	1177·5	421	1446·8	435	1707·3	449	1959·6
380	630·9	394	919·0	408	1197·1	422	1465·7	436	1725·6	450	1977·3
381	651·8	395	939·2	409	1216·6	423	1484·6	437	1743·8	451	1994·9
382	672·7	396	959·3	410	1236·0	424	1503·4	438	1762·1	452	2012·6
383	693·5	397	979·4	411	1255·4	425	1522·2	439	1780·3	453	2030·2

TABLE B. (*continued.*)

Mil.	Metr.	Mil.	Metr.	Mil.	Metr.	Mil.	Metr.	Mil.	Metr.	Mil.	Metr.
454	2047·8	502	2848·1	550	3575·3	598	4241·6	646	4856·4	694	5427·2
455	2065·3	503	2864·0	551	3589·8	599	4254·9	647	4868·7	695	5438·7
456	2082·8	504	2879·8	552	3604·2	600	4268·2	648	4881·0	696	5450·1
457	2100·2	505	2895·6	553	3618·6	601	4281·4	649	4893·3	697	5461·5
458	2117·6	506	2911·3	554	3633·0	602	4294·7	650	4905·6	698	5472·9
459	2135·0	507	2927·0	555	3647·4	603	4307·9	651	4917·8	699	5484·3
460	2152·3	508	2942·7	556	3661·7	604	4321·1	652	4930·0	700	5495·7
461	2169·6	509	2958·4	557	3676·0	605	4334·3	653	4942·2	701	5507·1
462	2186·9	510	2974·0	558	3690·3	606	4347·4	654	4954·4	702	5518·4
463	2204·1	511	2989·6	559	3704·6	607	4360·5	655	4966·6	703	5529·8
464	2221·3	512	3005·2	560	3718·8	608	4373·7	656	4978·7	704	5541·1
465	2238·4	513	3020·7	561	3733·0	609	4386·7	657	4990·9	705	5552·4
466	2255·5	514	3036·2	562	3747·2	610	4399·8	658	5003·0	706	5563·7
467	2272·6	515	3051·7	563	3761·3	611	4412·8	659	5015·1	707	5575·0
468	2280·6	516	3067·2	564	3775·4	612	4425·9	660	5027·2	708	5586·2
469	2306·6	517	3082·6	565	3789·5	613	4438·9	661	5039·2	709	5597·5
470	2323·6	518	3097·9	566	3803·6	614	4451·9	662	5051·2	710	5608·7
471	2340·5	519	3113·3	567	3817·7	615	4464·8	663	5063·3	711	5619·9
472	2357·4	520	3128·6	568	3831·7	616	4477·7	664	5075·3	712	5631·1
473	2374·2	521	3143·9	569	3845·7	617	4490·7	665	5087·2	713	5642·2
474	2391·1	522	3159·2	570	3859·7	618	4503·6	666	5099·2	714	5653·4
475	2407·9	523	3174·4	571	3873·7	619	4516·4	667	5111·2	715	5664·6
476	2424·6	524	3189·7	572	3887·6	620	4529·3	668	5123·1	716	5675·7
477	2441·3	525	3204·9	573	3901·5	621	4542·1	669	5135·0	717	5686·8
478	2458·0	526	3220·0	574	3915·4	622	4554·9	670	5146·9	718	5697·9
479	2474·6	527	3235·1	575	3929·3	623	4567·7	671	5158·8	719	5709·0
480	2491·3	528	3250·2	576	3943·1	624	4580·5	672	5170·6	720	5720·1
481	2507·9	529	3265·3	577	3956·9	625	4593·2	673	5182·5	721	5731·1
482	2524·3	530	3280·3	578	3970·7	626	4606·0	674	5194·3	722	5742·1
483	2540·8	531	3295·3	579	3984·5	627	4618·7	675	5206·1	723	5753·1
484	2557·3	532	3310·3	580	3998·2	628	4631·4	676	5217·9	724	5764·2
485	2573·7	533	3325·3	581	4011·9	629	4644·0	677	5229·7	725	5775·1
486	2590·2	534	3340·2	582	4025·6	630	4656·7	678	5241·4	726	5786·1
487	2506·6	535	3355·1	583	4039·3	631	4669·3	679	5253·2	727	5797·1
488	2622·9	536	3370·0	584	4052·9	632	4682·0	680	5264·9	728	5808·0
489	2639·2	537	3384·8	585	4066·6	633	4694·5	681	5276·6	729	5819·0
490	2655·4	538	3399·6	586	4080·2	634	4707·1	682	5288·3	730	5829·9
491	2671·6	539	3414·4	587	4093·8	635	4719·7	683	5300·0	731	5840·8
492	2687·9	540	3429·2	588	4107·3	636	4732·2	684	5311·6	732	5851·7
493	2704·1	541	3443·9	589	4120·8	637	4744·7	685	5323·2	733	5862·5
494	2720·2	542	3458·6	590	4134·3	638	4757·2	686	5334·8	734	5873·4
495	2736·3	543	3473·3	591	4147·8	639	4769·7	687	5346·4	735	5884·2
496	2752·3	544	3487·9	592	4161·3	640	4782·1	688	5358·0	736	5895·1
497	2768·3	545	3502·5	593	4174·7	641	4794·6	689	5369·6	737	5905·9
498	2784·4	546	3517·2	594	4188·1	642	4807·0	690	5381·1	738	5916·7
499	2800·4	547	3531·8	595	4201·5	643	4819·4	691	5392·7	739	5927·5
500	2816·3	548	3546·3	596	4214·9	644	4831·7	692	5404·2	740	5938·2
501	2832·2	549	3560·8	597	4228·2	645	4844·1	693	5415·7	741	5949·0

TABLE B. (*continued.*)

Mil.	Metr.	Mil.	Metr.	Mil.	Metr.	Mil.	Metr.	Mil.	Metr.	Mil.	Metr.
742	5959·7	751	6055·7	759	6140·1	767	6223·6	775	6306·2	783	6388·0
743	5970·4	752	6066·3	760	6150·6	768	6234·0	776	6316·5	784	6398·2
744	5981·2	753	6076·9	761	6161·1	769	6244·4	777	6326·7	785	6408·3
745	5991·9	754	6087·5	762	6171·5	770	6254·7	778	6337·0	786	6418·5
746	6002·5	755	6098·0	763	6182·0	771	6265·0	779	6347·2	787	6428·6
747	6013·2	756	6108·6	764	6192·4	772	6275·4	780	6357·4	788	6438·7
748	6023·8	757	6119·1	765	6202·8	773	6285·7	781	6367·6	789	6448·8
749	6034·4	758	6129·6	766	6213·2	774	6296·0	782	6377·8	790	6458·9
750	6045·1										

TABLE C.

Deg.	Metre.	Deg.	Metre.	Deg.	Metre.	Deg.	Metre.	Deg.	Metre.
0·2	0·3	4·2	6·2	8·2	12·1	12·2	17·9	16·2	23·8
0·4	0·6	4·4	6·5	8·4	12·4	12·4	18·2	16·4	24·1
0·6	0·9	4·6	6·8	8·6	12·6	12·6	18·5	16·6	24·4
0·8	1·2	4·8	7·1	8·8	12·9	12·8	18·8	16·8	24·7
1·0	1·5	5·0	7·4	9·0	13·2	13·0	19·1	17·0	25·0
1·2	1·8	5·2	7·6	9·2	13·5	13·2	19·4	17·2	25·3
1·4	2·1	5·4	7·9	9·4	13·8	13·4	19·7	17·4	25·6
1·6	2·3	5·6	8·2	9·6	14·1	13·6	20·0	17·6	25·9
1·8	2·6	5·8	8·5	9·8	14·4	13·8	20·3	17·8	26·2
2·0	2·9	6·0	8·8	10·0	14·7	14·0	20·6	18·0	26·5
2·2	3·2	6·2	9·1	10·2	15·0	14·2	20·9	18·2	26·8
2·4	3·5	6·4	9·4	10·4	15·3	14·4	21·2	18·4	27·1
2·6	3·8	6·6	9·7	10·6	15·6	14·6	21·5	18·6	27·4
2·8	4·1	6·8	10·0	10·8	15·9	14·8	21·8	18·8	27·7
3·0	4·4	7·0	10·3	11·0	16·2	15·0	22·1	19·0	28·0
3·2	4·7	7·2	10·6	11·2	16·5	15·2	22·4	19·2	28·2
3·4	5·0	7·4	10·9	11·4	16·8	15·4	22·7	19·4	28·5
3·6	5·3	7·6	11·2	11·6	17·1	15·6	22·9	19·6	28·8
3·8	5·6	7·8	11·5	11·8	17·4	15·8	23·2	19·8	29·1
4·0	5·9	8·0	11·8	12·0	17·6	16·0	23·5		

TABLE D.

Appr. Ht.	0°	5°	10°	15°	20°	25°	Appr. Ht.	30°	35°	40°	45°	50°	55°
	m.	m.	m.	m.	m.	m.		m.	m.	m.	m.	m.	m.
200	1·2	1·2	1·2	1·0	1·0	1·0	200	0·8	0·8	0·6	0·6	0·6	0·4
400	2·4	2·4	2·4	2·2	2·0	2·0	400	1·8	1·7	1·4	1·2	1·0	0·8
600	3·4	3·4	3·4	3·2	3·0	2·8	600	2·6	2·4	2·0	1·8	1·6	1·2
800	4·5	4·5	4·5	4·3	4·1	3·8	800	3·5	3·1	2·8	2·4	2·0	1·7
1000	5·7	5·7	5·7	5·3	5·1	4·8	1000	4·3	3·8	3·4	3·1	2·6	2·2
1200	7·0	7·0	6·8	6·4	6·0	5·8	1200	5·1	4·6	4·2	3·6	3·1	2·6
1400	8·2	8·2	8·0	7·6	7·1	6·7	1400	6·1	5·4	4·8	4·2	3·6	3·0
1600	9·2	9·2	9·0	8·8	8·2	7·6	1600	7·0	6·2	5·6	4·8	4·1	3·4
1800	10·4	10·4	10·2	9·8	9·4	8·6	1800	8·0	7·0	6·3	5·4	4·6	3·8
2000	11·6	11·5	11·3	11·0	10·4	9·6	2000	8·8	7·8	7·0	6·0	5·1	4·2
2200	12·8	12·6	12·6	12·1	11·4	10·6	2200	9·7	8·6	7·6	6·6	5·6	4·6
2400	14·0	14·0	13·8	13·3	12·5	11·6	2400	10·6	9·4	8·4	7·2	6·1	5·1
2600	15·2	15·2	15·0	14·4	13·6	12·6	2600	11·6	10·5	9·2	8·0	6·8	5·6
2800	16·6	16·5	16·4	15·6	14·8	13·6	2800	12·6	11·4	10·0	8·8	7·4	6·2
3000	17·9	17·7	17·6	16·8	15·8	14·6	3000	13·6	12·2	10·8	9·4	8·0	6·6
3200	19·1	18·9	18·7	18·0	17·0	15·7	3200	14·6	13·1	11·5	10·1	8·6	7·0
3400	20·5	20·3	20·1	19·3	18·4	16·9	3400	15·7	14·1	12·4	10·9	9·2	7·7
3600	21·8	21·7	21·4	20·4	19·6	18·0	3600	16·7	15·0	13·4	11·6	9·8	8·2
3800	23·1	22·9	22·6	21·6	20·6	19·1	3800	17·7	15·9	14·3	12·4	10·5	8·7
4000	24·6	24·4	24·0	22·9	21·9	20·3	4000	18·7	17·0	15·1	13·1	11·2	9·4
4200	25·9	25·7	25·3	24·3	23·0	21·6	4200	19·9	18·0	15·9	14·0	12·0	10·1
4400	27·5	27·3	26·8	25·8	24·3	23·0	4400	21·1	19·1	16·9	15·0	12·9	10·8
4600	28·9	28·7	28·2	27·1	25·6	24·3	4600	22·3	20·3	18·0	15·9	13·6	11·5
4800	30·4	30·2	29·6	28·4	27·0	25·5	4800	23·4	21·3	19·0	16·7	14·3	12·1
5000	31·8	31·6	30·9	29·8	28·4	26·7	5000	24·6	22·3	19·9	17·4	15·0	12·7
5200	33·0	32·8	32·1	31·0	29·7	28·0	5200	25·7	23·3	20·8	18·2	15·7	13·3
5400	34·3	34·1	33·5	32·4	30·8	29·2	5400	26·7	24·3	21·7	19·1	16·4	13·9
5600	35·7	35·5	34·8	33·7	32·1	30·2	5600	27·8	25·3	22·6	19·9	17·2	14·5
5800	37·1	36·9	36·1	35·0	33·2	31·3	5800	28·9	26·3	23·6	20·7	17·8	15·1
6000	38·5	38·3	37·5	36·3	34·3	32·3	6000	30·0	27·3	24·6	21·5	18·5	15·7

TABLE E.

Reduction of Metres into English Feet and Inches.

Metr.	Feet.	Inches.	Metr.	Feet.	Inches.	Metr.	Feet.	Inches.
1	3	3·370	50	164	0·514	900	2952	9·261
2	6	6·740	60	196	10·217	1000	3280	10·290
3	9	10·111	70	229	7·920	2000	6561	8·58
4	13	1·481	80	262	5·623	3000	9842	6·87
5	16	4·851	90	295	3·326	4000	13123	5·16
6	19	8·222	100	328	1·029	5000	16404	3·45
7	22	11·592	200	656	2·058	6000	19685	1·74
8	26	2·963	300	984	3·087	7000	22966	0·03
9	29	6·333	400	1312	4·116	8000	26246	10·32
10	32	9·702	500	1640	5·145	9000	29527	8·61
20	65	7·405	600	1968	6·174	10000	32808	6·90
30	98	5·108	700	2296	7·203			
40	131	2·811	800	2624	8·232			

Reduction of Decimetres, Centimetres, and Millimetres, to English Inches.

Dec.	Inches.	Cent.	Inches.	Milli.	Inches.
1	3·937	1	0·393	1	0·039
2	7·874	2	0·787	2	0·078
3	11·811	3	1·181	3	0·118
4	15·748	4	1·574	4	0·157
5	19·685	5	1·968	5	0·196
6	23·622	6	2·362	6	0·236
7	27·559	7	2·755	7	0·275
8	31·496	8	3·149	8	0·314
9	35·433	9	3·543	9	0·354
10	39·370	10	3·937	10	0·393

F. *Comparison of English and French Measures.*
(From Baily's Astronomical Tables.)

Eng. Inches.		Fr. Metres.
French Metre.....=39·37079	French Toise.....=1·949036
——— Toise.....=76·739400	——— Foot.....=0·324839
——— Foot.....=12·789900	——— Inch.....=0·027070
——— Inch.....= 1·065825	English Foot.....=0·304794
——— Line.....= 0·088819	——— Inch.....=0·025399

Constant logarithms (always additive) for converting

French Toises into Met.0·2898200 French Ft. into Eng. Ft.0·0276860
——— Feet into Metr. 9·5116687 ——— Met. into Eng. Ft.0·5159929
——— T. into Eng. Ft.0·8058372 Millimet. into Eng. In. 8·5951741

INDEX.

ABEL, Dr. Clarke, on the bank thrown up at the Cape of Good Hope, 86.

Aix in Provence, supracretaceous rocks of, 247.

Alps, supracretaceous rocks of, 231, 237; cretaceous rocks of, 290, 292; oolitic rocks of, 329.

Animal life, early, on the surface of the globe, remarks on the, 429.

Arago, M., on the temperature of the earth's surface, 7.

Arctic circle, numerous remains of elephants and rhinoceroses within the, 199.

Argillaceous, or clay slate, 433.

Arkose, 324.

Artesian wells, temperature of, 12.

Atchafalaya, raft of, 71; section of the alluvial banks of, *ib.*

Attraction, local, of the magnetic needle, renders the determination of currents doubtful, 109.

Auvergne, erosive action of rivers in, 57; extinct volcanos and supracretaceous rocks of, 271.

Ava, organic remains found in the kingdom of, 268.

Baculite limestone, of Normandy, 296.

Bagshot sands, 263.

Baku, inflammable gaseous exhalations of, 154; naphtha and petroleum springs of, 163.

Banda, Isle of, gradual rise of a promontory at, 127.

Barometer, tables for calculating heights by, 610.

Bars of rivers, 89.

Basalt, 137; chemical composition of, 452; columnar structure of, 470.

Basins, rock, 46.

Basterot, M. de, on the supracretaceous rocks of Bordeaux, 250.

Beach, raised, of Plymouth, 172; in Cornwall, 174; of Jura (Hebrides),

175; of Uddevalla, Sweden, 176; of St. Hospice, Nice, *ib.*; in South America, 178.

Beaches, shingle, 79; travel in the direction of the prevalent winds, *ib.*; bar up the valleys, forming lakes, 81; sandy, 84.

Beaufort, Capt., on the indurated beach, Selinty, Karamania, *note*, 85; on undercurrents in the Mediterranean, 114; on the gaseous exhalation of the Yanar, 152.

Beaumont, M. Elie de, on the erratic blocks of the Alps, 194; on the gravels of the Lyonais, Dauphiné, and Provence, 222; on the supracretaceous rocks of the Pertuis de Mirabeau, 246; on the oolitic rocks of Burgundy, 317; on the lias of the Alps, 327; on the Grès de Vosges, 356; on the elevation of mountains, 483.

Beechey, Capt., on the temperature of the sea, 24.

Belemnites, association of, with coal measure plants in the Alps, 328.

Bertrand de Doue, M., on the volcanic rocks and ossiferous beds of the Velay, 272.

Bertrand-Geslin, M., on the Val d'Arno, 245.

Beudant, M., on the volcanic rocks of Hungary, 278.

Bigsby, Dr., on the erratic blocks of North America, 192.

Black Head (Babbacombe Bay, Devon), relations of limestone and trappean rocks at, 464.

Boase, Dr., on the submarine forest, Mount's Bay, Cornwall, 170.

Boblaye, M., on the passage of the Meuse through the Ardennes, 58; on the shore-lines upon the limestones of Greece, 178; on the oolitic rocks of the North of France, 316.

2 s

THE END.

LONDON:

PRINTED BY RICHARD TAYLOR,

RED LION COURT, FLEET STREET.

ALERE   FLAMMAM.

Printed in the United States
By Bookmasters